PETROLEUM PROCESSING HANDBOOK

PETROLEUM PROCESSING HANDBOOK

edited by
John J. McKetta
The University of Texas at Austin
Austin, Texas

Marcel Dekker, Inc. **New York • Basel • Hong Kong**

Library of Congress Cataloging-in-Publication Data

Petroleum processing handbook / edited by John J. McKetta.
    p.   cm.
  "The contents of this volume were originally published in
Encyclopedia of chemical processing and design, edited by J.J.
McKetta and W.A. Cunningham"-T.p. verso.
  Includes bibliographical refernces and index.
  ISBN 0-8247-8681-5 (alk. paper)
  1. Petroleum--Refining--Handbooks, manuals, etc.   I. McKetta,
John J.   II. Encyclopedia of chemical processing and design.
TP690.P4723   1992
665.5'3--dc20   92-4374
                                                                                            CIP

The contents of this volume were originally published in *Encyclopedia of Chemical Processing and Design*, edited by J. J. McKetta and W. A. Cunningham. © 1979, 1981, 1982, 1987, 1988, 1990 by Marcel Dekker, Inc.

This book is printed on acid-free paper.

Copyright © 1992 by Marcel Dekker, Inc. All Rights Reserved.

Neither this book nor any part may be reproduced or transmitted in any form or by any means, electronic or mechanical, including photocopying, microfilming, and recording, or by any information storage and retrieval system, without permission in writing from the publisher.

Marcel Dekker, Inc.
270 Madison Avenue, New York, New York 10016

Current printing (last digit):
10 9 8 7 6 5 4 3 2 1

PRINTED IN THE UNITED STATES OF AMERICA

Preface

It is time that many of the petroleum processes currently in use be presented in a well-organized, easy-to-read and understandable manner. This handbook fulfills this need by covering up-to-date processing operations. Each chapter is written by a world expert in that particular area, in such a manner that it is easily understood and applied. Each professional practicing engineer or industrial chemist involved in petroleum processing should have a copy of this book on his or her working shelf.

The handbook is conveniently divided into four sections: products, refining, manufacturing processes, and treating processes. Each of the processing chapters contain information on plant design as well as significant chemical reactions. Wherever possible, shortcut methods of calculations are included along with nomographic methods of solution. In the front of the book are two convenient sections that will be very helpful to the reader. These are (1) conversion to and from SI units, and (2) cost indexes that will enable the reader to update any cost information.

As Editor, I am grateful for all the help I have received from the great number of authors who have contributed to this book. I am also grateful to the huge number of readers who have written to me with suggestions of topics to be included.

JOHN J. McKETTA

Contents

Preface	iii
Contributors	vii
Conversion to SI Units	xi
Bringing Costs up to Date	xiii

1 Products

Petroleum Products	2
Harold L. Hoffman	
Petroleum Products, Production Costs	13
Fabio Bernasconi	
Octane Boosting	25
John J. Lipinski and Jack R. Wilcox	
Octane Catalysts	31
John S. Magee, Bruce R. Mitchell, and James W. Moore	
Octane Options	50
Joseph A. Weiszmann, James H. D'Auria, Frederick G. McWilliams, and Frederick M. Hibbs	

2 Refining

Petroleum Processing	67
Harold L. Hoffman and John J. McKetta	
Petroleum Refinery of the Future	108
D. B. Bartholic, A. M. Center, Brian R. Christian, and A. J. Suchanek	
Petroleum Processes, Catalyst Usage	130
Richard A. Corbett	
Petroleum Processing Economics, Catalysts	155
Mattheus M. van Kessel, R. H. van Dongen, and G. M. A. Chevalier	
Petroleum Refinery Yields Improvement	170
Dale R. Simbeck and Frank E. Biasca	
Hazardous Waste Regulations	179
David Olschewsky and Alice Megna	
Petroleum Waste Toxicity, Prevention	190
Raymond C. Loehr	
Petroleum Refining Processes, United States Capacities	199
Debra A. Gwyn	
Petroleum Refining Processes, Worldwide Capacities	214
Debra A. Gwyn	

3 Manufacturing Processes

Coking, Petroleum (Delayed and Fluid) *J. D. McKinney*	245
Coking, Petroleum (Fluid) *D. E. Blaser*	253
Cracking, Thermal *W. P. Ballard, G. I. Cottington, and T. A. Cooper*	281
Cracking, Catalytic *E. C. Luckenbach, A. C. Worley, A. D. Reichle, and E. M. Gladrow*	349
Heavy Oil Cracking *Guy E. Weismantel*	480
Cracking, Catalytic, Optimization and Control *J. A. Feldman, B. E. Lutter, and R. L. Hair*	516
Deasphalting *Carl Pei-Chi Chang and James R. Murphy*	527
Dehydrogenation *Hervey H. Voge*	544
Dewaxing, Catalytic *J. D. Hargrove*	558
Dewaxing, Solvent *G. G. Scholten*	565
Dewaxing, Urea *G. G. Scholten*	583
Hydrocracking *Guy E. Weismantel*	592
Lubricating Oils: Manufacturing Processes *Avilino Sequeira, Jr.*	635

4 Treating Processes

Desalting, Crude Oil *Donald R. Burris*	666
Demetallization/Desulfurization of High Metal Content Petroleum Feedstocks *Richard A. Baussell, John Caspers, Kenneth E. Hastings, John D. Potts, and Roger P. Van Driesen*	677
Desulfurization, Liquids, Petroleum Fractions *Robert J. Campagna, James A. Frayer, and Raynor T. Sebulsky*	697
Desulfurizing Cracked Gasoline and Other Hydrocarbon Liquids by Caustic Soda Treating *K. E. Clonts and Ralph E. Maple*	727
Doctor Sweetening *Kenneth M. Brown*	736

Index   759

Contributors

W. P. Ballard Manager, Port Arthur Research Laboratories (Retired), Texaco, Inc., Port Arthur, Texas

D. B. Bartholic Engelhard Corporation, Specialty Chemicals Division, Menlo Park, Edison, New Jersey

Richard A. Bausell Safety Services Manager, Cities Service Research and Development Company, Tulsa, Oklahoma

Fabio Bernasconi, Ph.D. Ambrosetti Group, Milan, Italy

Frank E. Biasca Manager, Process Technology, SFA Pacific, Inc., Mountain View, California

D. E. Blaser Engineering Associate, Exxon Engineering Petroleum Department, Exxon Research and Engineering Company, Florham Park, New Jersey

Kenneth M. Brown Director, Treating Services (Retired), UOP Process Division, Des Plaines, Illinois

Donald R. Burris Manager, Technical Advisory Division, C-E Natco Combustion Engineering, Inc., Denver, Colorado

Robert J. Campagna Gulf Science and Technology Company, Pittsburgh, Pennsylvania

John Caspers Manager, LC-Fining Design, C-E Lummus Company, Bloomfield, New Jersey

A. M. Center Engelhard Corporation, Specialty Chemicals Division, Menlo Park, Edison, New Jersey

Carl Pei-Chi Chang Process Manager, Refinery Process Division, Pullman Kellogg, Houston, Texas

G. M. A. Chevalier Shell Internationale Petroleum, Maatschappij BV, The Hague, The Netherlands

Brian R. Christian Engelhard Corporation, Specialty Chemicals Division, Menlo Park, Edison, New Jersey

K. E. Clonts Vice President, Technical, Merichem Company, Houston, Texas

T. A. Cooper Staff Coordinator-Strategic Planning, Texaco, Inc., White Plains, New York

Richard A. Corbett, P.E. Refining/Petrochemical Editor, *Oil & Gas Journal*, Houston, Texas

G. I. Cottington Technologist, Port Arthur Research Laboratories, Texaco, Inc., Port Arthur, Texas

James H. D'Auria Director, Process Development, UOP Inc., Des Plaines, Illinois

J. A. Feldman Senior Process Analysis Engineer (Retired), Applied Automation, Inc., Bartlesville, Oklahoma

James A. Frayer Technical Consultant, Gulf Science and Technology Company, Pittsburgh, Pennsylvania

E. M. Gladrow Senior Research Associate, Exxon Research and Development Laboratories, Baton Rouge, Louisiana

Debra A. Gwyn Director of Editorial Surveys, *Oil & Gas Journal*, Tulsa, Oklahoma

R. L. Hair Information Technology Planner, Phillips Petroleum Company, Bartlesville, Oklahoma

J. D. Hargrove The British Petroleum Company Limited, Sunbury-on-Thames, Middlesex, England

Kenneth E. Hastings Vice President and Director of Research, Cities Service Research and Development Company, Tulsa, Oklahoma

Frederick M. Hibbs UOP Inc., Des Plaines, Illinois

Harold L. Hoffman Editor, *Hydrocarbon Processing*, Houston, Texas

John J. Lipinski Coastal Eagle Point Oil Company, Westville, New Jersey

Raymond C. Loehr, Ph.D. H. M. Acharty Centennial Chair and Professor, Environmental Engineering Program, University of Texas at Austin, Austin, Texas

E. C. Luckenbach E. & R. Luckenbach and Co., Mountainside, New Jersey

B. E. Lutter Engineering Director, Automation Group, Applied Automation/Hartman and Braun, Bartlesville, Oklahoma

John S. Magee, Ph.D. Technical Director, Katalistiks International, a unit of UOP, Inc., Baltimore, Maryland

Ralph E. Maple Assistant General Manager, Process Technology Division, Merichem Company, Houston, Texas

John J. McKetta, Ph.D., P.E. The Joe C. Walter Professor of Chemical Engineering, The University of Texas at Austin, Austin, Texas

J. D. McKinney Gulf Research and Development Company, Pittsburgh, Pennsylvania

Frederick G. McWilliams UOP Inc., Des Plaines, Illinois

Alice Megna Project Manager, ERT Inc., Dallas, Texas

Contributors

Bruce R. Mitchell (deceased)   Katalistiks International, a unit of UOP, Inc., Baltimore, Maryland

James W. Moore   Senior Research Supervisor, Katalistiks International, a unit of UOP, Inc., Baltimore, Maryland

James R. Murphy   Pullman Kellogg, Houston, Texas

David Olschewsky   Project Manager, ERT Inc., Dallas, Texas

John D. Potts   Manager of Research Staff, Cities Service Research and Development Company, Tulsa, Oklahoma

A. D. Reichle   Engineering Advisor, Exxon Research and Development Laboratories, Baton Rouge, Louisiana

G. G. Scholten   Managing Director, Edeleanu GmbH, Frankfurt am Main, Germany

Raynor T. Sebulsky   General Manager-Products, Refining & Products Division, Gulf Science and Technology Company, Pittsburgh, Pennsylvania

Avilino Sequeira, Jr., P. E.   Senior Technologies, Texaco, Inc., Port Arthur, Texas

Dale R. Simbeck   Vice President Technology, SFA Pacific, Inc., Mountain View, California

A. J. Suchanek   Engelhard Corporation, Specialty Chemicals Division, Menlo Park, Edison, New Jersey

R. H. van Dongen   Shell Internationale Petroleum, Maatschappij BV, The Hague, The Netherlands

Roger P. Van Driesen   Manager, Petroleum and Coal Process Marketing, C-E Lummus Company, Bloomfield, New Jersey

Mattheus M. van Kessel   Product Manager, Refinery Catalysts, SICC, London, United Kingdom

Hervey H. Voge (deceased)   Sebastopol, California

Guy E. Weismantel   President, Weismantel International, Kingwood, Texas

Joseph A. Weiszmann   Marketing Manager, Western U.S., UOP, Inc., Des Plaines, Illinois

Jack R. Wilcox   Harshaw/Filtrol Partnership, Los Angeles, California

A. C. Worley   Senior Engineering Associate, Exxon Research and Engineering Company, Florham Park, New Jersey

Conversion to SI Units

To convert from	To	Multiply by
acre	square meter (m²)	4.046×10^3
angstrom	meter (m)	1.0×10^{-10}
are	square meter (m²)	1.0×10^2
atmosphere	newton/square meter (N/m²)	1.013×10^5
bar	newton/square meter (N/m²)	1.0×10^5
barrel (42 gallon)	cubic meter (m³)	0.159
Btu (International Steam Table)	joule (J)	1.055×10^3
Btu (mean)	joule (J)	1.056×10^3
Btu (thermochemical)	joule (J)	1.054×10^3
bushel	cubic meter (m³)	3.52×10^{-2}
calorie (International Steam Table)	joule (J)	4.187
calorie (mean)	joule (J)	4.190
calorie (thermochemical)	joule (J)	4.184
centimeter of mercury	newton/square meter (N/m²)	1.333×10^3
centimeter of water	newton/square meter (N/m²)	98.06
cubit	meter (m)	0.457
degree (angle)	radian (rad)	1.745×10^{-2}
denier (international)	kilogram/meter (kg/m)	1.0×10^{-7}
dram (avoirdupois)	kilogram (kg)	1.772×10^{-3}
dram (troy)	kilogram (kg)	3.888×10^{-3}
dram (U.S. fluid)	cubic meter (m³)	3.697×10^{-6}
dyne	newton (N)	1.0×10^{-5}
electron volt	joule (J)	1.60×10^{-19}
erg	joule (J)	1.0×10^{-7}
fluid ounce (U.S.)	cubic meter (m³)	2.96×10^{-5}
foot	meter (m)	0.305
furlong	meter (m)	2.01×10^2
gallon (U.S. dry)	cubic meter (m³)	4.404×10^{-3}
gallon (U.S. liquid)	cubic meter (m³)	3.785×10^{-3}
gill (U.S.)	cubic meter (m³)	1.183×10^{-4}
grain	kilogram (kg)	6.48×10^{-5}
gram	kilogram (kg)	1.0×10^{-3}
horsepower	watt (W)	7.457×10^2
horsepower (boiler)	watt (W)	9.81×10^3
horsepower (electric)	watt (W)	7.46×10^2
hundred weight (long)	kilogram (kg)	50.80
hundred weight (short)	kilogram (kg)	45.36
inch	meter (m)	2.54×10^{-2}
inch mercury	newton/square meter (N/m²)	3.386×10^3
inch water	newton/square meter (N/m²)	2.49×10^2
kilogram force	newton (N)	9.806

Conversion to SI Units

To convert from	To	Multiply by
kip	newton (N)	4.45×10^3
knot (international)	meter/second (m/s)	0.5144
league (British nautical)	meter (m)	5.559×10^3
league (statute)	meter (m)	4.83×10^3
light year	meter (m)	9.46×10^{15}
liter	cubic meter (m³)	0.001
micron	meter (m)	1.0×10^{-6}
mil	meter (m)	2.54×10^{-6}
mile (U.S. nautical)	meter (m)	1.852×10^3
mile (U.S. statute)	meter (m)	1.609×10^3
millibar	newton/square meter (N/m²)	100.0
millimeter mercury	newton/square meter (N/m²)	1.333×10^2
oersted	ampere/meter (A/m)	79.58
ounce force (avoirdupois)	newton (N)	0.278
ounce mass (avoirdupois)	kilogram (kg)	2.835×10^{-2}
ounce mass (troy)	kilogram (kg)	3.11×10^{-2}
ounce (U.S. fluid)	cubic meter (m³)	2.96×10^{-5}
pascal	newton/square meter (N/m²)	1.0
peck (U.S.)	cubic meter (m³)	8.81×10^{-3}
pennyweight	kilogram (kg)	1.555×10^{-3}
pint (U.S. dry)	cubic meter (m³)	5.506×10^{-4}
pint (U.S. liquid)	cubic meter (m³)	4.732×10^{-4}
poise	newton second/square meter (N · s/m²)	0.10
pound force (avoirdupois)	newton (N)	4.448
pound mass (avoirdupois)	kilogram (kg)	0.4536
pound mass (troy)	kilogram (kg)	0.373
poundal	newton (N)	0.138
quart (U.S. dry)	cubic meter (m³)	1.10×10^{-3}
quart (U.S. liquid)	cubic meter (m³)	9.46×10^{-4}
rod	meter (m)	5.03
roentgen	coulomb/kilogram (c/kg)	2.579×10^{-4}
second (angle)	radian (rad)	4.85×10^{-6}
section	square meter (m²)	2.59×10^6
slug	kilogram (kg)	14.59
span	meter (m)	0.229
stoke	square meter/second (m²/s)	1.0×10^{-4}
ton (long)	kilogram (kg)	1.016×10^3
ton (metric)	kilogram (kg)	1.0×10^3
ton (short, 2000 pounds)	kilogram (kg)	9.072×10^2
torr	newton/square meter (N/m²)	1.333×10^2
yard	meter (m)	0.914

Bringing Costs up to Date

Cost escalation via inflation bears critically on estimates of plant costs. Historical costs of process plants are updated by means of an escalation factor. Several published cost indexes are widely used in the chemical process industries:

Nelson Cost Indexes (*Oil and Gas J.*), quarterly
Marshall and Swift (M&S) Equipment Cost Index, updated monthly
CE Plant Cost Index (*Chemical Engineering*), updated monthly
ENR Construction Cost Index (*Engineering News-Record*), updated weekly

All of these indexes were developed with various elements such as material availability and labor productivity taken into account. However, the proportion allotted to each element differs with each index. The differences in overall results of each index are due to uneven price changes for each element. In other words, the total escalation derived by each index will vary because different bases are used. The engineer should become familiar with each index and its limitations before using it.

Table 1 compares the CE Plant Index with the M&S Equipment Cost

TABLE 1 *Chemical Engineering* and Marshall and Swift Plant and Equipment Cost Indexes since 1950

Year	CE Index	M&S Index	Year	CE Index	M&S Index
1950	73.9	167.9	1971	132.3	321.3
1951	80.4	180.3	1972	137.2	332.0
1952	81.3	180.5	1973	144.1	344.1
1953	84.7	182.5	1974	165.4	398.4
1954	86.1	184.6	1975	182.4	444.3
1955	88.3	190.6	1976	192.1	472.1
1956	93.9	208.8	1977	204.1	505.4
1957	98.5	225.1	1978	218.8	545.3
1958	99.7	229.2	1979	238.7	599.4
1959	101.8	234.5	1980	261.2	659.6
1960	102.0	237.7	1981	297.0	721.3
1961	101.5	237.2	1982	314.0	745.6
1962	102.0	238.5	1983	316.9	760.8
1963	102.4	239.2	1984	322.7	780.4
1964	103.3	241.8	1985	325.3	789.6
1965	104.2	244.9	1986	318.4	797.6
1966	107.2	252.5	1987	323.8	813.6
1967	109.7	262.9	1988	342.5	852.0
1968	113.6	273.1	1989	355.4	895.1
1969	119.0	285.0	1990	357.6	915.1
1970	125.7	303.3			

TABLE 2 Nelson Inflation Refinery Construction Indexes since 1946 (1946 = 100)

Date	Materials Component	Labor Component	Miscellaneous Equipment	Nelson Inflation Index
1946	100.0	100.0	100.0	100.0
1947	122.4	113.5	114.2	117.0
1948	139.5	128.0	122.1	132.5
1949	143.6	137.1	121.6	139.7
1950	149.5	144.0	126.2	146.2
1951	164.0	152.5	145.0	157.2
1952	164.3	163.1	153.1	163.6
1953	172.4	174.2	158.8	173.5
1954	174.6	183.3	160.7	179.8
1955	176.1	189.6	161.5	184.2
1956	190.4	198.2	180.5	195.3
1957	201.9	208.6	192.1	205.9
1958	204.1	220.4	192.4	213.9
1959	207.8	231.6	196.1	222.1
1960	207.6	241.9	200.0	228.1
1961	207.7	249.4	199.5	232.7
1962	205.9	258.8	198.8	237.6
1963	206.3	268.4	201.4	243.6
1964	209.6	280.5	206.8	252.1
1965	212.0	294.4	211.6	261.4
1966	216.2	310.9	220.9	273.0
1967	219.7	331.3	226.1	286.7
1968	224.1	357.4	228.8	304.1
1969	234.9	391.8	239.3	329.0
1970	250.5	441.1	254.3	364.9
1971	265.2	499.9	268.7	406.0
1972	277.8	545.6	278.0	438.5
1973	292.3	585.2	291.4	468.0
1974	373.3	623.6	361.8	522.7
1975	421.0	678.5	415.9	575.5
1976	445.2	729.4	423.8	615.7
1977	471.3	774.1	438.2	653.0
1978	516.7	824.1	474.1	701.1
1979	573.1	879.0	515.4	756.6
1980	629.2	951.9	578.1	822.8
1981	693.2	1044.2	647.9	903.8
1982	707.6	1154.2	622.8	976.9
1983	712.4	1234.8	656.8	1025.8
1984	735.3	1278.1	665.6	1061.0
1985	739.6	1297.6	673.4	1074.4
1986	730.0	1330.0	684.4	1089.9
1987	748.9	1370.0	703.1	1121.5
1988	802.8	1405.6	732.5	1164.5
1989	829.2	1440.4	769.9	1195.9
1990	832.8	1487.7	795.5	1225.7

Bringing Costs up to Date

Index. Table 2 shows the Nelson Inflation Petroleum Refinery Construction Indexes since 1946. It is recommeded that the CE Index be used for updating total plant costs, and the M&S Index or Nelson Index for updating equipment costs. The Nelson Indexes are better suited for petroleum refinery materials, labor, equipment, and general refinery inflation.

Since

$$C_B = C_A(B/A)^n \qquad (1)$$

Here, A = the size of units for which the cost is known, expressed in terms of capacity, throughput, or volume; B = the size of unit for which a cost is required, expressed in the units of A; $n = 0.6$ (i.e., the six-tenths exponent); C_A = actual cost of unit A; and C_B = the cost for B being sought for the same time period as cost C_A.

To approximate a current cost, multiply the old cost by the ratio of the current index value to the index at the date of the old cost:

$$C_B = C_A I_B / I_A \qquad (2)$$

Here, C_A = old cost; I_B = current index value; and I_A = index value at the date of old cost.

Combining Eqs. (1) and (2):

$$C_B = C_A(B/A)^n (I_B/I_A) \qquad (3)$$

For example, if the total investment cost of Plant A was \$25,000,000 for 200-million-lb/yr capacity in 1974, find the cost of Plant B at a throughput of 300 million lb/yr on the same basis for 1986. Let the sizing exponent, n, be equal to 0.6.

From Table 1, the CE Index for 1986 was 318.4, and for 1974 it was 165.4.
Via Eq. (3):

$$C_B = C_A(B/A)^n (I_B/I_A)$$

$$= 25.0(300/200)^{0.6}(318.4/165.4)$$

$$= \$61,200,000$$

JOHN J. McKETTA

1
Products

Petroleum Products

Petroleum products are made from petroleum crude oil and natural gas. Similar products are made from other natural resources such as coal, peat, lignite, shale oil, and tar sands. Products from these other sources are frequently called "synthetic," even though their properties can be indistinguishable from crude oil derived products. Here the term "synthetic" is intended to denote the products came from a raw material other than the more common sources, crude oil or natural gas.

A list of the principal classes of products made from petroleum crude oil is given in Table 1. As an example of the relative product volume for each class, the average percentages are for United States crude oil refiners typical of the mid-1980s.

Fuels are the major class. Common uses for these products are: to burn in furnaces to supply heat, to aspirate into internal combustion engines to supply mechanical power, or to inject into jet engines to create thrust. In some cases the fuel is a gas, like natural gas or the lighter hydrocarbons from crude oil. In other cases the fuel is a clear or very pale orange tinted liquid—often with dyes added for product identity. And in still other cases the fuel is a heavy, dark liquid or semisolid, unable to flow until heated.

Building materials are also among petroleum products. For example, petroleum asphalt is used for roofing and road coverings. Petroleum waxes are used for waterproofing. After special chemical transformations, some petroleum fractions supply a wide range of plastics, elastomers, and other resins for construction uses.

Chemicals derived from petroleum are identified in Table 1 as simply "petrochemical feeds." The term "petrochemicals" was coined in an attempt to retain the identity of some chemicals as coming from petroleum. However, most manufacturing statistics do not use this distinction. So petrochemical production is often combined with chemicals derived from other sources within a single chemical class.

Take note that a highly industrialized economy, like that of the United States, diverts no more then about 7% of all petroleum products (feedstocks plus fuels) to the manufacture of petrochemicals. Yet these petrochemicals have a great variety of uses as shown by the partial listing of Table 2.

World Consumption

The trend in petroleum product usage is indicated by the growth in crude oil consumption. Table 3 gives world crude oil consumption in millions of barrels per day. The distribution among various areas reflect the high consumption within industrialized areas like North America (USA and Canada), Western Europe, and the USSR.

TABLE 1 Product Yields from U.S. Refineries, Mid-1980s Basis[a]

Product	Vol.%
Still gas	4.9
Liquefied gas	3.2
Gasoline, motor	45.8
Gasoline, aviation	0.2
Jet fuel	9.8
Kerosene	0.7
Special naphtha	0.4
Petrochemical feeds	3.1
Distillates	21.2
Lubricants	1.2
Waxes	0.1
Coke	3.8
Asphalt/road oil	3.1
Residuals	6.7
Miscellaneous	0.5
Total	104.7[b]

[a]Source: U.S. Energy Information Administration, *Petroleum Supply Annual 1986*, DOE/EIA-0340(86)1, published May 1987.
[b]100 wt.%. Volume gain because most products are lighter than original feed.

TABLE 2 Partial List of Petrochemical Uses

Absorbents	De-emulsifiers	Hair conditioners	Pipe
Activators	Desiccants	Heat transfer fluids	Plasticizers
Adhesives	Detergents	Herbicides	Preservatives
Adsorbents	Drugs	Hoses	Refrigerants
Analgesics	Drying oils	Humectants	Resins
Anesthetics	Dyes	Inks	Rigid foams
Antifreezes	Elastomers	Insecticides	Rust inhibitors
Antiknocks	Emulsifiers	Insulations	Safety glass
Beltings	Explosives	Lacquers	Scavengers
Biocides	Fertilizers	Laxatives	Stabilizers
Bleaches	Fibers	Odorants	Soldering flux
Catalysts	Films	Oxidation inhibitors	Solvents
Chelating agents	Finish removers	Packagings	Surfactants
Cleaners	Fire-proofers	Paints	Sweeteners
Coatings	Flavors	Paper sizings	Synthetic rubber
Containers	Food supplements	Perfumes	Textile sizings
Corrosion inhibitors	Fumigants	Pesticides	Tire cord
Cosmetics	Fungacides	Pharmaceuticals	
Cushions	Gaskets	Photographic chemicals	

TABLE 3 Crude Oil Consumption[a]

Area	Millions of barrels per day[b]			
	1970	1975	1980	1985
USA and Canada	15.9	17.6	18.3	16.7
Other Western Hemisphere	2.6	3.5	4.4	4.4
Western Europe	12.5	13.2	13.6	11.9
USSR	5.3	7.5	8.8	8.9
China	0.6	1.4	1.8	1.8
Other CPE countries	1.5	2.2	2.7	2.5
Africa	0.9	1.1	1.5	1.7
Asia and Middle East	2.6	3.5	4.8	5.6
Japan	4.0	5.0	4.9	4.4
Australasia	0.6	0.7	0.7	0.7
Total	46.5	55.7	61.5	58.6

[a]Source: *BP Statistical Review of World Energy*, issued annually.
[b]Barrel = 42 US gallons.

The drop between 1980 and 1985 is the result of a large increase in crude oil price set by oil producing countries. A fourfold price increase of Middle Eastern oil occurred in 1973. Other increases followed. By 1982, the price increase for the period was 12-fold. Because of the resulting increase in fuel price, many conservation measures were taken—especially with regard to fuels used in consuming countries. Later, a drop in crude oil price failed to return oil consumption to its earlier highs. By mid-1987, oil prices were about half of their earlier peak. Then consumption again began to increase, although at a much reduced pace—forecasted at about 2% annually.

Product Identity

Petroleum products are hydrocarbons—compounds with various combinations of hydrogen and carbon. Because there is an almost inconceivable number of hydrogen-carbon combinations, petroleum products take many forms, limited only by the imagination and ingenuity of the people who work with them. Many of the combinations exist naturally in the original raw materials. Other combinations are created by an ever-growing number of commercial processes for altering one combination to another (see *Petroleum Processing*). Each combination has its own unique set of chemical and physical properties. As a consequence, petroleum products are found in a wide variety of industrial and consumer products.

Many of these products are substitutes for earlier products from non-petroleum sources. For example: illuminating oil to replace sperm oil from whales; synthetic rubber to replace natural rubber from trees; man-made fibers to replace textiles from animals and vegetation. Each new use often imposes additional specifications on the new product. Also, product specifications tend to evolve to stay abreast of advances in both product application and manufacturing methods.

In earlier times there were two significant product specifications: density and boiling range. From these two physical properties, most other properties—both physical and chemical—were implied. Even today, with many sophisticated analytical tests available, these two specifications of density and boiling range are retained.

Density is determined relative to water at 60°F. But instead of using the units of specific density or specific gravity, the common unit of measure is one specified by the American Petroleum Institute. For this reason, the results are called degrees API gravity. The relation of this term to specific gravity is

$$°API = (141.5/\text{sp gr}) - 131.5$$

Thus, a petroleum product with the same specific gravity as water, 1.0, has an API gravity of 10. Products with densities less than water have API gravities larger than 10. For example, automotive gasoline generally has an API gravity of between 50 and 70, with the winter grades slightly lighter (greater API gravity) than the summer grades.

Boiling range of a petroleum product is reported in several ways. If the product were a single pure hydrocarbon, it would have a single boiling point. But most petroleum products are groups of hydrocarbons—each with its own normal boiling point as well as an influence on the vaporizing tendencies of neighboring hydrocarbons. The apparatus usually used for measuring boiling range is constructed according to standards specified by the American Society for Testing and Materials. The results are then called ASTM distillation temperatures. Some common terms used with boiling range are as follows:

Initial (IBP)—the temperature at which the first drop of condensate is formed from vaporizing a sample.
Percent distilled—temperatures associated with the recovery of various quantities of condensation from a vaporizing sample; e.g., a 10% ASTM temperature.
End (EP)—the highest temperature reached by the vapor during a distillation test.
Volatility—a term sometimes related to the distillation test; e.g., reported as volume % vaporized at specific temperatures. Volatility is used in a general way at other times to denote a product's overall vaporizing characteristics.

Characterization factor is a term that combines both density and boiling range. A popular term is the Watson characterization factor defined as follows:

$$\text{Watson } K = (T_B)^{1/3}/(\text{sp gr})$$

where T_B is the average of five temperatures (10, 30, 50, 70, and 90% vaporized) in degrees Rankin, and sp gr is specific gravity compared to water at 60°F.

To show how the characterization factor is related to chemical composition, consider several variations of six-carbon hydrocarbons. The paraffinic hydrocarbon hexane, C_6H_{14}, with its boiling point of 155.7°F (615.7°R) and its specific gravity of 0.664, has a Watson K factor of 12.8. The two isoparaffins are 12.8 and 12.6.

At the other end of the scale, the six-carbon aromatic hydrocarbon benzene, C_6H_6, has a boiling point of 176.2°F (636.2°R) and a specific gravity of 0.884, giving a Watson K factor of 9.7. Cyclohexane is 10.6 and five variations of monoolefins are in the 12.3–12.5 range.

While much better ways now exist for chemical analysis of petroleum products, the characterization factor is still an important criterion for buying and selling crude oil raw material.

Discretionary specifications exist for most petroleum products. Product specifications set minimum and maximum boundaries on a product's properties. At the discretion of the manufacturer, a product may be made to excell in one property or another, thereby commanding a higher price in a competitive market. In the sections to follow, the more popular fuel products will be described and their discretionary specifications identified.

Standards for fuel specifications in the United States are set by the American Society for Testing and Materials. This group has many committees dealing with various products. Committee D is concerned with fuels and related products. Specific specifications are numbered with a suffix denoting the year when a specification was updated. For example, the specifications for automotive gasoline are contained in the standard ASTM D 439-79.

Other countries have similar standard-setting organizations. In West Germany, it is Deutsches Institute fuer Normung, and the specifications are identified with DIN numbers. In the United Kingdom, it is the Institute of Petroleum which uses IP numbers, and in Japan, the Ministry of International Trade and Industry uses MITI numbers. Most of these groups cross-reference each other's numbers for easy comparison.

Gaseous Products

Fuels with four or less carbons in the hydrogen-carbon combination have boiling points less than normal room temperature. Therefore, these products are normally gases. Common classifications for these products are as follows.

Natural gas is methane denoted by the chemical structure CH_4, the lightest and least complex of all hydrocarbons. Yet, natural gas from an underground reservoir, when brought to the surface, can contain other heavier hydrocarbon vapors. Such a mixture is called a "wet" gas. Wet gas is usually processed to remove the entrained hydrocarbons heavier than methane. When isolated, the heavier hydrocarbons sometime liquefy and are called natural gas condensate.

Still gas is a broad classification for light hydrocarbon mixtures. "Still" is an abbreviation for distillation. Still gas is the lightest fraction created when crude oil is processed. If the distillation unit is separating light hydrocarbon fractions, the still gas will be almost entirely methane (C_1) with only traces of ethane and ethylene (C_2's). If the distillation unit is handling heavier fractions, the still gas might also contain propanes (C_3's) and butanes (C_4's).

Fuel gas and still gas are terms often used interchangably. Yet fuel gas is intended to denote the product's destination—to be used as a fuel for boilers, furnaces, or heaters.

LPG is an abbreviation for liquefied petroleum gas. It is composed of propane (C_3) and butane (C_4). LPG is stored under pressure in order to keep these hydrocarbons liquefied at normal atmospheric temperatures. Before LPG is burned, it passes through a pressure relief valve. The reduction in pressure causes the LPG to vaporize (gasify). Winter-grade LPG is mostly propane, the lighter of the two gases and easier to vaporize at lower temperatures. Summer-grade LPG is mostly butane. The better grades of LPG strive for reduced content of unsaturated hydrocarbons (propylene and butylene) because these hydrocarbons do not burn as cleanly as do saturated hydrocarbons.

Specifications for LPG are given in Table 4. Note the common use of metric units. While temperature conversion is fairly common, these other conversion factors might be helpful:

Pressure in kilopascals multiplied by 7.5 gives millimeters of mercury.
Heat in joules multiplied by 0.24 gives calories.

GASOLINE, or motor fuel, is intended for most spark-ignition engines such as those used in passenger cars, light duty trucks, tractors, motorboats, and engine-driven implements. Gasoline is a mixture of hydrocarbons with

TABLE 4 Specifications for Liquefied Petroleum Gases (ASTM D 1835-76)

Property	Special-Duty Propane	Commercial Propane	Commercial Butane	Propane–Butane Mixture
Distillation, 95% point, max, °C	−38.3	−38.3	2.2	2.2
Vapor pressure at 37.8°C, max, kPa	1430	1430	485	
Propylene, max, vol.%	5.0			
Butane and heavier, max, vol.%	2.5	2.5		
Pentane and heavier, max, vol.%			2.0	2.0
Sulfur at 15.6°C, 101 kPa, max, mg/m^3	229	343	343	343
Residue on evaporation 100 mL, max, mL	0.05	0.05	0.05	0.05
Oil stain observation	Pass	Pass	Pass	Pass
Corrosion, copper strip, max, no.		1		
Hydrogen sulfide content	Pass			
Moisture content	Pass	Pass		
Free-water content			None	None

boiling points in the approximate range of 100 to 400°F. Marketing specifications imposed on this fuel are intended to satisfy requirements of smooth and clean burning, easy ignition in cold weather, minimal evaporation in hot weather, and stability during long storage periods. These specifications are listed in Table 5.

Regular and premium grades are relative classifications for the octane numbers of gasolines. An octane number is a measure of gasoline's ability to resist spontaneous detonation. It is critical that detonation be at a precise time for a gasoline–air mixture in a spark-ignition engine. That time is determined by the electrical spark system. After ignition, the course of the detonation should progress smoothly, with a flame front moving across the combustion chamber.

If the fuel has a low octane number, the temperature and pressure wave caused by the spark-timed flame front can cause the remaining fuel–air mixture to ignite spontaneously. This secondary explosion causes an extra pressure pulse heard as knock. Then, more of the fuel's energy is lost as heat, and the engine delivers less motive power.

Octane numbers are measured in a single-cylinder laboratory engine. As in any spark-ignition engine, the combustion chamber of the laboratory engine is characterized by two volumes: the larger one determined when the piston is farthest from the cylinder head, the smaller one determined when the piston is closest to the head. The ratio of these two volumes is the engine's compression ratio. Engines with higher compression ratios require fuels with higher octane numbers if knocking is to be avoided.

TABLE 5 Specifications for Automotive Gasoline (ASTM D 439-79)

Property	Volatility Class				
	A	B	C	D	E
Octane number	No limit specified				
Distillation temperature, °C:					
10% evaporated, max	70	65	60	55	50
50% evaporated, min	77	77	77	77	77
50% evaporated, max	121	118	116	113	110
90% evaporated, max	190	190	185	185	185
End point, max	225	225	225	225	225
Temperature for vapor–liquid ratio of 20, min, °C	60	56	51	47	41
Vapor pressure, max, kPa	62	69	79	93	103
Lead content, max, g/L:					
Unleaded grade	0.013	0.013	0.013	0.013	0.013
Conventional grade	1.1	1.1	1.1	1.1	1.1
Corrosion, copper strip, max, no.	1	1	1	1	1
Gum, existent, max, mg/100 mL	5	5	5	5	5
Sulfur, max, wt.%:					
Unleaded grade	0.10	0.10	0.10	0.10	0.10
Conventional grade	0.15	0.15	0.15	0.15	0.15
Oxidation stability, min	240	240	240	240	240

For common gasoline engines the compression ratio is fixed. But the laboratory test engine has an adjustable head. Moving the head toward the piston increases the compression ratio and increases the tendency of a fuel to knock. To test a sample fuel, the compression ratio of the laboratory engine is set for track knock—determined by a bouncing-pin pressure gauge.

To relate the final engine conditions to an octane number, the knocking tendencies of two pure hydrocarbons are used as references. One of these hydrocarbons, isooctane (2,2,4-trimethyl pentane), is assigned an octane number of 100. The other, normal heptane, is assigned an octane number of zero. Mixtures of these reference hydrocarbons are assigned octane numbers equal to the volume percent of isooctane in the mixture. Then to complete the octane rating test, it is required to find the reference mixture that gives the same knock intensity in the test engine as that from the sample fuel. An exact match is not necessary, since the pressure gauge can be used to interpolate between near matches.

Research and Motor octane numbers identify other test engine variables—engine speed and intake air temperature. When the engine runs at 600 r/min and with 125°F intake air temperature, the rating is called a Research octane number, abbreviated RON or simply R. When the engine is operated at 900 r/min and with 300°F intake air temperature, the rating is a Motor octane number, MON or M. Both ratings are important to a multicylinder automobile engine because it usually operates over a wide range of conditions. For marketing purposes, a compromise is used: the arithmetic average of the Research and Motor octane numbers, abbreviated (R+M)/2.

Leaded and unleaded grades denote whether the gasoline mix includes lead additives. Lead compounds such as tetraethyl lead and tetramethyl lead are inexpensive additives to improve the octane rating of a gasoline mix. However, with the introduction of the catalytic muffler to automobile engines (to reduce exhaust emissions), unleaded gasolines were needed. Leaded fuel deactivated the catalyst in the muffler. The transition from an almost totally leaded gasoline market to an unleaded gasoline market is shown in Table 5.

Volatility is the third most popular marketing quality of a gasoline. Basically, volatility is a measure of a fuel's ability to vaporize. It attempts to combine vapor pressures and boiling points for the many components of a gasoline blend. Volatility is related to the following engine performance parameters: easy starting, quick warm-up, freedom from carburetor icing, rapid acceleration, freedom from vapor lock, good manifold distribution, and minimum crankcase dilution.

In short, volatility compromises two extreme properties: enough low boiling hydrocarbons to vaporize easily in cold weather and enough high boiling hydrocarbons to remain a liquid in an engine's fuel supply system during hotter periods. There must also be enough midboiling components to hold the mixture together. Several of the specifications in Table 5 relate to a fuel's volatility: distillation temperatures (used to determine volume percent vaporized for various conditions), temperature to achieve a 20 vapor–liquid ratio, and Reid vapor pressure. The five volatility classes (A through E) are tied to seasonal temperatures and locations by a matrix (not shown here) based on typical weather conditions.

TABLE 6 Specifications for Aviation Gasolines (ASTM D 910-79)

Property	Grade 80	Grade 100	Grade 100LL
Octane number, min:			
Lean	80	100	100
Rich	87	130	130
Performance number, min	87	130	130
Color	Red	Green	Blue
Tetraethyl lead, max, mL/L	0.13	1.06	0.53
Distillation temperature, °C:			
10% evaporated, max	75	75	75
40% evaporated, min	75	75	75
50% evaporated, max	105	105	105
90% evaporated, max	135	135	135
Final boiling point, max	170	170	170
Vapor pressure, max, kPa	48	48	48
Net heat of combustion, min, kJ/kg	43,520	43,520	43,520
Corrosion, copper strip, max, no.	1	1	1
Gum, potential (5 h), max, mg/100 mL	6	6	6
Lead precipitate, visible, max, mg/100 mL	3	3	3
Sulfur, max, wt.%	0.05	0.05	0.05
Freezing point, max, °C	−58	−58	−58
Water reaction, max, mL change	±2	±2	±2
Antioxidants, max, mg/L	12	12	12

Aviation fuels are of two types: gasolines and jet fuels. The specifications for aviation gasolines are given in Table 6 and have many of the same considerations as those for automotive gasolines. Among the added specifications for aviation gasoline are those for freezing point, water content, and storage stability.

Jet fuels are classified as "aviation turbine fuels," and their specifications are given in Table 7. In this case, ratings relative to octane number are replaced with properties concerned with the ability of the fuel to burn cleanly. These properties are discussed in the following section on diesel fuels.

DIESEL FUELS are distillates that are slightly heavier than gasoline. Diesel engines rely on compression-induced ignition. This means diesel fuel must self-ignite easily. In other words, a diesel fuel needs a property that is the opposite of the antiknock property of gasoline. That opposite property is called cetane rating. The term is derived from the reference fuel, normal cetane, which is easily ignited by compression with air and is the basis for comparing diesel fuel blends. This and other specifications for diesel fuels are given in Table 8. Grade 1-D is a lighter material suitable for winter or low temperature operation. Grade 2-D is suitable for warmer temperature operation. Both of these grades have low sulfur content. Grade 4-D is for special situations where the fuel's flow properties and sulfur content are not critical.

TABLE 7 Specifications for Aviation Turbine Fuels (ASTM D 1655-80a)

Properties	Jet A or A-1	Jet B
Density at 15°C, g/cm^3	0.7750–0.8394	0.7504–0.8013
Distillation temperature, °C:		
10% recovered, max	204.4	
20% recovered, max		143.3
50% recovered, max		187.8
90% recovered, max		243.3
Final boiling point, max	300	
Vapor pressure, max, kPa		20.7
Flash point, min, °C	37.8	
Freezing point, max, °C	−40 Jet A	−50
	−47 Jet A-1	
Viscosity at −20°C, max, mm^2/s (= cSt)	8	
Acidity, total, max, mg KOH/g	0.1	
Net heat of combustion, min, kJ/kg	42,780	42,780
Aromatic compounds, max, vol%	20	20
Sulfur, max, wt.%:		
Mercaptan	0.003	0.003
Total	0.3	0.3
Corrosion, copper strip after 2 h		
at 100°C, max, no.	1	1
Gum, existent, max, mg/100 mL	7	7
Water separation rating, max	2	2
Water interface rating, max	1	1

Sulfur has become one of the more important specifications for diesel fuels. The move is an attempt to further reduce sulfur in the atmosphere. The sulfur specification was not mentioned earlier when discussing gasoline, because gasoline seldom has significant amounts of sulfur. Usually the degree of processing for gasoline stock is so extensive that most sulfur compounds are removed incidental to the gasoline's other specifications, but diesel fuels and home heating oils are another matter.

TABLE 8 Specifications for Diesel Fuel Oils (ASTM 975-78)

	Grade		
Property	1-D	2-D	4-D
Distillation (90%) point, °C	288 max	282–338	
Flash point, min, °C	38	52	55
Water and sediment, max, vol.%	0.05	0.05	0.05
Carbon residue on 10% bottom, max, %	0.15	0.35	
Ash, max, wt.%	0.01	0.01	0.01
Viscosity at 40°C, kinematic, mm^2/s (= cSt)	1.3–2.4	1.9–4.1	5.5–24.0
Sulfur, max, wt.%	0.50	0.50	2.0
Corrosion, copper strip, max, no.	3	3	
Cetane number, min	40	40	30

TABLE 9 Specifications for Fuel Oils (ASTM D 396-79)

Property	No. 1	No. 2	No. 4 Light	No. 4	No. 5 Light	No. 5 Heavy	No. 6
Density at 15°C, max, g/cm^3	0.8495	0.8757	0.8757				
Distillation point, °C:							
10%, max	215						
90%, min		282					
90%, max	288	338					
Flash point, min, °C	38	38	38	55	55	55	60
Pour point, max, °C	−18	−6	−6	−6			
Water and sediment, max, vol.%	0.05	0.05	0.50	0.50	1.00	1.00	1.00
Carbon residue on 10% bottom, max, %	0.15	0.35					
Ash, max, wt.%				0.10	0.10	0.10	
Viscosity at 38°C, kinematic, mm^2/s (= cSt):							
min	1.4	2.0	2.0	5.8	26.4	65	
max	2.2	3.6	5.8	26.4	65	194	
Corrosion, copper strip, max, no.	3	3					
Sulfur, max, wt.%	0.5	0.5					

The sulfur is chemically combined with the hydrocarbons. There are many possible combinations, but chemical analyses show sulfur compounds are in greater concentration in the heavier fractions of crude oil. When a sulfur-containing fuel is burned, the sulfur compounds are decomposed and the sulfur portion becomes sulfur oxides; usually sulfur dioxide and sulfur trioxide.

Typically, sulfur content is determined by burning a sample of the fuel in a specified apparatus. The combustion products are absorbed in a solution which can later be analyzed for sulfur, usually with a solution that forms barium sulfate. Results are reported as weight percent sulfur.

An indicative test for corrosive sulfur is the copper strip test. In this case a brightly polished copper strip is immersed in a fuel sample. During the test, any sulfur in the sample will tarnish the copper. The degree of tarnish is measured against three standards designated 1, 2, and 3; with the larger numbers representing a greater amount of corrosive sulfur.

FUEL OILS are also called heating oils and are distillates that cover a broad range of properties. For this reason, their specifications are not so strict, as can be seen from the listing in Table 9. Some of their properties overlap specifications for the fuels mentioned earlier. Here are their grade definitions:

> *No. 1 fuel oil*—Very similar to kerosene or range oil (fuels used in stoves for cooking). This grade is defined as a distillate intended for vaporizing in pot-type burners and other burners where a clean flame is required.

No. 2 fuel oil—Often called domestic heating oil and having properties similar to diesel and heavier jet fuels. It is defined as a distillate for general purpose heating in which the burners do not require the fuel to be completely vaporized before burning.

No. 4 fuel oil—A light industrial heating oil. It is intended where preheating is not required for handling or burning. There are two grades of No. 4 oil, differing primarily in safety (flash) and flow (viscosity) properties.

No. 5 fuel oil—A heavy industrial oil. Preheating may be required for burning and, in cold climates, may be required for handling.

No. 6 fuel oil—A heavy residue oil sometimes called Bunker C when used to fuel ocean-going vessels. Preheating is required for both handling and burning this grade oil.

NONFUEL PRODUCTS account for about 10% of the materials made from petroleum crude oil. These include lubricants, waxes, asphalt, and petrochemical feedstocks. These topics are covered in this *Encyclopedia* in the Lubricating Oil articles, Asphalt articles, Petrochemical articles, and Wax articles.

HAROLD L. HOFFMAN

Petroleum Products, Production Costs

Production costs of refined petroleum products can be determined from a simple and accurate method that enables the user to calculate the break even value (BEV) to the refiner of all finished products from different grades of gasoline and middle distillate to specialties such as bitumens and lube basestocks.

The method, developed by the author, can also assess the effect on BEV of refinery configuration, mode of operation, and type of crude oil processes.

The BEV's are determined as a function of crude and residual fuel oil prices and are not influenced by the market price of the white products.

The ability to determine the production cost of a given product is a most valuable aid to the refiner, not just an academic exercise. Two important applications are "make or buy" decisions to meet needs for incremental amounts of product above the refinery baseload and the possibility of focusing refining operation on products that show the highest contribution to overall margins.

For these reasons the determination of the production costs of petroleum products is a subject which is attracting increasing interest. However, while it is relatively easy to calculate the overall cost of refining crude, it is not feasible to break this cost down by the various products produced. In fact, attempts to do so would be completely arbitrary.

On the other hand, by applying the incremental analysis technique to refining economics, it is possible to calculate the cost of producing an incremental amount of a specific product. In the refining industry this value is often assumed to be representative of the cost of production of that product, especially if in the calculation the "total" rather than the "incremental" yields and variable costs for the various processing units are taken into account.

It should, however, be noted that even with this latter adjustment, the value calculated remains intrinsically the "incremental" and not the "total" cost of production. In other words, for a given refinery the weighted average of the calculated production costs for the various components of the production slate does not necessarily coincide with the total cost of refining crude oil though, in general, it is not too different from this value.

Methodology

The assessment of incremental production costs is generally done via linear programming (LP). The increasing availability of computer facilities has led to the construction of very large and sophisticated LP models that, if used properly, enable the refiner to achieve accurate results and analyze different operational situations with little effort. The drawback of these models is that the inexperienced or even the average user quickly loses control, as the model complexity increases, of the factors which most influence the optimization process, and is, therefore, unable to interpret and criticize the results.

Misuse of the models in these circumstances can easily occur. Hence, frustration and distrust often accompany the use of these very powerful computer programs.

A way around this problem is to take advantage of simplified incremental analysis techniques (significantly more user friendly than the LP itself). They offer the advantage of clearly showing how the results are obtained. For general applications, the accuracy offered by these systems is quite satisfactory.

The methodology is based on a simplified version of the linear programming technique developed by the author, and is particularly suitable for analyzing refining operations. This methodology considers all the important factors that determine the value of the various products to the refiner, but uses only a limited number of basic steps to arrive at the final result.

While maintaining a good overall accuracy, it offers the main advantage of an easy understanding. Another important feature of the methodology is its flexibility. It can be applied to calculate the production costs of both major products and specialties reflecting all conceivable refinery configurations and mode of operations, and processing of any type of crude oil or alternative feedstock.

The methodology assesses the incremental variable cost of an oil product for a given refinery configuration by analyzing the effect on overall operations of producing a marginal amount of that product above the refinery base load. It is assumed that in this process the production of the other white products should remain unchanged, which on an incremental basis is possible provided that the refinery runs on "balanced" operation.

The only two degrees of freedom left to the system are the crude oil requirement and the residual fuel oil production. It is therefore possible to obtain BEV formulas that express the incremental value of a product or stream as a function of the crude oil and residual fuel oil prices:

$$P = cC - fF + V \qquad (1)$$

where P = product BEV, \$/metric ton
C = crude oil cost, \$/metric ton
F = residual fuel oil price, \$/metric ton
V = incremental variable costs, \$/metric ton of product
c = crude requirement (in weight) to produce one incremental ton of the product P
f = incremental residual fuel oil production (in weight) associated with production of one incremental ton of the product P

The price of residual fuel oil, in turn, depends on its sulfur level, and it is therefore important to calculate, for each formula, the sulfur level of the incremental fuel oil produced. The value of this incremental fuel oil can be obtained by linear interpolation between the market price of a high- and a low-sulfur grade of fuel oil.

The BEV formulas also take into account the energy requirement for producing a given product which, reflecting actual operations, is subtracted directly from the total residual fuel oil production.

The energy requirement can be estimated from the BEV formulas by subtracting the product and fuel oil coefficients from the crude coefficient. The BEV formulas also show the incremental variable costs (primarily catalyst, chemicals, and power requirements) associated with the production of a given product.

The BEV formulas express the variable incremental cost of production only. Fixed costs such as labor, overheads, and depreciation, which cannot be quantified by incremental analysis, have to be added in order to assess the full production costs.

It is noteworthy that the BEV formulas are based entirely on a material balance. Hence, their validity is not affected by the actual crude oil price.

Figure 1 shows the relative production cost (BEV) of the major petroleum products and specialties calculated for two refinery configurations. In order to explain more in detail the principle on which the proposed methodology works, the BEV calculation for a few specific applications is discussed in the following examples.

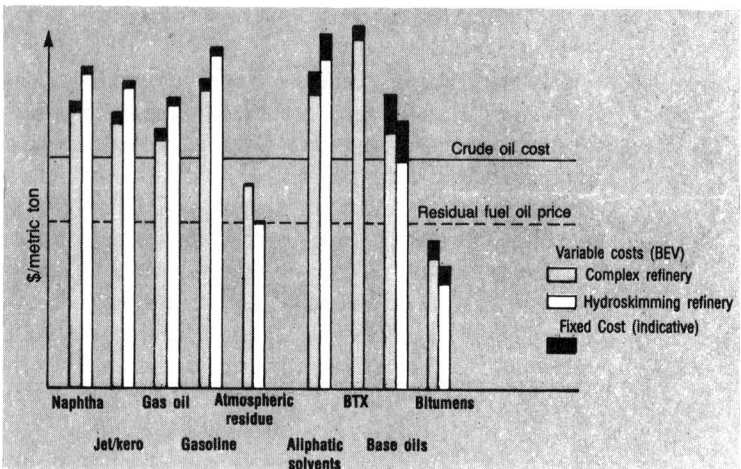

FIG. 1. Relative BEV of major oil products and specialties.

White Distillate

The term "white distillate" is applied to all the refinery streams with a distillation range between approximately 80 and 360°C (at atmospheric pressure), and with properties similar to the corresponding straight-run distillate from atmospheric crude distillation. Light distillate products, i.e., naphthas, kerosene, jet fuels, diesel fuels, and heating oils, are all manufactured by appropriate blending of white distillate streams.

The value to the refiner of these products is therefore directly related to the value of the white distillate. The only difference is some additional hydrotreating cost, if required.

In a refinery the incremental value of the various white distillate streams suitable for blending into a finished light distillate product is, on a volume basis, the same. It is in fact possible, on an incremental basis, to shift naphtha into middle distillate, or vice versa, by adjusting the cut points of the crude distillation tower.

There is also no difference, as far as incremental value is concerned, between a straight run and a cracked stream, as long as both can be blended in the same finished product without exceeding a quality specification constraint.

The determination of the BEV of the white distillate pool in a refinery is the basic step for determining the value of any refined product. All the white products, besides the distillate products, are in fact produced by further processing of specific white distillate streams.

The cost evaluation of "black" products is also significantly affected by the white distillate value. These latter evaluations involve a comparison with the alternative of blending a heavy residue stream in the residual fuel oil pool, the cost of which is dependent on the value of the cutter stock (a white distillate) required to make a marketable heavy fuel oil.

The data needed to derive the white distillate BEV formulas for two typical refinery configurations, i.e., hydroskimming and complex, are shown in Tables 1 and 2. In the complex configuration, vacuum distillation of the atmospheric residue is added to the hydroskimming base in order to recover a vacuum distillate stream which is then fed to a fluid catalytic cracker (FCC) with associated alkylation facilities. The vacuum residue undergoes a mild thermal upgrading in a visbreaker. In both cases the feedstock processed is a typical Middle East crude with gravity of 34°API.

The BEV calculation for the hydroskimming refinery is very simple, reflecting the simplicity of this configuration. It is based on two steps only. Crude is split into the relevant side streams in the crude atmospheric distillation tower, then a fuel oil balance is performed to calculate the amount of marketable residual fuel oil produced.

The total by row shows the incremental production of white distillate (light naphtha plus heavy naphtha plus middle distillate) and residual fuel oil obtainable by processing incrementally 100 tons of the given crude. It also shows the total variable operating cost involved with this operation.

The complex refinery case is conceptually more difficult. The atmospheric residue from distillation is sent to the vacuum unit where the vacuum distillate cut is recovered from the residue and then fed to the upgrading facility—the FCC plus alkylation block. The vacuum residue goes to the visbreaker.

TABLE 1 Incremental Formula Crude to White Distillate for Hydroskimming Refinery

	Atmospheric Distillation (metric tons)	Fuel Oil Balance (metric tons)	Total by Row (metric tons)
Crude, 34°API	(100.00)		(100.00)
Gases/LPG	1.32	(1.32)	
Light naphtha	5.24		5.24
Heavy naphtha	16.22		16.22
Middle distillate	30.62	(0.83)	29.79
			51.25 white distillate
Atmospheric residue	46.60	(46.60)	
Residual fuel oil		45.94	45.94
Refinery fuel:			
Direct operation	2.75	(2.75)	
Hydrotreating	0.23	(0.23)	
Variable operating costs, $/metric ton:			77
Direct operation	60		
Hydrotreating	17		
Residual fuel oil sulfur, wt.%		2.4	2.4

TABLE 2 Incremental Formula Crude to White Distillate for Complex Refinery

	Atmospheric Distillation (metric tons)	Vacuum Distillation (metric tons)	Visbreaking (metric tons)	FCC Alkylation (metric tons)	Naphtha Reforming (metric tons)	Fuel Oil Balance (metric tons)	Total by Row (metric tons)
Crude, 34° API	(100.00)						(100.00)
Gases/LPG	1.32		0.41	2.74	(2.24)	(2.23)	
Light naphtha	5.24		0.13				5.37
Heavy naphtha	16.22		0.19		18.78		35.19
Middle distillate	30.62		0.55	4.33		(3.61)	31.89
							72.45
							white distillate
Atmospheric residue	46.60	(46.60)					
Vacuum distillate		27.21		(27.71)			
Vacuum residue		18.89	(18.89)				
Visbreaking residue			17.61			(17.61)	
Cat naphtha				14.16	(14.16)		
Alkylate				2.38	(2.38)		
Decanted oil				1.69		(1.69)	
FCC coke				2.41		(2.41)	
Residual fuel oil						23.06	23.06
Refinery fuel:							
Direct operation	2.75	1.05	0.30	1.08	(0.94)	(4.24)	
Hydrotreating	0.23		0.01	0.05	(0.23)	(0.06)	
Variable operating costs, $/metric ton:							
Direct operation	60	26	9	64	(19)		144
Hydrotreating	17		1	4	(18)		
Residual fuel oil sulfur, wt.%						3.7	3.7

Among the upgraded streams, LCO and the thermally cracked light distillate can be assumed to be equivalent, on an incremental basis, to straight-run white distillate. However, this assumption is not valid for the cat naphtha and alkylate streams, which both have an intrinsically higher value than naphtha due to their relatively high octane characteristic.

In a real operation, if incremental amounts of gasoline-blending stocks from upgrading facilities are released into the gasoline pool and the total gasoline production has to remain unchanged, the most logical refiner's action would be to back out some reformate from the pool. Following this logic, alkylate and cat naphtha are incrementally exchanged with an equivalent amount of reformate which is then run backward through the reformer to get straight-run naphtha.

The reforming yield in this step reflects a severity of operation to produce a reformate with the same octane quality as the blend of the cat naphtha and alkylate streams. Note than on an incremental basis, it is perfectly feasible to run a processing unit "backwards."

The last step is the fuel oil balance. It determines the amount of cutter stock required to produce a marketable fuel oil. This step also takes into account that a portion of the fuel oil pool has to be burned in the refinery to meet the processing units' energy requirements. The total by rows also shows that, for this configuration, it is possible to run incremental crude to produce only white distillate residual fuel oil.

The BEV formulas derived from the calculations for white distillate are shown by Table 3, which also summarizes the formulas for the other products considered.

TABLE 3 Examples of BEV Formulas

White distillate:	
Hydroskimming	WD = $1.9444C - 0.8933F + 1.5$
Complex	WD = $1.3803C - 0.3182F + 2.0$
Naphtha:	
Hydroskimming	N = $2.0383C - 0.9365F + 1.6$
Complex	N = $1.4242C - 0.3284F + 2.1$
Gas oil:	
Hydroskimming	GO = $1.8352C - 0.8431F + 1.4$
Complex	GO = $1.2762C - 0.2943F + 1.8$
Gasoline from incremental crude, 91 (R+M)/2:	
Hydroskimming	G = $2.2620C - 1.1520F + 3.6$
Complex	G = $1.6148C - 0.4848F + 4.8$
Atmospheric residue:	
Complex	A = $0.6032C + 0.3959F - 0.6$

where WD, GO, N, G, A = BEV in $/ton for the product considered
    C = crude oil price in $/ton (34°API Middle East type)
    F = residual fuel oil price in $/ton (the sulfur level for the hydroskimming and complex cases is, respectively, 2.4 and 3.7 wt.%)

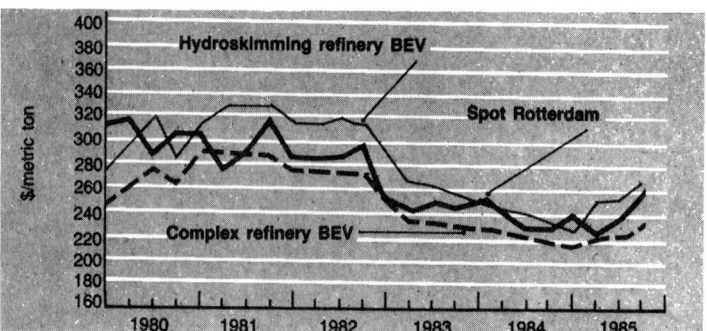

FIG. 2. Comparison of gas oil BEV with spot price.

Naphtha and Gas Oil

The BEV's of naphtha and gas oil can be derived directly from the white distillate BEV formulas by taking into account the specific gravity effect. For naphtha, a further correction should be applied to reflect the volumetric crude expansion that follows the atmospheric distillation as a result of the nonideal behavior of the various hydrocarbons in the crude oil, and that occurs primarily in the naphtha and lighter fractions.

It is of interest to compare the calculated production costs with the spot market prices (Figs. 2 and 3). Note how closely in Fig. 2 the production cost of gas oil follows the spot price development, which confirms the accuracy of the BEV formulas.

As expected, the complex refinery configuration enables the refiner to obtain production costs significantly lower than in the hydroskimming case. In the naphtha case, Fig. 3 compares the spot price ratio of naphtha and gas oil with the corresponding BEV ratio. The market price appears to be quite in line with production economics.

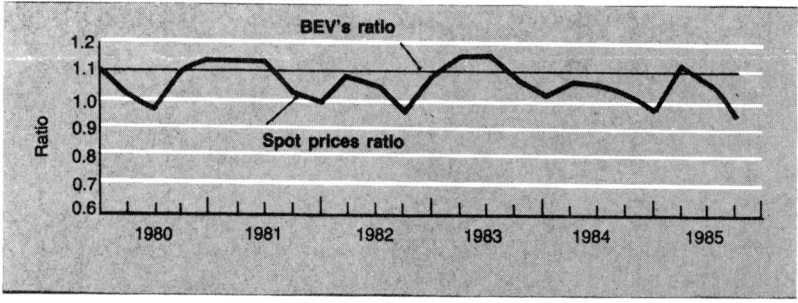

FIG. 3. Naphtha to gas oil BEV's and spot prices.

Petroleum Products, Production Costs

Gasoline

Gasoline production costs are very dependent on the manufacturing route and the quality specifications of the finished product. Typical production cases are:

Production from incremental crude, where gasoline is made by blending together all the suitable light streams, both straight run and cracked, produced by processing an incremental amount of crude. In this case the incremental gasoline blend reflects the average composition of the gasoline pool for the refinery configuration considered.

Production from naphtha reforming, where the incremental gasoline demand is met by reforming an incremental amount of straight run naphtha. This situation reflects primarily hydroskimming operations but, under particular circumstances, could apply also to conversion refineries.

Production of cracked gasoline. This is a typical situation for complex refineries processing atmospheric residue as an incremental feedstock alternative to crude oil.

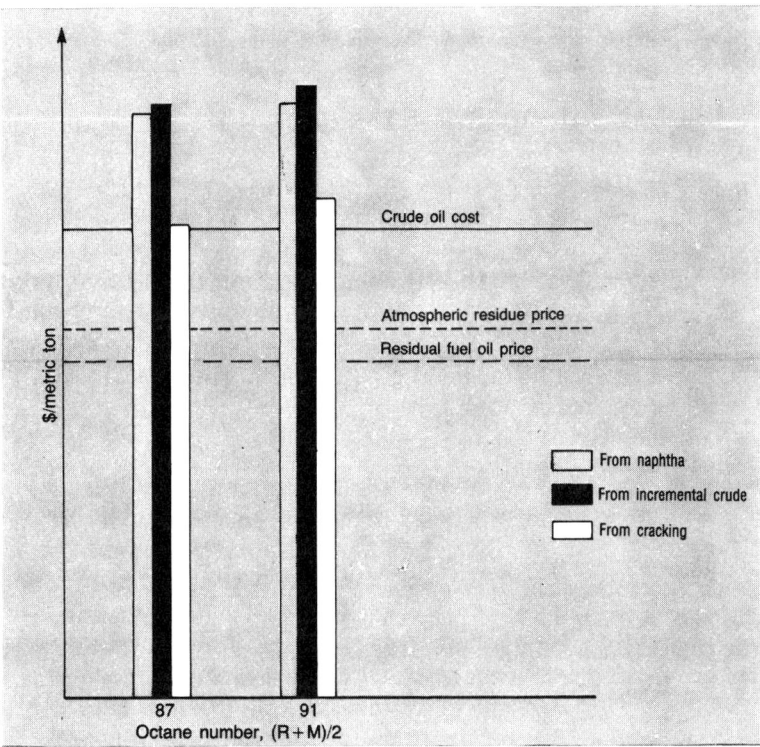

FIG. 4. Effect of gasoline production on gasoline BEV (complex refinery, 34°API gravity crude).

The histrogram in Fig. 4 shows the difference in production costs resulting from the above three manufacturing cases for two gasoline grades, i.e., 81 and 91 (R+M)/2, which are generally representative of leaded premium (0.4 g/L), and the common grade of unleaded gasoline to be introduced in Western Europe.

The gasoline from naphtha (with naphtha produced from crude) and gasoline from incremental crude have very similar BEV's, with the former method showing some cost advantages. The difference increases with the octane level of the gasoline, reflecting the increasing operational difficulties in making gasoline from crude and in absorbing light naphtha and cat naphtha in the pool.

Addition of a light naphtha isomerization unit would greatly reduce this difference. The gasoline from the cracking operation appears to offer significantly lower production costs.

This benefit is due primarily to the advantage of starting from a relatively cheaper raw material, i.e., atmospheric residue as an alternative to crude, and not to an intrinsically cheaper processing route. The BEV of cracked gasoline is particularly sensitive to the pool octane quality, and increases sharply at high octane values due to the need for reforming an increasing portion of the FCC gasoline prior to blending.

Atmospheric Residue

The evaluation of atmospheric residue, also known as straight-run residual fuel oil, has attracted much interest since the early 1980s. This product has developed a large market as an upgrading feedstock alternative to crude oil. As a result, very little atmospheric residue is blended directly to residual fuel oil today. The evaluation of atmospheric residue, which is of interest primarily for upgrading refineries, is generally aimed at producing the following information:

- Determination of the amount of white distillate which can be obtained from the processing of the atmospheric residue. This approach implies that the atmospheric residue is equivalent to a heavy crude, as indeed it is in the case of a refinery equipped with residue upgrading facilities.
- Determination of the BEV of the atmospheric residue as a function of crude and residual fuel oil values. This can be achieved by backing out from the refinery feedstock pool the amount of crude that would produce the same amount of white distillate as the atmospheric residue.

The BEV calculations are illustrated by Table 4. It is interesting to compare the developments of the spot price of high sulfur residual fuel oil with the calculated BEV of atmospheric residue (Fig. 5).

TABLE 4 BEV Calculation for Atmospheric Residue from 34° API Crude[a]

	Vacuum Distillation (metric tons)	FCC Alkylation (metric tons)	Visbreaking (metric tons)	Naphtha Reforming (metric tons)	Fuel Oil Balance (metric tons)	White Distillate to Crude[b] (metric tons)	Total by Row (metric tons)
Atmospheric residue	(100.00)						(100.00)
Vacuum distillate	59.46	(59.46)					
Vacuum residue	40.54		(40.54)				
Gases/LPG		5.88	0.88	(4.81)	(1.95)		
Light naphtha			0.28			(0.28)	
Heavy naphtha			0.41	40.30		(40.71)	
Middle distillate			1.18		(7.76)	(2.71)	
Cat naphtha		9.29		(30.38)			
Alkylate		30.38					
Decanted oil		5.11		(5.11)			
		3.63			(3.63)		
FCC coke		5.17			(5.17)		
Residual fuel oil					52.69	(13.10)	39.59
Crude, 34° API						60.32	60.32
Visbroken residue			37.79		(37.79)		
Refinery fuel:							
Direct operation	2.25	2.32	0.64	(2.02)	(3.19)		
Hydrotreating		0.11	0.02	(0.49)	0.36		
Variable operating cost, $/metric ton:							
Direct operation	55	137	19	(41)		(87)	55
Hydrotreating		9	2	(39)			
Residual fuel oil sulfur, wt.%					3.7	3.7	3.7

[a] Complex refinery.
[b] Based on the result of the calculations in Table 2.

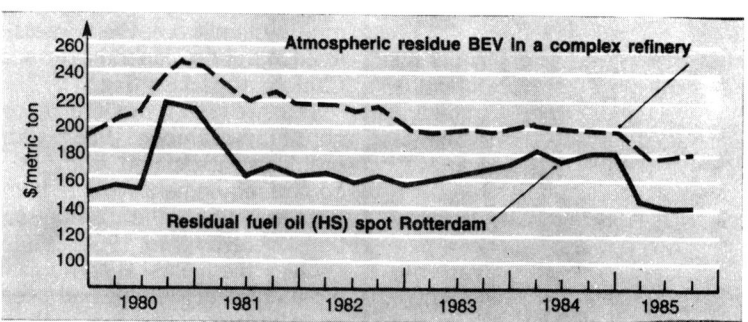

FIG. 5. Atmospheric resid BEV and HSFO spot prices.

Considering that the premium commanded by this product over fuel oil fluctuated until the end of 1985 at around $15/ton, it is clear that processing this feedstock as an alternative to crude is an attractive proposition. It is also important to note how the differential between the atmospheric residue BEV and the spot residual fuel oil price narrowed during 1984 due to the unusually strong fuel oil price relative to the crude price.

The methodology has been tested extensively under different refining operating environments and has always provided reliable results. It should be noted that it does not replace linear programming, but simply complements it.

In fact, our methodology performs an economic analysis based on a given mode of operation without questioning whether this mode reflects the optimal way of processing crude. The optimization process is left to the LP and the refiner's experience.

In a sense, the methodology carries out a sensitivity analysis around the optimal LP solution. The advantage is that this task is achieved in a significantly more efficient and clear way than resorting to LP simulation.

The methodology is particularly suitable for implementation on a personal computer using available spreadsheet software. It has the potential to become a power tool for the refiner to support "make or by" policies, planning decisions, and manufacturing-margins evaluations.

This material appeared in *Oil & Gas Journal*, pp. 27–32, February 9, 1987, copyright © 1987 by Pennwell Publishing Co., Tulsa, Oklahoma 74121, and is reprinted by special permission.

FABIO BERNASCONI

Octane Boosting

To meet additional octane requirements prompted by lead-phasedown rules, Coastal Eagle Point Oil Co. chose an octane-enhancing catalyst for its Westville, New Jersey, refinery. But octane enhancement was not the only benefit needed from the catalyst.

Coastal also had to increase the amount of resid charged to the fluid catalytic cracker (FCC) because of the nature of the crude oils processed in the refinery, and because of the operating nature of the crude and vacuum units. The combination of octane enhancement and increased resid-charging capability was necessary to achieve acceptable economics for the overall operation.

The goal of improving gasoline octane while charging poor-quality feedstocks was achieved by switching to an ultrastable Y (USY) type catalyst, Filtrol ROC-1DY.

Background

Because of its key position in most refineries, the FCC unit must have the flexibility to process a wide variety of feeds. The operation of the cat cracker must also be adjusted and maintained at optimum conditions for maximum profitability.

Some older cat crackers can be mechanically revamped to enable processing a wider range of feedstocks.

For example, upgrading regenerator and reactor internals, and the addition of heat-removal equipment, will allow for adding significant quantities of residual feeds to FCC combined feed. However, in many cases the use of the proper catalyst system, combined with appropriate unit operation, will also provide the refiner with a wider selection of feedstocks for the FCCU.

When faced with the economic reality of having to process heavier and poorer-quality feeds with a tighter capital budget, both refiners and catalyst manufacturers need to respond with innovative techniques to improve profitability. Generally, octane-improvement options are constrained by particular unit design and operating parameters. The most important parameters are:

Octane enhancing catalysts
Operating severity
FCC gasoline end point
Feedstock selection

Ultrastable Zeolites

It had long been theorized that ultrastable Y zeolites (USY) could enhance both gasoline research and motor octane levels without a significant negative effect on gasoline selectivity. Using a series of laboratory-synthesized FCC catalysts, Pine et al. of Exxon demonstrated the effect of USY unit cell size on catalytic octane performance and activity. As the unit cell size of steam-deactivated, USY-containing catalysts decreased, there was a steady and significant increase in both research and motor octane numbers, accompanied by a decreased microactivity test (MAT) activity and a strong tendency to make light C_2^- gas.

Also of considerable importance has been the USY's positive impact on coke selectivity.

Filtrol's research concentrated on the development of a zeolite system, taking advantage of the attributes of the conventional USY while controlling the light gas production and maintaining acceptable activity levels. The result of this research effort was the development of dealuminated Y (DY) zeolite. Filtrol's dealuminated Y catalysts retain all the benefits of regular USY while maintaining better activity and producing less light gas. In addition, Filtrol combines USY technology with a cost-effective matrix in its catalysts for upgrading heavy gas oils and resids.

The effects of feed contaminants, particularly metals, has been well catalyst. These metals tend to nonselectively catalyze undesirable dehydrogenation reactions while also causing a loss of surface areas and zeolite activity. For units processing large quantities of resid containing high levels of nickel and vanadium, it is critical that a catalyst tolerate significant metals loading while continuing to provide good performance.

Coastal's Application

Coastal's Eagle Point refinery has improved FCC octanes without yield penalties while charging poor-quality feedstocks to the unit. This was accomplished primarily by switching to Filtrol ROC-1DY. This catalyst combines octane-enhancing dealuminated zeolites with an active matrix to provide octane improvements together with the flexibility to process heavily contaminated poorer-quality feedstocks.

Octane Boosting

Coastal Eagle Point primarily processes heavy paraffinic crudes from Indonesia and China. While such crudes generate typically excellent FCC feedstock, the naphtha produced from cracking is of low octane with high sensitivity (RON-MON spread).

Additionally, paraffinic crudes produce poor-quality, catalytic-reformer feedstocks which restrict the octane production from the reformer. At Eagle Point these factors combined to severely strain the refinery's ability to produce motor gasolines from indigenous blendstocks at the lower levels mandated by the EPA.

From a cursory review, an immediate switch to a USY catalyst appeared to be the appropriate step. However, the Coastal FCC processes significant quantities of residual oil, and any change of catalyst would have to be consistent with a severe high-temperature, high-metal-on-catalyst operation.

In addition, because of the somewhat unorthodox configuration of the Eagle Point crude unit, crude running was very often restricted by the amount of resid which could be processed in the FCC.

The crude unit consists of three columns: atmospheric, gas oil, and vacuum, with the vacuum column being the unit bottleneck. As the crude charge became heavier in recent times, the limit of vacuum-column throughput has become more pronounced.

While resid cracking first started in the late 1970s, to maintain crude runs it had been necessary to route greater volumes of gas oil tower bottoms (vacuum tower charge) to the FCC, thus causing the critical link between the FCC and crude unit. Selection of a stable, coke-selective octane catalyst was of utmost importance.

Prior to switching to ROC-1DY catalyst, Eagle Point utilized a conventional high rare-earth-content Y zeolite catalyst. Residual content (1000°F + equivalent) in FCC feed averaged approximately 17 vol.%, limited primarily by coke burning and regenerator temperature limitations.

The addition of ROC-1DY catalyst provided the necessary coke selectivity to allow additional resid cracking and overall higher unit throughput. Resid content in feed now typically ranges between 20 and 27 vol.%, and an overall increase of approximately 15 vol.% in unit charge has been observed.

FCC-feedstock qualities are outlined in Table 1. The increase in charge was accomplished without deterioration of overall unit yields.

TABLE 1 Resid Feed Quantity Boosted

Feed Property	Base Operation	Current Operation
Catalyst	REY	ROC-1DY
FCCU charge rate	Base	Base + 15%
Vacuum resid, vol.%	13–19	20–27
Gravity, °API	30	30
UOP K	12.05	12.0
Molecular weight	330	370
Nickel, wt. ppm	3	5
Vanadium, wt. ppm	0.5	0.5

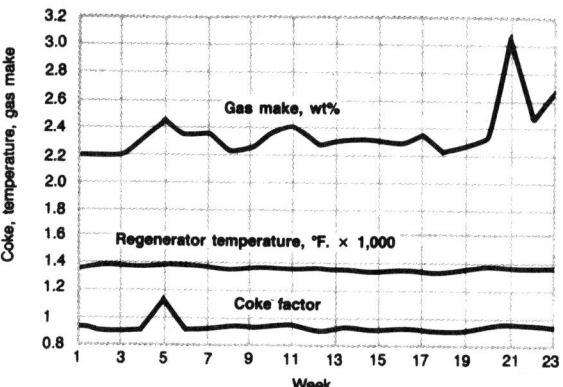

FIG. 1. Gas and coke selectivities.

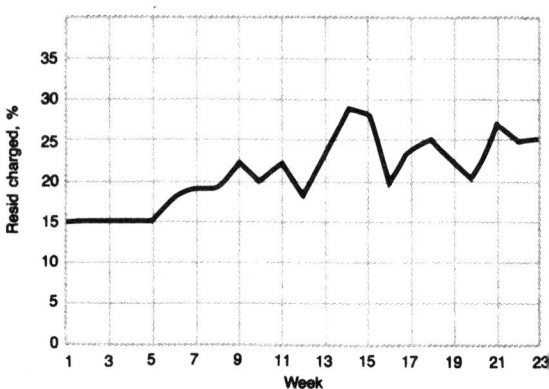

FIG. 2. Resid charge increased.

The new catalyst provides increased gasoline octane with significant coke reduction, remains stable, and limits dry-gas production. The attrition resistance and thermal stability of the catalyst has been excellent.

The effect on coke selectivity, dry gas, and regenerator temperature is shown in Fig. 1. The effect on resid in FCC feed is shown in Fig. 2.

Since the addition of ROC-1DY catalyst, the FCCU has been operated in both a maximum fuel oil (LCO) and maximum gasoline mode. To maximize the yield of middle distillate boiling range product, several steps were taken.

Reaction severity was reduced by lowering reactor temperature. Feed-preheat temperature was increased, and the quantity of heavy-cycle-oil recycle to the riser was also increased.

TABLE 2 Feed and Yields Compared

	Base Operation	LCO Mode	Gasoline Mode
Charge rate	Base	Base + 15%	Base + 15%
Vacuum resid, vol.%	13–19	20–27	20–27
Combined feed ratio	1.10	1.20	1.12
Riser outlet temperature, °F	955	945	965
Catalyst type	REY	ROC-1DY	ROC-1DY
Catalyst activity, MAT	68	63	64
Yields:			
C_2 and lighter, wt.%	2.40	2.22	2.29
$C_3 + C_4$, vol.%	20.9	17.6	18.8
C_5 + gasoline, vol.%	60.3[a]	48.8[b]	59.8[c]
Light cycle oil, vol.%	17.0	29.8	22.5
Decant oil, vol.%	10.3	11.0	10.6
Coke, wt.%	4.7	3.6	4.1
Gasoline: RONC	89.6	91.8	92.2
MONC	76.4	78.4	78.8
(R + M)/2	83	85.1	85.5
Ni + V on catalyst, wt. ppm	4050	5010	5260

[a] 430°F ASTM end point.
[b] 350° F ASTM end point.
[c] 430° F ASTM end point.

Octane Boosting

In addition to these adjustments, gasoline product was cut to 350°F. ASTM end point, adding to the total middle distillate yield. A sumary of this middle distillate operation is shown in Table 2.

In addition to providing the coke selectivity, which allowed processing the increased quantity of resid and total charge, the ROC-1DY contributed to the greater than two-number increase in both clear research (RON, 2.6) and motor octane (MON, 2.4) of the FCC gasoline. A second feature, the active matrix of this catalyst, has provided several additional benefits.

The most notable effect is the upgrading of heavy components into lighter boiling range products. During the LCO operation, which included additional nondistillable (1000°F+) material in the feed as well as reduced cracking severity, the total yield of decant oil increased only very slightly over the base operation, while the yield of LCO increased more than 10 vol.%. This upgrading contributed to the significant improvement in gasoline MON clear.

As the demand for gasoline increased, the FCCU operation was adjusted to produce more naphtha-boiling-range product. The HCO recycle was reduced, the reactor temperature was increased, and the gasoline end point was raised to 420°F. The resulting yield distribution is shown in Table 2.

Even at the higher cracking severity required to produce maximum gasoline yield, the total feed rate was not reduced from the previous operation, also maintaining the 20–27 vol.% vacuum resid in the combined FCC charge.

The coke and dry gas selectivities provided by the DY zeolite catalyst did not decline at this increased-severity operation.

The improvement in RON clear, normally associated with increased reactor temperature, was offset with the increase in gasoline end point for this operation.

It must be noted that the LCO is cut to a cloud-point specification rather than end point because of the paraffinic nature of both the feed and cycle oils.

With other catalysts it had been virtually impossible to raise naphtha end point above 400°F without adversely affecting LCO cloud point and the LCO/decant oil split.

In essence, the heavy gasoline acted as a cloud point cutter stock and economically could not be removed. Under the current operation, LCO cloud points have been reduced enough to allow production of FCC naphtha to any desired end point, constrained only by gasoline pool distillation specifications and the economics of upgrading the relatively low-octane, heavy FCC gasoline.

At 30¢/octane-barrel, the octane improvement alone from switching catalysts translated into a revenue improvement in excess of $10,000/day. The octane improvement experienced during operation with ROC-1DY is shown in Fig. 3.

The coke and dry gas selectivities, combined with the matrix properties, have provided a tool for controlling the two most limited factors, namely, reactor/regenerator heat balance and feed contaminant effects.

The coke and dry gas selectivities were demonstrated by increasing the feed rate and resid content in the feed while not exceeding the existing metallurgical and mechanical limitations of the unit.

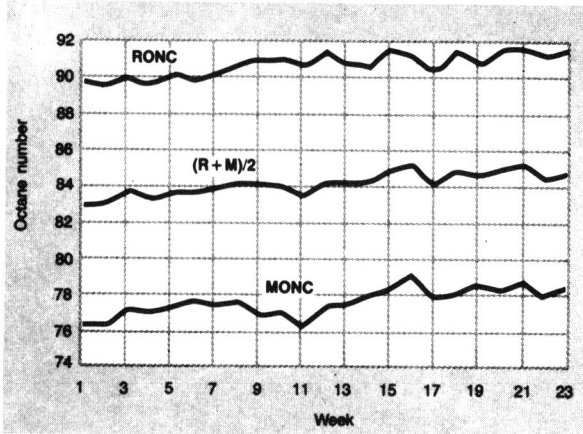

FIG. 3. Gasoline octane increased.

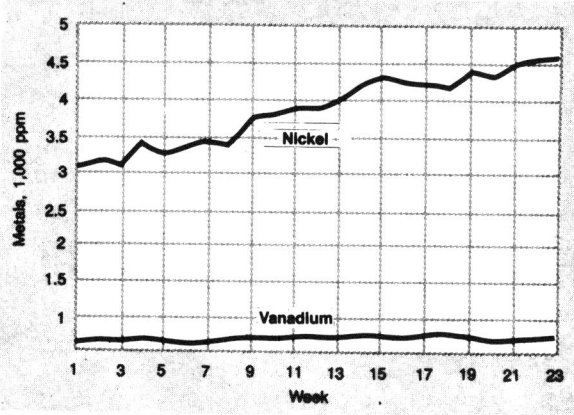

FIG. 4. Catalyst handles increased metals.

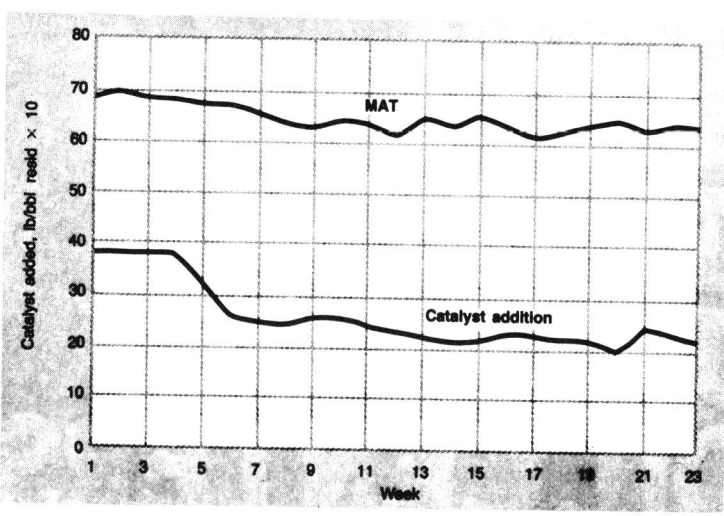

FIG. 5. Fresh catalyst addition minimized.

Eagle Point uses a metals passivator to control production of hydrogen. Nickel activity is Eagle Point's primary concern, because nickel accounts for approximatley 85% of the metal on equilibrium catalyst. The effect of increasing nickel content on catalyst is shown in Fig. 4.

Eagle Point was able to take advantage of the increased metal tolerance of ROC-1DY to minimize addition of fresh catalyst while maintaining the desired activity level, as shown in Fig. 5. Catalyst additions increased in rough proportion to the amount of resid cracking to maintain reasonable levels of metal contaminants.

Equilibrium catalysts are also added at times for metals control. Equilibrium additions typically average less than 25% of total catalyst makeup.

This material appeared in *Oil and Gas Journal*, November 24, 1986, and is reprinted with the permission of the editor.

JOHN J. LIPINSKI
JACK R. WILCOX

Octane Catalysts

Introduction

The ability of zeolite cracking catalysts to moderate the research octane number and, to a lesser extent, the motor octane number of FCC gasoline has been known for well over 10 years [1]. Considerable effort has been made during this period to understand the mechanism of the catalytic reactions involved. In the recent past a number of publications have dealt with the catalyst properties involved in controlling the organic reactions responsible for octane enhancement of the FCC gasoline [2–4]. Catalyst formulations are now widely available that control the amount of the all-important hydrogen transfer reactions taking place under cracking conditions. The principal properties of the catalysts which influence this octane-controlling reaction are the silica/alumina ratio of the Y zeolite present, the zeolite sodium content, and the degree of rare-earth exchange of the zeolite. Increased SiO_2/Al_2O_3 in the zeolite framework reduces the number of acid sites available for cracking and hydrogen transfer but increases their strength. X-ray determination of the framework unit cell constant, a_0, relates directly to the framework SiO_2/Al_2O_3 and has also been shown to relate to catalyst selectivity and octane enhancement [2]. Data have clearly illustrated that as the silica/alumina ratio is increased, octane enhancement occurs as a result of a reduced amount of hydrogen-transfer reactions relative to cracking. The reverse is observed as rare-earth exchange is increased; hydrogen transfer increases and octane enhancement decreases.* Since both catalyst activity and stability are increased as rare-earth exchange level is increased at low zeolite soda level, this valuable tool of catalyst performance control cannot be ignored. Thus a complete portfolio of octane catalysts to cover the wide variety of refinery needs must contain both hydrogen and hydrogen/rare-earth-type catalysts.

The petroleum industry is required to produce maximum octane barrel from the cat cracker, and the catalyst supplier must respond to this demand (currently 50+% of the United States and Canadian market = ~300 tons/day) with premium octane catalyst products—those containing high silica/alumina ratio zeolite Y frameworks. Direct synthesis of products in the silica/alumina range to properly control hydrogen transfer, i.e., from 5.5 to 10, has been unsuccessful, and indirect preparative procedures are currently widely used. These procedures, however, lead to the formation of zeolite crystal defect structures containing amorphous "debris" within the zeolite crystal which enhances nonselective cracking reactions, lowering gasoline yield and increasing coke make. Such defect structures also lack the hydrothermal stability required by today's high temperature regenerators and residual feed operations.

Union Carbide Corporation and its Katalistiks subsidiary have now completed the successful development and early commercial testing of a series of "High Stability Zone Catalysts" (HSZ) which contain derivatives of zeolite Y, identified as LZ-210, prepared by a novel "framework silicon enrichment process." Zeolites formed by this process greatly reduce the formation of catalytically harmful silica/alumina debris and defect structure instability. This procedure has been adopted to produce virtually any silica/alumina ratio desired. Y zeolite crystals with ratios of 10, 20, 30, and above can be readily produced, and they afford a high level of octane-barrel flexibility. This catalytic flexibility is the result of proper active site placement and strength made possible by using the "framework silicon enrichment process" for SiO_2/Al_2O_3 ratio control. This article describes the preparation and pilot-plant evaluation of two catalyst series, ALPHA and BETA, derived from the new "framework silicon enrichment process."

I. The Chemistry and Characteristics of Framework Silicon-Enriched Zeolites

A. Chemistry

The chemistry and functional role of Y zeolite in FCC catalysts has been extensively reviewed for over two decades [2, 5–8]. The topic of most papers

---

*Sodium on zeolite exchange sites also increases the amount of hydrogen transfer relative to the amount of cracking, thus reducing octane enhancement. Sodium levels smaller than 1% on zeolite must be achieved on premium octane catalysts for maximum octane enhancement to occur.

TABLE 1 High Stability Zone Boundaries and Framework Silicon Enriched Y

SiO_2/Al_2O_3	a_0 (Å)	Description	Potential Active Sites/Unit Cell
5.3 minimum	24.65 maximum	High quality NaY	55
6.0+	24.60	Beginning of zone	48
7–9	24.56–24.52	USY range (octane, coke selectivity)	43 to 35
6–20	24.60–24.35	LZ-210 range (octane barrels, coke selectivity)	48 to 17

is the acidic form of Y zeolite with SiO_2/Al_2O_3 ratios of about 5. Many investigations centered on modified Y zeolites, obtained from HY by post-synthesis treatments such as steaming or aluminum extraction by chemical agents [9–11]. The differences between the as-synthesized, cation-exchanged and steamed, and Al-extracted Y zeolites have been described in terms of crystal structure, chemistry, and catalytic performance. The morphological difference between the as-synthesized "perfect" crystals and the steamed or extracted "defect" crystals has received less attention.

Framework-silicon-enriched type Y (LZ-210) used in Katalistiks HSZ Catalysts are materials which vary in SiO_2/Al_2O_3 ratio in the range from 5 to 30 and beyond. Typically this range overlaps the catalytically familiar NaY and ultrastable Y materials which have been in use for many years. Table 1 contrasts various important catalytic properties of selected members of the range. Importantly, only LZ-210 covers the entire range and can be varied in active site concentration from 17 to 48/unit cell compared with the limited USY variation from 35 to 43.

As described below, the chemical reaction used for the silicon enrichment of the Y crystal framework (LZ-210) results in a substantially defect-free Y crystal lattice. Framework-silicon-enriched Y materials exhibit superior thermal and hydrothermal stability compared with reference high quality ammonium-exchanged type Y (Union Carbide Corp. LZ-Y62), steam-stabilized Y (LZ-Y82), and steamed and aluminum-extracted Y derivatives. The replacement of zeolite framework Al atoms by Si atoms can be achieved by the application of certain aqueous solutions of soluble silicon enriching agents at mild reaction temperatures [12]. A favored reaction scheme applied to Y zeolite is given as follows:

$$Na^+\text{-}O_4Al + (SiX_6)^{2-} \longrightarrow O_4Si + (AlX_5)^{2-} + NaX$$

Solid   Solution   Solid   Solution

The overall reaction seems to proceed through two consecutive steps. First, framework Al atoms are removed from the Y zeolite by the hydrolyzed $(SiX_6)^{2-}$ solution. Subsequently, silicon atoms are inserted in a slower reaction step. Substantial dealumination *without* silicon insertion may cause crystal collapse while the silicon-enriched product becomes more stable relative to the starting Y zeolite. It is also clear from the equation that exchanged sodium is a reaction product (as NaX) when the silicon-enriched product is formed. This favors the preparation of low Na$_2$O zeolites for octane catalysts—a well-known advantage for octane enhancement [2].

A variety of chemical, physical, and crystallographic evidence demonstrates that the silicon atoms are indeed inserted into framework sites vacated by aluminum atoms. For example, material balance of the reaction system shows perfect stoichiometry between aluminum removed and inserted silicon. Crystallographic information and particularly the gradual changes in lattice parameter, a_0, observed with incremental reagent additions indicate good distribution of the inserted silicon atoms throughout the Y zeolite crystal. According to Skeels, up to 60% replacement of aluminum by silicon is readily achieved while maintaining nearly full Y crystal integrity [12]. Measurement of the framework aluminum remaining after silicon insertion shows its tetrahedral coordination (shown in Table 2 as M$^+$/Al).

B. Hydrothermal Properties

LZ-210 zeolites exhibit 100% crystallinity retention after the preparative change and very low Na$_2$O contents compared with the untreated LZ-Y62 starting material except LZ-210 19.7 which loses 10% crystallinity. Significantly, the properties of aluminum-depleted Y products prepared by ethylenediaminetetraacetic acid (EDTA) extraction are quite different. By removing aluminum with EDTA to achieve even a modest increase in SiO$_2$/Al$_2$O$_3$ ratio (from 5 to 9.5), the product crystallinity is reduced to 68% relative to the untreated Y [12].

The chemical and physical characteristics of LZ-210 products with SiO$_2$/Al$_2$O$_3$ ratios from 6.4 to 19.7 are shown in Table 2 and follow closely the characteristics reported earlier by Skeels [12]. The crystal retention of the products after both chemical and steam treatment is reported using two independent measurements: surface area and oxygen adsorption capacity. In order to arrive at the best estimate, the average of these two determinations was calculated in terms of percent crystal retention relative to untreated reference high Na$_2$O·NH$_4$Y, LZ-Y62 (Na$_2$O = 2.25%). Also shown in Table 2 are data on aluminum-deficient zeolites made by alternate processes: one by steam stabilization, the second by EDTA aluminum extraction, and the third by steam stabilization followed by hydrochloric acid extraction (designed to remove alumina "debris"). All materials were subsequently NH$_4^+$ exchanged to reduce soda content to below 0.45 wt.% and were then steam treated at 1600°F, 23% steam, 5 h, and 15 lb/in.^{2}abs to determine hydrothermal stability. The characteristics of all samples before and after steaming are given in Table 2.

All treated zeolites showed an increased differential thermal analysis (DTA) collapse temperature as a result of the higher SiO$_2$/Al$_2$O$_3$ ratio of their crystal framework. However, the DTA crystal collapse temperatures show

TABLE 2 Comparative Properties of Aluminum-Deficient Y Zeolites Versus Framework Silicon Enriched Zeolite Y after Severe Hydrothermal Treatment[a]

Zeolite[b]	SiO$_2$/Al$_2$O$_3$[c]	Na$_2$O	M$^+$/Al[d]	O$_2$ Captured[e]	SA (m^2/g)[f]	a$_0$[g]	DATA[h]	% Relative Retention[i] O$_2$ Captured	SA	Av
NH$_4$Y	5.0	0.36	0.98	33.0	935	24.79	1565	25.8	12.5	19.2
Y82/USY	5.8	0.17	0.38	27.8	734	24.55	1850	59.4	59.1	59.2
EDTA Extracted	8.1	0.42	0.93	31.2	812	24.62	1800	51.4	56.1	53.7
USY, HCl Extracted	9.1	0.33	0.26	19.9	527	24.30	1985	23.2	30.3	26.7
LZ-210 Zeolites	6.4	0.38	0.94	35.1	949	24.69	1725	57.4	59.0	58.2
	8.4	0.05	0.98	33.8	923	24.58	1890	71.9	77.2	74.6
	11.7	0.05	0.88	30.7	863	24.42	2020	77.7	77.2	77.4
	19.7[j]	0.18	0.99	30.1	767	24.41	2150	76.8	80.4	78.6

[a] 1600°F, 23% steam, 5 h, 15 lb/in.^{2}abs.
[b] Zeolites used, all samples ammonium exchanged to indicated Na$_2$O level except LZ-210 19.7 (see Footnote j). NH$_4$Y: Union Carbide Corp. Linde Y62, repeatedly NH$_4^+$ exchanged to reduce Na$_2$O below 0.5 wt.%. Y82/USY: Typical of high quality ultrastable products at 0.2% Na$_2$O on zeolite. EDTA Extracted: Y62 extracted with EDTA (method of Refs. 10 and 11). USY, HCl Extracted: Y72 extracted with HCl (method of Pellet and Rabo). LZ-210 Zeolites: Y62 treatment as described in this article.
[c] Chemically determined.
[d] Number of equivalents of cation/number of equivalents of (AlO$_2^-$).
[e] Wt.% O$_2$ adsorbed at −183°C, 100 torr.
[f] Three-point nitrogen BET.
[g] X-ray diffraction d-spacing determination.
[h] Differential thermal analyzer collapse temperature, °F.
[i] Percentage relative retention after steaming versus unsteamed NH$_4$Y standard (Y62: 5.1 SiO$_2$/AL$_2$O$_3$; 2.2% Na$_2$O; O$_2$ capacity, 34.1; SA, 875 m^2/g). Relative retentions corrected for slight ($\approx$ 10%) reduction in crystallinity following preparative treatment.
[j] Used as "high" Na$_2$O control in present study.

very large increases in thermal stability for the LZ-210 crystals relative to both the low soda Y62 reference (2150°F for LZ-210 19.7 vs 1565°F for Y62) and the aluminum-depleted Y zeolites (1890°F for LZ-210 8.4 vs 1800°F for EDTA 8.1).

Characterization of the aluminum-deficient materials prepared by aluminum extraction and other preparative techniques described above showed that each product lost considerable crystallinity as a result of these overall processing steps. From Table 2 it is seen that following the steam treatment described earlier, their crystal retention decreased in the following order: steam-stabilized Y (Y-82) > EDTA-extracted Y > acid-extracted steam stabilized Y. The EDTA and acid-extracted products show lower crystallinity retention to any of the LZ-210 products. The best of the three, steam-stabilized Y (Y82), was about equivalent to 6.4 ratio LZ-210 even though LZ-210 had *not* received prior steam stabilization.

In light of these considerations, the LZ-210 zeolites are *expected* to show benefits as cracking components in FCC catalysts. These materials have less aluminum and more silicon atoms in their crystal at the start. Therefore, upon steaming and concurrent hydrolysis of the framework aluminum, they contain less defect sites. Because of this, the LZ-210 crystal is more stable under the hydrothermal conditions encountered in the FCC process compared to steamed reference Y zeolites. Higher intrinsic crystal stability and the resulting higher crystal retention result in enhanced cracking activity retention in service. The intrinsically higher crystal stability and the reduced initial aluminum content should reduce the formation of alumina and silica-alumina debris created within the Y crystal upon loss of crystallinity following exposure to steam at high temperatures. With a higher fraction of catalytically selective zeolite acid sites and less 'debris" acid sites, noted for low selectivity, better gasoline and coke selectivity is expected relative to alternative H-Y products. Additionally, the lower Na_2O content obtained by this preparative technique should result in improved vanadium tolerance [13, 14]. A description of catalytic data obtained in both bench and pilot plant studies with catalysts containing LZ-210 zeolites is given in the following sections.

II. Evaluation of High Stability Zone Catalysts Containing LZ-210 Zeolites

Two new series of catalysts, the ALPHA and BETA series, have been prepared and evaluated under a variety of conditions in units ranging from microactivity through pilot plants and in initial trial stages in three commercial units. The ALPHA and BETA series differ primarily in their degree of hydrogen transfer ability but are made, nevertheless, in a variety of activity levels. In order to supply some perspective for judgment of the new materials, a number of well-known competitive premium octane catalysts were evaluated

TABLE 3 Description of Premium High Octane Catalysts

Katalistiks grades containing framework enriched zeolites:
 ALPHA 500: Maximum octane barrel, high octane
 ALPHA 540: Maximum octane barrel containing rare-earth exchanged zeolite, RE
 BETA 500: Octane enhancement oriented, high octane barrels, premium octane
 BETA 540: Premium octane containing RE
 BETA 500M: Premium octane, commercial production
Competitors grades:
 D1: Octane oriented, standard steam dealuminated Y plus Si/Al debris, no RE
 D2: Octane barrel oriented, standard steam dealuminated Y plus Si/Al debris, contains RE
 E1: Octane barrel oriented, with RE
 E2: Octane oriented, steam dealuminated Y plus Si/Al debris, no RE
 F1: Octane oriented dealuminated Y with reduced framework Si/Al debris, contains RE
 F2: Octane barrel oriented, partially dealuminated Y, framework debris present, high RE

using identical deactivation procedures. A listing of the catalysts used in the comparative study is given in Table 3.

A. Microactivity and Pilot Plant Evaluation

1. Experimental

Catalyst Pretreatment. Prior to evaluation in the pilot plant, all catalysts were steam treated to simulate equilibrium activity. Two different steam severities were used to reflect hydrothermal stability under both moderate and severe conditions. The moderate steaming conditions were 100% steam, 1350°F, 14 h at atmospheric pressure. For simulation of high severity operations, several catalysts were steamed at 1500°F for 5 h (100% steam, atmospheric pressure). For microactivity tests (MAT), catalysts were steamed at either 1400 or 1500°F for 5 h at atmospheric pressure prior to testing. Vanadium tolerance data were generated by MAT tests on artificially poisoned catalysts prepared by a method described elsewhere with slight modifications [15].

Catalyst Evaluation. MAT testing conditions were the same as defined by ASTM D-3907-80 except as noted elsewhere. Pilot plant tests were made in a cyclic fixed fluidized bed unit over a range of conditions. The catalyst-to-oil ratio was varied from 3 to 5, and WHSV was varied from 32 to 53, inversely. Reactor temperature was held constant at 975°F. Feedstock inspections are shown in Table 4. The pilot-plant feed is a moderately paraffinic charge stock with a high boiling end point.

TABLE 4 Feedstock Properties

	Microactivity	Pilot Plant
Gravity, °API	27.5	25.9
Sulfur, wt.%	0.29	0.53
Nitrogen, wt.%	—	0.10
Carbon residue, Rams wt.%	0.22	0.59
Aniline point, °F	185.0	196.0
Pour point, °F	80	95.0
Metals, ppm:		
Nickel	0.4	0.4
Vanadium	<0.1	2.0
Molecular weight	330	391
UOPK factor	11.9	12.0
Distillation (D1160), °F:		
5 vol.%	548	658
10	601	700
30	697	782
50	750	845
70	820	918
90	955	—
Composition, wt.%:		
Aromatics	—	17.4
Naphthenes	—	18.9
Paraffins	—	63.7

2. Bench-Scale Evaluations

As noted earlier, the LZ-210 zeolites have greater thermal and hydrothermal stability than dealuminated or steamed defect zeolites with comparable SiO_2/Al_2O_3 ratios or unit cell sizes. This translates to higher stability and activity for HSZ zeolite catalysts. For example, the relative hydrothermal stability of ALPHA 540 is compared to several commercially available premium octane catalysts in Fig. 1. These results indicate that the competitive catalysts would require higher rare-earth levels to match the stability of ALPHA 540. This in turn would result in enhanced hydrogen transfer reactions and lower octane.

MAT results of ALPHA and BETA catalysts are compared to two competitive catalysts, F1 and D2, in Table 5. These results clearly show that at constant conversion, ALPHA and BETA catalysts have a distinct advantage in gasoline and coke selectivity. The F1 and D2 catalysts produce more coke and C_4's at the expense of gasoline. This is believed to be due to the nonselective reactions catalyzed by residual Al_2O_3 and SiO_2/Al_2O_3 "debris" remaining in their zeolite structure after preparation and deactivation.

Metals tolerance is of concern to refiners processing heavy residual feedstocks. Nickel contamination is of less concern at the present time due to the quite effective antimony process licensed by Phillips Petroleum Co. Of more concern is vanadium contamination which leads to an *irreversible*

Octane Catalysts

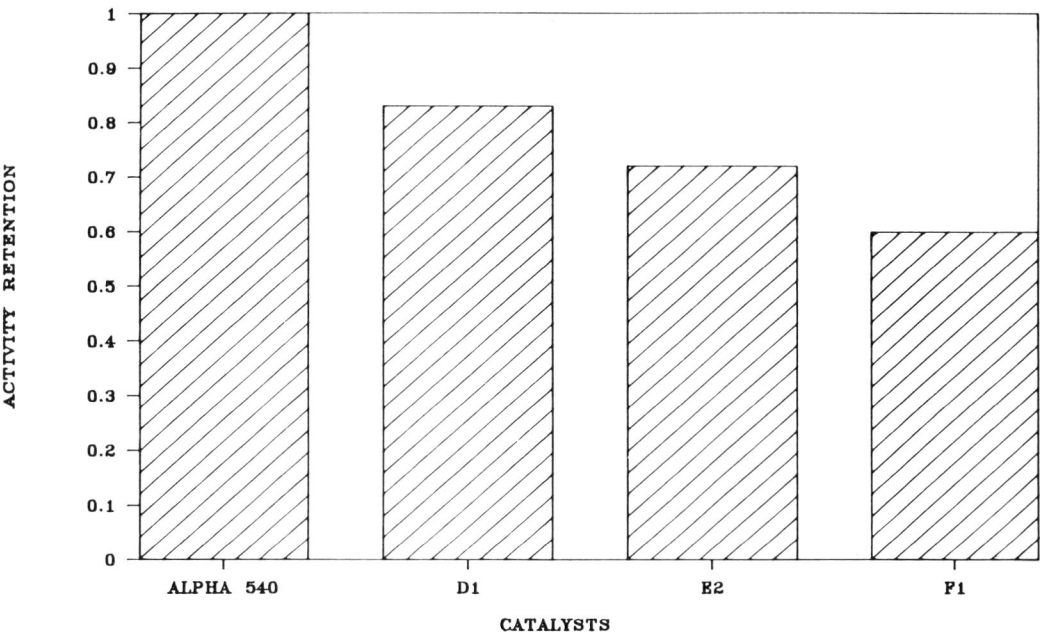

FIG. 1. Hydrothermal stability. Relative retention = 1/(microactivity at 1400°F steaming – microactivity at 1500°F steaming)/retention of ALPHA 540.

structural collapse of the zeolite. The ALPHA and BETA catalysts show very good vanadium tolerance relative to commercial catalysts of similar zeolite level (see Fig. 2). Comparing the slopes of the curves, it is apparent that ALPHA 540 is as tolerant as either competitive catalyst. BETA 540, containing an ultra low Na_2O LZ-210, shows a distinct advantage at higher vanadium levels.

TABLE 5 Microactivity Comparisons of ALPHA and BETA to Competitive Octane Catalysts at Constant Conversion

Catalyst[a]	F1	D2	ALPHA 500	ALPHA 540	BETA 500	BETA 540
Conversion			75			
Gasoline	53.0	56.3	59.4	58.5	57.6	58.0
Coke	4.5	3.4	2.5	2.5	2.7	2.6
TotC	11.4	9.9	8.4	8.9	9.0	8.9
$TotC_3^4$	4.8	4.3	3.8	4.1	4.5	4.4
C_2^-	1.3	1.1	1.0	1.0	1.2	1.1
LCO	19.0	18.9	19.1	19.2	19.2	19.5

[a]Steamed at 1400°F for 5 h, 100% steam, 15 lb/in.^{2}abs.

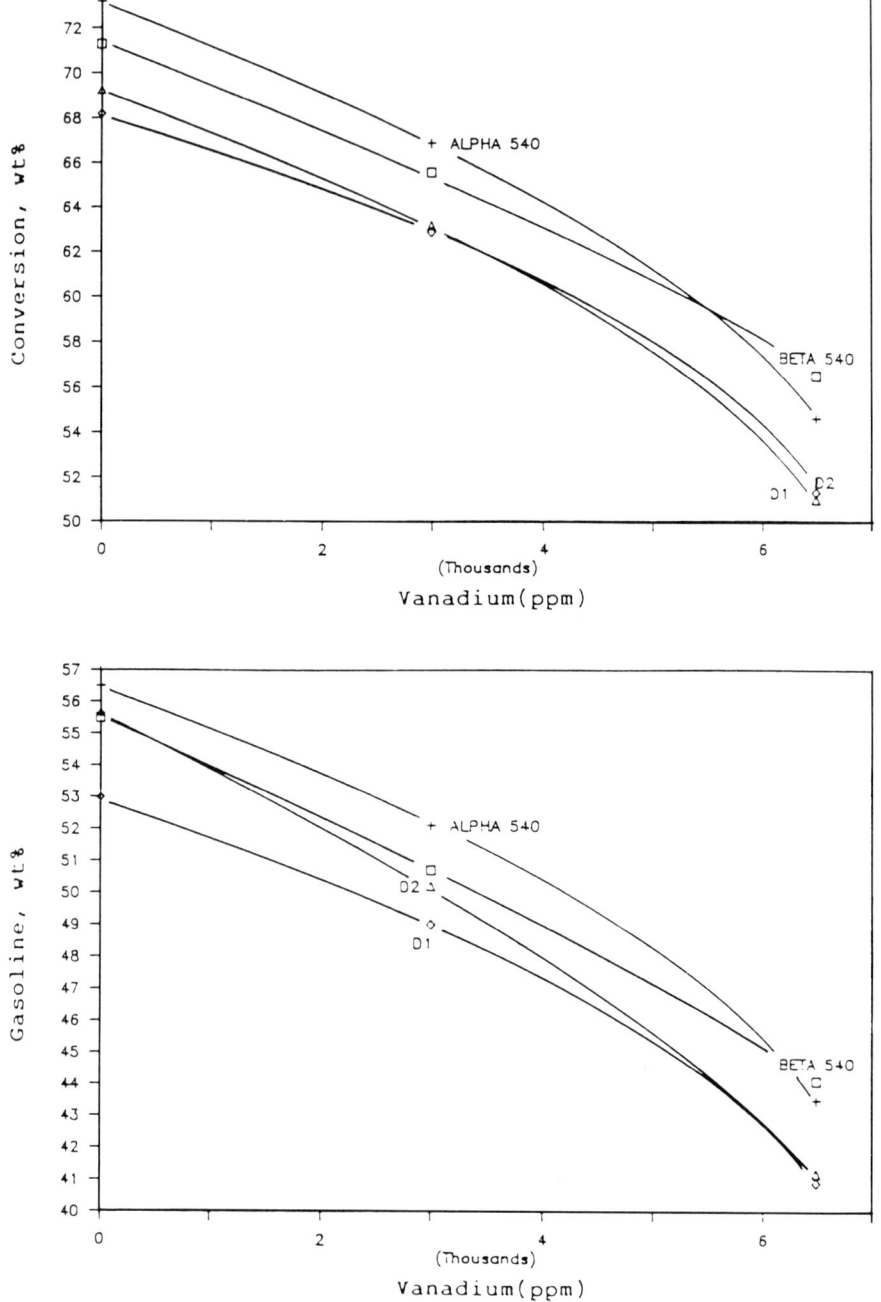

FIG. 2. Vanadium tolerance. MAT conditions: 32 WHSV, 3 cat/oil, 900°F. Deactivation: 1350°F, 14 h, 100% steam, atmospheric pressure.

Octane Catalysts

TABLE 6 Pilot-Plant Data. Comparison of ALPHA 500 and BETA 500 to a USY Catalyst at Constant Conversion[a]

	Catalyst		
Description	ALPHA 500	BETA 500	D1
Conversion: Vol.% FF	←	79.0	→
Product yields: Vol.% FF:			
Total C_3's	9.0	9.8	9.8
$C_3^=$	6.7	7.8	6.9
Total C_4's	15.8	18.0	17.2
iC_4	8.9	8.3	10.1
$C_4^=$'s	5.4	7.3	5.2
C_5–430°F gasoline	66.0	63.8	62.6
LCO 430–650°F	13.8	13.1	14.4
HCO 650°F+	7.2	7.9	6.6
Total C_3^+ liquid	111.8	112.9	110.6
Product yields: Wt.% FF			
C_2^-	1.6	1.8	1.7
Coke	3.9	3.4	4.3
Gasoline blend stock,			
FCC + C_3/C_4 alk.: vol.%	87.3	90.4	83.9
Gasoline product inspec.:			
RON: clear	91.8	92.5	92.3
MON: clear	81.3	81.3	81.7
Octane bbl:			
Research	6013	5902	5778
Motor	5366	5187	5114
Conversion at 4 C/O	76.1	76.4	75.9

[a]Pilot-plant C/O and WHSV varied at constant reactor temperature of 975°F. Steamed at 1350°F for 14 h, 100% steam, 15 lb/in.^{2}abs.

3. Pilot Plant Evaluations

(a) Moderate Steaming Severity

(1) Nonrare-Earth Catalysts. Pilot-plant results at a constant 79 vol.% conversion for ALPHA 500 (octane barrel oriented), BETA 500 (premium octane oriented), and a competitive octane catalyst, D1, are presented in Table 6. Catalysts were steam deactivated at 1350°F, 100% atmospheric steam for 14 h. The HSZ ALPHA and BETA catalysts show significant improvements in gasoline and in coke selectivity (+1.2 vol.% and +3.4 vol.% gasoline and −0.9 wt.% and −0.4 wt.% coke for BETA 500 and ALPHA 500, respectively) when compared to the conventional USY catalyst D1. This results in a 2.1% increase in octane barrels for BETA 500 and 4.1% for ALPHA 500 when compared to D1. In addition, the C_3's and C_4's produced have become more olefinic, which increases the amount of high octane alkylate which can be produced.

Gasoline yield and research octane barrels are plotted versus volume

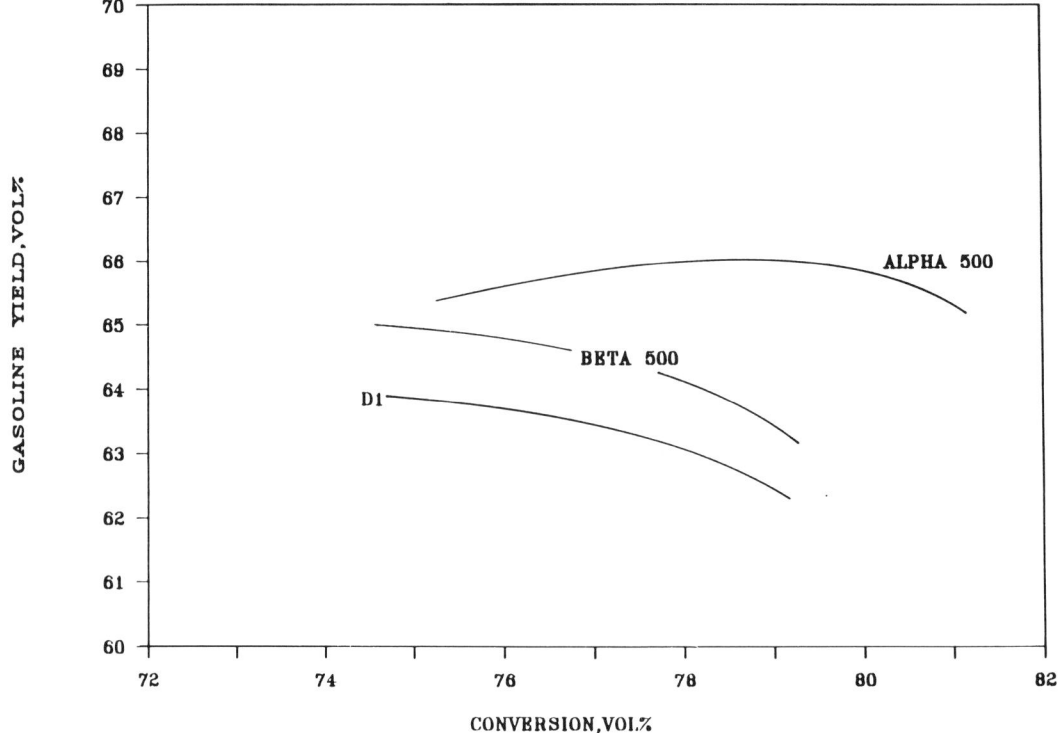

FIG. 3. Gasoline yield as a function of conversion of nonrare-earth catalysts. Steamed at 1350°F, 14 h, 100% steam.

percent conversion in Figs. 3 and 4. These plots show that the aforementioned improvements are evident over the entire range of pilot-plant operating severities covered in these tests. The slightly less olefinic nature of the gasoline produced by ALPHA 500 increases the conversion level at which overcracking occurs, which further increases the gasoline yield advantage at higher conversion levels.

For refiners regenerator temperature or air blower limited, results at constant coke yield can provide more important information than those at constant conversion (Table 7). At a constant coke yield, ALPHA 500 gives a 2.4 vol.% higher gasoline yield with an equivalent octane when compared to D1. This results in a 3.6% increase in octane barrels. BETA 500 gives a 1.3 number increase in research octane along with a 0.4 vol.% increase in gasoline yield compared to D1 at constant coke yield, resulting in a 2.1% increase in octane barrels. As in the constant conversion comparison, the HSZ catalysts produced more propylene and butylene than D1, increasing the amount of high octane alkylate which can be produced.

(2) Hydrogen/Rare-Earth-Type Catalysts—Octane Barrel Oriented Catalysts. Pilot plant results at a constant 75 vol.% conversion for ALPHA 540, BETA 540, and competitive catalysts E1 and F1 steam deactivated at 1350°F

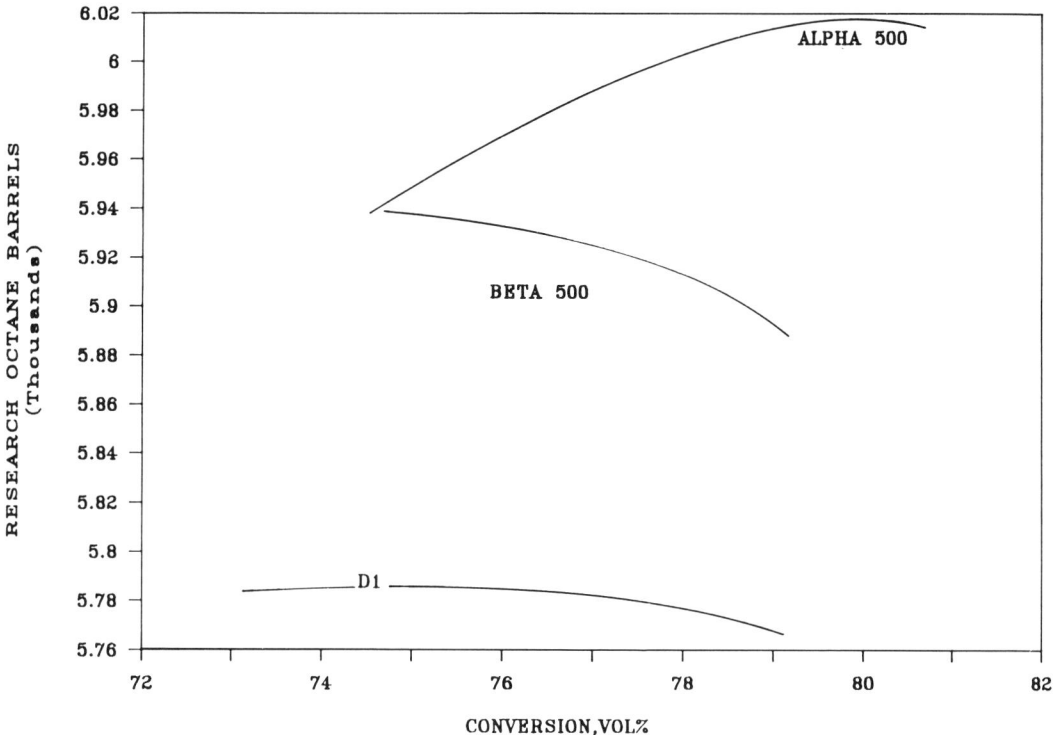

FIG. 4. Research Octane barrels as a function of conversion of nonrare-earth catalysts. Steamed at 1350°F, 14 h, 100% steam.

for 14 h with 100% atmospheric steam are presented in Table 8. ALPHA 540 shows substantially higher gasoline yields than all of the other catalysts (+3 vol.% compared to BETA 540 and E1 and +8.6 vol.% compared to F1) with comparable octanes and a slightly lower coke yield than the competitive catalysts. BETA 540 gives approximately a one number octane increase with equivalent or better gasoline and coke selectivity compared to the competitive catalysts. The dealuminated Y catalyst F1 is significantly less stable than the HSZ catalysts, giving approximately a 10 vol.% lower conversion at constant pilot-plant operating conditions and equivalent steam deactivation. Gasoline yield and research octane barrels are plotted versus volume percent conversion in Figs. 5 and 6.

(b) *Severe Steaming Severity.* BETA 500, BETA 540, D1, and F1 catalysts were also tested in the pilot plant after a severe steaming at 1500°F (see Section II-A-1) to determine relative stabilities. The purpose of this experiment was to establish relative performances in FCCU's operating under severe conditions. A direct comparison of the yield distributions was not possible due to the very low conversions obtained with D1 and F1 catalysts. To provide a constant conversion comparison, operating severities would have been so varied that the results would have been biased and would also have

TABLE 7 Pilot-Plant Data. Comparison of ALPHA 500 and BETA 500 to a USY Catalyst at Constant Coke[a]

Description	Catalyst		
	ALPHA 500	BETA 500	D1
Conversion: Vol.% FF	77.6	79.0	76.6
Product yields: Vol.% FF:			
Total C_3's	8.8	9.8	9.0
$C_3^=$	6.7	7.8	6.7
Total C_4's	14.9	18.0	15.9
iC_4	8.1	8.3	8.6
$C_4^=$'s	5.5	7.3	5.7
C_5–430°F gasoline	65.8	63.8	63.4
LCO 430–650°F	14.5	13.1	15.5
HCO 650°F+	7.9	7.9	7.9
Total C_3^+ liquid	111.9	112.6	111.7
Product yields: Wt.% FF:			
C_2^-	1.5	1.8	1.5
Coke		←—— 3.4 ——→	
Gasoline blend stock,			
FCC + C_3/C_4 alk.: vol.%	87.3	90.4	85.2
Gasoline product inspec.			
RON: clear	91.0	92.5	91.2
MON: clear	81.1	81.3	81.9
Octane bbl:			
Research	5988	5902	5782
Motor	5336	5187	5192

[a]Pilot-plant C/O and WHSV varied at constant reactor temperature of 975°F. Steamed at 1350°F for 14 h, 100% steam, 15 lb/in.^{2}abs.

presented a problem with pilot-plant constraints. Therefore, only runs at constant operating conditions were made. The incremental effects (1500 vs 1350°F steaming) upon conversion are shown in Table 9. It is apparent that considerable conversion loss occurs with D1 and F1 catalysts. On the other hand, the BETA catalysts, containing the high stability LZ-210 zeolites, show much better maintenance of conversion relative to their counterparts. These results point out the benefits HSZ catalysts can provide and reaffirm the advantage of the LZ-210 process in preparing "debris"-free zeolites of low soda content.

(c) Fractional Octanes. To determine the fractional octane enhancement of HSZ catalysts relative to a conventional hydrogen–rare earth Y catalyst, full range gasoline products (430°F end point) from SIGMA 400 and BETA 500 were cut into six fractions and rated for research octane. Gasoline products were generated at constant conversion. Figure 7 shows the research octane number as a function of boiling range of the gasoline products from these two catalysts. It is apparent that the major octane increase with BETA 500 occurs in the medium and heavy end of the full boiling range

TABLE 8 Pilot-Plant Data. Comparison of ALPHA 540 and BETA 540 to USY and Dealuminated Catalysts at Constant Conversion[a]

Description	Catalyst			
	ALPHA 540	El	Fl	BETA 540
Conversion: Vol.% FF	←―――― 75 ――――→			
Product yields: Vol.% FF:				
Total C_3's	6.6	7.2	7.6	7.5
$C_3^=$	4.9	5.5	5.8	6.0
Total C_4's	10.8	12.1	13.7	11.2
iC_4	5.6	5.4	7.5	5.2
$C_4^=$'s	4.3	5.0	4.5	4.8
C_5–430°F gasoline	68.8	65.7	60.2	65.6
LCO 430–650°F	15.8	15.8	15.8	15.8
HCO 650°F+	9.2	9.2	9.2	9.2
Total C_3^+ liquid	111.2	110.0	106.5	109.3
Product yields: Wt.% FF:				
C_2^-	1.2	1.4	1.5	1.4
Coke	3.6	3.8	4.0	4.1
Gasoline blend stock, FCC + C_3/C_4 alk.: vol.%	85.0	84.2	78.3	84.6
Gasoline product inspec.:				
RON: clear	88.5	88.9	88.4	89.7
MON: clear	79.4	79.4	78.2	79.7
Octane bbl:				
Research	6089	5840	5298	5884
Motor	5463	5217	4708	5228
Conversion at 4 C/O	80.7	77.5	69.4	78.0

[a]Pilot-plant C/O and WHSV varied at constant reactor temperature of 975°F. Steamed at 1350°F for 14 h, 100% steam, 15 lb/in.^{2}abs.

gasoline. This is due to an increase in both the olefin and aromatic levels in these fractions. This allows the refiner greater flexibility in increasing octane rating by slight increases in gasoline end point and/or reforming the heart cut of the FCC gasoline.

4. Commercial Trials of ALPHA and BETA Series Catalysts

Three commercial trials of various ALPHA and BETA catalysts have been recently started, one in the United States and two in Europe, and early data from the first trial with 20% BETA 500M in inventory are given in Table 10. The data are from a Central European refinery processing approximately 22,500 bbl/d (3500 T/d) of a predominately paraffinic feedstock, 24.5 API gravity, Ca 12.7 wt.%, and C_n/C_p 0.56. BETA 500M was added at rates from 1 to 1.6%/d on an inventory of 100% DELTA 400, Katalistiks premium octane catalyst. Equilibrium unit microactivity increased from 58 to 63, brought about by increased addition rate and increased catalyst activity. Increased

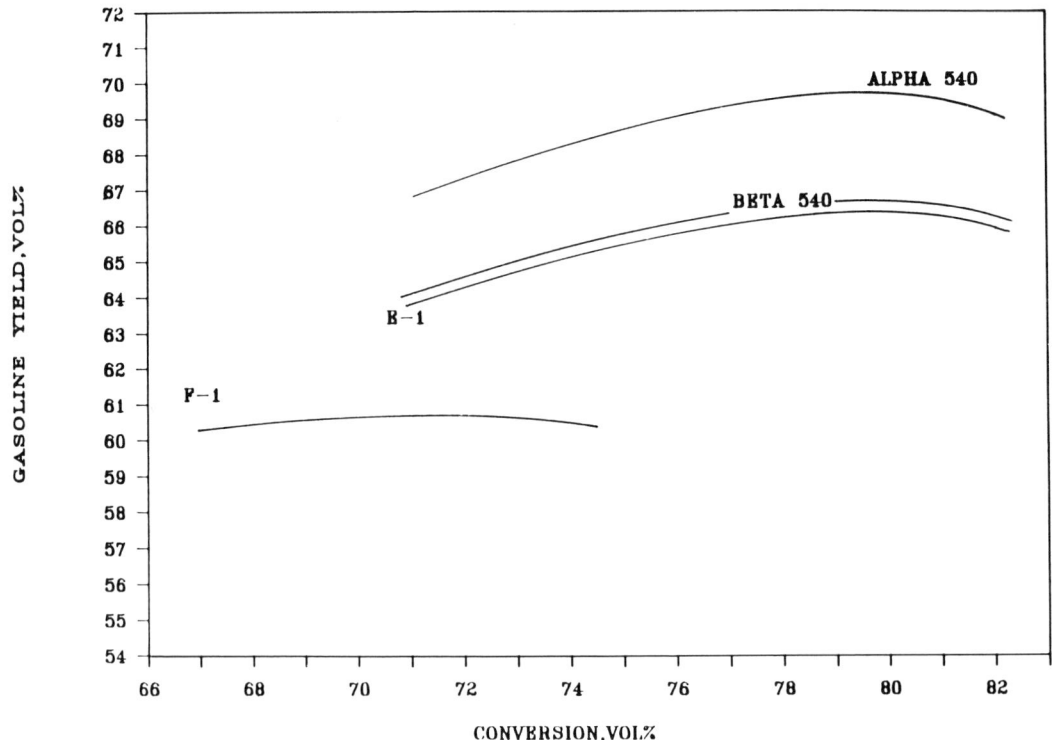

FIG. 5. Gasoline yield as a function of conversion of rare-earth catalysts. Steamed at 1350°F, 14 h, 100% steam.

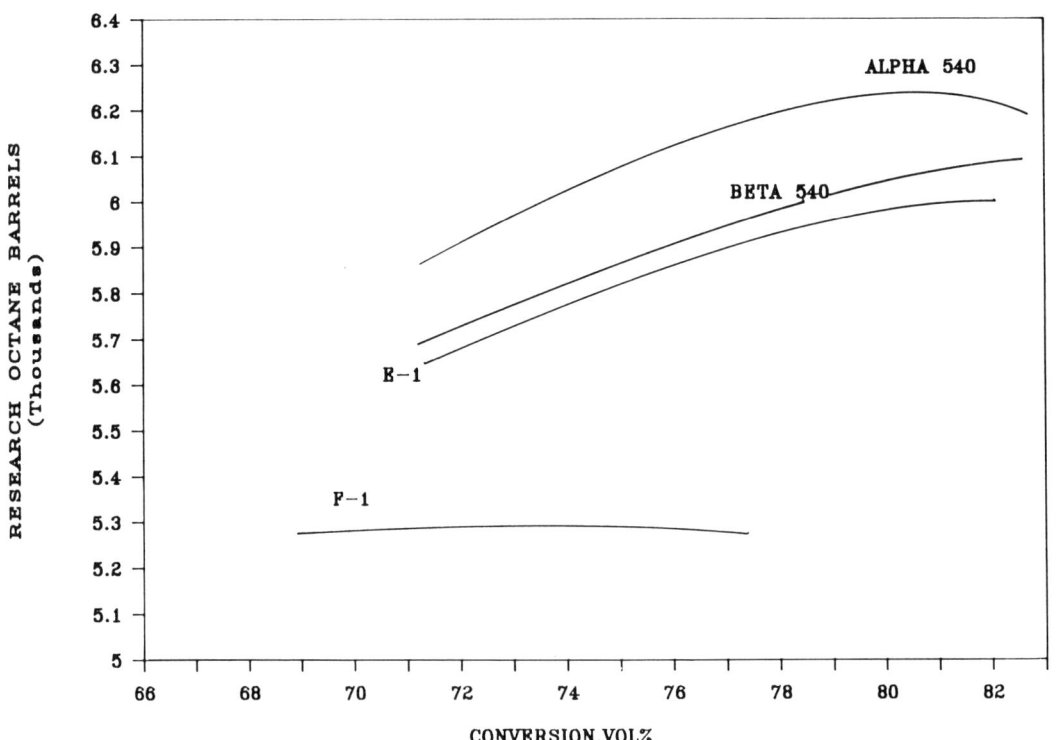

FIG. 6. Research Octane barrels as a function of conversion of rare-earth catalysts. Steamed at 1350°F, 14 h, 100% steam.

TABLE 9 Incremental Pilot-Plant Yield Losses at 1500°F Steaming Relative to 1350°F Steaming[a]

Catalyst	Conversion (vol.%)[b]
BETA 540	−0.9
Fl	−11.8
BETA 500	−3.7
Dl	−8.1

[a]1350°F, 14 h, 100% atmospheric steam; 1500°F, 5 h, 100% atmospheric steam.
[b]Compared at constant pilot-plant operating conditions.

conversion and gasoline yield were observed along with a RON increase of 0.8 on top of an already high 92.9 RON for the DELTA operation. The Motor response of 0.4 number was also significantly higher than expected for this relatively paraffinic feed. Some coke increase was observed, partly due to the lower combined feed temperature used during this part of the trial and a problem with stripping and feed dispersion steam. At the time the data were taken, there was a heat exchange problem which contributed to this lower

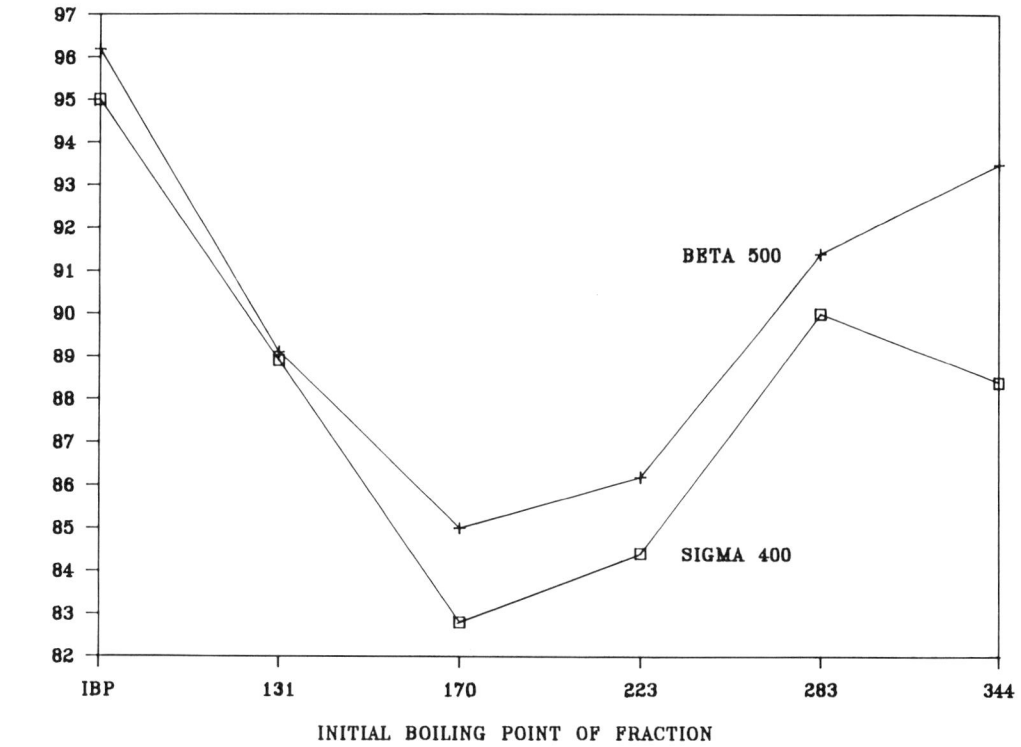

FIG. 7. Fractional Octane Number comparisons of SIGMA 400 and BETA 500.

TABLE 10 Commercial Trial Results

	DELTA 400 (100%)	BETA 500M (~20%)
Catalyst:		
Addition, %/d	1.2	1.6
MAT activity	58	63
Feedstock:		
API gravity	24.5	24.5
Ca, wt.%	12.7	12.5
C_n/C_p	0.56	0.59
Operating conditions:		
Reactor temperature, °F (°C)	966 (519)	964 (518)
Feed inlet temperature, °F (°C)	527 (275)	457 (236)
C/O ratio	6.0	5.8
Product yields, wt.%:		
Conversion	72.6	74.8
C_4 and lighter	14.3	14.7
C_5^{4+} gasoline	53.7	55.0
LCO	20.9	19.8
HCO	6.5	5.4
Coke	4.6	5.1[a]
Gasoline octanes:		
RONC	92.9	93.7
MONC	80.7	81.1

[a]The refinery was experiencing operational problems which limited the stripping and feed dispersion steam at this time.

feed temperature. BETA appeared to more selectively crack bottoms (6.5 wt.% in the DELTA operation versus 5.4 wt.% with BETA), an obvious advantage which, however, may also have contributed to the higher coke make.

At this stage of the commercial trial, detailed inspection of each product yield is not warranted. However, the major yield changes follow a pattern suggested by pilot unit performance:

Increased activity
Increased gasoline yield
Increased RON and MON

Summary and Conclusions

Framework silicon-enriched Y zeolites (LZ-210) when compared with aluminum-deficient faujasites exhibit both increased thermal and hydrothermal stability. Collapse temperatures as measured by DTA are at least 165°F higher for the enriched framework materials.

The increased stability afforded by the LZ-210 process is believed to be due to the drastically reduced number of defects compared with dealuminated structures, their variable but high SiO_2/Al_2O_3 ratio, and their inherently low Na_2O content.

The LZ-210 preparative technique produces zeolites of very low Na_2O level (as low as 0.05 wt.%) which in turn results in an inherently higher ratio of cracking to H-transfer and increased gasoline octane number.

High stability zone zeolite catalysts containing LZ-210 exhibit less non-selective cracking, resulting in both improved gasoline and coke selectivity while maintaining octane levels at least as high as dealuminated species.

The improved selectivity is believed due to the much reduced silica/alumina and alumina "debris" present in the LZ-210 compared with dealuminated structures.

Low Na_2O zeolites are also more vanadium tolerant since vanadium and sodium are virtually additive in crystal destabilization ability.

This article has described two high octane catalyst series which are available with a wide range of activities. The ALPHA series provides maximum octane barrels via maximum gasoline yield with high octane. BETA catalysts provide high octane barrels with premium octane. In comparison with well-known competitive octane catalysts, one or both of the series shows superior cracking characteristics. This provides the refiner with a wide selection of catalysts to meet individual needs and a portfolio of catalysts which can be custom-made as required.

Three commercial trials of ALPHA and BETA catalysts are in the early stages. A trial in Europe with a BETA catalyst at only about 20% changeover shows some increase in conversion, gasoline, and octane relative to our premium octane catalyst, DELTA 400.

This article was first presented at the Katalistiks' 7th Annual Fluid Cat Cracking Symposium, Venice, Italy, May 12, 1986, and is used with special permission.

References

1. J. S. Magee and R. E. Ritter, *Symposium on Octane in the 1980s*, ACS Division of Petroleum Chemistry, September 10–15, 1978.
2. L. A. Pine, P. J. Maher, and W. A. Wachter, *J. Catal.*, 85, 466–476 (1984).
3. W. J. Reagan, G. M. Woltermann, and S. M. Brown, *Symposium on Advances in Catalytic Cracking*, ACS Division of Petroleum Chemistry, August 28–September 2, 1983, p. 884.
4. J. S. Magee, W. E. Cormier, and G. M. Woltermann, *Oil Gas J.*, May 27, 1985.

5. J. S. Magee and J. J. Blazek, *Preparation and Performance of Zeolite Cracking Catalysts*, in ACS Monograph 171, 1976, pp. 615–679.
6. P. B. Venuto and E. T. Habib Jr., *Fluid Catalytic Cracking with Zeolite Catalysts*, Dekker, New York, 1979.
7. M. L. Poutsma, *Mechanistic Consideration of Hydrocarbon Transformation Catalyzed by Zeolite*, in ACS Monograph 117, 1976, pp. 437–551.
8. J. A. Rabo, *Zeolite Chemistry and Catalysis*, ACS Monograph 171, 1976.
9. C. V. McDaniel et al., *Zeolite Stability and Ultrastable Zeolite*, in ACS Monograph 171, 1976, pp. 285–331.
10. G. T. Kerr, *J. Phys. Chem.*, *72*, 2594 (1968).
11. G. T. Kerr, *J. Phys. Chem.*, *73*, 2780 (1969).
12. G. W. Skeels and D. W. Breck, *Zeolite Chemistry V*, Sixth International Zeolite Conference, 1984, pp. 87–96.
13. R. Pompe et al., *Appl. Catal.*, *13*, 171–179 (1984).
14. Y. Nishimura et al., *Symposium on Processing Heavy Oils and Residua*, ACS Division of Petroleum Chemistry, March 20–25, 1983, p. 707.
15. B. R. Mitchell, *Ind. Eng. Chem., Prod. Res. Dev., 19*, 209 (1980).

JOHN S. MAGEE
BRUCE R. MITCHELL
JAMES W. MOORE

Octane Options

Planning to increase octane requires examining the gasoline pool and determining how best to produce new components. Refiners all face higher octane requirements due to lead phaseout and engine performance demands.

The United States is expected to essentially eliminate allowable lead usage by the early 1990s while increasing the percent of unleaded premium. The result will be to increase the pool average from 86.8 to 88.3 $(R + M)/2$.

TABLE 1 Typical Gasoline Pool Composition

	Pool (%)
Butanes	0–10
Light naphtha	10–20
Reformate	20–60
FCC gasoline	25–50
Alkylate/poly gasoline	0–15
Others (cracked naphthas, oxygenates, purchased aromatics)	0–10

Western Europe plans to eliminate use of lead, although the timing is uncertain. Their pool octane number should increase from 92.2/82.1 to 94.6/84.7 RON/MON as a result.

Japan, which now uses no lead, is expected to increase pool octane from 91 to 92 RON.

A large number of options are available for providing additional octane to the typical gasoline pool (Table 1). Each has been evaluated for impact on pool octane and cost (¢/octane-barrel added) sensitivity to: feedstock, direct and indirect operating expenses, by-product credits and debits, RVP credits and debits, and investment requirement (using a 35% capital charge). Feedstock, product, and utility values used are shown in Table 2.

Results fall into groups based on relative attractiveness: those found that significantly upgrade the pool octane at lower cost; those that have either a smaller impact on the pool or cost more to implement; and those that require a longer-term capital commitment or need a price structure change to look attractive.

TABLE 2

Feedstock and Products

	Price, $/unit	
	U.S. gallon	MT
Propane	0.32	166
Mixed butanes	0.43	197
FCC C_3's	0.40	207
FCC C_4's	0.47	215
96 RON reformate	0.72	238
Methanol	0.45	135
Ethanol	1.40	469
TBA/methanol	0.66	227
MTBE	0.90	318
Gasoline at 87 (R + M)/2	0.61	207
Benzene	1.10	329
Toluene	0.95	288

Utilities

	Price, $/unit
Hydrogen, Mscf (MT)	2.40 (996)
Fuel gas, 10^6 Btu (GJ)	3.00 (2.84)
600 lb/in.^{2} gauge steam, Mlb (MT)	4.50 (9.90)
150 lb/in.^{2} gauge steam, Mlb (MT)	3.40 (7.48)
50 lb/in.^{2} gauge steam, Mlb (MT)	3.00 (6.60)
Boiler H_2O, Mlb (MT)	0.80 (1.78)
Cooling H_2O, M U.S. gal (m^3)	0.05 (0.013)
Power, kWh	0.04 —

Best Choices

Octane upgrading options judged most attractive are:

Once-through isomerization
Increasing reformer severity to produce one or two higher octane number reformate
Changing from semiregenerative to continuous reformer operation
Use of a high octane FCC catalyst
Conversion of LPG to aromatics (Cyclar Process) when hydrogen carries a high value
Use of a 50/50 isopropyl alcohol (IPA) and methanol blend
Addition of ethanol in the United States (because of current tax concessions totaling about 60¢/gal and the RVP waiver)

Next Best

Those options that impart a lesser pool octane impact or are more expensive are:

Recycle isomerization. This gives a larger octane increase than once-through isomerization, but costs slightly more
Revamp reformers and FCC units to allow for octane increases above those attainable by catalyst change alone
Selective hydrogenation (SHP) of FCC olefin feed to an HF alkylation unit
Production of MTBE from FCC butenes
Propylene dimerization (Hexall Process) via revamp of catalytic polymerization unit
Purchase of 50/50 *tert*-butyl alcohol (TBA) and methanol blend. Although this approach is attractive, material availability greatly limits its potential

Future Choices

Several options require a longer-term capital commitment or a change in pricing to be immediately interesting.

Addition of a new HF alkylation unit to use available FCC olefins. This is less cost-effective than other FCC olefin processing options

Octane Options

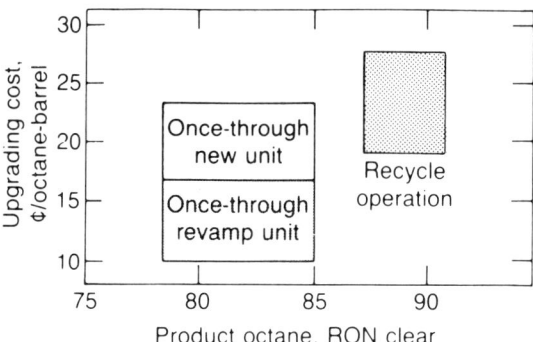

FIG. 1. Cost for upgrading octane by isomerization (feedstock: C_5/C_6 LSR, 68 RON, 30 ppm S).

Dehydrogenation of field butanes for use as isobutylene in preparation of MTBE or TBA. Significant capital investment is required but the octane-barrel cost is reasonable

Aromatization of propane and butanes (Cyclar Process) when hydrogen is low value

Dimerization of available FCC propylenes (Hexall Process) via a new unit

Reforming low octane portion of FCC gasoline after hydrotreating. This gives a sizable pool octane increase, particularly in MON, but the cost per octane-barrel is high

Outside purchase of toluene or MTBE carries a very high octane-barrel cost

The details and cost sensitivity of these options are evaluated according to functional groupings.

Light Naphtha Upgrading

Much information on light naphtha isomerization [1, 4, 5] shows significant octane improvement potential with a variety of options. Product octanes and upgrading costs associated with normal C_5/C_6 isomerization (the Penex design) are summarized in Fig. 1. All of the costs shown account for hydrogen consumption during isomerization and hydrotreating (where applicable). Once-through isomerization can give octanes in the range of 79 to 81 RON with a zeolitic catalyst that is very tolerant of contaminants and can process most feeds without hydrotreating. High activity alumina catalysts require feed hydrotreating but provide octanes of 83 to 86 RON in once-through operation. Higher octanes, in excess of 90 RON, can be obtained with a recycle isomerization/separation system (Penex/Molex) with costs around

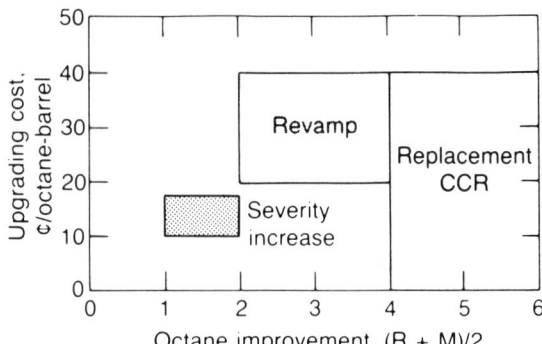

FIG. 2. Reformate octane upgrading.

25¢/octane-barrel. Light naphtha isomerization is clearly one of the more attractive options, particularly when existing equipment is available for reuse.

Reformate Upgrading

Upgrading reformate often is the most important option considered for meeting requirements of lead phasedown because of its quantity and the potential for improvement. The primary reforming alternatives investigated are severity increases via catalyst changes, unit revamps, and use of continuous regeneration reforming [2]. Figure 2 summarizes potential octane increases and costs associated with these alternatives. Use of newer high activity, stable catalysts allows reformer severity increases of 1.0 to 2.0 numbers depending on equipment limitations. Costs are attributable to yield

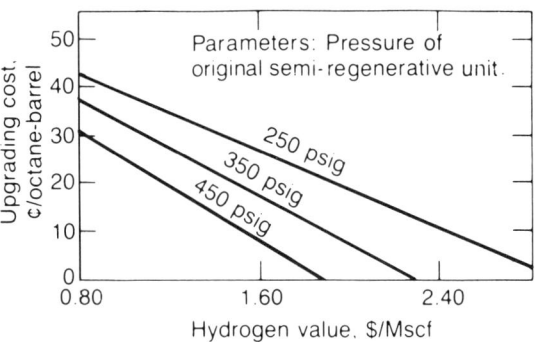

FIG. 3. Sensitivity of replacement continuous Platformer upgrading cost to hydrogen value (basis: 20,000 BPSD feed rate).

loss, higher utility consumption, and catalyst change out. Revamping an existing reformer, in conjunction with use of a high-activity catalyst, can produce as much as 4.0 (R + M)/2 improvement. Cost of this option is primarily due to capital charges for revamp that are unit specific. Replacing an existing unit with a continuous Platformer can increase reformate octane by as much as 6.0 numbers. The cost of replacement is quite sensitive to the pressure of the existing unit and the value of hydrogen (Fig. 3). This option looks extremely attractive when the existing unit is medium to high pressure. With high-cost hydrogen, a refiner can even justify replacing a low-pressure semiregenerative unit. Using the old reforming equipment to install isomerization further enhances the attractiveness of this option [8].

FCC Gasoline Upgrading

FCC gasoline is clearly one of the most frequently examined options for octane improvement, not because of any dramatic improvement potential, but because it is generally the largest component of the gasoline pool. Each FCC unit has design and operating parameters that dictate what octane improvement options can be implemented. The most important of these options are:

High octane catalyst
Severity increase via revamp
Reduction in gasoline end point

The octane improvement–cost relationship for these options is shown in Fig. 4.

Many United States FCC operators use octane catalysts that gain 1.0–1.5

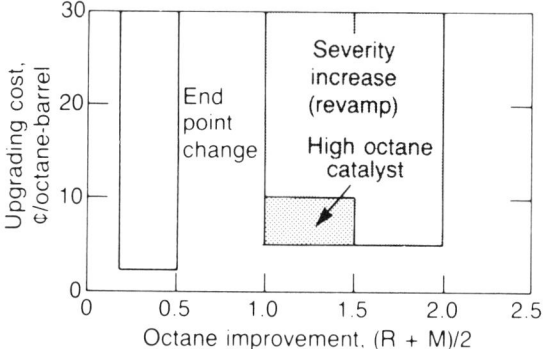

FIG. 4. FCC gasoline upgrading costs.

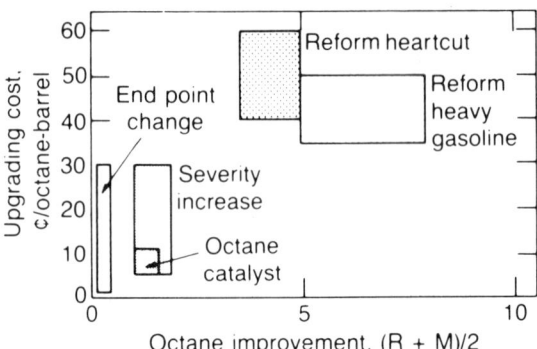

FIG. 5. Additional FCC gasoline upgrading costs.

numbers in FCC gasoline. The cost is 5 to 10¢/octane-barrel depending on catalyst price and consumption. The complexity involved in revamping an FCC unit to higher severity operation which is very dependent on unit specifics may cost from $1.0 to $5.0+ million for a 30,000 BPSD unit. The net cost for a 1 or 2 octane number increase, primarily attributable to the capital cost of the revamp, can range from 5 to 30¢/octane-barrel. Dropping the last 5 to 10% of FCC gasoline into the light cycle oil can increase octane by 0.2 to 0.5 (R + M)/2 at a cost that is essentially a function of the difference in price between FCC gasoline and light cycle oil.

Improving FCC gasoline octane can also be done by reforming the lowest octane portion of FCC gasoline. Depending on FCC operation, either a heartcut or a heavy FCC naphtha would be considered for reforming. These are compared with other FCC upgrading options in Fig. 5. While a very significant octane improvement can be gained by these steps, the high costs due to many processing steps make these less attractive options.

Alkylate and Poly Gasoline Upgrading

Most HF alkylation units are pushed to capacity when FCC units operate at higher severities. Under such circumstances there is little flexibility to increase alkylation gasoline octane. However, one step can be taken: add a new reactor in series. Improved heat removal capability keeps the alkylate octane as high as possible. Increasing isobutane-to-olefin ratio also helps increase the octane; however, for many units there is little room to increase this ratio. Addition of a selective hydrogenation process (SHP) unit on the alkylate feed stream can increase alkylate octane by 1.0 RON or more at a cost of 15 to 20¢/octane-barrel.

Existing catalytic condensation units are not very flexible at achieving increased octane with a given feedstock. Fixed bed propylene dimerization [3, 7] can be applied to existing catalytic condensation units via revamping.

Such a revamp gives improved product blending octane and product yield at a cost of 13 to 20¢/octane-barrel.

Oxygenates

The most widely discussed new components to the gasoline pool are oxygenated hydrocarbons, either methanol or ethanol based.

Methanol, which cannot be added directly to the pool due to phase separation problems in the presence of water, must be blended with higher alcohols such as IPA or TBA, or reacted with isobutylene to form MTBE. Ethanol requires no higher alcohol cosolvents; however, production cost makes ethanol unattractive without subsidies or tax concessions.

Table 3 summarizes the gasoline properties of MTBE, 50/50 IPA/methanol and TBA/methanol blends, and ethanol. Process technology is available for the production of MTBE, IPA, and TBA in the refinery [3]. In addition, MTBE, TBA/methanol blends, methanol, and ethanol may be purchased, although the quantity available for the first two is quite limited. For a hypothetical 100,000 BPSD FCC refinery, producing 45,000 BPSD of gasoline, the octane-barrel costs and percentage of pool that could be supplied by various oxygenate sources are shown in Table 4. The quantity of MTBE available from outside purchase or produced from the FCC C_4 stream is limited by isobutylene availability. The primary source of TBA is as a byproduct from propylene oxide production and is severely limited. The concentration of isobutylene in refinery C_4 streams is not high enough for TBA production. These feedstock limitations can be removed by installing isomerization and dehydrogenation processes to produce isobutylene from field butanes. Dehydrogenation can also be applied to refinery propane in order to increase IPA production, if desired.

IPA/methanol option, as shown in Table 4, is the lowest cost oxygenate option. It can be produced in significant volume from available FCC propylene. With existing United States tax allowance, ethanol is a marginally attractive option; without allowance it would be prohibitively expensive.

TABLE 3 Oxygenates Gasoline Properties

	Blend RON, Clear	Blend MON, Clear	Blend, RVP
MTBE	115–120	98–101	~10
IPA/methanol[a]	115–120	97–100	25–30[b]
TBA/methanol[a]	112–116	92–97	25–30[b]
Ethanol	120–130	98–104	~20

[a]50/50 blend.
[b]At 3.5% oxygen in gasoline.

TABLE 4[a] Oxygenate Cost for Gasoline Upgrading

Option	Percentage of Gasoline Pool	Cost ¢/Octane-barrel	Cost ¢/MON-barrel
MTBE purchase	Limited	40–60	45–70
MTBE via FCC butenes	1.8	20–30	35–45
MTBE via refinery C_4's	11.0[e]	30–45	45–55
IPA/methanol[d] via FCC propylene	7.0	20–25	30–40
TBA/methanol[d] purchase	Limited	25–30	15–25
TBA/methanol[d] via refinery C_4's	9.6[e]	35–50	65–75
Ethanol purchase	As required	30–40[b] >$2.00[c]	— —

[a] Basis: 100,000 BPSD FCC refinery, 45,000 BPSD gasoline.
[b] With 60¢/gallon tax allowance (U.S.).
[c] Without any tax allowance.
[d] 50/50 blend.
[e] Maximum allowable (U.S.).

MTBE production from FCC isobutylene is also cost attractive, but total potential United States production from this source is limited to about 70,000 BPSD. Addition of dehydrogenation technology makes significant quantities available but at a higher cost. Outside purchase of MTBE is not attractive at current United States prices. Purchase of TBA/methanol, on the other hand, is similar in cost to MTBE production, but quantities are severely limited. Dehydrogenation technology can provide sufficient TBA at a higher cost.

The octane upgrading costs for oxygenates (Figs. 6–10) can change dramatically with shifts in feedstock costs and gasoline prices.

Oxygenates will be restricted to use primarily in premium grade gasoline because of their high octane value and limited availability. Where MON is the controlling specification (Western Europe), the large sensitivity of oxygenates

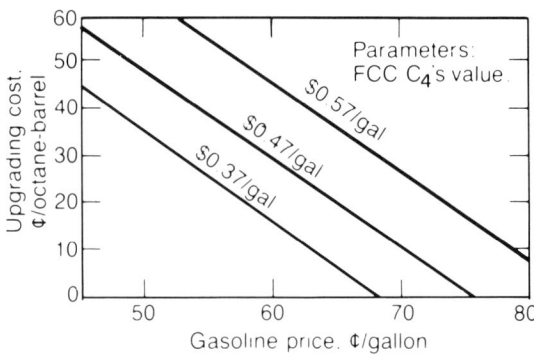

FIG. 6. MTBE upgrading cost sensitivity to gasoline price (basis: 820 BPSD MTBE).

Octane Options

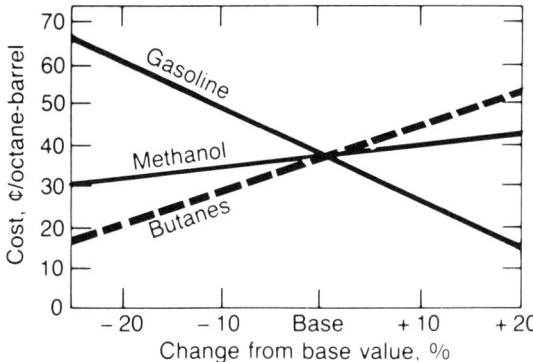

FIG. 7. MTBE dehydrogenation (Oleflex), sensitivity of production cost to gasoline, methanol, and butane values (basis: 5300 BPSD MTBE). Base values: gasoline, 61¢/gal; methanol, 45¢/gal; butanes, 43¢/gal.

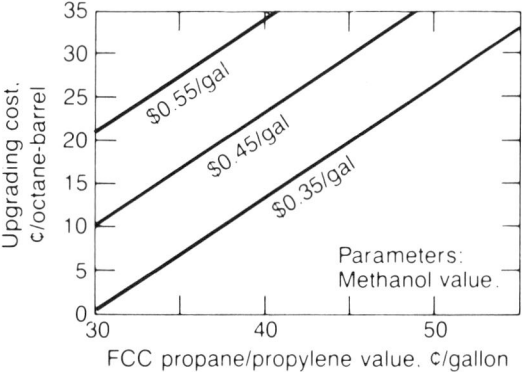

FIG. 8. IPA/methanol octane upgrading cost sensitivity to FCC C_3's and methanol values (basis: 3150 BPSD 50/50 IPA/methanol).

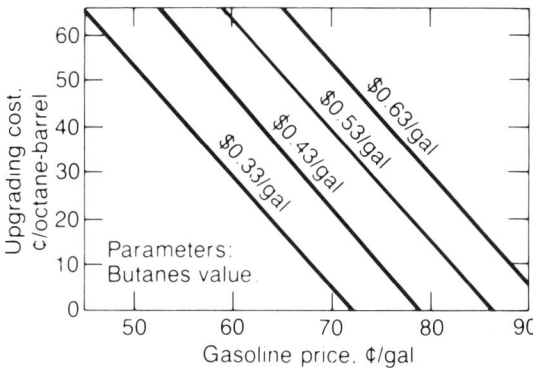

FIG. 9. TBA/methanol via dehydrogenation (Oleflex), sensitivity of upgrading cost to gasoline and butane price (basis: 7400 BPSD TBA/methanol and 620 BPSD MTBE).

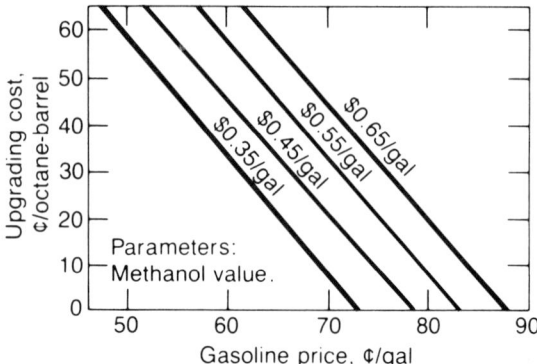

FIG. 10. TBA/methanol via dehydrogenation (Oleflex), sensitivity of upgrading cost to gasoline and methanol price (basis: 7400 BPSD TBA/methanol and 620 BPSD MTBE).

makes them a less attractive upgrading option. Furthermore, in the future it is quite possible that a debit will be incurred by oxygenates because of their low energy value per unit volume. Finally, alcohols are the subject of considerable debate regarding engine performance problems. Resolution of this debate will have a significant impact on the attractiveness of alcohols in future gasoline blends.

Light Hydrocarbons

HF alkylation is likely to be of some importance in Europe because of the large MON benefit it affords among traditional processes for converting light hydrocarbons. An HF alkylation unit in a 100,000 BPSD FCC refinery produces about 3700 BPSD of allkylate and has an upgrading cost of about 45¢/octane-barrel or 30¢/MON-barrel.

A number of new processes have been developed to convert C_3/C_4 hydrocarbons to high octane gasoline including those discussed earlier. Two other processes, aromatization of C_3/C_4 and solid bed dimerization of propylene to high octane hexene, are available also.

Aromatization of C_3/C_4 converts LPG to aromatics, offering a high-octane product as well as creating a large potential source of high-purity hydrogen [6]. Typical product blend octanes are 106 for the C_6+ product and 108 for the C_7+ product, which is about 70% of the liquid product. The C_7+ product is the potential gasoline blending component in situations where there is a benzene limitation or benzene has a high petrochemical value. In addition to very high octane, the product has favorable vapor pressure and distillation characteristics.

The octane-barrel upgrading cost for this process is quite sensitive to feedstock and product values (Figs. 11 and 12). When hydrogen is required for

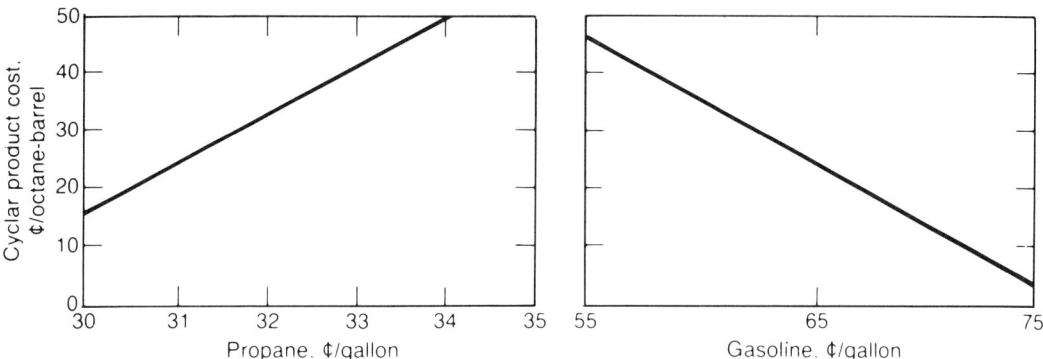

FIG. 11. Cyclar product cost vs propane and gasoline price.

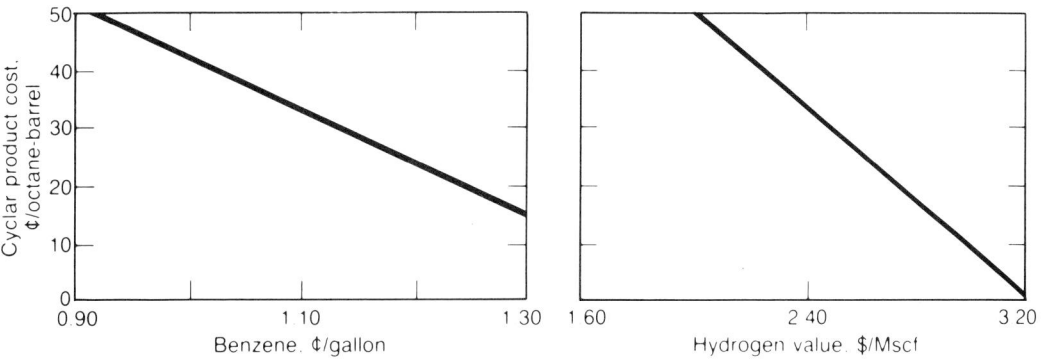

FIG. 12. Cyclar product cost vs benzene and hydrogen price.

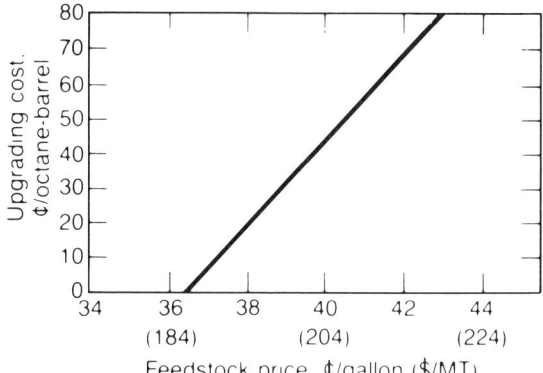

FIG. 13. Sensitivity of Hexall upgrading cost to feedstock price (basis: 5000 BPSD of 76% propylene/24% propane feed).

TABLE 5 Gasoline Octane Upgrading Costs [(R + M)/2 basis]

Stream	Percent of Gasoline Pool	Potential (R + M)/2 Impact	Cost ¢/Octane-barrel
Light naphtha:			
Once-through Penex (revamp)	10	1.0–1.6	10–15
Once-through Penex (new unit)	10	1.0–1.6	15–20
Recycle Penex/Molex	10	1.8–2.2	18–25
Recycle Penex/DIH	10	1.8–2.2	20–25
Reformate:			
Increased octane-catalyst change	30	0.4–0.6	10–20
Increased octane-revamp	30	0.5–1.0	20–40
New continuous Platformer	30	1.3–2.0	Negative–40
FCC gasoline:			
High octane catalyst	40	0.4–0.6	5–7
High severity revamp	40	0.4–0.8	5–30
Gasoline end point change	40	0.1–0.2	0–30
Reform FCC heart cut	12	1.4–1.8	40–50
Reform FCC heavy cut	24	1.9–3.0	35–50
Alkylate/cat poly:			
Selective hydrogenation (SHP)	10	0.1	15–20
Poly revamp to Hexall	2–5	0.1	0–20
Use of oxygenates:			
MTBE purchase	2–5	0.4–1.1	40–60
MTBE via FCC butenes	2–5	0.4–1.1	25–30
MTBE via Oleflex	5–10	1.1–2.2	30–45
IPA/methanol	3–7	0.6–1.4	20–25
TBA/methanol purchase	2–5	0.3–0.8	25–30
TBA/methanol via Oleflex	5–10	0.8–1.6	35–50
Ethanol (with 60¢/gal tax credit)	10 maximum	2.5 maximum	30–40
Light hydrocarbons:			
HF alkylation—new unit	5–10	0.4–0.8	40–50
Cyclar	5–10	1.0–2.0	10–60
Hexall—new unit	~4	~0.2	80–100
Other primary option:			
Toluene purchase	2–5	0.4–1.1	40–80

hydroprocessing and a reasonably priced feedstock is available, this approach is very attractive.

If refinery C_3 olefins are available, their conversion to hexenes by propylene dimerization represents an attractive option for augmenting gasoline pool octane [7]. The investment for a fixed bed unit to process 5000 BPSD of FCC C_3's is $12.0 million to produce 2768 BPSD of product with a blending octane of about 93 (R + M)/2 (101 RON and 85 MON). The octane upgrading cost for this product is a function of feedstock price (Fig. 13).

The potential impact on the gasoline pool and the costs associated with each improvement option are shown in Table 5. Ranges given are intended to account for variations in gasoline pool composition and in current feedstock, product, and utility values.

TABLE 6 Gasoline Upgrading Costs (MON basis)

Stream	Percent of Gasoline Pool	MON Impact	Cost ¢/Octane-barrel
Light naphtha:			
Once-through Penex (revamp)	10	1.0–1.6	10–15
Once-through Penex (new unit)	10	1.0–1.6	15–20
Recycle Penex/Molex	10	1.8–2.2	18–25
Recycle Penex/DIH	10	1.8–2.2	20–25
Reformate:			
Increased octane-catalyst change	50	0.5–0.8	12–25
Increased octane-revamp	50	0.6–1.4	24–45
New continuous Platformer	50	1.9–2.7	Negative–40
FCC gasoline:			
High octane catalyst	40	0.3–0.6	7–10
High severity revamp	40	0.3–0.6	7–10
Reform FCC heart cut	12	1.4–1.8	40–50
Reform FCC heavy cut	24	1.9–3.0	35–50
Alkylate/cat poly:			
Selective hydrogenation (SHP)	10	0.1	15–20
Poly revamp to Hexall	2–5	0.1	30–40
Use of oxygenates:			
MTBE purchase	2–5	0.4–1.0	45–70
MTBE via FCC butenes	2–5	0.3–0.9	35–45
MTBE via Oleflex	5–10	0.8–1.8	45–55
IPA/methanol	3–7	0.4–1.2	30–40
TBA/methanol purchase	2–5	0.2–0.6	15–25
TBA/methanol via Oleflex	4	0.6–1.4	65–75
Light hydrocarbons:			
HF alkylation—new unit	5–10	0.5–1.2	25–35
Cyclar	5–10	0.8–1.7	15–70
Other primary option:			
Toluene	2–5	0.4–1.0	45–90

Western Europe has a particular concern with gasoline sensitivity, causing those options with a high potential (R + M)/2 improvement to have reduced attractiveness when the pool MON improvement is low. Table 6 shows the potential MON impact of the options for use in Western Europe. The largest potential pool MON gain is obtained by fractionating out the low octane portion of the FCC gasoline stream and upgrading it by reforming. However, this is one of the most expensive options examined and will probably remain a last resort. Beyond FCC gasoline reforming, the largest potential MON gains can be obtained via isomerization of the light naphtha stream, especially with recycle, or upgrading of the reformate stream. Since these options are considerably lower in cost, they most likely will find wide application.

Overall, the United States, Western Europe, and Japan will use a variety of options (Table 7). With the diversity of individual refinery situations and available options, no single solution can be identified as the best.

TABLE 7 Geographical Distribution of Octane Upgrading Options

Option	United States (%)	Western Europe (%)	Japan (%)
Upgrading existing process units (Reforming, FCC, alkylation)	30–40	20–30	45–55
New process units (Penex, CCR Platformer, Cyclar)	40–50	50–60	45–50
Use of oxygenates	15–25	15–25	0–5

This material appeared in *Hydrocarbon Processing*, pp. 41–45, June 1986, and is reprinted with the permission of the editor.

References

1. T. Wheeler, S. H. Hobbs, and A. P. Krueding, *The Penex Process—Providing Future Octane Needs*, Presented at the 1985 UOP Technology Conference.
2. R. L. Peer, J. A. Weiszmann, and R. W. Bennett, *Platforming: Its Flexibility Meets Today's Changing Requirements*, Presented at the 1985 UOP Technology Conference.
3. J. A. Johnson, J. R. Mowry, and R. F. Anderson, *LPG Processing Options*, Presented at the 1985 UOP Technology Conference.
4. A. P. Krueding, J. A. Johnson, and S. W. Pappas, *New UOP Catalysts Plus Revamps to Meet Your Octane Goals*, Presented at the National Petroleum Refiners Association Annual Meeting, March 24–26, 1985.
5. R. J. Schmidt, J. A. Weiszmann, and J. A. Johnson, "Catalysts—Key to Low Cost Isomerization," *Oil Gas J.*, *83*(21) (1985).
6. R. F. Anderson, J. A. Johnson, and J. R. Mowry, *Cyclar: One Step Processing of LPG to Aromatics and Hydrogen*, Presented at the American Institute of Chemical Engineers Spring Meeting, March 24–28, 1985.
7. D. J. Ward, R. H. Friedlander, R. Frame, and T. Imai, *Increase Gasoline Yields with a New Propylene Dimerization Process*, Presented at the National Petroleum Refiners Association Annual Meeting, March 24–26, 1985.
8. J. H. D'Auria, B. L. Schaefer, and R. J. Schmidt, *A Two Step Approach to Increasing Octane and Yield*, Presented at the NPRA Annual Meeting, March 23–25, 1986.

JOSEPH A. WEISZMANN
JAMES H. D'AURIA
FREDERICK G. McWILLIAMS
FREDERICK M. HIBBS

2
Refining

Petroleum Processing

The world presently gets most of its energy from crude oil and natural gas. Petroleum is the major source of fuel used in transportation, manufacturing, and home heating.

Primary energy sources are defined as those coming from natural raw materials. The primary energy sources for the world during 1979 are reported [1] in Fig. 1. Oil and gas together furnished 65% of the total world energy usage for that year.

Note electricity is missing from this representation. Electricity is a secondary energy source because it is generated by consuming some of the other natural resources shown in Fig. 1. Thus, electricity should not appear in an energy balance unless the fuel from which it is generated is omitted from the other totals. Too often, this correction is not made and an inflated energy supply results.

World proven crude oil reserves determined for several earlier years [2] are shown in Table 1. Also included in this table are the annual production rates [3] and the ratio of the reserves to annual production.

What is disturbing at present is the fact that new oil sources are not being found fast enough to keep up with rapidly growing consumption rates. Furthermore, when oil is found, it is generally at greater depths or under seabeds farther from shore than was the case before. Reports proliferate now regarding the ultimate reserves likely to be found in the world.

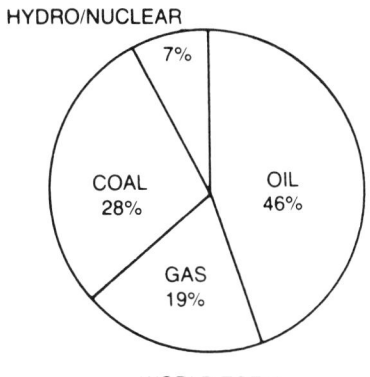

FIG. 1. World energy consumption, 1979. World total equals 140×10^6 daily barrels oil equivalent.

TABLE 1 World Proven Crude Oil Reserves Compared with Production Rates (10^9 barrels)

Year	Proven Reserves, First of Year [2]	Annual Production [3]	Ratio, Reserves Divided by Production, Years
1935	23.4	1.65	14.2
1940	34.9	2.15	16.2
1945	51.2	2.59	19.8
1950	76.4	3.80	20.1
1955	153.7	5.63	27.3
1960	255.8	7.67	33.4
1965	341.7	11.06	30.9
1970	516.7	16.69	31.0
1975	569.1	19.50	29.2
1980	625.8	22.4	28.4
1985	696.3	20.36	34.11
1986	692.8	20.15	34.38

Ultimate recoverable petroleum reserves have been estimated with wide variations, the consensus being between 1.5 to 2.5 trillion barrels (10^{12} bbl). Most estimators go to great length to explain the basis of their estimates. By comparison, proven reserves are in the neighborhood of 600 billion barrels. Proven reserves are generally taken to mean: "the oil remaining in the ground which geological and engineering information indicate with reasonable certainty to be recoverable in the future from known reservoirs under existing economic and operating conditions" [2].

Alternate feedstocks for refineries are being sought from coal, tar sands, and shale oil. Synthetic crude oil will be made from these raw materials so that conventional refining units can continue to be used to make consumer products.

In the meantime, top priority will be given to use crude oil to make liquid transportation fuels (because of their convenience) and petrochemical materials (because of their diversity of uses). It's a good bet that many of the gross conversion methods now applied to crude oil processing will be replaced in the future by more specific conversion. It is in this transition that knowledge of organic chemistry guides the development of new refining processes.

From Well to Refinery

A country-by-country listing [1] of crude oil production and consumption shows the importance of petroleum movement around the world. Production and demand rates of crude oil for various countries are shown in Table 2.

TABLE 2 Worldwide Petroleum Production and Demand for 1985 (thousands of barrels per day)[a]

Continent[b] and Country[c]	Production[d]	Demand	Production–Demand Ratio
North America, total	14,892	16,640	0.89
Canada	1,568	1,470	1.07
Mexico	2,703	e	—
United States	10,621	15,170	0.70
South America, total	3,623	4,430	0.82
Argentina	460	e	—
Brazil	563	e	—
Colombia	176	e	—
Trinidad and Tobago	185	e	—
Venezuela**	1,693	e	—
Other South America	546	e	—
Eastern Europe, total*	12,344	11,610	1.06
USSR	11,982	9,060	1.32
Other Eastern Europe	362	2,550	0.14
Western Europe, total	3,731	11,950	0.31
France	53	1,785	0.03
Germany, West	81	2,410	0.03
Italy	44	1,745	0.03
Netherlands	72	640	0.11
Norway	803	190	4.23
United Kingdom	2,444	1,635	1.49
Other Western Europe	234	3,545	0.07
Africa, total	5,284	1,715	3.08
Algeria**	980	e	—
Egypt	895	e	—
Libya**	1,050	e	—
Nigeria**	1,480	e	—
South Africa	e	e	—
Other Africa	879	e	—
Middle East, total	10,580	1,980	5.34
Iran**	2,258	e	—
Iraq**	1,436	e	—
Kuwait**	939	e	—
Saudi Arabia**	3,255	e	—
Other Middle East	2,692	e	—
Far East, total	6,958	10,140	0.69
Australia and New Zealand	597	670	0.89
Brunei and Malaysia	1,830	e	—
China*	2,498	1,760	1.42
India	606	e	—
Indonesia**	1,315	e	—
Japan	10	4,320	0.00
Other Far East	102	e	—
Total world	57,412	58,465	0.98
Communist controlled countries*	14,842	13,370	1.11
OPEC countries**	14,406		—

[a] Authority: United States Department of Energy; BP Statistical Review of World Energy, June 1985; World Oil; Statistics Canada.
[b] Continent totals may include countries not identified.
[c] Communist controlled countries are identified by a single asterisk; OPEC countries by a double asterisk.
[d] When available, figures include natural gas liquids production.
[e] Data not available.

The growth of world refining capacity attempts to keep up with the growing demand for petroleum products. A measure of this growth is shown in Fig. 2. The upper curve shows total refining capacity, while the lower curve shows the amount of crude oil run through the refineries.

One might wonder why refining capacity continued to surge ahead when crude throughput took a dip in the period from 1974 through 1976. For one thing, the amount of crude oil available for processing is subject to the whims of international trade. Since so much crude oil comes from some countries (notably, Middle Eastern countries) to be refined in other countries, international relations between countries are a strong factor that determines how much crude oil feedstock is available.

Another factor in refining growth is the time required to construct processing units. In highly industrialized countries like the United States, Japan, and Western European countries, there are mounting restrictions on new refinery sites. Thus, a decision to build a refinery and the actual completion of that refining capacity will take several years in order to fulfill local and governmental requirements. Then one to four years of actual construction activities are required before a new refinery will start processing feedstocks.

Refineries are located mostly in the countries consuming refined products. It is easier to transport crude oil to major refining centers than to transport separately the many individual products. The distribution of refining capacity [1] by areas for 1980 is depicted in Fig. 3.

The variety of ways crude oils are delivered to refineries is indicated by using United States refineries as an example. United States refineries get their feedstock via pipelines, tank trucks, barges, and ocean-going vessels [4]. The amount received by each of these routes is shown in Table 3. The large quantity coming by water explains why many refineries are located near oceans and why they own or lease such as large fleet of barges and ocean-going vessels.

Product Names

The distinction between refined products and petrochemicals is often a subtle one. In general, when the product is a fraction from crude oil that includes a fairly large group of hydrocarbons, the fraction is classified as a refined product. Examples of refined products are: gasoline, diesel fuel, heating oils, lubricants, waxes, asphalts, and petroleum coke.

By contrast, when the product from crude oil is limited to only one or two specific hydrocarbons of fairly high purity, the fraction is called a petrochemical. Examples of petrochemicals are: ethylene, propylene, benzene, toluene, and styrene—to name only a few.

There are many more identifiable petrochemical products than there are refined products. There are many specific hydrocarbons that can be derived from petroleum. However, these hydrocarbons lose individual identity when they are grouped into a refined product.

Petroleum Processing

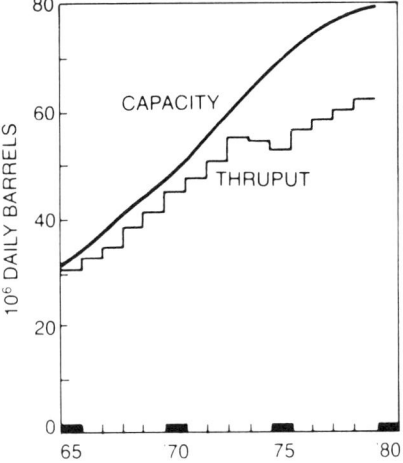

FIG. 2. World refining capacity and crude oil throughputs.

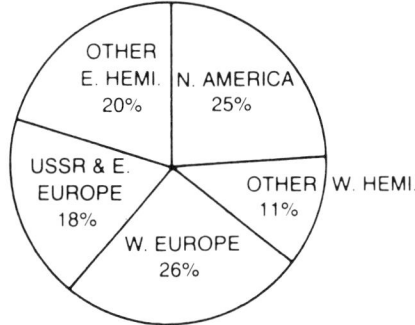

WORLD TOTAL = 80.37 × 10^6 DAILY BARRELS

FIG. 3. World refining capacity in early 1980.

TABLE 3 Method of Transportation for Crude Oil Received by United States Refineries, 1979 [4]

Transportation Method	Vol.%
Domestic crude oil:	
Pipelines	42.9
Tank cars and trucks	2.1
Tankers and barges	11.0
Subtotal	56.0
Foreign crude oil:	
Pipelines	12.7
Tankers and barges	31.3
Subtotal	44.0
Total receipts	100.0

Refined Products

Most refined products at the consumer level are blends of several refinery streams. Product specifications determine which streams are suitable for a specific blend. Part of the difficulty of learning about refining lies in the industry's use of stream names that are different from the names of the consumer products.

Consider the listing in Table 4. The names in the last column should be familiar because they are used at the consumer level. Yet within a refinery, these products will be blended from portions of crude oil fractions having the names shown in the first column. To make matters worse, specifications and statistics for the industry are often reported under yet another set of names—those shown in the middle column of Table 4.

Gasoline at the consumer level, for example, may be called benzol or petrol, depending on the country where it is sold. In the early stages of crude oil processing, most gasoline components are called naphthas. Kerosene is another example. It may be called coal oil to denote that it replaces stove oil (or range oil) once derived from coal. Kerosene's historical significance was first as an illuminating oil for lamps that once burned sperm oil taken from whales. But today, kerosene fractions go mostly into transportation fuels such as jet fuel and high quality No. 1 heating oil.

Product Specifications

Product application and customer acceptance set detailed specifications for various product properties. In the United States, the American Society for Testing and Materials (ASTM) and the American Petroleum Institute (API) are recognized for establishing specifications on both products and methods for testing. Other countries have similar referee organizations. For example,

TABLE 4 Several Names for the Same Material

Crude Oil Cuts	Refinery Blends	Consumer Products
Gases	Still gases	Fuel gas
	Propane/butane	Liquefied petroleum gas (LPG)
Light/heavy naphtha	Motor fuel	Gasoline
	Aviation turbine, Jet-B	Jet fuel (naphtha type)
Kerosene	Aviation turbine, Jet-A	Jet fuel (kerosene type)
	No. 1 fuel oil	Kerosene (range oil)
Light gas oil	Diesel	Auto and tractor diesel
	No. 2 fuel oil	Home heating oil
Heavy gas oil	No. 4 fuel oil	Commercial heating oil
	No. 5 fuel oil	Industrial heating oil
	Bright stock	Lubricants
Residuals	No. 6 fuel oil	Bunker C oil
	Heavy residual	Asphalt
	Coke	Coke

in the United Kingdom, it is the Institute of Petroleum (IP). In West Germany, it is Deutsches Institute fuer Normung (DIN). In Japan, it is the Ministry of International Trade and Industry (MITI).

Boiling range is the major distinction among refined products, and many other properties are related directly to the products in these boiling ranges. A summary of ASTM specifications for fuel boiling ranges [5] is given in Table 5.

TABLE 5 Major Petroleum Products and Their Specified Boiling Range [5]

Product Designation	ASTM Designation	Specified Temperature for vol.% Distilled at 1 atm (°F)		
		10	50	90
Liquefied petroleum gas (LPG):	D 1835			
Commercial propane		—[a]		—[b]
Commercial butane		—[a]		—[c]
Aviation gasoline (Avgas)	D 910	158 max	221 max	275 max[d]
Automotive gasoline:	D 439			
Volatility class A		158 max	170–250	374 max[e]
Volatility class B		149 max	170–245	374 max[e]
Volatility class C		140 max	170–240	365 max[e]
Volatility class D		131 max	170–235	365 max[e]
Volatility class E		122 max	170–230	365 max[e]
Aviation turbine fuel:	D 1655			
Jet A or A-1		400 max		—[f]
Jet B		—[g]	370 max	470 max
Diesel fuel oil:	D 975			
Grade 1-D				550 max
Grade 2-D				540–640
Grade 4-D		—	Not specified	—
Gas turbine fuel oil:	D 2880			
No. 0-GT		—	Not specified	—
No. 1-GT				550 max
No. 2-GT				540–640
No. 3-GT		—	Not specified	—
No. 4-GT		—	Not specified	—
Fuel oil:	D 396			
Grade No. 1		420 max		550 max
Grade No. 2		—[h]		540–640
Grade No. 4		—	Not specified	—
Grade No. 5		—	Not specified	—
Grade No. 6		—	Not specified	—

[a] Vapor pressure specified instead of front end distillation.
[b] 95% point, −37°F max.
[c] 95% point, 36°F max.
[d] Final point, 338°F max.
[e] Final point, all classes, 437° F max.
[f] Final point, 572°F max.
[g] 20% point, 290°F max.
[h] Flash point specified instead of front end distillation.

Boiling range also is used to identify individual refinery streams—as an example will show in a later section concerning crude oil distillation. The temperature that separates one fraction from an adjacent fraction will differ from refinery to refinery. Factors influencing the choice of cut point temperatures includes the following: type of crude oil feed, kind and size of downstream processes, and relative market demand among products.

Other specifications can involve either physical or chemical properties. Generally these specifications are stated as minimum or maximum quantities. Once a product qualifies to be in a certain group, it may receive a premium price by virtue of exceeding minimum specifications or by being below maximum specifications. Yet all too often, the only advantage for being better than specifications is an increase in the volume of sales in a competitive market.

The evolution of product specifications will, at times, appear sadly behind recent developments in more sophisticated analytical techniques. Certainly the ultimate specifications should be based on how well a product performs in use. Yet the industry has grown comfortable with certain comparisons, and these standards are retained for easier comparison with earlier products. Thus, it is not uncommon to find petroleum products sold under an array of tests and specifications—some seemingly measuring similar properties.

It is behind the scenes that sophisticated analytical techniques prove their worth. These techniques are used to identify specific hydrocarbons responsible for one property or another. Then suitable refining processes are devised to accomplish a desired chemical reaction that will increase the production of specific types of hydrocarbons.

In the discussion on refining schemes, major specifications will be identified for each product category. It will be left to the reader to remember that a wide variety of other specifications also must be met.

Product Yields

As changes occur in relative demand for refined products, refiners turn their attention to ways that will alter internal refinery streams. The big problem here is that the increase in volume of one fraction of crude oil will deprive some other product of that same fraction. This point is often overlooked when the question arises: "How much of a specific product can a refinery make?" Such a question should always be followed by a second question: "What other products will be penalized?"

Envision, for example, what would happen if the refining industry were to make all the gasoline it possibly could with today's present technology. The result would be to rob many other petroleum products. A vehicle which needs gasoline for fuel also needs such products as industrial fuels to fabricate the vehicle, lubricants for the engine's operation, asphalt for roads upon which the vehicle is to move, and petrochemical plastics and fibers for the vehicle's interior. Until adequate substitutes are found for these other petroleum products, it would be unwise to make only one product, even though sufficient technology may exist to offer this option.

TABLE 6 Product Yields from United States Refineries, 1986 [4]

Product	Vol.% of Refinery Input
Still gas	4.0
Ethane/ethylene	0.1
Liquefied gas	2.4
Gasoline	42.1
Jet fuel	7.4
Kerosene	1.6
Special naphtha	0.8
Petrochemical feed	4.4
Distillates	20.9
Lubricants	1.2
Waxes	0.1
Coke	2.3
Asphalt	3.0
Road oil	0.1
Residuals	11.8
Miscellaneous	0.8
Total	110.4[a]

[a]Includes 10.4% vol.% gain because most products are of less density than original feedstock.

This is not to say that substitutes will not be found nor that these substitutes will not be better than petroleum products. In fact, many forecasts suggest that petroleum will ultimately be allocated only to transportation fuels and petrochemical feedstocks. It appears that these uses are the most suitable options for petroleum crude oil.

In the United States, the relative portions of refined products [4] made from crude oil are shown in Table 6. This distribution of products is the result of a long-standing trend to convert the heavier less valuable fractions into lighter, more valuable fractions. The ways this can be done will be discussed in the section on refinery schemes.

Petrochemicals

The portion of crude oil going to petrochemicals may appear small compared to fuels, but the variety of petrochemicals is huge. The listing in Table 7 will give some idea of the range of petrochemical applications.

Despite their variety, all commercially manufactured petrochemicals account for the consumption of only a small part of the total crude oil processed. In the United States, where petrochemicals have grown swiftly, the total petrochemical output at the beginning of 1980 was little more than 6½ vol.% of all petroleum feedstocks. An estimated 2½ vol.% went to energy of

TABLE 7 Petrochemical Applications

Absorbents	De-emulsifiers	Hair conditioners	Pipe
Activators	Desiccants	Heat transfer fluids	Plasticizers
Adhesives	Detergents	Herbicides	Preservatives
Adsorbents	Drugs	Hoses	Refrigerants
Analgesics	Drying oils	Humectants	Resins
Anesthetics	Dyes	Inks	Rigid foams
Antifreezes	Elastomers	Insecticides	Rust inhibitors
Antiknocks	Emulsifiers	Insulations	Safety glass
Beltings	Explosives	Lacquers	Scavengers
Biocides	Fertilizers	Laxatives	Stabilizers
Bleaches	Fibers	Odorants	Soldering flux
Catalysts	Films	Oxidation inhibitors	Solvents
Chelating agents	Finish removers	Packagings	Surfactants
Cleaners	Fire-proofers	Paints	Sweeteners
Coatings	Flavors	Paper sizings	Synthetic rubber
Containers	Food supplements	Perfumes	Textile sizings
Corrosion inhibitors	Fumigants	Pesticides	Tire cord
Cosmetics	Fungacides	Pharmaceuticals	
Cushions	Gaskets	Photographic chemicals	

conversion, leaving 4 vol.% represented as petrochemical products. Even so, this quantity has been sufficient for petrochemical-based materials to replace many products once made from such raw materials as coal, lumber, metal ores, and so forth.

Refining Schemes

A refinery is a massive network of vessels, equipment, and pipes. The total scheme can be divided into a number of unit processes. In the discussion to follow, only major flow streams will be shown, and each unit will be depicted by a single block on a simplified flow diagram. Details will be discussed later.

Refined products establish the order in which each refining unit will be introduced. Only one or two key product specifications are used to explain the purpose of each unit. Nevertheless, the reader is reminded that the choice from among several types of units and the size of these units are complicated economic decisions. The trade-off among product types, quantity, and quality will be mentioned to the extent that they influence the choice of one kind of process unit over another.

Feedstock Identification

Each refinery has its own range of preferred crude oil feedstock for which a desired distribution of products is obtained. The crude oil usually is identified by its source country, underground reservoir, or some distinguishing physical or chemical property. The three most frequently specified properties are density, chemical characterization, and sulfur content.

API gravity is a contrived measure of density [6]. The relation of API gravity to specific gravity is given by

$$°API = \frac{141.5}{sp\ gr} - 131.5$$

where sp gr is the specific gravity, or the ratio of the weight of a given volume of oil to the weight of the same volume of water at a standard temperature, usually 60°F.

An oil with a density the same as that of water, or with a specific gravity of 1.0, would then be 10°API oil. Oils with higher than 10°API gravity are lighter than water. Since lighter crude oil fractions are usually more valuable, a crude oil with a higher °API gravity will bring a premium price in the market place.

Heavier crude oils are getting renewed attention as supplies of lighter crude oil dwindle. In 1967 the U.S. Bureau of Mines (now part of the U.S. Department of Energy) defined heavy crudes as those of 25°API or less. More recently, the American Petroleum Institute proposed to use 20°API or less as the distinction for heavy crude oils.

A characterization factor was introduced by Watson and Nelson [7] to use as an index of the chemical character of a crude oil or its fractions. The Watson characterization factor is defined as follows:

$$\text{Watson } K = (T_B)^{1/3}/sp\ gr$$

where T_B is the absolute boiling point in degrees Rankin and sp gr is specific gravity compared to water at 60°F. For a wide boiling range material like crude oil, the boiling point is taken as an average of the five temperatures at which 10, 30, 50, 70, and 90% is vaporized.

A highly paraffinic crude oil might have a characterization factor as high as 13 while a highly naphthenic crude oil could be as low as about 10.5. Highly paraffinic crude oils can also contain heavy waxes which make it difficult for the oil to flow. Thus, another test for paraffin content is to measure how cold a crude oil can be before it fails to flow under specific test conditions. The higher the pour point temperature, the greater the paraffin content for a given boiling range.

Sour and sweet are terms referring to a crude oil's approximate sulfur content. In early days, these terms designated smell. A crude oil with a high sulfur content usually contains hydrogen sulfide—the gas associated with rotten eggs. Then the crude oil was called sour. Without this disagreeable odor, the crude oil was judged sweet. Today, the distinction between sour and sweet is based on total sulfur content. A sour crude oil is one with more than 0.5 wt.% sulfur, whereas a sweet crude oil has 0.5 wt.% or less sulfur. It has been estimated that 58% of United States crude oil reserves are sour. More importantly, an estimated 81% of world crude oil reserves are sour [8].

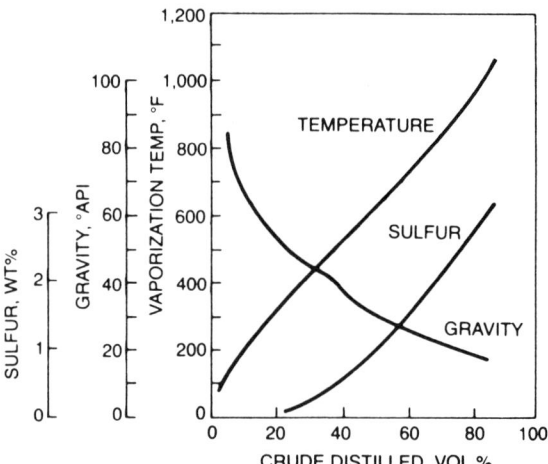

FIG. 4. Analysis of Light Arabian crude oil.

ASTM distillation is a test prescribed by the American Society for Testing and Materials to measure the volume percent distilled at various temperatures [5]. The results are often reported the other way around: the temperatures at which given volume percents vaporize [9]. These data indicate the quantity of conventional boiling range products occurring naturally in the crude oil. Analytical tests on each fraction indicate the kind of processing that may be needed to make specification products. A plot of boiling point, sulfur content, and API gravity for fractions of Light Arabian crude oil are shown in Fig. 4. This crude oil is among the ones most traded in the international crude oil market.

In effect, Fig. 4 shows that the material in the mid-volume range of Light Arabian crude oil has a boiling point of approximately 600°F, a liquid density of approximately 30°API, and an approximate sulfur content of 1.0 wt.%. These data are an average of eight samples of Light Arabian crude oil. More precise values would be obtained on a specific crude oil if the data were to be used in design work.

Since a refinery stream spans a fairly wide boiling range, the crude oil analysis data would be accumulated throughout that range to give fraction properties. The intent here is to show an example of the relation between volume distilled, boiling point, liquid density, and sulfur content.

Crude Oil Pretreatment

Crude oil comes from the ground admixed with a variety of substances: gases, water, and dirt (minerals). The technical literature devoted to petroleum refining often omits crude oil clean-up steps. It is likely presumed that the reader wishing to compare refining schemes will understand that the crude has already been through these clean-up steps. Yet cleanup is important if the

Petroleum Processing

crude oil is to be transported effectively and to be processed without causing fouling and corrosion. Cleanup takes place in two ways: field separation and crude desalting.

Field separation is the first attempt to remove the gases, water, and dirt that accompany crude oil coming from the ground. As the term implies, field separation is located in the field near the site of the oil wells. The field separator is often no more than a large vessel which gives a quieting zone to permit gravity separation of three phases: gases, crude oil, and water (with entrained dirt).

The crude oil is lighter than water but heavier than the gases. Therefore, crude oil appears within the field separator as a middle layer. The water is withdrawn from the bottom to be disposed of at the well site. The gases are withdrawn from the top to be piped to a natural gas processing plant or are pumped back into the oil well to maintain well pressure. The crude oil from the middle layer is pumped to a refinery or to storage awaiting transportation by other means.

Crude desalting is a water-washing operation performed at the refinery site to get additional crude oil cleanup [10]. The crude oil coming from field separators will continue to have some water and dirt entrained with it. Water washing removes much of the water-soluble minerals and entrained solids.

If these crude oil contaminants were not removed, they would cause operating problems during refinery processing. The solids (dirt and silt) would plug equipment. Some of the solids, being minerals, would dissociate at high temperature and corrode equipment. Still others would deactivate catalysts used in some refining processes.

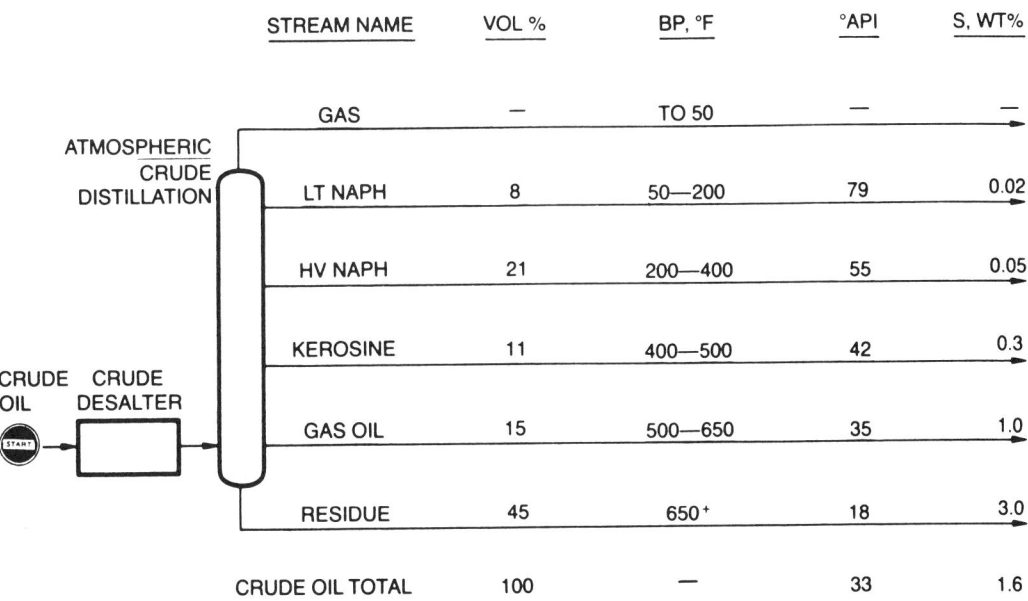

FIG. 5. Separating desalted crude oil into fractions.

Crude Oil Fractions

The importance of boiling range for petroleum products has already been discussed in connection with the earlier Table 5. The simplest form of refining would isolate crude oil into fractions having boiling ranges that would coincide with the temperature ranges for consumer products. Some treating steps might be added to remove or alter undesirable components, and a very small quantity of various chemical additives would be included to enhance final properties.

Crude oil distillation separates the desalted crude oil into fractions of differing boiling ranges. Instead of trying to match final product boiling ranges at this point, the fractions are defined by the number and kind of downstream processes.

The desalting and distillation units are depicted in Fig. 5 to show the usual fractions coming from crude oil distillation units. The discussion in the following paragraphs shows the relationships between some finished products and downstream processing steps.

Gasoline

The light and heavy naphtha fractions from crude oil distillation are ultimately combined to make gasoline. The two streams are isolated early in the refining scheme so that each can be refined separately for optimum blending in order to achieve required specifications—of which only volatility, sulfur content, and octane number will be discussed.

Volatility

A gasoline's boiling range is important during its aspiration into the combustion chamber of a gasoline-powered engine. Vapor pressure, a function of the fuel's boiling range, is also important. Boiling range and vapor pressure are lumped into one concept, volatility [11].

Lighter components in the gasoline blend are established as a compromise between two extremes: enough light components are needed to give adequate vaporization of the fuel-air mixture for easy engine starting in cold weather, but too much of the light components can cause the fuel to vaporize within the fuel pump and result in vapor lock.

Heavier components are a trade-off between fuel volume and combustion chamber deposits. Heavier components extend the yield of gasoline that can be made from a given volume of crude oil. But heavier components also contribute to combustion chamber deposits and spark plug fouling. Thus, an upper limit is set on gasoline's boiling range to give a clean-burning fuel.

Sulfur Content

Sulfur compounds are corrosive and foul smelling. When burned in an engine, these compounds result in sulfur dioxide exhaust. Should the engine be equipped with a catalytic muffler, as is the case for many modern automobile engines, the sulfur is exhausted from the muffler as sulfur trioxide, or sulfuric acid mist.

Caustic wash or some other enhanced solvent washing technique is usually sufficient to remove sulfur from light naphtha. The sulfur compounds in light naphtha are mercaptans and organic sulfides that are removed readily by these washing processes.

Heavy naphtha is harder to desulfurize. The sulfur compounds are in greater concentration and are of more complicated molecular structure. A more severe desulfurization method is needed to break these structures and release the sulfur. One such process is hydrotreating.

Hydrotreating is a catalytic process that converts sulfur-containing hydrocarbons into low-sulfur liquids and hydrogen sulfide [12]. The process is operated under a hydrogen-rich blanket at elevated temperature and pressure. A separate supply of hydrogen is needed to compensate for the amount of hydrogen required to occupy the vacant hydrocarbon site once held by the sulfur. Also, hydrogen is consumed to convert the sulfur to hydrogen sulfide gas.

Nitrogen and oxygen compounds are also dissociated by hydrotreating. The beauty of the process is that molecules are split at the points where these contaminants are attached. For nitrogen and oxygen compounds, the products of hydrotreating are ammonia and water, respectively. Thus, the contaminants will appear in the off-gases and are easily removed by conventional gas treating processes.

Octane Number (see articles on Octane)

Another condition to keep gasoline engines running smoothly is that the fuel–air mixture start burning at a precise time in the combustion cycle. An electrical spark starts the ignition. The remainder of the fuel–air mix should be consumed by a flame front moving out from the initial spark.

Under some conditions, a portion of fuel–air mixture will ignite spontaneously instead of waiting for the flame front from the carefully timed spark. The extra pressure pulses resulting from spontaneous combustion are usually audible above the normal sounds of a running engine and give rise to the phenomenon called "knock." Some special attributes of the knocking phenomenon are called pinging and rumble. All of these forms of knock are undesirable because they waste some of the available power of an otherwise smooth-running engine.

Octane number is a measure of a fuel's ability to avoid knocking. The octane number of a gasoline is determined in a special single cylinder engine where various combustion conditions can be controlled [5]. The test engine is adjusted to give trace knock from the fuel to be rated. Then various mixtures of isooctane (2,2,4-trimethyl pentane) and normal heptane are used to find the ratio of the two reference fuels that will give the same intensity of knock as that from the unknown fuel. Defining isooctane as 100 octane number and normal heptane as 0 octane number, the volumetric percentage of isooctane in heptane that matches knock from the unknown fuel is reported as the octane number of the fuel. For example, 90 vol.% isooctane and 10 vol.% normal heptane establishes a 90 octane number reference fuel.

Two kinds of octane number ratings are specified, although other methods are often used for engine and fuel development. Both methods use the same reference fuels and essentially the same test engine. Engine operating conditions are the difference. In one, called the *Research method*, the spark advance is fixed, the air inlet temperature is 125°F, and engine speed in 600 r/min. The other, called the *Motor method*, uses variable spark timing, a higher mixture temperature (300°F), and a faster engine speed (900 r/min).

The more severe conditions of the Motor method have a greater influence on commercial blends than they do on the reference fuels. Thus, a Motor octane number of a commercial blend tends to be lower than the Research octane number. Recently, it has become the practice to label gasoline with an arithmetic average of both ratings, abbreviated (R+M)/2.

Catalytic reforming is the principal process for improving the octane number of a naphtha for gasoline blending [10]. The process gets its name from its ability to re-form or re-shape the molecular structure of a feedstock. The transformation that accounts for the improvement in octane number is the conversion of paraffins and naphthenes to aromatics. The aromatics have better octane numbers than their paraffin or naphthene homologs. The greater octane number increase for the heavier molecules explains why catalytic reforming is usually applied to the heavy naphtha fractions.

Catalysts for reforming typically contain platinum or a mixture of platinum and other metal promoters on a silica-alumina support. Only a small concentration of platinum is used, averaging about 0.4 wt.%. The need to sustain catalyst activity and the expense of the platinum make it common practice to pretreat the reformer's feedstock to remove catalyst poisons.

Hydrotreating, already discussed, is an effective process to pretreat reforming feedstocks. The two processes go together well for another reason. The reformer is a net producer of hydrogen by virtue of its cyclization and dehydrogenation reactions. Thus, the reformer can supply the hydrogen needed by the hydrotreating reactions. A rough rule-of-thumb is that a catalytic reformer produces 800–1200 SCF of hydrogen per barrel of feed, while the hydrotreater consumes about 100–200 SCF/bbl for naphtha treating. The excess hydrogen is available for hydrotreating other fractions in separate hydrotreaters [13].

Distillates

Jet fuel, kerosene (range oil), No. 1 fuel oil, No. 2 fuel oil, and diesel fuel are all popular distillate products coming from 400 to 600°F fractions of crude oil.

One grade of jet fuel uses the heavy naphtha fraction, but the kerosene fraction supplies the more popular heavier grade of jet fuel, with smaller amounts sold as burner fuel (range oil) or No. 1 heating oil.

Some heating oil (generally No. 2 heating oil) and diesel fuel are very similar and are sometimes substitutes for each other. The home heating oil is intended to be burned within a furnace for space heating. The diesel fuel is intended for compression-ignition engines.

Hydrotreating improves the properties of all these distillate products. The process not only reduces the sulfur content of the distillates to a low level but also hydrogenates unsaturated hydrocarbons so that they will not contribute to smoke and particulate emissions—whether the fuel is burned in a furnace or used in an engine.

Residuals

Crude oil is seldom distilled at temperatures above about 650°F. At higher temperatures, coke will form and plug the lower section of the crude oil distillation tower. Therefore, the portion with a boiling point above 650°F is not vaporized—or at least not with the processing units introduced so far. This residual liquid is disposed of as industrial fuel oils, road oils, etc. The residual is sometimes called reduced crude because the lighter fractions have been removed.

Producing More Light Products

The refining scheme evolved to this point is shown in Fig. 6. It is typical of a low investment refinery designed to make products of modern quality. Yet the relative amounts of products are dictated by the boiling range of the crude oil feed. For Light Arabian crude oil reported earlier (see Fig. 4), all distillate fuel oils and lighter products (those boiling below 650°F) would comprise only about 55 vol.% of the crude oil feed rate.

For industrialized areas where the principal demand is for transportation fuels or high quality heating oils, a refining scheme of the type shown in Fig. 6 would need to dispose of almost half of the crude oil as low quality, less desirable, residual products. Moreover, the price obtained for these residual products is not only much lower than revenues from lighter products but also lower than the cost of the original crude oil. Thus, there are economic incentives to convert much of the residual portions into lighter products of suitable properties.

Relative volumes of petroleum product deliveries in the United States and the portions existing in Light Arabian crude oil are compared by boiling ranges in Fig. 7. Note that 80–85 vol.% of all United States petroleum products are lighter than the boiling temperature of 650°F compared to the 55 vol.% existing in the crude oil. Furthermore, half of all United States products

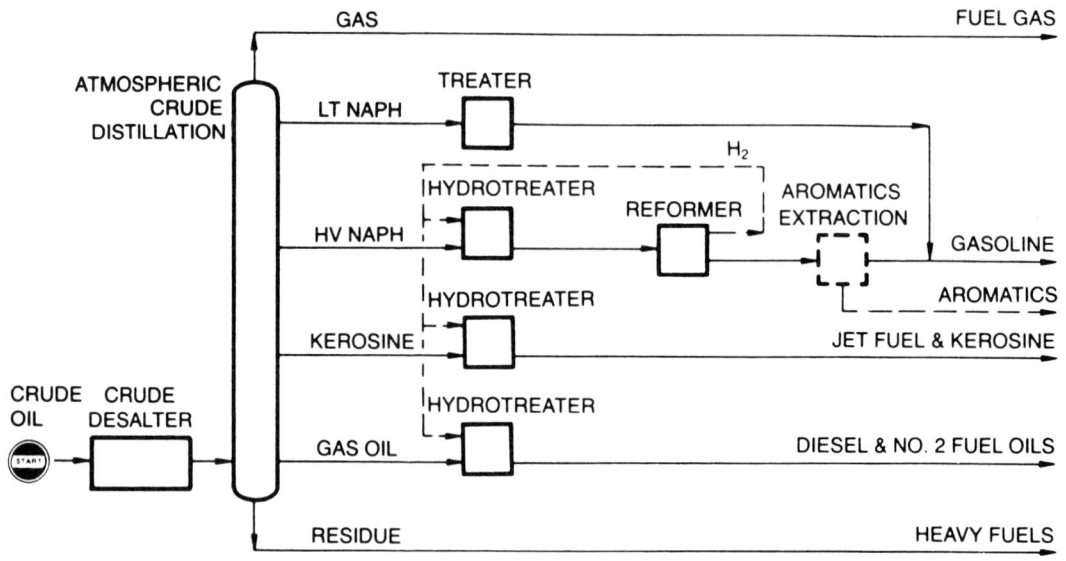

FIG. 6. Low investment route to modern products.

are gasoline and lighter distillates (boiling temperatures less than 400°F) compared to 29 vol.% in the crude oil.

This comparison can appear unfair since an array of products obtained easily from one crude oil would be difficult to obtain from another crude oil feed. Total product deliveries in the United States come from a variety of different crude oils, processed in a variety of different refining schemes. But the comparison serves to emphasize a long term trend in the refining industry—to convert heavy, less desirable fractions into lighter, more valuable products. The comparison also lays the foundation for the next group of processes to be discussed. The choice and arrangement of processes hereafter are intended to depict a breadth of refining technology, rather than suggest a commercial scheme for handling the example crude oil.

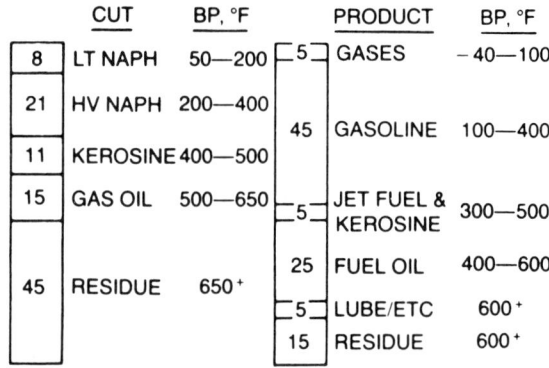

FIG. 7. Light Arabian crude oil compared to 1979 United States deliveries.

Cracking (see articles on Cracking)

These processes cause hydrocarbon molecules to break apart into two or more smaller molecules. Thermal cracking uses high temperature (above 650°F) and long residence time to accomplish the molecular split. Catalytic cracking accomplishes the split much faster and at lower temperatures because of the presence of a cracking catalyst.

Catalytic cracking involves not only some of the biggest units, with their large catalyst reactor-separators and regenerators, but it is also among the more profitable operations with its effective conversion of heavy feeds to light products. Gasoline from catalytic cracking has a higher octane number than thermally cracked gasoline. Yields include less gas and coke than thermal cracking; i.e., more useful liquid products are made. The distribution of products between gasoline and heating oils can be varied by different choices for catalysts and varying operating conditions.

The best feeds for catalytic crackers are determined by a number of factors. The feed should be heavy enough to justify conversion. This usually sets a lower boiling point of about 650°F. The feed should not be so heavy that it contains undue amounts of metal-bearing compounds nor carbon-forming materials. Either of these substances is more prevalent in heavier fractions and can cause the catalyst to lose activity more quickly.

Visbreaking is basically a mild, once-through thermal-cracking process. It is used to get just sufficient cracking of resid so that fuel oil specifications can be met. Although some gasoline and light distillates are made, this is not the purpose of the visbreaker.

Coking is another matter. It is a severe form of thermal cracking in which coke formation is tolerated to get additional lighter liquids from the heavier, dirtier fractions of crude oil. Here, the metals that would otherwise foul a catalytic process are laid down with the coke. The coke settles out in large coke drums that are removed from service frequently (about one a day) to have the coke cut out with high-pressure water lances. To make the process continuous, multiple coke drums are used so that some drums can be on-stream while others are being unloaded.

Hydrocracking achieves cracking with a rugged catalyst to withstand resid contaminants and with a hydrogen atmosphere to minimize coking. Hydrocracking combines hydrotreating and catalytic-cracking goals, but a hydrocracker is much more expensive than either of the other two. The pressure is so high (up to 3000 lb/in.^{2}) that very thick walled vessels must be used for reactors (up to 9 in. thick). The products from a hydrocracker will be clean (desulfurized, denitrified, and demetalized) and will contain isomerized hydrocarbons in greater amount than in conventional catalytic cracking. A significant part of the expense of operating a hydrocracker is for the hydrogen that it consumes.

Vacuum Distillation

Among the greater variety of products made from crude oil, some of the products (lubricating oils, for example) have boiling ranges that exceed

650°F—the general vicinity where cracking would occur in atmospheric distillation. Thus, by operating a second distillation unit under vacuum, the heavier parts of the crude oil can continue to be divided into specific products. Furthermore, some of the fractions distilled from vacuum units are better than atmospheric residue for cracking because the metal-bearing compounds and carbon-forming materials are more highly concentrated in the vacuum residue.

Reconstituting Gases

Cracking processes to convert heavy liquids to lighter liquids also make gases. Another way to make more liquid products is to combine gaseous hydrocarbons. A few small molecules of a gas can be combined to make one bigger molecule with fairly specific properties. Here, a gas separation unit is added to the refinery scheme to isolate the individual types of gases. When catalytic cracking is also part of the refining scheme, there will be a greater supply of olefins—ethylene, propylene, and butylene. Two routes for reconstituting these gaseous olefins into gasoline blending stocks are described below.

Polymerization ties two or more olefins together to make polymer gasoline. The double bond in only one olefin is changed to a single bond during each link between two olefins. This means the product will still have a double bond. For gasoline, these polymer stocks are good for blending because olefins tend to have higher octane numbers than their paraffin homologs.

However, the olefinic nature of polymer gasoline can also be a drawback. During long storage in warmer climates, the olefins can continue to link up to form bigger molecules of gum and sludge. This effect, though, is seldom important when the gasoline goes through ordinary distribution systems.

Alkylation combines an olefin and isobutane when gasoline is desired. The product is mostly isomers. If the olefin were butylene, the product would contain a high concentration of 2,2,4-trimethyl pentane. The reader is reminded that this is the standard compound that defines 100 on the octane number scale. Alkylates are high quality gasoline-blending compounds, having good stability as well as high octane numbers.

A Modern Refinery

A refining scheme incorporating the processes discussed so far is shown in Fig. 8. The variations are quite numerous, though. Types of crude oil available, local product demands, and competitive quality goals are just a few of the factors that are weighed to decide a specific scheme.

Many other processes play an important role in the final scheme. A partial list of these other processes would have the following goals: dewaxing lubricating oils, deoiling waxes, deasphalting heavy fractions, manufacturing specific compounds for gasoline blending (alcohols, ethers, etc.), and isolating specific fractions for use as petrochemical feedstocks.

Petroleum Processing

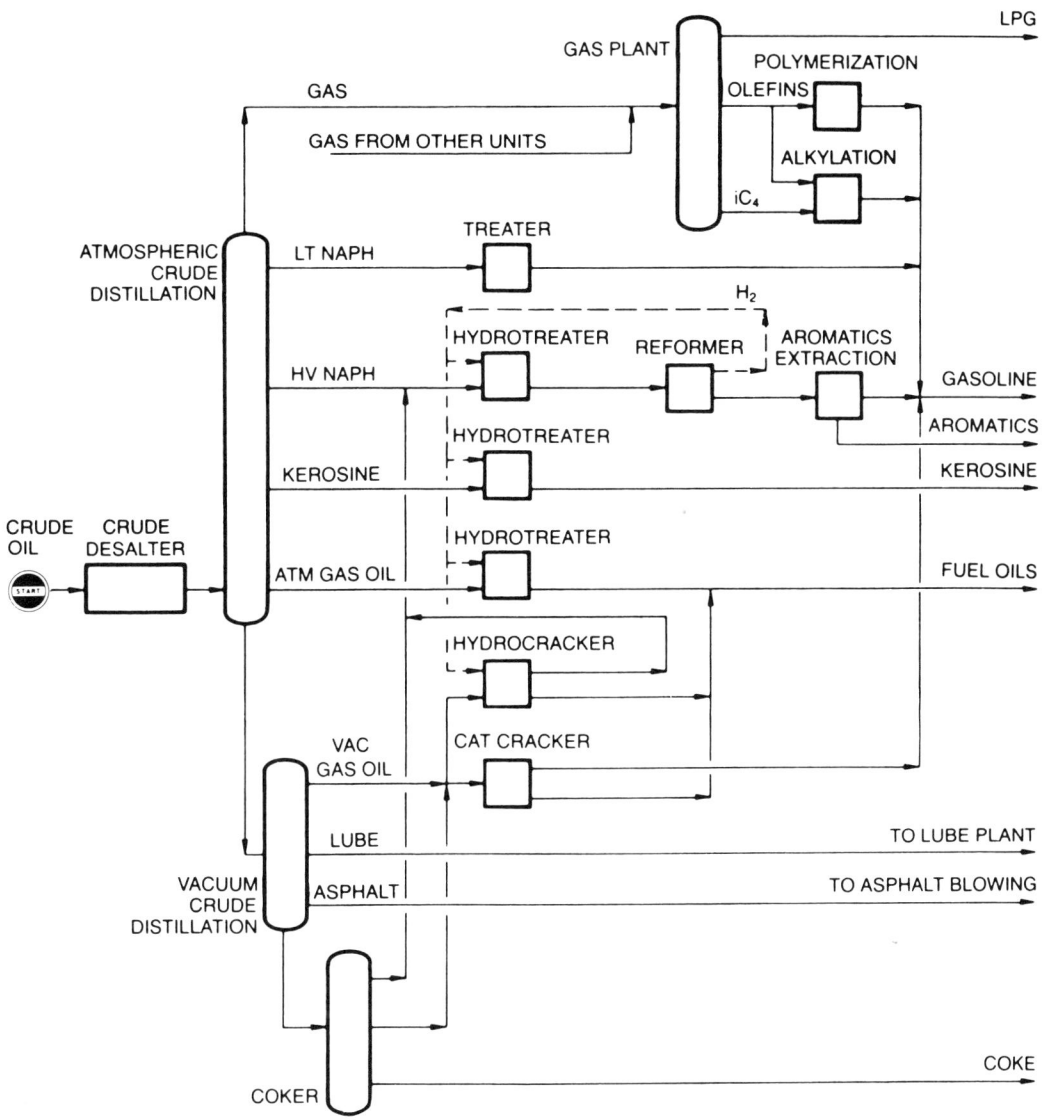

FIG. 8. High conversion refinery.

Petrochemicals

It has already been mentioned that petrochemicals account for only a little more than 6½ vol.% of all petroleum feedstocks. Earlier, Table 7 gave the vast array of the applications of these petrochemicals. Olefins and aromatics make up a big part of the total.

Ethylene is one of the most important olefins. It is usually made by cracking gases—ethane, propane, butane, or a mixture of these as might exist in a refinery's off-gases. When gas feedstock is scarce or expensive, naphthas and even whole crude oil have been used in specially designed ethylene crackers. The heavier feeds also give significant quantities of higher molecular weight olefins and aromatics.

Aromatics (see articles on Benzene), as has been pointed out, are in high concentration in the product from a catalytic reformer. When aromatics are needed for petrochemical manufacture, they are extracted from the reformer's product using solvents such as glycols (the Udex process, for example) and sulfolane, to name two popular ones.

The mixed aromatics are called BTX as an abbreviation for benzene, toluene, and xylene. The first two are isolated by distillation and the isomers of the third are separated by partial crystallization. Benzene is the starting material for styrene, phenol, and a number of fibers and plastics. Toluene is used to make a number of chemicals, but most of it is blended into gasoline. Xylene use depends on the isomer, p-xylene going into polyester and o-xylene going into phthalic anhydride. Both are involved in a wide variety of consumer products.

Process Details

So far, refining units have been described as they relate to other units and to final product specifications. Now, typical flow diagrams of some major processes will be presented to highlight individual features. In many cases the specific design shown herein is an arbitrary choice from among several equally qualified designers.

Crude Desalting (see *Desalting, Crude Oil*)

Basically a water-washing process, the crude desalter must accomplish intimate mixing between the crude oil and water, then separate them sufficiently so that water will not enter subsequent crude-oil distillation heaters.

A typical flow diagram is shown in Fig. 9. The unrefined crude oil is heated to 100–300°F for suitable fluid properties. The operating pressure is 40 $lb/in.^2$ gauge or more. Elevated temperatures reduce oil viscosity for better mixing, and elevated pressure suppresses vaporization. The washwater can be added either before or after heating.

Mixing between the water and crude oil is assured by passing the mixture through a throttling valve or emulsifier orifice. Trace quantities of caustic, acid, or other chemicals are sometimes added to promote treating. Then the water-in-oil emulsion is introduced into a high voltage electrostatic field inside a gravity settler. The electrostatic field helps the water droplets to agglomerate for easier settling.

Salts, minerals, and other water-soluble impurities in the crude oil are carried off with the water discharged from the settler. Clean desalted crude oil flows from the top of the settler and is ready for subsequent refining.

Petroleum Processing

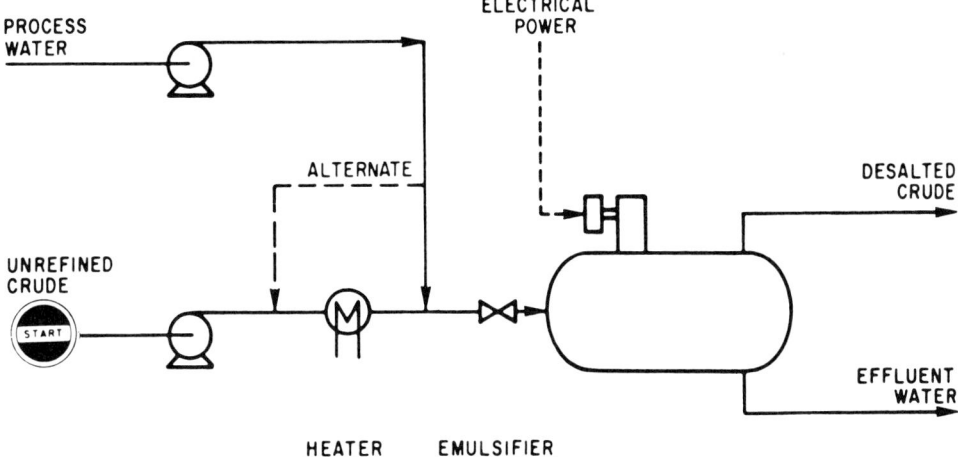

FIG. 9. Crude desalting. (Reprinted from Ref. 10 with permission.)

Additional stages can be used in series to get additional reduction in salt content of the crude oil. Two stages are typical, but some installations use three stages. The increased investment cost for multiple stages is offset by reduced corrosion, plugging, and catalyst poisoning in downstream equipment by virtue of lower salt content.

Crude Distillation

Single or multiple distillation columns are used to separate crude oil into fractions determined by their boiling range. Common identification of these fractions was discussed in connection with Fig. 5, but these should only be considered as a guide since a variety of refining schemes call for altering the type of separation made at this point.

A typical flow diagram of a three-stage crude distillation system [14] is shown in Fig. 10. The crude oil is heated by exchange with various hot products coming from the system before it passes through a fired heater. The temperature of the crude oil entering the first column is 600–700°F, or high enough to vaporize the heavy gas oil and all lighter fractions. The first column is depicted as having a larger diameter in the lower section because the quantity of vapors and liquids passing through this section require more cross-sectional area to avoid high pressure drop across individual contacting trays or to prevent high-velocity vapors from blowing the liquid up the column rather than giving good mixing as the liquids follow their normal path down the column. The final design depends upon the quantities of individual fractions and permitted tower loadings.

Since light products must pass from the feed point up to their respective drawoff point, any intermediate stream will contain some of these lighter materials. Stream stripping (note the group of steam strippers beside the first

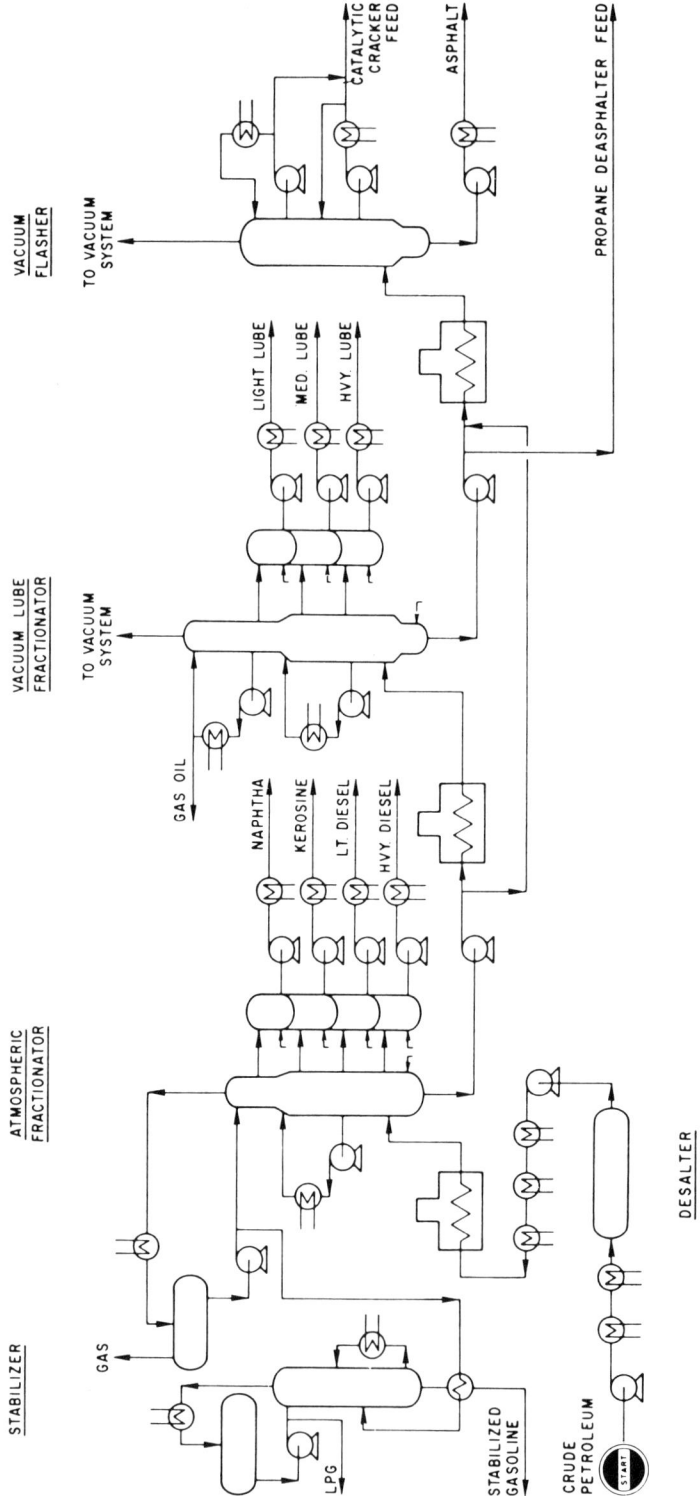

FIG. 10. Crude distillation—*Foster Wheeler Corp.* (Reprinted from Ref. 14 with permission.)

column in Fig. 10) is a way to reintroduce these light materials back into the tower to continue their passage up through the column.

The next two fractionating columns of Fig. 10 are operated under vacuum. Steam jet ejectors are used to create the vacuum so that the absolute pressure can be as low as 30–40 mmHg. The vacuum permits hydrocarbons to be vaporized at temperatures much below their normal boiling points. Thus, fractions with boiling points above 650°F can be separated with vacuum distillation without causing thermal cracking. Steam is often added to vacuum units to reduce the partial pressure of the hydrocarbons even further. If the steam is added ahead of the furnace associated with the vacuum columns, fluid velocity through the furnace tubes is increased and coke formation is minimized.

Lately, a popular addition to a crude distillation system has been a preflash column ahead of the two stages shown in Fig. 10. The preflash tower strips out the lighter portions of a crude oil before the remainder enters the atmospheric column. It is the lighter portions that set the vapor loading in the atmospheric column which, in turn, determines the diameter of the upper section of the column.

Incidentally, total refining capacity of a facility is reported in terms of its crude-oil handling capacity. Thus, the size of the first distillation column, whether a preflash or an atmospheric distillation column, sets the reported size of the entire refinery. Ratings in barrels per stream day (BPSD) will be greater than barrels per calendar day (BPCD). Processing units must be shut down on occasion for maintenance, repairs, and equipment replacement. The ratio of operating days to total days (or BPCD divided by BPSD) is called an "on-stream factor" or "operating factor." The ratio will be expressed either as a percent or a decimal. For example, if a refinery unit undergoes one shutdown period of one month during a three year duration, its operating factor is $(36 - 1)/36$, or 0.972, or 97.2%.

Outside the United States, refining capacity is normally given in metric tons per year. Precise conversion from one unit of measure to the other depends upon the specific gravity of the crude oil, but the approximate relation is one barrel per day equals 50 tons per year.

Hydrotreating

This is a catalytic hydrogenation process that reduces the concentration of sulfur, nitrogen, oxygen, metals, and other contaminants in a hydrocarbon feed. In more severe forms, hydrotreating saturates olefins and aromatics.

A typical flow diagram is shown in Fig. 11. The feed is pumped to operating pressure and mixed with a hydrogen-rich gas, either before or after being heated to the proper reactor inlet temperature. The heated mixture passes through a fixed bed of catalyst where exothermic hydrogenation reactions occur. The effluent from the reactor is then cooled and sent through two separation stages. In the first, the high-pressure separator, unreacted hydrogen is taken overhead to be scrubbed for hydrogen sulfide removal; the cleaned hydrogen is then recycled. In the second, the lower-pressure separator

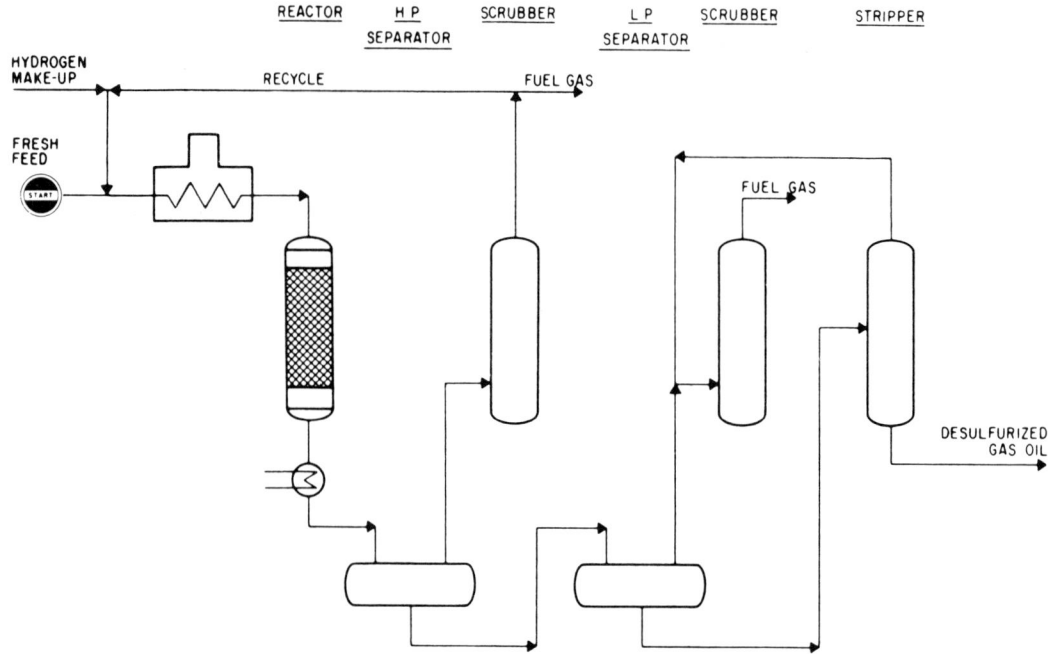

FIG. 11. Gulfining—Gulf R&D Co. and Houdry Division of Air Products and Chemicals, Inc. (Reprinted from Ref. 10 with permission.)

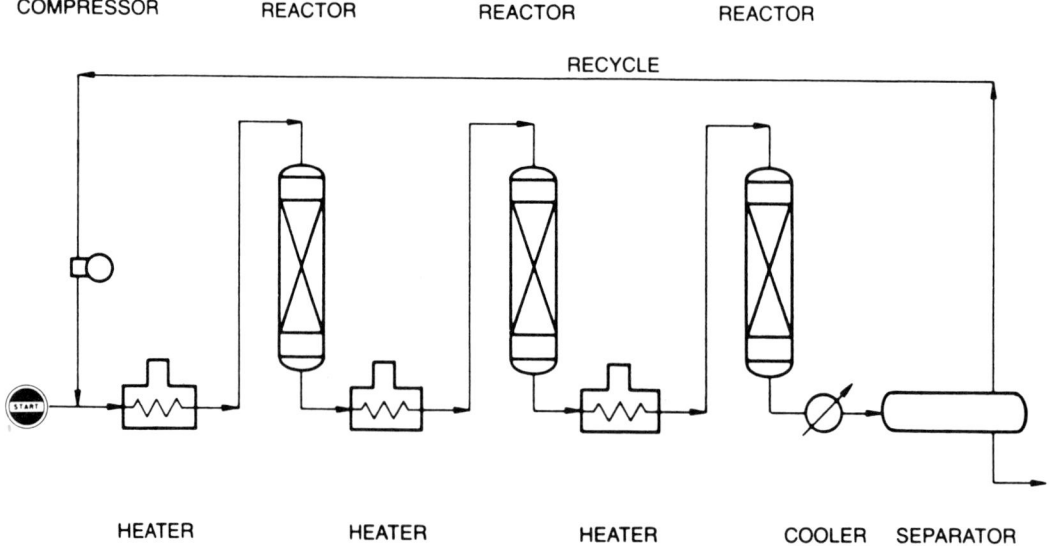

FIG. 12. Catalytic reforming, generalized flow diagram.

takes off the remaining gases and light hydrocarbons from the liquid product. If the feed is a wide-boiling range material from which several blending stocks are to be made, the second separator can be a distillation column for removing gases and light hydrocarbons, with sidestreams used for each of the liquid products.

The feed for hydrotreating can be a variety of different boiling range materials extending from light naphtha to vacuum residues. Generally, each fraction is treated separately to permit optimum conditions—the higher boiling materials requiring more severe treating conditions. For example, naphtha hydrotreating can be carried out at 200–500 lb/in.^{2}abs and at 500–650°F, with a hydrogen consumption of 10–50 SCF/bbl of feed. On the other hand, a residue hydrotreating process can operate at 1000–2000 lb/in.^{2}abs and at 650–800°F, with a hydrogen consumption of 600–1200 SCF/bbl [13]. Nevertheless, hydrotreating is such a desirable clean-up step that it can justify its own hydrogen manufacturing facilities, although the hydrogen-rich stream obtained as a by-product from catalytic reforming usually is sufficient for most operations.

Catalyst formulations constitute a significant difference among hydrotreating processes. Each catalyst is designed to be better suited to one type of feed or one type of treating goal [15]. When hydrotreating is done for sulfur removal, the process is called hydrodesulfurization and the catalyst generally is cobalt and molybdenum oxides on alumina. A catalyst of nickel-molybdenum compounds on alumina can be used for denitrogenation and cracked-stock saturation.

Catalytic Reforming

Some confusion comes from the literature when the term "naphtha reforming" is used to designate processes to make synthesis gas—a mixture containing predominantly carbon monoxide and hydrogen. However, naphtha reforming has another meaning and is the one intended here—production of an aromatic-rich liquid for use in gasoline blending.

A typical flow diagram is shown in Fig. 12. The feed is pumped to operating pressure and mixed with a hydrogen-rich gas before heating to reaction temperatures. Actually, hydrogen is a by-product of the dehydrogenation and cyclization reactions, but by sustaining a hydrogen atmosphere, cracking and coke formation are minimized.

The feed for catalytic reforming is mostly in the boiling range of gasoline to start with. The intent is to convert the paraffin and naphthene portions to aromatics. As an example, a 180–310°F fraction of Light Arabian crude oil was reported to have 8 vol.% aromatics before catalytic reforming, but was 68 vol.% aromatics afterwards. The feed paraffin content (69 vol.%) was reduced to less than half, and the feed naphthene content (23 vol.%) was almost completely absent in the product [16].

The extent of octane number change with changes in molecular configuration is shown in Table 8 where normal paraffins and naphthenes are compared with their aromatic homologs.

TABLE 8 Aromatics Have Higher Octane Numbers [6]

		Octane Number, Clear	
Hydrocarbon Homologs		Motor	Research
C_7 hydrocarbons:			
$\quad$ n-Paraffin	C_7H_{16} (n-heptane)	0.0	0.0
$\quad$ Naphthene	C_7H_{14} (cycloheptane)	40.2	38.8
	C_7H_{14} (methylcyclohexane)	71.1	74.8
$\quad$ Aromatic	C_7H_8 (toluene)	103.5	120.1
C_8 hydrocarbons:			
$\quad$ n-Paraffin	C_8H_{18} (n-octane)	-15^a	-19^a
$\quad$ Naphthene	$C_{18}H_{16}$ (cyclooctane)	58.2	71.0
	C_8H_{16} (ethylcyclohexane)	40.8	45.6
$\quad$ Aromatic	C_8H_{10} (ethylbenzene)	97.9	107.4
	C_8H_{10} (o-xylene)	100.0	120^a
	C_8H_{10} (m-xylene)	115.0	117.5
	C_8H_{10} (p-xylene)	109.6	116.4

[a] Blending value at 20 vol.% in 60 octane number reference fuel.

If the naphthenes are condensed (multirings or indanes), they tend to deactivate the reforming catalyst quickly. Control of the end point of the feed will exclude these deactivating compounds.

Catalysts which promote reforming reactions can give side-reactions. Isomerization is acceptable but hydrocracking gives unwanted saturates and gases. Therefore, higher operating pressures are used to suppress hydrocracking. This remedy has disadvantages. Higher pressures suppress reforming reactions too, although to a lesser extent. Generally, a compromise is made between desired reforming and undesired hydrocracking. The effects of operating conditions on competing reactions [17] are shown in Table 9.

In the late 1960s it was discovered that the addition of certain promoters, such as rhenium, germanium, or tin, to the platinum-containing catalyst would reduce cracking and coke formation. The resulting catalysts are

TABLE 9 Favored Operating Conditions for Desired Reaction Rates [17]

Feed	Reaction	Product	Desired Rate	To Get Desired Rate	
				Pressure	Temperature
Paraffins	Isomerization	Isoparaffins	Increase	Increase	Increase
	Dehydrocyclization	Naphthenes	Increase	Decrease	Increase
	Hydrocracking	Lower molecular weight	Decrease	Decrease	Decrease
Naphthenes	Dehydrogenation	Aromatics	Increase	Decrease	Increase
	Isomerization	Isoparaffins	Increase	Increase	Increase
	Hydrocracking	Lower molecular weight	Decrease	Decrease	Increase
Aromatics	Hydrodealkylation	Lower molecular weight	Decrease	Decrease	Decrease

referred to as bimetallic catalysts. These newer catalysts permit the process to enjoy the better reforming conditions of lower pressure without being unduly penalized by hydrocracking. Earlier pressures of 500 lb/in.^{2}gauge are now down to 150 lb/in.^{2}gauge.

Operating temperatures are important, too. The reactions are endothermic. Best yields would come from isothermal reaction zones, but this is difficult to achieve. Instead, the reaction beds are separated into a number of adiabatic zones operating at 500–1000°F with heaters between stages to supply the necessary heat of reaction and hold the overall train near a constant temperature. Three or four reactor zones are commonly used when it is desired to have a product with high octane numbers.

In the recent push to make gasoline with high octane numbers but without the use of antiknock additives, high severity catalytic reforming is the prime route. The big disadvantage is a yield loss. Newer catalysts make the loss less dramatic, but the penalty remains [18], as can be seen from Fig. 13.

Catalytic Cracking (see Articles on Cracking)

A typical diagram of a fluid catalytic cracker is shown in Fig. 14. The unit is characterized by two huge vessels, one to react the feed with hot catalyst and the other to regenerate the spent catalyst by burning off carbon with air. The activity of the newer molecular-sieve catalysts is so great that the contact time

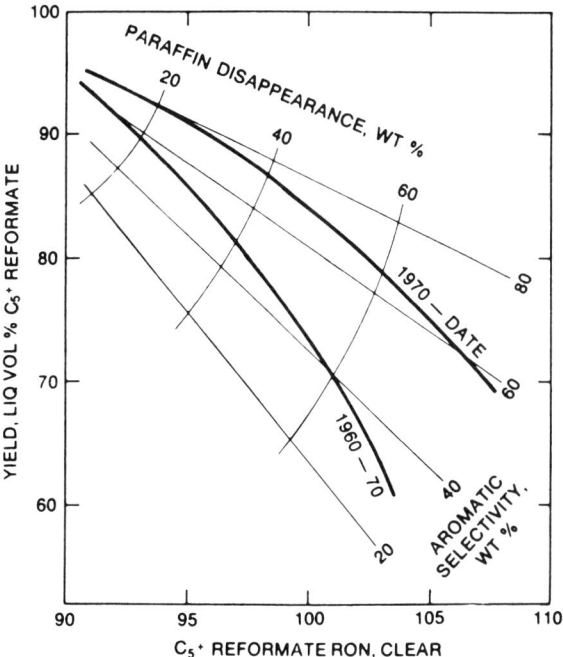

FIG. 13. Better octane numbers, less yield (Kuwait naphtha). (Reprinted from Ref. 18 with permission.)

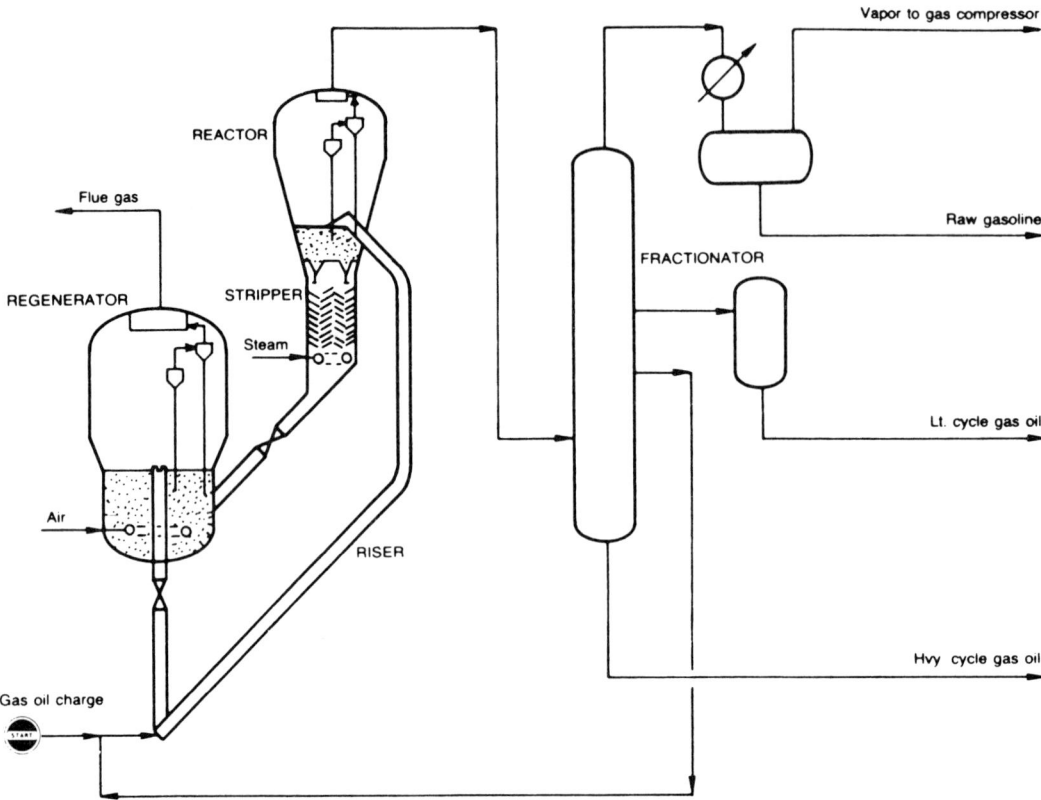

FIG. 14. Fluid catalytic cracking—Texaco Development Corp. (Reprinted from Ref. 10 with permission.)

between feed and catalyst is reduced drastically. If not, the oil will overcrack to give unwanted gases and coke. The short contact time is accomplished using a transfer line between the regenerator and reactor vessels. In fact, the major portion of the reaction occurs in this piece of pipe or riser and the products are taken quickly overhead. The main reactor vessels then are used to hold cyclone separators to remove the catalyst from the vapor products and to give additional space for cracking the heavier portions of the feed.

There are several configurations of reactors and regenerators. In some designs, one vessel is stacked on top of the other. All are big structures (150–200 ft high).

Riser cracking, as the short-time contacting is called, has a number of advantages. It is easier to design and operate. It can be operated at higher temperatures to give more gasoline olefins. It minimizes the destruction of any aromatics formed during cracking. The net effect can be the production of gasoline having octane numbers 2 or 3 numbers higher than earlier designs would give.

Better regeneration of the spent catalyst is obtained by operating at higher temperatures (1300–1400°F) [19]. The coke that is deposited on the catalyst is more completely burned away by higher temperature air blowing. The newer catalysts are rugged enough to withstand the extra heat, and newer metallurgy gives the regenerator vessel the strength it needs at higher temperatures.

Heavier feedstocks can be put into catalytic crackers. The nickel, vanadium, and iron in these heavier fractions do not deactivate the catalysts as fast as they once did because passivators are available now to add to the catalysts [20]. The extra sulfur that comes with heavier feeds can be prevented from exhausting into the atmosphere during regeneration because of catalysts that hold on to the sulfur compounds until the catalysts get into the reactor [21]. Then the sulfur compounds are cracked to light gases and leave the unit with the cracked products. Ordinary gas treating methods are used to capture the hydrogen sulfide coming from the sulfur in the feedstock.

Coking (see articles on Coking)

Coking is an extreme form of thermal cracking. The process converts residual materials that might not easily be converted by the more popular catalytic cracking process. Coking is also a less expensive process for getting more light stocks from residual fractions. In the coking process, the coke is considered a by-product that is tolerated in the interest of more complete conversion of residues to lighter liquids.

A typical flow diagram of a delayed coker is shown in Fig. 15. There are several possible configurations, but in this one the feed goes directly into the product fractionator in order to pick up heavier products to be recycled to the cracking operation. The term "delayed coker" signifies that the heat of cracking is added by the furnace and the cracking occurs during the longer residence time in the following coke drums. Furnace outlet temperatures are in the range of 900 to 950°F while the coke drum pressures are in the range of 20 to 60 lb/in.^{2} gauge.

The coke accumulates in the coke drum and the remaining products go overhead as vapors to be fractionated into various products. In this case, the products are gas, naphtha, light gas oil, heavy gas oil, and coke. When a coke drum is to be emptied, a large drilling structure mounted on top of the drum is used to make a center hole in the coke formation. The drill is equipped with high-pressure water jets (3000 lb/in.^{2} gauge or more) to cut the coke from the drum so that it can fall out a bottom hatch into a coke pit. From there, belt conveyors and bucket cranes move the coke to storage or to market [22].

Fluid Coking is a proprietary name given to a different type of coking process in which the coke is suspended as particles in fluids flowing from a reactor to a heater and back again. When part of the coke is gasified, the process is called *Flexicoking*. Both Fluid Coking and Flexicoking are proprietary processes of Exxon Research and Engineering Co.

A flow diagram for Flexicoking is shown in Fig. 16. The first two vessels are typical of Fluid Coking in which part of the coke is burned in the heater in order to have hot coke nuclei to contact the feed in the reactor vessel. The cracked products are quenched in an overhead scrubber where entrained coke is returned to the reactor. Coke from the reactor circulates to the heater

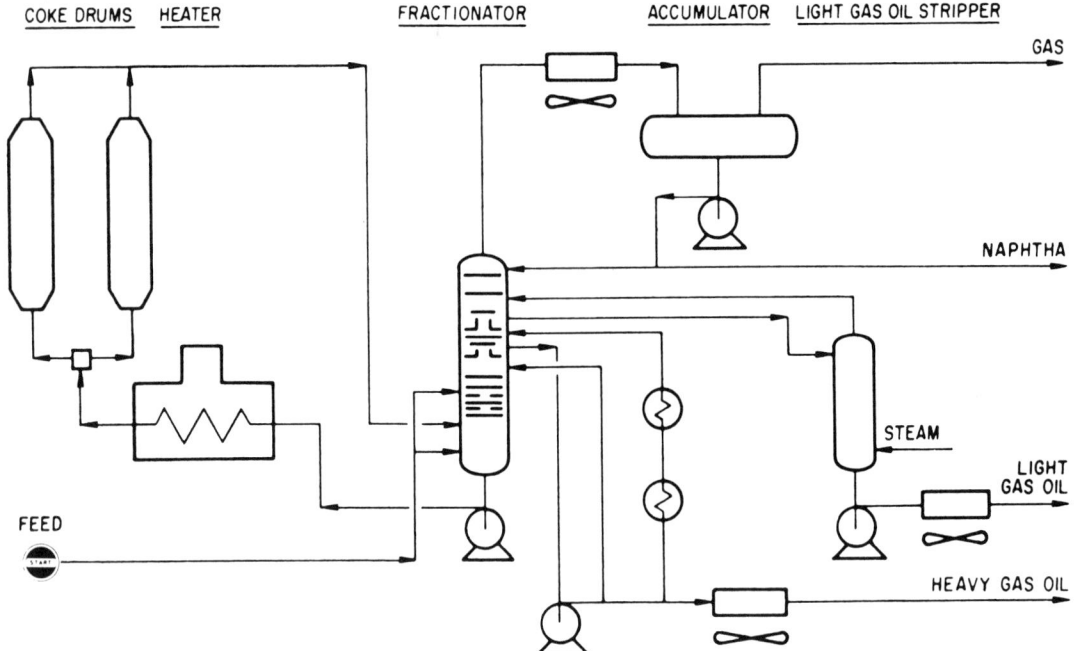

FIG. 15. Delayed coking—*Foster Wheeler Energy Corp.* (Reprinted from Ref. 10 with permission.)

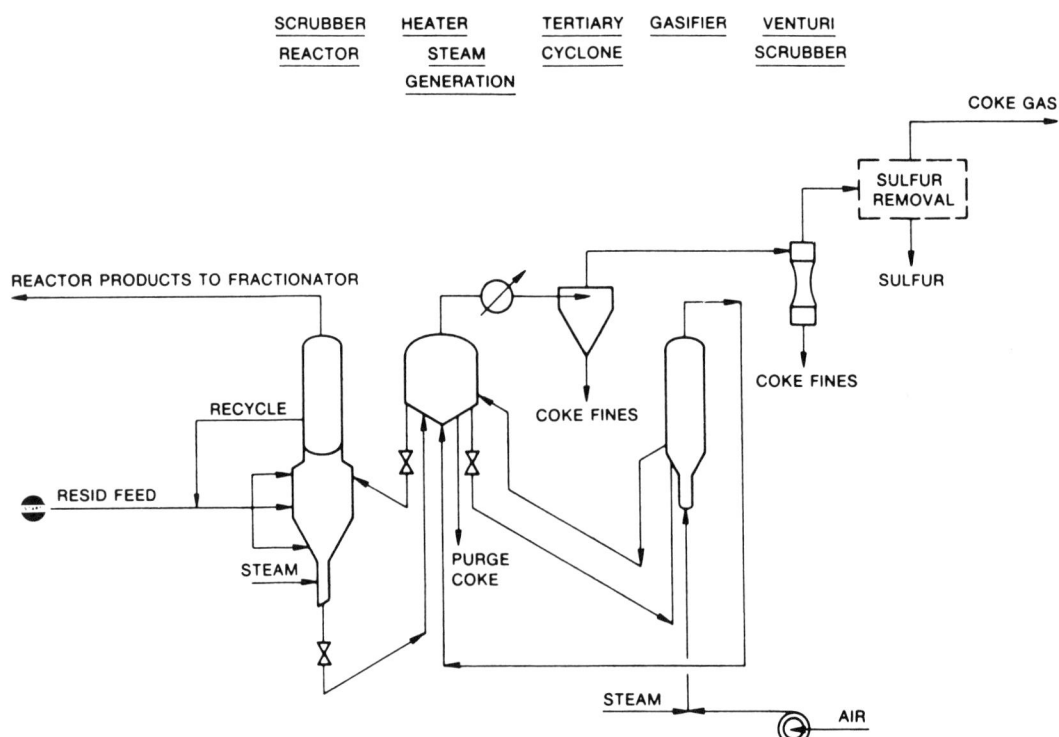

FIG. 16. Flexicoking—Exxon Research and Engineering Co. (Reprinted from Ref. 10 with permission.)

where it is devolatilized to yield a light hydrocarbon gas and residual coke. A sidestream of coke is circulated to the gasifier where, for most feedstocks, 95% or more of the gross coke product from the reactor is gasified at elevated temperature with steam and air. Sulfur that enters the unit with the feedstock eventually becomes hydrogen sulfide exiting the gasifier and is recovered by a sulfur removal step.

Hydrocracking (see *Hydrocracking*)

Before the late 1960s, most hydrogen used in processing crude oil was for pretreating catalytic reformer feed naphtha and for desulfurizing middle-distillate products. Soon thereafter, requirements to lower sulfur content in most fuels became an important consideration. The heavier fractions of crude oil were the hardest to treat. Moreover, these fractions were the ones offering additional sources of light products. This situation set the stage for the introduction of hydrocracking.

A typical flow diagram for hydrocracking is shown in Fig. 17. Process flow is similar to hydrotreating in that feed is pumped to operating pressure, mixed with a hydrogen-rich gas, heated, passed through a catalytic reactor, and distributed among various fractions. Yet the hydrocracking process is unlike hydrotreating in several important ways. Operating pressures are very high, 2000–3000 lb/in.^{2} gauge. Hydrogen consumption also is high, 1200–1600 SCF of hydrogen per barrel of feed depending on the extent of the cracking [13]. In fact, it is not uncommon to see hydrocrackers built with their own hydrogen manufacturing facilities nearby.

The catalysts for hydrocracking have a dual function. They give both hydrogenation and dehydrogenation reactions and have a highly acidic support to foster cracking. The hydrogenation-dehydrogenation component of the catalysts are metals such as cobalt, nickel, tungsten, vanadium,

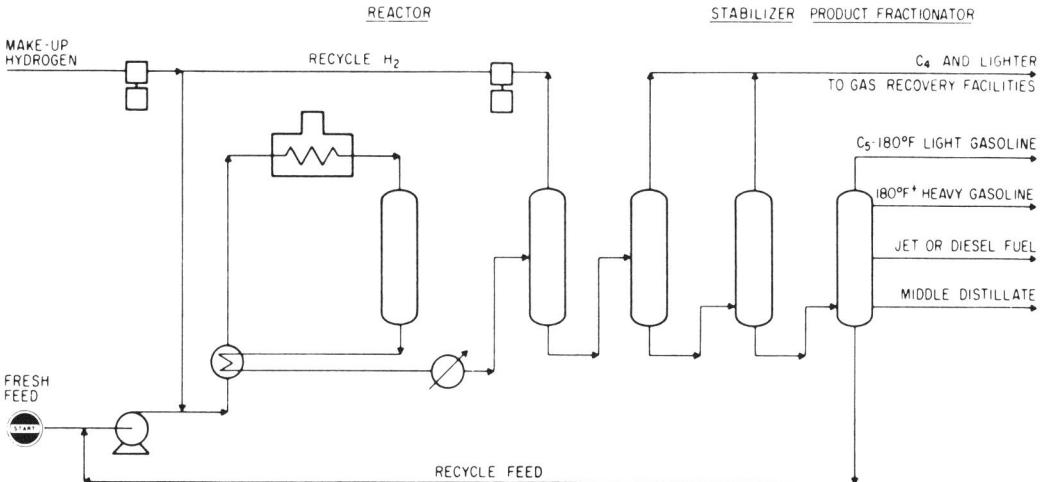

FIG. 17. Isocracking—Chevron Research Co. (Reprinted from Ref. 10 with permission.)

molybdenum, platinum, palladium, or a combination of these metals. The acidic support can be silica-alumina, silica-zirconia, silica-magnesia, alumina-boria, silica-titania, acid-treated clays, acidic-metal phosphates, or alumina, to name some given in the literature [23].

Great flexibility is attributed to most hydrocracking processes. Under mild conditions the process can function as a hydrotreater. Under more severe conditions of cracking, the process produces a varying ratio of motor fuels and middle distillates, depending on the feedstock and operating variables. Even greater flexibility is possible for the process during design stages when it can be tailored to change naphthas into liquefied petroleum gases or convert heavy residues into lighter products.

Because the hydrocracker is viewed as both a cracker and a treater, it can appear in refining process schemes in a number of different places. As a cracker, it is used to convert feeds that are too heavy or too contaminant-laden to go to catalytic cracking. As a treater, it is used to handle heating-oil fractions that need to be saturated to give good burning quality. But it is the trend to heavier feeds and lighter high-quality fuels that causes hydrocracking to offer advantages to future refining, even though the hydrocracking units are much more expensive to build and to operate.

The principle of an ebulliating catalyst bed is embodied in some proprietary designs, in contrast with the fixed-catalyst beds used in other versions of hydrocracking. The H-Oil process of Hydrocarbon Research, Inc. and the LC-Fining process jointly licensed by C-E Lummus and Cities Service Research and Development Co. are examples of hydrocracking processes that use a mixed-reaction bed instead of a fixed bed of catalyst.

Polymerization

This process usually is associated with the manufacture of plastic films and fibers from light hydrocarbon olefins with products like polyethylene and polypropylene. As a gasoline manufacturing process, the polymerization of light olefins emphasizes a combination of only two or three molecules so that the resulting liquid will be in the gasoline boiling range.

For early polymerization units the catalyst was phosphoric acid on a quartz or kieselguhr support. Many of these units were shut down when the demand for gasoline with increased octane numbers prompted the diversion of the olefin feeds to alkylation units that gave higher octane number products. Yet some refinery balances have more propylene than alkylation can handle, so a newer version of polymerization was introduced [24]. It is the Dimersol process of the Institut Francais du Petrole for which the flow diagram is shown in Fig. 18.

The Dimersol process uses a soluble catalytic complex injected into the feed before it enters the reactor. The heat of reaction is taken away by circulating a portion of the bottoms back to the reactor after passing through a cooling water exchanger. The product goes through a neutralizing system that uses caustic to destroy the catalyst so that the resulting polymer is clean and stable. Typical octane number ratings for the product are 81 Motor and 96.5 Research, unleaded.

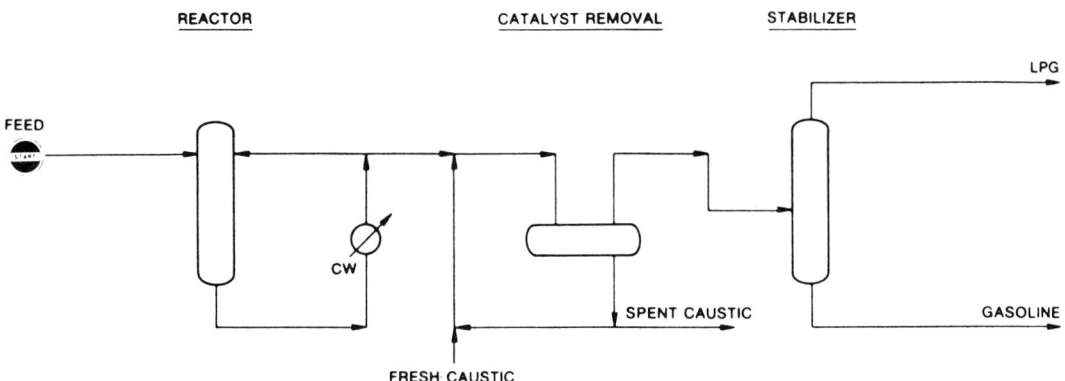

FIG. 18. Dimersol—Institut Francais du Petrole. (Reprinted from Ref. 10 with permission.)

Alkylation

This is another process that increases the total yield of gasoline by combining some of the gaseous light hydrocarbons to form bigger molecules boiling in the gasoline range. Alkylation combines isobutane with a light olefin, typically propylene and butylene. A flow diagram for an alkylation unit using hydrofluoric acid as a catalyst is shown in Fig. 19.

Common catalysts for gasoline alkylation are hydrofluoric acid or sulfuric acid. The reaction is favored by higher temperatures, but competing reactions among the olefins to give polymers prevent high quality yields. Thus, alkylation usually is carried out at low temperatures in order to make the alkylation reaction predominate over the polymerization reactions. Temperatures for hydrofluoric acid catalyzed reactions are approximately 100°F and for sulfuric acid they are approximately 50°F. Since the sulfuric-acid catalyzed reactions are carried out at below normal atmospheric temperatures, refrigeration facilities are included.

Alkylate product has a high concentration of 2,2,4-trimethyl pentane, the standard for the 100 rating of the octane number scale. Other compounds in the alkylate are higher or lower in octane number, but the lower octane number materials predominate so that alkylate has a Research octane number in the range of 92 to 99. Developments are underway to slant the reactions in favor of the higher octane materials [25]. Random samples of alkylate quality reported in the literature [26] are summarized in Table 10.

TABLE 10 Typical Alkylate Octane Numbers [25]

	Feed Olefin			
	C_2	C_3	$C_3 + C_4$	C_4
Research octane number, clear	101.5	90.5	93	96.5
Motor octane number, clear	93	89	91	95.5

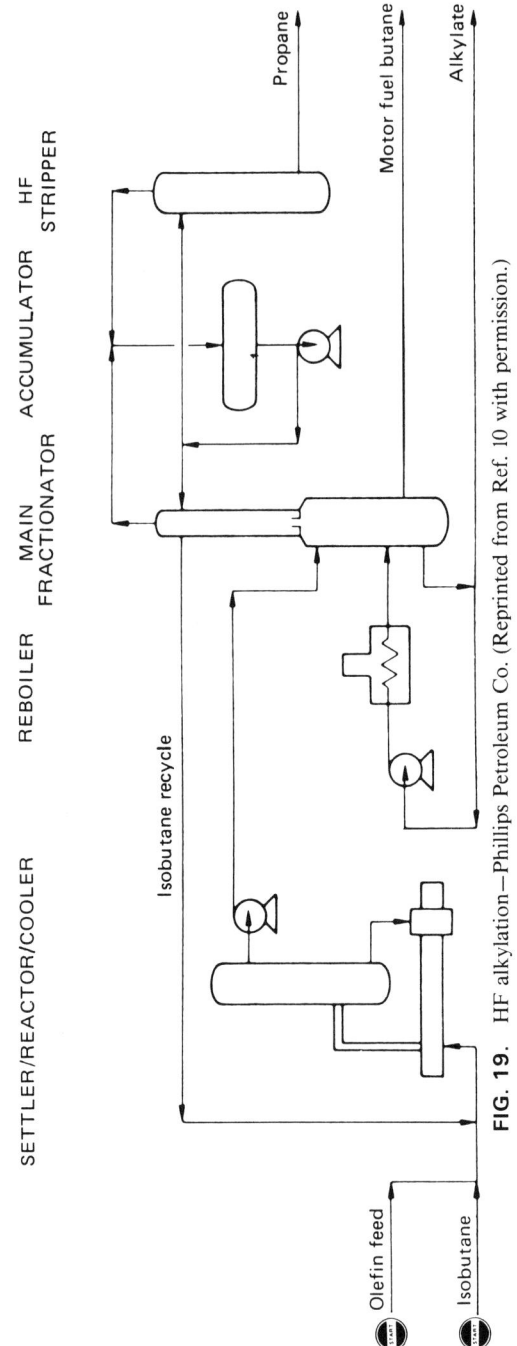

FIG. 19. HF alkylation—Phillips Petroleum Co. (Reprinted from Ref. 10 with permission.)

Ethylene

The evolution of cracking to make ethylene has progressed along two lines. In one, the ethylene is a by-product of fuel manufacturing, with the feedstock sometimes being a less desirable fuel material and at other times being a heart cut from some very desirable fuel material like naphtha. In the other line of progression, ethylene is pursued as a growing business of its own, with heavier by-product liquids being treated for use as gasoline blending stocks.

A popular starting material for ethylene cracking is ethane or propane. Some forecasts [27] suggest that these light hydrocarbon feeds may not be available in the growing volumes needed to keep up with a predicted 4.5–5% per year growth in ethylene demand. Thus many recent ethylene-cracking processes are tailored to handle heavier feedstocks. A flow diagram of an ethylene cracker [28] is shown in Fig. 20.

The feedstock is preheated and mixed with steam to be cracked in a tubular pyrolysis furnace. The products leave the furnace at 1400–1600°F and are rapidly quenched in an exchanger and sent to a gasoline fractionator where heavy fractions are removed. The gaseous products go to a quench

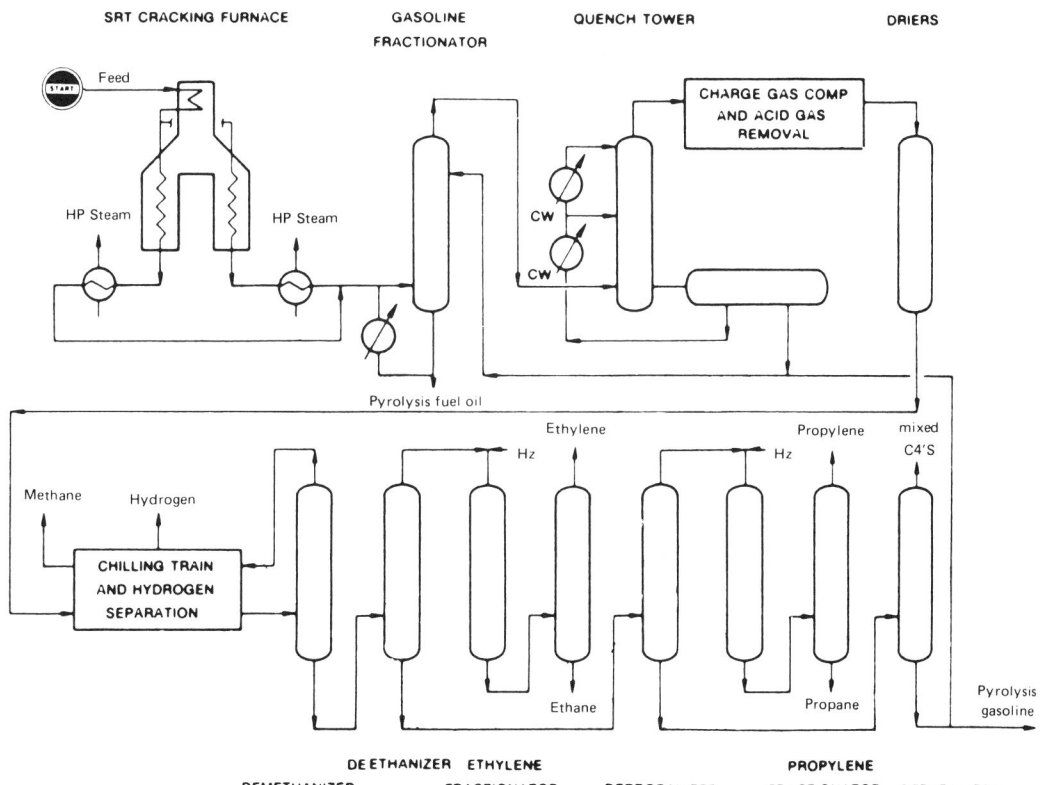

FIG. 20. Ethylene—C-E Lummus. (Reprinted from Ref. 28 with permission.)

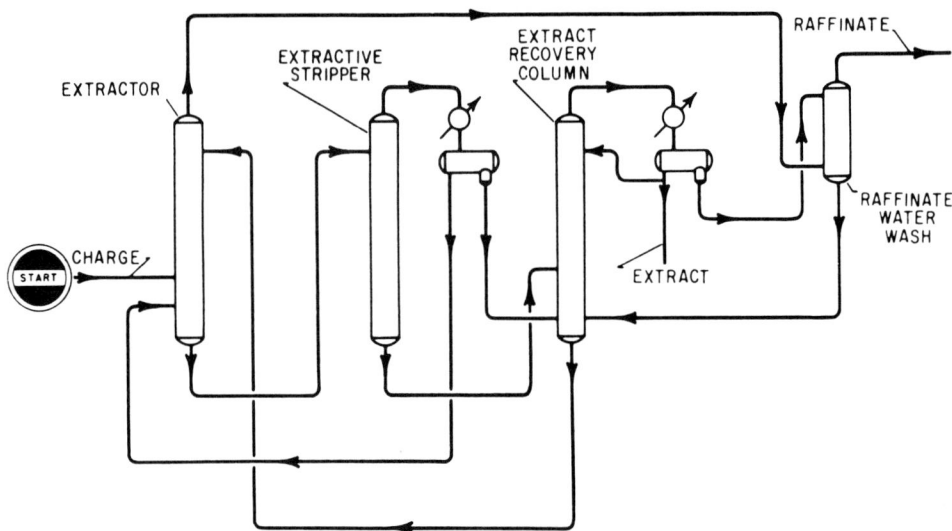

FIG. 21. Sulfone extraction—UOP Process Division of UOP, Inc., for process originally developed by Royal Dutch/Shell group. (Reprinted from Ref. 10 with permission.)

tower where direct water quench stops any further reaction. The flow diagram shows that the remainder of the process is intended to get light gas separation of the product.

The main derivatives of ethylene are polyvinyl chloride (PVC), ethylene glycol, and polyethylene.

Aromatics

Benzene, toluene, xylene, and ethylbenzene are made mostly from catalytic reforming of naphthas with units like those already discussed. As a gross mixture, these aromatics are the backbone of gasoline blending for high octane numbers. However, there are many chemicals derived from these same aromatics [29]. Thus many aromatic petrochemicals have their beginning by selective extraction from naphtha or gas-oil reformate.

A typical extraction process is the Sulfolane extraction process [10] shown in Fig. 21. The reformate is fed to a contactor for the countercurrent extraction of aromatic components with sulfolane solvent. The solvent, rich in aromatics, goes to a stripper and recovery column where the aromatics are separated from the solvent. Aromatic recoveries and purities are varied by choosing suitable operating parameters.

Refinery Size and Cost

The few, truly huge refineries give the impression that most refineries are big. The bigger world refineries have a capacity of more than 600,000 BPD [30]. Yet one should not be misled by a so-called "average" refinery designed for

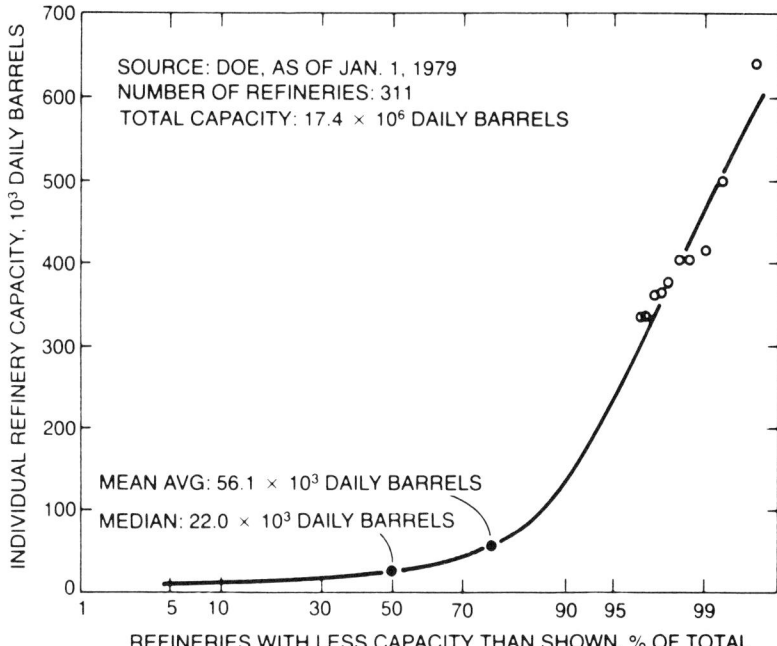

FIG. 22. United States refinery size distribution as of January 1, 1979.

100,000 BPD. Actually, the mean average size of the 240 refineries in the United States operating as of the first of 1988 was 58,250 BPD, computed from U.S. Department of Energy data [31]. Moreover, half of these refineries are only about 22,000 BPD (the median) or less. A distribution plot for United States refinery sizes is shown in Fig. 22 (as of 1980).

TABLE 11 Distribution of Unit Cost[a]

Process Unit	Relative Capacity (vol.%)	Unit[b] Cost ($/BPD)	Cost as Part of Total Refinery[b]	
			$/BPD	%
Crude distillation, atmospheric	100	275	275	8
Crude distillation, vacuum	35	320	110	3
Catalytic cracking, fresh feed	28	1,400	400	11
Hydrocracking	4	2,000	80	2
Coking	6	1,200	75	2
Thermal cracking	3	700	20	1
Catalytic reforming	21	1,100	230	7
Alkylation, product basis	5	2,000	100	3
Hydrotreating	46	400	190	5
Others and offsites	100	2,000	2,000	57
Total refinery	100	11,395	3,480	100

[a]Basis: Total United States refining capacity, first of 1987. Cost: 1987 basis.
[b]See the *Petroleum Processes, Cost Indexes* article to update these costs.

Costs are harder to nail down. Some reasonable estimates are given in Table 11. The relative capacity for each type of unit was fixed to match the relative capacity in existence in the United States at the first of 1987. The costs are also on a 1987 basis, although inflation has been changing these values rapidly.

An important item to notice is that more than half of the cost of a refinery is for materials and equipment other than those directly associated with specific petroleum processing. Some of these other items include storage, utilities, and environmental control systems. Land costs are excluded.

It should not be inferred that future construction will have the same cost distribution as shown in Table 11. The cost distribution in Table 11 is based on existing unit capacity. Future product demands will likely slant future construction in favor of one or another process with different relative unit costs based on future technology.

This article is an updating of Chapter 14 of *Riegel's Handbook of Industrial Chemistry*, 8th ed. (James A. Kent, editor), copyright © 1983 by Van Nostrand Reinhold, New York.

References

1. British Petroleum Co., *BP Statistical Review of the World Oil Industry 1987*, published annually.
2. "International Outlook Issue," *World Oil*, August 15, annually.
3. U.S. Department of the Interior, *Minerals Yearbook, Metals, Minerals and Fuels*, annually.
4. U.S. Department of Energy, "Petroleum Statement, Monthly," DOE/EIA-0109 (yr/month).
5. American Society for Testing and Materials, *1986 Annual Book of ASTM Standards*, Parts 23 and 24.
6. American Petroleum Institute, *Technical Data Book—Petroleum Refining*, 3rd ed., 1985.
7. K. M. Watson and E. F. Nelson, "Improved Methods for Approximating Critical and Thermal Properties of Petroleum Fractions," *Ind. Eng. Chem., 25*, 880 (1933).
8. H. L. Hoffman, "Sour Crude Limits Refining Output," *Hydrocarbon Process., 52*(9), 107–110 (1973).
9. E. P. Ferrero and D. T. Nichols, *Analyses of 169 Crude Oils from 122 Foreign Oilfields*, U.S. Department of the Interior, Bureau of Mines, Information Circular 8542, 1972.
10. "1980 Refining Process Handbook," *Hydrocarbon Process., 59*(9), 93–220 (1980).
11. G. H. Unzelman and E. J. Forster, "How to Blend for Volatility," *Pet. Refiner, 39*(9), 109–140 (1960).
12. H. Kay, "What Hydrogen Treating Can Do," *Pet. Refiner, 35*(9), 306–318 (1956).
13. H. G. Corneil and F. J. Heinzelmann, "Hydrogen for Future Refining," *Hydrocarbon Process., 59*(8), 85–90 (1980).

14. Foster Wheeler Corp., "Crude Distillation," *Hydrocarbon Process., 53*(9), 106 (1974).
15. T. F. Kellett et al., "How to Select Hydrotreating Catalyst," *Hydrocarbon Process., 59*(5), 139–142 (1980).
16. T. R. Hughes et al., "To Save Energy When Reforming," *Hydrocarbon Process., 55*(5), 75–80 (1976).
17. J. H. Jenkins and J. W. Stephens, "Kinetics of Cat Reforming," *Hydrocarbon Process., 59*(11), 163–167 (1980).
18. G. W. G. McDonald, "To Judge Reformer Performance," *Hydrocarbon Process., 56*(6), 147–150 (1977).
19. F. D. Hartzell and A. W. Chester, "FCCU Gets a Catalyst Promoter," *Hydrocarbon Process., 58*(7), 137–140 (1979).
20. G. H. Dale and D. L. McKay, "Passivate Metals in FCC Feeds," *Hydrocarbon Process., 56*(9), 97–102 (1977).
21. J. S. Magee et al., "A Look at FCC Catalyst Advances," *Hydrocarbon Process., 58*(9), 123–130 (1979).
22. R. C. Howell and R. C. Kerr, "Moving Coke? What to Expect," *Hydrocarbon Process., 60*(3), 107–111 (1981).
23. R. F. Sullivan and J. A. Meyer, *Catalyst Effects on Yields and Product Properties in Hydrocracking*, Presented at the American Chemical Society Philadelphia Meeting, April 6–11, 1975.
24. W. J. Benedek and J-L. Mauleon, "How First Dimersol Is Working," *Hydrocarbon Process., 59*(5), 143–149 (1980).
25. R. M. Heck et al., "Better Use of Butenes for High-Octane Gasoline," *Hydrocarbon Process., 59*(4), 185–191 (1980).
26. R. S. Logan and R. L. Banks, "Disproportionate Propylene to Make More and Better Alkylate," *Hydrocarbon Process., 47*(6), 135–138 (1968).
27. R. A. Klein, "Olefins Shift to Heavy Liquids," *Hydrocarbon Process., 59*(10), 113–115 (1980).
28. C-E Lummus, "Ethylene," *Hydrocarbon Process., 58*(11), 160 (1979).
29. J. R. Dosher, "Toluene: Octanes or Chemicals?" *Hydrocarbon Process., 58*(5), 123–126 (1979).
30. "World Refineries, 1980," *Int. Pet. Times, 84*(2119), 13–32 (March 15, 1980).
31. U.S. Department of Energy, Energy Information Administration, *Petroleum Refineries in the United States and U.S. Territories, January 1, 1988*, DOE/EIA-0111/88, released June 28, 1988.

HAROLD L. HOFFMAN
JOHN J. McKETTA

Petroleum Refinery of the Future

Introduction

The current lull in petroleum refinery modernization activity is certain to end soon. When it does, the industry will be looking for low cost routes to maximum yield of transportation fuels and maximum crude and product flexibility. There are advanced process technology and the refining expertise necessary to meet these needs, via either revamp or new construction. These processes are discussed, along with an advanced fluid bed design technology called TTR. Integration of these processes and design technology into existing refineries is demonstrated.

It is evident to most people in the oil refining business that there is very little refinery upgrading activity at the present time. In fact, because of the present market uncertainty, it might be correct to say that a paralysis has hit the oil business. However, since the survival of our oil industry depends on our providing an affordable energy supply for the world, this paralysis cannot last. In the writers' opinion, the start of a new era in refinery upgrading projects will take place in the not too distant future.

However, we do not believe that these upgrading projects will be the large mega-dollar (astro-yen) projects of the early 1980s. Instead, we believe they will be projects that minimize investment cost while maximizing the use of existing equipment to:

1. Maximize yield of transportation fuels.
2. Maximize refinery crude oil and product flexibility.
3. Minimize operating costs.

With the above in mind, Engelhard has developed what we believe to be the refinery of the future. It is a refining scheme that is applicable both to new refineries and to the revamping of existing refineries. This refining scheme can eliminate crude and vacuum units and their associated investment and operating costs. In addition, this refining scheme yields greater than 90 vol% transportation fuels on a wide range of crude oils and allows considerable flexibility in the product distillate to gasoline ratio.

Figure 1 is a schematic of our idea of the refinery of the future. With the exception of the ART and ESR* Processes, all of the units shown are

*ART and ESR are service marks of Engelhard Corporation for professional services related to selective vaporization processes for removing contaminants from petroleum feedstocks and to catalytic processes for removing SO_x/NO_x from gas streams.

Petroleum Refinery of the Future

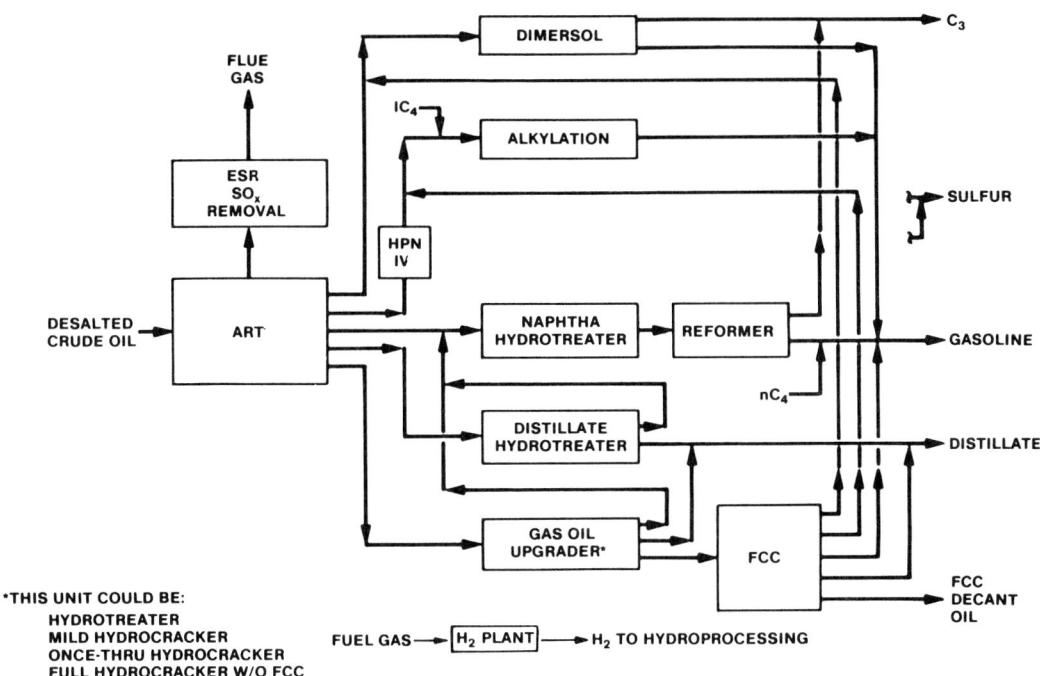

FIG. 1. Whole crude ART refinery.

conventional refining processes with which the industry is quite familiar. The capabilities of the ART Process are well documented [5, 7, 8] and thoroughly demonstrated, both commercially and in pilot plants. The ESR Process is new Engelhard technology for the removal of SO_x/NO_x from combustion flue gases. This process is explained in greater detail at the end of this article. Since the ART Unit serves as the crude and vacuum unit, by proper sizing of the ART Process combustor and ESR system, this refinery can be designed for complete crude flexibility. By proper selection and design of the gas oil hydroprocessing and FCC units, this refinery can achieve considerable flexibility in the distillate to gasoline product ratio. To illustrate the crude and product flexibility of this scheme, a study was made using an LP (linear program) model of the refinery in Fig. 1. Four crudes (Kuwait, Arab Heavy, Maya, and Bachaquero) were used with four different gas oil hydroprocessing severities (hydrotreating, and mild, once-through, and full hydrocracking). The results of this study are used throughout this paper to emphasize various points. For a 4.8-MM MT/Y (100,000 BPD) Maya crude case, the ART and ESR Processes would cost about $120 MM with operating costs of about $6.90/MT ($1.00/bbl), including the costs of the ESR system and ARTCAT fluidizable contact material.

There is also a great potential for revamping existing refineries to obtain the same crude and product slate flexibility at the lowest capital and operating costs. Engelhard has assembled a team of experts in refining technology that have the capability of applying the most modern and up-to-date fluid process technology and hydrogen addition technology to any refinery system to meet the crude and product requirements of the refiner. Using Table 1 as our guide, we will discuss application of this technology and expertise to new and existing refineries.

Minimize the Investment Costs

One way to minimize the capital requirements in a refinery modernization is to minimize operating pressure and hydrogen requirements in hydroprocessing units. The combination of the new ART and ESR technologies reduces the cost of heavy oil hydrogen processing by removing sulfur, nitrogen, and metals from the feedstock. This is accomplished by converting the asphaltenes to hydrocarbon products or fuel for the process, while not changing the molecular structure of the lower boiling (<540°C, <1000°F) virgin material in the feed. Thus, the downstream processing units see 30–50% less sulfur, 50–80% less nitrogen, and >95% fewer asphaltenes and metals. This greatly reduces the catalyst volume, hydrogen partial pressure, and hydrogen makeup and recycle rates required to achieve the desired effect and, therefore, reduces the overall capital costs.

In addition, since the entire 345°C+ (650°F+) ART product can be processed in downstream units, the ART Process (with its main column) can effectively replace the atmospheric and vacuum distillation columns. This represents considerable savings in capital investment for new refineries.

The best approach to minimizing the capital costs of refinery modernization may be the blending of new and revamped process units.

TABLE 1

A) Minimize the investment cost
B) Maximize the use of existing equipment
C) Maximize refinery flexibility, not only in crude type, but distillate-to-gasoline ratio
D) Maximize yield of high quality products, mainly gasoline and distillates
E) Minimize operating costs

Maximize the Use of Existing Equipment

This area will be divided into detailed discussions of an advanced fluid process technology and its application to FCC and coker revamps and the adaptation of existing hydrogen addition units to an ART refinery.

The key to optimizing ART integration into an existing refinery is determining the most advantageous place for the ART Process Unit in the refining scheme. The ART Process can be used anywhere up the refining ladder from wellhead upgrading [2, 6] to vacuum resid processing. Since the ART Process is really a selective vaporization process, it can replace the existing crude unit furnace so that the existing crude column can be used as the ART Process main column. This whole crude scheme has the advantages shown in Table 2. All the ART liquid products [naphtha, light gas oil (LGO), and heavy gas oil (HGO)] can be processed in existing conventional process equipment to further refine them into finished products: gasoline, jet, diesel fuel, and fuel oil.

TABLE 2 Benefits of Whole Crude ART Processing

I. Capital
   Eliminates need for crude and vacuum units
   Reduces hydrogen pressure and recycle requirements in downstream hydroprocessing units
   ART Unit investment is essentially the same on whole crude or atmospheric or vacuum resid since coke yield for ART Processing is 0.8–1.0 times the carbon residue
   Total refinery yields on ART Processing of whole crude compared to ART Processing of atmospheric or vacuum resid is essentially the same since feed material boiling below 540°C (1000°F) goes through the ART Process essentially unreacted
II. Operating Costs
   Eliminates fuel and operating costs associated with crude unit; savings of $3.50–$10.00/MT ($0.50–$1.50/bbl)
   Eliminates fuel and operating costs associated with vacuum unit; savings of $2.00–$3.50/MT ($0.30–$0.50/bbl)
   80%+ of the heat released in the ART combustor (coke burn) can be recovered as usable energy for operation of associated units and equipment
   Reduces requirements for auxiliary steam generation in the refinery
   Reduces the requirements for imported fuel gas
   High naphthenic acid crudes can be processed without concern for their corrosive effects
   Thermally unstable crudes can be processed without concern for the additional gas make

In optimizing the downstream processing of ART products, one must understand the refiner's ultimate needs. The selection of the ART Process with proper sizing of the combustor and ESR system can assure the desired crude flexibility. It is now necessary to determine the desired product slate. If one wants to produce ethylene and associated petrochemical products inexpensively from the 565°C+ (1050°F+) material in the crude, the ART Process can be adapted to this end [4]. However, most refiners are more concerned with the gasoline-jet-diesel distribution. Greater transportation fuel flexibility requires making better use of conventional downstream gas oil conversion processes. Engelhard possesses the revamp technology and expertise to incorporate many of these gas oil processes into our refinery of the future.

FCCU/Coker Revamps

The revamp schemes proposed in the following discussion are derived from an advanced fast fluid design technology that Engelhard Corporation has developed. A refiner may use this revamp technology to debottleneck an existing fluid catalytic cracking unit (FCCU), increasing its coke-burning capacity as much as threefold. This gives the refiner the option of increasing octane barrels by replacing LSR naphtha with FCC gasoline and alkylate in either of two ways:

1. Making the FCCU feed heavier by addition of residuum, while decreasing the refinery throughput.
2. Making the FCCU feed heavier by addition of residuum, while charging a heavier crude to the refinery.

This revamp scheme also allows the refiner to convert an existing FCCU to an ART unit. Other circulating fluid solids processes such as fluid cokers can be revamped in a manner similar to the revamp of an FCCU. In addition, a scheme has been developed for the revamp of delayed cokers to ART Units using the Engelhard fast fluid design technology. The conversion of a delayed or fluid coker to an ART Unit can increase the refinery liquid yield up to 10 LV % on crude and produce better quality products, requiring less hydrogen addition in downstream hydrotreaters.

Refiners with two existing FCCU's have the attractive option of revamping both FCCU's for up to two to three times their original designed coke-burning capacity while converting one of them to an ART Unit. Each FCCU can be revamped for a cost of $4 to $16 per MT/Y ($200 to $800 per BPSD), depending on the condition and utility of existing equipment. It is possible to convert fluid cokers and delayed cokers to ART Units for a relatively low revamp cost of $4 to $16 per MT/Y for fluid coker conversion and $10 to $16 per MT/Y for delayed coker conversion. In doing so, refiners can realize increased percentage yield of liquid products, higher quality products, lower operating costs, increased refinery energy efficiency, and frequently an increase in refinery capacity.

As all revamps will vary with the application, it is best to simply describe the major features of the Engelhard fast fluid design technology. The techniques for adapting these features to existing equipment is discussed in greater detail in another presentation [1].

Advanced FCC-ART Design Technology

Engelhard's advanced fluidized circulating unit design technology was first made public at the NPRA Annual Meeting in Los Angeles in March 1986 [3]. It was developed by Engelhard Corporation for both FCC and ART applications and is called the "Twin Transport Riser" system. The Twin Transport Riser (TTR)* design is a radical departure from the conventional FCC in use today. Figure 2 illustrates the new Engelhard proprietary design for the case of FCC operation.

The outstanding features of the TTR design are the following:

1. *Riser Regenerator.* A true "hollow pipe" riser regenerator that operates in the fast bed region of the fluidization phase diagram will be used. This is a

---

*TTR is a trademark of Engelhard Corporation.

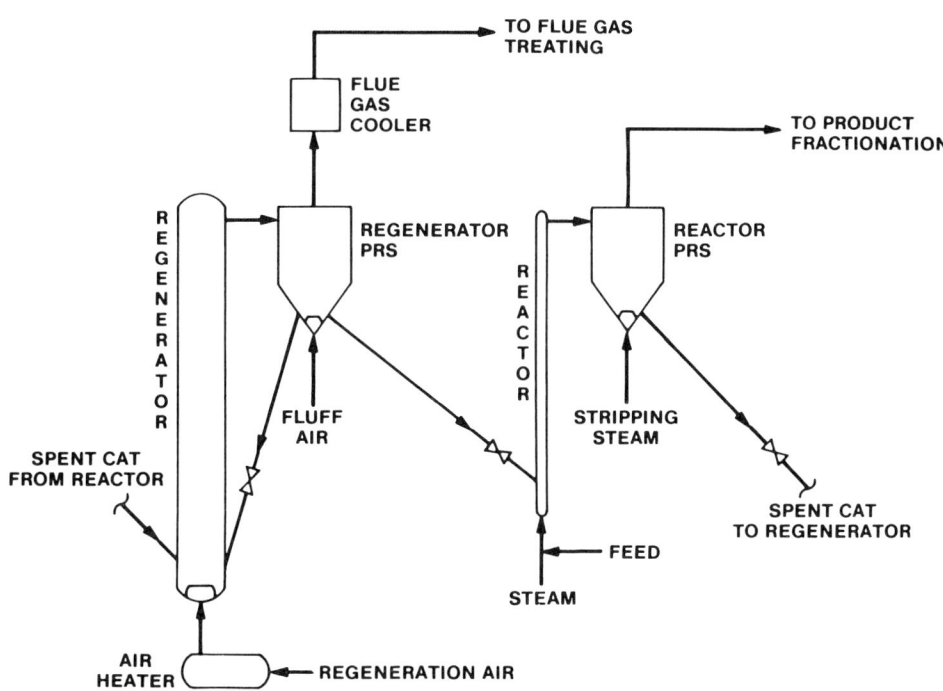

FIG. 2. Twin transport riser system.

significant leap from the bubbling or turbulent regimes of conventional designs. The benefits of the fast bed mode of operation are mechanical simplicity, reduced catalyst inventory, and reduced catalyst residence time in the regenerator. Smaller vessels reduce capital investment.

Lower residence time reduces hydrothermal deactivation of the catalyst at elevated temperatures. The mechanical simplicity and reduced inventory shorten the downtime for turnaround by making the cooldown and unloading times shorter and by virtually eliminating the need for scaffolding to inspect and repair internals.

2. *No Disengager*. The reactor riser discharges directly into an external cyclone system. This precludes the need for a disengager. In the conventional designs of today, the disengager is nothing more than a containment vessel for the cyclones. In fact, disengagers may often be an impediment to efficient operation. Disengagers tend to produce overcracking (more fuel gas and excessive coke) because of the slow disengaging of catalyst and hydrocarbon vapor in the large disengager volume. Much effort and money have been spent in developing countless mechanical schemes for rapid disengaging of catalyst and vapor. In the new design, rapid disengaging will occur in the proprietary external cyclone systems. Also, the elimination of the disengager reduces capital and maintenance costs.

3. *External Cyclones*. External proprietary Particulate Removal Systems (PRS)* that give essentially complete recovery of particles larger than 7 μm will be employed for both the reactor and regenerator. The high PRS efficiency allows the use of regenerator flue gas power recovery equipment without further fines removal. On the reactor side, it reduces the BS&W of the heavy product. Overall, the PRS units on the reactor and regenerator outlets reduce catalyst losses and also reduce the average particle size of the catalyst inventory. This will tend to improve catalyst fluidization and also increase the surface area per unit weight of catalyst (activity).

4. *Catalyst Cooler*. Catalyst coolers are not new to the industry, but they are not as yet common. Although not indicated in Fig. 2, they will be considered for installation on all ART Process Units and even FCCU's if one wishes to crack heavy oil. In the ART Process, cat coolers allow a wide variety of feedstock qualities to be used while maintaining the combustor (regenerator) temperature below its metallurgical limit. In FCCU's, cat coolers allow direct regenerator temperature control, which gives an added degree of control over cat/oil and, hence, yields. In addition, cat coolers can provide high pressure steam for turbine drivers, power generation, or heat exchange.

5. *Few Vessel Internals*. Other than the PRS units, all vessels in the new design contain very few internals. The few necessary internals are two air grids in the regenerator and regenerator PRS, one steam distributor in the reactor PRS, and one feed distributor in the reactor.

---

*PRS is a trademark of Engelhard Corporation.

The end result is a very simple design that is less expensive to build and maintain and easier to operate than conventional FCCU's. Most importantly, the mechanical simplicity and considerably reduced overall vessel volume of the TTR system give it an estimated grass roots construction cost 15 to 25% less than conventional designs of comparable feed quality and capacity. In addition, this new system can be constructed on a modular basis for units up to 1.2 MM MT/Y (25,000 BPD). This simplicity is also what makes it an ideal design for revamping existing equipment.

Hydrotreater/Hydrocracker Integration

As discussed previously, hydrogen addition to the 345°C+ (650°F+) gas oil is much more economical after ART processing because the ART Process removes the typical heavy oil hydroprocessing catalyst poisons. This results in lower operating pressures and less hydrogen circulation and consumption when compared to hydroprocessing of virgin residual oils or thermally degraded products from even the most efficient carbon rejection processes. Therefore, desulfurization of ART 345°C+ (650°F+) product (HGO) for FCC feed or different levels of hydrocracking of ART HGO product to produce high quality diesels or jet fuel are viable options in achieving product slate flexibility. The important implication of this is that existing hydrogen addition capacity in most refineries is adequate to handle the HGO and lighter ART products. Only minimal modifications should be necessary in adapting existing hydrotreaters/hydrocrackers for ART product processing. Based on the results of pilot-plant work performed by Shell Chemical, a 1986 NPRA paper defined the hydrotreating requirements for the products from ART Processing of whole Maya crude [3]. The work clearly shows that only conventional vacuum gas oil (VGO) hydrotreating conditions are required for ART HGO. The paper further shows that greater than 90% yield of transportation fuels can be obtained from an HGO hydrotreating/FCC refinery processing Maya crude.

There are two approaches to residue upgrading which have been used in refinery designs—the indirect desulfurization approach (carbon-out) or the direct desulfurization approach (hydrogen-in). The hydrotreaters constructed for each of these schemes can be readily adapted to the processing of ART products. This will be further explained for both refinery designs in the following discussion.

The indirect desulfurization or "carbon-out" approach is illustrated in Fig. 3. This strategy is generally used with heavier, less desirable crudes. It includes elimination of the poor quality vacuum residue (carbon-out) as fuel oil, coke, or pitch. Depending on the way the residue is handled, the penalty to the plant's ultimate liquid yield is substantial, ranging from 8 to greater than 20 vol.%.

The VGO hydrotreaters for these refineries were generally designed to process heavy vacuum gas oil plus deasphalted oil or coker gas oil. The distillate and naphtha hydrotreaters were designed to process a mix of virgin and cracked stocks. Thus, in the plant designed for carbon-out processing, the naphtha and distillate hydrotreaters should be more than adequate for the

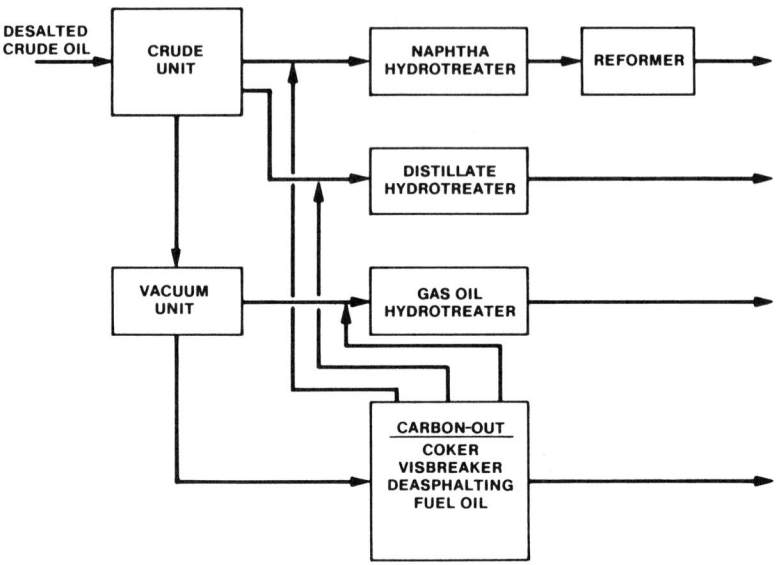

FIG. 3. Indirect desulfurization (carbon-out).

hydrotreating of the ART naphtha and light gas oil (LGO) [3]. The VGO hydrotreater can easily hydrotreat the ART HGO with today's modern catalyst. Further, depending on the pressure, LHSV, available hydrogen, and fractionator capabilities, the unit could achieve 25–30 wt.% conversion with mild hydrocracking. In fact, advances in catalysis have allowed the conversion of many existing VGO hydrotreaters to mild hydrocracking operation to achieve increased distillate production.

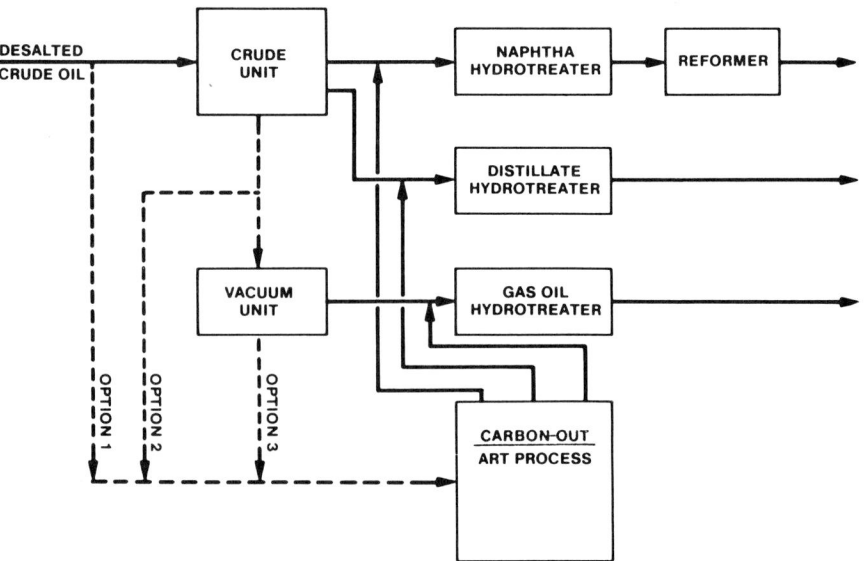

FIG. 4. Integration of ART into existing indirect desulfurization scheme.

Petroleum Refinery of the Future

Figure 4 shows the carbon-out scheme coupled with ART Processing, with three feed options [whole crude, atmospheric tower bottoms (ATB), or vacuum tower bottoms (VTB)]. The feedstock flexibility of the ART Process gives the refiner many options in choosing his revamp scheme, including the shutdown of his crude and vacuum units. Note that the ART Process eliminates the fuel oil, coke, or pitch product.

The direct desulfurization or "hydrogen-in" approach is illustrated in Fig. 5. The design crude for these refineries was usually good quality such as Kuwait, Light Arabian, or, in the worst case, Heavy Arabian. The atmospheric residue hydroprocessing unit (ARDS) is the pivotal process in this kind of plant. It is also the unit which restricts the feedstocks which can be processed in the refinery. As the metals and asphaltene content of the residue increase beyond the design basis, the catalyst performance becomes unacceptable. Because of the low conversion in residue hydrodesulfurization, increases in distillate can only be achieved by processing more crude. The ARDS unit is a bottleneck to refinery capacity and flexibility.

In combination with ART Processing, both crude and product flexibility can be achieved by conversion of the ARDS unit to ART heavy gas oil hydrotreating, mild hydrocracking, once-through hydrocracking, or full hydrocracking. Residue hydrotreating and gas oil hydrocracking operating conditions are virtually the same. Typical design conditions for the resid hydrotreater are also shown in Fig. 5 for a 15% conversion end-of-run case. When processing resid, the unit produces low sulfur fuel oil or heavy oil FCC feedstock with sulfur levels ranging between 0.1 and 1.0 wt.%. The hydrogen consumption ranges between 100 and 160 NM^3/M^3 (600 and 1000 SCF/bbl). The unit is high pressure, contains large volumes of catalyst, and multiple reactors and/or multiple reactor trains. The important point is that the ARDS unit was designed to process a full residue and has hydrogen circulation and

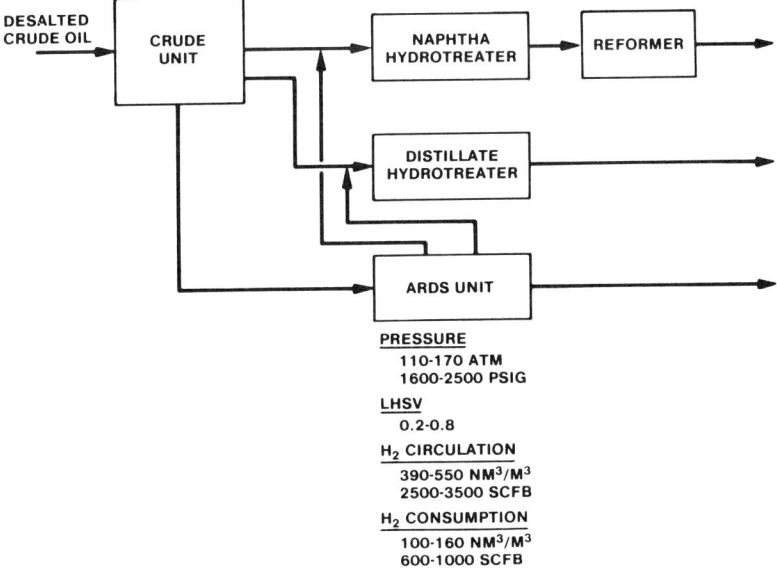

FIG. 5. Direct desulfurization (hydrogen-in).

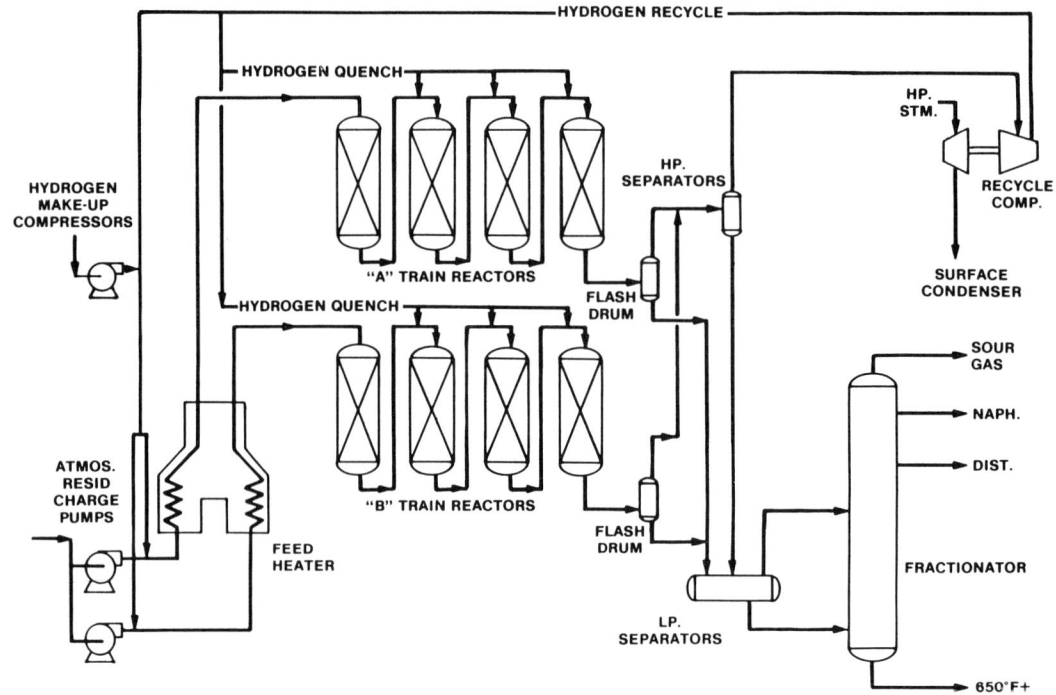

FIG. 6. ARDS Unit process flow diagram.

makeup capacity designed for that feed. Figure 6 shows the complexity of a plant designed for deep desulfurization of an atmospheric residue to produce feed to a heavy oil FCC operation. The two trains can easily be converted to an ART HGO hydrocracker or hydrotreater, with the possible requirement of additional hydrogen quench in the hydrocracking case.

Figure 7 shows the hydrogen-in refinery coupled with ART Processing of the whole crude or ATB. ART Processing, by efficiently reacting the asphaltenes, has changed the crude residue to heavy gas oil with 60–70 vol.% yield, similar in quality to a virgin heavy vacuum gas oil that has been blended with DAO. Now, because the asphaltene has been converted to lighter products and the ART HGO is only 60–70% of the ARDS design throughput, the hydrogen availability on both makeup and circulation will increase by the ratio of the feed rates. With proper catalyst changes, the fixed-bed residue hydrotreater can now be used for hydrotreating FCC feed or for mild hydrocracking, once-through hydrocracking, or full liquid hydrocracking as limited by hydrogen availability or pressure requirements. The ease of hydrotreating and mild hydrocracking ART HGO has already been illustrated [3]. Pilot-plant work on once-through and full liquid hydrocracking is now being done in conjunction with Shell Chemical.

There is an important point to make here. Table 3 compares the ART HGO from each of the base crudes with its corresponding atmospheric residue. As can be seen, the difference is dramatic because of the asphaltene removal from the ART HGO. In fact, as pointed out in the 1986 NPRA paper [3], a residue hydrotreater converted to hydrotreat ART HGO would

Petroleum Refinery of the Future

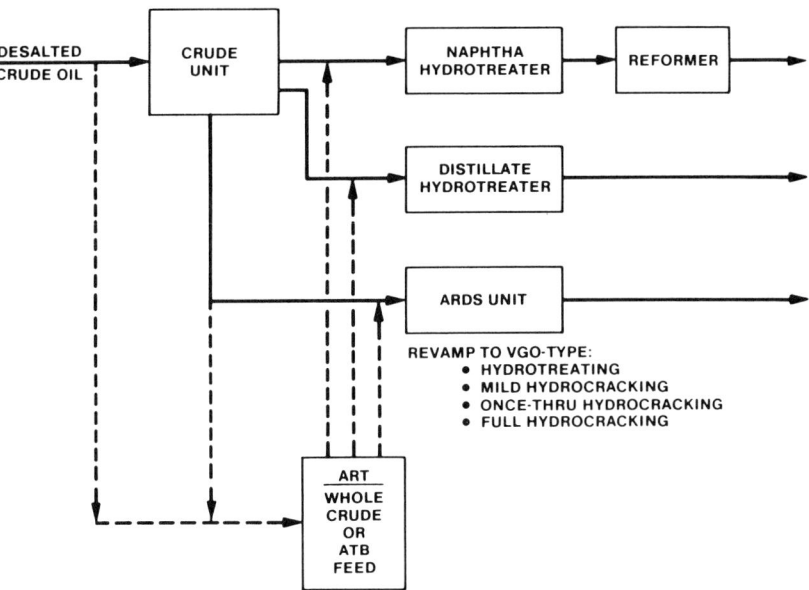

FIG. 7. Integration of ART into existing direct desulfurization scheme.

TABLE 3 Comparison of ATB and ART HGO

	Kuwait	Heavy Arabian	Maya	Bachaquero
ART HGO				
Yield, vol.%	30.1	33.2	34.9	45.2
Specific gravity	0.953	0.959	0.953	0.959
°API	17	16	17	16
S, wt.%	3.8	4.4	3.2	2.5
N, ppm wt.	1900	1750	2450	4400
Ni + V, ppm wt.	2	2	21	19
C_7 insols., wt.%	0.35[a]	0.25	0.2	0.5[a]
RCR, wt.%	4.0	4.1	4.8	4.5
Atmospheric residue				
Yield, vol.%	48.5	52.9	58.1	70.9
Specific gravity	0.971	0.983	1.000	0.997
°API	14.3	12.5	10.0	10.5
S, wt.%	4.2	4.5	3.75	2.7
N, ppm, wt.	2300	2200	5050	5800
Ni + V, ppm, wt.	60	65	303	580
C_7 insols., wt.%	11.1[b]	7.5	17.5	15.2[b]
RCR, wt.%	11.1	11.9	15.2	15.2

[a] Typical C_7 insols. removal at 98°.
[b] C_7 insols. assumed equal to RBC.

require less than half the pressure, less than 10% of the catalyst, and less than 50% of the hydrogen makeup and circulation. Clearly, there must come a time when we stop putting hydrogen into the asphaltene molecule. Further, if the existing ARDS unit were used only to hydrotreat the ART HGO without changing the catalyst, the catalyst life would be in the range of 5–10 years depending on the operating conditions.

Another important point to make here concerns the fines content (BS&W) of the ART heavy gas oil product. Refiners may anticipate that catalyst particles will cause bed plugging in downstream residual oil fixed-bed hydroprocessing units. These units are typically loaded with 1 mm ($^1/_{32}$ in.) extrudates or small-lobed catalysts, which tend to filter the fines and result in plugging. However, because the ART HGO is similar to VGO, it can be hydroprocessed with a larger 1.5 or 3 mm ($^1/_{16}$ or $^1/_8$ in.) catalyst. The larger catalyst is much less susceptible to plugging. In addition, the proprietary particle separation technology (PRS) which Engelhard possesses, and plans to install in all new and revamped ART and FCC Units, removes essentially all of the particles larger than 7 μm. This is significantly better than the 25-μm residual oil feed filters that many refiners use to alleviate the plugging problem on these fixed-bed residual oil hydroprocessors.

To complete the picture on ART product hydrotreating, Table 4 details the quality of the naphtha produced from whole crude ART processing of each of the base crudes in the LP study. Table 5 repeats this for the light distillates. Incremental hydrogen consumptions for hydrotreating these products are also shown.

Maximize Refinery Flexibility

The LP study indicates very clearly the wide variety of crude oils which can be ART processed and the flexibility in product distribution which the ART HGO processability allows. Indeed, the ART Unit has demonstrated its

TABLE 4 Naphthas from ART Processing Whole Crude [80–190°C (180–375°F) boiling range]

	Kuwait	Heavy Arabian	Maya	Bachaquero
Yield, % by vol.	22.4	19.0	15.1	11.0
Specific gravity	0.763	0.751	0.755	0.755
°API	54	57	56	56
Sulfur, wt.%	0.1	0.25	0.3	0.25
Nitrogen, ppm wt.	10	30	25	50
Olefins, wt.%	11	8	9	20
Consumption,[a] NM^3/M^3	17	16	19	32
SCFB	110	100	120	200

[a]To produce reformer feed quality.

TABLE 5 LGO from ART Processing Whole Crude [190–340°C (375–650°F) boiling range]

	Kuwait	Heavy Arabian	Maya	Bachaquero
Yield, % by vol.	29.5	30.8	36.4	30.1
Specific gravity	0.850	0.850	0.845	0.876
°API	35	35	36	30
S, wt.%	1.4	1.35	1.9	1.1
N, ppm wt.	250	275	300	550
Bromine no., mg/g	6	9	12	17
Cetane index	47	47	49	40
H_2 consumption, NM^3/M^3	43	47	68	76
SCFB	270	300	430	480

ability to process even tar sand bitumen [6]. The four crude oils used in the LP study are described in Table 6. They vary from 0.955 to 0.871 specific gravity (16.7 to 30.9°API), 6.0 to 11.3 wt.% RBC, and 30 to 430 ppm metals. As mentioned previously, the same ART Unit can be designed to handle all four crudes if the ART combustor is designed for the highest RBC crude and the ESR system is designed for the most sulfurous crude.

For each of the crudes in Table 6, the refinery scheme in Fig. 1 yields a product slate which can be shifted from maximum gasoline to maximum distillate. Trading hydrocracking severity for FCC capacity increases the distillate to gasoline ratio, as detailed in Tables 7a through 7d. Among the four crude cases, the distillate to gasoline ratio varies from 0.7 for the hydrotreating case to 1.8 for the full hydrocracking case.

TABLE 6 Crude Qualities

	Kuwait	Ah-Safaniya	Maya	Bachaquero
Specific gravity	0.871	0.889	0.913	0.955
°API	30.9	27.7	23.5	16.7
Sulfur, wt.%	2.5	2.9	3.0	2.3
Nitrogen, ppm wt.	1300	1300	3000	4300
RBC, wt.%	6.0	7.0	9.7	11.3
Nickel, ppm wt.	10	10	24	45
Vanadium, ppm wt.	22	28	240	385
TBP distillation, vol.%:				
C_4^-	2.5	1.9	0.9	0.5
15–80°C (60–180°F)	6.4	5.6	3.8	1.6
80–190°C (180–375°F)	17.6	15.6	15.3	6.4
190–340°C (375–650°F)	25.0	24.0	21.9	20.6
340°C+ (650°F+)	48.5	52.9	58.1	70.9

TABLE 7a 100,000 BPD ART Process Refinery (gas oil conversion process: hydrotreater and FCC)

	Kuwait	Ah-Safaniya	Maya	Bachaquero
Refinery inputs (BPSD):				
Crude	100,000	100,000	100,000	100,000
i-Butane	—	170	1,890	1,460
n-Butane	900	1,540	3,930	5,040
Purchased fuel oil	—	670	—	630
Total inputs	100,900	102,380	105,820	107,130
Refinery products (BPSD):				
Propane LPG	3,370	3,140	2,860	2,170
Butane LPG	—	—	—	—
Gasoline [87(R+M)/2 and 11 RVP]	54,250	52,550	50,120	54,390
Distillate (0.3% S)	37,850	41,180	45,640	41,620
FCC decant oil	1,940	2,170	2,460	3,620
Total liquid products	97,410	99,040	101,080	101,800
Vol.% recovery on total feed	96.5	96.7	95.5	95.0
Sulfur LT/d	325	385	410	330
Distillate cetane index	49	50	47	40

TABLE 7b 100,000 BPD ART Process Refinery [gas oil conversion process: mild hydrocracker (25% conversion) and FCC]

	Kuwait	Ah-Safaniya	Maya	Bachaquero
Refinery inputs (BPSD):				
Crude	100,000	100,000	100,000	100,000
i-Butane	—	—	1,390	770
n-Butane	250	1,260	3,800	4,690
Purchased fuel oil	850	1,580	970	1,450
Total inputs	101,100	102,840	106,160	106,910
Refinery products (BPSD):				
Propane LPG	3,470	3,250	2,980	2,330
Butane LPG	—	—	—	—
Gasoline [87(R+M)/2 and 11 RVP]	54,060	53,480	49,810	52,780
Distillate (0.3% S)	38,580	41,100	47,460	44,870
FCC decant oil	1,720	1,990	2,140	3,040
Total liquid products	97,830	99,820	102,390	103,020
Vol.% recovery on total feed	96.8	97.1	96.5	96.4
Sulfur LT/d	325	385	410	330
Distillate cetane index	50	50	48	41

TABLE 7c 100,000 BPD ART Process Refinery [gas oil conversion process: once-through hydrocracker (50% conversion) and FCC]

	Kuwait	Ah-Safaniya	Maya	Bachaquero
Refinery inputs (BPSD):				
Crude	100,000	100,000	100,000	100,000
i-Butane	—	—	470	—
n-Butane	—	—	3,020	3,060
Purchased fuel oil	1,680	2,600	1,870	2,860
Total inputs	101,680	102,600	105,360	105,920
Refinery products (BPSD):				
Propane LPG	3,660	3,470	3,210	2,640
Butane LPG	1,050	490	—	—
Gasoline [87(R+M)/2 and 11 RVP]	50,070	48,790	44,740	46,320
Distillate (0.3% S)	42,400	45,210	51,800	50,420
FCC decant oil	1,130	1,310	1,400	1,990
Total liquid products	98,310	99,270	101,150	101,370
Vol.% recovery on total feed	96.7	96.8	96.0	95.7
Sulfur LT/d	325	385	410	330
Distillate cetane index	51	51	49	44

TABLE 7d 100,000 BPD ART Process Refinery (gas oil conversion process: diesel mode hydrocracker)

	Kuwait	Ah-Safaniya	Maya	Bachaquero
Refinery inputs (BPSD):				
Crude	100,000	100,000	100,000	100,000
i-Butane	—	—	—	—
n-Butane	—	—	160	1,570
Purchased fuel oil	1,490	2,570	3,790	3,580
Total inputs	101,490	102,570	103,950	105,150
Refinery products (BPSD):				
Propane LPG	4,020	3,830	1,910	—
Butane LPG	3,060	470	—	—
Gasoline [87(R+M)/2 and 11 RVP]	43,200	40,910	36,200	34,760
Distillate (0.3% S)	49,770	53,900	60,410	62,220
FCC decant oil	—	—	—	—
Total liquid products	100,050	99,110	98,520	96,980
Vol.% recovery on total feed	98.6	96.6	94.8	92.2
Sulfur LT/d	325	385	410	330
Distillate cetane index	53	52	51	48

Maximize Yield of Transportation Fuels

The results of the LP study indicate, for all 16 cases (4 crudes × 4 schemes), greater than 90 vol.% yield of transportation fuels on total refinery feed, which includes crude, isobutane, n-butane, and purchased fuel oil. The maximum yield of transportation fuels is 93 vol.%, and this is in the Maya full hydrocracking case. Additional products include propane LPG, butane LPG, FCC decant oil, and sulfur.

In all cases the hydrogen demand is greater than the catalytic naphtha reformer supply. The net hydrogen requirement is supplied by steam reforming of refinery gases and LPG.

Reforming severity is governed by the 87 (R+M)/2 requirement of the gasoline pool. The distillate cetane index meets the ASTM minimum specification of 40 in all cases. Only in the Bachaquero crude hydrotreating and mild hydrocracking cases (Tables 7a and 7b) is the cetane index below 44, running 44 to 53 in all other cases.

The purchased fuel oil is for process heaters and does not include boiler fuel. The FCC decant oil could be used as fuel oil, but it is reported separately because it usually has a higher value than fuel oil. Two cases, the Kuwait and Maya crudes in the hydrotreating scheme (Table 7a), are in fuel balance. The fuel gas produced in the refinery just meets the steam reformer and fuel requirements. All other cases require some fuel oil import for process heaters, primarily due to the higher fuel requirements of higher severity levels of hydroprocessing of the ART HGO. Ultimately, it is the excellent conversion of fuel oil grade materials that allows this refining scheme to run in fuel oil deficit while yielding greater than 90 vol.% transportation fuels.

Minimize Operating Costs

As we have shown, the integration of the ART Process into existing refineries results in increased profitability and flexibility and a reduction in the energy required to process a barrel of oil. This energy reduction is a result of eliminating the need to operate a crude and vacuum unit and by lowering the pressure and hydrogen requirements for downstream processing. Note that these are the same factors which reduce the ART refinery investment costs. At the same time, one is using the low value energy component of the crude (low hydrogen content asphaltenes) as fuel to vaporize the crude and produce steam for operation of rotating equipment. The effective integration of this produced energy into an existing refinery is of major importance when considering the ART Process.

The ART Process converts about 80% of the Ramsbottom carbon to coke. Therefore, the quantity of coke made per barrel of whole crude is the same whether the whole crude, atmospheric tower bottoms (ATB), or vacuum tower bottoms (VTB) is processed. The coke is burned in the combustor to

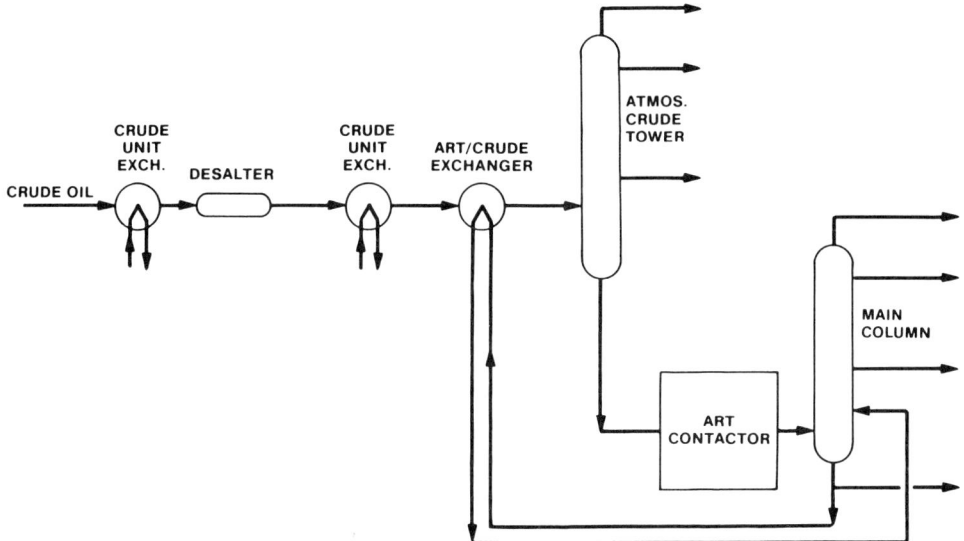

FIG. 8. ART and crude unit integration.

produce the energy to vaporize the contactor feed and produce steam in the flue gas cooler. Figure 1 shows the ART unit with the main column replacing the crude unit furnace and tower. Alternatively, the ART process can be integrated into an existing refinery as shown in Fig. 8. In this case the crude tower continues to operate and the ART Unit feed is ATB rather than whole crude. Both of these processing schemes ultimately utilize the heat of combustion of the coke to vaporize the crude and reduce or eliminate the need for the crude unit furnace. The thermal efficiency in the whole crude ART scheme is higher due to the direct exchange of the combustion heat in the circulating ARTCAT with the crude oil in the ART contactor, and therefore would have lower fuel costs. However, some refiners may prefer the option in Fig. 8 of keeping the virgin distillate separate from the ART product distillate.

The integration of the ART energy into the Fig. 1 refining scheme is demonstrated with a utility balance for Bachaquero crude in the hydrotreating case. Of the 16 cases studied, this case has the highest ART coke make and the lowest utility requirement and should therefore be the most difficult to integrate. Because the energy demands of the various units in the refinery LP are available on an electricity basis, the utility balance, detailed in Table 8, assumes that only the ART and FCC main process drivers are steam-driven and that all other process drivers in the refinery are electric-driven. Therefore, the letdown of HP steam to MP and LP steam and the condensing of the excess HP steam is assumed to be done across turbines to generate electricity. From Table 8, there is a net electricity demand of 14,200 kW (45% of the total utility requirement). In other words, integration of the ART Process into this refinery scheme has reduced the outside utility requirements by 55%. The remaining 45% utility requirement is a "flywheel" control on the refinery. This flywheel control will generally be in the form of an auxiliary boiler from

TABLE 8 Utility Balance [5 MM MT/Y (100,000 BPD) Bachaquero crude hydrotreater–FCC case]

	Power Demand	
Process plants		27,000 kW
Offsite		4,300 kW
Total		31,300 kW

HP Steam	kg/h (lb/h)	kW Generated
Generated	435,000 (959,000)	
Internal usage	240,000 (530,000)	
MP letdown	36,000 (80,000)	600
LP letdown	100,000 (220,000)	4,000
Condensing	57,000 (125,000)	12,500
		17,000 (55% of demand)
Net kW required		14,200 (45% of demand)

which individual turbine drivers can be supplied. A further reduction in this flywheel load could be accomplished by installing a power recovery system on the FCC flue gas.

This is further demonstration of the energy economy attainable with the ART Process. Even in this high coke make, low utility demand case, all the recovered heat energy from coke burning is readily utilized. The ART Process supplies roughly half of the refinery's energy demand, still allowing the refiner the utility control that he requires.

Engelhard ESR Process

When one considers the ART Process, it is normal to consider treating ART combustor flue gas for SO_x removal. Up to now the systems employed for flue gas treating have had limitations in that they all have some sort of disposal problem.

For this reason Engelhard has developed ESR, an SO_x removal system that generates sulfur as the only "by-product" and therefore eliminates any disposal problems. The ESR Process can be applied to the flue gas from the wide range of processes in Table 9 in order to realize the advantages listed in Table 10.

The ESR Process is based on a commercially demonstrated FCC catalyst technology which Engelhard began developing over 10 years ago. At that time it was recognized that a proprietary Engelhard catalyst was uniquely effective in reducing the SO_x concentration in FCC regenerator flue gas. Since that

TABLE 9 Engelhard ESR Process Technology Applications
Flue gas source:
Power boilers
Smelters
FCC units
ART units
Sulfur plants

TABLE 10 Engelhard ESR Process Technology Benefits
No waste by-products
Only product is sulfur, which is salable
Utilizes commercially demonstrated technology
Low operating costs
Low capital investment in existing refinery applications
System can be used upstream of power recovery systems in lieu of 3rd stage separator

time, Engelhard has emerged as a leader in catalytic technology for the reduction of SO_x emissions from FCC flue gases. The chemical mechanism upon which both the FCC SO_x technology and ESR Process are based is illustrated in Table 11.

The primary goal of an FCC catalyst is to produce transportation fuels, not reduce SO_x emissions. As a result, SO_x removal catalysts developed for use in FCC units are limited in their effectiveness. Engelhard has melded its expertise in fluid system designs with its catalyst technology and experience in SO_x/NO_x abatement in order to develop the ESR process. Use of ESR technology can provide greater than 95% SO_x removal from flue gases and generate sulfur as the only by-product, while utilizing existing refinery equipment as shown in Fig. 9. In the ESR Process, SO_x in flue gas streams is converted to H_2S, which may be removed in conventional amine systems with subsequent conversion to sulfur in a Claus unit.

Again, we are able to integrate this technology into existing systems to reduce investment. In many high SO_x concentration flue gas applications, low investment and operating costs allow the ESR system to show a healthy ROI from sales of the sulfur product.

TABLE 11 FCC SO_x Technology
Regenerator:
Oxidation:
$Coke + O_2 \longrightarrow CO_2 + H_2O + SO_x$
$MeO + SO_x \longrightarrow MeSO_4$
Reactor:
Reduction:
$MeSO_4 + 4H_2 \longrightarrow MeS + 4H_2O$
$MeS + H_2O \longrightarrow MeO + H_2O$
Amine system captures H_2S and releases as sulfur plant feed where:
$SO_2 + 2H_2S \longrightarrow 3S + 2H_2O$

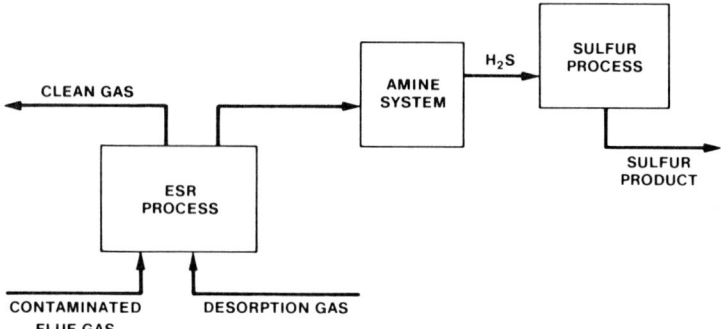

FIG. 9. ESR process.

Summary

The four crude oils discussed in this article—Kuwait, Arabian Heavy, Maya, and Bachaquero—cover a wide range of refinery feedstock quality. If one designs the ART combustor to handle the Bachaquero coke production and the ESR system to process the Maya flue gas SO_x, any of the four crudes can be charged to the same ART refinery. This crude oil flexibility can result in a more profitable operation.

Product flexibility is now also achievable due to the increased processability that the ART Process has given to the heavy gas oil fraction. By proper selection and combination of ART heavy gas oil hydroprocessing and fluid catalytic cracking, considerable control over the distillate-to-gasoline ratio is possible. In the 16 cases discussed in this paper (4 crudes × 4 schemes), the distillate-to-gasoline ratio varied from 0.7 to 1.8. And in all cases the total volume yield of transportation fuels was greater than 90% on crude oil.

The feedstock and product flexibility that the ART Process allows has been proven since 1983, when a 2.8-MM MT/Y (55,000 BPD) ART Unit was successfully integrated into Ashland Oil Company's Catlettsburg, Kentucky refinery complex. The ART Unit's feedstock has ranged from 1.006 specific gravity (9.2°API) and 17.5 wt.% RBC (100% VTB) to 0.924 specific gravity (21.7°API) and 5.5 wt.% RBC. At times it has processed portions of oxidized asphalt as well as material from an existing "tar pit." The product flexibility of the Ashland complex is demonstrated by its ability to meet large seasonal fluctuations in product demands. To meet its relatively large demand for asphalt in the summer months, Ashland bypasses the ART Unit. The remainder of the year, the ART Unit is operated to meet the refinery's quota of both gasoline and lube oils.

Application of the advanced TTR and ESR technologies in combination with modern hydrogen addition technology can substantially reduce the capital and operating costs in refinery modernization projects. This low-cost approach to feedstock and product flexibility is Engelhard's concept of the refinery of the future.

This article was first presented at the Japan Petroleum Institute, Tokyo, October 27, 1986, and is used here with permission.

References

1. B. R. Christian and D. F. Barger, *New Fluid Bed Design Expands Refinery Flexibility through Revamp*, Presented at the 1986 AIChE Annual Meeting, November 2-7, 1986.
2. A. K. Logwinuk and H. D. Sloan, *Upgrading Offshore California Crudes with ART*, Presented at the 1986 API Spring Meeting, May 15, 1986.
3. A. J. Suchanek and A. S. Moore, *Modern Residue Upgrading—By ART*, Presented at the 1986 NPRA Annual Meeting, March 23-25, 1986.
4. A. K. Logwinuk and H. D. Sloan, *Residuum Upgrading for Petrochemicals*, Presented at the 1986 NPRA Annual Meeting, March 23-25, 1986.
5. D. B. Bartholic and A. K. Logwinuk, *Upgrading of Crude Oils Utilizing the ART Process*, Presented at the Hungarian Conference on Crude Oil Upgrading, September 25-27, 1985.
6. P. M. Geren and A. M. Center, *Upgrading and Enhanced Recovery of Tar Sand Bitumen and Heavy Crude Using the ART Process*, Presented at the 1985 UNITAR Heavy Crude and Tar Sands Conference, July 22-31, 1985.
7. D. B. Bartholic, D. F. Barger, and A. M. Center, *The Applicability of the ART Process to Whole Maya Crude*, Presented at the 1985 NPRA Annual Meeting, March 24-26, 1985.
8. D. B. Bartholic, A. K. Logwinuk, H. B. Boyd, and C. P. Chang, *The ART Process*, Presented at the 1985 AIChE Houston National Meeting, March 24-28, 1985.

D. B. BARTHOLIC
A. M. CENTER
BRIAN R. CHRISTIAN
A. J. SUCHANEK

Petroleum Processes, Catalyst Usage

Worldwide catalyst manufacturers and suppliers in 1987 continue to add new petroleum-refining industry catalysts to the more than 900 available, but at a slightly slower pace compared to previous years. According to the latest survey, there were 94 new catalysts added during 1987 compared to more than 150 added in each of the two previous annual surveys.

This doesn't mean reduced activity in the industry. Rather, suppliers are concentrating more on effective application of catalysts previously introduced and on providing more customized formulations. Fluid catalytic cracking (FCC) catalysts continue to get the bulk of the attention of manufacturers with 45 new additions to that process category. Many of these are octane-enhancing catalysts in response to continued octane production needs fostered by the final stages of phaseout of lead in gasoline.

Hydrocracking and mild hydrocracking catalysts scored the second largest increase with 21 new additions to these categories.

Close scrutiny of these and other process categories shows that several of the new catalysts are incremental modifications of existing catalysts that make them highly specific to finely tuned processes. This supports an apparent trend toward customized specialty catalysts, rather than toward bulk, general purpose types, according to many in industry. Suppliers also eliminated about 33 catalysts that are no longer used or that have been replaced by improved catalysts.

Scope, Purpose

Oil & Gas Journal's International Refining Catalyst Compilation lists catalysts from essentially all of the world's manufacturers and suppliers, for practically all catalytic refining processes.

The compilation is intended to meet two main objectives:

1. To list all catalysts by their specific designations in their specific area of process application, for instance, catalytic reforming and hydrotreating.
2. To differentiate in some way each catalyst from another, i.e., to explain why one has a different designation from another.

The characteristics of the listed catalysts are furnished by the manufacturers or suppliers of the catalysts. If a supplier offers two or more families of catalysts in a single process area, the objective is to describe how the families differ on the basis of physical, chemical, or functional characteristics. And if there are several members in the family, how each differs from the others.

Petroleum Processes, Catalyst Usage

Abbreviations for catalyst compilation

ABD -- Apparent bulk density
AGO -- Atmospheric gas oil
Aro -- Aromatics
HC -- Hydrocracking or hydrocracked
Cet. -- Cetane
Cyl. -- Cylinder
DAO -- Deasphalted oil
Dist. -- Distillate
DO -- Diesel oil
Ext. -- Extrudate
GASO--Gasoline
HAGO -- Heavy atmospheric gas oil
HDM -- Hydrodemetallization
HDN -- Hyurodenitrogenation
HDS -- Hydrodesulfurization
HDT -- Hydrotreating or hydrotreated
HFO -- Heavy fuel oil
HGO -- Heavy gas oil

HVGO -- Heavy vacuum gas oil
Jet -- Jet fuel
Kero. -- Kerosine
LCO -- Light cycle oil
LGO -- Light gas oil
LSFO -- Low sulfur fuel oil
LSR -- Light straight-run
MS -- Microsphere
Oct. -- Octane
'P' -- Proprietary
RE -- Rare earth
RDS -- Resid desulfurization
SA -- Surface area
Sph. -- Sphere
Sh. ext -- Shaped extrudate
VGO -- Vacuum gas oil
ZEO. -- Zeolite

OGJ international refining-catalyst compilation—1987

Catalyst designation	Primary differentiating characteristics	Application (feedstock)	Application (product)	Form	Bulk density compacted, lb/cu ft,* g/cc	Carrier, support	Active agents	Availability
\multicolumn{9}{c}{Catalytic naphtha reforming}							
CRITERION CATALYSTS Co. LP.								
AEROFORM® PHF-5 & PHF-5A	Monometallic; differing particle size.	Naphtha	Gaso. or BTX	Cyl.	42	Al$_2$O$_3$	"P"	√
PRHF-30 PRHF-37 PRHF-50 PRHF-58	Bimetallic, different amounts of active metals	Naphtha	Gaso. or BTX	Cyl.	"P"	"	"P"	L
KX-120, KX-130 KX-160	(See Exxon Research & Engineering listing)							L
TRILOBE® KX-130	(See Exxon Research & Engineering listing)			Sh. ext.				L
AEROFORM® PHF-4 & PHF-4A	Monometallic; different particle sizes.	Naphtha	Gaso. or BTX	Cyl.	42	Al$_2$O$_3$	PtCl	√
TRILOBE® P-8 & P-8A	Monometallic, high particle density, different particle sizes	Naphtha	Gaso. or BTX	Sh. ext.	44	"	PtCl	√
AEROFORM® PR-6 & PR-6A	Bimetallic; different particle sizes.	Naphtha	Gaso. or BTX	Cyl.	42	"	Pt/ReCl	√
TRILOBE® PR-8, PR-8A & PR-8E	Bimetallic, high particle density catalyst; different particle sizes	Naphtha	Gaso. or BTX	Sh. ext.	44	"	Pt/ReCl	√
TRILOBE® PR-18 & PR-18A	Bimetallic, high particle density shaped skewed metals levels; different particle sizes	Low sulfur naphtha	Gaso. or BTX	Sh. ext.	44	"	Pt/ReCl	√
IMP-RNA-2	(See Instituto Mexicano del Petroleo Listing)			Sh. ext.				L
CHEVRON RESEARCH CO.								
Rheniforming Type F	Latest version, long cycle life stable product yields	Napthas, SR & cracked coal naphthas	Gas & Aro.	Cyl.	40	"P"	Pt/Re	L
CYANAMID KETJEN								
CK-303 CK-304 CK-305 CK-306	All monometallics, different active metal contents	Naphtha	Gaso. or BTX	Cyl.	44	Al$_2$O$_3$	PtCl	√
CK-433	Bimetallic	Naphtha	Gaso. or BTX	Cyl.	44	"	Pt/ReCl	√
CK-522 TRILOBE®	Bimetallic, high particle density	Naphtha	Gaso. or BTX	Sh. ext.	45	"	"	√
CK-542 TRILOBE	High-Re high particle density, shaped catalyst	Low S-Naphtha	"	"	"	"	"	√

(continued)

Catalytic naphtha reforming—continued

Catalyst designation	Primary differentiating characteristics	Application (feedstock)	Application (product)	Form	Bulk density compacted, lb/cu ft,* g/cc	Carrier, support	Active agents	Availability
PHRF-30	Multimetallic	Naphtha	Gaso. or BTX	Cyl.	"P"	"	"P"	L
CK-473	High-Re catalyst	Naphtha	Gaso. or BTX	Cyl.	44	"	Pt/ReCl	√
KX-120, KX-130, KX-160	See Exxon Research & Engineering							L
ENGELHARD CORP.								
KX-120	(See Exxon Research Engineering cat. naphtha reforming)							
E-301	Moderate stability	Naphthas, S < 5 ppm wt	Gaso. or Aro.	Cyl.	45	Al_2O_3	Pt	√
E-302	More stable than E-301	Naphthas, S < 10 ppm wt	Gaso. or Aro.	Cyl.	45	"	"	√
E-311	Maximize LPG production	Naphthas, S < 5 ppm wt	LPG gaso.	Cyl.	45	"	"	√
E-601	High stability	Naphthas, S < 1 ppm wt	Gaso. or Aro.	Cyl.	45	"	Rt, Re	√
E-603	High stability. Lower metal loading than E-601.	Naphthas, S < 1 ppm wt	Gaso. or Aro.	Cyl.	45	"	"	√
E-611	Very high stability	Naphthas S = 0.1 ppm wt	Gaso. or Aro.	Cyl.	45	Al_2O_3	Pt/Re	√
(Note: Preceeding catalysts available in higher density at 49.5 lb/cu ft. They are designated HD, for example E – 301 HD).								
E-801	Stability of E-603 at higher activity	Napathas S <1 ppm wt	Motor fuel or aromatics	"	51	"	"	√
E-802	Stability of E-611 at higher activity	Naphthas, S ≤ 0.5 ppm wt	"	"	51	"	"	√
E-803	Stability of E-611 at higher activity, optimized metals loading	"	"	"	51	"	"	√
E-804	Stability & activity of E-801, optimized metals loading.	"	"	"	51	"	"	√
EXXON RESEARCH & ENGINEERING CO.								
KX-120	Multimetallic semiregen. or cyclic units; different promotor	Virgin/cracked naphthas	High oct. gaso & aro.	—	—	—	Pt, etc.	L
KX-130	", different promotor	"	"	—	—	—	—	L
KX-160	", different promotor	"	"	—	—	—	—	L
INSTITUTO MEXICANO DEL PETROLEO								
IMP-RNA-1	High activity bimetallic	Desulfurized naphtha, aromatics production	BTX aro. high oct. gas	Cyl.	0.70*	Al_2O_3	Pt/Re	L
IMP-RNA-2 TRILOBE*	"	Desulfurized naphtha, aromatic production	BTX aro. high octane gas.	Exts.	0.70*	Al_2O_3	Pt/Re	L
KATALYSATORENWERKE HÜLS GmbH								
H 2440		Fraction 35-195 °C.	Gaso., BTX	Cyl. exts.	0.8*	Al_2O_3	Pt, 0.3-0.8% monometallic	√
H 6413		"	"	"	0.6*	"	"	√
H 2462		"	"	"	0.6*	"	Pt, 0.3-0,8% + "P"	√
PROCATALYSE								
RG 412	H.P. reformer	Naphtha	Gaso.	Cyl.		"	Pt 0.35%	√
RG 432	M.P. reformer	Naphtha	Gaso.	Cyl.		"	Pt 0.35% + Ir	√
RG 482	M.P. & L.P. reformer	Naphtha	Gaso.	Cyl.		"	Pt 0.3% + Re	√
RG 492	M.P. & L.P. reformer	Naphtha	Gaso.	Cyl.		"	Pt/Re	√
RG 442	Reformer	Naphtha	Gaso. & LPG	cyl.	"	Pt + Ir + promotor		
CR 201	Cont. regen. reformer	Naphtha	Gaso.	Sph.		"	Pt Sn	√
AR 401	Cont. regen. reformer (Aromizing-IFP)	Naphtha	Aro.	Sph.		"	Pt/promotor	L
AR403	Cont. regen. reformer (Aromizing IFP)	Naphtha	Aro.	Sph.		"	Pt/Promotor	L
UOP								
R-11†	Higher activity monometallic	Semi-regen. or swing	Gaso. aro.	Sph.	32	Al_2O_3	Pt	Contact UOP
R-12†	Monometallic, higher stability than R-11	Semi-regen.	Gaso., aro.	Sph.	32	"	Pt	"
R-55†	Monometallic, higher activity and stability than R-12,	Semi-regen. or swing	Gaso. aro.	Ext.	Dense-loaded 52	Al_2O_3	Pt	Contact UOP
R-15†	Monometallic	Semi-regen.	LPG, gaso.	Sph.	32	SiO_2/Al_2O_3	Pt	"
R-16H†	Bimetallic, higher stability than monometallics	High severity, semi-regen.	Gaso. aro.	Sph.	32	Al_2O_3	Pt/Re	"
R-16F†	Bimetallic, low Pt, higher stability than monometallics, reduced and sulfided	High severity, semi-regen.	Gaso. aro.	Sph.	32	"	Pt/Re	"
R16G†	Bimetallic, higher activity and stability than R-16F and H,	High severity, semi-regen.	Gaso. aro.	Sph.	32	"	Pt/Re	"
R-18†	Bimetallic, higher activity and stability than R-16 series,	High severity, semi-regen.	Gaso. aro.	Sph.	37	"	Pt/Re	Contact UOP

(continued)

Catalytic naphtha reforming—continued

Catalyst designation	Primary differentiating characteristics	Application (feedstock)	Application (product)	Form	Bulk density compacted, lb/cu ft,* g/cc	Carrier, support	Active agents	Availability
R-22	Bimetallic, higher yields, reduced state	CCR or semi, regen.	Higher production of oct. bbl & aro.	Sph.	32	"	"P"	"
R-30	Bimetallic, higher yields and activity reduced state.	CCR or swing reactor	"	Sph.	35	"	"P"	"
R-32	Bimetallic, higher yields and activity reduced state	CCR	"	Sph.	35	"	"P"	"
R-50†	Bimetallic, higher activity and stability than R-16 series,	High severity, semi-regen.	Gaso. aro.	Ext.	Dense loaded 52	"	Pt/Re	"
R-51	"	"	"	"	43	"	Pt/Re	"
R-60†	Bimetallic, higher stability than R-50, high activity	High severity, semi-regen.	Gaso. aro.	Sph.	45	"	Pt/Re	"
R-62	Bimetallic, Lower Pt than R-60, higher stability than R-50, high activity	High severity, semi-regen.	Gaso., aro.	Sph.	44	"	Pt/Re	"

†Reduced and sulfided.

Dimerization catalysts

KALI-CHEMIE AG								
Beadcat PK Series	Mechanical and thermal stability	C_3/C_4 olefins	Gaso.	Bead	0.80*	SiO_2/Al_2O_3	Acid sites Al_2O_3	√
PROCATALYSE								
LC 1252		Propylene/ butylene	Gaso./hexenes/ octenes	Liquid	—	—	Ni	L

Isomerization C_4 catalysts

BP								
BP Isom	High activity, regenerable			Cyl.		Al_2O_3	Pt Cl	L
UOP								
I-8	Higher activity for n-butane isom to iso-butane			Ext.	48	Al_2O_3	Pt	Contact UOP
ENGELHARD CORP.								
RD-290C	Regenerable, high activity			Cyl.	48	Al_2O_3	Pt cl	L

Isomerization (C_5 & C_6) catalysts

BP								
BP Isom	High activity, L.P. operation, regenerable	C_5, C_6 fractions		Cyl.		Al_2O_3	Pt Cl	L
ENGELHARD CORP.								
RD-290C	Regenerable, high activity	C_5, C_6 paraffins		Cyl.	48	Al_2O_3	PtCl	L
PROCATALYSE								
IS632	Non-chlorinated, high activity	C_5/C_6 paraffins	High oct. gaso. or isomers	Ext.	0.50*	Special	Pt	√
UNION CARBIDE CORP.								
HS-10 HS-12	Non-chloride activated, high tolerance for feed impurities	C_5/C_6 Isom.	Octane upgrade	Cyl.	46	Mole sieve Al_2O_3	Pt	L
UOP								
I-7	High activity, sulfur and water tolerant	SR or HDT C_5/C_6 streams		Ext.	36	"P"	Pt	Contact UOP
I-8	High activity, L.P. application	"		Cyl.	48	Al_2O_3	Pt	"

Isomerization (xylenes) catalysts

ENGELHARD CORP.								
O-750*	Xylene isom.	C_8 aromatics	P, o-xylene	Cyl.	"P"	"P"	Pt	L
KALI-CHEMIE AG								
Beadcat PK series	Xylenes isom.	xylenes	p, o-xylene	Bead	0.80*	SiO_2/Al_2O_3	Acid sites	√
UOP								
I-9	High activity & high selectivity.	Mixed xylenes, ethylbenzene	p, o-xylene	Sph.	37	"P"	"P"	Contact UOP

Fluid catalytic cracking

AKZO CHEMICALS INC.								
Ketjen Lowcoke® 80/800 series KLC-80/800	Low delta coke, Low regn. temp., Gaso. selective	FCC feed, Lt. resid	Gaso., mid dist., alky. feed	MS	0.78*-0.82*		Low RE zeo.	√
Ketjen Lowcoke® 80-800 series KLC-80/800+	Extra low Δ coke, high oct. & oct. bbl	VGO Lt. resid	Gaso.-mid-dist	"	0.76*-0.82*		Stab. low RE Zeo.	√

(continued)

Petroleum Processes, Catalyst Usage

Fluid catalytic cracking—continued

Catalyst designation	Primary differentiating characteristics	Application (feedstock)	Application (product)	Form	Bulk density compacted, lb/cu ft,* g/cc	Carrier, support	Active agents	Availability
Ketjen Multicrack® 20/200 Series KMC-20/200	Low coke, gas make, int. oct. gaso. selective	VGO, resid	Gaso., oct. LCO, cet.	"	0.76*–0.80*	Active matrix	Low RE zeo.	√
Ketjen Multicrack® 50/500 series KMC-50/500	Max. bot. conv., int. oct., Gaso. selective	VGO, resid, HDT feeds	"	"	0.74*–0.78*	"	Low RE Zeo.	√
Ketjen Multicrack® 20/200+ Series KMC-20+/200+	High oct. & oct. bbl, low coke, gas make	VGO, resid	"	"	0.75*–0.79*	"	Stab. Low RE Zeo.	
Ketjen Multicrack® 50/500+ series KMC-50+/500+	High Oct. & oct. bbl, max. bott. upgrade	VGO, resid, HDT feeds	"	"	0.72*–0.78*	"	Stab. Low RE Zeo.	√
Octaboost® 600 series KOB-600	High oct., low coke, gas, Bott. upgrade	VGO, resid	Alky feed, gaso. oct, LCO, cet.	"	0.72*–0.80*	"	USY	√
Octaboost® 610 Series KOB-610	Low soda, attrition resist., High oct., bott. upgrade	"	"	"	0.72*–0.80*	"	USY	√
Octaboost® 620 Series KOB-620	High oct. & oct. bbl, bott. upgrade	"	"	"	0.72*–0.80*	"	USY	√
Octaboost® 600/610 Low coke series KOBN-600/610 LC	Max. oct., extra low Δ coke	"	"	"	0.72*–0.80*	"	USY	√
Ketjen Metal Resist® 900 Series KMR-900	Metal resist., bottom conv., low Δ coke	High metals, hvy. resid	Gaso.; oct., LCO	"	0.75*–0.82*	V+Ni trap	Low RE Zeo.	√
Ketjen Metal Resist® 920 Series KMR-920	Metals resist, bott. conv., low Δ Ni sensitivity	VGO, resid	Gaso., oct., Diesel	"	0.75*–0.80*	Selective active matrix	Stab. low RE Zeo.	√
Ketjen Metal Resist® Additive MR	Metals resist.	High metals feed	Gaso., oct, LCO	Incorporated into any ketjen FCC catalyst		Vi+Ni trap		√
CATALYSTS & CHEMICALS INDUSTRIES CO., LTD.								
MRZ-210	Low activity, high liquid yield	FCC feed	Gaso. & LCO	MS	0.74*	SiO_2/Al_2O_3	Zeo.	√
MRZ-220	High liquid yield	"	"	"	"	"	"	√
MRZ-230	"	"	"	"	"	"	"	√
MRZ-203	High octane metals tolerance	FCC feed	Gaso.	MS	0.74*	SiO_2/Al_2O_3	Zeo.	√
MRZ-204	"	n	"	"	"	"	"	√
MRZ-206	"	"	"	"	0.72*	"	"	√
MRZ-203S	High oct. metal tolerance, high hydrothermal stability	FCC feed, high metals feed	Gaso.	MS	0.74*	"	"	√
MRZ-204S	"	"	"	"	"	"	"	√
MRZ-206S	"	"	"	"	0.70*	"	"	√
MRZ-208	High gasol. yield & oct.	FCC feed	Gaso.	MS	0.68*	"	"	√
MZR-208S	"	"	"	"	"	"	"	√
MRZ-230	High gaso. yield, high metal feed tolerance	FCC feed, high metals feed	Gaso.	MS	0.74*	SiO_2/Al_2O_3	Zeo.	√
MRZ-240	"	"	"	"	0.74*	"	"	√
MRZ-260	"	"	"	"	0.70*	"	"	√
Hylic-30	High liquid	FCC feed	"	"	0.80*	"	"	√
-40	"	"	"	"	"	"	"	√
Hycon-20	Hy oct. and metals tolerance	FCC feed	Gaso.	"	0.78*	SiO_2/Al_2O_3	Zeo.	√
-30	"	"	"	"	"	"	"	√
-40	"	"	"	"	"	"	"	√
Valec-50	"	"	"	"	0.70*	"	"	√
-60	"	"	"	"	0.68*	"	"	√
-70	"	"	"	"	0.65*	"	"	√
RCZ-540	High oct., low coke, high metal tolerance	FCC feed	Gaso.	MS	0.75*	SiO_2/Al_2O_3	Zeo.	√
RCZ-550	"	"	"	"	"	"	"	
RCZ-560	"	"	"	"	"	"	"	√
Octup	Oct. additive	FCC feed	Gaso.	MS	0.70*	SiO_2/Al_2O_3	ZSM-5	√
CROSFIELD CATALYSTS								
XL Series	Bottoms crack, and high gaso. and oct. selectivity	Gas oils	Bottoms upgrade, high liquid product. oct. and cet.	MS		SiO_2/Al_2O_3	Zeo. matrix	√
XL 60	"	"	Low activity	"	0.87*	"	"	
XL 80	"	"	Medium activity	"	0.86*	"	"	√
XL 100	"	"	High activity	"	0.84*	"	"	√
SLS Extra Series	High oct. gaso. liquid product	Gas oils, resids	Gaso, LCO,	MS		SiO_2/Al_2O_3	Zeo. matrix	√
SLS 60 Extra	"	"	Low activity	MS	0.75*	SiO_2/Al_2O_3	"	√
SLS 80 Extra	"	"	Med. activity	MS	0.74*	"	"	√
SLS 100 Extra	"	"	High activity	MS	0.72	"	"	√
SLS Series	Coke selectivity, gaso. mid. dist. oct., max. oct bbl	Gas oils and resids	Max. gaso. oct. bbl, selective bottom cracking	MS		SiO_2/Al_2O_3	Zeo. matrix	
SLS-60	"	"	"	"	0.78* ABD	"	"	√
SLS-80	"	"	"	"	0.76*	"	"	√
SLS-100	"	"	"	"	0.76*	"	"	√

(continued)

Petroleum Processes, Catalyst Usage

Fluid catalytic cracking—continued

Catalyst designation	Primary differentiating characteristics	Application (feedstock)	Application (product)	Form	Bulk density compacted, lb/cu ft,* g/cc	Carrier, support	Active agents	Availability
GLS Series	Max. gaso liq. product, low coke, gas select	Gas oils, resids	Gaso. LCO, act.	MS		SiO_2/Al_2O_3	Zen. CMT	√
GLS 60	"	"	Low activity	"	0.75*	"	"	√
GLS 80	"	"	Med. activity	"	0.74*	"	"	√
GLS 100	"	"	High activity	"	0.72*	"	"	√
Quantum M Series	Max. oct., Oct. bbl, low Coke, gas select	Gas oils, resids	Oct., oct. bbls	MS		SiO_2/Al_2O_3	Zeo. CMT	√
Quantum 80 M	"	"	Med. activity	"	0.72*	"	"	√
Quantum 100 M	"	"	Low activity	"	0.70*	"	"	√

Vanadium trap (VT), FLuidization additive (CFA), and combustion promoter (CP) can be added to all catalysts

DAVISON CHEMICAL DIVISION, W. R. GRACE & CO. (Each catalyst group arranged in decreasing order of activity)

Catalyst designation	Primary differentiating characteristics	Application (feedstock)	Application (product)	Form	Bulk density lb/cu ft,* g/cc	Carrier, support	Active agents	Availability
CBZ-1	General purpose, increased LCO yield	Gas oil, mild resid	Gaso. & dist.	MS	0.52* ABD	"P"	Zeo. matrix	√
CBZ-4		"	"	"	0.58*	"P"	"	√
DA-440	Max. unit retention, high stability, bot. cracking	Gas oil, resid	Gaso. & dist.	MS	0.82* ABD	"P"	Zeo. matrix	√
DA-400	"	"	"	"	0.83*	"	"	√
DA-300	"	"	"	"	0.82*	"	"	√
DA-250	"	"	"	"	0.85*	"	"	√
DA-200	"	"	"	"	0.85*	"	"	√
DZ-8	General purpose, bot. cracking, low ABD	Gas oil, mild resid	Gaso. & dist.	MS	0.52* ABD	"P"	Zeo. matrix	√
GX-30	Bot. cracking, nitrogen, metals tolerance	Gas oil, resid, coker gas oil high N-feeds	Gaso. & dist.	MS	0.74* ABD	"P"	Zeo. matrix	√
GX-20	"	"	"	"	0.76*	"	"	√
GX-10	"	"	"	"	0.78*	"	"	√
GXO-41	Oct. enhancement, bot. coker cracking, coke selectivity, stability	Gas oil, resid, gas oils, high N-feed	Gaso. & dist.	MS	0.78* ABD	"P"	Zeo. matrix	√
GXO-40	"	"	"	"	0.75*	"	"	√
GXO-27	"	"	"	"	0.77*	"	"	√
GXO-28*	Max. stability, oct., coke selectivity, bot. upgrade, metals tolerance	Gas oil, resid, coker gas oils, high N-feed	Gas. & dist.	MS	0.78* ABD	"P"	Z-14US zeo.	√
GXO-25+	"	"	"	"	0.77*	"	"	√
GXO-25	"	"	"	"	0.77*	"	"	√
GXO-31+	"	"	"	"	0.72*	"	"	√
GXO-31	"	"	"	"	0.72*	"	"	√
GXO-25S	"	"	"	"	0.76*	"	"	√

*Catalysts arranged in decreasing order of activity

Catalyst designation	Primary differentiating characteristics	Application (feedstock)	Application (product)	Form	Bulk density	Carrier, support	Active agents	Availability
Super Nova D†	Coke selectivity, oct. stability max. unit retention, bott. cracking	Gas oil, resid, coker gas oils, high N-feeds	Gaso. & dist.	MS	0.75* ABD	"P"	Z-14US zeo.	√
Super Nova D	"	"	"	"	0.75*	"	"	√
Nova D	"	"	"	"	0.83*	"	"	√
Nova D-2	"	"	"	"	0.81*	"	"	√
Nova-SX*	Coke, gas selectivity mod. cost	Gas oil, resid	Gaso. & dist.	MS	0.78* ABD	"P"	Zeo.	√
Nova-S	"	"	"	"	0.78*	"	"	√
Octacat+*	Max. oct. stability, min. coke	Gas oil, resid, coker gas oils	Gaso. & dist.	MS	0.70* ABD	"P"	Z-14US zeo.	√
Octacat	Maximum octane enhancement and stability, minimum coke make	Gas oil, resid, coker gas oils	Gaso. & dist.	"	0.70* ABD (ABD)	"P"	Z-14US zeo.	√
Octacat-2	"	"	"	"	0.73*	"	"	√
Octacat-3	"	"	"	"	0.76*	"	"	√
Octacat B	Max. oct., stability, bott. cracking	"	"	"	0.70*	"	"	√
Octacat-D*	Max. retention, with improved bott. cracking	Gas oil, resid, coker gas oils, high N-feeds	Gaso. & dist.	MS	0.75* ABD	"P"	Z-14US zeo.	√
Octacat-D2	"	"	"	"	0.76*	"	"	√
Octacat-D3	"	"	"	"	0.78*	"	"	√
RC-25	High liquid product selectivity, isobutane yield, metal tolerance	Gas oil, resid	Gaso., dist.	MS	0.73* ABD (ABD)	"P"	Zeo. matrix	√
Super-D Magnum	"	"	"	"	0.73*	"	"	√
Super-D Extra	"	"	"	"	0.77*	"	"	√
Super-D	"	"	"	"	0.79*	"	"	√
Super-D2	"	"	"	"	0.80*	"	"	√
Super-D3	"	"	"	"	0.80*	"	"	√
Super-D4	"	"	"	"	0.81*	"	"	√
Gencat	High activity, mod. cost flush, coke & gas make selectivity	Gas oil, resid	Gaso. & dist.	MS	0.80*	"P"	Zeo.	√
XP-60	Bott. cracking, oct., coke selectivity, retention.	Gas oil, resid	Max. bott. reduction	MS	0.80*	"P"	Z-14 US matrix	√
XP-45	"	"	"	"	0.80*	"	"	√

(continued)

Fluid catalytic cracking—continued

Catalyst designation	Primary differentiating characteristics	Application (feedstock)	Application (product)	Form	Bulk density compacted, lb/cu ft,* g/cc	Carrier, support	Active agents	Availability
HP-75+	Bott. cracking, oct., coke selectivity, retention	Gas oil, resid	Max. bott. reduction	MS	0.80*	"P"	Z-14 US zeo. matrix	√
XP-75		"	"	"	0.80*	"	"	√
CP-3	Combustion promotion		Additive	MS	0.85* ABD	"P"	"P"	√
CP-5	High activity, stability		"	MS	"P"	"P"	"P"	√
Additive O	Gaso. oct.		Additive	MS	0.80* ABD	"P"	"P"	√
DVT	Vanadium passivation		Additive	MS	0.65* ABD	"P"	"P"	√
Additive R	SO_x		Additive	MS	0.98* "	"P"	"P"	√
ENGELHARD CORP.								
ULTRASIV*-280 ULTRASIV*-260 ULTRASIV*-250 ULTRASIV*-240 ULTRASIV*-230 ULTRASIV*-220	US-280 highest activity, vandidium tolerance, low gas, attrition resist.	Gas oils, gas oil-resid mixtures, resid	Incremental resid processing	MS	1.1*	In situ SiO_2/Al_2O_3	Zeo. mod. RE, mod. SA	√
HEZ-55™ EPZ*-55 HEZ-54™ HEZ-53™ HEZ-52™ HEZ-50™	HEZ-55 highest activity, EPZ designates O promoted designates CO promoted	Gas oils, high N-gas oils, Calif. gas oils	Bot. upgrade oct. bbl, cet.	MS	1.0*	"	Zeo., moder., RE high SA	√
MAGNASIV*-380 MAGNASIV*-370 MAGNASIV*-360	MAG-380 highest activity, low coke, gas make, metal tolerance, hydrothermal stability, bott. upgrade	Gas oils, high N-gas oils. Calif. gas	Max. liquid yield	MS	1.05*	"	Zeo., RE mod. SA,	√
Octasiv Plus* 680, 670 660	680 highest activity, stability metals tolerance, hydro thermal stab., bott. upgrade	Gas oils, gas oil-resid, resid	Max. oct. bbl	MS	1.0*	"	Ultrastable zeo., low RE, low matrix SA	√
Octasiv Plus* 580, 570, 560	580 highest activity, stability, metals tolerance, hydrothermal stab.	Gas oils, gas oil resid, resid	Max. olefin production	MS	1.0*	in situ SiO_2/Al_2O_3	Ultrastable zeo., mod. matrix SA	√
Octasiv Plus* 160, 150, 140	160 highest activity, oct. adv. over REO	Gas oils, gas oil resid, resid	Bott. upgrading, and metals tolerance.	MS	1.0*	In situ SiO_2/Al_2O_3	Ultrastable zeo., low RE mod. matrix SA	√
ULTRASOX*	Available in a variety of cracking & sulfur reduction activities	High gas oils, gas oil-resid mix	SOx reduction	MS	1.1*	In situ SiO_2/Al_2O_3	Zeo./Prop. SOx reduction	√
HFZ*-33 HFZ*-20 HFZ*-23 HFZ*-25	HFZ-33 highest activity	Gas oils	Olefins for chemical feed, high oct. cet.	MS	1.0*	In situ / SiO_2/Al_2O_3	Zeo. No RE, High SA	√
Ultradyne* 980/ 980, 970, 960 950, 940, 930	UD980 is highest activity low matrix SA	Gas oil	Low coke production, low gaso. bromine number, high gaso. yield, low dry gas yield	MS	1.0*	In situ SiO_2/Al_2O alumina	Zeolite	√
Ultradyne* Plus 1580, 1560, 1550, 1540, 153	UDP 1580 highest activity. low matrix SA oct. cat.	Gas oil, resid	Oct. bbl, low coke, dry gas	MS	1.0*	"	Ultrastable zeo. low RE, low matrix SA	√
Dynacat* 1080, 1070, 1060, 1050, 1040, 1030	DC 1080 highest activity vanadium tolerant metal	Resid, high metals gas oil	High activity. maintenance under contaminated situations	MS	"	"	"	√
Dynasiv* 780, 770, 760, 750 740, 730	DS 780 is highest activity	Gas oil, mild resid	Mod. oct. bott. upgrade, high gaso. selectivity	"	"	"	"	√
Dynasiv Plus* 880, 870, 860 850, 840, 830	DSP880 is highest activity	Gas oil/ resid	Oct. gaso. selectivity, bott. upgrade	"	"	"	Ultrastabilized zeo.	√

(continued)

Petroleum Processes, Catalyst Usage

Fluid catalytic cracking—continued

Catalyst designation	Primary differentiating characteristics	Application (feedstock)	Application (product)	Form	Bulk density compacted, lb/cu ft,* g/cc	Carrier, support	Active agents	Availability
Octidyne* 1170, 1160, 1150, 1140, 1130	OP1170 is highest activity, RON, MON imp., low coke	Gas oil/ resid	Max. oct. max. gaso. selectivity, reduced regen. temp	"	"	"	"	✓
DynaSox* 1380, 1340, 1370, 1360, 1350, 1330	DSX 1380 is highest activity	Gas oil/ resid	Mod. SOx reduction built in Gaso. selectivity selection and bott. upgrade	"	"	"	Zeo.	✓
Nitrodyne* 1490 1480, 1470	ND1490 extremely high activity	High N feed, coker gas oils	High in-unit activity, converversion bot. upgrade	MS	1.0*	SiO_2/Al_2O_3	Zeo.	✓
Z-100*		Gas oil, resid	RON, MON, Imp., definicity	MS	1.0*	SiO_2/Al_2O_3	XSM-5	✓
HARSHAW/FILTROL								
F/X-450, 460 470, 480	480 highest act. oct. cat	Gas oils through resid	Max. oct., min. coke, bot. upgrade	MS	0.80*	Active Al_2O_3	Low RE, low sodium, dealum (DY) Zeo., active stable crystalline alumina matrice	✓
FSS-1 DY 2 DY 3 DY	Very stable, all purpose. "3" highest activity	Gas oils, resids, high N-feed	High gaso. yields, bott. upgrading, severe operations	MS	0.80*	Active Al_2O_3	High RE, Dy zeo. low soda, active stable crystal-line alumina matrix	✓
HRO-550 560 570 580	580 highest act. Max. oct. bbl	Gas oil through resid	Oct. appr. F/X, stab. appr. FSS-DY	MS	0.80*	Active Al_2O_3	"	✓
(TEX-)1 DY -2 DY -3 DY	Stable high oct. gaso. "3" highest activity	Gas oils or resid	High gaso. yields & O.N. low coke, high LCO cet. grading	MS	0.80*	Active Al_2O_3	Moderate RE DY zeo. active stable crystalline alumina matrix.	✓
ROC-1DY 2DY	Resists metals, Low coke make. Highest activity, 2DY	Heavy gas oil, high metals resids.	"	MS	0.80*	Active Al_2O_3	Mod. to high RE DY zeo. active stable crystaline alumina matrix	✓
SOC-1 DY 2 DY	Very stable, low coke make high activity. "2" highest activity	HGO, high metals resids.	High gaso. yield & O.N., low coke, high LCO, cet. bott. upgrading, severe operations	MS	0.80*	Active Al_2O_3	Low to mod. RE, DY, zeo., low soda, active crystalline alumina matrix	✓
FLEX-230 -440 -650 -860	860 highest act., gen. purpose	Gas oils through resid	Max. gaso. select. min. gas make,or units not requiring max. stab.	MS	0.80*	Active Al_2O_3	High RE, DY Zeo., active stable crystalline alumina matrix	✓
EXSEL-620 -630 -640	640 highest act., gen purpose	Gas oil, high met resid. (blend cat.)	High gaso. select. bott. upgrade gas, metals control	MS	0.80*	Active Al_2O_3	Low RE, DY Zeo. active stable crystalline alumina matrix	✓
DOC-1 DY -2 DY	Diesel oil production	Gas oils resids.	Max. LCO yield, high cet. bott. upgrading	"	"	"	Low RE DY zeo. active crystalline alumina matrix	✓
FOCUS-1 -2	High oct. gaso., low gas and coke make.	Gas oils through resids.	Highest ON gaso., low coke, gas, make, bott. upgrade	MS	0.80* (ABD)	"	"	✓
S grade	Reduced matrix act.	Gas oils	Red. activity of any filtrol FCC cat.	MS	0.80*	"	Active stable crystalline alumnina matrix.	✓
BCA	Bott. crack additive	High met. resid	Bott. crack., metals flush	MS	0.80*	"	"P"	✓
Intercat Corp Z-cat	Octane additive	FCC feeds	High oct. gaso.	MS	50	"P"	ZSM-5	✓

(continued)

Fluid catalytic cracking—continued

Catalyst designation	Primary differentiating characteristics	Application (feedstock)	Application (product)	Form	Bulk density compacted, lb/cu ft,* g/cc	Carrier, support	Active agents	Availability
KATALISTIKS INTERNATIONAL INC.								
Alpha-540	High activity oct.	Gas oils, resids, reduced crudes	Max. oct. bbl, high selectivities	MS		SiO_2/Al_2O_3	LZ-210, active matrix	✓
Alpha-640	"	"	"	"		"	"	
Beta-540	Octane	Gas oils, resids, reduced crudes	Max. octane	MS		SiO_2/Al_2O_3	LZ-210, active matrix	✓
Beta-640	"	"	"	"		"	"	
Beta-600	Highest octane	Gas oils, resids, reduced crudes	Max. octane	MS		SiO_2/Al_2O_3	LZ-210, active matrix	✓
Delta-400	Max. RON, MON without yield penalty	Gas oil, resid, reduced crude	High liquid, max. mid-bbl yield. Low 650+ yield.	MS	0.74* ABD	SiO_2/Al_2O_3	USY Zeo., & low SA matrix matrix	✓
Delta-444	High RON, MON, increased gasoline	"	High gaso. yields with high oct.	MS		SiO_2/Al_2O_3	USY Zeo., RE & active matrix	
Delta-454	"	"	"	"		"	"	✓
Delta-455	"	"	"	"		"	"	✓
Delta-464	"	"	"	"		"	"	✓
Gamma-419	High RON, MON, OCt. bbl, high stability, coke selectivity selectivity	Gas oil, resid, reduced crude	High Liquid mid bbl yields, low gas, coke, 630+ yields	MS	0.74*	SiO_2/Al_2O_3	High SiO_2 zeo. matrix matrix	✓
Gamma-419+	"	"	"	"	"	"	" + low RE	✓
Gamma-450	"	"	"	"	"	"	"	✓
Gamma-550	"	"	"	"	0.73*	"	"	✓
Gamma-620	"	"	"	"	0.72*	"	"	✓
Sigma-200 Plus	High oct. bbl, good good coke, gas selectivity. 5 activity levels for dist. gas. operations	Gas oils, resids, reduced crudes	High liquid, max. mid-bbl yield, low coke gas make.	MS	0.76* ABD	SiO_2/Al_2O_3	High silica, low RE zeo. matrix	✓
Sigma-300 Plus	"	"	"	"	0.75*	"	"	✓
Sigma-400 Plus	"	"	"	"	0.74*	"	"	✓
Sigma-500 Plus	"	"	"	"	"	"	"	✓
Sigma-600 Plus	"	"	"	"	0.72*	"	"	✓
Sigma-200	General purpose, high stability, 4 activity levels.	Gas oil, resid, reduced crude high N, coker and gas oil	High liquid max. mid-bbl yield. low 650 + yield, low gas, attrition resistance, low bromine number, balanced butane. yields.	MS	0.74 ABD	SiO_2/Al_2O_3	Mixed zeo. low SA matrix	✓
Sigma-300	"	"	"	"	0.75* "	"	"	✓
Sigma-400	"	"	"	"	0.74* "	"	"	✓
Sigma-600	"	"	"	"	0.72*	"	"	✓
EKZ-2	General purpose, 3 activity levels	Gas, dirty feeds, coker gas oils	High liquid yield, low gas, coke	MS	0.79* ABD	SiO_2/Al_2O_3	Fully ReO exchanged zeo.	✓
EKZ-3	"	"	"	"	0.78*	"	"	✓
EKZ-4	"	"	"	"	0.76*	"	"	✓
BMZ-200	Max. dist. or max gas.	Gas oil, resid, reduced crude, high N coker gas oil	High liquid, max. mid-bbl yield. Low 650+ yield, low gas, attrition resistance. low bromine number, high isobutane yield.	MS	0.76* ABD	SiO_2/Al_2O_3	High silica, fully RE, zeo. SA matrix	✓
BMZ-300	"	"	"	"	0.75*	"	"	✓
BMZ-400	"	"	"	"	0.74*	"	"	✓
BMZ-600	"	"	"	"	0.72*	"	"	✓
Ekonokat-1	General purpose, 2 activity levels.	Gas oils, dirty feeds, coker gas oils	High liquid yield, low gas, coke low 650+ yield.	MS	0.76* ABD	SiO_2/Al_2O_3	Mixed zeo., low SA matrix	✓
Ekonokat-2	"	"	"	"	0.72* ABD	"	"	✓

W. R. GRACE CATALYSTS PTY. LTD. (see complete listing of Davison Chemical Division, W.R. Grace & Co.)

Hydrocracking

AKZO CHEMICALS INC.								
Ketjenfine, LC-fining	(See Lummus Crest HC listing)	Resid HC		Cyl.		"P"	Co Mo	L
Ketjenfine H-oil, (HRI Texaco)		Resid HC		Cyl.		"P"	CoMo & NiMo	L
Ketjenfine HC-143	(See BP HC listing)	VGO HC		Cyl.		"P"	"P"	L
KC-2000	NiMo		Gaso.	Cyl.		"P"	"P"	
KC-2100	Precious metal		"	"		"	"	
KC-2200	NiW		Mid-dist.	"		"	"	
CRITERION CATALYSTS Co. LP.								
IPB-2		AGO, VGO	Gaso. & dist	Cyl.		"P"	"P"	L
HC-102	(See Unocal HC listing)							L

(continued)

Petroleum Processes, Catalyst Usage

Hydrocracking—continued

Catalyst designation	Primary differentiating characteristics	Application (feedstock)	Application (product)	Form	Bulk density compacted, lb/cu ft,* g/cc	Carrier, support	Active agents	Availability
MD-2	(See BP HC Listing)							L
HDS-1442B	Ebullating bed	Resid HDT		Cyl.	34	Al_2O_3, MoO_3	C_0O/	L
HDS-1443B	"	"		"	36	"	NiO/MoO_3	L
GC-30-36	(See Chevron Research HC listing)							L
ARCO TECHNOLOGIES INC.								
ARCO H-H Cat		Raw lube stock	High quality lube products	Cyl.	48			L
BP								
MD-2		VGO	Mid. Dist.	Cyl.		"P"	"P"	L
BASF AG								
M 8-80	Raffination, HC	HGO, VGO	Mid. dist. steam-cracker feed	Cyl.	0.70*	Amorph. alumina-silicates	NiO MoO_3	L
M 8-81	HC	"	"	"	0.68*	"	"	√
CHEVRON RESEARCH CO.								
ICR 106	High HDN, high cracking activity,	VGO, CGO, LCO, AGO	Diesel, jet, naphtha, Isocracker 2nd stage feed	Cyl.	55-63	"P"	"P"	L
ICR 120	Max. diesel yield	"	"	Cyl.	50-58	"P"	"P"	L
ICR 126	Longer life	"	"	Cyl.	55-63	"P"	"P"	L
ICR 113	Lower cost	" DAO	"	Cyl.	45-53	"P"	"P"	L
ICR 117	Naphtha yield	VGO, CGO, LCO, AGO	Naphtha, jet, diesel, Isocracker 2nd stage feed	Cyl.	50-58	"P"	"P"	L
ICR 202	High activity, selectivity	Hydrofined VGO, CGO, LCO	Reformer feed, lt naphtha, jet fuel, isobutane	Cyl.	53-61	"P"	"P"	L
ICR 204	Long cycle life	"	"	Cyl.	44-52	"P"	"P"	L
GC-36	Max. gaso. HC	Dist. lt gas oils	Gaso., naphtha kero. jet	Cyl.	50-60	SiO_2/Al_2O_3	"P"	√
GC-30	Max. dist. HC	Dist. H, H, Hvy. VGO	Furnace oil, diesel, jet, kero.	Cyl.	50-60	SiO_2/Al_2O	"P"	L
KATALCO DIVISION, ICI AMERICAS								
KAT-4000	Active metal	Atm. & Vac. resid		Cyl.	34	Al_2O_3	CoMo	√
KAT-5000	Active metal	Amt. & vac. resid		Cyl.	34	"	Ni Mo	√
LUMMUS CREST INC.								
LC-Fining	Heavy feed, metals tolerance	Atm. bott., vac. bott. hvy crudes, tars, etc.	Dist., low S fuel	Cyl.	Low	Al_2O_3	CoMo/NiMo	L
PROCATALYSE								
HYC 642	High cracking activity	Gas oil, VGO	Gaso. dist	Ext.		Special	NiMo	√
CRITERION CATALYSTS Co. LP.								
S 324	HDN & 1 stage HC	Coal oil HC,		Cyl.	0.86*	Al_2O_3	Ni Mo	√
S 324T	"	"		Sh. ext.	0.80*	"	"	√
S 424	HDN, 1 stage HC	1 stage HC		Sh. ext.	0.82*	Al_2O_3	NiMo	√
S 411				"	0.85*			√
S 653	2nd stage HC	VGO to naphtha, middle dist.		Ext.	0.85*	SiO_2/Al_2O_3	NiW	√
S-753	"	"		"	0.70*	Crystalline SiO_2/Al_2O_3		L
S 354	Heavy dist., HC			Ext.	1.02*	Al_2O_3	NiW	√
S 454	"			Ext.	1.05*	Al_2O_3	NiW	√
UOP								
DHC-2	Bifunctional, strong hydrogenation good cracking activity	VGO, DAO, CGO other non-SR gas oils	High selectivity to mid. dist.	Sph.	41	Amorphous	"P"	
DHC-6	Bifunctional, good hydro. very good cracking activity	"	"	Sph.	44	Amorphous	"P"	
DHC-8	Bifunctional, good hydro. very good cracking activity, imp. temp. stability	"	"	Sph.	44	Amorphous	"P"	
DHC-100	Bifunctional excellent, hydro. function, improved cracking activity, greater temp. stability	"	"	Ext.	"P"	"P"	"P"	
HC-8	Second stage high activity Very active, stable	"	High naphtha selectivity	Cyl.	42	Zeolite	"P"	
HC-100	Second stage high activity, much more active than HC-8	"	"	"	48	Zeolite	"P"	
HC-101	High activity, strong hydrog.	"	High LPG selectivity	"	50	Zeolite	"P"	
RCM-3	High activity, ebullating bed resid conversion units, high strength spheres	Vac. resid	Naphtha, dist.-lates, gas oils	Sph.	41	"P"	"P"	

(continued)

Hydrocracking—continued

Catalyst designation	Primary differentiating characteristics	Application (feedstock)	Application (product)	Form	Bulk density compacted, lb/cu ft,* g/cc	Carrier, support	Active agents	Availability
RCM-4	"	"	"	Sph.	39	"P"	"P"	
UNION CARBIDE CORP.								
HC-14	(See Unocal's HC listing)							
HC-16	"							
HC-18	"							
HC-22	"							
HC-24	"							
HC-28	"							
HC-30	"							
UNOCAL CORP.								
HC-14	Flexibility for fuels	Virgin cracked dist. VGO	Gaso. kero., jet	Cyl.	"P"	Mole sieve	Nonnoble metals	L
HC-16	Flexibility for fuels	"	Kero. jet, diesel	Cyl.	"P"	"	"	L
HC-18	Max. gaso., LPG gaso. naphtha	"	LPG, gaso. petrochemical naphtha	Cyl.	"P"	"	Noble metals	L
HC-22	Max. diesel, jet	"	Diesel, jet	Cyl.	"P"	"	Nonnoble metals	L
HC-28	Max. LPG, gaso. petrochemical naphtha	"	LPG, Gaso., petrochemical naphtha	Cyl.	"P"	"	Noble metal	L
HC-30	Selective HC for pour point reduction	Petroleum and shale oil fractions	Low pour point dist., lube stocks syncrudes	Cyl.	"P"	"	Nonnoble metal	L
HC-80	"	"	"	"	"	"	"	L
HC-102	Max. diesel	Virgin and cracked dist. VGO	Diesel	Sh. ext.	"P"	Amorphous	Nonnoble metal	L

Mild hydrocracking

Catalyst designation	Primary differentiating characteristics	Application (feedstock)	Application (product)	Form	Bulk density compacted, lb/cu ft,* g/cc	Carrier, support	Active agents	Availability
AKZO CHMICALS INC.								
KF-742	CoMo	VGO Conv.	Dist. boiling range mat.	AQ	45	Al_2O_3		✓
KF-840	"	"	"	"	50	"		✓
KF-1001	CoMo	"	"	Cyl.	51-54	"		✓
KF-1002	Nickel moly.	"	"	"	48	"		✓
KF-1011	"	"	"	"	51	"		✓
KF-1012	Co Mo	"	"	"	42	"		✓
KF-1014	Co Mo	"	"	"	41	"		✓
KF-1015	NiMo	"	"	"	42	"		✓
TLF-141 (Elf)	"P"	Pour pt. protection		Cycl.	"P"	"P"		L
CRITERION CATALYSTS Co. LP.								
HCM-40	(See Unocal MHC listing)						"P"	L
HCM-110	"						"	L
BASF AG								
M 8-80	Raffination, MHC	HGO, VGO	Mid. dist.	Cyl.	0.70*	Amorph. alumina-silicates	NiO MoO_3	L
M 8-81	MHC combination with M 8-26	"	"	"	0.68*	"	"	L
BP								
MH 128	Mid. dist. selectivity	VGO	Mid. dist.	Sh. ext.	45	"P"	"P"	✓
CATALYST & CHEMICALS INDUSTRIES CO., LTD.								
HT-D3	High mid. dist. yield	VGO	Kero., diesel &	Cyl.	0.76*	Al_2O_3		✓
HT-D5	High mid. dist. yield	VGO	desulf.	Cyl.	0.78*	"		✓
CHEVRON RESEARCH CO.								
ICR 106	High cracking activity, selective for mid-dist.	VGO, CGO, LCO, AGO	Diesel, FCC feed, LSFO	Cyl.	55-63	"P"	"P"	L
ICR 126	Longer life, higher conversion	"	Diesel, naphtha, FCC feed, LSFO	Cyl.	55-634	"P"	"P"	L
ICR 114	Lower cost	"	Diesel, FCC feed, LSFO	Cyl., sh.	42-54	"P"	"P"	L
ENGLEHARD CORP.								
HPN*-IVB	Select. hydrog. trace butadiene, isom butene to butene 2	C_3, C_4	Dimersol, alky feed	Sph.	47	"P"	"P"	L
EXXON RESEARCH & ENGINEERING CO.								
RT-2, RT-228, RT-621	See also ER&E "Hydrorefining"							L
RT-3, RTNX-1	High activity, regenerability VGO's	Virgin cracked	FCC feed prep; naphtha, mid-dist.	—	—	"	CoMo, NiMo	L
HALDOR TOPSOE								
TK-551	High stability	VGO range	Mid. dist. treated VGO	Cyl.	46	Al_2O_3	Ni Mo	✓
TK-581	"	"	"	"	46	"	"	✓
HARSHAW/FILTROL								
HPC-50	High activity N removal, polynuclear, aro. sat., high S removal activity, severe operations	HVGO, VGO, LCO CGO, coker dist., lube stocks	FCC, HC feed pretreat, lubes	Kloverleaf® cyl. ext. (4-lobed)	51	Al_2O_3	NiO, MoO_3	✓

(continued)

Petroleum Processes, Catalyst Usage

Hydrocracking—continued

Catalyst designation	Primary differentiating characteristics	Application (feedstock)	Application (product)	Form	Bulk density compacted, lb/cu ft,* g/cc	Carrier, support	Active agents	Availability
KATALCO, DIVISION, ICI AMERICAS								
477	% of active metals	FCC feed, blends		Cyl.	45	Al_2O_3	CoMo	✓
479	"	"		Cyl.	51	"	"	✓
497	"	"		Katform	45	"	"	✓
504K	"	FCC feed, gas oil blends, HC feed, cycle oils, coker dist., tillates	FCC feed, HC feed	Cyl.	45	Al_2O_3	Ni Mo	✓
506	"			"	36	"	"	✓
594	% of active metals, shape	"		Katform	52	"	"	✓
596	"	"		"	57	"	"	✓
PROCATALYSE								
HTH 544	Good stability, high mid. dist. selectivity	HVGO		Cyl.		SiO_2/Al_2O_3	Ni Mo	✓
HTH 546	"	"		SiO_2/Al_2O + promotor		"	"	✓
HTH 548	"	"		"		"	"	✓
CRITERION CATALYSTS Co. LP.								
MHC-1	Combined conv. & HDS activity	VGO	Mid. dist. plus treated VGO	Cyl.	0.86*	Al_2O_3	Ni Mo	✓
S-424	"	"	"	Sh. ext.	0.82*	"	"	✓
MHC-200	Low pressure MHC	"	"	"	0.76*	"	Co Mo	✓
UNOCAL CORP.								
HCM-40	High stability MHC	Virgin & cracked dist., VGO	Dist. + hot feed hydrotreated	Sh. ext.	"P"	"P"	"P"	L
HCM-110	"	"	"	"	"P"	"	"	L
UNITED CATALYSTS INC.								
C20-7-02	Composition, physical properties	Lube oil, HVGO HVGO	Hydrog. of olefins, cracked prodct	Ext.	44	Al_2O_3	Ni Mo	✓
C20-7-02CDS	"	"	"	"		"	"	✓
C20-7-04	"	"	"	"	50	"	"	✓
C20-7-06CDS	"	"	"	Ext.	44	"	"	✓
UOP								
DHC-2	Bifunctional clean-up, hydrog. good cracking activity	VGO, CGO	Dist., HDS, lubes	Sph.	41	"P"	"P"	✓
MHC-10	Bifunctional, high hydro. cracking activity	"	"	Cyl.	42	"P"	"P"	✓

Hydrotreating/hydrogenation, saturation catalysts

Catalyst designation	Primary differentiating characteristics	Application (feedstock)	Application (product)	Form	Bulk density compacted, lb/cu ft,* g/cc	Carrier, support	Active agents	Availability
AIR PRODUCTS & CHEMICALS INC.								
HR 836	Pretreat	Py gaso.	Sat. hydrocarbons	Cyl.		Al_2O_3	Co Mo	L
AKZO CHEMICALS INC.								
Active supports	Used to support bed, for feed distribution in reactor top, relative low activity, reduced coking and pressure drop, sat. of diolefins in cracked feeds (AQ = assymetric quadralobe)							
A-540	Ni Mo			AQ	36	Al_2O_3		✓
A-541	Co Mo			AQ	36	"		✓
KAS 082-5B	Ni Mo			Sph.	52	"		✓
AS-20	Ni Mo			AQ		"		✓
KF-H-OIL	Co Mo	Resid, hvy. crude, tons, etc.	Dist, LSFO	Cyl.	"P"	Al_2O_3		L
KF-H-OIL	Ni Mo	"	"	"	"	"		✓
KFR-10	Ni Mo	HDM of resid		AQ		"		✓
KFR-30	"	HDM/HDS of resid		AQ		"		✓
KFR-50	"	HDS of resid		AQ		"		✓
KF INT-RI	"	HDM of resid		AQ	36	"		✓
KF 153/153S	Ni Mo	SR, cracked napthas, dist., VGO, LCO, lube		Cyl.	48	Al_2O_3		✓
KF 842	"	SR, cracked napthas, dist., VGO		AQ	42-45	"		✓
KF 844	"	Coker naphtha		AQ	42-45	"		✓
KF 840	High Act. Ni Mo	SR, cracked naphtha, dist. VGO		AQ	50	"		✓
KF 843	High act, Aro sat. Ni Mo	LCO, VGO, HC lube		AQ	56	"		✓
KF 603	Ni, Co Mo	HDM, HDS, resid		Cyl.	42	"		✓
KF 643	"	"		AQ	39	"		✓
KF 604	Ni Mo	"		Cyl.	45	"		✓
KF 644	"	"		AQ	41	"		✓
KF 609	Ni, Co Mo	"		Cyl.	37	"		✓
KF 649	"	"		AQ	37	"		✓
KF 330	Ni W	H, Pi lube, diolefin Sat.		Cyl.	52	"		✓
KF 124 LD	Co Mo	SR naphthas, dist. gas oil, VGO		Cyl.	38	Al_2O_3		✓
KF 124 HD	"	Naphtha, kero., gas oil, VGO, lube		"	48	"		✓

(continued)

Hydrotreating/hydrogenation, saturation catalysts—continued

Catalyst designation	Primary differentiating characteristics	Application (feedstock)	Application (product)	Form	Bulk density compacted, lb/cu ft,* g/cc	Carrier, support	Active agents	Availability
KF 165	"	SR, cracked naphtha, dist., VGO, LCO, lube		Cyl.	49	Al_2O_3		√
KF SC-1315	Shell oil catalyst							√
KF 640	Cobalt Co Mo, high act.	Naphtha, kero. gasoil, VGO		AQ	41	Al_2O_3		√
KF 742	Co Mo, high act.	SR, & cracked, naphtha dist. VGO, lube py. naphtha		AQ	45	"		√
KF 702	"	SR, cracked naphtha, dist. VGO; Lube, py.		AQ		"		√
KF 707	"	"		AQ	48	"		√
KF 742	"	SR, & cracked naphtha, dist. VGO; lube, py. naphtha		AQ	45	"		√
CRITERION CATALYSTS Co. LP.								
HDS-2(1.6) HDS-2(3.2)	Different particle sizes	Naphthas through LGO		Cyl.	36	Al_2O_3	CoO MoO_3	√
HDS-3(1.6) HDS-3(3.2) TRILOBE* HDS-3(1.6) TRILOBE* HDS-3(3.2)	Different particle size & shapes " "	Naphthas through LGO " "		Cyl. sh. ext. " "	44-47 " "	Al_2O_3 " "	NiO MoO_3 " "	√ √ √ √
HDS-9(1.6) HDS-9(3.2) TRILOBE* HDS-9(1.6) TRILOBE* HDS-9(3.2)	Different particle size & shapes " "	Naphthas through HGO " "		Cyl. or sh. ext. " "	49-52 " "	Al_2O_3 " "	NiO MoO_3 " "	√ √ √ √
TRILOBE* HDS-20(1.6) HDS-20(3.2)	Different particle sizes	Naphthas through HGO		Sh. ext.	~ 46	Al_2O_3	CoO MoO_3	√
TRILOBE* HDN-30(1.6) HDN-30(3.2)	Different particle sizes	Naphthas through HGO		Sh. ext.	~ 52	Al_2O_3	NiO MoO_3	√
TRILOBE* HDS-35(1.6) HDS-35(3.2)	Different particle sizes	Naphthas through LGO		Sh. ext.	~ 46		NiO, CoO, MoO_3	√
AS-100	(See Unocal listing)							L
TRILOBE* HDN-60(1.6) HDN-60(3.2)	Different particle sizes	Naphthas through HGO		Sh. ext.	~ 54	Al_2O_3	NiO, MoO_3	√
RT-3	(See Exxon R&E listing)							L
IMP-DSD-3	(See Instituto del Mexicano Petroleo Listing)							L
TRILOBE* HDS-24 (1.6) HDS-24 (3.2)	Different particle sizes	Naphthas through HGO		Sh. ext.	~ 35	Al_2O_3	CoO MoO_3	√
TRILOBE* HDN-34(1.6) HDN-34(3.2)	"	"		"	~ 35	"	NiO, MoO_3	√
RIFLED* TRILOBE* HDN-60(1.6) HDN-60(3.2)	"	"		"	~ 50	"	NiO, MoO_3	√
CSM-481, CSM-482		"		"	"P"	"P"	"P"	L
TRILOBE* HDS-22 (1.6) HDS-22 (3.2)	Different particles	"		"	~ 43	Al_2O_3	CoO, MoO_3	√
BASF AG								
M 8-12		LPG, nat. gas SR fractions, blends with cracked oils, py gaso.	Steam reforming feed, cat. reformer feed, jet diesel feed for aro. extract.	Cyl.	0.65*	$\gamma - Al_2O_3$	CoO, MoO_3	√
M 8-14	More active for hvy. fractions	HGO, VGO, SR, blends, with cracked oils, paraffins, lube	Jet, diesel, HC feed, tech. white lube oil	Star shaped	0.60*	$\gamma - Al_2O_3$	"	√
M 8-24		Blends of SR fractions with cracked oils	Cat. reformer feed	Cyl.	0.70*	$\gamma - Al_2O_3$	NiO, MoO_3	√

(continued)

Hydrotreating/hydrogenation, saturation catalysts—continued

Catalyst designation	Primary differentiating characteristics	Application (feedstock)	Application (product)	Form	Bulk density compacted, lb/cu ft,* g/cc	Carrier, support	Active agents	Availability
M 8-24	More active for hvy fractions	Blend of HGO, or VGO with cracked oils, lube	Gas oil, FCC feed, lube oil	Star shaped	0.65*	γ–Al$_2$O$_3$	NiO, MoO$_3$	√
M-8-26	High HON, HDS act., olefin hydrog.	Naphtha, kero, GO, VGO, HGO	HC pretreat	Starshaped & cyl.	0.82*	Special	NiO, MoO$_3$	√
H 0-11		Selective hydro. of acteylene acetylene in ethylene	Ethylene	"	0.46*	SiO$_2$	Pd	√
H 0-12		Selective hydro. of acetylenes in C$_3$- and C$_4$-		Cyl.	1.1*	Al$_2$O$_3$	Pd	√
H 0-22		Selective hydro. of diolefins & unsat aromatics	Py. gaso.	Cyl. sph.	0.65*-0.80	"	Pd	√
H 1-80		Hydro. of aromatics aromatics	Food-grade white oil	Cyl.	1.0*	SiO$_2$	Ni	L
EXXON RESEARCH & ENGINEERING CO.								
RT-3 RT-NX-1	High activity, regenerability	Virgin cracked VGO's, distillates	Feed prep. for cat. ref. FCC, S, N removal etc	—	—	—	Co/Ni Mo	L
HALDOR TOPSOE								
TK-250/251		Nat. gas, refinery gas. SR naphtha	Deep HDS, steam reforming cat. reforming HDS olefin sat. of refinery gases	Ring	32	Al$_2$O$_3$	Co Mo(250) Ni Mo (251)	√
TK-450/451	Naphtha to LGO	Deep HDS	"	Cyl.	34	Al$_2$O$_3$	Co Mo (450)/ Ni,Mo (451)	√
TK-551		SR & cracked naphthas HVGO	Deep HDS, HDN, aro.	Ring cyl.	46	"	Ni Mo	√
TK-550	Lower selectivity for HDN, aro. sat. TK-551, lower H$_2$ consumption	"	Deep HDS	Ring cyl.	44	"	Co Mo	√
TK-561	Extra high HDN act.	"	"	Cyl.	50	"	NiMo	√
HARSHAW/FILTROL								
HPC-8	Low density, high pore vol.	Naphtha through VGO	HOS	Kloveleaf®, or Cyl.	33	Al$_2$O$_3$	CoO, MoO$_3$	√
HPC-6B	Low density	"	HDS	"	39	"	"	√
HPC-40B	HDS, HDN	CGO, HVGO, LCO, CGO, coke dist., naphtha, lube	FCC feed, gaso. jet	"	47	"	Nio, MoO$_3$	√
HPC-50	HDS, HDN, PNA sat.	"	FCC feed lube, gaso. jet	"	51	"	"	√
HPC-70	High act. HDS, HON, PNA sat. severe operations	"	FCC Feed, HC pretreat., lube	"	51	"	"	√
HPC-66	HDS	HS HGO, heating oil, naphtha	GO, heating oil, gaso.	"	44	"	CoO, MoO$_3$	√
HPC-60	High act. HDS	"	"	"	46	"	"	√
INSTITUTO MEXICANO DEL PETROLEO								
IMP-DSD-3	High act. trilobe	Petro. fract.	Naphtha for reforming, high diesel act. FCC feed	Sh. ext	0.74*	γ-Al$_2$O$_3$	Mo, Ni, P	L
IMP-DSD-4	High act. sphere	"	"	Sph.	0.55*	γ-Al$_2$O$_3$	Mo, Ni	L
IMP-DSD-5	Good activity sphere	Gasolines to diesel HDS	Naphtha for reforming good quality diesel	Sph.	0.55*	"	Mo, Ni, P	L
IMP-DSD-1	Good activity sphere	"	"	Sph.	0.86*	γ-Al$_2$O$_3$	Mo, Ni	L
KATALCO DIVISION, ICI AMERICAS								
477	% of active metals, shape	SR naphtha through gas oil/resid blends		Cyl.	46	Al$_2$O$_3$		√
477S				Sph.	36		"	√
479				Cyl.	52		"	√
497	"	"		Katform	45	"	"	√
499	"	"		"	51	"	"	√
482	Active metal	Sour gases, Lt. hydro. carbon		Cyl.	32	"	"	√
502	% of active metals, shape	SR naphtha through gas oil/resid blends		Cyl.	46	Al$_2$O$_3$,	NiMo	√
502S				Sph.	36	"	"	√
504K				Cyl.	56	"	"	√
506		"		Cyl.	58	"	"	√
592	"	"		Katform	45	"	"	√
594	"	"		Katform	52	"	"	√
596				Katform	57			√
550	Shape	Cracked stocks/cycle oil, jet, lubes		Cyl.	50	Al$_2$O$_3$,	NiW	√
550S				Sph.	37	"	"	√
PROCATALYSE								
LD 271	Selective hydrog. of propylene rich cut from steam cracking, also butene-1 rich cut	Steam cracker propylene rich, butene-1 rich cut	Propylene, butene 1	Sph.		Al$_2$O$_3$	Pd promoter	√

(continued)

Hydrotreating/hydrogenation, saturation catalysts—continued

Catalyst designation	Primary differentiating characteristics	Application (feedstock)	Application (product)	Form	Bulk density compacted, lb/cu ft,* g/cc	Carrier, support	Active agents	Availability
LD 275	Selective hydrog. of pyr. gaso. fractions	Pyr. gaso.	Gaso.	Sph.		Al_2O_3	Pd promoter	✓
LD 277	Acetylene removal from butadiene rich steam cracker cut	C_4 steam cracker cut	Butadiene rich C_4 cut	"		Al_2O_3	Pd promoter	✓
LD 2773	Selective hydrogen. of butadiene in butene rich cuts from FCC	FCC C_4 cut	C_4 cut	"		"	"	✓
LD 277	MAPD removal from FCC propylene cuts	FCC C_3 cut	C_3 cut	"		"	"	✓
Ni PS_2	Total hydrog. of benzene into cyclohexane	Benzene	Cyclohexane	Powder		Ni	Ni	✓
LT 261	Selective hydrog. of ethylene ethylene	Ethylene cut (vapor phase)	Ethylene	Sph.		Al_2O_3	Pd	
LD 145	Selective hydrogen. of py. gaso. and 180 400° C. fraction	Py. gaso.	Gaso.	"		Al_2O_3	Ni Mo	
LD 155	"	Py. gaso. hvy. gaso.	Gaso., special fuel	"		"	Ni W	
LD 241	Selective hydrogenation of pyrolysis gasoline	Py. gaso. line	Sweet gaso.	Sph.		Al_2O_3	Ni	
LD 265	Selective hydrogen. of pyrolysis propylene cut, butylene cut, and gaso.	Pyrolysis propylene cut, butylene cut, gaso.	Propylene, butylene, gaso.	"		"	Pd	
HC 102	Total hydrogenation	Benzene	Cyclohexane	Liquid			Ni	✓
Cu 640	Hydrogenation	Aldehydes	Alcohols	Cyl.		SiO_2	Cu	✓
Ni 563	Hydrogenation	Aldehydes, unsat. drocarbons	Alcohols, solvents	Pellets		"	Ni	✓
CRITERION CATALYSTS Co. LP.								
S 214	Multi purpose HDT	Naphtha to HVGOs		Sh. ext.	0.71*	Al_2O_3	Ni Mo	✓
S 324	High performance, HDS, HDN	Coker naphtha and py. gaso. to HVGOs		Cyl.	0.86*	"	"	✓
S 324T	"	"		Sh. ext.	0.80*	"	"	
S 424	High activity HDS, HDN, sat.	Pretreat. cracker feed, 1st stage HC		Sh. ext.	0.82*	Al_2O_3	Ni Mo	✓
S 411	"	"		"	0.85*	"	"	✓
S 514	Support function-replaces inert ceramic material.	All HDT applications-preferentially cracked feeds		Sph.	0.80*	Al_2O_3	Ni Mo	✓
S 344	Heavy duty HDS	Wide variety of stocks, HVGO		Cyl.	0.78*	Al_2O_3	Co Mo	✓
S 344T	"	"		Sh. ext.	0.73*	"	"	
S 444	High activity HDS	VGO, LGO, HGO Cracked GO		Cyl.	0.77*	Al_2O_3	Co Mo	✓
S 444T	"	"		Sh. ext.	0.72*	"	"	
S 447	Deep HDS	VGO, resid. LGO, LGO, HGO		Sh. ext.	0.71*	"	"	✓
S 544	Support function. Can replace all ceramic material.			Sph.	0.80*	Al_2O_3	Co Mo	✓
S 127	Initial residue HDS	Heavy resid. with moderate metal content.		Cyl.	0.70*	Al_2O_3	Ni Mo	✓
S 227	Tail end deep residue HDS	Heavy resid. with moderate metal content		"	0.70*	"	"	✓
S 147	Initial residue HDS	Heavy resid. with modererate metal content.		"	0.70*	Al_2O_3	Co Mo	✓
S 247	Tail end deep residue HDS	Heavy resid. with moderte metal content.		Cyl.	0.70*	Al_2O_3	Co Mo	✓
S 311	Hydrocracker tail end	Hydrocracked feed		Sh. ext.	0.76*	"	Ni Mo	
S 614	Hydrog. of aromatics	Kero. specialty oils		Cyl.	0.41*	SiO_2/Al_2O_3	Pt	✓
S 204	Hydrogenation of di-olefins	Pyrolysis gaso.			0.72*	Al_2O_3	Ni	✓
S 504	"	Py. gaso. line	Support balls, replaces inert ceramic material	Sph.	0.80*	"	"	✓
UNOCAL CORP.								
HC-B, HD-D, HC-F, HC-H, HC-K	(See Unocal Hydrorefining listing)							
AS-100	Saturation of aromatics	Naphtha, kero, jet fuel	Low aro. naphthas, high smoke point kero., jet	Cyl.	"P"	"P"	Noble metal	L
N-21, N-22, N-23, N-30, N-100	(See Unocal Hydrorefining listing)							

(continued)

Petroleum Processes, Catalyst Usage

Hydrotreating/hydrogenation, saturation catalysts—continued

Catalyst designation	Primary differentiating characteristics	Application (feedstock)	Application (product)	Form	Bulk density compacted, lb/cu ft,* g/cc	Carrier, support	Active agents	Availability
UNITED CATALYSTS INC.								
C20-5-01		SR, coker naphtha, Kedro. diesel	HDS, removal of metal contaminants imp. color, odor, stability	Tablet or ext.	50	Al_2O_3	Co Mo	√
C20-6-01	Composition, physical properties	Lt. naphthas through heavy residues	HDS hydrog. of olefins	Ext. 1/8-in. or 1/16-in.	38	Al_2O_3	Co Mo	√
C20-6-03	"	"	"	"	44	"	"	√
C20-6-04	"	"	"	"	51	"	"	√
C20-7-02	Composition physical properties	Cracked stocks, high olefin N content feed	HDS, HDN, hydrog. of olefins, de-wax	"	44	"	Ni Mo	√
C20-7-03	"	"	"	"	38	"	"	√
C20-7-04	Composition physical properties	Cracked stock, high olefin and nitrogen content feedstocks	HDS, HDN, hydrog. of olefins, de-wax	Ext.	50	"	"	√
C20-6-04 CDS	High HDS activity	VGO, LGO, HGO		Ext.	37-43	Al_2O_3	CoMo	√
C20-7-07 CDS	High HDS, denitrification activity	Pretreat. FCC feed	FCC feed	Ext.	47-53	"	NiMo	√
G-87	Total hydrogen. of Benzene to Cyclohexane	Benzene	Cyclohexane	Sph.	52-58		Ni	√
G-98B	"	Bz with S present	"	Tab	48-52	SiO_2	NiCuCo	√
C46-7-03	High hydrogen. activity	Aro. sat		Ext.	40-48	SiO_2	Ni	√
C46-8-03	"	"		Ext.	40-48	Al_2O_3	Ni	√
G-68A	Hydrogen. of C_5 dienes	Isoprene		Tab	32-38	"	Pd	√
G-68C	Selective hydrogen. DPG-diolefins	Py. gaso.	Gaso.	Sph.	42-48	"	"	√
G-68D	Selectivity hydrogen. trace butadiene & isomerize B1 to B2	C_3 - C_5 stream	Alky. feed	Tab	64-69	"	"	√
C38-1	Selective hydrogen. trace butadiene with isomerization	C_3 - C_5 stream with S	"	Ext.	52-58	Al_2O_3	Ni	√
C20-7-06CDS	"	"	"	Ext.	44	"	Ni Mo	√
UOP								
S-12	High HDS, HDN	Naphtha through VGO, cracked stocks, high severity	HDS, HDN	Cyl.	49	Al_2O_3	Co Mo	√
PF-4	Selective hydrogenation	Alphamethylstyrene in Cumene. Diolefin Sat.-	Cumene, py. gaso.	Sph.	32	Al_2O_3	Pd	
H-3	High hydrogenation activity, S tolerant	Aro. sat (usually benzene)	High-purity cycloparaffins (usually cyclohexane)	Sph.	32	Al_2O_3	Pt	
H-4	High hydrogenation activity	" plus selective diolefin sat., meth.	"plus upgrade alky feed"	Sph.	44	"	Ni	
H-8	"	Aro sat. dist. Sat. of residual aromatics in N-Paraffin streams	Improved burning or ignition qualities. High purity N-Paraffins	Sph.	32	"	"	

Hydrorefining

Catalyst designation	Primary differentiating characteristics	Application (feedstock)	Application (product)	Form	Bulk density compacted, lb/cu ft,* g/cc	Carrier, support	Active agents	Availability
AKZO CHEMICALS INC.								
HC-K	Co Mo, high act.	SR, cracked naphthas, dist. VGO, LCO		Cyl.	48	Al_2O_3		L
RF-100, 200, 220	(See Unocal Hydrorefining listing)			AQ				L
RT-2, RT-3, RT-621, RT-228, RT-NX-1, TN-8	(See Exxon Research & Engineering Hydrorefining)							
CRITERION CATALYSTS Co. LP.								
HC-B, HC-D, TRILOBE* HC-F, HC-H	(See Unocal HF listing)			Cyl. or Sh. ext.				L
TRILOBE* RF-11, RF-25	(See Unocal HF listing)			Sh. ext.				L
RT-2	(See Exxon R&E HF listing)							L
RT-3	(See Exxon R&E HF listing)							L

(continued)

Petroleum Processes, Catalyst Usage

Hydrorefining—continued

Catalyst designation	Primary differentiating characteristics	Application (feedstock)	Application (product)	Form	Bulk density compacted, lb/cu ft,* g/cc	Carrier, support	Active agents	Availability
CSM-471		Resid HDT		Sh. ext.	"P"	"P"	"P"	L
HDS-1442A HDS-1442B	Different particle sizes	Resid or VGO HDT		Cyl.	~ 34	Al_2O_3	CoO MoO_3	√
HDS-1443-A HDS-1443-B	Different particle sizes	"		Cyl.	~ 36	"	NiO MoO_3	√
HDS-9(1.6) HDS-9(3.2) HDS-9(1.6) HDS-9(3.2)	Different particle sizes and shapes	HGO/cat. feed HDT	Fuel oil/FCC feed	Cyl. or sh. ext.	49-52	"	"	√
TRILOBE* HDS-20(1.6) HDS-20(3.2)	Different particle sizes	"	Fuel oil/FCC feed	Sh. ext.	~ 46	"	"	√
TRILOBE* HDN-30(1.6) HDN-30(3.2)	"	"	Fuel oil/FCC feed	Sh. ext.	~ 52	"	NiO MoO_3	√
GC-100, 101, 102, 105, 106, 107	(See Chevron Research HF listing)							L
TRILOBE* HDN-60(1.6) HDN-60(3.2)	Different particle sizes	HGO/cat. feed HDT	Fuel oil/FCC feed	Sh. ext.	~ 54	Al_2O_3	NiO MoO_3	√
GC 405, 36, 30, 26	(See Chevron Research HF listing)							L
N-21, N-22, N-23 N-30	(See Unocal HF listing)							L
RIFLED® TRILOBE® HDN-60(1.6) HDN-60(3.2)	Different particle sizes	HGO/cat. HDT		Sh. ext.	~50	Al_2O_3	NiO MoO_3	√
CSM-481 CSM-482		HGO thru resid		Sh. ext. ext.	"P"	"P"	"P"	L
CDM-152		"		"	"	"	"	L
TRILOBE* HDS-22 (1.6) HDS-22 (3.2)	Diff. particle size	Fuel oil, FCC feed		Sh. ext.	~ 43	Al_2O_3	CoO, MoO_3	√
BASF AG								
M 8-21		Raw lube feed	Lube oil, white oil	Cyl.	Ca. 72*	Al_2O_3	Ni Mo	
M8-23	Raffination	Aro. Hydrog.	Food grade	Cyl.	0.82	$\gamma\, Al_2O_3$	NiO, MoO_3	
H 1-80		Aro. Hydrog.	"	"	1.0*	SiO_2	Ni	√
CATALYST & CHEMICALS INDUSTRIES CO., LTD.								
CDS-R2	High HDS, low reaction temp.	Atm. & vac. resid	Desulfurized oil	Cyl.	0.68*	Al_2O_3	—	√
CDS-R9	High HDS & demet.	Atm. & vac. resid	Desulfurized oil	Cyl.	0.66*	"	—	√
CDS-DM-1	High demet.	"	Demetalized oil	Cyl.	0.58*	"	—	√
CDS-DM5	High deme. with desulfurization ability	"	"	"	0.62*	"		
CDS-D5	High HDS activity	Naphtha & kero.	Desulfurized naphtha, kero.	Cyl.	0.68*	—	—	√
CDS-D9	"	Naphtha, kero. gas oil	Desulfurized Kero, VGO		0.72*	Al_2O_3		
CDS-D11	"	Kero. diesel, VGO	Desulfurized, kero. diesel, VGO	Cyl.	0.73*	—	—	√
CDS-D13	Superior high HDS activity	"	"	"	0.78*			√
CDS-DN1	High HDS and HDN activity	Naphtha	Desulfurized naphtha	"	0.77*	"		
CDS-DR1	High HDS activity	20-30% Atm. resid, and 80-70% VGO	FCC feed, desul- furized oil	"	0.70*	"		
CHEVRON RESEARCH CO.								
ICR 112	High HDS activity, low H_2 consumption	VGO, CGO	LSFO, FCC feed	Cyl.	53	"P"	"P"	L
ICR 114	High HDN activity	VGO, CGO	FCC feed	"P"	57	"P"	"P"	L
ICR 105	Designed for resid HDT	DAO, topped crudes, atm. vac. resids	LSFO, FCC feed, coker feed	Cyl.	49	"P"	"P"	L
ICR 121	Resid feeds	"	"	"P"	43	"P"	"P"	L
ICR 122	Resid HDM	"	LSFO, FCC feed, resid HDS feed, coker feed	"P"	30	"P"	"P"	L
GC-101 GC-106	High metals resist.	HDS, atm., vac., resid	0.5-1.0% S HFO	Cyl.	30-40	Al_2O_3	"P"	L
GC-102 GC-107	High coke deactivation resist. HDS cat.	HDS amt., vac. resids amt. & vac.	0.2-0.6% sulfur HFO, resid FCC feed prep.	Cyl. Sh. "	33-41	Al_2O_3	"P"	L
GC-100 GC-105	High severity HDS resid HDS	"	0.05-0.2% S HFO	Cyl. sh.	33-42	"	"P"	L
GC-125 GC-130	Metals resist.	High metals, amt., vac. resids	Demet./HDS	Cyl. Sh. ext.	30-37	"	"P"	L

(continued)

Petroleum Processes, Catalyst Usage

Hydrorefining—continued

Catalyst designation	Primary differentiating characteristics	Application (feedstock)	Application (product)	Form	Bulk density compacted, lb/cu ft,* g/cc	Carrier, support	Active agents	Availability
GC-405	High activity, hydrogenation, selective hydrocracking	Conventional refined, virgin lube feed	Lube, base oils	Cyl. Sph.	40-48(c) 32-38 (s)	Al₂O₃	"P"	L
GC-36	High activity hydro., selective HC	Poor quality lube feed	"	Cyl.	50-55	SiO₂, Al₂O₃	"P"	L
GC-30	Higher activity.	"	"	Cyl.	50-60	"	"P"	L
GC-26	Selective hydro.	Semi-refined lube, specialty oil product feed-	Lube industrial oils, specialty products	Cyl.	60-70	Al₂O₃	"P"	L

EXXON RESEARCH & ENGINEERING CO.

Catalyst designation	Primary differentiating characteristics	Application (feedstock)	Application (product)	Form	Bulk density	Carrier, support	Active agents	Availability
RT-2, RT-228, RT-621	Contaminant tolerant for resid processing	Atm. vac. resids	Feed prep for FCC, coking LSFO, Lt. dist.	—	—	Tailored pore size supports	Co Mo., Ni Mo	L
RT-3, RT-NX-1	High activity, regenerability	Virgin/cracked VGO's dist.	Feed prep. for cat. reforming, FCC, S, N removal	—	—	—	Co/Ni Mo	L

HALDOR TOPSOE

Catalyst	Primary differentiating	Application (feedstock)	Application (product)	Form	Bulk density	Carrier	Active agents	Availability
TK-550	Smaller pore for lighter feeds	AGO, VGO	HDS	Ring	37/46	Al₂O₃	Co Mo	✓
TK-551	"	VGO, FCC feed, CGO	HDS, HDN, PNA	Ring, Cyl.	37/46	"	Ni Mo	✓
TK-581	Larger pore, cat. heavier feeds	"	"	Ring	37/46		Ni Mo	✓
TK-709	Low HDS activity, high HDM selectivity, high metals	Atm resid, vac. resid, DAO	Front end HDS, HDM	Ring	25 33	Al₂O₃	Mo	✓
TK-710/711	Slightly higher HDS act. high HDM selectivity, high metals	"	"	Ring	31/40	"	Co Mo (710) Ni Mo (711)	✓
TK-750/751	Medium HDS activity, HDM selectivity metals tolerance	"	Intermediate in HDS, HDM	Ring	31/40	"	Co Mo (750) Ni Mo (751)	✓
TK-770/771	High HDS activity, low HDM selectivity, limited metals tolerance	"	Final cat. in HDS, HDM	Cyl.	45	"	Co Mo (770) Ni Mo (771)	✓

INSTITUTO MEXICANO DEL PETROLEO

Catalyst	Characteristics	Feedstock	Product	Form	Bulk density	Carrier	Active agents	Availability
IMP-HDW-10	Zeolitic for dewaxing lube oil cuts	Wide range of lube	Dewaxed lube-	Cyl.	0.80*	Zeolite Al₂O₃	Pt/Zeolite	L

KATALCO DIVISION, ICI AMERICAS

Catalyst								
477, 477S, 479, 497, 499	(See also Katalco Corp. Hydrotreating, etc.)							
502, 504, 506, 592, 594, 596	"							

UNOCAL CORP.

Catalyst	Characteristics	Feedstock	Product	Form	Bulk density	Carrier	Active agents	Availability
N-30	High activity, HDS	Virgin, cracked dist. naphthas, VGOS	Desulfurized products	Sh. ext.	"P"	"P"	Co Mo	L
N-100	"	"	"	Cyl.	"P"	"P"	"	L
N-23	"	"	"	Cyl.	"P"	"P"	"	L
N-22	HDS, HDN	"	Low S, low N products	Cyl.	"P"	"P"	Ni Mo	L
N-21	"	"	"	Cyl.	"P"	"P"	"	L
HC-H	HDS, HDN	Virgin, cracked dist. VGO	HC, & FCC Feed	Sh. ext.	"P"	"P"	Ni Mo	L
HC-K	"	"	"	"	"P"	"P"	"	L
HC-F	"	"	Dist. products, HC FCC feed	"	"P"	"P"	"	L
HC-D	"	"	"	Cyl.	"P"	"P"	"	L
HC-B	"	"	"	Cyl.	"P"	"P"	Ni Mo	L
RF-11	HDS	Atm. resid., whole crudes	Fuel oil, FCC coker feed	Sh. ext.	"P"	"P"	"P"	L
RF-25	High activity bifunctional demet.	"	"	"	"P"	"P"	"P"	L
RF-100	HDS	"	"	"	"P"	"P"	"P"	L
RF-200	High activity bifunctional demet.	"	"	"	"P"	"P"	"P"	L
RF-220	"	"	"	"	"P"	"P"	"P"	L

UOP

Catalyst	Characteristics	Feedstock	Product	Form	Bulk density	Carrier	Active agents	Availability
RCD-5	Metals tolerance, HDS	Atm. or vac. resid.	LSFO, cracking, coker feed	Sph.	41	Al₂O₃	Co Mo	Contact UOP
RCD-5A	High activity for HDS, mod. metal tolerance	"	"	Sph.	46	"	"	"
RCD-7	High activity for S and Con. Carbon Removal	Demetallized oil, deasphalted resid	LSFO, cracking feed	Cyl.	53	"	"	"
RCD-8	High metals tolerance, high demet. mod. HDS moderate HDS activity	Atm. vac. resid., deasphalted resid.	LSFO, cracking coker feed	Cyl.	36	"P"	"P"	"
RCD-9	High metals tolerance, high deme. high HDS activity	"	"	Cly.	37	"P"	"P"	"

(continued)

Hydrorefining—continued

Catalyst designation	Primary differentiating characteristics	Application (feedstock)	Application (product)	Form	Bulk density compacted, lb/cu ft,* g/cc	Carrier, support	Active agents	Availability
PROCATALYSE								
HR 306	HDS, HDN	Naphtha through VGO		Cyl.		Al_2O_3	Co Mo	√
HT 308	HDT	Residues		"	0.67*	"	Co Mo	√
HR 346	HDS, HDN, hydro.	Residues		Cyl.	"	"	Ni Mo	√
HR 348	HDS, HDN, hydro. aromatics	Naphtha through VGO	Desulfurized N-free, hydro. product	Cyl.	0.77*	"	"	√
HR 354	HDS, HDN, aro. hydrog.	Naphtha through VGO	N-free, hydro product	Cyl.		Al_2O_3	W Ni	√
HMC 841	HDT, HDM	GO, VGO, atm., vac. resid.		Sph.		Al_2O_3	NiMO	√
HMC 849	HDM	"		"		"	NiMO	√
HMC 845	HDT, HDM	"		"		"	NiMO	√
Actispheres 923	Catalyst protection and support	Naphtha through atm. resid		"		"	Ni, Co, Mo	√

Polymerization

Catalyst designation	Primary differentiating characteristics	Application (feedstock)	Application (product)	Form	Bulk density compacted, lb/cu ft,* g/cc	Carrier, support	Active agents	Availability
INSTITUTO MEXICANO DEL PETROLERO								
IMP-TPC-1	High selective extrudate	Propylene	Propylene tetramer, high oct. gaso.	Cyl.	0.88*	Kieselgurgh	H_3PO_4	L
UNITED CATALYSTS INC.								
C84-1-01 & C-84-02-04	Particle size	Propylene, benzene, butenes	Cumene polygaso. higher olefins	Sph.	55	Diatomaceous earth	H_3PO_4	√
C84-3-01 & C84-3-02	Form, particle size	"	"	Ext.	"	"	"	√
UOP								
SPA-1	High act. & select. larger size chamber reactor	Propylene, butenes, pentenes, benzene, toluene	Petrochemicals, gaso., tetramer, cumene, cymene	Cyl.	58	SiO_2	H_3PO_4	Contact UOP
SPA-2	Small size for tubular reactor	"	"	"	58	"	"	Contact UOP

Sulfur (elemental) recovery

Catalyst designation	Primary differentiating characteristics	Application (feedstock)	Application (product)	Form	Bulk density compacted, lb/cu ft,* g/cc	Carrier, support	Active agents	Availability
ALCOA								
S-100 1/4-in	Lower pressure drop	Claus feed hydrocarbons	Withstands pressure drop due to coking and ammonia. All converters and subdew point tail gas processes.	Sph.	45	Activated Al_2O_3	Activated Al_2O_3	√
S-100 3/16-in.	Higher activity	Refinery or nat. gas	All converters and subdewpoint tail-gas processes	"	45	"	"	√
S-100 5/16-in.	Ext. low pressure drop	"	Withstands extreme coking and ammonia.	Sph.	45	"	"	√
SP-100 3/16 in.	Promoted, lower pressure drop	High COS and CS_2 conversion	COS and CS_2 conversion. Withstanding fouling due to cracking of aromatics.	"	47	Activated Al_2O_3	"P"	√
SP-100	Promoted	High COS and CS_2 conversion	"	"	47	"	"P"	√
SRU 1/2-in. 3/8-in.	Active bed support, high physical strength	All claus plants	High strength bed support and hold down with claus activity	Sph.	50	Activated Al_2O_3	Activated Al_2O_3	√
ARI TECHNOLOGIES								
ARI-310	Active ingredient in chelated aqueous solution	Refinery gas, sour water stripper gas, amine tail gas, nat. gas	Elemental S	Liquid	74.8	None	Fe	L
CATALYST & CHEMICALS INDUSTRIES CO. LTD.								
CSR-2	High Claus activity	Claus reaction feed feed gas.	S	Sph.	0.84*	Al_2O_3	—	√
CSR-3	High hydrolysis of organic sulfide	"	"	"	0.87*	"		
CSR-7	High prevention of sulfation	Claus reaction feed	S	Sph.	0.80*	"	—	√
ENGELHARD CORP.								
SuReCat® 2-4, 3-6	High act., low attr. promoted sulfur resist.	Gas with 2:1 H_2S/SO_2	Purified gas, S	Sph.	50-52	Al_2O_3	Al, Ti, Fe oxide	√
SRC 2-4	High Claus activity	"	Purified gas, S	Granules	53	Al_2O_3	"	√
SRC-4-8	Higher than SRC 2-4 activity	"	"	"	55	"	"	√

(continued)

Petroleum Processes, Catalyst Usage

Sulfur (elemental) recovery—continued

Catalyst designation	Primary differentiating characteristics	Application (feedstock)	Application (product)	Form	Bulk density compacted, lb/cu ft,* g/cc	Carrier, support	Active agents	Availability
LPD 5-16	Modest activity. Bed support	"	"	Ext.	52	Al_2O_3	Al, Ti, Fe oxide	√
LPD 5-8	Modest activity, bed support only	"	"	Ext.	52			√
KAISER CHEMICALS								
S-301 3 × 6, 1/2 × 1/4	High porosity, high activity, lowest attrition	H_2S, CO_2 conv. conv. COS, CS_2 conversion	S	Sph.	45	Activated Al_2O_3	Activated Al_2O_3	√
S-201 3 × 6, 1/2 × 1/4	Standard high porosity	H_2S and SO_2 con., COS, CS_2 con.	"	"	45	" alumina	" alumina	√
S-200 3 × 6, 1/2 × 1/4	Low attrition, mod. activity CS_2 conv.	H_2S and SO_2 converstion	"	"	48	"	"	√
S-501 3 × 6, 1/2 × 1/4	Promoted, sulfation resist.	Improved COS, CS_2 conv.	"	"	52	"	"P"	√
S-701	Plus thermal stability	Very high CS_2, COS, conv. and H_2S conver.	"	Kloverleaf®	52	"	"P"	√
RHONE POULENC								
CR 4-6	High pore volume	H_2S	S	Sph.	0.67*	Al_2O_3		√
DR 5-10	First generation Claus	"	"	"	0.75*	"		√
AM 4-6	Sulfation protection	"	"	"	0.75*	"	Fe	√
CRS 21	High COS, & CS_2 conv.	"	"	"	0.71*	"	TiO_2	√
CRS 31	High COS, CS_2 conv. Unsulfatable	"	"	Cyl	0.95*	TiO_2	TiO_2	√
CT 739	Catalytic incineration	H_2S & SO_2	SO_2	Sph.	0.60*	SiO_2	Fe	√
CRITERION CATALYSTS Co. LP.								
S 949	High activity Claus, high COS, CS_2 hydrolysis		S	Sph.	0.53*	SiO_2	—	√
UNOCAL CORP.								
Selectox-32	Solid for oxidation of H_2S to S	Acid gas from refineries, nat. gas plants, or waste treatment plants.	S, no SO_3	Cyl.	"P"	"P"	Non-noble metals	L
Selectox-33	"	"	"	"	"	"	"	L

Steam hydrocarbon reforming catalysts

Catalyst designation	Primary differentiating characteristics	Application (feedstock)	Application (product)	Form	Bulk density compacted, lb/cu ft,* g/cc	Carrier, support	Active agents	Availability
BASF AG								
M 8-12	Hydrogen. of organic S compounds	Nat. refinery gas, LPG, Lt. naphtha	Steam reformer feed	Ext.	0.72*	Al_2O_3	Co Mo	√
M 8-21		" high CO_2	"	"	0.72*	"	Ni/Mo	√
R 5-10	Removal of H_2S	S containing gas	"	"	1.2*	—	ZnO	√
G 1-25	Steam reforming (gasification)	Nat. & refinery gas	Process gas, CO-shift	Rings	1.1-1.25*	Al_2O_3	Ni	√
G 1-25 S	"	"	"	"	0.9*	" High therm. stab.	Ni	√
G 1-50	"	LPG naphtha	"	Rings	1.1*	—	Nickel, alkalized on special carrier	√
G 1-80	Autothermal steam reforming	Naphtha/LPG	Process gas, tubular reformer	Tablets	"P"	"P"	Ni	L
K 6-10	High temp. shift conv.	Process gas, ex. steam reforming/partial oxidation	Process gas for H_2 prod.	Tablets	1.0*	—	Fe, Cr oxide	√
K 8-11	Shift conversion	Process gas ex. partial oxidation	"	Ext.	"P"	"P"	Co Mo	√
K 3-110	Low temp. shift conv.	Process gas, ex. HTS	"	Tablets	"P"	ZnO, Al_2O_3	Cu	√
R 1-10	Methanation	Methanation of CO_2, CO	H_2	Tablets	0.9*	Al_2O_3	Ni	√
DYCAT INTERNATIONAL								
Dycat 140	S removal	Gas, naphtha	S free hydrocarbons	Sph.	70	None	Zno	√
Dycat 873	High act. steam reforming	Nat. gas LPG, refinery gas, lt. naphtha	H_2, CO, CO_2	Rings	68-70	α Al_2O_3	Ni, La	√
Dycat 873F	"	"		Enhanced Rings	64-66	"	"	√
Dycat 890	"	Naphtha, gas CO, CO_2	H_2, CO, CO_2	Rings	68-70	"	"	√
Dycat 804	Methanation			Rings, cyl. sph.	54-56	Refractory	"	√
Dycat 812		Secondary reforming	Process gas, NH_3 prod.	Rings	72	α Al_2O_3	"	√
HALDER TOPSOE								
RKS-1	Medium act. steam reforming	Lt. to hvy. NG and ref. gas.	H_2, CO, CO_2	Rings	62-65	$MgAl_2O_4$	Ni	√

(continued)

Steam hydrocarbon reforming catalysts—continued

Catalyst designation	Primary differentiating characteristics	Application (feedstock)	Application (product)	Form	Bulk density compacted, lb/cu ft,* g/cc	Carrier, support	Active agents	Availability
R-67	High act. steam reforming	C_1-C_4 hydrocarbon, ref. gas	"	"	65-68	Mg Al_2O_4	Ni	✓
R-67-7H	"	"	"	Cyl. with holes	53-56	"	"	✓
RKNR	"	CH_4, LPG, naphtha, lt., hvy. offgas	"	Rings	57-60	MgO, Al_2O_3	"	L
HTZ-3	Fine desulf. of steam reforming, feed	CH_4, LPG, naphtha	S free hydrocarbon ref. gas	Ext.	78	—	ZnO	✓
HTZ-5	"	"	"	"	62	—	"	✓
SK-12	High temperature CO conv.	Steam ref. effluent	Gas, 2-3% CO	Cyl.	65	—	Fe, Cr oxides	✓
LSK	Low temp. CO conv., regen. Cl_2 resist.	HT-shift effluent	Gas, 0.1-0.3% CO	Cyl.	66	—	Cu, Zn, Cr oxides	✓
LK-801	Low temp. CO conv. S resistant	HT-shift effluent	Gas with 0.1-0.3% CO	Cyl.	67	—	Cu, Zn, Al oxides	✓
PK-5	Methanation	H_2, 2% CO + CO_2	H_2 w/o CO, CO_2 O_2 containing contaminants.	Rings	39	Al_2O_3	Ni	✓
ICI CATALYSTS								
41-6	HDS	Gas/naphtha		Ext.	0.6*	Al_2O_3	Co Mo	✓
61-1	"	"		"	0.67*	"	Ni Mo	✓
32-4	S removal	"		Sph.	1.08*	None	ZnO	✓
25-3	"	Heavy gas		"	1.0*	"	Ni K_2O	✓
59-3	Chloride removal	"		"	0.85*	Al_2O_3	Alkali	✓
46-1	Steam reforming	Naphtha		Ring	1.15*	"	Ni, K_2O	✓
46-4	"	"		"	1.0*	Ca, Al_2O_3	Ni	✓
57-3	"	Nat. gas, ref gas		"	1.0*	"	"	✓
KATALCO CORP. DIVISION, ICI AMERICAS								
23-1, 23-3	Steam reforming	Nat., refin. gas		Ring	66/75	Al_2O_3	Ni	✓
23-4, 23-4M	"	"		4 hole-ring	66/75	"	Ni	✓
25 Series	"	Hvy nat. gas		Rachig & 4 hole ring	66	Ca Aluminate	Ni/Potash	✓
46-9	"	" & LPGs		Ring	66	"	Ni/Potash	✓
46-1	"	Naphtha		Ring	72	"	Ni/Potash	✓
46-4	"	Naphtha		Ring	62	"	Ni	✓
71-1	High temp. shift	Reformer product gas		Cyl.	77	None	Fe, Cr oxides	✓
52-2	Low temp. shift	HT shift product gas		Cyl.	62	None	Cu, Zn, Al_2O_3	✓
53-1	"	"		Cyl.	62	None	"	✓
11-3	Methanation	CO_2 removal plant gas		Cyl.	70	Ca/Aluminate	Ni	✓
11-4	"	"		Cyl.	75	"	Ni	✓
59-3	Chloride removal	Any gas stream		Sph.	55	Alumina	Promoted Al_2O_3	✓
32-4	S removal	"		"	77	None	ZnO	✓
UNITED CATALYSTS INC.								
C11-2, C11-2S	Steam hydrocarbon reforming	Nat. gas-C_3H_3 feed	H_2, 65-72%	Raschig rings	50-55†	Ca Aluminate	Ni	✓
C11-9-01, 02, 03, 04, 06	Composition differs	"	"	Raschig rings	70-95†	Refractory Al_2O_3	Ni	✓
C11-9-09 HGS	"	"	"	"Wheels"	70	"	Ni	✓
G90A, & B, & C	"	"	"	Raschig rings	50-60†	"	Ni	✓
G56A, B, H	"	"	"	Raschig rings	55-60	Ca Aluminate	Ni	✓
C11-9-061	Composition differs	C_4H_{10}-naphtha feeds	H_2, 65-72%	Raschig rings	80-85†	Refractory Al_2O_3	Ni	✓
C11-9-062	"	"	"	Raschig or HGSA	80-85 70 (HGS)	"	Ni	✓
C11-N/C11-NK (split loading)	"	C_4H_{10}-naphtha	"	Raschig ring	67	Refractory	Ni	✓
† Bulk density depends on size of rings and Ni content.								
C7-2	Desulfurization, metal content	NG steam reformer feed	S free streams	Ext.	70	Binder	ZnO	✓
C7-4					80			
G-72D	"	"	"	"	70	"	"	✓
G-1	Conv. of S compounds to H_2S, absorption of H_2S	NG streams	"	Tablets	"P"	"P"	"P"	✓
C12-3-05	High temperature shift conv.	Process gas ex. reformers	CO free gas	Tablets	65	—	Fe/Cr	✓
G3	"	"	"	"	70	—	Fe/Cr	✓
C18 HC	Low temperature shift conv.	Process gas	"	Tablets	80-85	—	Cu	✓
C18-HCS	"	"	"	"	70-80	—	Cu	✓
G66A	"	"	"	Tablets	80	—	Cu	✓
G66B	"	"	"	Tablets	90/85	—	"	✓
C13-3	Methanation (differ in form & metal content)	Process gas	H_2 w/o CO, CO_2	Tablets	65	Refractory	Ni	✓
C13-4		"	"	Sph.	55	"		
C13-5		"	"	"	60	"		

(continued)

OGJ international refining-catalyst compilation—continued

Catalyst designation	Primary differentiating characteristics	Application (feedstock)	Application (product)	Form	Bulk density compacted, lb/cu ft,* g/cc	Carrier, support	Active agents	Availability
C25-2-02	Co/H_2O shift, H_2S containing gas	Process gas, partial oxidation		Ext., tabs	40-47	Al_2O_3	Co Mo	
G65	Co/H_2O shift H_2S containing gas	Process gas, partial oxidation	"	Tab.	Tablets: 65(T)	Al_2O_3	Co Mo	√
G65RS	Withstands elevated temp.			Ext., tab.	58(ex) 60(T)			
G87	Methanation	"	"	Ext.	60 (ex)	"	"	√
G87RS	"			Sph.	55 (S)			
PROCATALYSE								
MT 15	CO Methanation		Pure H_2	Spheres		Al_2O_3	Ni	√

Sweetening catalysts

Catalyst designation	Primary differentiating characteristics	Application (feedstock)	Application (product)	Form	Bulk density	Carrier, support	Active agents	Availability
INSTITUTO MEXICANA DEL PETROLEO								
IMP OM-1	Sweet in homogenous phase	LPG, gaso.		Powder	—	—	Co	L
IMP-OM-2	Sweet in heterogenous phase	Gaso., turbine fuel		"	—	—	"	L
UOP								
MEROX FB*	Active agent dispersed in water	Lt. nap. to diesel	Merox*, fixed bed	Liquid slurry			"P"	L
MEROX WS*	Active agent dissolved in water	C_1 through LSR, thermal, gaso.	Merox*, liq./liq.	Liquid solution			"P"	L
MEROX No. 10*	Active agent supported on activated carbon	Lt. nap. to diesel	Merox*	Granular		Activated carbon	"P"	L
MEROX No. 8*	"	"	"	Granular		Activated carbon	"P"	L
ARI TECHNOLOGIES								
ARI-100, 100 W.S.	Powder (WS-in water soluble bags)	LPG and gaso.	Merox*, Mericat* liq./liq.	Powder		None	"P"	√
ARI-100L	"	"	"	Liquid suspen.		None	"P"	√
ARI-100EXL	"	"	"	Liquid		None	"P"	√
ARI-120	"	LPG, gaso., kero jet	Merox*, fixed bed	Powder		None		√
ARI-120L	"	"	"	Liquid slurry		None		√

Claus unit tail gas treatment catalysts

Catalyst designation	Primary differentiating characteristics	Application (feedstock)	Application (product)	Form	Bulk density	Carrier, support	Active agents	Availability
AKZO CHEMICALS INC.								
KF-124 LD	Co Mo			Cyl.	38	Al_2O_3		√
KF-124 HD	"			Cyl.	48	"		√
KF 165	"			Cyl.	48	"		√
ALCOA								
S-100 1/8-in.	Subdewpoint Claus tail-gas			Sph.	46	—	None	√
CRITERION CATALYSTS Co. LP.								
N-25, N-39	(See Unocal "Claus unit listing")							L
HDS-2		Tail gas		Cyl.	~ 36	Al_2O_3	CoO MoO_3	√
Selectox-33	(See Unocal Claus unit listing)							L
ARI TECHNOLOGIES								
ARI-310	Active ingredient in chelated aqueous solution	Claus tail gas, amine acid gas	Elemental S	Liquid	74.8	None	Iron	L
KAISER CHEMICALS								
S-301, 3 × 6 5 × 8	High porosity, low attrition	Sub-dew point Claus tail gas		Sph.	45	Activated Al_2O_3	Activated Al_2O_3	√
S-201, 3×6, 5×8	Standard high porosity alumina	Sub-dewpoint Claus tail gas		Sph.	45	Activated Al_2O_3	Activated Al_2O_3	
S-701	Promoted sulfur recovery	High CS_2 and COS conv.	TGCU process	Kloverleaf*	52	"P"	"P"	√
PROCATALYSE								
TG 103	Total hydrog. of S compounds to H_2S	Claus tail gas	H_2S	Sph.	0.75*	Al_2O_3	CoMo	√
TG 105	"	"	"	"	"	"	"	√
RHONE POULENC								
A 2.5	Catalysts for Sulfreen	Claus effluent		Sph.	0.77*	Al_2O_3	—	√
CRS 31	Oxidation of H_2S to S	"		Ext.	0.95*	TiO_2	—	√
AM 2.5	Catalyst protection (Sulfreen units)	"		Sph.	0.82*	Al_2O_3	Fe	√
CRITERION CATALYSTS Co. LP.								
S 099	Incineration of Claus, SCOT tailgas	Claus, SCOT tailgas		Sph.	0.78*	Al_2O_3	—	√
S 599	Support function, replaces inert ceramic material.	"		Sph.	0.78*	"	—	√
S 234	Conv. of S compounds† H_2S	Claus tailgas		Cyl.	0.53*	"	Co Mo	√
S 534	"	"		Sph.	0.77*	"	"	√

(continued)

Petroleum Processes, Catalyst Usage

Steam hydrocarbon reforming catalysts—continued

Catalyst designation	Primary differentiating characteristics	Application (feedstock)	Application (product)	Form	Bulk density compacted, lb/cu ft,* g/cc	Carrier, support	Active agents	Availability
UNOCAL CORP.								
Selectox-33	Solid for oxidation of H_2S to S, catalytic incineration of SO_2	Claus tail gas, acid		Cyl.	"P"	"P"	Non-noble metals	L
Selectox-32	"	"		Cyl.	"P"	"P"	"	L
N-25	Hydro., hydrolysis of gaseous S compounds	"		Cyl.	"P"	"P"	"P"	L
N-39	"	"		"	"	"	"	L
N-239	"	"	H_2S	"	"	"	Co Mo	L
UNITED CATALYSTS INC.								
C29-2-02		Claus tail gas		Ext.	37	Al_2O_3	Co Mo	√
C29-2-03	Stabilized carrier (SC)	"		Ext.	"	"	"	√
C29-2-04	SC plus lower metals	"		"	"	"	"	√
C29-3-03	SC plus spherical form (SF)	"		Sph.	55	"	"	√
C29-3-04	SC, SF, plus lower metals	"		"	50	"	"	√
C29-4-03	Composition, SC, SF	"		"	55	"	Ni Mo	√
N-239	(See Unocal Corp., Claus unit Listing)							L
G-41P	COS hydrolysis	High CS_2, COS conv.		Ext.	34-38	Al_2O_3	"P"	√

Other refining catalysts

Catalyst designation	Primary differentiating characteristics	Application (feedstock)	Application (product)	Form	Bulk density compacted, lb/cu ft,* g/cc	Carrier, support	Active agents	Availability
CRITERION CATALYSTS Co. LP.								
TSR-10	(See Unocal "other listing")							
SOAR-100	"							
Bayer AG OC-1038	Hydrog. & etnerification, condensation, dehydrolization	Methanol & C_5 cracked fraction acetone	Tame MIBK	Beads	1.2*	Polystyrene DVB copolymerisate	Sulfonic acid groups, Pd	L
AC 10FT	Oligomerization	Isobutene	Di & tri-isobutene	Powder	1.3*	"	"	√
BP								
CDW-12	Producing specialty and lube oils	Dist. raffinates	High quality specialty, lube oil	Cyl.		Zeolite	Noble metal	L
CDW 14	Improving low temp. properties of mid. dist.	Kero. gas oils oils	Jet, kero. diesel, heating oils	Cyl.		Zeolite	Noble metal	L
FF-62	Lube oil color improvement	Solvent refined lube oils	Finished lube oils	Sph.		"P"	"P"	L
CHEVRON RESEARCH CO.								
GC-500	Selective arsine removal	Lt. HC gas	Arsine free feed	Cyl.	34-35	Al_2O_3	"P"	L
GC-501	"	"	"	"	"	"	"	L
ENGELHARD CORP.								
HZ-1*	Gas oil cracking, Houdriflow TTC. HZ-1 is octane catalyst	HGO	Olefins, LPG, LCO	Ext.	0.95*	In situ zeolites	Zeo./No RE	√
HZ-PLUS*	High attrition resistant	"	"	Ext.	0.95*	"	"	√
EMCAT*-100	RE exchanged version, max. liquid yield. HCC, TOC xylenes plus	HGO; toluene, mixed C_9 transalky	LPG, gaso. LCO xylene, benezene	Ext.	0.95*	In situ zeolites	Zeo. RE exchanged	√
EMCAT*-80	Lower activity version	"	"	Ext.	0.95*	"	"	√
HZ-PLUS P*, EMCAT*-100P and	Prepromoted products for max. CO burning.	HGO cracking	LPG, gaso. LCO	Ext.	0.95*	In situ zeolites	Zeo. No RE HZ-P, RE exch. (EMCAT)	√
HRD-264	Vapor phase S removal	Naphtha reformer	S-feed naphtha	Ext.	70	Al_2O_3	"P"	√
HALDOR TOPSOE								
HTZ-3	Fine desulfur. of steam ref. feed reforming feed stocks	Methane, LPG, naphtha, ref. gas	S-free hydrocarbon	Ext.	82		ZnO	√
HTZ-5	" low temp., low S	"	"		62		"	√
SK-12	High temp. CO conv.	Steam ref. effluent	Gas with 2-3% CO	Cyl.	65		Fe/Cr oxides	√
LSK	Low temp. CO conv., regenerable, chlorine resistant	HT-shift effluent	Gas with 0.1-0.3% CO	Cyl.	66		Zn, CR oxides	√
LK-801	Low temp. CO conv., S resistant	"	"	Cyl.	64		Cu, Zn, Al oxides	√
PK-5	Methanation	H_2 w/o CO, CO_2 containing H_2 with up to 2% contaminants		Rings	39	Al_2O_3	Ni	√
ICI Catalysts								
32-4	Desulf. of steam ref. feed	CH_4, LPG naphtha ref. gas	S-free hydrocarbons	Sph.	1.08*	None	Zno	√
75-1	Cold desulf. of hydrocarbons	NG, LPG, naphtha, kero., H_2 off gas, cat. ref	"	Sph.	0.85*	None	Zno	√
59-3	HCl removal	NG, H_2 off gas, cat. ref.	Chloride-free hydrocarbons	Sph.	0.85*	Al_2O_3	Activated	√

(continued)

Petroleum Processes, Catalyst Usage

Other refining catalysts—continued

Catalyst designation	Primary differentiating characteristics	Application (feedstock)	Application (product)	Form	Bulk density compacted, lb/cu ft,* g/cc	Carrier, support	Active agents	Availability
INSTITUTO MEXICANO DEL PETROLEO								
IMP-TPC-1	Cumene production	Benzene, propane propylene	Cumene	Cyl.	0.86-0.95*	—	Fe, K, oxides	L
IMP-ES-1	Styrene production	Ethylbenzene water vapor	Styrene	Cyl.	1.0-1.4*	—	Fe, K, oxides	L
KALI-CHEMIE AG								
KC PERL-KATOR® series	Moving bed cat. cracking in TCC, Houdriflow	Gas oil cracking		Beads		SiO_2/Al_2O_3	Acid sites	L
D 1	Hvy. feeds, gaso. yields secondary	HGO/VGO	C_3/C_4 olefins	Bead	0.80*	SiO_2, Al_2O_3	Amorphous aluminosilicate	√
D 8	Improved yield selectivity more gasoline, less avg gas & coke	"	Gaso. dist.	"	0.90*	"	Stab. zeolites	√
D 9A	High act. version imp. stability	"	"	"	"	"	"	√
D 10A	Max. Co burning in kiln gas	"	"	"	"	"	Stab. zeolites, CO combustion promoter	√
D 11	High octane	HGO/VGO	High oct. gaso.	"	"	"	Stab. zeo.	√
D 12	High gaso. selectivity	"	High oct. gaso., low low coke yields	"	"	"	"	√
D 14	"	"	High oct. gaso.	"	"	"	"	√
XP-catalyst	Xylenes-plus process	Transalkylation, disproportionation toluene	Xylenes/benzene	"	"	"	RE exchanged zeo.	L
Beadcat PK series	Xylenes Isom.	Xylenes	p, o-xylene	"	0.8*	"	Acid sites	√
PROCATALYSE								
RS 103	Oxidation of S compounds	$S-H_2S$, COS-CS_2, RSH	SO_2	Sph.		Al_2O_3	Metallic oxides	√
DN 281	NOx reduction	Industrial gas	N_2	Sph.		"	Metallic oxides	√
FF 62	Lube oil hydrofinishing	Extracted lube	Finished base oil	Sph.		"		L
HR 346	Wax hydrotreating	Slack wax	Food-grade wax	Cyl.		"	Ni-Mo	√
CRITERION CATALYSTS Co. LP.								
S 538	CO/H_2O shift reactions	Treatment of syngas		Sph.	0.82*	Al_2O_3	Co Mo	√
UNOCAL CORP.								
TSR-10	Trace S removal	Naphtha	Cat. ref. feed	Sh. ext.	"P"	"P"	"P"	L
TSR-11	"	"	"	"	"P"	"P"	"P"	L
SOAR-100	Arsenic removal	Shale oil, coal liq.	Low arsenic prod.	Sh. Ext.	"P"	"P"	"P"	L
UNITED CATALYSTS								
C53-2-01	COS hydrolysis	FCC P-P gas		Ext.	28-32	Al_2O_3	PTS	√
G-132	Arsine removal	"		Tab	75-85		Cu	√
C28-1-01	S removal	Naphthas, cat. ref. feed		Ext.	45-50	"P"	"P"	√
TSR-11	(See Unocal "other" listing)							L
C125-1-01	HCl removal	H_2 offgas cat. ref.		Ext.	47-53	Zn	Ca	√
G-92	"	"		Sph.	49-55	Al_2O_3	Na	√
UOP								
TA-4	Transalkylation	Toluene and C_9 aro.	Benzene, xylenes	Cyl.	45	"P"	"P"	Contact UOP
DEH-5	Paraffin dehydro.	C_6-C_{20} paraffins	C_6-C_{20} monoolefins	Sph.	21	Al_2O_3	Pt	"
DEH-6	Light paraffin dehydrog.-	C_3-C_5 paraffins	C_3-C_5 monoolefins	"	21	"	"P"	"
DEH-7	Paraffin dehydro.	C_6-C_{20} n-paraffins	C_6-C_{20} monoolefins	"	20	"	"P"	"
A-2	Alkyl. of benzene with ethylene	Ethylene, benzene	Ethylbenzene	Sph.	35	Al_2O_3	"P"	"

Methyl tertiary butyl ether (MTBE) catalysts

Catalyst designation	Primary differentiating characteristics	Application (feedstock)	Application (product)	Form	Bulk density compacted, lb/cu ft,* g/cc	Carrier, support	Active agents	Availability
BAYER AG								
SPC 118	Etherification	Methanol, C_4 raffinates		Beads	1.2*	Polystyrene, DVB CO-Polymerisate	Sulfanic acid groups	L

Combustion Promotors (FCC)

Catalyst designation	Primary distinguishing characteristics	Application	Additive to or component of catalyst	Form	Carrier, support	Active agents	Available
AKZO CHEMICALS INC.							
Ketjen Oxycat Series							
KOC-50	High oct., stab. attrit. resist.	CO combustion	Add. or blend	Powder	Al_2O_3	Pt	√
KOC-10	"	"	"	"	"	"	√
Ketjen Desox-3							
KD SOx-3	Attrition resist.	SOx-trans.	Additive	"	"	Pt	√

(continued)

Combustion Promotors (FCC)—continued

Catalyst designation	Primary distinguishing characteristics	Application	Additive to or component of catalyst	Form	Carrier, support	Active agents	Available
AMBUR CHEMICAL CO. INC.							
CCA Brand	Combination of noble metals on active alumina	CO combustion	Additive	MS	High purity Al_2O_3	Pt, Pd	
CCA-1#	Highest activity	Partial, complete CO comb.	"	"	"	"	√
CCA-5+*	High activity	"	"	"	"	"	√
CCA-6#	High activity/improved support	"	MS	"	"	Al_2O_3	√
CCA-8#	High act. imp. support	"	"	"	0.97*	Al_2O_3	√
CATALYSTS & CHEMICALS INDUSTRIES CO. LTD.							
SP-10S	CO combustion solid promoter	FCC feed	Additive	Sph.	Al_2O_3	"P"	√
DAVISON CHEMICAL DIVISION, W.R. GRACE & CO.							
CP-3	High activity, stability	Partial and complete CO combustion	Additive	MS	"P"	"P"	√
ENGELHARD CORP.							
COCAT*-1	Low activity	Partial CO comb.	Additive	MS	SiO_2/Al_2O_3	Precious metal	√
COCAT*-5/ PROCAT 500	High activity	Full CO comb.	Additive	"	"	"	√
COCAT*-7/ PROCAT 700	High act., imp. support	Full CO combustion	Additive	"	"	"	√
INTERCAT CORP.							
COP-850	Pt. content	CO combustion	Additive	MS	Al_2O_3	Pt	√
COP-550	"	"	"	"	"	"	
KATALISTIKS INTERNATIONAL INC.							
KCP-8	High activity comb. CO	Partial, full CO comb. CO Add. or blend		MS	"P"	Precious metals	√
KCP-5	Medium activity	"	"	0.80 (ABD)	"P"	"	√
UOP							
UNICAT* Cl-3	Active metal	CO combustion	Additive	Powder	Al_2O_3, Al_2O_3/SiO_2	Pt	√

W.R. GRACE CATALYSTS PTY. LTD. (See Davison Chemical Division Co. combustion listing)

Sulfur oxides reduction catalyst (FCC)

Catalyst designation	Characteristics	Form	How introduced to FCC system	
CHEVRON RESEARCH CO.				
Transcat	Fully fluidizable powder	Added to FCC cat.	Separate add. to FCC	
DAVISON CHEMICAL DIV. W. R. GRACE				
Additive R	Additive to reduce regenerator flue gas sox	Additive to FCC cat.	Via separate addition	
ENGELHARD CORP.				
ULTRASOX* MAGNASIV* 370S	Typical FCC particle	Typical FCC catalyst	As fresh makeup	√
W.R. GRACE CATALYSTS PTY. LTD. (See Davison Chemical Div. Sox listing)				
KATALISTIKS INTERNATIONAL INC.				
DESOX	Removes SOx from FCC flue gas, releases as H_2S	MS	Added or blended with FCC catalyst.	

The information shown is the supplier's best effort to meet the objectives of the compilation. The blanks that appear in parts of the table reflect the supplier's response, and are not there because the information requested is not applicable or available. The symbol "P" appears occasionally to signify some proprietary aspect of a catalyst that the supplier chose not to divulge.

The scope of the compilation includes all catalysts that are available by sale or license to refiners. Those that are available for general sale are denoted by a check mark in the far right column. Those that are only available through licensing arrangements are denoted by an L in the column. Custom catalysts for proprietary processes, unavailable to the general industry by themselves, are not listed.

The compilation is broken down alphabetically into the various refining processes, and further alphabetically by supplier. The nature of some catalyst types causes them to appear in more than one processing catagory. And in some cases, more than one supplier may list the same catalyst because of licensing or sales agreements.

Liberal abbreviations (see page 87) are used to try to include as much information as possible in the limited space of the compilation.

The compilation is designed to provide a ready reference for both refiners and manufacturers and to sort out the sometimes-confusing nomenclature used to designate or label them and to reveal the current catalysts available for various processes and for specific objectives.

This material appeared in *Oil & Gas Journal*, pp. 41 ff., October 5, 1987, copyright © 1987 by Pennwell Publishing Co., Tulsa, Oklahoma 74121, and is reprinted by special permission. The *Oil & Gas Journal* updates this information each year (occasionally every second year). The reader should contact the *Oil & Gas Journal* for the latest information.

RICHARD A. CORBETT

Petroleum Processing Economics, Catalysts

Catalyst expenditures are a minute fraction of the total operating costs of a refinery. But the essential role catalysts play in refining gives them high leverage on operating profits.

Therefore, catalysts should be selected primarily on a performance basis. To assess the performance, an integrative analysis of the benefits and costs should be done, especially in conversion processes with a high upgrading margin. These processes play a key role in the refinery blending schemes.

General hydrotreating catalysts should be selected on a price/performance basis, with special attention paid to mechanical strength and regenerability.

Refiners have faced a turbulent era characterized by overcapacity, strong competition, and refinery closures. Recent large swings in crude prices have added to the economic difficulties.

Catalyst manufacturers have faced similar competitive market conditions. Catalyst production overcapacity, caused by overly optimistic consumption projections in the late 1970s, has led to fierce competition and price erosion.

This increased economic pressure on both refiners and catalyst manufacturers has driven the industry to greater cost effectiveness. As a result, the search for better, more cost-efficient catalysts and processes has been stimulated.

Shell has examined the effects of catalyst selection on refinery economics. Illustrated here are developments in catalyst performance and highlights of the scope of economic improvements in current refinery operations achievable through the judicious choice of catalysts.

Examples given have been selected on the basis that no major capital investments are required. To depict realistic economics of an average-size European refinery of today, we have assumed the recent base product prices shown in Table 1. These data are used throughout this article to achieve consistency.

Catalyst Functions and Costs

Essential catalyst functions in the refinery comprise pretreatment and upgrading of various oil fractions, pollution abatement, and energy conservation.

TABLE 1 Product Price Bases[a]

	$/metric ton
Premium mogas	175
Regular mogas	157
Naphtha	138
Kerosene	182
Gas oil	165
Fuel oil, 1% sulfur	93
Fuel oil, 3.5% sulfur	78

[a]Source: Platts barges, Rotterdam, average 1st half 1986.

Petroleum Processing Economics, Catalysts

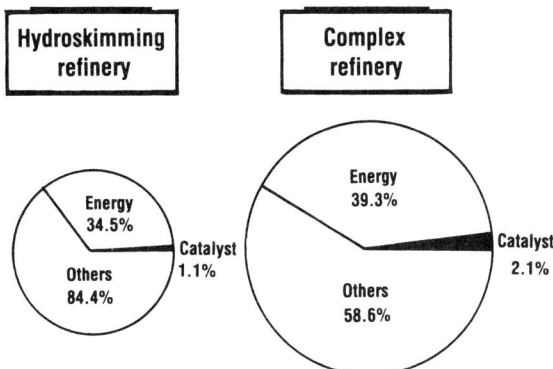

FIG. 1. Cost of catalyst in refining. Crude distillation capacity 15,000 metric tons/SD, 70% utilization, refinery fuel $80/metric ton, excluding depreciation.

It is common practice to attribute added value in refining only to conversion processes. Although pretreatment processes (hydrotreating) are crucial, they are generally considered cost centers, and, as such, fall into the same category as sulfur recovery and pollution abatement.

Cases in this article refer mainly to the main money-making catalytic processes such as catalytic cracking, hydrocracking, and residue conversion.

In great contrast to the crucial role of catalysts in the refinery, the catalyst costs involved are relatively small, if not minute. For a typical European hydroskimming refinery, catalyst expenditures are estimated at 1% of the operating costs (excluding depreciation), or about $0.5 million/year. Fuel costs, assuming a fuel price of $80/metric ton of refinery fuel, amount to about one-third of the operating costs. The remaining two-thirds relate to manpower, fixed costs, chemicals, and overheads (Fig. 1).

However, in a complex refinery, FCC and/or HC operating costs are appreciably higher. A relative shift to higher catalyst and energy costs is a matter of course.

The preceding cost perspective suggests that, in economic terms, catalyst performance can have a high leverage on refinery performance as a whole. In particular, this must be the case for upgrading/conversion processes, yielding high-added-value products.

Catalytic Reforming

Table 2 reveals the advances in catalytic reforming catalysts, using a 3000 metric ton/day, semiregenerative unit as the basis for comparison.

The results at 95° research octane (RON) indicate a substantial gain for the low-density bimetallic catalyst. With the high-density catalyst, the higher product yield is offset by a higher product density, and the increased catalyst costs result in a small overall loss.

Thus, the use of high-density bimetallic catalyst to produce a 95 RON product appears to be overkill.

TABLE 2 Advantages of Modern Catalysts in Reforming

Basis: Semiregenerative unit
No equipment limitations
Constant fuel consumption for constant RON-0

Intake:	3 000 metric ton/day
Catalyst volume:	110 m^3
Cycle length:	1 year
Cat replacement:	Once every 7 years
Naphtha quality:	P/N/A: 50/35/15 wt.%[a]
C$_5$+ yield:	Middle of cycle

	Catalyst Type		
	Monometallic	Bimetallic Low Density	Bimetallic High Density
Year	1965	1975	1985
Relative stability, stream-days	1	2	4
Catalyst replacement cost, $/metric ton of feed	0.157	0.160	0.281
RON clear = 95			
C$_5$+ yield, wt.% on feed	83.6	85.5	86.5
Density, D15/4	0.782	0.788	0.792
Gross process margin, $/metric ton	Base	+0.45	+0.37
RON clear = 100			
C$_5$+ yield, wt.% on feed	—	78.8	81.5
Density, D15/4	—	0.797	0.803
Gross process margin, $/metric ton	—	Base	+1.30

[a]Paraffins, naphthenes, aromatics.

For 100 RON, however, this type of catalyst is a clear winner with an annual extra margin of about $1.3 million. By spending an additional $0.12 for the catalyst, an extra margin of $1.3/metric ton of feed may be cashed in. This represents a leverage of more than ten to one. It may be noted that the use of monometallic catalyst for the highest octane product is not practical.

This example illustrates that, by proper catalyst selection, substantial gains can be made. Comparing the additional upgrading margin with the extra catalyst costs for the 100 RON case, it would seem economically foolhardy to persist in the use of any other than the best-performing catalyst.

The use of margin differentials may give the impression that catalyst costs are high in catalytic reforming. For a good understanding, it should be noted that the catalyst costs in the examples amount to only 0.6–1.0% of the gross processing margin.

Fluid Catalytic Cracking

Currently, all FCC plants use zeolite catalysts, with some exceptions in lesser-developed countries.

Petroleum Processing Economics, Catalysts

TABLE 3 Developments in Fluid Cat Cracking Performance

	Catalyst Type and Year			
	Low Alumina, 1950s	High Alumina, 1960s	General Purpose	
			Zeolitic, 1970s	Catalysts, 1980s
Yields, wt.% on feed				
C_1–C_2	4	3	3	2
C_3–C_4	6	10	12	13
Gasoline	33	35	45	53
Light cycle oil	22	21	20	21
Heavy cycle oil/slurry	28	24	14	5
Coke	7	7	6	6
Gasoline RON	93	93	91	92
Fuel consumption, % on feed	6.9	7.2	7.5	7.9
Catalyst consumption, metric ton/1000 metric ton feed	1.6	1.4	0.7	0.65
Catalyst costs (Europe 1986), $/metric ton	1000	1000	1000	1000
Catalyst costs, $/metric ton of feed	1.60	1.4	0.70	0.65
Gross process margin, $/metric ton of feed	Base	+1.4	+6.0	+12.8

The continuous improvement in FCC catalysts can best be demonstrated by the milestones achieved in their development.

The second example (Table 3) reviews the developments in FCC performance, and the economic consequences from a catalyst perspective. It goes without saying that with the advent of zeolite catalysts displaying an order of magnitude higher activity than amorphous materials, changes in reactor design were called for.

The improvements indicated in the example reflect the combined effects of superior catalysts and reactor optimizations tailored around them. The data refer to a typical FCC unit of 4000 metric ton/SD, running on waxy distillate feedstock with a boiling range of 360–540°C, and a specific gravity of 0.91.

A dramatic improvement in gasoline yield and concurrently higher selectivities (lower gas make and lower heavy cycle oil yields) are evident. The zeolitic catalysts may be operated at a considerably lower coke make than the amorphous catalysts. Alternatively, heavier feeds may be processed while achieving the same coke make.

To allow a straightforward comparison in our example, the feed quality and coke figures have been kept constant. A dramatic decrease in catalyst consumption can be noted.

The gasoline quality tends to be somewhat lower with the general-purpose zeolite catalysts. Recent developments, not included here, feature octane-enhancing ZSM-5 or USY-type zeolite catalysts.

These catalysts produce gasolines richer in olefins and, to some extent, aromatics. This may result in RON improvements of several points, while a modest increase in motor octane number (MON) may also be achieved [2]. However, some yield loss may be incurred with such catalysts.

Fuel consumption figures indicate a small increase with time. This is mainly related to the higher energy losses in the product fractionation section through higher gasoline production. It should be noted that the fuel consumption figures (Table 3) do not reflect any energy-saving measures which have been implemented in most units in recent years.

It can be concluded that catalyst unit costs have decreased substantially over the years through the application of more active and stable catalysts. Furthermore, maturing of the FCC process technology has resulted in greatly improved gross process margins.

Without going into details, further improvements in the overall economics may be achieved by increasing the feed heaviness, for instance, by inclusion of residual material while accepting higher catalyst consumptions [3].

Companies often regard the purchase of catalyst as an out-of-pocket expense, isolated from the technical and economic benefits it may or may not imply. This often goes hand in hand with a conservative approach to catalyst change-outs.

Clearly, the implication of catalyst selection in conversion processes may be enormous, and an integrated approach is necessary to assess costs and benefits rather than opting for the cheapest or lowest risk solution. An evaluation of potential plant constraints, such as in reactor design, reactor section heat balance, and possible consequences in product blending schemes, is normally required.

The major responsibility for making such analyses is vested in the refiner, and, if need be, assisted by process specialists. On the other hand, a catalyst company with the proper resources can stimulate the application of better products by emphasizing the integrative approach, and the overall economic benefit. The next example illustrates such an integrated approach.

Mild Hydrocracking

Mild hydrocracking (MHC) of vacuum gas oil (VGO) produces lighter products, mainly in the kerosine and diesel range, as well as minor amounts of gasoline and gas. As such, MHC has a much higher selectivity toward middle distillates than FCC. The quality of the various MHC product fractions makes them valuable blending components in a refinery.

Process requirements for MHC are typically within the range of existing VGO desulfurization units and FCC feed pretreaters. The main differences are in catalyst choice and operating temperatures.

MHC is favored by application of maximum operating temperatures throughout the catalyst life cycle. The other processes are normally run in a temperature-programmed way to maximize the catalyst life cycle and to achieve the process targets in terms of sulfur and nitrogen specifications of the product.

Often, MHC can be carried out in existing units with minor modifications in the product work-up section. This renders MHC a very attractive process for the production of additional middle distillates of good quality, with a short pay-out time.

Usually profits are maximized by applying the most severe operating conditions feasible within the equipment constraints in reactor and product fractionating section. Commercial experience indicates that the increased catalyst turnover in such operations is readily paid back.

Much less considered benefits of MHC are the attractiveness of the unconverted MHC bottoms as FCC feed, and the flexibility it can achieve in tandem with an FCC unit with regard to product slates. To illustrate the benefits, two process configurations are compared in Fig. 2.

Scheme 1 (Fig. 2, top) represents the cat feed hydrotreatment (CFH) mode of operation. In the CFH unit, the VGO feed is hydrotreated to improve the FCC feed quality, and the complete product is transferred into the FCC unit. In Scheme 2 (Fig. 2, bottom) an extra distillation step is included downstream of the CFH/MHC unit to increase the middle distillates yield.

The comparison is taken from an actual pilot-plant study on Shell S-424 which was later applied as an MHC catalyst in a commercial unit running on a mixed VGO of Middle East origin. MHC conditions, as well as properties of feed and MHC 370°C+ product fractions, are given in Table 4 for various operating severities.

A large reduction in the nitrogen content, the density, the viscosity, and the Ramsbottom carbon test (RCT) is found. These improvements give rise to a better FCC operation.

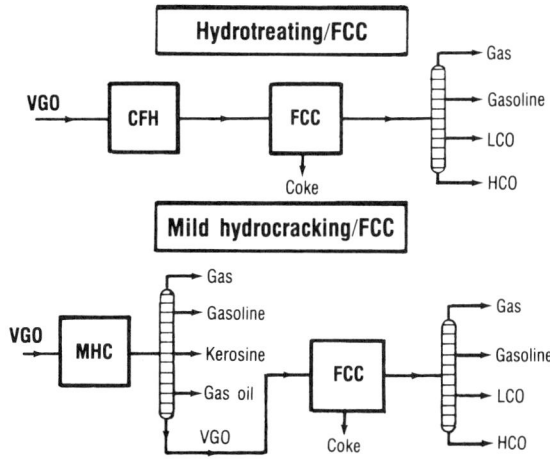

FIG. 2. Process flow schemes.

TABLE 4 Properties of VGO Fractions via MHC with Shell 424 Catalyst

Product	Feed	A	B	C
Conditions:				
Temperature, °C		T	$T+14$	$T+29$
Pressure, bar		62	62	62
Weight hourly space velocity, h^{-1}			Constant	
H_2/feed, Nm^3/t		325	325	325
Yield of 370°C+ product, wt.% on feed	78.5	66.8	59.5	53.1
Net conversion to 370°C−, wt.%		15.1	24.3	32.5
Physical properties:				
Sulfur, wt.%	1.99	0.07	0.05	0.03
Nitrogen (basic), ppm wt	493	34	39	11
Density 70/4, kg/L	0.888	0.855	0.853	0.852
Viscosity, 60°C, cSt	32.1	21.4	19.1	16.7
RCT, wt.%	0.28	0.11	0.12	0.27
Aniline point, °C	94.2	95.8	96.0	94.8

At constant coke make in the FCC unit, 6% on intake has been assumed, the gasoline yield increases by more than 10 percentage points on FCC intake going from untreated VGO to the MHC bottom products for condition A. At the higher MHC conversion levels (B and C), the improvement is slightly less, owing to the increase in RCT with higher MHC temperatures. The integrated yield picture of the MHC-FCC combination is depicted in Fig. 3.

For the gas yield, the C_4-fractions produced in both MHC and FCC units have been taken together. Gasoline is a composite of C_5 to 220°C material from the FCC and C_5 to 180°C from the MHC unit.

The middle distillates are built up from the 180–370°C fraction from the MHC and light cycle oil (220–370°C) from the FCC unit. Heavy cycle oil (HCO) is the oil fraction boiling above 370°C from the FCC unit.

Figure 3 indicates that, by increasing the conversion level in MHC, the total gasoline production can be reduced, while the middle distillate production increases. The loss in gasoline yield from FCC is mainly caused by the lower amount of 370°C+ feed from the MHC unit, and could be compensated for by processing a deeper-cut VGO or coprocessing of some residual material.

By adjustment of the conversion level in MHC, the product slate of the refinery can be steered to meet the requirements at any period in time, allowing great flexibility in operation. Clearly, mild hydrocracking with optimized catalysts and conditions provides a greater bonus and more flexibility to the refiner than is apparent from a study of the MHC process alone.

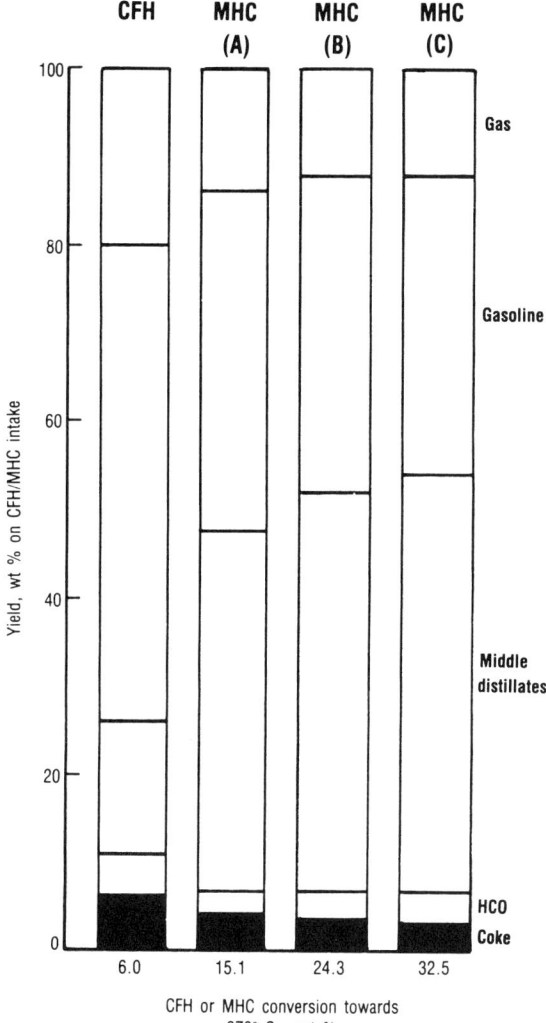

FIG. 3. Integrated yield of MHC-FCC combination.

Hydrocracking

Hydrocracking is being increasingly applied as a conversion process because of its product flexibility and the ability to produce high-quality middle distillates. Current trends are the use of increasingly heavy feedstocks and the desire to maximize middle distillate yield.

By using judicious combinations of zeolitic and amorphous cracking functions, new high-performance hydrocracking catalysts have been designed. The improvements in selectivity that are possible with the new materials are illustrated in Table 5.

TABLE 5 Improvements in Hydrocracking Selectivity

	Catalyst	
	A	B
Yields: wt.% on feed		
H_2 consumption	3.2	3.2
Gas (C_1–C_2)	2.6	1.4
LPG (C_3–C_4)	8.4	4.6
Gasoline:	16	10
RON	84	82
Naphtha:	30	20
PNA, wt.%	39/53/8	40/53/7
Density	0.75	0.75
Kerosene:	43	64
Density	0.79	0.79
Smoke point, mm	22	24
Catalyst costs, $1000/year	600	900
Upgrading margin, $ million/year	Base	Base + 9.2

The data relate to a two-stage Shell hydrocracker operated on Middle East VGO and aimed to produce maximum kerosene yield. A marked increase in kerosene output is evident going from Catalyst A to B, while production of gas, gasoline, and naphtha is reduced accordingly.

With the product values indicated in Table 1, a net upgrading margin differential of $9.30/metric ton results. For a medium-size hydrocracker of 3000 metric tons/day, this represents an increase in upgrading margin of the formidable sum of $9.2 million/year. The increased catalyst costs are included in this figure. Additional benefits such as the slight reduction in hydrogen and fuel consumption are not included.

It is important to note that the higher purchase cost of the sophisticated MD-selective Catalyst B represents only 3% of the added upgrading margin. This clearly illustrates the high leverage of catalysts on the overall performance of the unit. Furthermore, it is another clear example indicating that the selection of a catalyst should be on a performance basis.

A further way to exploit the high-performance HC Catalyst B is to make use of its higher activity. This quality, together with the substantial decrease in gas make, may render possible a substantially higher throughput of an existing hydrocracker. Those knowledgeable in the field will appreciate the large economic benefits this may offer.

In recent years there have been large seasonal fluctuations in product demands and prices (Fig. 4). With a modern high-performance catalyst, the hydrocracker product out-turn can be adjusted readily to accommodate demands by adapting the process conditions.

The range of yield flexibilities that is feasible without changing the catalyst is indicated in Table 6. The naphtha yield may be varied from 25 to 71 wt.% on feed, and the kerosene plus gas oil yield from zero to 61 wt.% on feed.

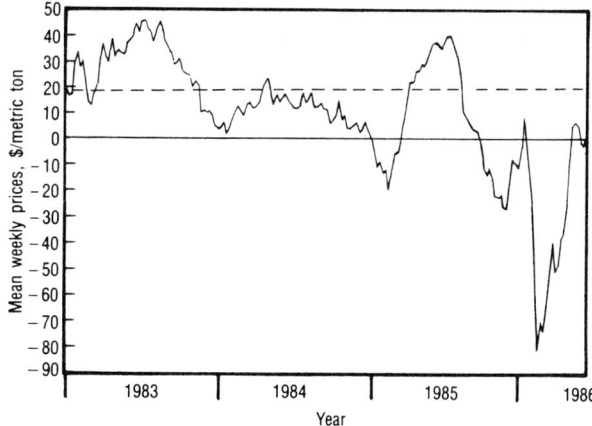

FIG. 4. Naphtha–gas oil differential. Platts barges, Rotterdam.

The economic incentives are illustrated in Fig. 5, which shows the differential upgrading margin, including hydrogen and fuel costs, between naphtha and distillate modes.

It is indicated that at a naphtha price of at least $18/metric ton above the gas oil price, it will be attractive to operate in the naphtha mode. Going back to Fig. 4, it can be seen that during the past few years this would have been the case in the summers of 1983 and 1985.

TABLE 6 Hydrocracking Catalyst Flexibility[a]

	Mode of Operation	
	Maximum Naphtha	Maximum Distillates
Yields, wt.% on feed:		
H_2 consumption	3.2	2.4
H_2S	1.6	1.6
C_{1+2}	0.4	0.3
C_{3+4}	12.1	5.7
Gasoline	18.2	8.0
Naphtha:	70.9	25.4
PNA, wt.%	39/52/9	40/52/8
Density	0.75	0.75
Kerosene:	—	49.9
Density		0.80
Smoke point, mm		24
Gas oil:	—	11.5
Density		83
Cetane number		52
Fuel consumption	6.3	6.0

[a] 3600 metric ton/SD, two-stage unit.

Product price levels are difficult to predict. Hence, the ability to adjust the product out-turn in the course of a run may create an interesting economic opportunity for the refinery.

It is this yield flexibility combined with high selectivity to the desired (low-sulfur) products and significant volumetric yield gain (up to 28%) that makes hydrocracking an important process in the modern refinery for upgrading of heavy oil fractions, and catalyst selection the key to optimum economics.

Residue Hydroprocessing

Some 10 years ago, when quite a number of residue-processing units were built, the prime objective was sulfur removal at minimum hydrogen consumption. Conversion was not considered of interest.

Over the years this has changed. Sulfur premium on low-S fuel has diminished in significance with the growing realization that the conversion accompanying HDS is the greater economic driving force. The higher conversion is achieved readily by running at higher temperatures and/or reducing the throughput within the unit constraints.

Higher severity generally results in a more rapid catalyst deactivation, especially with metal-containing residual feeds. As a result of the higher reaction temperatures, the rate of metal deposition on the catalyst is increased in the front end of the reactor train. Metals deposition on the catalyst in the tail end of the train tends to be less at the higher severities simply because they have already been removed and deposited, to a large extent, upstream.

Because poisoning by metals is reduced in the tail end of the former HDS unit, catalyst deactivation by coke laydown has become the predominant factor. Thus, we see that the front end gets more metals and the tail end more coke.

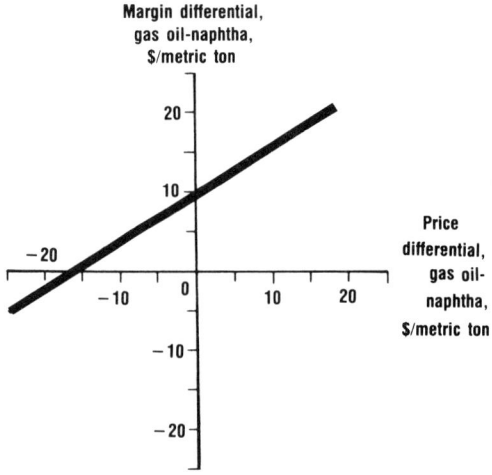

FIG. 5. HC upgrading differential in the naphtha/gas oil mode.

TABLE 7 Catalyst Comparison for HDS[a]

	Catalysts	
	Desulfurization Mode Combination A	Conversion Mode Combination B
Sulfur removal, %	88	91
Yields, wt.% on long residue:		
H_2 consumed	1.2	1.4
C_5–350°C distillate	13	19
350–550°C FCC feed	62	59
550°C+ residue	22	18
Upgrading margin, $/metric ton	Base	+3.1

[a]Feed: 4000 metric ton/day Kuwait long residue, 1-year cycle length.

This shift in duties calls for a new combination of catalytic functions to achieve an economically attractive run length of, say, 1 year. During this period, every catalyst in the system should be exploited to its maximum capability.

Shell has developed two new catalysts in the series of residue hydroconversion catalysts that reflect these needs. They feature improved hydrogenation activity and coke resistance and combine perfectly with the Shell hydrodemetallization catalyst up front.

The new combination allows deeper conversion at practically the same purchase price as the former combination. A comparison of yield data with the old and new catalyst systems on an annual average basis is given in Table 7.

In the conversion mode, some 4 wt.% less unconverted vacuum residue remains. Taking into account the fuel consumption, an extra upgrading of $3/metric tons of atmospheric residue feed results. This corresponds to a net benefit of $4 million/year for a 4000 metric ton/day unit.

This is another illustration of the fact that, through the application of optimum catalysts, substantial profits can be obtained, provided the catalyst system is fully exploited rather than spared.

General HT, HDS

Amorphous Al_2O_3-based hydrotreating (HT) and hydrodesulfurization (HDS) catalysts should also be mentioned. The main qualities sought in these workhorses, present in every refinery, are high activity/stability, sturdiness, and regenerability. Heavier burdens are put on these catalysts with ever-increasing quantities of cracked materials in their feed diet.

It is interesting to note that, contrary to the doubling in activity during the 1970s, the present price in real terms has dropped slightly below that of 1975

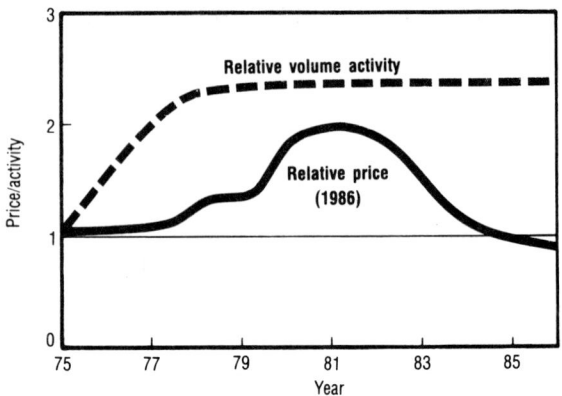

FIG. 6. Price and activity developments. Co/Mo/A 1203 catalysts.

(Fig. 6). The price increase toward the end of the 1970s was mainly due to increased metal costs.

Despite continuing high metal costs, a clear erosion of catalyst prices is evident, illustrating the strong competition between the catalyst suppliers. Substantial further price reductions should not be expected for the near future.

The competitive state of the general HT/HDS catalyst business, at present, may tempt refineries into a predominantly price buying policy. Caution is recommended, however, since catalyst quality varies with regard to mechanical strength and regenerability.

The best approach is to buy a top-grade catalyst from a renowned catalyst supplier at a competitive strength, based on regenerability of the catalyst and technical service capabilities of the supplier.

The workhorse thus selected may start its active life in a high-performance duty and be reused several times, while cascading it into less severe duties in the refinery when it is growing in age. Various companies offer high-quality, off-site regeneration services which enable the refinery to carry through repeated use of a top-grade, fresh material.

Future Developments

Catalyst manufacture is developing from an art into a science of making high-quality, value-adding, cost-efficient catalysts. At present, much catalyst research is carried out in areas of high-margin conversion processes.

A major future role is envisaged for zeolitic catalysts which, through their structure, render possible a large (shape) selectivity effect unprecedented in amorphous catalyst carriers. The number of formulations possible on zeolite-based materials is extremely large.

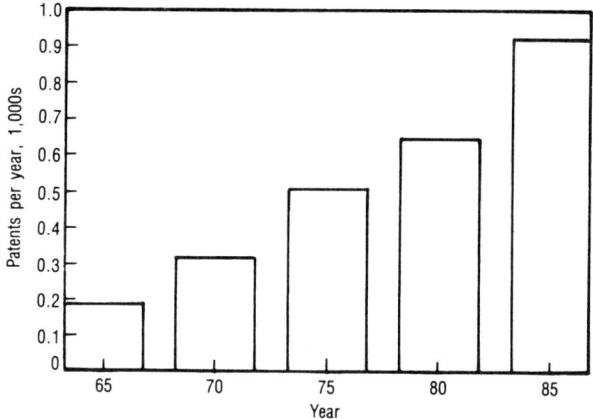

FIG. 7. Zeolite application patents. First or primary patent application.

Moreover, the research effort to develop new materials has been stepped up considerably in recent years, as illustrated by the growth in patent applications presented in Fig. 7. The trend toward zeolite-containing catalysts that is evident in processes such as FCC and HC will continue.

This material appeared in *Oil & Gas Journal*, pp. 55 ff., February 16, 1987, copyright © 1987 by Pennwell Publishing Co., Tulsa, Oklahoma 74121, and is reprinted by special permission.

References

1. Paper AM-86-58, National Petroleum Refiners Association Annual Meeting, Los Angeles, March 23–24, 1986.
2. R. A. Corbett, "More Refining Catalysts Available," *Oil Gas J.*, p. 127 (November 18, 1985).
3. R. E. Wrench and J. W. Wilson, "FCCU Upgrade to Heavy Oil Cracking Improves Margins," *Oil Gas J.*, p. 53 (October 6, 1986).

MATTHEUS M. van KESSEL
R. H. van DONGEN
G. M. A. CHEVALIER

Petroleum Refinery Yields Improvement

This article addresses the use of coke and coal to supplement refinery fuel and utility requirements, thereby increasing refineries liquid yield per barrel of crude oil. The impending deregulation of natural gas and the increased processing of heavy low quality crude oil makes refinery gas a valuable product which must be utilized in only premium applications. Coke and coal can be effectively utilized in direct combustion processes to generate refinery utilities and process heat. Coke and coal can also be effectively utilized in gasification processes to generate fuel gas and synthesis gas to supplement refinery gas.

Introduction

The refining industry is still adjusting to the drastic changes of the last 10 years. Consider the modern refinery of the early 1970s; that refinery processed light high quality (low sulfur and metals) crude at $3 per barrel. There were growing markets for major products, especially gasoline, and there were usually reasonably priced markets for residual fuel. Natural gas could be purchased at $0.20 per million Btu or $1.15 per barrel crude oil equivalent (COE). Therefore, natural gas was the most economical fuel to supplement refinery gas in meeting the refinery fuel and utility requirements. Furthermore, the use of natural gas rather than liquid fuels increased refinery liquid yield per barrel of crude oil.

Today's refinery operates under much different conditions. Many refineries now process crude oil priced at almost $30 per barrel, including increasing amounts of heavy low quality crude. Markets for gasoline are stagnant while residual fuel markets are drying up. These changes in crude quality and product markets have led to large investments in heavy crude oil and residual upgrading. The use of higher severity process technology has generally increased refinery utility and fuel requirements. Refinery gas remains the primary plant fuel. Natural gas price, thanks to federal regulations, can vary by an order of magnitude. However, the future price of natural gas is generally considered to be headed toward the price of low sulfur residual fuel which is over $30 per barrel. At that price the economics of continued use of natural gas to supplement the refinery fuel and utility requirements become subject to question.

This article addresses the use of low cost coke and coal to supplement the refinery fuel and utility requirements and increase the refinery liquid yield per barrel of crude oil.

Better Refinery Gas Utilization

Refinery gas is mostly methane and hydrogen produced as a by-product in the many refining processes. Traditionally, refinery gas is utilized as plant fuel gas to fire process heaters and to generate process steam and power. In the past there were no economical alternative uses for this gas because its market value was controlled by the price of natural gas which was very low. Today, however, natural gas prices are moving toward low sulfur fuel value. Therefore, refinery gas is quickly becoming a valuable product which must be utilized in only premium applications.

The refinery gas analysis, yield, variability, and best utilization varies greatly with every specific refinery. However, as natural gas proceeds toward deregulation, a general forced ranking of the best to the worst refinery gas uses is beneficial. We consider the following applications the best use of refinery gas in decreasing order:

1. Hydrogen for hydroprocessing
2. Synthesis gas for chemicals
3. Fuel gas for radiant heating
4. Natural gas for export sales
5. Fuel gas for combustion turbine shaft power
6. Fuel gas for steam generation

The most valuable use of refinery gas is for hydrogen production. The decline in crude quality and changing refinery product markets is increasing the hydrogen requirements of oil refineries. Hydrogen can be effectively and economically recovered from the refinery gas by cryogenics, pressure swing absorption, or semipermeable membranes. The methane-enriched refinery gas from hydrogen separation can then be steam reformed to meet any additional hydrogen requirements. Our detailed technical and economic analyses indicate that steam methane reforming (SMR) is the lowest cost source of hydrogen even at relatively high methane values. SMR is a cheaper hydrogen source than even low value alternative feedstocks such as high sulfur residue, high sulfur coke, or coal [1].

The second most valuable use of refinery gas is the production of synthesis gas (hydrogen and carbon monoxide) for the manufacture of chemicals such as methanol. The potential for methanol as an octane enhancer indirectly or directly is very real as demonstrated by the use of methyl tertiary butyl ether (MTBE) and low level blending. Large scale testing of methanol as a primary engine fuel continues, with impressive results [2]. Important research and development advances are being made on the chemistry and catalysis of synthesis gas reactions for the manufacture of other chemicals as illustrated in Fig. 1. Essentially all these chemicals have a much higher value than gasoline. However, they must meet the potential competition from products currently made via ethylene or from imports produced from low priced remote natural gas.

The third best use of refinery gas is as fuel gas to fire radiant-type process furnaces. This type of heat transfer is commonly used in oil refinery processes to meet the high temperature requirements of most processes, plus the necessary high heat transfer rate and heat flux control, and for ease of decoking. The use of refinery gas has a number of technical and operability advantages over direct combustion of coal or coke for radiant process heating. Efficient use of refinery gas for this application required recovery of the lower level heat by convection-type heat transfer. Therefore, the process feed, combustion air, and fuel gas should be preheated as much as possible before considering steam generation in the convection section.

The fourth best use of refinery gas is purification and sale as natural gas, if the market value of the gas (on an energy basis) is near that of low sulfur fuel oil. This situation already exists in Europe and Japan. After natural gas deregulation, this situation may also develop in the United States. Purification of the refinery gas to produce substitute natural gas is relatively easy, especially if hydrogen is already being recovered from the raw refinery gas.

A lower value use of refinery gas is to generate shaft power. Purchased electricity is usually a lower cost option if utilities in the area generate most of their electricity from coal. If the electricity price is high enough to warrant the consideration of making shaft power from a premium fuel gas like refinery gas, it is likely that the local utility is making much of the electricity from premium fuels such as natural gas or low sulfur fuel oil. In most cases the combustion of coal or coke to generate steam will be the first alternative to purchased electricity. Nevertheless, refinery gas might be effective for shaft power if utilized efficiently. Specifically, the gas should be used in combustion turbines to generate the shaft power. The flue gas leaving the combustion turbine exhaust is about 1000°F and still contains 12–15 vol.% oxygen. This gas should be effectively utilized as preheated oxidant to steam methane reforming furnaces or to other large radiant-type process furnaces. Steam generation from this hot flue gas is a last resort option.

The lowest value use of refinery gas is direct combustion, specifically to cogenerate steam and power. It is usually more economical to generate steam from coal or coke because of their much lower value relative to refinery gas. It must be noted, however, that refinery gas usually generates a certain amount of steam in the convection section of refinery gas-fired radiant furnaces.

Petroleum Refinery Yields Improvement

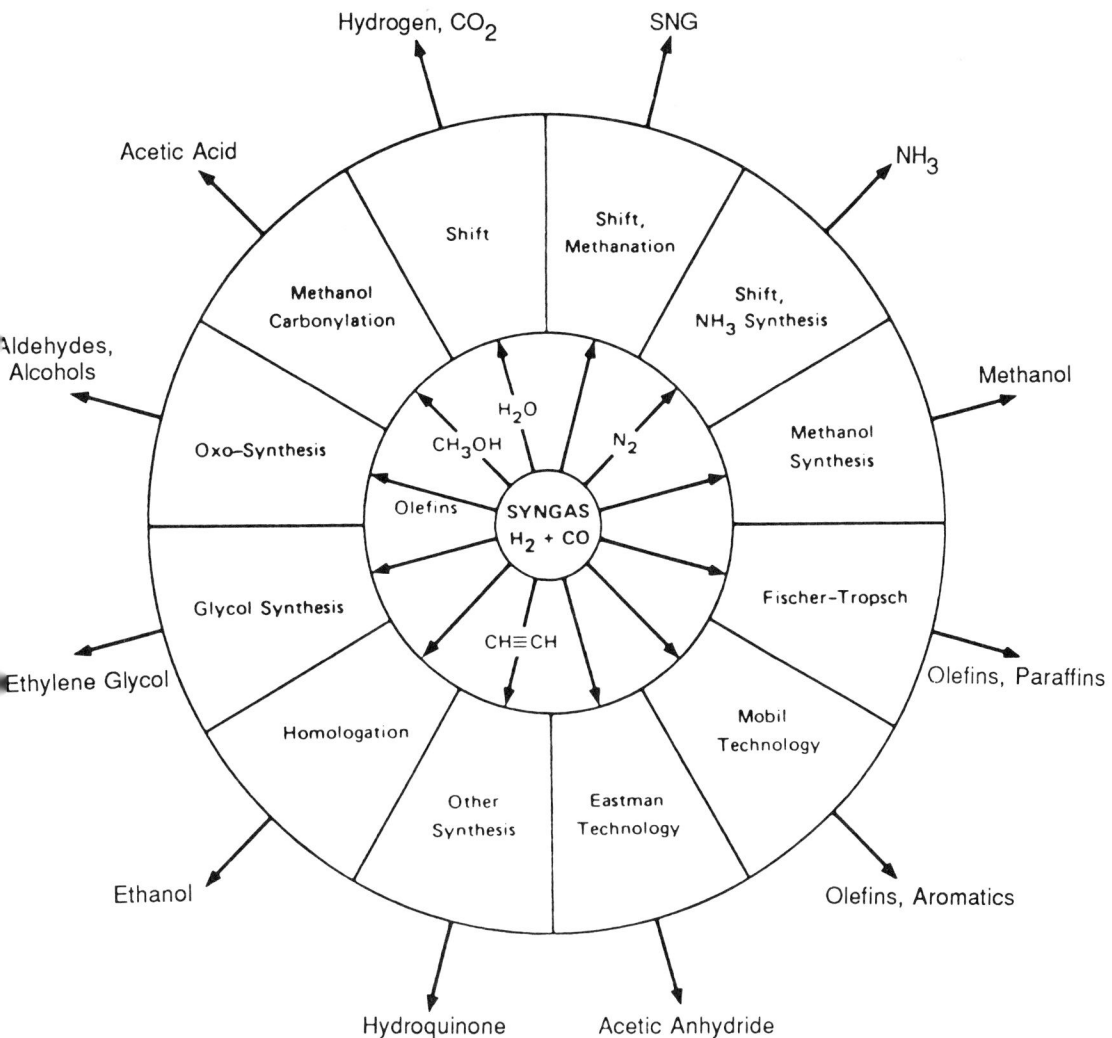

FIG. 1. Synthesis gas technology: commercial and developmental.

Direct Combustion of Coal or Coke

Refineries and chemical plants normally have a multitude of process heaters, some rather small, interspersed throughout the processing area. The use of coal in process heaters in existing refineries and chemical plants is therefore complicated by limited space availability for replacing small, compact gas-fired furnaces with larger coal-fired furnaces and for providing for coal receiving and pulverizing, particulate and ash recovery, and flue gas desulfurization. In addition, some process heaters such as reforming furnaces and cracking furnaces have high temperature radiant sections to provide rapid, controlled heating. The use of coal in these applications has been restricted because of the potential metallurgical attack on furnace tubes by the sulfur and ash constituents in the hot flue gas.

Utilities, steam, and in some cases power are normally generated in central facilities where facilities for handling solid fuels could be concentrated. Before the use of low cost natural gas became widespread in 1950, coal and coke were commonly burned in oil refineries to generate steam and utilities. More recently, Shell Oil burned coal and delayed coke in their Wood River, Illinois, Refinery, and Tosco burned fluid coke in their Avon, California, Refinery. The largest coke burning operation in a refinery today is that of the Suncor plant at Fort McMurray, Alberta, where approximately 2200 tons/day of delayed coke is burned in 3 boilers, each rated at 750,000 lb/h of steam. Today, electric utilities effectively operate large coal-fired boilers which must meet environmental standards that may be more stringent than those for smaller industrial coal-fired boilers. Utilities also do this with personnel who have substantially lower wage rates than oil refinery workers. Coal and coke are the normal fuels for steam production in most energy-intensive industries other than the oil and chemical industries. The chemical industry already is investing in coal-based utilities, and some refiners have installed boiler fireboxes which could accommodate solid fuels at a later date.

The conventional technology for generating steam from coal or coke is the pulverized boiler followed by electrostatic precipitation for particulates control and wet limestone scrubbing for flue gas desulfurization (FGD). However, we believe that the development of fluidized bed combustion (FBC) technology will greatly improve the potential for coal and coke combustion in oil refineries [3]. FBC has a number of advantages over pulverized combustion for refinery application including

Better feedstock flexibility
Lower NO_x
Less space requirements
Produces a dry solid waste, not a sludge

Chevron, Conoco, Gulf Oil, and Shell (Royal Dutch) already have operating FBC units while several other oil companies currently have units under construction or own companies involved in FBC development.

Direct combustion of coal or coke can be utilized to replace some of the higher temperature process heating which currently uses gas-fired radiant heaters. Oil or other higher temperature heat transfer media, such as inorganic salt eutectic mixtures, can be heated by direct combustion of solid fuels.

Alternatively, coke or coal can be used directly in fired process heaters. This technology is more complex than that for gas-fired heating; however, it is not impossible. For example, the Getty Oil refinery at Delaware City has a crude oil heater designed to fire a mixture of coke and refinery gas.

Innovative development work is being done on advanced solid fuel combustors. This work involves combustion of low ash coal/water slurry in advanced combustors which remove most of the ash (in the combustor) by gas velocity impingement or cyclonic motion. The work is being conducted by all major combustion turbine manufacturers in the United States under DOE funding. The current research is directed toward coal-fired combustion turbines. However, this research could lead to development of coal or coke fuel-combustors which could be utilized for firing process heaters.

Synthetic Fuels

In 1980, a synthetic fuels group was told that the lowest cost coal liquids were not made by direct or indirect coal liquefaction but by the use of coal (or coke) in oil refining to increase liquid yield [4]. We called these coal liquids "phantom coal liquids." This involves first the direct combustion of coal or coke for steam and utilities, as already discussed, and after this is done, gasification of coal or coke to produce fuel gas for use in radiant process furnaces. The coal- or coke-derived fuel gas would replace refinery gas which could be better utilized for production of hydrogen, synthesis gas, or natural gas. Finally, coal or coke gasification could ultimately be utilized to supplement hydrogen or synthesis gas requirements.

High sulfur green coke has advantages over coal for these synthetic fuel applications in refineries. This is due to increasing supplies of low quality coke which are being generated in oil refining. The traditional market for low quality coke is direct combustion by utilities and other industries. However, the increasingly higher sulfur levels in coke, together with the poor flame characteristics, put coke at a disadvantage relative to coal for direct combustion. High sulfur coke is essentially the only low value refinery product since its market value is controlled by steam coal prices. Coke gasification, on the other hand, has important advantages over coal for certain gasification technologies. Slagging gasifiers are favored for coke gasification due to coke's low reactivity [5].

The British Gas Corporation (BGC)/Lurgi Slagger is well suited for the manufacture of fuel gas or synthesis gas from delayed coke. A large-scale (700 ton/day) demonstration plant has begun operation in Westfield, Scotland, and petroleum coke has been tested in the BGC/Lurgi pilot plant, also at Westfield. Fines are an important consideration for the BGC/Lurgi. Delayed cokers can produce as low as 15 wt.% minus 1/4 in. fines by proper control of the coker operation, the high pressure water cutting of the coke in the drum, and correct crushing. The fines can be effectively burned in coal- or coke-fired boilers or injected in the tuyeres of the gasifier. The low volatiles content of the coke greatly reduce the liquid hydrocarbon production for this slagging fixed-bed gasifier.

The Texaco process is also well suited for the manufacture of hydrogen or synthesis gas from fluid or delayed coke. Two 900 ton/day Texaco gasifiers without gas coolers are successfully operating at the Tennessee Eastman chemical plant in Kingsport, Tennessee. The Cool Water project at Daggett, California, will soon be testing a large-scale Texaco gasifier with gas coolers. Also, Ube Ammonia will soon be operating several large-scale Texaco gasifiers on petroleum coke at their chemical plant in Ube City, Japan. The low ash content of coke assures a high ratio of moisture and ash free (MAF) feedstock to water for this slurry feed-entrained flow slagging gasifier. Compared to the Texaco gasifier feeding coal, the higher MAF feed-to-water ratio of coke reduces the oxygen requirements and gas cooler duty while increasing the yield of synthesis gas. The Texaco gasification process has the unique advantage of feedstock flexibility. Specifically, a Texaco gasification plant designed to gasify coke could easily be utilized to gasify other attractive feedstocks such as low quality residual or coal. This is very important from the standpoint of future market/price uncertainty for residual fuel and maintaining the freedom to only operate the coker when the refinery LP model indicates it is most economical.

The GKT (also Koppers-Totzek), Shell, and Dow gasification processes are also well suited for coke gasification as all three are entrained flow slagging technologies. A large-scale test of petroleum coke was done by a GKT gasifier in Spain in 1975. The commercially proven GKT has the disadvantage of low pressure operation. The Shell process still requires a large-scale demonstration. The Dow process is currently operating a large-scale demonstration plant at Plaquemine, Louisiana.

Limitations

There are a number of limitations and uncertainties which are hindering the move to coal and coke utilization in the oil industry. These include the following:

Low refinery utilization
Lack of growth in refinery product markets
High capital costs for solids combustion or gasification
Lack of experience in solids processing
Natural gas deregulation uncertainty
Uncertainty in environmental regulation

Poor refinery utilization and lack of market growth is causing oil companies to consolidate refineries. This situation also limits refinery expenditures to only those projects with the best return on investment, which are generally heavy crude oil and residual upgrading technology [6].

Solids combustion or gasification in existing refineries is a difficult investment to justify. The economics of burning undervalued refinery gas in "paid-off" equipment can easily look better in the short term than solids combustion or gasification processes which require large investments. Furthermore, most refinery operations personnel have no experience in solids combustion or gasification. Therefore, it is natural that they would be reluctant to accept technologies they know little about, and in addition, are likely to be more difficult to operate than gas-fired equipment.

Natural gas deregulation continues to be a political issue which must be resolved sooner or later. Until complete natural gas deregulation is accomplished, the use of coal and coke in refineries could be limited to steam generation. However, eventually natural gas will very likely approach low sulfur fuel oil in value. At that time, coal or coke gasification could be an economical alternative for refinery fuel gas.

Environmental uncertainty tends to promote "status quo." However, the acid rain issue will eventually force new environmental regulations which will ultimately affect oil refineries. Our analysis of acid rain control technologies indicates advantages for gasification in meeting stringent environmental standards [7]. This will become an important issue once the start-up and operation of the Cool Water coal gasification combined cycle demonstration plant is successful.

Summary

Oil refineries are currently faced with a situation similar to that of steel producers 10 years ago. At that time the steel industry knew that in the long term it was best to produce steel with basic oxygen furnaces and continuous casting machines. However, some companies saw there were only small short-term returns on the large investments associated with these new technologies. These investments appeared to be unattractive, especially when compared to producing steel in old existing open hearth furnaces and hot rolling mills which were completely amortized. As we are all aware, this type of fascination with only short-term return on investment led to the demise of many steel producers. Today, refiners realize that in the long term it is better to generate refinery utilities from solid fuels than from valuable crude oil by-products. However, the large investments in solids combustion or gasification generate only small short-term returns at today's fuel prices, especially when compared to the current refineries which are generating utilities in amortized equipment with undervalued regulated gas.

The refining industry has to respond to a number of drastic changes over the next 10 years. Oil refinery investments are currently directed at heavy crude oil and residual oil upgrading. Once bottom-of-the-barrel processing is optimized, refineries must begin optimizing "top-of-the-barrel" processing. In the near future, refinery gas and natural gas will be too valuable to be utilized for plant fuel. The profitable refineries of the future will utilize refinery gas for only the highest value applications such as hydrogen, synthesis gas, and even natural gas sales. Coal or coke will be combusted and gasified to meet the refineries fuel and utilities requirements, thereby greatly increasing the refinery yield of valuable products per barrel of crude oil.

Much of this is taken from an article presented at the API Meeting, New Orleans, Louisiana May 15, 1984 and is used with the permission of American Petroleum Institute, Washington, D.C. 20005.

References

1. SFA Pacific, Inc., *Synthesis Gas for Chemicals and Fuels*, multisponsored report, Mountain View, California, October 1983.
2. R. L. Dickenson et al., *Methanol Fuel—Status and Outlook*, Presented at the 183rd National Meeting of the American Chemical Society, Las Vegas, Nevada, March 30, 1982.
3. SFA Pacific, Inc., *The SFA Quarterly Report*, multisponsored report, Mountain View, California, January 1984.
4. D. R. Simbeck et al., *Coal Liquefaction: Direct Versus Indirect*, Presented at Coal Technology '80, Houston, Texas, November 19, 1980; *Oil Gas J.,*, pp. 254–268 (May 4, 1981).
5. SFA Pacific Inc., *Coal Gasification Systems: A Guide to Status, Applications and Economics*, Electric Power Research Institute, Palo Alto, California, Report No. AP-3109, June 1983.
6. SFA Pacific, Inc., *Upgrading Heavy Crude Oils and Residues to Transportation Fuels: Technology, Economics and Outlook*, multisponsored project in progress.
7. SFA Pacific, Inc., *The SFA Quarterly Report*, multisponsored report, Mountain View, California, March 1984.

DALE R. SIMBECK
FRANK E. BIASCA

Hazardous Waste Regulations

Petroleum refinery operators face more stringent regulation of the treatment, storage, and disposal of hazardous wastes. Under recent regulations, a larger number of compounds have been, and are being, studied. Long-time methods of disposal, such as land farming of refinery waste, are being phased out.

As a result, many refineries are changing their waste management practices. An informal survey of nine refineries showed that eight were planning to close land treatment units because of the uncertainty of continuing the practice.

New regulations are becoming even more stringent, and they encompass a broader range of chemical constituents and processes. Continued pressure from the U.S. Congress has led to more explicit laws allowing little leeway for industry, the U.S. Environmental Protection Agency (EPA), or state agencies.

A summary of the current regulations and what they mean to refiners is given in the following. It is hoped that the summary will clarify the sometimes confusing rules, and help refiners deal with a situation that could entail major costs for the industry in the years to come.

Regulatory Background

The hazardous waste regulatory program, as we know it today, began with the Resource Conservation and Recovery Act (RCRA) in 1976. The Used Oil Recycling Act of 1980 and Hazardous and Solid Waste Amendments of 1984 (HSWA) were the major amendments to the original law.

RCRA provides for the tracking of hazardous waste from the time it is generated, through storage and transportation, to the treatment or disposal sites. RCRA and its amendments are aimed at preventing the disposal problems that lead to a need for the Comprehensive Environmental Response Compensation and Liability Act (Cercla), or Superfund, as it is known.

Subtitle C of the original RCRA lists the requirements for the management of hazardous waste. This includes the EPA criteria for identifying hazardous waste, and the standards for generators, transporters, and companies which treat, store, or dispose of the waste. The RCRA regulations also provide standards for design and operation of such facilities.

Requirements

The first step to be taken by a generator of waste is to determine whether that waste is hazardous. Waste may be hazardous by being listed in the regulations, or by meeting any of the four characteristics: ignitability, corrosivity, reactivity, and extraction procedure (EP) toxicity.

Generally:

If the material has a flash point less than 140°F, it is considered ignitable.
If the waste has a pH less than 2.0 or above 12.5, it is considered corrosive. It may also be considered corrosive if it corrodes stainless steel at a certain rate.
A waste is considered reactive if it is unstable and produces toxic materials, or it is a cyanide or sulfide-bearing waste which generates toxic gases or fumes.
A waste which is analyzed for EP toxicity and fails is also considered a hazardous waste. This procedure subjects a sample of the waste to an acidic environment. After an appropriate time has elapsed, the liquid portion of the sample (or the sample itself if the waste is liquid) is analyzed for certain metals and pesticides. Limits for allowable concentrations are given in the regulations.

The specific analytical parameters and procedures for these tests are referred to in 40CRF 261.

The 1980 RCRA regulations apply to companies generating more than 1000 kg of hazardous waste per month. In 1984, HSWA amended Subtitle C (hazardous waste management) to include standards for generators of lesser quantities.

When these regulations took effect on September 22, 1986, companies generating greater than 100 but less than 1000 kg of hazardous waste per month had to comply with RCRA. This added more than 100,000 generators of hazardous waste to the RCRA program.

The 1984 amendments also brought the owners and operators of underground storage tanks into the RCRA fold. This can have a significant effect on refineries which store product in underground tanks. Now, in addition to the hazardous waste being controlled, petroleum products and other hazardous substances (raw materials, etc.) are regulated by RCRA Subtitle I.

A facility that treats, stores, or disposes of hazardous waste on site (a TSDF) is required to obtain a permit for these activities. EPA realized that facilities in operation when the regulations became effective would require some time to comply with those regulations. Therefore, the permit application process was divided into two parts, A and B.

The Part A application was submitted on a form provided by EPA which supplied general information regarding the waste handled at the facility and the means by which it was managed. The Part B application, on the other hand, is a detailed document thoroughly describing how the facility complies with the requirements for security, spill response, etc., as well as detailed engineering information on the hazardous waste management units.

Submittals

Submission of Part A permit applications for existing facilities prior to November 19, 1980, qualified the refinery for interim status. This means that the refinery was allowed to continue operation according to certain regulations during the permitting process. These interim status regulations are less stringent than the regulations for a permitted facility.

When revisions of the regulations first cause a facility to be regulated, that facility may also obtain interim status by submitting the Part A application within 180 days of the effective date. The Part B applications were then to be "called in" by EPA at later times.

Although the legislators recognized that the EPA would not be able to meet its permitting demands overnight, neither did they expect that it would take 6 years before permits were issued. This delay was one of the concerns the 1984 legislature addressed by putting time limits on EPA actions.

Deadlines were established for submission of the Part B applications. Existing land disposal facilities had to submit applications by November 8, 1985, existing incinerators by November 8, 1986, and all other existing hazardous waste facilities by November 8, 1988. Owners or operators of facilities or units that are to be built must submit both Parts A and B at the time of application. No construction may commence until the permit is issued.

Waste Disposal

The Hazardous and Solid Waste Amendments (HSWA) were signed into law on November 8, 1984. This legislation represented a strong bias against land disposal of hazardous waste.

In addition to the Part B permit application deadlines mentioned, other deadlines were placed on EPA and the regulated community. These deadlines require EPA to promulgate regulations regarding certain issues by a certain period in time. If no regulations are developed by that time, the law goes into effect as Congress wrote it.

These are referred to as "hammer" provisions. Some of the hammer provisions which affect refineries are:

A ban on the disposal of bulk or noncontainerized liquids in landfills went into effect on May 8, 1985. The prohibition also bans solidification of liquids using absorbent material including absorbents used for spill cleanup. Contaminated soils resulting from a spill are not included in this prohibition.

On November 8, 1886, certain spent halogenated and nonhalogenated solvents were banned. HSWA banned the "California list" of hazardous chemicals on July 8, 1987, from land disposal units. These include liquid hazardous wastes which contain certain metals, free cyanides, polychlorinated, biphenyls, corrosives, and liquid or nonliquid halongenated organic compounds.

By August 8, 1988, five hazardous wastes from specific sources in the refining industry are scheduled for disposal prohibition and/or treatment standards. These five are: dissolved air flotation (DAF) float, slop oil emulsion solids, heat exchanger bundle cleaning sludge, API separator sludge, and leaded tank bottoms (EPA ID Nos. K047–K051).

Petroleum refineries often use surface impoundments for hazardous waste management. These must be either retrofitted with double liners and leak detection systems by November 8, 1988, or closed.

With HSWA, Congress mandated EPA to restrict disposal or hazardous waste in land disposal units. These units include landfills, surface impoundments, land treatment units, and waste piles. EPA is currently investigating every hazardous waste to determine whether it can continue to be disposed of in land disposal units. Final decisions for listed refinery wastes were to be made by August 8, 1986.

Waste Cleanup

Under RCRA, EPA has the authority to require a facility to clean up releases of hazardous waste or waste constituents. The 1980 regulation provided for cleanup of hazardous waste released from active treatment, storage, and disposal facilities occurring after November 19, 1980. Superfund was expected to handle contamination which had occurred before that date.

Now, under the provision of HSWA, the agency can mandate cleanup before a permit is issued. This specific enforcement authority is granted to EPA under Section 3008(h) of HSWA, or during the permitting process under the authority of 3004(u) and (v).

Congress also mandated that EPA look at specific types of compounds to determine if they should be listed as hazardous waste. The two industrial categories specifically named for these studies are the inorganic chemicals industry and refineries.

As part of the review process, EPA is to expand its EP toxicity test to include a larger number of compounds. Toward that end, EPA scheduled implementation of the Toxic Characteristic Leaching Procedure (TCLP) test by December 1987.

The TCLP adds an additional 38 compounds to be considered in determining whether a waste is hazardous. The proposed allowable level for many of these organic compounds is in the parts per billion range.

For example, the proposed level for benzene is 0.07 mg/L or 7.0 ppb. A further expansion of the TCLP constituents list is expected in 1988. While the test has not yet replaced the EP toxicity test, the TCLP is currently used to determine if the waste will be subject to the land disposal ban.

Regulatory Authority

Originally, RCRA directed the EPA to authorize state government to handle the hazardous waste program. The states were to establish regulations that were substantially equivalent, or more stringent, than EPA rules. Amendments passed by Congress in 1984 required EPA to once again become involved with the hazardous waste permitting process through oversight responsibility. As a result of this involvement, refiners now have to answer to a dual authority: EPA and the state agency. EPA's responsibility is to review portions of the application or permits that are a result of the HSWA legislation. The program is extremely complex and involves significant company/agency interaction.

Cutbacks in federal funds have also had an effect on the programs. For example, an insufficient number of personnel at EPA and state agencies seriously delays the review of the permit application and preparation of the permit.

Additionally, when a Notice of Deficiency (NOD) is issued, it may no longer spell out in detail what is expected of the company to meet the requirements. It may only tell the company to meet the requirements. It may only tell the company that the submission is deficient without further explanation of what the agency expects.

Furthermore, when permitting a land treatment unit, only one NOD is allowed. If insufficient information is presented in response to the NOD, EPA will move toward denial of the permit.

Denial of a hazardous waste treatment or disposal permit requires the refinery to transport its waste off-site. This adds additional cost for disposal of the waste.

Management of Wastes

Before a refinery can determine if its waste is hazardous, it must first determine that the waste is indeed a solid waste. In 40 CFR 261.2, the definition of solid waste can be found. If a waste material is considered a solids waste, it may be a hazardous waste in accordance with 40 CFR 261.3.

There are two ways to determine whether a waste is hazardous. These are to see if the waste is listed in the regulations or to test the waste to see if it exhibits one of the characteristics. The lists and the characteristics can be found in 40 CRF 261.

There are four lists of hazardous wastes in the regulations. These are wastes from nonspecific sources (F list), wastes from specific sources (K list), acutely toxic wastes (P list), and toxic wastes (U list). And there are the four characteristics mentioned before: ignitability, corrosivity, reactivity, and extraction procedure toxicity.

Exempted Wastes

Certain waste materials are excluded from regulation under RCRA. The various definitions and situations which allow waste to be exempted can be confusing and difficult to interpret. One such case is the interpretation of the "mixture" and "derived-from" rules.

According to the mixture rule, mixtures of solid waste and listed hazardous wastes are, by definition, considered hazardous. Likewise, the derived-from rule defines solid waste resulting from the management of hazardous waste to be hazardous. These definitions can be found in 40 CFR 261.3(a) and (c), respectively.

There are five specific listed hazardous wastes (K list) generated in refineries. These are K048–K052. Additional listed wastes, those from nonspecific sources (F list) and those from the commercial chemical product lists (P and U), may also be generated at refineries.

Because of the mixture and derived-from rules, special care must be taken to ensure that hazardous wastes do not "contaminate" nonhazardous waste. Under the mixture rule, adding one drop of hazardous waste in a container of nonhazardous materials makes the entire container contents a hazardous waste.

API Separator Sludge

As an example of the problems such mixing can cause, consider the case with API separator sludges. The sludges are a listed hazardous waste, K051. EPA determined that wastewater from a properly operating API separator is not hazardous unless it exhibits one of the characteristics of a hazardous waste. That is, the derived-from rule does not apply to the wastewater.

However, if the API separator is not functioning properly, solids carryover in the wastewater can occur. In this case, the wastewater contains a listed hazardous waste, the solids from the API sludge. The wastewater would be considered a hazardous waste regulated under RCRA because it is a mixture of a nonhazardous waste and a hazardous waste.

This wastewater is often further cleaned by other treatment systems (filters, impoundments, etc.). The solids separating in these systems continue to be API separator sludge, a listed hazardous waste. Therefore, all downstream wastewater treatment systems are receiving and treating a hazardous waste. These downstream systems are considered hazardous waste management units subject to regulation under RCRA and possibly requiring an RCRA permit.

Oily wastewater is often treated or stored in unlined wastewater treatment ponds in refineries. These wastes appear to be similar to API separator waste. EPA has expressed concern over management of this material in this manner.

On November 12, 1980, EPA proposed a regulation that would include, as hazardous waste, sludge generated in oil/water/solid separators which are used prior to biological wastewater treatment. Much debate has occurred over the years, with EPA being pressured from opposing sides by the petroleum industry and environmental groups. Comment on the proposal was reopened on February 11, 1985. Final regulation is expected soon.

Once a refinery understands the regulatory background faced under RCRA and HSWA, it can begin analyzing refinery units to determine which are regulated under RCRA.

Land Treatment and Storage

Provisions of the regulations focused on the units used to treat, store, or dispose of hazardous waste. Hazardous waste surface impoundments and land treatment units (LTU) are under specific regulatory scrutiny. Changes in these regulations will affect the refinery industry because of the dependence on these types of waste management units.

Surface Impoundments

HSWA Section 3005(j)(l) requires surface impoundments existing as of November 8, 1984, to be retrofitted with double liner and leak detection systems. An owner or operator using surface impoundments to manage hazardous waste has three options: retrofit the surface impoundment, apply for an exemption, or stop receiving hazardous waste and close the unit.

One of these options must be implemented by November 8, 1988. Furthermore, Congress has allowed four exemptions to the retrofitting requirements on surface impoundments:

1. Interim status impoundments having at least one liner that is not leaking, are one-fourth of a mile from an underground source of drinking water, and are in compliance with groundwater monitoring requirements for the permitted facilities.
2. Impoundments that are part of a secondary or later phase of aggressive biological treatment. These units also must comply with Clean Water Act and groundwater monitoring requirements.
3. Units where the owner or operator can prove that there will be no future migration of waste.
4. Impoundments subject to a consent order decree or agreement with EPA, and corrective action meeting minimum technological standards.

Facilities denied exemptions will have one year from November 8, 1987, to either close the facility or retrofit the impoundment.

Land Treatment

The land treatment issue comes under extreme fire because of HSWA. The significant issue for land treatment is the "no-migration provision." Congress' intent is to provide safe hazardous waste disposal, and to prevent the spread of hazardous waste or hazardous waste constituents in the groundwater, surface water, soils, or into the air.

RCRA Section 1002 states that, "reliance on land disposal should be minimized or eliminated, and land disposal, particularly landfills and surface impoundments, should be the least favored method for managing hazardous waste."

EPA will be aggressive in its review of RCRA permits involving land treatment. A Part B hazardous waste permit application can range from moderately complex to extremely complex. The applications to permit land treatment facilities are the most complex.

The permitting process involves submission of the application and review by the governing agency. If the application is found to be incomplete, a notice of deficiency (NOD) is given to the permittee spelling out those items that need to be more completely addressed.

The permittee works closely with the agency to satisfy the NOD. There may be several communications between the permittee and the agency, with several NOD's, before the agency is satisfied with the application.

However, in review of land treatment applications, generally one NOD is effected. If insufficient information is presented, EPA will move toward denial of the permit. This places an extreme burden on the permittee to ensure that all items are thoroughly presented in the initial application.

Land Disposal Ban

Congress has allowed that land disposal could continue if there were no migration of hazardous waste or hazardous waste constituents to groundwater, surface water, soil, or air. To minimize such spread of contamination, certain categories of waste have been targeted to be banned from land disposal. One of these is liquids.

Liquids

No bulk hazardous waste liquid or any such waste containing free liquids can be placed in landfills. No nonhazardous liquids can be placed in a landfill requiring an RCRA permit. This includes liquids absorbed on materials that would release those liquids if compressed or biodegraded, and absorbents used during spill cleanups. EPA has allowed that this does not include contaminated soil resulting from a spill.

A test was developed to determine whether a waste is considered a liquid or contains free liquids. This is the paint filter liquids test. A predetermined amount of waste is put into the paint filter. If any portion of the waste passes through the filter within 5 min, the waste is considered liquid. It cannot be landfilled without further treatment to solidify the waste.

EPA will develop a confined pressure test to squeeze a waste/absorbent mixture to see if any liquid is released. Certain pressure conditions will be set to duplicate compacting in a landfill.

If the waste material releases liquid under those conditions, it cannot be landfilled without further treatment. EPA's current thinking is that absorbent with greater than 1% carbon in its makeup is biodegradable. Wastes absorbed on such materials would also be banned from landfills. The agency proposed these regulations on December 24, 1986, and final regulations were expected to be issued in November 1987.

According to the EPA, liquid materials must be stabilized before placement in a landfill. Therefore, the stabilization process cannot be accomplished within the landfill cell.

Containerized liquids are being regulated as well. Disposal of containerized liquids will be minimized, however, if a bulk liquid is found to contain absorbents, an owner or operator can containerize the materials and landfill them in compliance with the containerized liquids requirements.

This means of disposal can be significantly more expensive than other methods. A refinery generating such waste would be subject to the regulations and added cost of treatment or containerization.

The first of the land-ban regulations for solvent and dioxins was final on November 8, 1986. Past disposal of these wastes remains unaffected by the new standards. The standards apply to disposal of certain of these wastes in landfills, surface impoundments, waste piles, LTU's, and injection wells (in 2 years).

Notification

As a generator of hazardous waste, a refiner must notify the owner or operator of a treatment or disposal facility, in writing, that the waste being sent there is restricted by the bans. Additional treatment is needed before land disposal.

If the generator finds that a waste can be land disposed without added treatment, a certification statement must be submitted with each shipment which contains the hazardous waste number, the manifest number, the applicable treatment standards, and the waste analysis data. This is only required when the generator is representing that the waste meets the treatment standards.

If the refinery is operating under an extension or petition, the specific extension or petition information must be submitted with each exempted shipment. The land disposal facility must ensure that its records include the required certifications, or that its own analysis proves the waste complies with these disposal regulations.

Air Toxicity

EPA took its first step toward an air toxicity characteristic on February 5, 1987, with a proposed rule and notice of public hearing. These proposed standards would limit emissions of volatile organics at some treatment, storage, and disposal facilities. To be regulated under these proposed rules, a treatment, storage, or disposal facility must have equipment, process vents, or accumulator vessels in compliance with the standards. These sources must contain hazardous waste or hazardous waste derivatives with concentrations of greater than 10% total organics. The facility is also required to obtain an RCRA permit for an independent reason.

Congress wishes to minimize migration of hazardous waste to various media, including the air. Therefore, landfarming of waste, which allows uncontrolled evaporation of volatile organics to the air, is a method which has a questionable future. This migration issue, along with those discussed above, further demonstrate the difficulty in permitting landfarm facilities.

The land-ban regulatory package allows for exemption for a treatment surface impoundment and case-by-case extension for those who have made a good faith effort to develop an alternative treatment technology.

Under HSWA, EPA has the option of developing appropriate treatment standards for particular wastes. If a refiner meets these requirements, that particular waste could be exempted from the land disposal restrictions.

Other options in HSWA allow an EPA extension of a prohibition deadline for up to 2 years if there are not sufficient alternative treatment technologies or disposal capacity to protect human health and the environment. These extensions will be granted on a national basis, not case-by-case.

Exemptions to Bans

Six provisions in the statute allow continued land disposal of certain hazardous waste in view of specific bans.

1. An owner or operator working under an extension provided by 40 CFR 268.5 or a successful petition under 40 CFR 268.6.
2. Land disposal which is a result of RCRA corrective action or Superfund response action.
3. A conditionally exempt small quantity generator (one who generates less than 100 kg per calendar month).
4. A small quantity generator generating between 100 and 1000 kg per calendar month is excluded for a 2-year period ending November 8, 1988.
5. Treatment of hazardous waste in a surface impoundment described by 40 CRF 268.4.
6. If the EPA provides a national extension under Subpart C due to a lack of disposal capacity for a particular waste for a period not greater than 2 years. (During the national extensions, restricted waste must be disposed of in a facility meeting the minimum technology requirements in HSWA 3004(0).)

The statutory and regulatory programs concerning the land disposal ban provide several options and waivers. However, the congressional-stated preference against land disposal makes the use of these options and waivers very difficult.

Refiners now face a new level of regulatory control and requirements. These result in additional analytical and treatment technologies in normal operating practices.

The refining industry, as well as other industries, will increasingly feel the effects of the land bans on their hazardous waste management practices. Current practices of land disposal must change along with management attitudes for waste handling.

Toxicity Measurement

Although the Toxic Characteristic Leaching Procedures (TCLP) test has yet to be approved as a replacement for the EP toxicity characteristic, a generator of waste must use the test to find out if the waste will be subject to the land disposal ban.

Corrective Action

The most significant overall impact of HSWA is the authority to require corrective action for releases of hazardous waste or hazardous waste constituents into the environment. These provisions will allow the EPA to deal with releases into any media of the environment.

The EPA has the authority to require corrective action under HSWA Sections 3004(u) and 3004(v) and the interim status corrective action order provision of 3008(h) for continuing releases at RCRA facilities. In an enforcement mode, Section 3008(h) provisions are invoked when there is indication of a release of hazardous waste to the environment.

The EPA's broad interpretation of these provisions allows them to expand control to the entire facility. This includes process areas where spills and improper management may be contributing to environmental contamination in addition to waste management units.

In a permitting mode, in addition to active operational hazardous waste units, the EPA must review all solid waste management units (SWMU) at a facility prior to issuing an RCRA permit. The process that EPA will use to define the need for corrective action is called RCRA facility assessment (RFA).

This review must identify any release of hazardous waste or hazardous waste constituents from both active and inactive units. Requirements for remedial actions will be established. The RFA is completed in several steps:

1. The preliminary review (PR)—This is a paperwork evaluation. It identifies the potential for any solid waste management units on the property.
2. Visual sight inspection (VSI)—EPA or state personnel physically inspect the property to establish the likelihood of releases to the environment. The governing agency can then require sampling and/or remedial investigation to determine the extent of migration of any contamination. Once that is identified, the proper corrective actions are developed and made part of the RCRA permit.

The way refineries handle their waste in the future depends largely on the ever-changing RCRA regulations. Waste management is the focus, not waste disposal. Reuse/recycle options must be explored to keep pace with the desire and intent of Congress under the RCRA program.

This material appeared in *Oil & Gas Journal*, pp. 39–44, January 4, 1988, and is reprinted by special permission. Copyright © 1988 by Pennwell Publishing Co., Tulsa, Oklahoma 74121.

DAVID OLSCHEWSKY
ALICE MEGNA

Petroleum Waste Toxicity, Prevention

Land treatment can protect human health and the environment from wastes as diverse as liquid municipal wastewater, liquid and dewatered municipal sludge, and petroleum industry wastes.

The basic technology also can be used for bioremediation of contaminated soils at Superfund and other sites. The use of land treatment under controlled and well-designed and operated conditions should continue to be available to the petroleum and other industries.

Land treatment is a managed technology that involves the controlled application of a waste on the soil surface and the incorporation of the waste into the upper soil. It is not the indiscriminate dumping of waste on land, and differs significantly from landfilling, waste piles, and surface impoundments. These methods store wastes in man-made or natural excavations and use a combination of liners and leachate collection systems to control the migration of the waste or resultant by-products.

Land treatment relies on the dynamic physical, chemical, and biological processes occurring in the soil. As a result, the constituents in the applied wastes are degraded, immobilized, or transformed to environmentally acceptable components.

The land-treatment process uses the assimilative capacity of the soil to degrade, immobilize, and contain the applied waste. The upper soil layer (the top 6–12 in.) in which the waste is incorporated is the zone of incorporation (ZOI).

This zone, in conjunction with the underlying soils where additional treatment and immobilization of the applied waste constituents occurs, is referred to as the treatment zone. For hazardous wastes, the treatment zone depth is defined as 5 ft. The transformations, biological oxidation, and immobilization occur primarily in the upper part of the treatment zone.

Land-treatment technology has been used successfully for the treatment and disposal of petroleum industry wastes; other industrial wastes such as from the wood preserving, pulp and paper, and food processing industries; and municipal wastewater and sewage sludges. It is also used for remediation of contaminated soil and is a suitable closure process at specific sites.

When the technology is designed and operated using proper procedures and methods, land treatment has been shown to be protective of human health and the environment.

The design and operation of a land-treatment facility is based on sound scientific and engineering principles as well as on extensive practical field experience. A land-treatment site is designed and operated to maximize waste degradation and immobilization, minimize release of dust and volatile compounds as well as percolation of water-soluble waste compounds, and control surface water runoff.

Land treatment can be a viable management practice for the treatment for disposal of nonhazardous wastes as well as some types of hazardous wastes. The applied wastes can be liquid, semisolid, or solid.

At some point, wastes may no longer be applied at a site and the site must be closed. A closure and postclosure plan must be prepared and approved by the appropriate regulatory agency. This plan can utilize degradation and immobilization as a closure mechanism [1].

Land treatment is an important technology for the treatment and disposal of oily wastes from United States refineries. About 100 land treatment facilities treating petroleum industry wastes, primarily from the refining sector, exist in the United States [2].

In addition, 26 of 38 Canadian refineries and at least 10 refineries in Europe use land treatment as an environmentally sound, cost-effective waste-disposal technology. At these sites the constituents in the applied wastes have been degraded and immobilized successfully. At many petroleum industry land-treatment sites, water balances indicate that there is little or no net percolation as a result of the small amount of water added with the applied waste, the low precipitation, and the water loss due to evaporation.

Land treatment of refinery wastes has been successfully practiced in all of the major climatic regions of the United States, Europe, and Canada under a wide range of hydrogeologic conditions. The data suggest that there are very few site limitations which cannot be overcome with the appropriate design and operation of land-treatment facilities.

Regulatory Framework

On November 8, 1984, the U.S. Congress passed amendments to the Resource Conservation and Recovery Act (RCRA) that specifically prohibit land disposal of hazardous wastes beyond specified dates unless the U.S. Environmental Protection Agency (EPA) determines that a land disposal method is protective of human health and the environment. The EPA was required to set a schedule for land restriction decisions for all hazardous waste listed as of November 8, 1984.

The schedule for the land disposal restriction decisions was published in the *Federal Register* on May 18, 1986. Decisions regarding petroleum refining industry-listed wastes such as K048 DAF float and K051 API separator sludge were made in August 8, 1988.

Under Section 3004(K) of RCRA, land disposal of hazardous waste is defined to include "...any placement of such hazardous waste in a...land treatment facility...." As a result, land-treatment facilities for hazardous wastes are subject to the banning decisions. Land treatment is "disposal" for purposes of 3004(K) restrictions, but it is also RCRA "treatment" as the term is defined under RCRA 1004(34). Land treatment of hazardous wastes involves the incorporation of the waste into the ZOI in order to degrade, transform, or immobilize hazardous constituents.

After this *in-situ* treatment occurs, the long-term disposal of the remaining residues occurs. This dual aspect—treatment and disposal—makes land treatment unique among land disposal alternatives. These unique aspects have been recognized by EPA (51 FR 1710, January 14, 1986).

The amendments specify that a method of land disposal may not be protective of human health and the environment unless upon application by an interested person it has been demonstrated to the administrator "that there will be no migration of hazardous constituents from the disposal unit or injection zone for as long as the wastes remain hazardous."

The statute also requires EPA to set "levels or methods of treatment, if any, which substantially diminish the toxicity of the waste or substantially reduce the likelihood of migration of hazardous constituents from the waste so that short-term and long-term threats to human health and the environment are minimized."

Properly designed and operated land-treatment facilities achieved this requirement. This results from the degradation, immobilization, and transformations that occur at a land-treatment site.

Wastes that meet the treatment standards established by EPA are not subject to the land disposal prohibitions. If EPA fails to set treatment standards (best demonstrated available technologies—BDAT) for any of the scheduled listed hazardous wastes by May 8, 1990, all such wastes are prohibited from land disposal unless EPA grants a case-by-case petition.

In the *Federal Register* of November 7, 1986, EPA identified procedures that it will use to establish treatment standards for hazardous wastes, evaluate petitions for a variance from the treatment standard, and evaluate petitions demonstrating that continued land disposal of hazardous waste is protective of human health and the environment [3]. These procedures indicate how BDAT will be considered for hazardous wastes such as those from the petroleum industry.

RCRA specified that BDAT may be expressed as either a performance standard or a method of treatment. EPA decided to establish BDAT treatment standards as performance standards rather than adopting an approach that would require the use of specific treatment methods because of a belief that performance standards offer the regulated community greater flexibility to develop and implement compliance strategies, provide an incentive to develop innovative treatment technologies, and ensure that a technology is operated properly.

Even though specific technologies are identified as being able to meet the indicated performance standards, it is not necessary that those technologies actually be used to meet the indicated performance standards [3].

The most economical technology, or combination of technologies, that can achieve the identified performance standards can be used. The advantage of meeting the promulgated performance standards is that the residues from such treatment are not subject to the land-disposal prohibitions and can be disposed of in an RCRA Subtitle C land disposal facility that has been fully permitted and that has an approved waste management plan.

The terms "demonstrated" and "available" were defined in the November 7 notice. To be considered a demonstrated treatment technology for the purposes of this final rule, a full-scale facility must be known to be in operation for the waste or similar wastes.

In order to be considered "available," a demonstrated technology must substantially diminish the toxicity of the waste or substantially reduce the likelihood of migration of hazardous constituents from the waste. This ensures that all wastes are adequately treated. Treatment will always be deemed substantial if it results in nondetectable levels of the hazardous constituents in the toxicity characteristic leaching procedure (TCLP) extract.

EPA recognizes that treatment standards need to incorporate a variability factor. This factor is intended to account for variations that arise from mechanical limitations in the equipment, from variations in waste characteristics, and from variations in analytical test methods. The variability factor was established as the ratio of the calculated 99th percentile concentration, C_{99}, to the mean concentration.

BDAT Examples

The BDAT standards for the petroleum refining industry listed hazardous wastes were promulgated on August 17, 1988 [4]. These wastes were: (1) K048-DAF float, (2) K049-slop oil emulsion solids, (3) K050-heat exchanger bundle cleaning sludge, (4) K051-API separator sludge, and (5) K052-leaded tank bottoms.

The standards for the organic constituents in these wastes were based on the results of the performance achievable by solvent extraction and/or incineration. Standards for metals in the residues of these processes (arsenic, total chromium, nickel, and selenium) were based on the performance of a chemical stabilization process. These technologies were judged to be demonstrated and available because they are commercially available and provide substantial reduction of the organic and metal hazardous constituents.

It is important to note that while the standards for organics are based on data from the use of solvent extraction and fluidized bed incineration, other treatment technologies such as rotary kiln incineration and biodegradation also may be able to achieve the BDAT standards and are not precluded. Industry may use any technology not otherwise prohibited, e.g., impermissible dilution, to meet the treatment standard and is not limited to only those technologies considered by EPA in determining BDAT.

Treatment standards for wastes for which destruction technologies are appropriate are based on total constituent analysis, typically mg of constituent/kg of waste or mg/L of wastewater. For those wastes, such as incinerator residues, for which stabilization or fixation is appropriate, treatment standards are based on concentrations in an extract developed by the Toxicity Characteristic Leaching Procedure (TCLP). BDAT standards for petroleum refinery nonwastewaters and for petroleum refinery wastewater are indicated in Tables 1 and 2, respectively.

It is very clear that dilution, such as simply mixing wastes with soil, will not be allowed as a method to comply with the restrictions and standards. This is a generic emphasis that applies to all media-surface waters, the atmosphere, and soil.

A new section, part 268.3, has been added to the federal regulations to cover this point [3]. This part states that, "No generator, transporter, handler, or owner or operator of a treatment, storage, or disposal facility shall in any

TABLE 1 BDAT Standards for Refinery Wastes (K048 to K052) (nonwastewaters) [4]

Total Composition (mg/kg)		TCLP Extract (mg/L)	
Anthracene (A)[a]	6.2	Arsenic (D)	0.004
Benzene (B)	9.5	Chromium (total) (D)	1.7
Benzo(a)anthracene (C)	1.4	Nickel (D)	0.048
Benzo(a)pyrene (D)	0.84	Selenium (D)	0.025
Bis(2-ethylhexyl) phthalate (E)	37		
o-Cresol (F)	2.2		
p-Cresol (F)	0.90		
Chrysene (G)	2.2		
Di-n-butyl phthalate (H)	4.2		
Ethylbenzene (B)	67		
Phenanthrene (B)	7.7		
Phenol (D)	2.7		
Pyrene (G)	2.0		
Toluene (B)	9.5		
Cyanides (total) (D)	1.8		

[a]Letters refer to the specific waste for which the constituent and standard are relevant: A = K049, K051; B = K048, K049, K051, K052; C = K051; D = K048, K049, K050, K051, K052; E = K048, K049, K051; F = K052; G = K048, K049, K051; H = K048, K051.

way dilute a restricted waste or the residual from treatment of a restricted waste as a substitute for adequate treatment."

An active treatment process is required to diminish the toxicity of a hazardous waste and substantially reduce the migration of hazardous constituents.

TABLE 2 BDAT Standards for Refinery Wastes (K048 to K052) [4]

Wastewaters, Total Composition (mg/L)			
Acenaphthene (A)[a]	0.050	Di-n-butyl phthalate (I)	0.060
Anthracene (B)	0.039	Ethylbenzene (C)	0.011
Benzene (C)	0.011	Fluorene (I)	0.050
Benzo(a)anthracene (A)	0.043	Naphthalene (C)	0.033
Benzo(a)pyrene (D)	0.047	Phenanthrene (C)	0.039
Bis(2-ethylhexyl) phthalate (E)	0.043	Phenol (D)	0.047
Carbon disulfide (F)	0.011	Pyrene (E)	0.045
Chrysene (G)	0.043	Toluene (C)	0.011
o-Cresol (H)	0.011	Xylenes (C)	0.011
p-Cresol (H)	0.011	Chromium (total) (D)	0.020
2,4-Dimethylphenol (B)	0.033	Lead (D)	0.037

[a]Letters refer to the specific waste for which the constituent and standard are relevant: A = K051; B = K049, K051; C = K048, K049, K051, K052; D = K048, K049, K050, K051, K052; E = K048, K049, K051; F = K049; G = K048, K049, K051; H = K052; I = K048, K051.

No-Migration Petitions

A petroleum refinery also can continue to use land treatment for its wastes if it is able to demonstrate that the wastes remain contained in the soil. This is accomplished by preparing a "no-migration petition" and having such petitions be accepted by the appropriate regulatory agency.

A no-migration variance is a decision by EPA to allow the land disposal, at a particular facility, of specific wastes not meeting the treatment (BDAT) standards established for those wastes. A no-migration petition must demonstrate "to a reasonable degree of certainty that there will be no migration of hazardous constituents from the disposal unit or injection zone for as long as the waste remains hazardous."

The term "to a reasonable degree of certainty" is interpreted to mean that hazardous constituents applied to a land disposal unit will not exceed human health-based or environmentally protective levels beyond the edge of the disposal unit which normally is defined as the limit of the engineered components. This definition does not allow for fate and transport of hazardous constituents above acceptable levels outside the boundary of the unit. The petition must demonstrate that the no-migration standard is met for all media. Many wastes that are currently land-treated contain significant amounts of organic constituents and/or metals. Therefore, the petition must demonstrate a sufficient level of degradation and/or immobilization of waste constituents and metals within the treatment zone. For the volatile constituents, the applicant must demonstrate that the health-based or environmental standards are not exceeded by air emissions.

Land Treatment and BDAT

Although no BDATs have yet been established for petroleum industry wastes, the general approach is for establishing such BDATs by EPA is clear, and example performance standards are available. This allows a comparison of land treatment as used in the petroleum industry to the approach being used by EPA to assure that land disposal methods are protective of human health and the environment. To be considered a demonstrated facility, full-scale facilities must be known to be in operation for a waste or similar waste. To be available, a demonstrated technology must substantially diminish the toxicity of a waste or substantially reduce the likelihood of migration of hazardous constituents from the waste.

Air stripping, biological treatment, and carbon adsorption are acceptable BDATs for solvents. Ninety-five percent reduction of volatile organics is judged an acceptable performance standard. An active treatment process is required.

Available data indicate that land treatment, as practiced by the petroleum industry, achieves the equivalent or greater performance of BDATs and in most cases also results in no migration of hazardous constituents. Land treatment for petroleum wastes has been used for several decades at location throughout the world, with a variety of wastes (Table 3) and for small and large facilities. Thus, petroleum waste land treatment is a demonstrated technology.

Land treatment is an active treatment process that accomplishes degradation of organics, immobilization of inorganics, and results in the transformation of hazardous constituents to less and nonhazardous constituents. Land treatment results in biological treatment of organics, such as occurs in a conventional waste treatment system; sorption of organics to the organic carbon in the soil, such as occurs in more conventional carbon adsorption processes; immobilization of metals in the soil; and small quantities of volatile emissions, such as occur in air stripping and biological treatment.

Land treatment of petroleum industry wastes substantially reduces the migration of hazardous constituents from the soil. Table 4 illustrates results from a detailed laboratory petroleum industry land-treatment study that evaluated average as well as extreme rainfall events [5].

Most of the compounds did not migrate below 0.5 ft. Leachate collected from these columns indicated that no PAH or volatile compounds leached during the study, nor did the leachate exhibit any toxicity. Similar results have been obtained in field studies as well as other laboratory studies. Land treatment of petroleum industry wastes reduces the toxicity of the applied wastes. Table 2 indicates that the soil samples below the zone of incorporation did not exhibit any toxicity using the standard "Microtox" procedure. Other examples of toxicity reduction are noted in the following paragraphs.

Mutagenicity determinations used to evaluate the detoxification of petroleum wastes indicated a reduction from mutagenic to nonmutagenic activity when API separator sludge slop oil emulsion solids were incubated with clay loam and sand loam soils, respectively [6]. When the land treatment of an oily waste was evaluated, waste application and soil tilling did reduce the number of earthworms at a field site [7]. However, no long-term toxicity resulted. As the waste constituents were degraded and immobilized, the soil biota were able to recover and return to background conditions.

TABLE 3 Types of Wastes Applied to Full-Scale Refinery Land-Treatment Facilities

Oil–water separator sludge
Dissolved air flotation sludge
Induced air flotation sludge
Tank bottoms
Slop oil emulsion solids
Oily sludge
Filter clays
Biological sludges
Cooling tower sludge
Lime sludge
Coker blowdown sludge

TABLE 4 Organic Compounds Detected in Soil Collected from Below ZOI of a Controlled Land-Treatment Study[a]

	Depth below ZOI (ft)				
	0–1.5		1.5–2.5		2.5–3.5
Compound	Column A	All Other Columns	Column A	All Other Columns	All Columns
Volatiles:					
Benzene	<0.91	<0.91	<0.18	<0.18	<0.18
Toluene	<0.90	<0.90	<0.18	<0.18	<0.18
Ethylbenzene	<0.90	<0.90	<0.18	<0.18	<0.18
p-Xylene	<0.89	<0.89	<0.18	<0.18	<0.18
m-Xylene	<0.90	<0.90	<0.18	<0.18	<0.18
o-Xylene	<0.91	<0.91	<0.18	<0.18	<0.18
Naphthalene	<0.78	<0.78	<0.16	<0.16	<0.16
2-Methylnaphthalene	0.77	<0.54	<0.11	<0.11	<0.11
1-Methylnaphthalene	4.68	<1.02	<0.20	<0.20	<0.20
PAH:					
Naphthalene	<5.0	<5.0	<5.0	<5.0	<5.0
Acenaphthylene	<7.5	<7.5	<7.5	<7.5	<7.5
Acenaphthene	<2.5	<2.5	18.40	<2.5	<2.5
Fluorene	<0.5	<0.5	<0.5	<0.5	<0.5
Phenanthrene	4.125	<0.5	<0.5	<0.5	<0.5
Anthracene	<0.5	<0.5	<0.5	<0.5	<0.5
Fluoranthene	4.6	<0.5	<0.5	<0.5	<0.5
Pyrene	110.0	<1.0	<1.0	<1.0	<1.0
Benzo(a)anthracene	<0.5	<0.5	<0.5	<0.5	<0.5
Chrysene	<0.5	<0.5	<0.5	<0.5	<0.5
Benzo(b)fluoranthene	<0.5	<0.5	<0.5	<0.5	<0.5
Benzo(k)fluoranthene	15.28	<0.5	<0.5	<0.5	<0.5
Benzo(a)pyrene	<0.5	<0.5	<0.5	<0.5	<0.5
Dibenzo(a,h)anthracene	<0.5	<0.5	<0.5	<0.5	<0.5
Benzo(g,h,i)perylene	15.58	<1.0	<1.0	<1.0	<1.0
Indeno(1,2,3,cd)pyrene	6.42	<0.5	<0.5	<0.5	<0.5
Oil and grease	13,600	<600	<600	<600	<600
Microtex (% luminescence)	NT[b]	NT	NT	NT	NT

[a]Source: API, 1987 [5]; data are from soil columns that had petroleum industry wastes applied at realistic land treatment rates to the ZOI; all units are of compound per kg of dry soil except for Microtox.
[b]NT, nontoxic.

Considerable data exist on the immobilization of metals at petroleum industry land-treatment facilities. The metals in the applied waste are immobilized in the treatment zone, primarily in the upper soil layers [2].

Soils from petroleum industry land-treatment facilities have been evaluated using the EPA extraction procedure (EP). The metal concentrations in the EP extract were significantly lower than the EP limits used to define a hazardous material [2]. Thus, although the metals are immobilized and do accumulate, they do not appear to be of environmental concern.

Industry studies have shown that air emissions do occur at petroleum industry land-treatment facilities [2]. However, there has been no indication that such emissions have caused adverse impacts on human health and the environment.

Available information demonstrates that land treatment is an active treatment process that differs from other hazardous waste land disposal methods. Land treatment is not merely dilution nor merely storage; it is a treatment process that accomplishes degradation, immobilization, and transformation.

Land treatment of petroleum industry wastes reduces the toxicity of the applied waste and significantly reduces the migration of hazardous constituents. Numerous studies have shown that such constituents do not migrate deep into the treatment zone and, therefore, do not migrate through the 5-ft treatment zone. Land treatment does reduce the risk to human health and the environment from constituents in petroleum industry wastes and achieves performance standards similar to those established for BDAT.

This material is an adaptation of an article by R. C. Loehr, J. R. Ryan, and J. E. Rucker that appeared in *Oil & Gas Journal*, pp. 40 ff., November 2, 1987, copyright © 1987 by Pennwell Publishing Co., Tulsa, Oklahoma 74121, and is reprinted by special permission.

References

1. *Guidance Manual on Hazardous Waste Land Treatment Closure/Post Closure*, 40 CFR Part 265, Interim Final, Environmental Protection Agency, Office of Solid Waste, Washington, D.C., April 1987.
2. *Land Treatment Practices in the Petroleum Industry*, American Petroleum Institute, Washington, D.C., June 1983.
3. Environmental Protection Agency, "Hazardous Waste Management System: Land Disposal Restrictions, Final Rule," *Fed. Regist.*, pp. 40571–40654 (November 7, 1986).
4. Environmental Protection Agency, "Land Disposal Restrictions for First Third Scheduled Wastes; Final Rule," *Fed. Regist.*, pp. 31188–31222 (August 17, 1988).
5. *The Land Treatability of Appendix VIII Constituents Present in Petroleum Refinery Waste: Laboratory and Modeling Studies*, American Petroleum Institute, Washington, D.C., May 1987.
6. R. C. Sims, J. L. Sims, D. L. Sorenson, W. J. Doucette, and L. L. Hastings, "Waste/Soil Treatability Studies for Four Complex Industrial Wastes: Methodologies and Results," in *Waste Loading Impacts on Soil Degradation, Transformation, and Immobilization*, Vol. 2, Utah State University; National Technical Information Service, U.S. Department of Commerce, Washington, D.C., PB 87-111746, October 1986.
7. R. C. Loehr, J. H. Martin, E. F. Neuhauser, R. A. Horton, and M. R. Malecki, *Land Treatment of an Oily Waste—Degradation, Immobilization and Bioaccumulation*, Cornell University; National Technical Information Service, U.S. Department of Commerce, Washington, D.C., PB 85-166353, February 1985.

RAYMOND C. LOEHR

Petroleum Refining Processes, United States Capacities*

For a better understanding of the survey, the following information may be of help.

Calendar-day figures reported in this survey are the average volume a refinery unit processes each day including downtime used for turnarounds. This is actual total volume for the year divided by 365.

Stream-day figures represent the amount a unit can process when running full capacity for short periods.

This survey includes only facilities charging whole crude, plus lube plants not charging whole crude.

Totals. When an asterisk (*) appears beside a figure in the crude capacity columns, it indicates that the figure was not reported but has been converted into either a calendar-day or stream-day figure using the crude conversion factor of 0.95.

Legend. Hydrocracking includes those processes where 50% or more of the feed is reduced in molecular size; for hydrorefining, 10% of the feed or less is reduced in molecular size; and for hydrotreating, essentially no reduction in molecular size of feed occurs.

Values for hydrogen volumes represent either generation or upgrading to 90+% purity.

Cat reforming definitions are:

Semiregenerative is characterized by shutdown to the reforming unit at specified intervals or at the operator's convenience for regeneration of catalyst *in situ*.
Cyclic is characterized by continuous or continual regeneration of catalyst *in situ* in any one of several reactors that can be isolated from and returned to the reforming operation. This is accomplished without changing feed rate or octane.
Other includes: Nonregenerative—catalyst replaced by fresh catalyst; continuous—continuous regeneration of a part of the catalyst in a special regenerator followed by continuous addition to the reactor; moving-bed catalyst systems.

---

*Editor's Note: These surveys are updated each March by the *Oil & Gas Journal* staff. All figures are in barrels per calendar day and are as of January 1, 1989.

Detailed United States breakdown by company, plant, and state are given in the tables. Included are:

United States crude capacity and downstream production by states (Table 1).
Detailed United States survey (Table 2).
Inactive refineries (Table 2).

The following numbers identify processes in Table 2:

Thermal Processes:
 1. Gas-oil cracking
 2. Thermal cracking
 3. Visbreaking
 4. Coking (fluid)
 5. Coking (delayed)
 6. Other

Cat Cracking:
 1. Fluid
 2. Other

Cat Reforming:
 Semiregenerative:
  1. Conventional catalyst
  2. Bimetallic catalyst
 Cyclic:
  3. Conventional catalyst
  4. Bimetallic catalyst
 Other:
  5. Conventional catalyst
  6. Bimetallic catalyst

Cat Hydrocracking:
 1. Distillate upgrading
 2. Residual upgrading
 3. Lube-oil manufacturing
 4. Other

Cat Hydrorefining:
 1. Residual desulfurizing
 2. Heavy gas-oil desulfurizing
 3. Cat-cracker and cycle-stock feed pretreatment
 4. Middle distillate
 5. Other

Cat Hydrotreating:
 1. Pretreating cat-reformer feeds
 2. Naphtha desulfurizing

3. Naphtha olefin or aromatics saturation
   4. Straight-run distillate
   5. Other distillate
   6. Lube-oil "polishing"
   7. Other

Alkylation:
   1. Sulfuric acid
   2. Hydrofluoric acid

Aromatics/Isomerization:
   1. BTX
   2. Hydrodealkylation
   3. Cyclohexane
   4. C_4 feed
   5. C_5 feed
   6. C_5 and C_6 feed

Hydrogen:
   1. Steam methane reforming
   2. Steam naphtha reforming
   3. Partial oxidation
   4. Cryogenic
   5. Other

This material appeared in *Oil & Gas Journal*, March 20, 1989, and is reprinted by special permission. Copyright © 1989 by Pennwell Publishing Co., Tulsa, Oklahoma 74121.

DEBRA A. GWYN

TABLE 1 Survey of Operating Refineries in the United States (state capacities as of January 1, 1989)

State	No. Plants	Crude Capacity b/cd	Crude Capacity b/sd	Vacuum Distillation	Thermal Operations	Cat Cracking Fresh Feed	Cat Cracking Recycle	Cat Reforming	Cat Hydro-cracking	Cat Hydro-refining	Cat Hydro-treating	Alky Poly.	Aromatics-Isomerization	Lubes	Asphalt	Hydrogen (MMcfd)	Coke (t/d)
Alabama	3	138,750	143,900	48,000	10,000	—	—	27,000	—	13,000	43,500	—	6,000	—	20,500	8.0	400
Alaska	6	217,000	234,437	6,000	14,000	—	—	12,000	9,000	—	12,000	—	6,500	—	8,000	12.8	—
Arizona	1	5,710	6,000	2,000	—	—	—	—	—	—	—	—	—	—	1,160	—	—
Arkansas	3	58,570	61,000	30,500	—	18,500	775	9,000	—	—	20,000	4,800	3,000	4,000	9,700	2.8	—
California	31	2,278,593	2,418,930	1,350,265	503,200	645,500	14,000	561,500	385,000	525,000	948,450	127,200	19,700	29,900	94,754	997.4	20,092
Colorado	2	72,000	76,000	23,000	—	23,000	1,000	16,500	—	—	26,500	3,800	—	—	7,400	—	—
Delaware	1	140,000	150,000	95,000	46,000	65,000	5,000	56,000	19,000	—	110,000	13,500	3,370	—	—	40.0	2,180
Georgia	2	36,800	40,000	5,000	—	—	—	—	—	—	—	—	—	—	27,500	—	—
Hawaii	2	129,500	135,000	68,000	13,000	22,000	—	12,000	16,000	—	14,500	5,625	1,500	—	1,300	19.5	—
Illinois	6	920,600	977,000	361,000	106,400	342,000	12,400	294,300	66,000	35,000	546,800	91,000	30,200	4,600	36,600	66.8	6,300
Indiana	4	422,900	441,000	232,200	28,500	159,000	4,000	99,000	—	80,000	152,800	33,900	36,000	6,400	43,500	—	1,550
Kansas	8	344,025	363,583	124,650	55,690	127,000	5,000	98,300	—	44,000	165,300	39,100	43,469	2,500	9,000	—	2,070
Kentucky	2	218,900	226,300	92,000	57,600	100,000	—	53,000	—	40,000	142,300	13,000	17,600	7,900	30,000	20.0	—
Louisiana	17	2,274,515	2,374,300	908,300	408,500	736,500	13,300	494,300	149,000	301,000	942,600	202,200	79,300	37,000	55,500	162.6	14,775
Michigan	4	120,100	127,095	30,000	—	45,500	1,300	33,500	—	19,800	40,000	9,600	6,000	—	10,000	—	—
Minnesota	2	285,600	299,220	192,000	58,000	78,000	1,000	55,500	—	86,500	127,000	19,150	23,300	—	49,000	20.0	2,800
Mississippi	5	358,600	378,879	285,600	70,000	72,000	2,000	95,800	68,000	194,000	59,800	19,700	5,500	5,000	40,600	217.5	3,450
Montana	4	137,200	143,000	55,250	7,700	50,600	7,700	35,500	4,900	14,000	100,200	11,670	5,800	—	23,500	19.3	435
Nevada	1	4,500	4,700	2,500	—	—	—	—	—	—	—	—	—	—	—	—	—
New Jersey	6	464,250	486,368	249,400	31,500	256,000	25,000	89,500	—	65,000	270,600	30,500	43,000	8,500	73,000	11.0	1,010
New Mexico	3	72,800	77,107	13,900	—	27,100	6,600	17,400	—	—	29,100	5,400	4,000	—	4,100	—	—
New York	1	42,750	45,000	27,000	—	—	—	—	—	—	—	—	—	—	17,000	—	—
North Dakota	1	58,000	60,000	—	—	26,000	5,200	12,100	—	—	16,600	5,100	4,000	—	—	—	—
Ohio	4	482,650	508,000	163,000	29,900	176,000	8,800	160,600	86,200	23,000	169,500	26,600	65,400	2,100	19,000	65.0	1,250
Oklahoma	6	387,500	404,316	141,400	20,500	136,500	5,840	91,000	5,000	20,000	137,000	36,100	29,400	9,500	18,100	10.0	1,040
Oregon	1	15,000	15,789	16,000	—	—	—	—	—	—	—	—	—	—	11,500	—	—
Pennsylvania	8	730,400	768,000	320,180	—	239,300	6,800	208,820	51,000	50,000	430,900	49,000	21,050	21,310	43,000	49.5	—
Tennessee	1	57,000	60,000	12,000	—	30,000	—	10,000	—	—	31,000	5,500	4,000	—	3,500	—	—
Texas	31	4,058,350	4,294,500	1,826,300	355,000	1,616,500	134,250	1,150,900	276,500	827,000	2,249,950	301,750	304,215	88,800	67,400	614.0	11,694
Utah	6	154,500	150,400	45,500	8,500	55,400	9,100	29,100	—	7,100	35,100	14,500	8,050	—	1,700	—	350
Virginia	1	51,000	53,000	29,000	13,500	27,500	2,000	9,750	—	—	25,500	2,400	—	—	—	—	750
Washington	7	469,675	489,517	220,500	70,000	102,500	13,000	119,800	52,000	25,500	180,000	27,300	4,250	—	16,600	80.0	3,500
West Virginia	2	15,500	16,000	10,850	—	—	—	4,900	4,500	—	3,900	—	—	5,559	—	1.2	—
Wisonsin	1	32,000	34,000	20,500	—	11,000	1,000	8,000	—	5,800	9,000	1,300	—	—	13,500	—	—
Wyoming	5	163,500	171,500	75,000	6,600	64,500	10,000	36,850	—	21,000	59,950	10,700	1,800	1,500	6,000	—	—
Total	188	15,418,738	16,243,841	7,082,395	1,914,090	5,252,900	295,065	3,901,920	1,192,100	2,396,700	7,099,850	1,110,395	776,404	234,569	762,414	2,417.4	73,646

TABLE 2 United States Refineries: Location, Capacities, Types of Processing

Company and Location	Crude Capacity b/cd	Crude Capacity b/sd	Vacuum Distillation	Thermal Operations	Cat Cracking Fresh Feed	Cat Cracking Recycle	Cat Reforming	Cat Hydrocracking	Cat Hydrorefining	Cat Hydrotreating	Alky. *Poly	Aromatics-Isomerization	Lubes	Asphalt	Hydrogen (MMcfd)	Coke (t/d)
Alabama:																
Belcher Refining Co.—Mobile Bay	*14,250	15,000	14,000	—	—	—	—	—	—	—	—	—	—	10,000	—	—
Hunt Refining Co.—Tuscaloosa	44,500	47,600	14,000	⁵10,000	—	—	²6,000	—	²9,000	⁶6,000 / ³1,500	—	—	—	10,500	⁸8.0	400
Louisiana Land & Exploration Co.—Saraland	80,000	81,300	20,000	—	—	—	²21,000	—	⁴4,000	¹21,000 / ⁷15,000	—	⁶6,000	—	—	—	—
Total	138,750	143,900	48,000	10,000	—	—	27,000	—	13,000	43,500	—	6,000	—	20,500	8.0	400
Alaska:																
ARCO Alaska Inc.—Kuparuk	12,000	12,000	—	—	—	—	—	—	—	—	—	—	—	—	—	—
Prudhoe Bay	14,000	18,000	—	⁶14,000	—	—	—	—	—	—	—	—	—	—	—	—
Chevron U.S.A. Inc.—Kenai	22,000	22,500	6,000	—	—	—	—	—	—	—	—	—	—	6,000	—	—
Mapco Alaska Petroleum—North Pole	*94,737	—	—	—	—	—	—	—	—	—	—	²2,500	—	2,000	—	—
Petro Star Inc.—North Pole	7,000	7,200	—	—	—	—	—	—	—	—	—	—	—	—	—	—
Tesoro Petroleum Corp.—Kenai	72,000	80,000	—	—	—	—	⁴12,000	⁹9,000	—	¹12,000	—	⁴4,000	—	—	¹12.8	—
Total	217,000	234,437	6,000	14,000	—	—	12,000	9,000	—	12,000	—	6,500	—	8,000	12.8	—
Arizona:																
Intermountain Refining Cl—Fredonia	5,710	6,000	2,000	—	—	—	—	—	—	—	—	—	—	1,160	—	—
Arkansas:																
Berry Petroleum Co.—Stevens	3,800	4,000	1,500	—	—	—	—	—	—	—	—	—	—	1,000	—	—
Cross Oil & Refining Co. of Arkansas—Smackover	6,770	7,000	4,000	—	—	—	—	—	—	⁴4,500	—	—	4,000	2,200	—	—
Lion Oil Co.—El Dorado[a]	48,000	50,000	25,000	—	¹18,500	775	⁵9,000	—	—	¹10,000 / ⁵5,000	¹4,800	⁶3,000	—	6,500	²2.8	—
Total	58,570	61,000	30,500	—	18,500	775	9,000	—	—	20,000	4,800	3,000	4,000	9,700	2.8	—
California:																
Anchor Refining Co. Inc.—McKittrick	10,000	11,000	7,000	—	—	—	—	—	—	—	—	—	—	—	—	—
Atlantic Richfield Co.—Carson	214,000	225,000	112,000	⁵56,000	⁸82,000	—	²48,000	¹22,000	—	¹40,000 / ⁴18,000 / ³8,000 / ⁷80,000	¹9,000 *3,000	—	—	—	¹70.0	2,500
Chemoil Refining Corp.—Signal Hill	14,400	16,000	—	—	—	—	—	—	—	—	—	—	—	—	—	—
Chevron U.S.A. Inc.—El Segundo	330,000	380,000	179,000	⁴54,000	⁸60,000	—	²60,000	¹45,000	²24,000 / ⁶60,000 / ⁴14,000	¹56,000 / ⁴18,000	⁸8,000	¹1,500	—	—	²112.0	2,900
Richmond[b]	270,000	280,000	175,000	—	⁸63,000	—	²68,000	¹45,000 / ²30,000 / ³30,500	⁴60,000 / ⁶65,000	¹63,000 / ⁶18,200	⁹9,200 *2,100	—	11,800	11,000	¹135.0	—
Conoco Inc.—Santa Maria	9,500	10,000	7,800	—	—	—	—	—	—	—	—	—	—	6,800	—	—

TABLE 2 (continued)

Company and Location	Crude Capacity		Vacuum Distillation	Thermal Operations	Cat Cracking		Cat Reforming	Cat Hydro-cracking	Cat Hydro-refining	Cat Hydro-treating	Alky. *Poly	Aromatics-Isomerization	Lubes	Asphalt	Hydrogen (MMcfd)	Coke (t/d)
	b/cd	b/sd			Fresh Feed	Recycle										
Edgington Oil Co. Inc.—Long Beach	41,600	44,730	21,165	⁴27,500	¹64,000	11,000	²28,000	¹29,500	—	—	—	—	—	15,554	—	—
Exxon Co.—Benicia	128,000	132,000	67,000	—					¹37,000	¹22,500 ¹22,000 ⁴12,500 ⁵12,500 ¹17,000	¹14,000 *2,000	—	—	—	¹104.0	1,100
Fletcher Oil & Refining Co.—Carson	26,500	27,000	—	—	¹10,500	1,000	⁴5,000	—	²11,500	⁵5,000	—	—	—	—	—	—
Golden Bear Division, Witco Chemical Corp.—Oildale	10,818	11,500	10,700	—	—	—	—	—	—	⁹950	—	—	4,800	4,000	—	—
Golden West Refining Co.—Santa Fe Springs	40,600	42,300	25,000	³13,800	¹13,500	—	²19,000	¹11,000	—	²12,000	³3,000	—	—	4,000	¹11.0	—
Huntway Refining Co.—Benicia	8,400	9,000	7,500	—	—	—	—	—	—	—	—	—	—	4,500	—	—
Wilmington	5,500	6,000	5,000	—	—	—	—	—	—	—	—	—	—	3,500	—	—
Kern Oil & Refining Co.—Bakersfield	20,000	22,000	—	—	—	—	²3,000	—	—	⁴4,500	—	—	—	—	—	—
Lunday-Thagard Co.—South Gate	7,700	8,100	4,200	—	—	—	—	—	—	—	—	—	—	2,400	—	—
Mobil Oil Corp.—Torrance	123,000	130,000	95,000	⁴48,000	⁶3,000	—	²36,000	¹21,700	⁶8,000	¹21,000 ³16,000 ⁴28,000	²17,000	—	—	—	¹137.0	2,900
Newhall Refining Co. Inc.—Newhall	18,800	20,000	10,000	—	—	—	²2,500	—	—	¹12,500 ⁴7,000	—	—	—	5,000	—	—
Oxnard Refinery—Oxnard	*4,275	4,500	—	—	—	—	—	—	—	—	—	—	—	3,000	—	—
Pacific Refining Co.—Hercules	⁵52,250	55,000	17,000	³12,000	—	—	²15,000	—	—	¹15,000	—	—	—	—	—	—
Paramount Petroleum Corp.—Paramount	42,700	46,500	29,000	—	—	—	²10,500	—	²11,000	²11,000 ⁴7,000	—	—	—	15,000	—	—
Powerine Oil Co.—Santa Fe Springs	*46,550	49,000	24,800	⁵10,400	¹12,500	—	¹8,500	¹8,000	¹13,500 ⁴6,000	¹8,500 ⁵5,000	²3,200	⁴1,800	—	4,000	¹19.0	470
San Joaquin Refining Cl—Bakersfield	18,000	20,000	14,000	³10,000	—	—	—	—	—	—	—	—	4,000	5,000	—	—
Shell Oil Co.—Martinez	140,100	143,000	96,500	⁴22,000	¹67,000	1,000	²28,000	¹27,000	⁵50,000	²17,000 ³16,000 ⁵19,000 ⁶6,300 ⁷15,000	¹8,000 *3,200	—	4,500	11,000	¹110.0	127
Wilmington	144,000	149,000	75,000	⁵53,000	¹42,000	—	²24,000	—	—	²25,000 ³14,000 ⁴34,000 ⁷32,000 ⁸1,500	¹8,600	—	—	—	¹36.0	2,500
Sunland Refining Corp.—Bakersfield	15,000	15,000	—	—	—	—	²1,500	—	—	—	—	—	—	—	—	—
Texaco Refining & Marketing Inc.—Bakersfield	47,000	48,000	23,000	⁵11,800	—	—	²23,000 ⁵3,000	¹14,300	²15,000	¹14,000	¹4,400	—	—	—	¹21.0	625
Wilmington	75,000	78,400	42,000	⁴48,000	¹28,000	—	²38,000	¹20,000	—	²18,000 ⁴12,000	—	—	—	—	⁴48.0	1,650
Tosco Corp.—Martinez	126,000	132,600	105,000	⁴42,000	¹55,000	1,000	²20,000 ²3,000	¹23,000	²50,000	¹13,500 ⁴28,000	¹12,000	—	—	—	¹80.0	1,500
Union Pacific Resources—Wilmington	68,000	72,000	42,000	²24,000	¹38,000	—	⁶14,500	—	⁴40,000	¹15,000 ³54,000	²10,500	⁴9,000	—	5,000	—	1,200
Unocal Corp.—Los Angeles	108,000	111,000	83,000	³20,000	²47,000	—	²52,000	¹22,000	—	⁵36,000	²10,000	—	—	—	⁴9.4	—
Rodeo	112,900	120,300	72,600	⁵50,700	—	—	²34,000	¹32,500	—	¹23,000 ⁶9,500 ⁶14,500	—	⁴7,400	4,800	—	¹65.0	2,620
Total	2,278,593	2,418,930	1,350,265	503,200	645,500	14,000	561,500	385,000	525,000	948,450	127,200	19,700	29,000	94,754	997.4	20,092

Company/Location	C1	C2	C3	C4	C5	C6	C7	C8	C9	C10	C11	C12	C13	C14	C15	C16
Colorado:																
Colorado Refining Co.—Commerce City	28,000	30,000	10,000	—	8,000	—	[2]9,000	—	—	[1]9,000 [1]7,500 [4]10,000	*1,200 *2,600	—	—	3,400 4,000	—	—
Conoco Inc.—Commerce City	44,000	46,000	13,000	—	[1]15,000	1,000	[2]7,500	—	—			—	—		—	—
Total	72,000	76,000	23,000	—	23,000	1,000	16,500	—	—	26,500	3,800	—	—	7,400	—	—
Delaware:																
Star Enterprise—Delaware City	140,000	150,000	95,000	[4]46,000	[1]65,000	5,000	[2]18,000 [5]38,000	[4]19,000	—	[1]55,000 [4]55,000	[1]8,000 *5,500	[2]2,900 [2]470	—	—	[1]40.0	2,180
Total	140,000	150,000	95,000	46,000	65,000	5,000	56,000	19,000	—	110,000	13,500	3,370	—	—	40.0	2,180
Georgia:																
Amoco Oil Co.—Savannah	28,000	30,000	—	—	—	—	—	—	—	—	—	—	—	22,500 5,000	—	—
Young Refining Corp.—Douglasville	8,800	10,000	5,000	—	—	—	—	—	—	—	—	—	—		—	—
Total	36,800	40,000	5,000	—	—	—	—	—	—	—	—	—	—	27,500	—	—
Hawaii:																
Chevron U.S.A. Inc.—Barber's Point	48,000	50,000	28,000	—	[1]22,000	—	—	—	—	[3]3,500	[4]4,500 1,125[c]	[4]1,500	—	1,300	[1]2.5	—
Hawaiian Independent Refinery Inc.—Ewa Beach	81,500	85,000	40,000	[3]13,000	—	—	[2]12,000	[1]16,000	—	[1]11,000	—	1,500	—	—	[2]17.0 19.5	—
Total	129,500	135,000	68,000	13,000	22,000	—	12,000	16,000	—	14,500	5,625	1,500	—	1,300	19.5	—
Illinois:																
Clark Oil & Refining Corp.—Blue Island Hartford	64,600 60,000	68,000 65,000	27,000 18,000	— [6]14,500	[1]25,000 [1]26,000	1,000 1,000	[2]30,500 [2]12,000	[1]9,500 —	— —	[1]20,500 [1]12,000 [2]4,000 [3]8,000	[6]6,000 [3]8,000	[4]4,000	— —	4,500 —	— [1]2.5	— 750
Marathon Petroleum Co.—Robinson	195,000	205,000	62,000	[4]4,000 [5]22,000	[1]41,000	400	[4]38,000 [4]41,000 [2]46,000	[1]23,000	[5]6,000	[1]65,000 [4]11,500 [1]72,000 [2]9,000 [3]75,000	[2]12,000 [3]25,000	[6]11,500	—	—	[1]25.0	1,200
Mobil Oil Corp.—Joliet	180,000	200,000	88,000	[5]38,000	[1]98,000	—		—	—			—	—	—	—	2,350
Shell Oil Co.—Wood River	274,000	286,000	108,000	—	[1]94,000	—	[2]22,000 [3]75,000	[1]33,500	[3]29,000	[1]64,000 [2]52,500 [6]10,500 [3]35,000	[1]22,000	[1]3,800	4,600	28,500	[1]28.3	—

(continued)

TABLE 2 (*continued*)

Company and Location	Crude Capacity		Vacuum Distillation	Thermal Operations	Cat Cracking		Cat Reforming	Cat Hydro-cracking	Cat Hydro-refining	Cat Hydro-treating	Alky. *Poly	Aromatics-Isomerization	Lubes	Asphalt	Hydrogen (MMcfd)	Coke (t/d)
	b/cd	b/sd			Fresh Feed	Recycle										
Unocal Corp.—Lemont	147,000	153,000	58,000	¹27,900	¹58,000	10,000	²29,800	—	—	¹29,800 ²22,200 ³6,400 ⁵39,000 ⁷2,100 ⁴4,300	²18,000	¹3,500 ⁶7,400	—	3,600	¹11.0	2,000
Total	920,600	977,000	361,000	106,400	342,000	12,400	294,300	66,000	35,000	546,800	91,000	30,200	4,600	36,600	66.8	6,300
Indiana:																
Amoco Oil Co.—Whiting	350,000	360,000	203,000	²28,500	¹140,000	4,000	⁸85,000	—	⁸80,000	¹87,000 ²42,000 ⁶4,300	¹26,000	¹13,000 ⁶21,000	6,400	40,000	—	1,550
Indiana Farm Bureau Cooperative Association Inc.—Mt. Vernon	20,600	21,500	7,200	—	—	—	²4,000	—	—	²6,000	²1,700	⁶2,000	—	3,500	—	—
Laketon Refining Corp.—Laketon	8,300	10,000	5,000	—	—	—	—	—	—	—	—	—	—	—	—	—
Rock Island Refining Corp.—Indianapolis⁵	44,000	49,500	17,000	—	¹19,000	—	²10,000	—	—	¹13,500	⁶6,200	—	—	—	—	—
Total	422,900	441,000	232,200	28,500	159,000	4,000	99,000	—	80,000	152,800	33,900	36,000	6,400	43,500	—	1,550
Kansas:																
Coastal Refining and Marketing Inc.—Augusta	*28,500	30,000	10,000	—	¹14,500	—	²10,000 ²4,500	—	—	¹13,000 ⁴4,200	—	⁶6,000	—	2,500	—	—
Derby Refining Co.—El Dorado⁴	*29,925	31,500	10,000	⁵5,500	¹19,000	—	²6,500	—	—	²7,000	¹2,800 ²2,800 ⁵3,500	—	—	—	—	210
Farmland Industries Inc.—Coffeyville	56,500	60,723	19,500	²12,000	²3,000	1,500	²16,000 ⁵5,300	—	—	²6,500 ⁵7,500	²6,000	⁶8,000	2,500	—	—	600
Phillipsburg	26,400	27,460	10,000	—	—	—	—	—	—	—	—	—	—	2,000	—	—
National Cooperative Refinery Association—McPherson	70,900	75,000	27,000	²22,000	¹20,000	1,000	²15,000	—	—	¹10,000 ²14,500 ⁴13,000	²6,000	⁴2,000 ⁹9,500	—	—	—	650
Texaco Refining & Marketing Inc.—El Dorado	78,200	82,900	32,000	¹13,000	¹31,500	2,500	²25,000	—	⁸44,000	¹20,000 ²20,000 ⁴4,800 ⁴9,000	²12,500	¹2,969 ⁶15,000	—	—	—	610
Total Petroleum Inc.—Arkansas City	53,600	56,000	16,150	³3,190	¹19,000	—	²16,000	—	—	¹16,000	²5,500	—	—	4,500	—	—
Total	344,025	363,583	124,650	55,690	127,000	5,000	98,300	—	44,000	165,300	39,100	43,469	2,500	9,000	—	2,070
Kentucky:																
Ashland Petroleum Co.—Catlettsburg	213,400	220,000	92,000	²2,600 ⁶55,000ᶜ	¹60,000 ²40,000ᶜ	—	²25,000 ⁵27,000	—	⁸40,000	¹60,000 ²6,000 ⁶6,000 ⁴40,000 ⁷5,000 ⁷12,000	²12,000 *1,000	¹5,400 ⁶12,000	7,900	30,000	¹20.0	—
Somerset Refinery Inc.—Somerset	5,500	6,300	—	—	—	—	⁵1,000	—	—	⁴1,300	—	⁶200	—	—	—	—
Total	218,900	226,300	92,000	57,600	100,000	—	53,000	—	40,000	142,300	13,000	17,600	7,900	30,000	20.0	—

Refinery														
Louisiana:														
Atlas Processing Co., Division of Pennzoil—Shreveport	46,200	50,000	24,300	—	—	—	²10,000	—	—	¹10,000 ⁶3,400 ⁷1,200 ⁵5,500	—	—	8,500	⁶6.1	—
Calcasieu Refining Co.—Lake Charles	12,000	13,500	—	—	—	—	—	—	—	—	—	—	—	—	—
Calumet Refining Co.—Princeton	4,100	4,400	8,000	—	—	—	—	—	—	—	—	—	3,000	—	—
Canal Refining Co.—Church Point	9,865	10,000	—	—	—	—	¹1,900	—	—	—	—	—	—	¹,000	—
Citgo Petroleum Corp.—Lake Charles	320,000	330,000	83,000	⁶63,000	¹150,000	—	²46,000 ⁴45,000	³37,000	⁴40,000	⁹1,000 ⁴14,000	²20,200	—	9,000	—	3,000
Conoco Inc.—Westlake	156,500	164,000	63,000	²12,000 ⁵53,000	¹40,000	—	¹5,900 ⁵11,500	—	—	³33,000 ⁵109,000 ⁷13,500	¹7,500 *2,100	—	—	—	3,400
Exxon Co.—Baton Rouge⁸	455,000	474,000	205,000	⁵90,000	¹155,000	—	⁴90,000	²24,000	—	⁵95,000 ¹2,500 ²55,000 ⁶17,000 ⁷2,000 ⁴43,000	¹33,200 ⁶8,000	⁵7,500	16,500	28,900	4,980
Hill Petroleum Co.—Krotz Springs	55,300	57,500	24,000	—	¹28,000	—	²12,500	—	—	¹12,500	*6,600	—	—	—	—
Kerr-McGee Refining Corp.—Cotton Valley	33,000	34,000	20,000	—	—	—	—	—	—	—	—	—	—	—	—
Marathon Petroleum Co.—Garyville	7,800	8,500	—	—	—	1,000	—	—	²71,000 ⁴38,000	⁴48,000 ⁴16,000	²26,000	⁴18,500 ⁶16,000	—	25,000	—
Mobil Oil Corp.—Chalmette	255,000	263,000	125,000	—	¹85,000	—	⁴48,000	¹18,000	⁴43,000 ⁴24,000	⁴45,000	²19,000	¹7,000	—	—	1,550
Murphy Oil USA Inc.—Meraux⁶	145,000	155,000	70,000	⁵33,000	¹51,000	—	²28,000 ³19,000	—	⁴15,000	²29,000	¹8,700	—	—	—	—
Placid Refining Co.—Port Allen¹	92,500	95,400	40,000	—	¹35,000	2,500	⁶23,000 ²10,000	—	—	¹10,000	²3,600	—	—	—	—
Shell Oil Co.—Norco	*47,500	50,000	20,000	—	¹18,500	2,500	²18,000 ¹38,000	¹35,000	¹70,000	²29,000 ²28,000	¹15,000 ⁶9,400	—	—	—	—
	215,000	220,000	78,000	⁶21,000 ⁶17,700 ⁶85,800	—	—							—	—	1,000
Sohio Oil Co. (BP America)—Belle Chasse	*194,750	205,000	73,000	¹21,000	¹89,000	2,300	²37,500	—	—	²42,000 ⁴24,000 ⁵22,000	¹28,400	¹24,000 ²6,300	—	—	845
Star Enterprise—Convent	225,000	240,000	75,000	³12,000	¹85,000	5,000	²40,000	²35,000	—	⁴40,000 ⁴37,000 ⁶65,000	¹14,500	—	—	—	—
Total	2,274,515	2,374,300	908,300	408,500	736,500	13,300	494,300	149,000	301,000	942,600	202,200	79,300	37,000	55,500	14,775
													162.6		

(continued)

TABLE 2 (continued)

Company and Location	Crude Capacity		Vacuum Distillation	Thermal Operations	Cat Cracking		Cat Reforming	Cat Hydro-cracking	Cat Hydro-refining	Cat Hydro-treating	Alky. *Poly	Aromatics-Isomerization	Lubes	Asphalt	Hydrogen (MMcfd)	Coke (t/d)
	b/cd	b/sd			Fresh Feed	Recycle										
Michigan:																
Crystal Refining Co.—Carson City	4,000	6,200	—	—	—	—	—	—	—	—	—	—	—	—	—	—
Lakeside Refining Co.—Kalamazoo	5,600	*5,895	—	—	—	—	—	—	—	—	—	—	—	—	—	—
Marathon Petroleum Co.—Detroit	68,500	71,000	30,000	—	¹27,000	1,300	¹1,000	—	¹14,000	¹21,000	¹4,000	—	—	10,000	—	—
Total Petroleum Inc.—Alma	42,000	44,000	—	—	²18,500	—	²18,500 ²14,000	—	²3,800 ²2,000	¹19,000	²4,600 *1,000	⁶6,000	—	—	—	—
Total	120,100	127,095	30,000	—	45,500	1,300	33,500	—	19,800	40,000	9,600	6,000	—	10,000	—	—
Minnesota:																
Ashland Petroleum Co.—St. Paul Park	67,100	69,220	32,000	—	¹23,000	—	¹23,500	—	¹23,000	¹24,500 ²7,200 ⁸8,300	²5,500 *350	⁸8,300	—	14,000	—	—
Koch Refining Co.—Rosemount	*218,500	230,000	160,000	⁵58,000	¹55,000	1,000	⁶26,000 ⁶6,000	—	¹63,500	¹26,000 ²15,000 ⁴6,000	⁸8,500 *1,100 ³3,700°	⁸15,000	—	35,000	²20.0	2,800
Total	285,600	299,220	192,000	58,000	78,000	1,000	55,500	—	86,500	127,000	19,150	23,300	—	49,000	20.0	2,800
Mississippi:																
Amerada-Hess Corp.—Purvis	30,000	*31,579	20,000	⁴8,000	²16,000	—	²5,800	—	—	²5,800 ⁴6,000	¹3,500	—	—	—	—	250
Chevron U.S.A. Inc.—Pascagoula	295,000	310,000	243,000	⁶62,000	¹56,000	2,000	²90,000	⁶68,000	¹96,000 ²63,000 ⁴30,000 ⁵5,000	¹48,000	¹16,200	⁵5,500	—	20,000	²15.0	3,200
Ergon Refining Inc.—Vicksburg	16,800	18,300	15,600	—	—	—	—	—	—	—	—	—	5,000	12,000	¹2.5	—
Southland Oil Co.—Lumberton	5,800	6,500	—	—	—	—	—	—	—	—	—	—	—	3,500	—	—
Sandersville	11,000	12,500	7,000	—	—	—	—	—	—	—	—	—	—	5,100	—	—
Total	358,600	378,879	285,600	70,000	72,000	2,000	95,800	68,000	194,000	59,800	19,700	5,500	5,000	40,600	217.5	3,450
Montana:																
Cenex—Laurel¹	40,400	42,500	14,000	—	¹12,000	3,000	²12,000	—	⁴14,000	¹15,000 ⁴38,000 ³3,500	²3,000	²2,000	—	6,000	—	—
Conoco Inc.—Billings*	48,500	50,000	20,000	—	¹15,500	1,000	²12,500	—	—		²5,000	³3,400	—	6,500	—	—
Exxon Co.—Billings	42,000	44,000	18,000	⁴⁷7,700	¹21,000	3,500	¹10,000	⁴4,900	—	¹15,500 ⁴10,000 ⁵10,000 ⁶6,000	²3,400	—	—	11,000	⁵19.3	435
Montana Refining Co.—Great Falls	6,300	6,500	3,250	—	²2,100	200	²1,000	—	—	¹1,000 ⁴1,200	*270	⁶400	—	—	—	—
Total	137,200	143,000	55,250	7,700	50,600	7,700	35,500	4,900	14,000	100,200	11,670	5,800	—	23,500	19.3	435
Nevada:																
Nevada Refining Co.—Tonopah	4,500	4,700	2,500	—	—	—	—	—	—	—	—	—	—	—	—	—
New Jersey:																
Amerada-Hess Corp.—Port Reading	—	—	—	—	¹50,000	—	—	—	—	—	⁴4,500 *5,000	—	—	—	—	—

Company—Location																
Chevron U.S.A. Inc.—Perth Amboy	80,000	85,000	46,000	—	—	—	—	—	—	—	—	—	—	35,000	—	—
Coastal Eagle Point Oil Co.—Westville	*109,250	115,000	45,000	³10,000	¹50,000	—	—	—	—	⁴40,000 ³30,000 ⁵17,000 ⁷4,800	¹3,000 *2,500	—	—	—	—	—
Exxon Co.—Linden	130,000	135,000	66,000	—	¹120,000	25,000	—	²38,000	²50,000	²29,000 ¹9,000 ⁶5,000	¹10,500	²25,000	—	38,000	—	—
Mantua Oil Co. LP—Thorofare	45,000	*47,368	30,000	—	—	—	—	—	—	—	—	—	8,500	—	—	—
Mobil Oil Corp.—Paulsboro¹	100,000	104,000	62,400	⁵21,500	¹36,000	—	²23,500	—	⁵15,000	¹23,500 ⁶42,000 ⁹300	²5,000	—	—	—	¹11.0	1,010
Total	464,250	486,368	249,400	31,500	256,000	25,000	89,500	—	65,000	270,600	30,500	43,000	8,500	73,000	11.0	1,010
New Mexico:																
Bloomfield Refining Co.—Bloomfield	16,800	18,107	—	—	⁵5,400	500	²2,800	—	—	¹2,800 ⁶6,800 ¹¹10,000 ⁴⁶,500 ⁵3,000	*2,000 ²1,400 ²2,000	⁴4,000	—	700	—	—
Giant Industries Inc.—Gallup	18,000	19,000	7,900	—	⁷7,200	3,600	⁶6,800	—	—							
Navajo Refining Co.—Artesia	*38,000	40,000	6,000	—	¹¹14,500	2,500	²7,800	—	—					3,400	—	—
Total	72,800	77,107	13,900	—	27,100	6,600	17,400	—	—	29,100	5,400	4,000	—	4,100	—	—
New York:																
Cibro Petroleum Products Co.—Albany	*42,750	45,000	27,000	—	—	—	—	—	—	—	—	—	—	17,000	—	—
North Dakota:																
Amoco Oil Co.—Mandan	58,000	60,000	—	—	¹26,000	5,200	⁴12,100	—	—	²16,600	²3,400 *1,700	⁴4,000	—	—	—	—
Total	58,000	60,000	—	—	26,000	5,200	12,100	—	—	16,600	5,100	4,000	—	—	—	—
Ohio:																
Ashland Petroleum Co.—Canton	66,000	68,000	33,000	—	¹25,000	—	⁵20,000	—	³23,000	¹20,000 ⁴7,000 ⁷6,500	²7,000 *500	⁶6,500	—	12,000	—	—
Sohio Oil Co. (BP America)—Lima	*171,000	180,000	51,000	⁵16,200	¹36,000	7,800	⁵53,000	⁴35,000	—	¹59,000	—	²24,400 ²6,500 ⁸16,600	2,100	—	—	620
Toledo	*120,650	127,000	49,000	⁵13,700	¹55,000	—	²³23,000 ⁴19,000	⁴35,000	—	¹37,000	²11,300	—	—	7,000	¹24.0	630
Sun Co. Inc.—Toledo⁵	125,000	133,000	30,000	—	¹60,000	1,000	²45,600	²28,200	—	¹40,000	²7,800	⁹19,000 ²2,400	—	—	⁴¹41.0	—
Total	482,650	508,000	163,000	29,900	176,000	8,800	160,600	86,200	23,000	169,500	26,600	65,400	2,100	19,000	65.0	1,250

(continued)

TABLE 2 (continued)

Company and Location	Crude Capacity (b/cd)	Crude Capacity (b/sd)	Vacuum Distillation	Thermal Operations	Cat Cracking Fresh Feed	Cat Cracking Recycle	Cat Reforming	Cat Hydro-cracking	Cat Hydro-refining	Cat Hydro-treating	Alky. *Poly	Aromatics-Isomerization	Lubes	Asphalt	Hydrogen (MMcfd)	Coke (t/d)
Oklahoma:																
Barrett Refining Corp.—Thomas	13,000	*13,684	—	—	—	—	—	—	—	—	—	—	2,000	—	—	—
Conoco Inc.—Ponca City	136,000	140,000	40,000	⁵20,500	¹46,000	—	²34,500	—	—	¹34,500 ²24,000	²12,000 *2,100	⁴4,500	—	—	—	740
Kerr-McGee Refining Corp.—Wynnewood⁰	43,000	45,000	13,000	—	¹20,000	—	²8,500	¹5,000	—	⁹,000 ¹12,000	²5,000	⁴4,000 ⁴500	—	5,000	¹10.0	—
Sinclair Oil Corp.—Tulsa	50,000	*52,632	26,500	—	¹18,000	5,000	²12,000	—	—	⁵,000	³3,000	⁶6,000	—	2,500	—	—
Sun Co. Inc.—Tulsa⁹	85,000	90,000	29,900	—	¹30,000	840	²24,000	—	—	²24,000 ⁶10,500	²7,000	²2,200 ²1,200 ³2,000 ⁴3,000	7,500	4,600	—	300
Total Petroleum Inc.—Ardmore	60,500	63,000	32,000	—	²22,500	—	⁶12,000	—	²20,000	¹18,000	²7,000	⁶6,000	—	6,000	—	—
Total	387,500	404,316	141,400	20,500	136,500	5,840	91,000	5,000	20,000	137,000	36,100	29,400	9,500	18,100	10.0	1,040
Oregon:																
Chevron U.S.A. Inc.—Portland	15,000	15,789	16,000	—	—	—	—	—	—	—	—	—	—	11,500	—	—
Pennsylvania:																
Atlantic Refining & Marketing Corp.—Philadelphia	125,000	130,000	83,000	—	¹29,000	—	²60,000	⁴30,000	—	¹54,000 ⁵50,000 ²24,000	²20,000	—	—	35,000	⁹40.0	—
Chevron U.S.A. Inc.—Philadelphia	174,100	180,000	80,000	—	⁵53,300	5,000	²34,000	—	—	²34,000 ⁴30,000	—	¹4,000 ²1,300	—	—	—	—
Kendall-Amalie Division																
Witco Chemical Co.—Bradford	8,500	11,500	—	—	—	—	²3,300	—	—	—	—	—	4,000	—	—	—
Pennzoil Products Co.—Rouseville	15,700	16,500	6,500	—	—	—	⁵5,820	—	—	¹6,500 ⁷7,800	—	⁶1,150	4,750	—	⁵3.5	—
Quaker State Oil Refining Corp.—Smethport⁹	6,500	7,000	2,680	—	—	—	²2,100	—	—	²2,900	—	⁶800	2,560	—	—	—
Sohio Oil Co. (BP America)—Marcus Hook	*171,000	180,000	75,000	—	¹50,000	1,600	⁶48,000	*21,000	³50,000	¹64,000 ⁴22,000 ²22,000	²12,000	—	—	—	—	—
Sun Co. Inc.—Marcus Hook	165,000	175,000	46,000	—	¹87,000	—	²39,600	—	—	¹54,300 ⁴14,400 ⁵13,000 ⁶6,000	¹12,000	¹7,000	10,000	—	⁶6.0	—
United Refining Co.—Warren	*64,600	68,000	27,000	—	¹20,000	200	²16,000	—	—	²20,000 ⁴6,000	¹3,000 *2,000	⁶6,800	—	8,000	—	—
Total	730,400	768,000	320,180	—	239,300	6,800	208,820	51,000	50,000	430,900	49,000	21,050	21,310	43,000	49.5	—
Tennessee:																
Mapco Petroleum Inc.—Memphis	*57,000	60,000	12,000	—	¹30,000	—	²10,000	—	—	¹10,000 ²5,000 ⁵16,000	²3,000 *2,500	⁶4,000	—	3,500	—	—
Total	57,000	60,000	12,000	—	30,000	—	10,000	—	—	31,000	5,500	4,000	—	3,500	—	—

Company												
Texas:												
Amoco Oil Co.—Texas City	420,000	440,000	195,000	¹41,000	¹195,000	43,000	⁴160,000	¹55,500 ²60,000	¹76,000	¹126,000 ²70,000 ³54,000	¹19,000 ²31,000 ³19,000	¹45,000 ⁶27,000 ⁵5,000 ²2,500	—	—	—	¹180.0	2,175
Champlin Refining Co.—Corpus Christi	150,000	160,000	80,000	⁵32,000	¹70,000	—	⁵52,000	—	²55,000 ³40,000	—	—	⁴3,000	—	5,500	—	—	1,800
Chevron U.S.A. Inc.—El Paso	66,000	68,000	54,000	—	¹22,000	—	²25,000 ²23,000 ⁴44,100	—	⁴18,000	¹25,000 ²67,100 ⁴138,000 ⁶13,900	¹5,500 ²16,900	¹7,095 ²2,905 ²2,500 ⁶7,200	—	—	—	—	—
Port Arthur	329,000	338,000	163,200	⁵34,000	¹110,000	6,000							10,000	—	—	—	1,840
Coastal Refining & Marketing Inc.—Corpus Christi	*90,250	95,000	53,000	³11,000 ⁵12,000	¹18,500	—	²11,000 ⁶17,500	¹10,000	—	¹30,000 ²20,000 ⁶25,000	²3,200 *1,750	¹11,000 ²7,000 ⁶5,300	—	10,000	¹24.0 ⁴15.0	650
Crown Central Petroleum Corp.—Houston	100,000	105,000	40,000	¹12,500	¹56,000	—	²14,000 ⁵22,000	—	⁴10,000	¹26,000	²13,000	¹12,000 ²2,000	—	—	—	350
Diamond Shamrock Corp.—Sunray¹	105,000	107,000	47,000	—	¹45,000	—	²29,000	⁴20,000	—	¹33,000	¹8,700 ⁴4,600	—	—	5,000	—	—
Three Rivers⁵	50,000	52,000	20,000	—	¹19,000	—	²11,000	—	—	¹11,000	²6,000	—	1,000	—	—	—
El Paso Refining Co. Ltd.—El Paso	21,000	21,000	—	⁵4,000	¹7,000	2,000	²5,600	—	—	²5,600	²2,500 *500	⁴500	—	—	—	100
Exxon Co. U.S.A.—Baytown¹	493,000	517,000	258,000	⁴28,000	¹155,000	15,000	⁴60,000 ⁴63,000	¹9,000	²95,000	¹139,000 ²23,500 ⁴143,000 ⁵80,000 ⁶44,100	²29,000	—	31,200	7,000	¹85.0	100
Fina Oil & Chemical Co.—Big Spring¹⁰	55,000	60,000	24,000	—	¹22,000	—	²20,000	—	²6,000	¹25,000 ²43,000 ⁵15,000	²5,000	¹600	—	—	—	—
Port Arthur¹	100,000	110,000	50,000	—	¹36,000	—	⁵34,000	—	²18,000 ³13,000	¹40,000 ⁶28,000	¹5,000	¹10,000 ⁶8,500	—	2,000	—	—
Hill Petroleum Co.—Houston⁴	67,500	70,000	27,000	—	¹55,000	—	¹11,500	—	—	¹15,000 ⁴22,500 ⁷7,500	¹5,500	²2,000	—	5,000	—	—
Texas City	119,600	130,000	54,000	³19,000	¹43,500	—	²11,000 ⁵12,000	—	⁴29,000	¹23,000	²6,200	—	—	—	—	—
Howell Hydrocarbons Inc.—San Antonio	2,750	2,900	—	—	—	—	¹1,000	—	—	—	—	¹1,000	—	—	—	—
Koch Refining Co.—Corpus Christi	125,000	130,000	42,000	⁵12,000	¹40,000	800	³15,000 ⁶33,500	—	—	¹49,500 ⁵8,000	²8,400	²7,600 ⁶3,200	—	—	—	375
LaGloria Oil & Gas Co.—Tyler	55,000	60,000	16,000	¹6,000	¹17,000	850	²4,000 ⁵11,700	—	—	²16,000	⁴4,800	⁶5,000	—	—	—	250
Liquid Energy Corp.—Bridgeport	10,000	10,800	—	—	—	—	—	—	—	—	—	—	6,000	—	—	—
Lyondell Petrochemical Co.—Houston¹	265,000	286,000	129,000	⁵40,000	¹90,000	—	²110,000	—	²48,000 ⁴46,000 ⁵35,000	¹110,000 ⁵10,000 ⁶43,000 ⁷7,000	¹14,000	¹11,000	—	—	—	2,650

(continued)

TABLE 2 (continued)

Company and Location	Crude Capacity (b/cd)	Crude Capacity (b/sd)	Vacuum Distillation	Thermal Operations	Cat Cracking Fresh Feed	Cat Cracking Recycle	Cat Reforming	Cat Hydro-cracking	Cat Hydro-refining	Cat Hydro-treating	Alky. *Poly	Aromatics-Isomerization	Lubes	Asphalt	Hydrogen (MMcfd)	Coke (t/d)
Marathon Petroleum Co.—Texas City	69,500	72,000	27,000	—	¹38,000	1,000	¹9,000	—	—	—	²11,000	²2,500	9,400	—	—	—
Mobil Oil Corp.—Beaumont	275,000	290,000	86,000	²29,500	¹102,000	—	²57,000 ³46,000	¹32,000	—	¹92,000 ⁶3,500 ⁷116,000 ⁷2,150	¹13,000	²20,000	—	—	⁶60.0	1,404
Phillips 66 Co.—Borger	105,000	110,000	—	—	¹60,000	10,400	²26,000	—	¹50,000 ⁴40,000	²26,500	²14,000	³3,060 ⁴11,000 ⁵12,100 ⁶12,500	—	—	⁵50.0	—
Sweeny	175,000	195,000	83,000	—	¹87,000	12,000	²36,000	—	¹75,000 ⁵50,000	²53,000	²10,500	¹5,575 ²7,630 ³9,100 ⁶7,800	—	—	⁴80.0	—
Pride Refining Inc.—Abilene	42,750	45,000	12,500	⁵55,000	—	—	—	—	—	—	—	—	—	—	—	—
Shell Oil Co.—Deer Park	215,900	227,000	89,500	⁶19,000	¹65,000	5,000	¹20,000 ²43,000	⁴65,000	¹45,000	²64,000 ³37,500 ⁶11,500 ⁷77,000	⁸8,100	¹20,000	10,700	5,900	¹65.0	—
Odessa	28,600	29,500	10,000	—	¹10,500	—	¹11,000	—	—	¹11,000	²3,300	¹350	—	—	—	—
Southwestern Refining Cl—Corpus Christi	104,000	108,000	36,000	—	¹50,000	—	²30,000	—	²18,000	²40,000 ⁴27,000	²4,000 ⁴4,200	⁶6,500	—	—	—	—
Star Enterprise—Port Arthur and Port Neches⁷	250,000	278,000	143,100	—	¹110,000	31,500	²40,000	¹15,000	—	²40,000 ⁴80,000 ⁶18,500	⁹9,000	—	17,400	14,000	—	—
Trifinery—Corpus Christi	28,500	29,500	20,000	—	—	—	—	—	—	—	—	—	—	10,500	—	—
Unocal Corp.—(Beaumont), Nederland	120,000	126,300	43,000	—	¹39,000	4,000	¹12,000 ²20,000	—	—	¹43,000 ³10,600	²4,200 ³1,400	⁶6,000 ³1,200	3,100	2,500	—	—
Valero Refining Co.—Corpus Christi	20,000	21,500	24,000	—	²54,000ᵃ	2,700	—	—	¹60,000	—	²9,500	—	—	—	⁵55.0	—
Total	4,058,350	4,294,500	1,826,300	355,000	1,616,500	134,250	1,150,900	276,500	827,000	2,249,950	301,750	304,215	88,800	67,400	614.0	11,694
Utah:																
Amoco Oil Co.—Salt Lake City	40,000	41,500	3,800	—	¹18,000	4,000	¹7,600	—	—	¹7,600	⁴4,000	³3,000	—	—	—	—
Big West Oil Co.—Salt Lake City	24,000	25,000	—	—	²5,000	1,000	²5,000	—	—	¹6,000	²1,500	⁶1,700	—	—	—	—
Chevron U.S.A.—Salt Lake City	45,000	46,000	35,500	⁸8,500	¹11,000 ⁷7,000	—	²5,500	—	⁵5,500	¹5,500	⁴4,300	⁴750	—	—	—	350
Crysen Refining Inc.—Woods Cross	12,500	13,400	3,000	—	—	1,000	²3,000	—	—	¹3,000	—	—	—	—	—	—
Pennzoil Products Co.—Roosevelt	8,000	8,500	—	—	¹6,000	500	²2,000	—	¹1,600	²2,000	²2,600	—	—	—	—	—
Phillips 66 Co.—Woods Crossᵇ	25,000	26,000	3,200	—	²8,400	2,600	⁴6,000	—	—	²11,000	²2,100	⁴2,600	—	1,700	—	—
Total	154,500	160,400	45,500	8,500	55,400	9,100	29,100	—	7,100	35,100	14,500	8,050	—	1,700	—	350
Virginia:																
Amoco Oil Co.—Yorktown	51,000	53,000	29,000	⁵13,500	¹27,500	2,000	²9,750	—	—	¹10,000 ⁵15,500	⁴2,400	—	—	—	—	750
Total	51,000	53,000	29,000	13,500	27,500	2,000	9,750	—	—	25,500	2,400	—	—	—	—	750
Washington:																
Atlantic Richfield Co.—Ferndale	162,000	170,000	95,000	⁵50,000	—	—	²56,000	¹52,000	⁴18,000	¹38,000	—	—	—	—	⁸80.0	2,500
Chevron U.S.A. Inc.—Seattle	5,000	*5,263	6,000	—	—	—	—	—	—	—	—	—	—	5,000	—	—

Company—Location																		
Shell Oil Co.—Anacortes	84,000	88,000	36,000	—	41,000	7,000	25,000	—	—	7,500	[2]32,000 [5,7]20,500 [4]13,500 [4]15,000	[1]10,900 [2]5,900 [*,1]1,200	[4]2,750	—	—	—	—	—
Sohio Oil Co. (BP America)—Ferndale	77,000	79,000	28,000	—	[2]25,500	2,000	[4]11,800	—	—	—	—	—	—	—	—	—	—	—
Sound Refining Inc.—Tacoma	11,900	12,754	6,000	—	—	—	—	—	—	—	—	—	—	—	3,600	—	—	—
Texaco Refining & Marketing Inc.—Anacortes[d]	97,000	100,000	30,000	[2]20,000	[1]36,000	4,000	[1]7,000 [2]14,000	—	—	—	[4]22,000 [1]13,000 [5]15,000	[8]8,000 [*,1]1,300	—	—	—	—	—	1,000
U.S. Oil & Refining Co.—Tacoma	*32,775	34,500	19,500	—	—	—	[2]6,000	—	—	—	[6]6,000 [2]1,000 [4]4,000	—	[6]1,500	—	8,000	—	—	—
Total	469,675	489,517	220,500	70,000	102,500	13,000	119,800	52,000	—	25,500	180,000	27,300	4,250	—	16,600	80.0	—	3,500
West Virginia:																		
Mid Atlantic Fuels Inc.—St. Mary's[9]	5,000	5,200	2,000	—	—	—	[1]1,500	—	—	—	—	—	—	1,119	—	—	—	—
Quaker State Oil Refining Corp.—Newell	10,500	10,800	8,850	—	—	—	[2]3,400	[3]4,500	—	—	[1]3,900	—	—	4,440	—	[1]1.2	—	—
Total	15,500	16,000	10,850	—	—	—	4,900	4,500	—	—	3,900	—	—	5,559	—	1.2	—	—
Wisconsin:																		
Murphy Oil USA Inc.—Superior	32,000	34,000	20,500	—	[1]11,000	1,000	[8]8,000	—	—	[4]5,800	[1]9,000	[2]1,300	—	—	13,500	—	—	—
Wyoming:																		
Amoco Oil Co.—Casper	40,000	41,000	17,000	—	[1]13,500	2,700	[4]7,000	—	—	—	[1]7,100 [2]7,200 [5]5,400	[1]2,500 [2]3,200	—	1,500	—	—	—	—
Frontier Oil & Refining Co.—Cheyenne	35,000	38,000	20,000	[5]6,600	[1]12,000	300	[2]6,600	—	—	—	[6]6,000 [2]3,750 [4]4,000	—	[4]1,800	—	—	—	—	—
Little America Refining Co.—Casper	22,000	24,000	8,600	—	[1]14,000	3,000	[6]6,000	—	—	—	[2]14,500 [4]12,000	—	—	—	1,000	—	—	—
Sinclair Oil Corp.—Sinclair	54,000	55,000	30,000	—	[1]21,000	1,000	[2]14,500	—	—	[3]21,000	[3]3,500 [*]700 [2]800	—	—	—	5,000	—	—	—
Wyoming Refining Co.—Newcastle	12,500	13,500	—	—	[2]4,000	3,000	[1]2,750	—	—	—	—	—	[1]1,800	—	—	—	—	—
Total	163,500	171,500	75,600	6,600	64,500	10,000	36,850	—	—	21,000	59,950	10,700	1,800	1,500	6,000	—	—	—

Inactive Refineries:
1. Mountaineer Refining Co. Inc., La Barge, Wyoming—300 b/cd

[a]ART.
[b]Reduced crude converter.
[c]Solvent extraction, 28,500 b/sd.
[d]Solvent extraction, 12,000 b/sd.
[1]Solvent extraction, 5,500 b/d.
[2]Solvent extraction, 50,000 b/sd.
[3]Dimersol.
[4]Solvent extraction, 4,500 b/sd.
[5]Solvent extraction, 6,000 b/sd.
[6]Solvent extraction, 4,000 b/sd.
[7]Solvent extraction, 8,500 b/sd (PDA).
[8]Solvent extraction, 9,500 b/sd.
[9]Cumene.
[10]Solvent extraction, 9,000 b/sd.
[11]Solvent extraction, 5,000 b/sd.
[12]Solvent extraction, 5,800 b/sd.
[13]Solvent extraction, 730 b/sd.
[14]Solvent extraction, 15,000 b/sd.
[15]Solvent extraction, 7,000 b/sd.
[16]Solvent extraction, 53,000 b/sd.
[17]Solvent extraction, 10,000 b/sd.
[18]Solvent extraction, 18,000 b/sd.
[19]Solvent extraction, 11,000 b/sd.
[20]Heavy oil cracker.
[21]Solvent extraction, 900 b/sd.

Petroleum Refining Processes, Worldwide Capacities*

Calendar figures reported in this survey (Tables 1 and 2) are the average volume a refinery unit processes each day including downtime used for turnarounds. This is the actual total volume for the year divided by 365. (Stream day figures represent the amount a unit can process when running full capacity for short periods.)

The following numbers identify processes in Table 2:

Thermal Processes:
1. Gas oil cracking
2. Thermal cracking
3. Vibrating
4. Fluid coking
5. Delayed coking
6. Other

Catalytic Cracking:
1. Fluid
2. Other

Catalytic Reforming
   Semiregenerative:
   1. Conventional catalyst
   2. Bimetallic catalyst
   Cyclic:
   3. Conventional catalyst
   4. Bimetallic catalyst
   Other:
   5. Conventional catalyst
   6. Bimetallic catalyst

Catalytic Hydrocracking:
1. Distillate upgrading
2. Residual upgrading
3. Lube-oil manufacturing
4. Other

*Editor's Note: These surveys are updated each December by the Oil & Gas Journal staff. All figures are in barrels per calendar day and are as of January 1, 1989.

Catalytic Hydrorefining:
1. Residual desulfurization
2. Heavy gas oil desulfurization
3. Cat cracker and cycle stock feed pretreatment
4. Mid distillate
5. Other

Catalytic Hydrotreating:
1. Pretreating cat reformer feeds
2. Naphtha desulfurizing
3. Naphtha olefin or aromatics saturation
4. Straight-run distillate
5. Other distillates
6. Lube oil "polishing"
7. Other

Alkylation:
1. Sulfuric acid
2. Hydrofluoric acid

Aromatics/Isomerization:
1. BTX
2. Hydrodealkylation
3. Cyclohexane
4. C_4 feed
5. C_5 feed
6. C_5 and C_6 feed

Hydrogen
1. Steam methane reforming
2. Steam naphtha reforming
3. Partial oxidation
4. Cryogenic
5. Other

This material appeared in *Oil & Gas Journal*, December 26, 1988, and is reprinted by special permission. Copyright © 1988 by Pennwell Publishing Co., Tulsa, Oklahoma 74121.

DEBRA A. GWYN

TABLE 1 Survey of Operating Refineries Worldwide (capacities as of January 1, 1989)

Country	No. Plants	Crude (b/cd)	Vacuum Distillation	Thermal Operations	Catalytic Cracking	Catalytic Reforming	Cat Hydro-cracking	Cat Hydro-refining	Cat Hydro-treating	Alkyla-tion	Aromatics-isomerization	Lubes	Asphalt	Hydrogen (MMcfd)	Coke (t/d)
Abu Dhabi	2	180,000	47,000	—	—	30,055	27,000	—	100,399	—	—	—	5,300	53.0	—
Algeria	4	464,700	15,000	—	—	55,600	—	—	23,600	—	1,900	2,400	—	—	—
Angola	1	32,100	1,900	—	—	1,900	—	—	3,800	—	—	—	950	—	—
Argentina	11	690,400	259,300	125,100	124,800	41,000	23,900	2,800	64,200	6,850	—	6,550	21,800	19.0	1,050
Australia	10	644,100	154,000	—	190,300	159,600	—	33,700	250,000	32,850	13,800	13,200	16,400	—	—
Austria	1	204,000	68,000	17,000	24,000	32,000	7,700	68,000	49,300	—	9,000	2,300	5,850	—	—
Bahrain	1	243,000	156,000	20,000	39,000	18,000	—	50,000	40,000	1,300	—	—	5,000	32.2	—
Bangladesh	1	31,200	3,530	—	—	1,650	—	2,430	1,940	—	—	—	1,200	—	—
Barbados	1	3,000	—	—	—	—	—	—	—	—	—	—	—	—	—
Belgium	4	630,500	298,000	62,800	102,000	80,300	—	190,100	182,700	7,710	—	—	25,300	46.0	—
Bolivia	3	57,500	2,961	—	—	14,900	—	—	15,379	—	—	377	—	—	—
Brazil	13	1,407,300	698,000	33,800	350,500	24,800	—	—	133,400	—	—	16,000	13,800	20.1	1,265
Brunei	1	10,000	—	—	—	—	—	—	—	—	—	—	—	—	—
Burma	2	26,300	4,800	1,700	—	7,000	—	—	—	—	—	—	—	—	40
Cameroon	1	42,000	—	—	—	—	—	10,500	9,500	—	—	—	—	—	—
Canada	27	1,855,800	657,850	88,250	387,600	369,600	206,350	51,000	771,600	71,145	55,870	17,360	112,050	309.2	756
Chile	3	146,800	74,800	20,000	36,670	10,000	—	—	6,000	1,000	—	—	1,070	—	—
China, Taiwan	2	570,000	117,325	13,500	22,500	56,250	18,000	167,400	—	3,150	43,200	—	—	100.0	800
Colombia	4	227,400	120,000	60,000	91,000	6,000	—	—	34,240	11,300	5,100	2,300	2,400	21.0	—
Congo	1	21,000	8,000	—	—	2,000	2,000	—	3,500	—	—	—	—	—	—
Costa Rica	1	16,200	600	—	—	1,200	—	—	3,400	—	—	—	300	—	—
Cyprus	1	17,100	2,600	6,500	—	4,250	—	—	7,750	—	—	—	1,200	—	—
Denmark	3	176,500	41,050	—	—	31,700	—	10,000	85,800	—	3,900	—	8,000	—	—
Dominican Republic	1	47,000	—	70,670	—	7,500	—	—	21,700	—	—	—	—	—	—
Ecuador	3	123,300	43,020	25,200	16,000	2,780	—	—	—	—	—	—	—	0.8	—
Egypt	8	489,203	46,725	16,470	—	30,940	—	—	84,386	804	1,452	3,848	6,633	—	541
El Salvador	1	17,000	1,900	—	—	3,800	—	—	12,100	—	—	—	—	5.5	—
Ethiopia	1	18,000	1,600	—	—	2,270	—	—	—	—	—	—	—	—	—
Finland	2	241,000	81,400	46,300	41,800	42,900	14,200	104,500	57,930	7,980	3,600	—	11,800	20.0	—
France	14	1,875,970	684,500	152,850	319,200	247,000	13,500	264,900	523,000	10,150	14,960	31,600	59,650	28.0	—
Gabon	1	24,000	—	7,200	—	1,400	—	—	3,600	—	—	—	—	—	—

Petroleum Refining Processes, Worldwide Capacities

Country															
Germany	15	1,518,200	608,700	289,500	179,400	275,100	96,200	296,400	624,200	11,300	62,290	9,000	75,840	156.5	2,860
Ghana	1	26,600	118,100	—	—	5,850	—	—	5,850	—	—	—	—	—	—
Greece	4	384,500	118,100	41,120	50,500	51,300	—	41,100	100,200	5,700	4,250	3,100	5,600	9.1	—
Guatemala	1	16,000	—	—	—	3,000	—	—	5,000	—	—	—	—	—	—
Honduras	1	14,000	—	—	—	1,800	—	—	5,000	—	—	—	—	—	—
Hungary	3	220,000	120,000	13,300	20,000	23,000	6,000	13,800	76,600	3,300	9,800	4,000	10,800	9.5	881
India	12	1,051,441	332,996	110,670	133,953	26,030	100,100	18,000	70,630	—	4,250	15,000	37,229	130.8	700
Indonesia	6	714,200	229,000	81,700	12,600	61,500	93,180	3,120	51,500	600	—	5,000	7,500	70.0	—
Iran	4	530,000	215,200	80,800	—	63,845	38,000	100,000	34,225	—	—	6,000	12,450	64.0	—
Iraq	8	318,500	82,650	—	—	43,500	—	—	13,000	—	—	10,168	6,715	—	—
Ireland	1	56,000	—	—	—	11,000	—	5,500	11,000	—	—	—	—	—	—
Israel	2	180,000	90,000	70,000	20,000	26,000	—	45,000	30,000	—	—	—	3,000	—	—
Italy	19	2,450,200	590,100	360,400	278,300	284,300	59,000	337,100	579,200	38,300	64,000	33,610	131,370	60.9	1,400
Ivory Coast	2	60,000	42,000	—	—	16,600	11,500	11,500	21,000	—	—	—	6,000	—	—
Jamaica	1	34,200	1,900	—	—	3,240	—	—	19,350	—	—	—	—	—	—
Japan	41	4,362,750	1,698,940	82,800	596,950	551,650	101,750	1,370,280	1,973,740	33,880	63,550	48,550	58,475	944.5	650
Jordan	1	100,000	16,900	—	4,410	8,640	4,800	—	17,100	—	—	—	—	7.9	—
Kenya	1	90,000	1,700	—	—	9,000	—	—	33,000	—	—	—	—	—	—
Korea	6	880,000	67,400	49,000	—	62,450	29,000	21,000	110,800	—	17,630	6,200	15,200	68.4	1,200
Kuwait	4	817,000	338,000	54,000	42,000	33,000	170,000	205,000	263,000	—	—	—	5,000	527.0	1,900
Lebanon	2	37,000	12,730	—	7,250	7,492	—	7,392	3,100	—	—	—	409	—	—
Liberia	1	15,000	1,000	—	—	2,000	—	—	3,300	—	—	—	200	—	—
Libya	3	329,400	3,432	—	—	13,982	—	—	37,126	—	—	635	1,716	—	—
Madagascar	1	16,350	1,200	7,000	—	2,600	—	3,600	5,500	—	—	—	—	—	—
Malaysia	4	209,300	7,150	—	—	21,980	—	—	60,850	—	—	—	1,700	—	—
Martinique	1	12,800	—	—	—	7,410	—	—	8,350	—	—	—	—	—	—
Mexico	9	1,354,000	600,700	82,000	267,000	157,800	18,000	250,000	207,000	3,860	10,594	15,500	15,000	70.0	260
Morocco	2	154,600	27,400	—	5,600	26,800	—	—	38,600	—	—	2,100	13,700	—	—
Netherlands	7	1,380,700	399,370	154,360	125,500	166,400	30,000	148,200	522,620	11,500	—	11,700	14,320	86.0	—
Netherlands Antilles	1	320,000	158,000	67,000	42,000	15,000	—	—	121,500	5,600	—	8,000	8,000	—	—
New Zealand	1	88,400	36,600	—	—	21,900	21,500	6,500	47,400	—	—	—	2,300	44.0	—
Nicaragua	1	14,500	2,000	—	—	2,800	—	—	10,000	—	—	—	800	—	—
Nigeria	4	414,500	122,290	—	84,000	71,000	—	—	110,686	9,294	3,901	3,878	14,850	—	—
Norway	3	239,400	4,000	77,300	—	30,100	—	—	79,700	—	3,600	—	4,000	—	480
Oman	1	76,932	—	—	—	15,386	—	—	19,506	—	—	—	—	—	—
Pakistan	3	130,050	14,200	—	—	5,450	—	—	24,800	—	1,250	3,815	4,050	0.1	—
Panama	1	100,000	14,000	—	—	7,500	—	—	30,000	—	—	—	5,000	—	—
Paraguay	1	7,500	—	—	—	—	—	—	—	—	—	—	—	—	—
Peru	6	172,438	36,915	—	22,540	1,680	—	—	—	—	—	1,080	1,530	—	—

TABLE 1 (continued)

Philippines	4	254,000	58,600	—	23,100	39,400	—	23,000	91,200	—	3,400	5,000	—	—
Portugal	3	313,300	55,300	—	10,100	52,200	10,200	49,800	76,200	—	3,000	5,800	—	—
Puerto Rico	2	123,000	59,300	—	12,000	5,800	15,000	12,000	—	—	9,000	900	18.0	—
Qatar	1	62,000	—	—	—	12,030	—	—	39,640	1,000	—	—	—	—
Saudi Arabia	7	1,375,000	280,655	32,000	76,000	180,900	70,100	40,000	340,000	12,500	—	30,500	201.0	—
Senegal	1	29,800	18,800	—	—	2,870	—	—	—	—	—	—	—	—
Sierra Leone	1	10,000	—	—	—	—	—	—	—	—	—	—	—	—
Singapore	5	852,000	256,100	137,700	—	58,300	48,100	80,600	243,400	—	12,800	12,000	79.0	—
Somalia	1	10,000	—	—	—	—	—	—	—	—	—	—	—	—
South Africa	4	433,500	104,000	63,500	84,400	62,800	18,000	77,300	142,150	7,000	3,000	6,140	34.0	—
Spain	10	1,285,000	388,800	173,000	164,000	171,600	15,000	192,100	278,900	1,700	8,400	40,220	59.2	550
Sri Lanka	1	50,000	2,400	12,500	—	3,750	—	2,100	15,300	—	—	1,000	—	—
Sudan	1	21,440	—	—	—	1,700	—	—	7,400	—	—	—	—	—
Sweden	5	426,500	122,000	67,000	25,000	71,000	—	90,500	118,000	3,500	2,500	28,500	1.9	—
Switzerland	2	132,000	24,000	20,000	—	26,000	—	8,000	49,200	—	—	5,200	—	—
Syria	2	243,744	61,346	36,384	—	18,994	—	18,684	47,755	—	—	8,490	9.9	487
Tanzania	1	13,500	—	—	—	3,000	—	—	5,000	—	—	—	—	—
Thailand	3	191,045	22,800	16,920	8,100	26,280	—	17,550	69,000	—	—	1,100	4.7	—
Trinidad	2	300,000	171,000	18,000	28,000	31,000	—	104,500	41,800	2,650	2,700	2,000	—	—
Tunisia	1	34,000	—	—	—	3,300	—	—	—	—	—	—	—	—
Turkey	5	724,654	159,376	4,500	34,128	63,127	—	15,096	153,524	—	4,000	29,617	0.4	—
United Kingdom	15	1,803,100	806,840	152,000	442,000	307,500	21,000	115,600	776,590	66,700	21,600	39,800	78.2	2,150
United States	188	15,418,738	7,082,395	1,914,090	5,252,900	3,901,920	1,192,100	2,396,700	7,099,850	1,110,395	244,569	762,414	2,417.4	73,646
Uruguay	1	33,000	13,000	—	4,100	2,500	—	—	3,700	—	—	—	—	—
Venezuela	6	1,201,100	560,600	138,000	193,100	6,000	—	296,100	—	61,250	7,590	39,800	108.9	100
Virgin Islands	1	545,000	215,000	80,000	—	125,000	—	130,000	290,000	—	—	—	—	—
Yemen, N.	1	10,000	—	—	—	2,500	—	—	—	—	—	—	—	—
Yemen, S.	1	161,500	9,500	—	—	10,800	—	2,700	—	—	—	2,000	—	—
Yugoslavia	7	609,135	178,759	35,630	52,050	74,612	7,200	21,970	107,878	3,400	4,900	27,325	14.9	137
Zaire	1	16,700	—	—	—	3,500	—	—	5,000	—	—	—	—	—
Zambia	1	20,594	2,170	—	—	5,070	—	—	8,250	—	—	—	—	—
Total	599	56,132,476	20,237,175	5,289,514	10,046,351	8,697,933	2,488,380	7,533,122	17,873,394	1,547,668	1,396,121	610,730	1,809,263	5,931.0	91,853

Petroleum Refining Processes, Worldwide Capacities

TABLE 2 Worldwide Refineries: Location, Capacities, Types of Processing

	Company and Refinery Location	Crude	Vacuum Distillation	Thermal Operations	Catalytic Cracking	Cat Reforming	Cat Hydro-cracking	Cat Hydro-refining	Cat Hydro-treating	Alky. *Poly.	Aromatics-Isomerization	Lubes	Asphalt	Hydrogen (MMcfd)	Coke (t/d)
Abu Dhabi	Abu Dhabi National Oil Co.—Ruwais	120,000	47,000	—	—	²19,150	¹27,000	—	²34,350 ⁴2,630	—	—	—	—	⁵3.0	—
	Umm Al-Nar 2	60,000	—	—	—	⁴10,905	—	—	²17,530 ⁴5,889	—	—	—	—	—	—
	Total	180,000	47,000	—	—	30,055	27,000	—	100,399	—	—	—	—	53.0	—
Albania	Government-owned refineries at Balish, Cerrik, Stalin.														
	Total	40,000													
Algeria	Sonatrach: Arzew	60,000	6,000	—	—	8,600	—	—	8,600	—	—	2,400	2,400	—	—
	Hassi Messaoud	23,700	—	—	—	2,000	—	—	—	—	—	—	—	—	—
	Maison Carree	58,000	—	—	—	15,000	—	—	15,000	—	1,900	—	2,900	—	—
	Skikda	323,000	9,000	—	—	30,000	—	—	—	—	—	—	5,300	—	—
	Total	464,700	15,000	—	—	55,600	—	—	23,600	—	1,900	2,400	—	—	—
Angola	Petrangol SARL—Luanda	32,100	1,900	—	—	¹1,900	—	²2,800	¹,²3,800	—	—	—	950	—	—
Argentina	Destileria Argentina de Petróleo SA—Lomas de Zamora	2,000	1,500	—	—	—	—	—	—	—	—	—	—	—	—
	Esso SAPA—Campana	94,500	51,200	²22,400	²29,700	²8,500	—	—	²8,300 ⁸8,700	—	—	1,000	—	—	1,050
	Galvan	18,600	—	²7,700	—	—	—	—	—	—	—	—	—	—	—
	Isaura SA—Bahia Blanca	12,000	—	—	—	—	—	—	—	—	—	—	—	—	—
	Shell Cia. Argentina de Petróleo SA—Buenos Aires	121,700	54,000	²27,000	²29,200	²12,000	—	—	²14,900 ⁴7,500	²1,700 *500	—	1,400	4,200	—	—
	Yacimientos Petroliferos Fiscales—Camo Duran	32,000	—	—	—	—	—	—	—	—	—	—	—	—	—
	Dock Sud	4,000	1,000	—	—	—	—	—	—	—	—	—	—	—	—
	La Plata	216,000	69,500	²22,000	⁴5,000	¹9,000	—	—	¹9,000 ⁴3,000 ⁵5,900	¹1,200 ²750 *2,700	—	4,150	17,600	—	—
	Lujan de Cuyo	129,000	66,000	²30,000	²20,900	¹9,000	²3,900	—	⁴4,400 ²2,500	—	—	—	—	—	—
	Plaza Huincul	23,000	—	²3,500	—	²2,500	—	—	—	—	—	—	—	—	—
	San Lorenzo	37,600	16,100	³12,500	—	—	—	—	—	—	—	—	—	—	—
	Total	690,400	259,300	125,100	124,800	41,000	23,900	—	64,200	6,850	—	6,550	21,800	19.0	1,050
Australia	Ampol Refineries Ltd.—Lytton	72,000	—	—	¹27,500	¹13,000	—	—	⁵5,000	¹4,000 *1,100	⁶5,800	—	—	—	—
	Australian Lubricating Oil Refinery Ltd.—Kurnell	—	6,900	—	—	—	—	—	—	—	—	3,400	—	—	—
	BP Australia—Brisbane	42,000	11,400	—	¹15,800	²8,500	—	—	⁶6,300	—	—	—	1,500	—	—
	Kwinana	100,000	43,000	—	²4,000	⁶6,400 ²16,400	—	⁸8,200	⁴7,600 ¹12,000	²1,850 ²2,400	—	2,600	1,700	19.0	—

TABLE 2 (continued)

Company and Refinery Location	Crude	Vacuum Distillation	Thermal Operations	Catalytic Cracking	Cat Reforming	Cat Hydro-cracking	Cat Hydro-refining	Cat Hydro-treating	Alky. *Poly.	Aromatics-Isomerization	Lubes	Asphalt	Hydrogen (MMcfd)	Coke (t/d)
Austria														
Caltex Refining CPL—Kurnell[a]	100,000	20,000	—	[1]39,000	[1]25,600	—	—	[1]25,600 [4,5]14,000	[1]2,600 *2,700	—	—	3,000	—	—
Mobil Oil Australia Ltd.—Adelaide	—	20,000	—	—	—	—	—	—	—	—	4,700	—	—	—
Petroleum Refineries Australia PL—														
Adelaide	46,500	—	—	—	[2]20,000	—	—	[2]42,500	—	—	—	1,200	—	—
Altona	98,600	22,000	—	[2]24,000	[2]28,000	—	—	[2]35,000 [4]10,000	[2]2,700	—	—	3,000	—	—
Shell Refining (Australia) PL—Clyde	75,000	21,000	—	[1]35,000	[2]13,700	—	[2]7,000	[4]37,000	[5]5,000 *3,000	—	—	3,000	—	—
Geelong	110,000	9,700	—	[2]25,000	[2]8,000 [6]20,000	—	[2]4,000 [3]11,000 [3]3,500	[3]55,000	[2]6,000 *1,500	[3]8,000	2,500	3,000	—	—
Total	644,100	154,000	—	190,300	159,600	—	33,700	250,000	32,850	13,800	13,200	16,400	—	—
Austria														
OeMV—Schwechat	204,000	68,000	[3]17,000	[2]24,000	[2]14,000 [4]18,000	[1]7,700	[2]35,000 [4]33,000	[3]37,000 [6]1,300 [7]11,000	—	[9]9,000	2,300	5,850	—	—
Total	204,000	68,000	17,000	24,000	32,000	7,700	68,000	49,300	—	9,000	2,300	5,850	—	—
Bahrain														
Bahrain Petroleum Co. BSC (Closed)—Sitra	243,000	156,000	[3]20,000	[1]39,000	[2]18,000	—	[1]50,000	[1]18,000 [4]22,000	*1,300	—	—	5,000	[3]32.2	—
Total	243,000	156,000	20,000	39,000	18,000	—	50,000	40,000	1,300	—	—	5,000	32.2	—
Bangladesh														
Eastern Refinery Ltd.—Chittagong	31,200	3,530	—	—	[1]1,650	—	[4]2,430	[2]1,940	—	—	—	1,200	—	—
Barbados														
Mobil Oil Barbados Ltd.—Bridgetown	3,000	—	—	—	—	—	—	—	—	—	—	—	—	—
Belgium														
Belgian Refining Corp. NV—Antwerp	91,500	45,000	[3]20,300	—	[2]12,000	—	[1,2]22,500 [2]25,000	—	—	—	—	—	—	—
Esso Inc.—Antwerp	218,000	108,300	—	[2]25,500	[4]38,100	—	[2]79,200	[2]44,400 [4]71,200	—	—	—	2,000	[1]46.0	—
Nynas Petroleum NV—Antwerp	15,000	12,500	—	—	—	—	—	—	—	—	—	12,000	—	—
Fina Raffinaderij Antwerpen—Antwerp	306,000	132,200	[2]42,500	[1]76,500	[2]30,200	—	[3,4]63,400	[2]67,100	[2]6150 *1,560	—	—	11,300	—	—
Total	630,500	298,000	62,800	102,000	80,300	—	190,100	182,700	7,710	—	—	25,300	46.0	—
Bolivia														
Yacimientos Petroliferos Fiscales Bolivianos—Cochabamba[b]	39,500	2,961	—	—	[1]8,500	—	—	[1]8,500 [6]479	—	—	377	—	—	—
Santa Cruz	15,000	—	—	—	[2]6,400	—	—	[6]6,400	—	—	—	—	—	—
Sucre	3,000	—	—	—	—	—	—	—	—	—	—	—	—	—
Total	57,500	2,961	—	—	14,900	—	—	15,379	—	—	377	—	—	—

Petroleum Refining Processes, Worldwide Capacities

Country	Company / Location														
Brazil	Petróleo Brasileiro SA—Araucária, Paraná	151,000	61,000	—	¹47,200	—	—	—	—	—	—	—	—	—	—
	Betim, Minas Gerais	144,600	68,400	—	¹34,100	—	—	—	—	—	—	—	—	—	—
	Canoas, Rio Grande do Sul	72,300	36,500	—	¹15,100	—	—	—	—	—	—	—	—	—	—
	Cubatão, São Paulo	163,600	81,100	⁵25,100	⁵53,400	¹11,000	—	²12,000	—	—	—	8,800	²5.0	—	—
	Duque de Caxias, Rio de Janeiro	226,600	112,500	—	¹47,200	¹13,800	—	⁵30,000	—	—	12,500	2,500	²5.0	—	1,265
	Fortaleza, Ceará	10,000	4,700	—	—	—	—	—	—	—	—	—	—	—	—
	Manaus, Amazonas	—	5,600	—	¹2,000	—	—	—	—	—	—	—	—	—	—
	Mataripe, Bahia	121,400	63,200	—	¹27,700	—	—	⁶3,300	—	—	3,500	2,500	¹1.1	—	—
	Mauá, Santo André, São Paulo	32,500	10,000	—	²18,200	—	—	⁷2,000	—	—	—	—	—	—	—
	Paulínia, São Paulo	302,000	150,000	—	¹44,000	—	—	²20,100	—	—	—	—	²9.0	—	—
	São José dos Campos, São Paulo	164,000	100,000	—	¹59,100	—	—	⁶66,000	—	—	—	—	—	—	—
	Refinaria de Petróleo Ipiranga SA—Rio Grande do Sul	9,300	5,000	—	¹2,500	—	—	—	—	—	—	—	—	—	—
	Refinaria de Petroleos de Manguinhos SA—Rio de Janeiro	10,000	—	—	—	—	—	—	—	—	—	—	—	—	—
	Total	1,407,300	698,000	33,800	350,500	24,800	—	133,400	—	—	16,000	13,800	20.1	—	1,265
Brunei	Brunei Shell Petroleum CL—Seria	10,000	—	—	—	—	—	—	—	—	—	—	—	—	—
Bulgaria	Government-owned refineries at Burgas, Pleven, Ruse.														
	Total	300,000													
Burma	Petrochemical Industries Corp.—Chauk	6,300	2,500	—	—	—	—	—	—	—	—	—	—	—	—
	Syriam	20,000	2,300	²1,700	—	—	—	—	—	—	—	—	—	—	40
	Total	26,300	4,800	1,700	—	—	—	—	—	—	—	—	—	—	40
Cameroon	Sonara-National Refining CL—Cape Limboh, Limbe	42,000	—	—	—	²7,000	²10,500	²9,500	—	—	—	—	—	—	—
Canada Alberta	Husky Oil Operations Ltd.—Lloydminster	23,500	15,000	—	¹47,000	⁴20,900	—	²51,000	⁴15,000	—	⁴6,700	12,000	—	—	—
	Imperial Oil Ltd.—Edmonton	164,900	66,100	—	—	—	²12,200	—	—	—	3,200	5,800	—	—	—
	Petro-Canada Products Inc.—Edmonton	115,500	27,000	⁵7,100	¹36,000	²12,600	⁴18,200	⁹9,500	—	⁶6,600 / ⁴6,600	—	—	¹30.8	—	431

(continued)

TABLE 2 (continued)

			Charge Capacity (b/cd)							Production Capacity (b/cd)					
	Company and Refinery Location	Crude	Vacuum Distillation	Thermal Operations	Catalytic Cracking	Cat Reforming	Cat Hydro-cracking	Cat Hydro-refining	Cat Hydro-treating	Alky. *Poly.	Aromatics-Isomerization	Lubes	Asphalt	Hydrogen (MMcfd)	Coke (t/d)
	Shell Canada Ltd.–Scotford	59,700	—	—	—	⁶21,500	¹40,500	—	²21,550 ²21,000	—	¹²5,300 ³270	—	—	⁶62.0	—
	Turbo Resources Ltd.–Balzac	27,500	6,950	—	¹8,800	²7,550	—	—	²7,550 ⁴7,550 ⁵15,300	*1,035	—	—	—	—	—
	Total	391,100	115,050	7,100	91,800	62,550	58,700	15,000	177,150	22,735	15,270	3,200	17,800	92.8	431
British Columbia	Chevron Canada Ltd.–North Burnaby	36,000	9,400	—	¹10,500	¹10,000	—	—	¹10,000	²2,000 *600	—	—	2,500	—	—
	Husky Oil Operations Ltd.–Prince George	9,500	3,800	—	¹3,300	²1,400	—	—	⁶6,900	—	—	—	1,300	—	—
	Imperial Oil Ltd.–Ioco	42,800	20,900	—	¹12,300	²6,700	—	—	²6,700 ⁴5,700 ⁵5,700	*1,200	—	—	700	—	—
	Petro-Canada Products Inc.–Port Moody	37,200	11,200	—	—	²8,600	—	—	¹15,400 ⁴9,400	—	—	—	—	—	—
	Taylor	17,400	1,600	—	¹6,300	²3,000	—	—	²4,000 ⁴7,200	²1,000	—	—	500	—	—
	Shell Canada Ltd.–Shellburn, Burnaby	24,000	6,650	—	¹5,400	²3,600	—	—	²5,400 ⁴9,900	*720	—	—	2,250	—	—
	Total	166,900	53,550	—	37,800	33,300	—	—	86,300	5,520	—	—	7,250	—	—
New Brunswick	Irving Oil Ltd.–St. John	237,500	61,750	³18,000	¹17,100	⁷34,650	¹29,700	—	⁴5,450 ⁴40,500	*1,710	⁹9,500	—	11,700	³40.0	—
	Total	237,500	61,750	18,000	17,100	34,650	29,700	—	85,950	1,710	9,500	—	11,700	40.0	—
Newfoundland	Newfoundland Processing Ltd.–Come By Chance	105,000	50,000	—	—	²26,000	¹35,000	—	²26,000 ⁵18,000	—	—	—	—	—	—
	Total	105,000	50,000	—	—	26,000	35,000	—	44,000	—	—	—	—	—	—
Northwest Territories	Imperial Oil Ltd.–Norman Wells	3,500	—	—	—	—	—	—	—	—	—	—	—	—	—
Nova Scotia	Imperial Oil Ltd.–Dartmouth	82,300	39,800	—	¹23,000	⁴8,600	—	—	¹8,600 ⁴25,700 ⁵8,600	*2,600	—	—	4,200	—	—
	Texaco Canada Ltd.–Halifax	20,000	8,600	—	¹7,200	²3,600	—	⁴5,200	¹3,600	—	—	—	—	¹1.7	—
	Total	102,300	48,400	—	30,200	12,200	—	5,200	46,500	2,600	—	—	4,200	1.7	—
Ontario	Imperial Oil Ltd.–Sarnia	122,600	28,500	⁴21,000	¹23,500	²13,300 ⁴14,300	¹10,400	—	²32,300 ⁴13,300 ⁵24,700	²6,800 *4,200	—	6,200	—	¹21.8	—

Petroleum Refining Processes, Worldwide Capacities

	Petro-Canada Products Inc.—Clarkson	41,500	35,560	—	—	⁹9,800		—	¹10,300	—	—	4,900	4,600	¹13.3	—
	Oakville	80,500	41,000	—	²5,400	⁵5,000 ²11,500		—	—	³3,100	—	—	9,500	—	—
	Shell Canada Ltd.—Sarnia	71,000	24,700	⁴4,050	¹14,400	²21,000	⁶6,750	—	—	*1,600	¹2,700	—	—	¹38.0	—
	Suncor Inc.—Sarnia	70,000	15,000	⁵5,200	²16,000	²26,000	²18,600	⁴3,700	¹15,000 ⁵6,300 ²25,600	⁴4,100	¹7,400	—	—	—	—
	Texaco Canada Ltd.—Nanticoke	95,000	31,000	—	²40,000	⁶24,000	—	—	²24,000	¹7,400	—	—	—	—	—
	Total	480,600	175,700	30,250	119,300	124,900	46,050	3,700	185,100	27,200	10,100	11,100	14,100	⁷73.1	—
Quebec	Petro-Canada Products Inc.—Montreal	87,400	40,700	¹12,600	¹17,200	²31,300	⁴14,400	—	¹26,600 ⁴9,000	¹2,300 *800	¹8,700 ²2,800	—	14,300	¹42.8	—
	Shell Canada Ltd.—Montreal	120,000	49,400	¹12,000	¹22,000	²20,700	¹11,700	—	²34,200 ⁴27,000	²2,500	⁶6,800	3,060	7,200	—	—
	Ultramar Canada Inc.—St. Romuald	103,000	33,000	—	¹34,600	²15,000	—	—	¹15,000 ⁴15,000	*3,900	—	—	30,000	—	—
	Total	310,400	123,100	24,600	73,800	67,000	26,100	—	126,800	9,500	18,300	3,060	51,500	42.8	—
Saskatchewan	Consumers' Cooperative Refineries Ltd.—Regina	45,200	23,000	⁸8,300	¹7,600	⁵9,000	¹10,800	²7,100	¹10,800 ²2,700 ⁵6,300	*1,880	²2,700	—	—	⁵58.8	325
	Petro-Canada Products Inc.—Moose Jaw	13,300	7,300	—	—	—	—	—	—	—	—	—	5,500	—	—
	Total	58,500	30,300	8,300	17,600	9,000	10,800	27,100	19,800	1,880	2,700	—	5,500	⁵58.8	325
	Total Canada	1,855,800	657,850	88,250	387,600	369,600	206,350	51,000	771,600	71,145	55,870	17,360	112,050	309.2	756
Chile	ENAP—Gregorio-Magallanes	8,800	—	—	—	—	—	—	—	—	—	—	5,500	—	—
	Petrox SA—Talcahuano	72,000	35,800	¹10,000	¹16,670	²4,000	—	—	—	—	—	—	70	—	—
	Refinería de Concón—Concón	66,000	39,000	¹10,000	²20,000	⁶6,000	—	—	²6,000	¹1,000	—	—	1,000	—	—
	Total	146,800	74,800	20,000	36,670	10,000	—	—	6,000	1,000	—	—	1,070	—	—
China	Government-owned refineries at Anshan, Beijing, Dalian, Daqing, Fushun, Hangzhou, Jinxi, Karamai-Dushanzi, Lanchow, Lenghu, Maoming, Nanchong, Nanjing, Shanghai, Shengli, Tianjin, Yumen.														
	Total	2,200,000													
China, Taiwan	Chinese Petroleum Corp.—Kaohsiung	446,500	83,125	⁵13,500	¹22,500	²36,000	¹18,000	¹27,000 ²40,500	—	³3,150	²29,700 ³13,500	—	—	⁵50.0	800
	Tao-Yuan	123,500	34,200	—	—	²20,250	—	⁴40,500 ¹27,000 ²18,000 ⁴14,400	—	—	—	—	—	⁵50.0	—
	Total	570,000	117,325	13,500	22,500	56,250	18,000	167,400	—	3,150	43,200	—	—	100.0	800

(*continued*)

TABLE 2 (continued)

Company and Refinery Location	Crude	Vacuum Distillation	Thermal Operations	Catalytic Cracking	Cat Reforming	Cat Hydro-cracking	Cat Hydro-refining	Cat Hydro-treating	Alky. *Poly.	Aromatics-Isomerization	Lubes	Asphalt	Hydrogen (MMcfd)	Coke (t/d)
Colombia														
Empresa Columbiana de Petróleos —Barrancabermeja-Santander¹	150,000	80,000	³40,000	²62,000	²6,000	—	—	⁶6,000 ³540 ²22,000 ⁴4,300 ¹1,400	¹3,000 ²500 •2,000	¹3,300 ²1,100 ³700	2,300	2,400	¹21.0	—
Cartagena, Bolívar	70,000	40,000	²20,000	²9,000	—	—	—	—	•5,800	—	—	—	—	—
Orito, Putumayo	2,400	—	—	—	—	—	—	—	—	—	—	—	—	—
Tibu, N. de Santander	5,000	—	—	—	—	—	—	—	—	—	—	—	—	—
Total	227,400	120,000	60,000	91,000	6,000	—	—	34,240	11,300	5,100	2,300	2,400	21.0	—
Congo														
Coraf–Pointe-Noire	21,000	8,000	—	—	²2,000	¹2,000	—	²3,500	—	—	—	—	—	—
Costa Rica														
Refinadora Costarricense de Petroleo SA–Limón	16,200	600	²6,500	—	²1,200	—	—	¹1,200 ²2,200	—	—	—	300	—	—
Total	16,200	600	6,500	—	1,200	—	—	3,400	—	—	—	300	—	—
Cuba														
Government-owned refineries at Cabaiguan, Havana, Santiago de Cuba.														
Total	160,000													
Cyprus														
Cyprus Petroleum Refinery Ltd.–Larnaca	17,100	2,600	—	—	²4,250	—	—	⁴7,750	—	—	—	1,200	—	—
Czechoslovakia														
Government owned refineries at Bratislava, Kolin, Kralupi, Pardubice, Strazke, Zaluzi, Zvolen.														
Total	455,000													
Denmark														
AS Dansk Shell–Fredericia	55,000	—	²22,000	—	²12,000	—	²10,000	²20,000 ¹9,200 ⁵11,400	—	²3,900	—	8,000	—	—
Statoil AS–Kalundborg	65,000	22,000	²28,000	—	²9,200	—	—	—	—	—	—	—	—	—
Kuwait Petroleum Refining (Danmark) A/S–Gulfhavn	56,500	19,050	²10,900 ³9,770	—	²10,500	—	—	²22,300 ⁴²22,900	—	—	—	—	—	—
Total	176,500	41,050	70,670	—	31,700	—	10,000	85,800	—	3,900	—	8,000	—	—
Dominican Republic														
Falconbridge Dominicana C por A–La Peguera	16,000	—	—	—	—	—	—	²6,200	—	—	—	—	²0.8	—
Refineria Dominicana de Petróleo SA–Haina	31,000	—	—	—	²7,500	—	—	⁴15,500	—	—	—	—	—	—
Total	47,000	—	—	—	7,500	—	—	21,700	—	—	—	—	0.8	—
Ecuador														
CEPE–Esmeraldas	90,000	43,020	²25,200	¹16,000	²2,780	—	—	—	—	—	—	—	—	—
Anglo Ecuadorian Oilfields Ltd.–Sta. Elena Peninsula	32,300	—	—	—	—	—	—	—	—	—	—	—	—	—
Texaco–Lago-Agrio	1,000	—	—	—	—	—	—	—	—	—	—	—	—	—
Total	123,300	43,020	25,200	16,000	2,780	—	—	—	—	—	—	—	—	—

Petroleum Refining Processes, Worldwide Capacities

Egypt	Alexandria Petroleum Co.—Alexandria (El-Mex)	99,250	22,500	—	—	—	—	⁶1,436 ⁷76	—	⁶603 ²81	1,436	2,010	¹1.0	—
	Cairo Oil Refining Co.—Mostorod	125,060	—	—	³9,000	—	—	¹11,700 ⁴14,400	—	—	—	—	—	—
	Tanta	21,800	—	—	—	—	—	—	—	—	—	—	—	—
	El Ameria Oil Refining Co.—Alexandria	60,340	14,725	—	³9,400	—	—	¹12,000 ⁴14,400 ⁶1,608	²804	¹86 ¹482	1,407	2,010	¹1.0	—
	El-Nasr Petroleum Co.—El-Suez Wadi-Feran	58,400 9,680	—	—	—	—	—	—	—	—	—	2,613	—	—
	Suez Oil Processing Co.—El-Suez	62,530	9,500	⁵16,470	—	—	²12,540	¹16,660 ⁴10,800 ⁶1,306	—	—	1,005	—	³3.5	¹540
	Assiout	52,143	—	—	—	—	—	—	—	—	—	—	—	—
	Total	489,203	46,725	16,470	30,940	—	—	84,386	804	1,452	3,848	6,633	5.5	541
El Salvador	Refineria Petrólera Acajutla SA—Acajutla	17,000	1,200	—	³3,200	—	—	⁵5,000 ⁷7,000	—	—	—	—	—	—
	Total	17,000	1,200	—	3,200	—	—	12,000	—	—	—	—	—	—
Ethiopia	Ethiopian Petroleum Corp.—Assab	18,000	1,600	—	²2,270	—	—	—	—	—	—	—	—	—
Finland	Neste Oy—Naantali	44,000	13,700	²7,100	²13,300	⁶,800	—	³7,100 ⁴14,700	*280	—	—	3,000	—	—
	Porvoo	197,000	67,700	³39,200	²28,500	³36,100	¹14,200	²51,000	²4,200 *3,500	³3,600	—	8,800	²20.0	—
	Total	241,000	81,400	46,300	41,800	42,900	14,200	57,930	7,980	3,600	—	11,800	20.0	—
France	CRD—Total France—Gonfreville L'Orcher[d]	309,000	73,000	³16,600	¹31,500	²45,400	—	³59,600 ³9,000 ⁵49,100 ⁵10,400 ³24,500	*1,150	¹1,360	7,400	20,250	—	—
	La Mède	136,000	47,500	²20,850	²29,700	²23,400	—	²71,600	²2,400	—	—	16,200	—	—
	Mardyck	122,000	41,600	—	²29,800	²20,500	—	—	—	—	—	—	—	—
	Cie. Rhénane de Raffinage—Reichstett-Vendenheim	85,000	36,000	²5,000 ¹11,000	¹14,000	²13,000	—	²37,500	—	—	—	4,500	—	—
	Elf France—Donges	200,000	99,000	¹44,000	²25,400	—	—	²21,000	—	—	—	—	—	—
	Feyzin	175,070	34,000	¹30,200	²8,400	—	—	¹35,600 ⁴⁵38,000 ¹8,700	²3,700	⁶4,300	—	—	—	—
	Grandpuits	92,000	46,900	³13,400	²28,800	²12,900	—	⁴⁵28,900 ²23,500 ⁴⁵26,100	²2,900	⁵5,000	—	—	—	—

(continued)

TABLE 2 (continued)

Company and Refinery Location	Crude	Vacuum Distillation	Thermal Operations	Catalytic Cracking	Cat Reforming	Cat Hydro-cracking	Cat Hydro-refining	Cat Hydro-treating	Alky. *Poly.	Aromatics-Isomerization	Lubes	Asphalt	Hydrogen (MMcfd)	Coke (t/d)
Esso SAF—Fos sur Mer	100,000	23,000	—	¹17,000	¹16,000	—	—	⁴48,000	—	—	7,000	—	—	—
Port Jerome	136,000	56,000	—	¹20,000	¹16,000	—	—	⁴36,000	—	—	—	—	—	—
Mobil Oil Francaise—														
Notre Dame de Gravenchon	57,000	36,500	—	—	²13,300	—	—	¹13,000	—	—	5,500	1,500	—	—
								⁴13,000						
Shell Francaise—Berre l'Etang	128,000	60,000	²23,000	¹38,000	¹19,000	—	²35,000	²33,000	—	—	—	5,500	—	—
Petit Couronne	154,000	113,000	⁵10,000	¹19,000	²28,000	—	²33,000	²48,000	—	—	7,500	9,000	—	—
Ste. Francaise des Petroles BP—														
Dunkirk	—	18,000	—	—	—	—	—	—	—	—	4,200	—	—	—
Lavera	181,900	—	²22,800	¹26,000	¹5,700	¹13,500	²32,300	²14,000	—	¹4,300	—	2,700	¹28.0	—
Total	1,875,970	684,500	152,850	319,200	247,000	13,500	264,900	523,000	10,150	14,960	31,600	59,650	28.0	—
Gabon														
Ste. Gabonaise de Raffinage—Port Gentil	24,000	—	²7,200	—	³1,400	—	—	⁴3,600	—	—	—	—	—	—
Germany, East														
Government-owned	470,000													
Germany, West														
Deutsche BP-Erdol Ingolstadt—Vohburg	102,000	35,300	—	¹17,200	¹17,200	—	²14,000	²28,000	—	—	—	5,500	—	—
Deutsche Marathon Petroleum GmbH—Burghausen	70,000	—	⁶26,000	—	—	—	—	¹18,000	—	¹2,200	—	—	—	—
Deutsche Shell AG—Godorf	170,000	73,000	²29,000	—	²17,000	¹19,000	²32,000	²52,000	—	¹4,000	—	12,000	—	—
			⁵7,000		⁶21,000			²34,000		²3,000				
								²12,000		⁶6,000				
Harburg-Grasbrook	86,000	45,000	²14,000	¹15,000	⁶16,000	—	²14,000	²31,000	—	—	6,100	7,000	—	—
							⁶6,000	²2,000						
Deutsche Texaco AG—Heide	80,000	34,000	²16,000	¹9,000	¹18,000	—	²29,000	²26,000	⁸300	¹7,000	—	3,500	—	—
Erdoel Raffinerie Neustadt GmbH—Neustadt-Donau	144,000	65,400	²15,300	¹22,700	¹17,700	—	²27,200	²22,500	—	—	—	5,400	—	—
Esso AG—Ingolstadt	95,000	31,400	—	²24,500	³12,800	—	—	²20,400	—	—	—	7,400	—	—
								⁴47,400						
Karlsruhe	150,000	46,700	²28,000	—	²27,900	—	—	²43,000	—	—	—	10,000	—	1,240
			²23,000					⁴73,000						
			⁵17,000											
			⁶1,900											
Mobil Oil AG—Worth	93,000	23,000	—	¹19,000	²15,900	—	—	²28,000	—	—	—	5,000	—	—
								⁴26,000						
Oberrheinische Mineralolwerke GmbH—Karlsruhe	142,000	102,000	²26,000	¹66,000	²11,400	—	²45,900	²37,400	⁸8,500	—	—	11,800	—	—
					²20,100		⁴28,000							
Ruhr Oel GmbH—Gelsenkirchen	215,400	48,200	²16,000	¹6,000	²11,900	¹30,000	²63,000	²53,000	⁸2,500	¹10,620	—	4,840	⁵25.0	1,200
			²28,000		²27,000					²2,150				
										³2,420				
Union Rheinische Braunkohlen Kraftstoff AG—Wesseling	103,000	71,000	²27,000	—	²14,000	²28,000	²16,000	²14,000	—	¹6,000	—	—	¹32.0	—
								²16,000					²42.0	
								³15,000					⁴17.0	

Petroleum Refining Processes, Worldwide Capacities

	Company—Location													
	Wintershall AG—Lingen	65,000	28,900	15,300	—	—	¹19,200	²21,300	¹15,000 ²2,500	—	¹500 ²900 ³2,200 ⁴2,650 ⁵2,650	—	—	3,400	¹40.0	420
	Mannheim	—	—	—	—	—	—	—	³5,500	—	—	—	—	—	—	—
	Salzbergen	2,800	4,800	—	—	—	—	—	⁶2,500	—	—	2,900	—	²0.5	—
	Total	1,518,200	608,700	289,500	179,400	275,100	96,200	296,400	624,200	11,300	62,290	9,000	75,840	156.5	2,860
Ghana	Ghanaian Italian Petroleum CL—Tema	26,600	—	—	—	²5,850	—	—	⁵5,850	—	—	—	—	—	—
Greece	Hellenic Aspropyrgos Refinery SA—Aspropyrgos	110,000	51,000	²21,600	²23,300	²14,700 ⁶20,000	—	²15,600 ³25,500	²24,200	⁴1,300	—	—	—	⁵9.1	—
	Motor Oil (Hellas) Corinth Refineries SA—Aghii Theodori	100,000	59,000	¹19,520	²27,200	²8,500	—	—	²11,800 ³25,800 ⁶3,100	²2,400 ⁴2,000	⁴4,250	3,100	2,800	—	—
	Petrola Hellas SA—Elefsis	108,000	—	—	—	—	—	—	—	—	—	—	—	—	—
	Thessaloniki Refining Co. AE—Thessaloniki	66,500	8,100	—	—	²8,100	—	—	²16,500 ⁴18,800	—	—	—	2,800	—	—
	Total	384,500	118,100	41,120	50,500	51,300	—	41,100	100,200	5,700	4,250	3,100	5,600	9.1	—
Guatemala	Texas Petroleum Co.—Escuintla	16,000	—	—	—	²3,000	—	—	³3,000 ⁴2,000	—	—	—	—	—	—
	Total	16,000	—	—	—	3,000	—	—	5,000	—	—	—	—	—	—
Honduras	Refineria Texas de Honduras SA—Puerto Cortes	14,000	—	—	—	²1,800	—	—	³3,000 ⁴2,000	—	—	—	—	—	—
	Total	14,000	—	—	—	1,800	—	—	5,000	—	—	—	—	—	—
Hungary	Dunai KV—Százhalombatta	150,000	84,000	¹13,300	²20,000	²23,000	—	—	²23,000 ³35,000 ⁶2,200 ⁴16,400	²3,300	¹7,100 ⁶2,700	4,000	6,300	—	—
	Tiszai KV—Leninváros	60,000	30,000	—	—	—	—	—	—	—	—	—	4,500	—	—
	Zalai KV—Zalaegerszeg	10,000	6,000	—	—	—	—	—	—	—	—	—	—	—	—
	Total	220,000	120,000	13,300	20,000	23,000	—	—	76,600	3,300	9,800	4,000	10,800	—	—
India	Bharat Petroleum CL—Mahul, Bombay	140,000	42,500	—	¹31,000	²5,200	—	—	—	—	²2,500 ¹375	—	1,000	—	—

(continued)

TABLE 2 (*continued*)

Company and Refinery Location	Crude	Charge Capacity (b/cd)						Production Capacity (b/cd)					Hydrogen (MMcfd)	Coke (t/d)
		Vacuum Distillation	Thermal Operations	Catalytic Cracking	Cat Reforming	Cat Hydro-cracking	Cat Hydro-refining	Cat Hydro-treating	Alky. *Poly.	Aromatics-Isomerization	Lubes	Asphalt		
Bongaigaon Refinery & Petrochemicals Ltd.—Bongaigaon, Assam	27,392	—	⁵10,400	—	²1,930	—	—	¹1,930	—	—	—	—	—	166
Cochin Refineries Ltd.—Kerala	90,000	34,200	³19,600	¹20,000	⁵5,000	—	—	⁵9,400 ⁷18,000	—	—	—	6,900	—	—
Hindustan Petroleum CL—														
Mahul, Bombay⁶	110,000	50,000	—	¹12,000	⁶18,000	⁵6,000	—	⁶6,000	—	—	6,000	6,031	—	—
Visakhapatnam	92,699	44,496	—	¹19,583		—	—	—	—	—	—	2,585	—	—
Indian Oil CL—Barauni	66,000	15,000	⁵20,170	—	—	—	—	—	—	—	—	—	—	525
Digboi	10,050	—	³800	—	—	—	—	—	—	—	1,200	513	—	40
Gauhati	60,400	—	⁵5,500	—	—	—	—	—	—	—	—	—	—	150
Gujarat	165,200	47,000	³18,900	¹19,800	³7,000	—	—	—	—	¹1,375	—	4,300	—	—
Haldia	55,000	20,000	³8,500	—	³4,500	—	⁴13,800	⁶3,800	—	—	3,800	3,500	—	—
Mathura	120,000	43,000	³18,900	¹19,800	—	—	—	—	—	—	—	8,600	—	—
Madras Refineries Ltd.—Madras	114,700	36,800	²4,900 ³3,000	¹11,770	²2,400	—	—	¹2,400 ²25,100 ⁴4,000	—	—	4,000	3,800	⁹9.5	—
Total	1,051,441	332,996	110,670	133,953	26,030	6,000	13,800	70,630	—	4,250	15,000	37,229	9.5	881
Indonesia														
Pertamina—Balikpapan, Kalimantan	221,500	84,800	—	—	⁶18,000 ¹12,000	¹49,700	—	¹18,000	—	—	—	—	⁶60.0	—
Cilacap, Central Java	271,200	30,700	³49,900	—	⁶18,000	—	⁴18,000	¹18,100	—	—	5,000	7,500	—	—
Dumai, Central Sumatra	99,400	83,500	⁵31,800	—	⁵5,600 ⁶7,900	⁵0,400	—	⁶6,300 ³9,100	—	—	—	—	⁷70.8	700
Musi, South Sumatra	81,400	30,000	—	¹12,600	—	—	—	—	¹300 *300	—	—	—	—	—
Pangakalan Brandan, North Sumatra	4,500	—	—	—	—	—	—	—	—	—	—	—	—	—
Sungai Pakning, Central Sumatra	36,200	—	—	—	—	—	—	—	—	—	—	—	—	—
Total	714,200	229,000	81,700	12,600	61,500	100,100	18,000	51,500	600	—	5,000	7,500	130.8	700
Iran⁸ National Iranian Oil Co.—Esfahan	200,000	96,800	³38,000	—	29,600	¹30,000	—	29,000	—	—	—	5,450	70.0	—
Shiraz	40,000	18,400	³8,800	—	6,245	⁵9,380	3,120	5,225	—	—	—	—	—	—
Tabriz	80,000	—	—	—	—	¹26,000	—	—	—	—	—	—	—	—
Tehran	210,000	100,000	⁵34,000	—	28,000	²7,800	—	—	—	—	6,000	7,000	—	—
Total	530,000	215,200	80,800	—	63,845	93,180	3,120	34,225	—	—	6,000	12,450	70.0	—
Iraq⁸ Oil Refineries Administration—Baji	150,000	65,000	—	—	22,000 16,500	38,000	31,000 28,000 41,000	—	—	—	—	—	64.0	—
Basra	70,000	17,650	—	—	—	—	—	13,000	—	—	2,008	5,450	—	—
Daura	71,000	—	—	—	5,000	—	—	—	—	—	8,160	1,815	—	—
K3-Haditha	7,000	—	—	—	—	—	—	—	—	—	—	—	—	—
Khanaqin	12,000	—	—	—	—	—	—	—	—	—	—	—	—	—
Muthia	4,500	—	—	—	—	—	—	—	—	—	—	4,900	—	—
Qaiyarah, Mosul	2,000	—	—	—	—	—	—	—	—	—	—	—	—	—
Iraqi Company for Oil Operations—Kirkuk	2,000	—	—	—	—	—	—	—	—	—	—	—	—	—
Total	318,500	82,650	—	—	43,500	38,000	100,000	13,000	—	—	10,168	6,715	64.0	—

Petroleum Refining Processes, Worldwide Capacities

Country	Company														
Ireland	Irish Refining plc—Whitegate	56,000	—	—	—	⁴11,000	—	⁴5,500	²11,000	—	—	—	—	—	—
Israel	Oil Refineries Ltd.—Ashdod	70,000	35,000	²25,000	—	²10,000	—	⁴25,000	¹15,000	—	—	—	—	—	—
	Haifa	110,000	55,000	⁴45,000	¹22,000	²16,000	—	²20,000	¹15,000	—	—	—	3,000	—	—
	Total	180,000	90,000	70,000	22,000	26,000	—	45,000	30,000	—	—	—	3,000	—	—
Italy	Agip Plus SpA—Livorno	100,000	36,000	—	—	²9,500	—	⁴31,700	¹20,000	—	—	8,800	3,500	—	—
	Agip Raffinazione—Rho, Milan	79,800	32,800	—	²14,300	⁴9,000	—	⁴15,400	²20,700	²1,600	⁶9,100	1,410	4,100	⁵14.3	—
	Sannazzaro, Pavia	210,000	—	—	¹27,000	²17,000	—	⁴41,000	⁴9,800	²3,200	—	—	—	²2.8	—
						²10,000			²27,000						
						⁵16,700			²24,000						
	Taranto	83,300	13,500	²18,150	—	²17,000	—	⁴13,700	¹16,000	—	—	—	—	²10.0	—
				³37,250				⁵20,500	⁵5,200						
	Venezia	90,000	38,000	²25,000	—	²16,000	—	⁴18,000	⁴4,000	—	⁵5,200	—	6,500	—	—
	Anonima Petroli Italiana—Falconara, Marittima	78,900	13,000	²4,500	—	²10,500	—	—	¹18,900	—	—	—	—	⁵3.8	—
				³1,000					⁴26,900						
	Arcola Petrolifera SpA—La Spezia	18,000	8,000	—	—	—	—	—	—	—	—	16,900	11,500	—	—
	Esso Italiana SpA—Augusta, Siracusa	180,000	94,000	—	¹45,300	²9,300	—	—	¹16,600	²9,100	⁴5,300				
						⁴7,300			⁴37,600						
									⁶23,000						
	Industrie Chimiche Italiane de Petrolio SpA—Frassino, Mantova	55,000	19,600	²8,000	—	²7,000	—	—	²14,000	—	⁶2,500	—	3,000	—	—
				³19,000					⁴15,000						
	Iplom SpA—Busalla	46,500	12,600	³13,100	—	—	—	—	—	—	—	6,500	12,500	—	—
	Isab—Priolo Gargallo	220,000	110,000	²15,000	—	⁴34,500	⁵9,000	⁴86,000	²59,000	—	⁶15,000	—	—	⁵18.0	—
				³9,000											
	Mobil Oil Italiana SpA—Naples	100,000	22,500	7,500	²18,400	²12,500	—	—	¹12,500	²3,500	—	—	650	—	—
									²12,000						
	Raffineria di Roma SpA—Rome	80,500	11,300	²28,400	—	²14,200	—	²17,000	²19,900	—	⁴5,200	—	9,620	—	—
	Raffineria Mediterranea SrL—Milazzo	160,000	72,000	—	¹28,000	²11,000	—	—	⁴10,000	¹3,600	⁵800	—	—	—	—
	Raffineria Siciliana SrL—Gela	80,000	32,000	²29,500	¹34,000	²10,000	—	⁴25,000	²7,000	⁴9,000	⁴4,000	—	—	—	1,400
	Saras SpA—Sarroch	285,000	—	²27,000	¹63,000	²27,000	—	—	¹27,000	²6,800	—	—	—	—	—
									⁴60,000						
	Sarpom—Trecate, Novara	248,200	24,800	—	¹16,300	²18,500	—	⁴38,000	⁴69,700	1,500	—	—	—	⁵12.0	—
						⁴8,100									
	Selm-Soc. Energia Montedison—Priolo	240,000	50,000	²25,000	¹32,000	²9,000	—	⁴18,000	¹9,000	—	¹2,000	—	80,000	—	—
											²1,300				
											⁶3,600				
	Tamoil Italia SpA—Cremona	95,000	—	³3,000	—	³3,800	—	⁴12,800	²14,400	—	—	—	—	—	—
						²6,400									
	Total	2,450,200	590,100	360,400	278,300	284,300	59,000	337,100	579,200	38,300	64,000	33,610	131,370	60.9	1,400

(continued)

Petroleum Refining Processes, Worldwide Capacities

TABLE 2 (continued)

	Company and Refinery Location	Crude	Vacuum Distillation	Thermal Operations	Catalytic Cracking	Cat Reforming	Cat Hydrocracking	Cat Hydrorefining	Cat Hydrotreating	Alky. *Poly.	Aromatics-Isomerization	Lubes	Asphalt	Hydrogen (MMcfd)	Coke (t/d)
Ivory Coast	Ste. Ivorienne de Raffinage—Abidjan	50,000	34,500	—	—	[2]16,600	[1]11,500	[1]1,500	[4]21,000	—	—	—	—	—	—
	Ste. Multinationale de Bitumes—Abidjan	10,000	7,500	—	—	—	—	—	—	—	—	—	6,000	—	—
	Total	60,000	42,000	—	—	16,600	11,500	11,500	21,000	—	—	—	6,000	—	—
Jamaica	Petrojam Ltd.—Kingston	34,200	1,900	—	—	[1]3,240	—	—	[1,3]5,850 [4]13,500	—	—	—	—	—	—
	Total	34,200	1,900	—	—	3,240	—	—	19,350	—	—	—	—	—	—
Japan	Asia Oil CL—Sakaide, Kagawa	13,500	36,000	—	[1]13,500	[2]9,000	—	[2]5,200 [4]8,780 [4]17,100	[3]9,000 [3]3,600	—	—	—	—	[5]32.0	—
	Cosmo Oil CL—Chiba	209,000	57,000	—	[2]6,600	[2]31,800	—	[1]57,000 [2]33,200	[1]31,800 [2]7,600 [4]60,800	—	—	—	—	[1]21.0 [2]62.0	—
	Sakai	104,500	32,300	—	[1]17,550	[2]6,650	—	[2]18,000	[1]17,650 [2]21,900	—	—	—	—	[5]10.6	—
	Yokkaichi City	166,300	70,300	—	[2]23,700	[2]5,400 [4]9,900	—	[2]31,500	[1]15,300 [4]61,800 [5]5,850	—	—	5,850	—	[2]12.0	—
	Fuji Oil CL—Sodegaura	126,600	79,600	[6]19,000	[2]14,000	[2]13,600	—	[2]52,400	[2]27,000 [4]41,600 [6]6,800	—	—	—	—	[2]27.9	—
	General Sekiyu Seisei KK—Sakai	140,400	63,000	—	[2]32,500	[2]26,100	—	[2]36,000	[2]36,000 [4]54,300	—	[1]17,300	—	—	[2]21.6	—
	Idemitsu Kosan CL—Chita, Aichi	123,500	—	—	[2]27,000	[5]16,200	—	[2]49,500	[2]23,400 [4]43,200	[1]5,400	—	—	—	[2]63.8 [4]13.2	—
	Himeji, Hyogo	104,500	—	—	—	[2]12,600	—	[2]36,000	[2]25,200 [4]31,500	—	—	—	—	[3]33.6	—
	Ichihara, Chiba	199,500	57,000	—	[1]37,800	[2]15,300	[9]9,450	[2]36,000 [3]35,100 [5]5,400	[2]38,700 [4]47,700	—	—	5,170	7,900	[2]64.9 [5]3.5	—
	Tokuyama, Yamaguchi	95,000	47,500	—	[1]20,700	—	—	[2]40,500	[1]13,500 [4]31,500	—	—	—	1,530	[2]18.0	—
	Tomakomai, Hokkaido	85,500	19,000	—	[2]16,200	—	[1]13,500	—	[1]24,300 [2]22,860	—	—	—	—	[4]2.4	—
	Kainan Petroleum Refining CL—Kainan	64,000	37,500	—	—	—	—	—	[2]21,500 [4]9,000	—	—	11,600	17,500	[2]4.0	—
	Kashima Oil CL—Kashima, Ibaragi	150,000	67,000	—	[1]17,500	[5]14,000	—	[1]20,000 [2]25,000 [3]35,000 [4]10,000	[1]14,500	—	—	—	—	[2]22.5	—
	Koa Oil CL—Marifu	110,000	49,000	—	[1]20,000	[2]9,700	—	—	[1]11,000 [4]29,000	—	—	—	1,700	[2]6.6	500
	Osaka	80,000	55,000	[5]22,800	[1]18,000	[2]9,000	—	[2]15,000	[1]11,000 [4]14,000	—	—	—	—	—	—
	Kyokuto Petroleum Ltd.—Chiba	120,000	55,000	—	—	[2]19,000	[2]28,000	—	[1]23,000 [4]20,000	—	—	—	—	[2]20.0	—

Petroleum Refining Processes, Worldwide Capacities

Kyushu Oil CL—Oita	130,000	—	¹16,500	²8,000	—	²33,000 ²27,000	¹9,000	—	—	—	—	—
Mitsubishi Oil CL—Kawasaki	55,000	5,800	—	²8,500	—	⁴17,000 ⁵45,000 ²30,000 ³11,000 ⁴25,000	—	—	—	1,800	—	—
Mizushima	220,000	97,000	¹34,000	²25,000 ⁶18,500	—	³14,000	*880	¹12,700	2,800	5,700	⁵14.9 ²60.0	—
Nansei Sekiyu KK—Nishihara, Okinawa	45,000	—	—	¹8,600	—	—	—	—	—	—	—	—
Nichimo Sekiyu Seisei KK—Kawasaki	76,000	—	—	¹7,600	—	²21,100 ⁴13,500	—	—	—	—	—	—
Nihonkai Oil CL—Toyama	50,000	—	—	4,000	—	²7,200 ²22,800	—	—	—	—	—	—
Nippon Mining CL—Chita	85,000	40,000	¹13,000	²22,000	—	5,000 ²43,000 ⁴46,000	—	¹11,000 ²3,000 ³1,450	—	—	¹23.7 ⁴9.2	—
Funakawa	6,000	2,740	—	—	—	²28,300 ⁴59,600 ⁶6,000	—	—	1,070	—	—	—
Mizushima	190,200	107,000	¹38,000	²20,000	—	⁵21,300	¹9,000	—	4,900	—	—	—
Nippon Oil CL—Niigata	26,000	2,000	²4,000	—	—		—	—	—	425	—	—
Nippon Petroleum Refining CL—Muroran	150,000	61,000	¹21,000	²21,000	⁴40,000	²32,000 ⁵48,000	—	—	—	4,160	⁵62.0	—
Nakagusuku, Okinawa	—	—	—	²2,500	—	¹2,500	—	—	—	—	—	—
Negishi	305,000	110,000	¹39,000	²20,000 ⁶30,000	—	⁵53,000 ⁶69,000	¹6,000	—	4,310	6,660	⁵22.0	—
Yokohama	—	18,500	—	²1,300	—	⁷3,300 ⁵7,300	—	—	1,250	—	—	—
Okinawa Sekiyu Seisei—Yonashiro, Okinawa	57,000	—	—	—	—	⁴18,000	—	—	—	—	—	—
Seibu Oil CL—Yamaguchi	110,100	42,200	¹18,100	²17,800	—	¹34,200 ²13,500 ⁴40,100 ⁴24,900	—	—	—	—	²42.7	—
Showa Shell Sekiyu KK—Kawasaki	136,800	27,500	—	²19,100	—	²7,600 ⁴18,700	—	⁴5,500	—	—	¹42.4	—
Niigata	27,500	—	—	²4,300	—	²11,600 ⁴2,000	—	—	—	—	—	—
Showa Yokkaichi Sekiyu CL—Yokkaichi	215,700	96,300	¹22,600	²38,100	—	²61,400 ⁴39,200	—	¹7,600 ⁶5,000	5,500	6,300	¹28.3	—
Taiyo Oil CL—Ehime	61,750	19,000	—	²5,400	¹10,800	²35,600 —	—	—	—	—	²54.7	—

(*continued*)

TABLE 2 (continued)

Company and Refinery Location	Crude	Charge Capacity (b/cd)						Production Capacity (b/cd)					
		Vacuum Distillation	Thermal Operations	Catalytic Cracking	Cat Reforming	Cat Hydro-cracking	Cat Hydro-refining	Cat Hydro-treating	Alky. *Poly.	Aromatics-Isomerization	Lubes	Asphalt	Hydrogen (MMcfd)	Coke (t/d)
Toa Nenryo Kogyo—Kawasaki	190,000	116,900	—	¹74,100	²15,200	—	⁵74,000	²32,300 ⁴31,400 ⁵11,400	*9,500	—	—	—	²39.0	—
Wakayama	157,700	58,600	—	¹31,600	²34,200	—	²3,800	²33,700 ⁴39,000 ⁵20,900	¹3,100	—	6,100	—	¹16.0	—
Toa Oil CL—Kawasaki	65,000	53,000	⁶21,000	—	²11,000	—	—	²17,000 ³3,500 ⁴22,000 ⁵46,000	—	—	—	—	¹34.7	150
Toho Oil CL—Owase	35,000	—	—	—	—	—	—	—	—	—	—	—	—	—
Tohoku Oil CL—Sendai	75,700	50,200	—	—	²11,300	—	²29,200	²11,200 ⁴17,300	—	—	—	4,800	—	—
Total	4,362,750	1,698,940	82,800	596,950	551,650	101,750	1,370,280	1,973,740	33,880	63,550	48,550	58,475	944.5	650
Jordan Jordan Petroleum Refinery—Zerka	100,000	16,900	—	⁴4,410	²8,640	¹4,800	—	²15,100 ⁴2,000	—	—	—	—	²7.9	—
Total	100,000	16,900	—	4,410	8,640	4,800	—	17,100	—	—	—	—	7.9	—
Kenya Kenya Petroleum Refineries Ltd.—Mombasa	90,000	1,700	—	—	¹3,800 ³5,200	—	—	⁴33,000	—	—	—	—	—	—
Total	90,000	1,700	—	—	9,000	—	—	33,000	—	—	—	—	—	—
Korea, North Government-owned, Unggi	42,000	—	—	—	—	—	—	—	—	—	—	—	—	—
Korea, South Honam Oil Refinery CL—Yosu	355,000	—	³30,000	—	²15,300	—	—	²22,500 ²2,000 ⁶1,200	—	—	—	3,600	—	—
Kukdong Oil CL—Busan[b]	10,000	7,000	—	—	—	—	—	⁶6,000	—	—	2,000	4,000	²0.4	—
Daesan	60,000	34,000	⁵19,000	—	²3,000	¹22,000	—	⁴12,000 ³3,600	—	—	—	—	²54.0	1,200
Kyung In Energy CL—Inchon	60,000	—	—	—	²3,600	—	—	⁴4,000	—	—	—	—	—	—
Ssangyong Oil Refining CL—Onsan[i]	90,000	22,000	—	—	²4,000	³7,000	⁴21,000	⁴4,000 ²1,100 ⁶2,400	—	—	4,200	5,000	²14.0	—
Yukong Ltd.—Ulsan	305,000	4,400	—	—	²17,650 ⁶18,900	—	—	²38,900 ⁴10,500 ⁶2,600	—	¹11,930 ²4,900 ³800	—	2,600	—	—
Total	880,000	67,400	49,000	—	62,450	29,000	21,000	110,800	—	17,630	6,200	15,200	68.4	1,200
Kuwait Getty Oil Co. (subsidiary of Texaco Inc.)—Mina Al-Zour	70,000	—	—	—	—	—	—	—	—	—	—	—	—	—
Kuwait National Petroleum Co.—Mina Abdulla	190,000	117,000	⁵54,000	—	—	¹34,000	¹88,000	²7,000 ⁴66,000	—	—	—	—	¹161.0	1,900

Petroleum Refining Processes, Worldwide Capacities

Country	Company—Location														
	Mina Al-Ahmadi	370,000	90,000	—	¹28,000	²33,000	¹34,000	¹117,000	¹34,000 ⁴66,000 ¹25,000 ⁴65,000	—	—	—	5,000	¹156.0	—
	Shuaiba	187,000	131,000	—	²14,000	—	¹60,000 ²42,000	—		—	—	—	—	²210.0	—
	Total	817,000	338,000	54,000	42,000	33,000	170,000	205,000	263,000	—	—	—	5,000	527.0	1,900
Lebanon	Tripoli Oil Installations—Tripoli	20,000	12,730	—	⁷7,250	²4,392 ²3,100	—	²7,392	¹3,100 —	—	—	—	409	—	—
	Mediterranean Refining Co.—Sidon	17,000	—	—	—		—	—		—	—	—	—	—	—
	Total	37,000	12,730	—	7,250	7,492	—	7,392	3,100	—	—	—	409	—	—
Liberia	Liberia Petroleum Refining—Monrovia	15,000	1,000	—	—	²2,000	—	—	²2,300 ⁴1,000	—	—	—	200	—	—
	Total	15,000	1,000	—	—	2,000	—	—	3,300	—	—	—	200	—	—
Libya	Azzawiya Oil Refining Co.—Azzawiya	120,000	3,432	—	—	²12,982	—	—	16,908 ⁵18,718	—	635	—	1,716	—	—
	Ras Lanuf Oil & Gas Processing Co.—Ras Lanuf	201,000	—	—	—	¹1,000	—	—	²1,500	—	—	—	—	—	—
	Sirte Oil Co.—Brega	8,400	—	—	—	—	—	—	—	—	—	—	—	—	—
	Total	329,400	3,432	—	—	13,982	—	—	37,126	—	635	—	1,716	—	—
Madagascar	Solima—Managareza, Tamatave	16,350	1,200	²7,000	—	²2,600	—	²3,600	⁵5,500	—	—	—	—	—	—
Malaysia	Esso Malaysia Berhad—Port Dickson	47,300	4,750	—	—	²7,980	—	—	¹4,250 ⁴13,600	—	—	—	—	—	—
	Petronas Penapisan Sdn. Berhad—Kerteh, Kemaman, Terengganu	27,000	—	—	—	—	—	—	—	—	—	—	—	—	—
	Sarawak Shell Berhad—Luton	45,000	—	—	—	—	—	—	—	—	—	—	—	—	—
	Shell Refining Co. Berhad—Port Dickson	90,000	2,400	—	—	⁴4,000 ¹0,000	—	—	²8,000 ⁴5,000	—	—	—	1,700	—	—
	Total	209,300	7,150	—	—	21,980	—	—	60,850	—	—	—	1,700	—	—
Martinique	Ste. Anonyme de la Raffinerie des Antilles—Fort-de-France	12,800	—	—	—	²7,410	—	—	¹4,850 ⁴3,500	—	—	—	—	—	—
	Total	12,800	—	—	—	7,410	—	—	8,350	—	—	—	—	—	—
Mexico	Petroleos Mexicanos—Azcapotzalco	110,000	48,000	¹20,000	¹24,000	—	—	—	—	²1,200 *430	—	—	—	—	—
	Cadereyta	235,000	137,000	—	⁴0,000	²0,000	—	⁵0,000	¹36,000	—	—	—	—	—	—

(continued)

TABLE 2 (continued)

	Company and Refinery Location	Crude	Charge Capacity (b/cd)						Production Capacity (b/cd)					
			Vacuum Distillation	Thermal Operations	Catalytic Cracking	Cat Reforming	Cat Hydro-cracking	Cat Hydro-refining	Cat Hydro-treating	Alky. *Poly.	Aromatics-Isomerization	Lubes	Asphalt	Hydrogen (MMcfd)	Coke (t/d)
	Ciudad Madero[j]	195,000	81,500	[3]8,000 [4]9,000	[1]43,000	[2]15,000	—	[4]37,000	[1]15,000	[2]1,200 *430	—	—	—	[1]10.0	260
	Minatitlan	200,000	83,000	—	[1]24,000 [2]16,000	[2]48,000	—	[4]37,000	[1]52,000	*600	[1]6,820 [2]1,400 [3]2,374	500	—	—	—
	Poza Rica	50,000	—	—	—	—	—	—	—	—	—	—	—	—	—
	Reynosa	9,000	—	—	—	—	—	—	—	—	—	—	—	—	—
	Salamanca	235,000	101,200	[3]4,000	[1]40,000	[2]24,800	[2]18,000	[2]26,000	[1]33,000 [6]10,000	—	—	15,000	15,000	[1]60.0	—
	Salina Cruz	165,000	75,000	—	[1]40,000	[2]20,000	—	[4]50,000	[1]25,000	—	—	—	—	—	—
	Tula, Hidalgo	155,000	75,000	[3]41,000	[2]40,000	[3]30,000	—	[4]50,000	[1]36,000	—	—	—	—	—	—
	Total	1,354,000	600,700	82,000	267,000	157,800	18,000	250,000	207,000	3,860	10,594	15,500	15,000	[1]70.0	260
Morocco	Samir—Mohammedia	129,000	20,000	—	—	[2]24,000	—	—	[1]24,000 [3]9,700 [6]2,100	—	—	2,100	13,700	—	—
	Ste. Cherifienne des Petroles—Sidi Kacem	25,600	7,400	—	[2]5,600	[1]2,800	—	—	[1]2,800	—	—	—	—	—	—
	Total	154,600	27,400	—	5,600	26,800	—	—	38,600	—	—	2,100	13,700	—	—
Netherlands	BP Raffinaderij Nederland NV—Rotterdam	437,000	73,000	[3]35,000	[1]47,500	[2]27,000	—	[4]33,200	[2]148,000 [3]25,200 [4]9,900 [6]61,200 [5]13,500	—	—	—	—	—	—
	Esso Nederland BV—Rotterdam	155,000	66,500	[6]31,500	—	[4]25,200	—	—		[2]4,700	—	—	—	[1]21.0	—
	Kuwait Petroleum Europoort BV—Rotterdam[k]	75,500	32,870	[2]9,460 [3]6,400	—	[2]8,200	—	—	[1]12,120 [4]28,400 [6]6,000	—	—	4,600	3,320	—	—
	Shell Nederland Raffinaderij BV—Pernis	348,000	148,000	[3]45,000	[1]78,000	[2]17,000 [6]42,000	—	[2]77,000	[2]138,000 [3]8,000 [6]300	[2]6,800	—	7,100	5,500	—	—
	Smid & Hollander Raffinaderij BV—Amsterdam	8,200	6,000	—	—	—	—	—	—	—	—	—	5,500	—	—
	Texaco Petroleum Mij. (Nederland) BV—Pernis	207,000	28,000	[3]27,000	—	[2]29,000	—	[4]38,000	[1]29,000	—	—	—	—	—	—
	Total Raffinaderij Nederland NV—Vlissingen	150,000	45,000	—	—	[6]18,000	[1]30,000	—	[1]18,000 [5]25,000	—	—	—	—	[1]65.0	—
	Total	1,380,700	399,370	154,360	125,500	166,400	30,000	148,200	522,620	11,500	—	11,700	14,320	86.0	—
Netherlands Antilles	Refineria Isla Curazao SA—Emmastad	320,000	158,000	[2]67,000	[4]42,000	[2]15,000	—	—	[2]60,000 [3]34,000 [5]25,000 [6]2,500	[2]5,000 *600	—	8,000	8,000	—	—
	Total	320,000	158,000	67,000	42,000	15,000	—	—	121,500	5,600	—	8,000	8,000	—	—

Petroleum Refining Processes, Worldwide Capacities

New Zealand	New Zealand Refining—Whangarei	88,400	36,600	—	—	—	²21,900	¹21,500	²6,500	³5,600 ⁴11,800	—	—	—	—	2,300	¹44.0	—
	Total	88,400	36,600	—	—	—	21,900	21,500	6,500	47,400	—	—	—	—	2,300	44.0	—
Nicaragua	Esso Standard Oil SA Ltd.—Managua	14,500	2,000	—	—	—	¹2,800	—	—	²4,800 ³5,200	—	—	—	—	800	—	—
	Total	14,500	2,000	—	—	—	2,800	—	—	10,000	—	—	—	—	800	—	—
Nigeria	Nigerian Petroleum Refining Cl.—Kaduna¹	104,500	36,290	—	—	¹18,000	²15,300	—	—	²1,600 ⁴15,750 ⁵36	—	²291	—	3,878	14,850	—	—
	Port Harcourt, Alesa Eleme	60,000	—	—	—	—	¹6,000	—	—	—	—	—	—	—	—	—	—
	Port Harcourt, Rivers State	150,000	54,000	—	—	¹40,000	⁶33,000	—	—	⁵33,000	⁷7,020 ²,274ᵐ	⁵3,610	—	—	—	—	—
	Warri	100,000	32,000	—	—	¹26,000	²16,700	—	—	⁵14,500 ¹16,700 ⁵8,600	—	—	—	—	—	—	—
	Total	414,500	122,290	—	—	84,000	71,000	—	—	110,686	9,294	3,901	—	3,878	14,850	—	—
Norway	Esso Norge AS—Slagen-Valloy	90,000	4,000	¹28,000	—	—	²10,000	—	—	²0,000 ⁵9,000	—	—	—	—	4,000	—	—
	Norske Shell AS—Sola	60,000	—	¹26,000	—	—	²10,000	—	—	¹16,000 ⁴9,000	—	—	—	—	—	—	—
	Rafinor AS—Mongstad	89,400	—	¹23,300	—	—	²10,100	—	—	²10,100 ⁵15,600	—	²3,600	—	—	—	—	480
	Total	239,400	4,000	77,300	—	—	30,100	—	—	79,700	—	3,600	—	—	4,000	—	480
Oman	Oman Refinery Co.—Mina Al Fahal	76,932	—	—	—	—	²15,386	—	—	²19,506	—	—	—	—	—	—	—
Pakistan	Attock Refinery Ltd.—Rawalpindi	36,000	1,000	—	—	—	—	—	—	—	—	—	—	115	100	—	—
	National Refinery Ltd.—Korangi, Karachi	47,750	13,200	—	—	—	²2,750	—	—	²2,800 ⁴4,000	—	¹1,250	—	3,700	3,950	⁷0.1	—
	Pakistan Refinery Ltd.—Karachi	46,300	—	—	—	—	¹2,700	—	—	⁷18,000	—	—	—	—	—	—	—
	Total	130,050	14,200	—	—	—	5,450	—	—	24,800	—	1,250	—	3,815	4,050	0.1	—
Panama	Refinería Panama SA—Las Minas	100,000	14,000	—	—	—	²7,500	—	—	¹10,000 ⁴20,000	—	—	—	—	5,000	—	—
	Total	100,000	14,000	—	—	—	7,500	—	—	30,000	—	—	—	—	5,000	—	—
Paraguay	Petroleos Paraguayos—Villa Elisa	7,500	—	—	—	—	—	—	—	—	—	—	—	—	—	—	—

(continued)

TABLE 2 (*continued*)

	Company and Refinery Location	Crude	Vacuum Distillation	Thermal Operations	Catalytic Cracking	Cat Reforming	Cat Hydro-cracking	Cat Hydro-refining	Cat Hydro-treating	Alky. *Poly.	Aromatics-Isomerization	Lubes	Asphalt	Hydrogen (MMcfd)	Coke (t/d)
Peru	Petróleos del Peru:														
	Conchan	5,670	3,915	—	—	—	—	—	—	—	—	—	—	—	—
	Iquitos, Loreto	9,450	—	—	—	—	—	—	—	—	—	—	—	—	—
	La Pampilla, Lima	93,900	11,400	—	[1]7,600	[1]1,680	—	—	—	—	—	—	—	—	—
	Marsella, Loreto	1,953	—	—	—	—	—	—	—	—	—	—	—	—	—
	Pucallpa	2,565	—	—	—	—	—	—	—	—	—	—	—	—	—
	Talara	58,900	21,600	—	[1]4,940	—	—	—	—	—	—	1,080	1,530	—	—
	Total	172,438	36,915	—	22,540	1,680	—	—	—	—	—	1,080	1,530	—	—
Philippines	Caltex (Phillippines) Inc.—Batangas	63,000	18,100	—	[1]8,100	[3]8,100	—	—	[1,2]10,800 [4]14,400	—	—	—	—	—	—
	Petron Corp.—Limay, Bataan	130,000	20,000	—	[2]15,000	[2]23,800	—	[4]18,000	[2]52,000	—	—	—	5,000	—	—
	Philippine Petroleum Corp.—Plililla	—	20,500	—	—	—	—	—	—	—	—	3,400	—	—	—
	Pilipinas Shell Petroleum—Tabangao	61,000	—	—	—	[1]7,500	—	[2]5,000	[4]14,000	—	—	—	—	—	—
	Total	254,000	58,600	—	23,100	39,400	—	23,000	91,200	—	—	3,400	5,000	—	—
Poland	Government-owned refineries at Czechowice, Gdansk, Glinik Mariampolski, Jasto, Jealicze, Kralaty, L. Warynski, Plock, Trzebinia. Of these, Plock is by far the largest, accounting for roughly 80% of Polish refinery production. Its crude capacity is nearly 300,000 b/d.														
	Total	385,000													
Portugal	Petrogal—Lisbon	—	18,900	—	[2]10,100	[2]14,200 [4]10,500	—	[1]3,400	[1]38,400 [4]7,300 [6]3,000	—	[1]8,200	3,000	700 2,700	—	—
	Leça da Palmeira, Porto	88,300	11,000	—	—	—	—	—	—	—	—	—	—	—	—
	Sines	225,000	25,400	—	—	[2]27,500	[4]10,200	[4]31,400 [5]15,000	[1]27,500	—	—	—	2,400	—	—
	Total	313,300	55,300	—	10,100	52,200	10,200	49,800	76,200	—	8,200	3,000	5,800	—	—
Puerto Rico	Caribbean Petroleum Corp.—Bayamón	38,000	19,300	—	[1]12,000	[2]5,800	[3]15,000	[4]12,000	—	*1,000	—	9,000	900	—	—
	Yabucoa Sun Oil Co.—Yabucoa"	85,000	40,000	—	—	—	—	—	—	—	—	—	—	—	—
	Total	123,000	59,300	—	12,000	5,800	15,000	12,000	—	1,000	—	9,000	900	—	—
Qatar	National Oil Distribution Co.—Umm Said	62,000	—	—	—	[2]2,200 [6]9,830	—	—	[2]17,240 [4]10,500 [5]11,900	—	—	—	—	—	—
	Total	62,000	—	—	—	12,030	—	—	39,640	—	—	—	—	—	—
Romania	Government-owned refineries at Bacau, Borzesti, Brazi, Brazov, Cimpina, Darmanesti, G. Gheorghiu Dej, Navodari, Onesti, Pitesti, Ploesti, Sulpacu, Telaejen.														
	Total	617,000													
Saudi Arabia	Arabian American Oil Co.—														
	Ras Tanura	450,000	30,000	—	—	[4]60,000	—	—	[1]60,000	—	—	—	12,500	—	—
	Arabian Oil CL—Ras Al Khafji	30,000	—	—	—	—	—	—	—	—	—	—	—	—	—

Petroleum Refining Processes, Worldwide Capacities

	Jeddah Oil Refinery—Jeddah	91,000	15,000	—	¹10,000	²2,900	—	—	²32,000	—	—	2,500	—	—	—
	Petromin-Mobil—Yanbu	250,000	106,000	—	¹66,000	⁴32,000	—	—	⁴40,000	²12,500	—	10,000	—	—	—
	Petromin-Shell—Al-Jubail⁰	250,000	77,000	³32,000	—	⁶17,000	²39,000	¹40,000	¹17,000	—	¹5,300	—	¹84.0	—	—
									⁵2,000		²6,100				
									⁴45,000						
	Riyadh Oil Refinery—Riyadh⁰	134,000	52,655	—	—	²8,600	²31,100	—	²30,900	—	8,000	5,500	117.0	—	—
						⁵5,400			⁴12,900						
						³5,000									
	Yanbu Petromin Refinery—Yanbu	170,000	—	—	—		—	—	¹35,000	—	—	—	—	—	—
									⁴15,200						
	Total	1,375,000	280,655	32,000	76,000	180,900	70,100	40,000	340,000	12,500	19,400	—	30,500	201.0	—
Senegal	Ste. Africaine de Raffinage—M'Bao (Dakar)	29,800	18,800	—	—	⁴2,870	—	—	—	—	—	—	—	—	—
Sierra Leone	Sierra Leone Petroleum Refining Co.—Freetown	10,000	—	—	—	—	—	—	—	—	—	—	—	—	—
Singapore	BP Refinery Singapore PL—Pasir Panjang	28,000	—	—	—	—	—	—	—	—	—	—	—	—	—
	Esso Singapore PL—Pulau Ayer Chawan	170,000	41,800	—	—	²8,600	—	—	²19,000	—	—	7,000	3,400	—	—
									⁴48,400						
	Mobil Oil Singapore PL—Jurong	193,000	70,000	³45,000	—	²18,000	—	—	¹22,000	—	—	—	—	—	—
									⁴36,000						
	Shell Eastern Petroleum Ltd.—Pulau Bukom	300,000	85,600	²64,700	—	²6,800	¹21,600	²44,600	²18,700	—	—	5,800	4,300	²40.0	—
						⁶13,900			⁴88,300						
	Singapore Refinery CPL⁰—Pulau Merlimau	161,000	58,700	¹28,000	—	⁴11,000	¹9,500	⁴36,000	¹11,000	—	—	—	4,300	²18.0	—
							⁴17,000							⁴14.0	
														⁵7.0	
	Total	852,000	256,100	137,700	—	58,300	48,100	80,600	243,400	—	—	12,800	12,000	79.0	—
Somalia	Iraqsoma Ref. Co.—Mogadishu	10,000	—	—	—	—	—	—	—	—	—	—	—	—	—
South Africa	Caltex Oil SA PL—Cape Town	90,000	44,000	³9,500	¹25,000	²10,500	—	⁴22,300	¹14,250	²2,200	—	—	1,140	—	—
	Mobil Oil Refining Co.—Southern Africa PL—Durban	65,000	32,000	³11,000	¹15,000	²15,000	—	—	¹21,000	—	⁴⁺⁶6,000	—	4,000	—	—
									⁵28,000						
	Durban⁰	—	—	—	—	—	—	—	—	—	—	3,000	—	—	—
	National Petroleum Refiners of South Africa PL—Sasolburg OFS	78,500	28,000	—	¹17,400	²11,300	¹8,600	—	¹11,300	²4,800	⁴3,400	—	1,000	¹25.0	—
							²9,400		⁴10,600					⁵9.0	

(continued)

TABLE 2 (*continued*)

Company and Refinery Location	Crude	Vacuum Distillation	Thermal Operations	Catalytic Cracking	Cat Reforming	Cat Hydrocracking	Cat Hydrorefining	Cat Hydrotreating	Alky. *Poly.	Aromatics-Isomerization	Lubes	Asphalt	Hydrogen (MMcfd)	Coke (t/d)
Shell and BP South Africa														
Petroleum Refineries PL—Durban	200,000	—	²43,000	²27,000	²26,000	—	²55,000	⁴57,000	—	—	—	—	—	—
Total	433,500	104,000	63,500	84,400	62,800	18,000	77,300	142,150	7,000	9,400	3,000	6,140	34.0	—
Spain														
Asfaltos Españoles SA—Tarragona	21,000	16,800	—	—	—	—	—	—	—	—	—	15,000	—	—
Cia. Española de Petróles—San Roque (Cádiz)	160,000	40,000	²35,000	¹36,000	²35,000	—	—	⁴24,000 ⁵15,000 ³20,000	—	¹12,500 ²1,800	—	—	—	—
Tenerife	130,000	12,000	²35,000	—	²13,000	—	—	—	—	—	—	8,000	—	—
Petróleos del Mediterraneo—Castellón de la Plana	112,000	24,000	—	¹14,000	²14,000	—	²11,000 ⁴25,000	²25,000	—	—	—	1,650	—	—
Petronor SA—Somorrostro, Vizcaya	220,000	90,000	²40,000	²40,000	²30,500	—	²35,000	²45,000 ⁴40,000	—	—	—	1,670	²12.0	—
Repsol Petroleo SA—Cartagena, Murcia	112,000	18,000	—	—	²25,000	—	²30,000	²29,000	—	—	3,500	3,500	—	550
La Coruña	135,000	29,000	²10,500 ³13,500	²28,000	²22,000	—	⁴17,000	²22,000	—	—	—	—	—	—
Puertollano, Ciudad Real	135,000	62,000	—	²30,000	²18,000	—	²9,500 ⁴22,000	²18,000	²1,700	—	3,000	3,500	²6.2	—
Tarragona	180,000	72,000	²30,000	²16,000	—	¹15,000	²22,000 ⁴20,600	²16,000	—	—	—	—	²10.0	—
Unión Explosivos Rio Tinto SA—La Rábida, Huelva	80,000	25,000	²9,000	—	²14,100	—	—	¹14,100 ⁴8,000 ³2,800	—	²2,420 ³3,500 ⁴1,800	1,900	6,900	²5.0 ³6.0	—
Total	1,285,000	388,800	173,000	164,000	171,600	15,000	192,100	278,900	1,700	22,020	8,400	40,220	59.2	550
Sri Lanka														
Ceylon Petroleum Corp.—Sapugaskanda	50,000	2,400	²12,500	—	²3,750	—	²2,100	⁴8,000 ⁴7,300	—	—	—	1,000	—	—
Total	50,000	2,400	12,500	—	3,750	—	2,100	15,300	—	—	—	1,000	—	—
Sudan														
Port Sudan Refinery Ltd.—Port Sudan	21,440	—	—	—	¹1,700	—	—	⁴7,400	—	—	—	—	—	—
Sweden														
AB Nynas Petroleum—Gothenburg	12,500	9,000	—	—	—	—	—	—	—	—	2,500	8,500	—	—
Nynäshamn	28,000	31,000	—	—	—	—	—	⁶2,500	—	—	—	20,000	²1.9	—
BP Raffinaderi—Gothenburg	106,000	—	—	—	²20,500	—	—	²45,000	—	—	—	—	—	—
Shell Raffinaderi BV—Gothenburg	80,000	17,000	²27,000	—	²16,000	—	²16,000	²22,000 ⁴5,000	—	⁴4,000	—	—	—	—
Skandinaviska Raffinaderi AB—Brofjorden-Lysekil	200,000	65,000	²40,000	²25,000	²34,500	—	²44,000 ⁴30,500	²43,500	⁴3,500	—	—	—	—	—
Total	426,500	122,000	67,000	25,000	71,000	—	90,500	118,000	3,500	4,000	2,500	28,500	1.9	—

Petroleum Refining Processes, Worldwide Capacities

Country	Company/Location	Crude	Vacuum	Thermal	Cat. Cracking	Hydrocracking	Cat. Reforming	—	Hydrotreating	Other Hydro	Alkyl.	Aromatics	Isom.	Lubes	Asphalt	Waxes	Coke
Switzerland	Raffinerie de Cressier SA—Cressier	60,000	24,000	—	—	—	[2]16,000	—	[8]8,000	[1]27,000 [4]5,500	—	[6]3,600	—	5,200	—	—	—
	Raffinerie du Sud-Ouest SA—Collombey	72,000	—	—	—	—	[2]10,000	—	—	[1]16,700	—	—	—	—	—	—	—
	Total	132,000	24,000	20,000	—	—	26,000	—	8,000	49,200	—	3,600	—	5,200	—	—	—
Syria	Banias Refining Co.—Banias	126,350	54,101	[2]4,099	—	—	[4]16,728	—	—	[1]16,769 [6]6,828 [14]14,266	—	—	—	4,016	—	—	—
	Homs Refinery Co.—Homs	117,394	7,245	[5]12,285	—	—	[2]2,266	—	[2]3,024 [4]15,660	[1]3,172 [4]6,720	—	—	—	4,474	[29]9.9	—	487
	Total	243,744	61,346	36,384	—	—	18,994	—	18,684	47,755	—	—	—	8,490	9.9	—	487
Tanzania	Tanzanian & Italian Petroleum Refining CL—Kigamboni, Dar es Salaam	13,500	—	—	—	—	[3]3,000	—	—	[4]5,000	—	—	—	—	—	—	—
Thailand	Esso Standard Thailand Ltd.—Sriracha	64,220	6,650	—	—	—	[5]7,200	—	[4]14,400	—	—	—	—	—	—	—	—
	Petroleum Authority of Thailand—Bangchak, Bangkok	61,750	—	—	—	—	[2]9,450	—	[5]3,150	[4]4,950 [5]5,850	—	—	—	—	[4]4.7	—	—
	Thai Oil CL—Sriracha	65,075	16,150	[3]16,920	[7]8,100	—	[5]9,630	—	—	[16]16,200 [32]32,000 [5]10,000	—	—	—	1,100	—	—	—
	Total	191,045	22,800	16,920	8,100	—	26,280	—	17,550	69,000	—	—	—	1,100	4.7	—	—
Trinidad	Trinidad and Tobago Oil CL—Pointe-a-Pierre	220,000	166,000	[3]18,000	[2]8,000	—	[4]25,000	—	[2]100,000	[5]18,000 [6]3,800 [4]20,000	[1]2,200 *450	[1]2,800	2,700	—	—	—	—
	Point Fortin	80,000	5,000	—	—	—	[3]6,000	—	[4]4,500	—	—	—	—	2,000	—	—	—
	Total	300,000	171,000	18,000	28,000	—	31,000	—	104,500	41,800	2,650	2,800	2,700	2,000	—	—	—
Tunisia	Ste. Tunisienne Industries des Raffinage—Bizerte	34,000	—	—	—	—	3,300	—	—	—	—	—	—	—	—	—	—
Turkey	Anadolu Tasfiyehanesi AS—Mersin	90,000	—	—	—	—	[2]10,500	—	—	[2]22,000 [4]11,500	—	—	—	—	—	—	—
	Turkish Petroleum Refineries Corp.: Aliaga-Izmir[a]	249,000	37,500	—	[1]14,000	—	[2]10,700	—	—	[1]12,000 [14]14,500	—	—	4,000	10,000	[3]0.4	—	—
	Batman, Siirt	24,350	4,140	[2]4,500	—	—	[2]1,300	—	—	[1]1,300 [20]20,285	—	—	—	4,640	—	—	—
	Izmit	248,455	88,060	—	[20]20,128	—	[20]20,285	—	—	[6]6,290 [5]36,482	—	—	—	8,800	—	—	—

(continued)

TABLE 2 (continued)

Company and Refinery Location	Crude	Vacuum Distillation	Thermal Operations	Catalytic Cracking	Charge Capacity (b/cd) Cat Reforming	Cat Hydrocracking	Cat Hydrorefining	Cat Hydrotreating	Production Capacity (b/cd) Alky. *Poly.	Aromatics-Isomerization	Lubes	Asphalt	Hydrogen (MMcfd)	Coke (t/d)
United Kingdom														
England														
Kinkkale	112,849	29,676	—	—	²20,342	—	²15,096	¹20,342 ²8,825	—	—	—	6,177	—	—
Total	724,654	159,376	4,500	34,128	63,127	—	15,096	153,524	—	—	4,000	29,617	0.4	—
Conoco Ltd.—South Killingholme	130,000	104,000	²40,000 ⁵60,000	¹38,000 ²50,000	—	—	—	⁴76,000 ⁵45,000 ⁷30,000	*3,500	³3,000 ⁶19,000	—	—	—	2,150
Eastham Refinery, Ltd.— Eastham, Cheshire	12,000	10,500	—	—	—	—	—	—	—	—	—	8,000	—	—
Esso Petroleum CL—Fawley	300,200	120,840	—	¹76,500	²27,900 ⁴22,500	—	—	¹²73,260 ⁴⁵104,130 ⁶7,200	*12,600	⁶16,150	7,200	7,650	—	—
Lindsey Oil Refinery Ltd.— Killingholme, South Humberside	190,000	80,000	²26,000	¹40,000	²33,600	—	⁴40,600	¹²45,400 ³3,600	²4,100 *1,900	—	—	7,100	—	—
Mobil Oil CL—Coryton, Essex	145,000	86,000	—	¹47,000	²27,000	—	—	²35,000 ⁵33,000	²14,500	⁴8,500	5,500	1,700	—	—
Phillips Imperial Petroleum Ltd.—Port Clarence	100,000	—	—	—	—	—	—	—	—	—	—	—	—	—
Shell U.K. Ltd.—Shell Haven	92,000	30,000	—	—	⁴34,000	²21,000	—	²34,000 ⁴29,000	—	—	—	3,000	—	—
Stanlow	262,000	90,000	—	¹54,000	²29,000 ⁶27,000	—	²41,000	²56,000 ²14,000	—	¹9,000 ²5,000	4,500	8,000	—	—
Total	1,231,200	521,340	126,000	305,500	201,000	21,000	81,600	585,590	36,600	60,650	17,200	35,450	—	2,150
Scotland														
BP Refinery Grangemouth Ltd.— Grangemouth	178,500	76,000	—	¹19,000	²32,000	—	—	⁴62,500	²3,800	—	—	4,350	¹77.4	—
William Briggs & Sons Ltd.—Dundee	6,400	6,000	—	—	—	—	—	—	—	—	—	—	—	—
Total	184,900	82,000	—	19,000	32,000	—	—	62,500	3,800	—	—	4,350	77.4	—
Wales														
Amoco UK Ltd.—Milford Haven	102,000	48,500	—	¹31,500	²17,500	—	—	²17,500 ⁴29,000	²3,100 1,200ᵐ	—	—	—	—	—
BP Refinery Llandarcy Ltd.— Llandarcy, Neath	—	27,000	—	—	—	—	—	—	—	—	4,400	—	—	—
Gulf Oil-GB—Milford Haven	105,000	37,000	—	—	²18,000	—	—	²29,000 ³14,000	—	—	—	—	—	—
Pembroke Cracking Co. (65% Texaco, 35% Gulf Oil-GB)	—	91,000	²26,000	¹86,000	⁶39,000	—	⁴34,000	¹39,000	²22,000	⁴10,000	—	—	²0.8	—
Texaco Ltd.—Pembroke, Dyfed	180,000	203,500	26,000	117,500	74,500	—	34,000	128,500	26,300	10,000	4,400	—	0.8	—
Total	387,000	806,840	152,000	442,000	307,500	21,000	115,600	776,500	66,700	70,650	21,600	39,800	78.2	2,150
Total United Kingdom	1,803,100													
Uruguay														
Ancap—La Teja, Montevideo	33,000	13,000	—	¹4,100	²2,500	—	—	⁴3,700	—	—	—	—	—	—
USSR														
Total	12,300,000													

Government-owned refineries at Achinsk, Angarsk, Baku, Batumi, Chimkent, Drogobych, Fergana, Gorki, Grozny, Guryev, Ishimbai, Khabarovsk, Kherson, Kirishi, Komsomolsk, Krasnovodsk, Kremenchug, Kuibyshev, Lisichansk, Mazheikiai, Moscow, Mozyr, Nadvornaya, Nizhnekamsk, Odessa, Omsk, Orsk, Pavlodar, Perm, Polotsk, Ryazan, Saratov, Syzran, Tuapse, Ufa, Ukhta, Vannovskiy, Volgograd, Yaroslavl.

Country	Refinery														
Venezuela	Corpoven—El Palito, Carabobo	105,000	63,200	—	—	⁴5,000	—	—	—	—	²0,500	—	—	—	—
	El Toreño, Barinas	4,500	—	—	—	—	—	—	—	—	—	—	—	—	—
	Puerto La Cruz, Anzoategui	195,100	—	—	¹11,400	—	—	—	—	—	²2,100	—	—	—	—
	San Roque, Anzoategui	5,200	1,700	—	—	—	—	—	—	—	—	—	—	—	—
	Lagoven—Judibana, Falcon	600,000	343,600	⁴9,700	¹76,100	—	—	²162,000 ⁴65,300 ²33,700 ³35,100	—	²16,000	¹8,400	1,890	39,800	¹82.9	100
	Maraven—Punta Cardon, Falcon	291,300	152,100	⁸88,300	¹60,600	—	—	—	—	²22,650	⁴9,900	5,700	—	¹26.0	—
	Total	1,201,100	560,600	138,000	193,100	6,000	—	296,100	—	61,250	18,300	7,590	39,800	108.9	100
Virgin Islands	Hess Oil Virgin Islands Corp. St. Croix	545,000	215,000	³80,000	—	²125,000	—	²130,000	²125,000 ⁴165,000	—	³30,000 ⁵15,000	—	—	—	—
	Total	545,000	215,000	80,000	—	125,000	—	130,000	290,000	—	45,000	—	—	—	—
Yemen, North	Yemen Hunt Oil Co.—Marib	10,000	—	—	—	¹2,500	—	—	—	—	—	—	—	—	—
Yemen, South	Aden Refinery Co.—Little Aden¹	161,500	9,500	—	—	¹10,800	—	⁴2,700	—	—	—	—	2,000	—	—
Yugoslavia	Bosanski Brod⁹	112,835	44,609	—	—	²7,612 ⁶13,500	⁷7,200	—	²20,200 ⁴13,438	—	—	—	13,310	¹14.0	—
	Lendava	14,600	—	—	—	—	—	—	—	—	—	1,500	4,000	—	—
	Novi Sad	63,000	30,000	—	—	⁵8,200	—	⁴4,600	²10,700 ⁴4,400 ⁴4,600	—	—	—	—	¹0.5	—
	Pancevo	108,000	44,850	¹21,700	²0,750	²8,800	—	⁴5,000 ⁹9,370 ⁴3,000	⁵8,800	²3,400	¹215	—	6,300	—	—
	Rijeka	160,000	43,000	⁸8,000	²21,300	²17,500	—	⁴3,000	²17,500 ⁶5,000	—	²2,800 ²1,535 ⁶9,000	2,200	—	¹0.4	—
	Sisak	150,000	15,700	⁵5,930	¹10,000	²19,000	—	—	²16,900 ³3,640 ⁵2,700	—	⁴4,900	—	3,715	—	137
	Zagreb	700	600	—	—	—	—	—	—	—	—	1,200	—	—	—
	Total	609,135	178,759	35,630	52,050	74,612	7,200	21,970	107,878	3,400	18,450	4,900	27,325	14.9	137
Zaire	Sozir—Muanda	16,700	—	—	—	⁵3,500	—	—	⁵5,000	—	—	—	—	—	—
Zambia	Indeni Petroleum Refinery CL— Bwana Nkubwa Area, Ndola	20,594	2,170	—	—	⁵5,070	—	—	⁸8,250	—	—	—	—	—	—

(continued)

TABLE 2 (continued)

Inactive Refineries
Germany
  1. Erdol Raffinerie Duisburg GmbH, Duisburg, 37,800 b/cd.
Italy
  2. ERG SpA, Genoa, San Quirico, 130,000 b/cd.
Japan
  3. Asia Oil CL, Yokohama, Kanagawa, no crude distillation. Cat reforming: 12,600 b/cd. Cat hydrorefining: 28,000 b/cd. Cat hydrotreating: 39,100 b/cd.
  4. Cosmo Oil CL. Matsuyama, no crude distillation. Cat reforming: 4,500 b/cd. Cat hydrotreating: 17,100 b/cd.
Togo
  5. Ste. Togolaise des Hydrocarbures, Lome, 20,000 b/cd.

[a]Solvent extraction, 4,800 b/cd.
[b]Solvent extraction, 166 b/cd.
[c]Solvent extraction, 35,000 b/cd.
[d]Solvent extraction, 9,800 b/cd.
[e]Operating on a semi-integrated basis.
[f]Solvent extraction, 10,000 b/cd.
[g]Present status uncertain due to Iran-Iraq war.
[h]Solvent extraction, 2,100 b/cd.
[i]Solvent extraction, 8,000 b/cd.
[j]Solvent extraction, 38,000 b/cd.
[k]Solvent extraction, 3,200 b/cd.
[l]Solvent extraction, 8,573 b/cd.
[m]Dimersol.
[n]Solvent extraction, 16,000 b/cd.
[o]Solvent extraction, 12,500 b/cd.
[p]Solvent extraction, 8,800 b/cd.
[q]Singapore Petroleum CPL shares percentages of refinery capacities.
[r]South African Oil Refinery PL is owner; Mobil is operator.
[s]Solvent extraction, 11,700 b/cd (furfural), 6,700 (propane deasphalting).
[t]Solvent extraction, 7,920 b/cd.
[u]Solvent extraction, 3,300 b/cd (deasphalting).

3
Manufacturing Processes

Coking, Petroleum (Delayed and Fluid)

Coking is a thermal cracking-type operation used to convert petroleum residua to coke, gas, and distillates. Two types of petroleum coking processes are presently operating: (1) Delayed coking, which uses multiple coking chambers to permit continuous feed processing wherein one drum is making coke and one drum is being decoked; and (2) fluid coking, which is a fully continuous process where product coke can be withdrawn as a fluidized solid. Installed capacities for coking in the United States are:

Coking Process	Fresh Feed (bbl/d)	Coke (tons/d)
Delayed	910,000	38,500
Fluid	135,000	4,500

Source: *Oil and Gas Journal*, April 7, 1975.

Coker Charge Stocks

Crude oil residua obtained from the vacuum distillation tower as a bottoms fraction are the usual charge stocks to coking. Longer residua, such as atmospheric tower bottoms, may be charged but it is generally not attractive to thermally degrade the gas-oil fraction contained in the longer residua. Other charge stocks to coking are solvent decarbonizer or deasphalter bottoms, bitumens such as Athabasca Tar and Gilsonite, shale oil bottoms, thermal tars, pyrolysis tar from ethylene plants, and decant oil from fluid catalytic cracking. The latter three stocks can be coked to provide a very high quality specialty coke used in graphite articles.

Delayed Coking

Delayed coking is the oldest, most widely used process and has changed very little over its 40 plus year history. As shown in Fig. 1, fresh feed is charged to the bottom of the fractionator and subsequently charged with some recycle, usually about 10%, through a preheater and then by upflow to the coke drum. Converted liquid and gaseous products pass to the fractionator for separation. Coke drums, used alternately to allow continuous processing, are operated on a cycle, typically 48 h, as shown in Table 1.

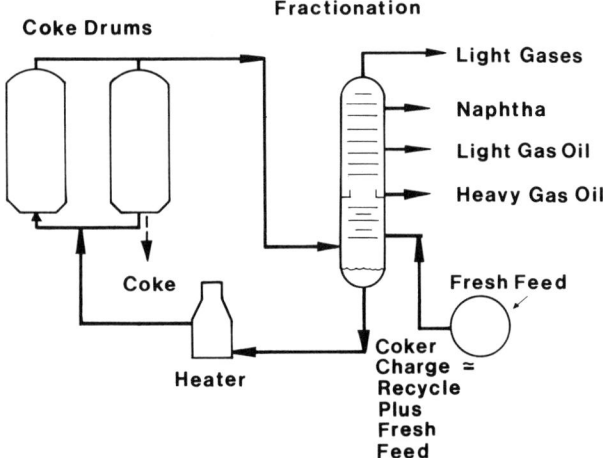

FIG. 1. Schematic of a delayed coker.

TABLE 1 Cycle of Coke Drums

Operation	Time (h)
Coking	24
Decoking:	24
Switch drums	0.5
Steam, cool	6.0
Drain, unhead, decoke	7.0
Rehead, warmup	9.0
Spare time, contingency	1.5

Coking Petroleum (Delayed and Fluid)

Delayed coking units fractionate the coke drum overhead effluent into fuel gas which is ethane and lighter, propane–propylene, butane–butene, naphthas, light gas oils, and heavy gas oils. Typically, a coker operates at 25 to 40 lb/in.^{2} gauge, heater outlet of 910 to 935°F, and a recycle ratio of 0.1 to 0.25. Coking reactions are endothermic, resulting in coke drum temperatures in the range of 810 to 840°F. Yields and product quality vary widely due to the broad range of feedstock types charged to delayed coking. The yields are well documented in the literature [1–3] for typical charge stocks. A contiguous set of yields for a moderate sulfur content vacuum residue and product properties are shown in Table 2. These were derived in a large-scale pilot plant and fully typify commercial operation.

Coker naphthas have boiling ranges up to 430°F, are olefinic, and must be

TABLE 2 Typical Yields and Product Quality from Delayed Coking Vacuum Residue[a]

Charge stock properties:			
Gravity: °API	12.3		
Sulfur: wt.%	0.68		
Carbon residue: wt.%	13.0		
Product yields, % by volume of fresh feed:			
Propane–propylene	3.4 (30% olefin)		
Butane–butenes	3.8 (40% olefin)		
Gasoline, C_5–400°F:	22.8		
Pentane–pentene	3.6 (35% olefin)		
Hexanes and heavier	19.2		
Furnace Oil	18.4		
Gas Oil	37.6		
Coke yield: wt.% of fresh feed	23.7		
Gas, C_2 and lighter: wt.% of fresh feed	3.5		
ft^3/bbl charged	257		

Product Inspections:

	Gasoline	Furnace oil	Gas oil
Gravity: °API	56.5	34	23
Sulfur: wt.%	0.2	0.3	0.6
Bromine no.	60	30	—
ASTM distillation:			
Over point: °F	120	400	
End point: °F	380	640	
10% at: °F	180	435	670
50%	265	500	760
90%	340	590	860

Coke inspections:		
Volatile matter: %		12
Sulfur: wt.%		1.3

[a]Source: Gulf Research and Development Company.

upgraded by hydrogen processing for removal of olefins, sulfur, and nitrogen. They are then used conventionally for reforming to gasoline or chemicals feedstock. Middle distillates, boiling in the range of 430 to 680°F, also are hydrogen treated for improved storage stability, sulfur removal, and nitrogen reduction. They can then be used for either diesel or burner fuel or further processed to gasoline. The gas-oil boiling up to about 950°F end point is low in metals content and may be charged to fluid catalytic cracking either directly or after hydrogen upgrading when low sulfur is a requirement. Technical advances in hydrotreating, reforming, fluid catalytic cracking, and hydrocracking make it economically feasible to coke residua and upgrade coker distillates.

Design Features

A well designed delayed coker will have an operating efficiency of better than 95%. Units are generally scheduled for shutdown for cleaning and repairs on 12 to 18 months schedules. Each operator must determine the most economical cycle for his plant since extended runs may result in expensive maintenance and longer turnaround times. The record delayed coking run [4] appears to be 931 days reported by American Oil, Yorktown.

The charge heater and the coke drums are the most critical parts of the delayed coking process. Of the two, run length is determined more by the heaters than the coke drum and its ancillary equipment. The remaining equipment is largely conventional.

Heater

The function of the heater or furnace is to preheat the charge quickly, to avoid preliminary decomposition, to the required temperature. Since coking is endothermic, the furnace outlet temperature must be about 100°F higher than the coke drum temperature to provide the necessary process heat. The heater run length is a function of coke laydown in heater tubes, and careful design is necessary to avoid premature shutdown. Cycle lengths should be at least 1 yr. When the charge stock is derived from crude distillation, double desalting is desirable since salt deposits will shorten heater cycles.

Factors influencing the coker heater design have been well summarized [1, 2, 5]. A critical zone of incipient cracking exists for each charge stock according to Mekler and Brooks [1]. To minimize coke laydown, it is necessary to provide turbulence in this zone by correct sizing of heater tubes. This is often accompanied by steam or condensate addition to increase velocity where insufficient feed vaporization occurs.

The coking heater does not have as broad an operating range as a thermal cracking or visbreaking heater where both contact time and temperature can be varied to achieve the conversion desired. The coker heater must reach a fixed outlet temperature for the required coke drum temperatures. Thus the coker heater requires a short residence time, high radient heat flux, and good control of heat distribution. For illustration, long run lengths have been reported for a heater with the following characteristics [6]:

1. Cold oil inlet velocity of 7 ft/s
2. Heat transfer section velocity of 60 to 70 ft/s
3. Total residence time of 250 s
4. Pressure differential of 140 to 150 lb/in.^{2}

In order to achieve good performance, the design must include analyses of velocity, pressure, vaporization, and temperature profiles over the entire heater.

Coke Drums and Ancillary Equipment

The function of the coke drum is to provide the residence time required for the coking reactions and to accumulate the coke made. In sizing coke drums [2], a superficial vapor velocity in the range of 0.3 to 0.5 ft/s is used. Drums with heights of 97 ft have been constructed and approach a practical limit for hydraulic coke cutting. Drum diameters up to 26 ft have been used, and larger drums do appear feasible [7]. Coke height in the drum during filling is monitored by radioactive level detectors which permit filling to within 7 to 8 ft of the upper tangent line of the drum. Generally, the design will allow for 10 ft outage where coke density can be predicted from known feed character with an additional safety factor for doubt in coke yield, density, etc.

Hydraulic cutters are used to remove coke from the drum [8]. The first step is to bore a vertical pilot hole through the coke. Cutting heads with horizontally directed nozzles then undercut the coke and drop it out of the bottom of the drum. Hydraulic pressures in the range of 3000 to 3600 lb/in.^{2} are used in the 26-ft diameter coking drums [7].

Fluid Coking

This is a fluid solids process that cracks feed thermally over heated coke particles in a reactor vessel to gas, liquid products, and coke. Heat for the process is supplied by partial combustion of the coke with the remaining coke being drawn as product. The new coke is deposited in a thin fresh layer (~ 5 μm) on the outside surface of the circulating coke particle, giving an onion skin effect.

Fluid coking is a licensed process of Exxon Research and Engineering Company and its first commercial [9] use was in late 1954. A schematic flow is shown in Fig. 2. The equipment is quite similar to that used in fluid cracking and follows comparable design concepts except that catalyst is replaced by the fluidized coke solids. Small particles of coke made in the process circulate in a fluidized state between the vessels and are the heat transfer medium. Thus the process requires no high temperature preheat furnace. Fluid coking is carried

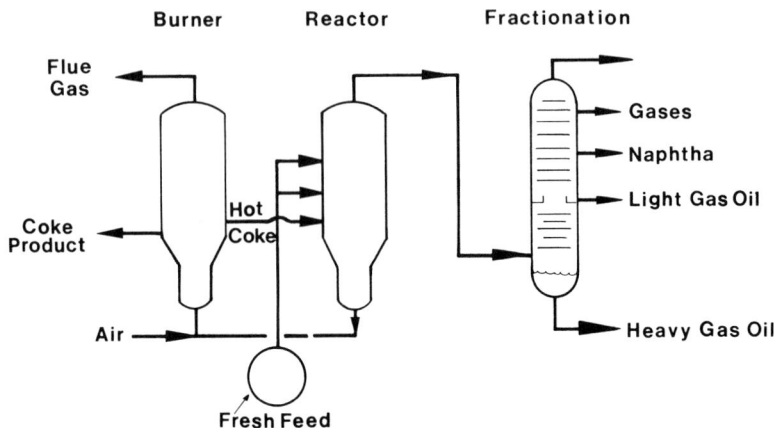

FIG. 2. Schematic of a fluid coker.

out at essentially atmospheric pressure and temperatures in the range of 900 to 1050°F with short residence times of the order of 15 to 30 s.

Due to the higher thermal cracking severity used in fluid coking compared to delayed coking, the products are somewhat more olefinic and slightly less desirable for downstream processing. Product yields taken from Ref. 9 are shown in Table 3. In general, products are handled for upgrading in a comparable manner from both coking processes.

The coke product from the fluidized process is a laminated sphere with an average particle size of 170 to 220 μm, readily handled by fluid transport

TABLE 3 Fluid Coking Yields and Product Quality

Feedstock:	South Louisiana vacuum pitch
Gravity, °API	11.6
Carbon residue: wt.%	13.0
Sulfur: wt.%	0.6
Yields based on fresh feed, wt.%:	
C_3 and lighter	8
C_4	2
C_5/430°F naphtha	16
430/1015°F gas oil	57
Coke (gross)	17
Product quality:	
Naphtha:	
Gravity, °API	56
Sulfur, wt.%	0.2
Research octane, clear	73
Gas oil:	
Gravity, °API	22
Sulfur: %	0.5

techniques. It is much harder and denser than delayed coke, and in general is not as desirable for manufacturing formed products. A modified process called Flexicoking [10–12] adds a gasifier for the coke to the basic fluid coking process. The first commercial use of this process was started up in September 1976 by Toa Oil Company of Japan.

The Flexicoking process produces a clean fuel gas with a heating value of about 90 Btu/scf. The coke gasification can be controlled to burn about 95% of the coke to maximize production of coke gas or at a reduced level to produce both gas and a coke which has been desulfurized by about 65%. This flexibility permits adjustment for coke market conditions over a considerable range of feedstock properties. Fluid coke is currently being used in power plant boilers.

Petroleum Coke Properties and Uses

Delayed coker drum conditions are typically controlled to produce a green (uncalcined) coke of 9 to 12% volatile content. A lower volatile content makes cutting coke from the drum more difficult and makes meeting the 24-h decoking cycle time more difficult. Ultimate coke use is generally governed by sulfur and metals contents and by coke structure. Cokes exceeding about $2\frac{1}{2}$% sulfur content and 200 ppm vanadium are mainly used for fuel or fuel additives. Total coke use in the United States in 1972 was 40% for fuel, 36% for aluminum anodes, 12% for graphite, and 12% for carbides and miscellaneous nonfuel applications.

Coke qualities required in nonfuel use include low sulfur, metals, and ash and its physical structure. Some typical coke properties are shown in Table 4.

TABLE 4 Petroleum Coke Qualities (typical ranges for cokes used in carbon and graphite manufacture)

	Green	Calcined
Sulfur, %	0.3–2.5	0.2–2.5
Vanadium, ppm	0–500	0–500
Volatile matter, %	9–13	Nil
Ash, total, %	0.1–0.8	
Silicon, %	0.01–0.08	
Iron, %	0.01–0.06	
Soluble salts, %	0.2–0.8	
Calcium, ppm	2.5–500	
Sodium, ppm	2.5–1000	
Moisture, %	[a]	
Electrical volume resistivity, $\Omega/\text{in.}^3$	—	0.045 max

[a] Not to freeze and stick coke.

Green petroleum coke is calcined between 2200 and 2600°F to reduce the volatile content to less than 1%. This calcined coke is an industrial carbon used as a raw material in the aluminum and graphite industries. Whereas coke was once a cheap by-product of a process to convert residual fuels to more valuable distillates, growing uses for coke and advances in technology make petroleum coke itself an inherently valuable product.

There are two basic mechanisms of coke formation [13, 14] which coexist for most of the charge stocks used in commercial coking operations. From virgin crude residua, coke is formed primarily from high molecular weight compounds, asphaltenes, and resins by dealkylation-precipitation reactions. These reactions produce a disordered cross-linked structure giving amorphous carbon plus asphalt which upon calcining to fix carbon leads to an isotropic structure. This type of coke is difficult to graphitize, and when low in impurities is the type used mainly in aluminum manufacture.

The second mechanism produces coke by condensation of polyaromatic compounds, and upon calcining has an acicular structure. This premium needle coke is more anisotropic, is readily graphitized, and commands a premium compared to other coke. Cokes used for graphite have high mechanical strength, low coefficient of thermal expansion (CTE), low porosity, and are low in sulfur, ash, and metallic impurities.

Premium cokes are generally made in the delayed coking process through careful selection of feedstocks. Fluid coke has an onionlike structure and exhibits very limited three-dimensional ordering when graphitized. It has found limited commercial importance in nonfuel uses.

References

1. V. Mekler and M. E. Brooks, "New Developments and Techniques in Delayed Coking," *Proc. Am. Pet. Inst.*, May 28, 1959; *Refining Eng.*, *31*(10), (September 1959).
2. K. E. Rose, "Delayed Coking—What You Should Know," *Hydrocarbon Process.*, July 1971.
3. "Petroleum Coke Takes New Luster," *Heat Engineering* (Foster Wheeler Corp.) May-June, 1970; reprinted in *Oil Gas J.*, September 14, 1970.
4. Anonymous, *Hydrocarbon Process.*, *48*(9), 129 (1964).
5. S. B. Heck, Jr., "Process Design of a Modern Delayed Coker," Preprint 46–72, American Petroleum Institute, May 11, 1972.
6. Transcript of NPRA Question and Answer Session on Refining Technology, 1969.
7. Transcript of NPRA Question and Answer Session on Refining Technology, 1971.
8. H. W. Nelson, "Petroleum Coke Handling Problems," *Ind. Eng. Chem., Prod. Res. Develop.*, September 1970.
9. J. McDonald and C. O. Rhys, Jr., "The Fluid Coking Process—Commercial Experience to Date," *Refining Eng.*, *31*(10), (September 1959).
10. J. P. Matula, H. N. Weinberg, and W. Weismann, "Flexicoking—An Advanced Fluid Coking Process," Preprint 45–72, American Petroleum Institute, May 11, 1972.

11. J. W. Brown, W. L. Schuette, and L. G. Sherman, "Latest Developments in Flexicoking," AIChE Paper, March 19, 1975.
12. J. P. Matula, B. V. Molstedt, and D. F. Ryan, "Flexicoker Prototype Demonstrates Successful Operation," Preprint 10-75, American Petroleum Institute, May 13, 1975.
13. R. R. Jakob, "Coke Quality and How To Make It," *Hydrocarbon Process.*, September 1971.
14. T. Reis, "To Coke, Desulfurize, and Calcine. Part 2: Coke Quality and Its Control," *Hydrocarbon Process.*, June 1975.

J. D. McKINNEY

Coking, Petroleum (Fluid)

Introduction

Fluid coking is a continuous fluid bed process for upgrading residual petroleum stocks to gas oil, naphtha, gas, and coke. Flexicoking is an extension of the fluid coking process, integrating a coke gasifier with fluid coking. About 99% of the products from Flexicoking are in the liquid and gaseous forms. Throughout this century, with only short-term exceptions, there has been a considerable economic driving force for upgrading residua. This has led to the development of processes to reduce residua yields such as thermal cracking, visbreaking, delayed coking. vacuum distillation, and deasphalting.

In the later 1940s and early 1950s there was a large incentive to develop a continuous process to convert heavy vacuum residua into lighter, more valuable products. During this period, Esso Laboratories, now Exxon Research and Development Laboratories, and Standard Oil Development Company, now Exxon Research and Engineering Company, developed fluid coking using fluid solids principles, and the Lummus Company developed contact coking, using the moving bed principle [21]. The first commercial fluid coker went on-stream in late 1954 [2, 5, 28]. A total of 13 fluid cokers had been built by 1978 (Table 1).

During the late 1960s, environmental considerations indicated that, in many areas, it would no longer be possible to utilize high sulfur coke as a boiler fuel. This and other environmental considerations resulted in the development of Flexicoking to convert this coke into clean fuel [3]. The process was developed by Exxon Research and Engineering Company and Exxon Research and Development Laboratories and was demonstrated in a 750 bbl/sd prototype unit during 1974 and 1975 [19]. The first commercial Flexicoker,

TABLE 1 Fluid Cokers

Company	Location	Design Feed Rate (bbl/d)	Initial Startup Date	Comments
Exxon	Baltimore, Maryland	10,000	October 1955	Shutdown 1957
Exxon	Baytown, Texas	750	February 1974	Prototype Flexicoker, shutdown July 1975
Exxon	Benicia, California	23,000	April 1969	
Exxon	Billings, Montana	3,800	December 1954	
Getty	Delaware City, Delaware	42,000	August 1957	
Gulf	Purvis, Mississippi	4,800	December 1957	
Imperial	Sarnia, Ontario	14,000	April 1968	Combination unit
Marathon	Detroit, Michigan	4,000	October 1956	Shutdown
Pemex	Madero, Tamaulipas, Mexico	10,000	February 1968	
Toscopetro	Martinez, California	42,000	June 1957	
Petrofina	Montreal, Quebec	3,800	August 1956	Shutdown 1976
Toscopetro	Bakersfield, California	4,000	April 1957	
Syncrude	Mildred Lake, Alberta	72,900	July 1978	
Syncrude	Mildred Lake, Alberta	72,900	October 1978	
Toa Sekiyu	Kawasaki, Japan	21,300	September 1976	Flexicoker

owned by the Toa Oil Company, went on-stream in 1976. Engineering was proceeding on two more commercial Flexicokers in 1978.

Extensive development work on processes similar to fluid coking and contact coking has been carried on in the Union of Soviet Socialist Republics [1]. In 1971 Soskind, Botnikov, and Voronova reported the operation of thermo contact cracking (fluid coking) pilot plants with throughputs of 320 to 520 t/d [27].

Fluid Coking Process

Figure 1 is a schematic flow plan of a fluid coker. Fluid coking uses two major fluid solids vessels, a reactor and a burner. The residuum feed is injected into the reactor where it is thermally cracked to vaporized products and coke which is deposited on fluid coke particles. Fluid coke circulates between the vessels to provide heat to the reactor. Steam is introduced into the bottom of the reactor to fluidize the coke particles and to strip residual hydrocarbon vapor from the coke. The volume of gas increases progressively up the reactor bed due to vapor formation by cracking.

Vapor products leave the bed and pass through cyclones which remove most of the entrained coke. The cyclones discharge the vapor into the bottom of a scrubber. The remaining coke dust is scrubbed out with a pumparound stream and the products are cooled to condense the heavy tar. The resulting slurry is

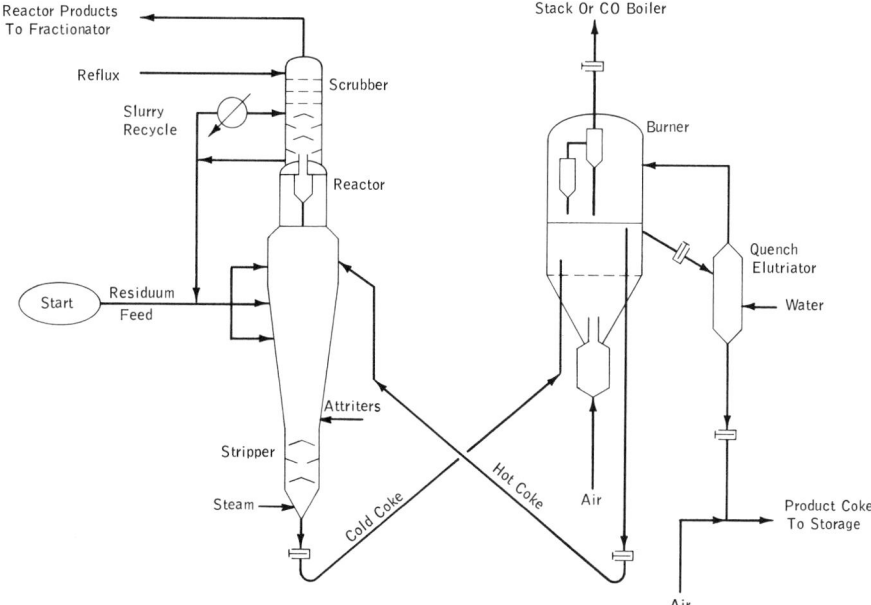

FIG. 1. Schematic flow diagram of a fluid coker.

recycled to the reactor. The scrubber overhead vapors are sent to a fractionator where they are separated into wet gas, naphtha, and various gas-oil fractions. The wet gas is compressed and further fractionated into the desired components.

In the reactor the coke particles flow down through the vessel into the stripping zone. The stripped coke then flows down a standpipe and through a slide valve which controls the reactor bed level. A riser carries the "cold coke" to the burner. Air is introduced to the burner to burn part of the coke to provide reactor heat. The hot coke from the burner flows down a standpipe through a slide valve which controls coke flow and thus the reactor-bed temperature. A riser carries the hot coke to the top of the reactor bed. Combustion products from the burner bed pass through two stages of cyclones to recover coke fines and return them to the burner bed. The combustion products pass through a variable orifice which controls the burner pressure. The burner gas is burned in a CO boiler or released to a stack.

Coke is withdrawn from the burner to keep the solids inventory constant. To aid in keeping the coke from becoming too coarse, large particles are selectively removed as product in a quench elutriator drum and coke fines are returned to the burner. The product coke is quenched with water in the quench elutriator drum and pneumatically transported to storage.

A simple jet attrition system in the reactor provides additional seed coke to maintain a constant particle size within the system.

Flexicoking Process

Figure 2 is a schematic flow plan of a Flexicoker. The reactor side and the hot and cold coke transfer lines are identical with those of a fluid coker. The burner vessel is replaced by a heater vessel, and a third major vessel, the gasifier, is used to gasify the coke production.

The cold coke from the reactor is partially devolatilized in the heater vessel. Some methane and hydrogen are produced. The devolatilized coke is circulated to the gasifier where it is reacted with steam and air at an elevated temperature to produce hydrogen, carbon monoxide, carbon dioxide, hydrogen sulfide, and nitrogen. This gas, together with entrained coke fines, is returned to the heater bed where some of its heat is picked up to provide a portion of the reactor heat. The additional heat required by the reactor is supplied by coke circulated from the gasifier to the heater.

The coke gas leaves the heater through two stages of cyclones, and additional heat is removed by steam generation. Additional coke fines are recovered in external tertiary cyclones and in a venturi scrubber. Hydrogen sulfide is removed from the gas and recovered as sulfur in commercially available processes such as Stretford [20]. The gas is used as fuel for boilers or process furnaces.

A small quantity of coke is removed from the heater to purge agglomerates from the system. The coke fines collected in the tertiary cyclones and venturi scrubber contain most of the ash from the residuum feedstock.

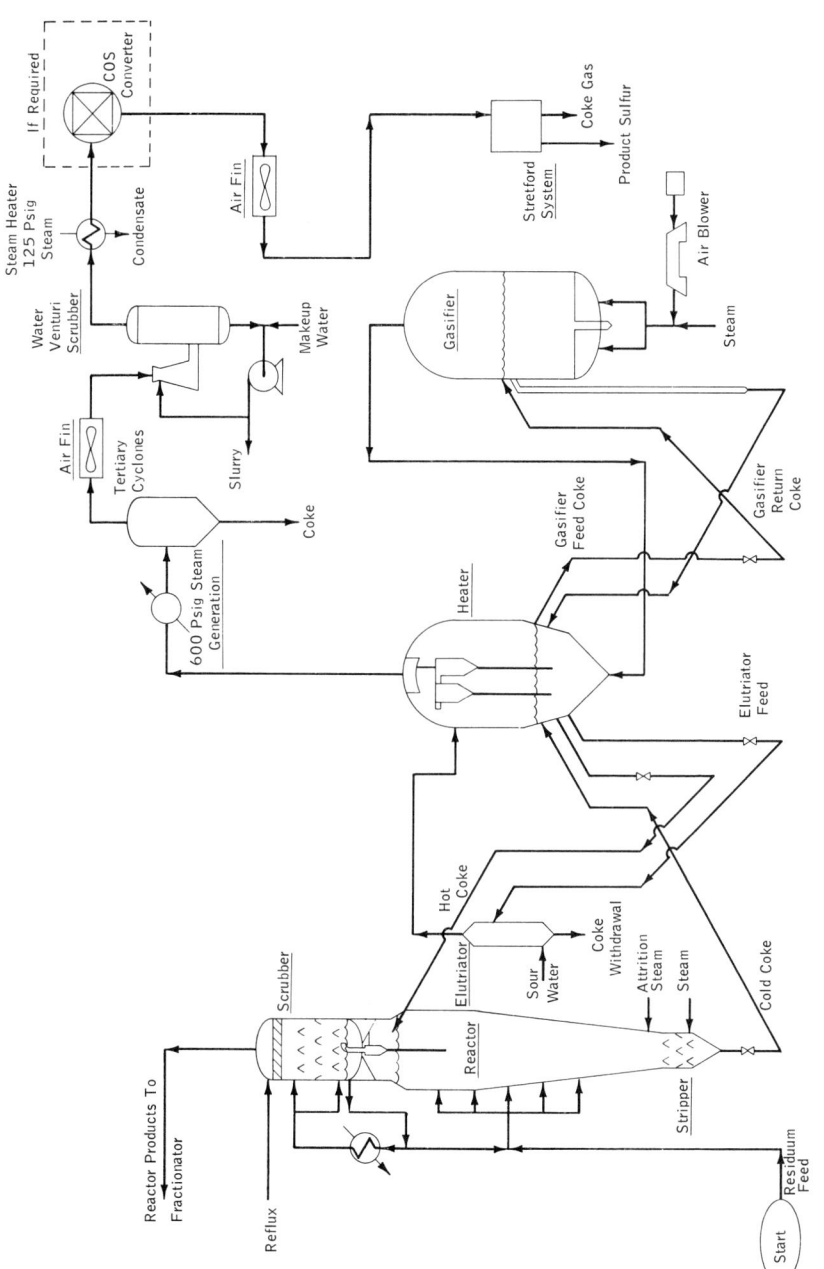

FIG. 2. Schematic flow diagram of the Flexicoker.

Application of the Fluid Coking Processes

Fluid coking and Flexicoking are versatile processes which are applicable to a wide range of heavy feedstocks and provide a variety of products. The feedstock should have a Conradson carbon residue of more than 5%. There is no upper limit on Conradson carbon residue. Suitable feed stocks include vacuum residua of all types, catalytic cracking polymer, asphalt, whole and reduced tar sands bitumen, visbreaker tar, and shale oil. Processing costs are insensitive to feed contaminants such as organic metals, ash, sulfur, and nitrogen [4].

The liquid products from the coker can, following cleanup via commercially available gas-oil hydrodesulfurization technology, provide large quantities of low sulfur fuel (less than 0.2 wt.% sulfur). Another major application for the processes is upgrading heavy low value crudes into lighter products. There are large reserves of heavy crudes such as those found in the Orinoco tar belt in Venezuela and the tar sands of Alberta. The third major application is the reduction or complete elimination of heavy fuel oil or asphalt from the refinery products. The incentive for fluid coking or Flexicoking increases relative to alternate processing, such as direct hydroprocessing, as feedstock quality (Conradson carbon, metals, sulfur, nitrogen, etc.) decreases.

Figure 3 shows a typical product distribution and possible disposition of the products. The gas oil produced can be used as catalytic cracker or hydrocracker feedstock or can be hydrodesulfurized for inclusion in various heating oil and fuel oil products. The naphtha is normally hydrotreated and reformed for motor gasoline. The light ends can be fractionated into cuts for alkylation, chemicals or hydrogen unit feedstocks, or used as fuel. Fluid coke is used in electrodes for aluminum manufacture, in silicon carbide manufacture, in ore sintering operations, and as fuel. The coke from a feedstock containing a large amount of contaminants may not be suitable for these uses, either from a product contamination or environmental standpoint. This problem is overcome by the Flexicoking process. The Flexicoker converts 95+% of the gross coke to a gas with a heating value of 100 to 120 Btu/scf which can be burned in process furnaces and boilers. Firing of the gas is discussed in Addendum C of a paper presented to the American Petroleum Institute [20]. The coke fines from a Flexicoker contain most of the metals in the feedstock and may be suitable for metals recovery [24]. For example, coke fines produced by Flexicoking from a high vanadium residuum contained 15% vanadium. The coke withdrawn from the bed has a sulfur content of 2 to 3% compared to 5 to 8% for normal fluid coke from a high sulfur residuum and could be used as a fuel in some locations.

Product Mix Flexibility

Table 2 indicates the change in yields and product quality which results from a change from a low cut point, high reactor temperature operation to a high cut point operation with a lower reactor temperature.

Table 3 illustrates a case where the refiner was interested in increasing butene yields. By increasing operating severity, the butene yield was increased 70%.

Coking, Petroleum (Fluid)

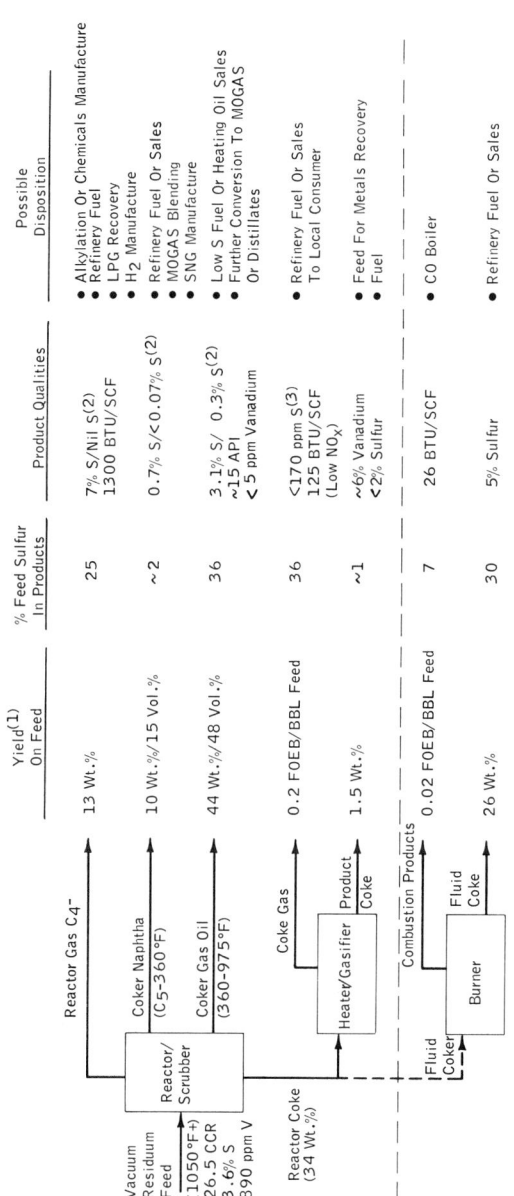

FIG. 3. Fluid and Flexicoker yield pattern, product qualities, and possible dispositions [8].

TABLE 2 Operating Flexibility Available [5]

Feed characteristics:	
Con carbon, wt.%	15.5
Gravity, °API	6.4
LV below 1000°F, %	8.0
Sulfur, wt.%	2.6
Nitrogen, wt.%	1.0
Nickel, ppm	283
Vanadium, ppm	126

	Low reactor temperature, high cut point maximum gas oil		High reactor temperature, low cut point low metals gas oil	
Yields:	Wt.%	LV%	Wt.%	LV%
Hydrogen sulfide	0.5		0.7	
Hydrogen	0.1		0.2	
C_1 to C_3	8.0		9.6	
C_4's	1.6	2.8	2.0	3.5
C_5 to 215°F (ASTM)	4.2	6.2	5.1	7.5
215 to 400°F	8.6	11.4	10.4	13.9
400° to end point	58.4	62.5	51.8	56.3
Gross coke	18.5		20.2	
Net coke	10.0		10.6	
Product qualities:				
C_5 to 215 naphtha				
Gravity, °API	71.6		71.6	
Sulfur, wt.%	0.41		0.43	
Nitrogen, wt.%	0.009		0.009	
215 to 400°F				
Gravity	52.3		52.2	
Sulfur, wt.%	0.66		0.69	
Nitrogen, wt.%	0.036		0.025	
Gas oil				
Gravity, °API	16.9		18.4	
Sulfur, wt.%	2.36		2.28	
Nitrogen, wt.%	0.73		0.63	
Nickel, ppm	2.7		0.43	
Vanadium, ppm	0.5		0.05	
Conradson carbon, wt.%	2.3		1.0	
Aniline point, °F	140		110	
Coke				
Sulfur, wt.%	3.4		3.4	
Nickel, ppm	1520		1400	
Vanadium, ppm	680		620	

TABLE 3 Operating for Higher Butene Yields [5]

	Low Severity		High Severity	
	Wt.%	LV%	Wt.%	LV%
Hydrogen sulfide	0.6		0.8	
Hydrogen	0.1		0.1	
Methane	2.8		3.8	
Ethylene	0.9		2.0	
Ethane	2.2		2.8	
Propylene	1.7		2.9	
Propane	1.6		2.7	
C_4's	1.9	3.4	2.9	5.0
$C_5/350$	9.3	13.1	12.5	17.8
350/685	18.9	22.6	18.6	22.5
685/CP	32.8	34.1	17.5	17.8
Gross coke	27.3		34.4	
Net coke	18.6		25.3	
C_4 Breakdown	Wt.%		Wt.%	
Butadiene	7		9	
cis-trans Butene-2	35		42	
Butene-1	17		17	
Isobutene	13		13	
Normal butane	23		14	
Isobutane	5		5	
	100		100	

TABLE 4 Effect of Flexicoking on Overall Refinery Yields [8]

	Crude Composition	Flexicoking Gas-Oil Hydrodesulfurization	Flexicoking Gas-Oil Hydrodesulfurization and Cat Cracking
Yields, LV% on crude			
Gas, FOEB	3	7	12
Naphtha	16	23	47
Middle distillate	26	31	35
Gas oil (LSFO)	25	34	
Residuum	30		
Total	100	95	94

The fluid coking processes can be used to produce a high yield of low sulfur fuel oil as well as to completely eliminate residual fuel and asphalt from the refinery product slate [4, 20]. This flexibility is shown on Table 4 which gives overall refinery yields on Heavy Arabian crude oil for two processing schemes, one producing low sulfur fuel oil and the other eliminating fuel oil. These different distributions are obtained by varying the fluid coker/Flexicoker operating conditions and changing the downstream processing of the coker reactor products.

Process Variations

There are many variations which can be used to adapt the process to particular refining situations. Once-through or partial recycle coking can be used where there is a small market for heavy fuel oil, or where a quantity of high sulfur material can be blended into the fuel oil pool.

A combination unit, wherein atmospheric tower residuum is fed directly into the coker scrubber, may eliminate the need for a vacuum tower upstream of the coker [5]. In this case a large portion of virgin gas oil is flashed off in the scrubber tower.

Another interesting variation is cracking of gas oil or naphtha in the hot

TABLE 5 On-Site Investment and Utility Requirements 30 kbbl/sd Arabian Heavy 1050+ Residuum

	Fluid Coker Including CO Boiler and Electrostatic Precipitator and Flue Gas Desulfurization	Flexicoker Including Stretford Unit
Total prime contract cost, 1974 M$	31	44
Without flue gas desulfurization, M$	26	—
Process manning, total men	28	36
Maintenance, % investment/yr	7	7
Low-pressure steam, net requirement, klb/h	16.3	78.4
High-pressure steam, net produced, klb/h	246.6	279
Boiler feed water, gal/min	818	872
Cooling water, gal/min	2,208	1,031
Process water, gal/min	55	223
Electricity, kW[a]	7,890	20,398
Instrument and refinery air, scf/min	350	350
Stretford solution, gal/d	—	3,335
Caustic, st/d	0.82	—
Sour water, gal/min	189	141

[a]Assumes electric motor driver on air blower. If steam turbine were used, electricity required would decrease to 2,950 kW for the fluid coker and 3,370 kW for the Flexicoker. Production of high-pressure steam would be reduced accordingly.

coke riser. The high temperature of the riser greatly increases the cracking severity and increases the yield of unsaturated gases such as ethylene. The products of riser cracking provide part or all of the lift gas in the hot coke riser, reducing the steam requirement.

A partial gasification operation can be used to produce low sulfur coke [22]. In this operation 15 to 25% of the gross coke can be withdrawn while producing a coke with 2 to 3% sulfur.

Investment and Operating Costs

Approximate on-site investment costs and operating cost factors for 30,000 bbl/sd cokers are given in Table 5. The investments are for 1974 at a U.S. Gulf Coast location. The estimates exclude gas compression, light ends, and coke handling, as these costs vary greatly with any particular refining situation.

Process Calculations

The process calculations for fluid coking are basically heat and material balance calculations which involve prediction of yields and setting operating conditions to match yields with refinery requirements.

Operating Conditions and Process Control

The reactor temperature is normally set at 950 to 1000°F. Low temperature favors high liquid yields and reduces the unsaturation of the gas, but increases the reactor holdup requirements. The burner temperature is normally 100 to 200°F above the reactor temperature. The reactor temperature is controlled by regulating the amount of coke sent to the reactor from the burner. Burner temperature is controlled by the air rate to the burner.

Low pressure is normally desired, as low pressure provides maximum gas-oil recycle cut point, minimizes steam requirements, and reduces air blower horsepower. Reactor pressure normally floats on the gas compressor suction pressure but is higher due to the pressure drop through the piping, the condenser, the fractionation tower, and the reactor cyclone. The burner pressure is set by the unit pressure balance required for coke circulation and is normally controlled at a fixed differential pressure relative to the reactor.

Reactor coke level is controlled by the cold coke slide valve on the transfer line from the reactor to the burner. Burner coke level is controlled by the coke withdrawal rate.

In Flexicoking the heater temperature is controlled by the rate of coke circulation between the heater and the gasifier. The unit inventory of coke is controlled by adjusting the air rate to the gasifier. Gasifier temperature is controlled by steam injection to the gasifier.

Reactor Yields and Product Qualities

In all coking processes the yields are a function of feed properties, the severity of the operation, and recycle cut point. Severity is a function of time and temperature. Low severity and high gas-oil cut point favor high liquid yields. High severity and low gas-oil cut point increase coke and gas yields. A large amount of yield and product quality information has been published [5, 10, 11, 13, 14, 25, 29]. Some yield and product quality data are shown in Tables 2, 3, 6, and 7. The yield data are plotted on Fig. 4. Data from these sources indicate that the gross coke yield is directly related to feedstock Conradson carbon residue. Coke quality data are shown in Table 8. Contaminant levels of the low heating value gas from a Flexicoker are shown in Table 9.

In most cases, high liquid yield and minimum coke and gas yields are required. A general discussion of cracking rates can be found in the literature. Theoretically, two cracking rates should be considered. The rate at which the liquid cracks and vaporizes after initially laying down on the coke particles should determine the reactor holdup required. Secondary vapor-phase cracking should determine the distribution of the products between gas, naphtha, and gas oil. The vapor residence time can be determined from the reactor volume and the volume flow of hydrocarbon vapor and steam, and can be divided into time in the fluid bed and time in the disperse phase. The former is a function of the coke holdup or weight space velocity (W/H/W) which is normally expressed as reciprocal hours. Since, for maximum liquid yield, the secondary cracking time should be kept at a minimum, it is normally desirable to design at the maximum operable W/H/W.

Reactor Holdup Requirement

The maximum rate at which feed can be injected into a fluid coker is limited by a condition known as bogging. If the feed injection rate exceeds the vaporization rate for an extended period of time, the thickness of the tacky oil film on the particles will increase until the particles rapidly agglomerate, causing the bed to lose fluidity. When fluidization is lost, the heat transfer rate is greatly reduced, further aggravating the condition. Coke circulation cannot be maintained due to the loss of reactor fluidization. The conditions required to avoid a bog are listed by Soskind, Botnikov, and Voronova [27].

1. The feedstock must be uniformly distributed over the entire surface of the heat transfer medium. . . .
2. The layer of feed material on the particles should not be too great . . . the thickness of the sticky plastic layer depends on the specific flow rate of feedstock, its coking factor and the recirculation rate of the heat-transfer medium. . . .
3. The bed temperature and the initial temperature of the heat transfer medium should be sufficiently high that the first stage of the process is completed in a short time. . . .
4. The heat-transfer medium should not consist of particles which are too fine.

Coking, Petroleum (Fluid)

TABLE 6 Typical Fluid Coking Yields and Product Quality from a Variety of Feeds [14]

	Crude Source of Feed									
	Los Angeles Basin			Panhandle	West Texas	Coleville	Ordonez	Hawkins	Elk Basin	South Louisiana	Kuwait
Feed to fluid coker	Vacuum pitch	Deasphalter pitch	Visbreaker tar	Vacuum pitch	Vacuum pitch	Vacuum pitch	Vacuum pitch	Vacuum pitch	Vacuum pitch	Vacuum pitch	Vacuum pitch
Feed inspections											
Gravity, °API	6.7	−0.2	−3.5	17.3	8.4	3.7	2.5	4.2	2.5	11.6	5.6
Conradson carbon, wt.%	17	33	41	11	19	21	26	24	30	13	21.8
Sulfur, wt.%	2.1	2.3	2.1	0.7	3.5	4.2	4.4	4.3	3.8	0.6	5.5
% distilled at 1000°F	25	0	5	0	10	20	10	10	0	5	0
Pour, °F	120+	120+	120+	85	120+	120+	120+	120+	120+	120+	120
Ultimate yields on coker feed											
C_3 and lighter, wt.%	8	10	12	7	9	9	10	10	11	8	10
C_4, vol.%	3	3	3	2	3	3	3	4	4	3	4
C_5-430°F naphtha, vol.%	17	18	14	21	21	20	20	19	18	21	21
430/1015°F gas oil, vol.%	62	45	32	69	58	49	49	52	45	61	48
Coke wt.% (gross)	21	36	48	12	20	30	30	27	34	17	28
Product quality											
Naphtha (C_3/430)											
Gravity, °API	55	54	55	59	56	54	57	56	56	56	56
Sulfur, wt.%	—	1.7	0.8	0.2	0.7	1.3	0.6	0.9	0.3	0.2	0.5
Research octane, clear	80	82	—	66	78	82	77	76	77	73	78
Gas oil (430/1015°F)											
Gravity, °API	17	14	15	29	20	15	17	17	17	22	15.5
Sulfur, wt.%	1.7	2.2	1.7	0.4	2.6	3.8	3.7	3.4	3.2	0.5	4.7
Watson K Factor	11.3	11.0	11.0	12.0	11.3	11.0	11.1	11.1	11.2	11.6	—

TABLE 7 Flexicoking Yields from Various Residua [11, 25]

Vacuum Residuum Properties	Arabian Light	Iranian Heavy	Arabian Heavy	Bachaquero	West Texas Sour Asphalt
°API gravity	6.5	5.1	4.4	2.6	−0.2
Conradson carbon, wt.%	19.2	21.4	24.4	26.5	34.0
Sulfur, wt.%	4.29	3.43	5.34	3.66	4.6
Nitrogen, wt.%	0.34	0.77	0.41	0.81	0.65
V + Ni, ppm	90	525	225	1040	137
Flexicoking yields on vacuum residuum					
C_3^- gas, wt.%	9.8	9.9	10.7	10.6	11.3
C_4 saturates, wt.%	0.6	0.6	0.6	0.7	0.7
C_4 unsaturates, wt.%	1.3	1.3	1.4	1.4	1.5
C_5/360°F naphtha, wt.%	11.2	11.0	10.6	10.4	9.2
LV%	15.5	15.4	15.0	14.8	13.4
360/975°F gas oil, wt.%	53.7	50.8	46.3	43.7	33.4
LV%	58.0	55.1	50.4	47.7	36.9
Gross coke, wt.%	23.4	26.4	30.4	33.2	43.9
Purge coke, wt.%	1.1	1.2	1.4	1.5	2.0
Coke gas, FOE vol.%	13.1	15.5	18.3	21.3	30.0

The heat reserve of the granules should be sufficient to cover the entire energy requirements in connection with heating the feedstock, supplying the energy for the endothermic cracking reaction and evaporating the decomposition products.

The article also states that "the limiting mass feed rate of the feedstock and mass loading per unit area of the bed depend on the fluidization rate, and increase appreciably with temperature."

A general approach to the problem of predicting the maximum operable feed rate to a fluid coker was proposed in a British patent [9]. The proposed equation is:

$$(W/H/W)_{max} = \frac{100 K_c \times 2^{\frac{T-1025}{38}}}{\% \text{ coke make}}$$

The equation is an integrated and highly modified form of the empirical Arrhenius equation for the effect of temperature on reaction rate. The maximum allowable feed rate will vary with the size of the unit, the feed temperature, the feed distribution into the reactor, the particle size of the solids, and the fluidization rate. K_c is generally between 0.15 and 0.35. Although the equation indicates the reaction rate doubles every 38°F, the temperature changes required to double the reaction rate varies from 30 to 50°F, depending upon feedstock. Data from Soskind et al. [27] and the equation from the British patent [9] are compared in Fig. 5. Similar effects are shown although the

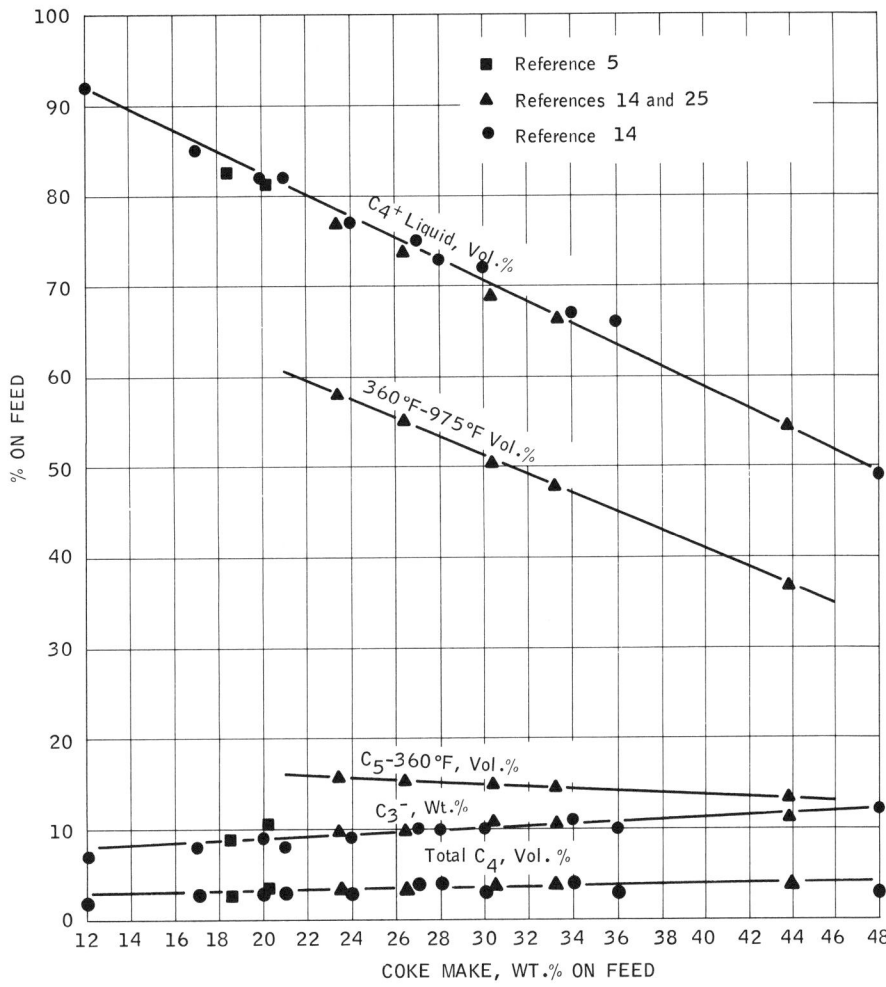

FIG. 4. Fluid coking yields [5, 14, 25].

TABLE 8 Typical Fluid and Flexicoke Properties [11]

	Flexicoke	Fluid Coke
Bulk density, lb/ft^3	50	60
Particle density, lb/ft^3	85	95
Surface area, m^2/g	70	<12
Average particle size, μm	120	170–240
Sulfur, wt.%	2.0	6.0

TABLE 9 Typical Flexicoker Coke Gas Contaminant Levels [11] (Dry Basis)

	After Particulate Removal	After Sulfur Removal
H_2S, vppm	7100	<10
COS, vppm	150	<15
NH_3, vppm[a]	<3	<3
HCN, vppm	<3	Nil
Solids, lb/Mscf	0.0042	Nil
Sulfur, wt.% FOE basis	9.7	<0.04

[a]Below detectable limit of 3 vppm.

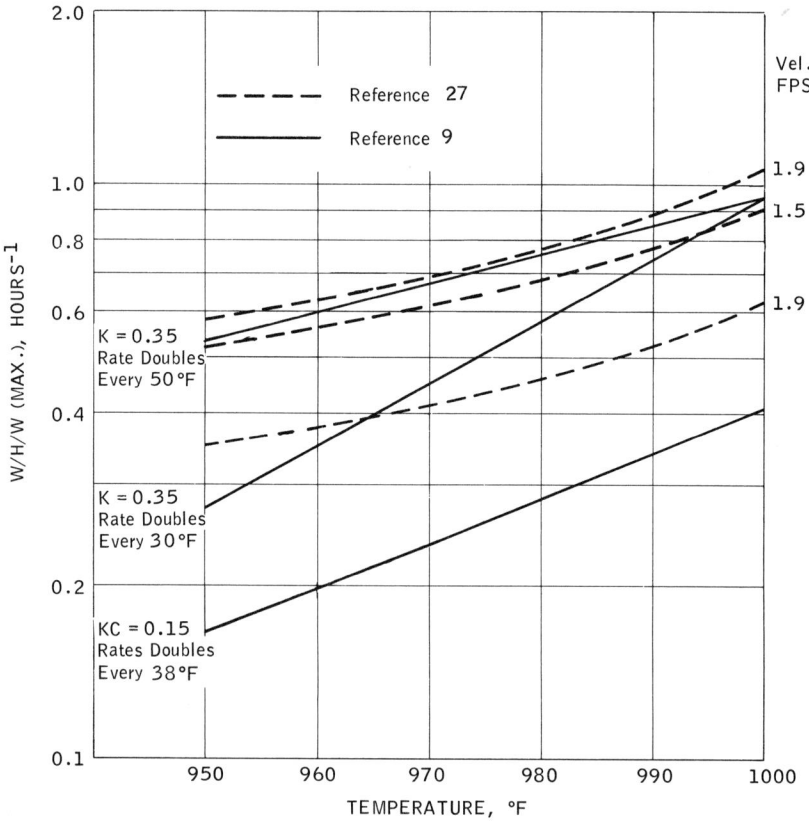

FIG. 5. Maximum weight space velocity [9, 27].

absolute values vary. This variation would be expected as the data from the Soviet Union are apparently from a single feedstock in a large pilot unit, and the data in the British patent were obtained from several feedstocks in a small laboratory unit. Additional information is obviously required to calculate reactor holdup, as the upper and lower lines in Fig. 6 represent a 2- to 3-fold difference in the coke holdup required in a reactor.

Fluid Coker Heat Balance

Air blower size, coke burning rate, and coke circulation rate are determined by heat balances. A reactor heat balance and an overall heat balance are required.
The reactor heat requirement is the sum of:

Sensible heat to the fresh feed
Latent heat to vaporize the fresh feed
Heat of cracking
Sensible heat to the recycle feed
Latent heat to vaporize the recycle feed
Heat to the reactor steam
Convection and radiation heat losses from the reactor and transfer lines

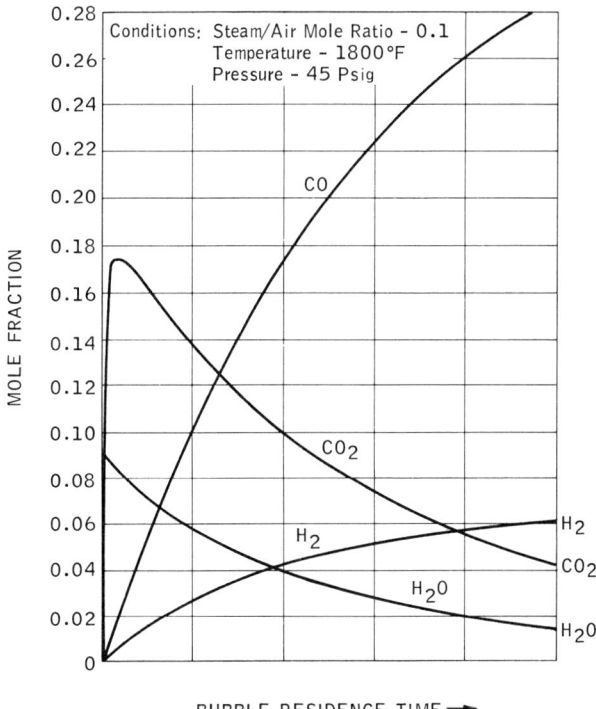

FIG. 6. Predicted concentration of gasification products and reactants [4, 20].

Circulation of hot coke from the burner supplies the required reactor heat. The total heat requirement is the sum of:

The reactor heat requirement
Heat to the burner and elutriator steam
Heat to the product coke
Radiation and convection heat losses from the burner and elutriator
Heat to combustion air

Coke must be burned to supply this heat. The heating value of coke is about 14,000 Btu/lb [28]. A substantial quantity of CO will be produced in the fuel-rich environment of the burner, so the heat realized in the burner will be less than 14,000 Btu/lb.

The required coke circulation rate can be calculated from the reactor heat requirement, the reactor and burner temperature, and the specific heat of coke.

Flexicoker Heat Balance

The reactor heat balance is the same for Flexicoking as for fluid coking. The overall heat balance is simple in theory and operation but complex mathematically.

The gasification reactions are given in Appendix A of Ref. 20.

$$C + O_2 \rightarrow CO_2 \quad \text{(very fast)} \quad (1)$$

$$C + \tfrac{1}{2}O_2 \rightarrow CO \quad \text{(very fast)} \quad (2)$$

$$C + CO_2 \rightarrow 2CO \quad \text{(slow)} \quad (3)$$

$$C + H_2O \rightarrow CO + H_2 \quad \text{(slow)} \quad (4)$$

$$CO + \tfrac{1}{2}O_2 \rightarrow CO_2 \quad \text{(fast)} \quad (5)$$

$$H_2O + CO \rightleftarrows CO_2 + H_2 \quad \text{(fast)} \quad (6)$$

Reactions (1), (2), and (5) are highly exothermic, very rapid, and primarily take place in the lower part of the coke bed. Reaction (6) reaches equilibrium rapidly at bed temperature and fixes the composition of the gas leaving the gasifier. Reactions (3) and (4) are both highly endothermic and very slow, and therefore set the bed depth and bed temperature. In addition to these reactions, the sulfur in the coke is rapidly converted to H_2S in the presence of carbon and steam [19, 26].

The air and steam rates must be set to consume the desired quantity of coke. The steam-to-air ratio, bed temperature, and bed depth must be set to balance the exothermic and endothermic reactions to provide the heat required for the process. The overall heat requirement is the sum of:

Reactor heat requirement
Heat to heater and gasifier steam

Radiation and convection losses from the heater and gasifier and transfer lines between these vessels

In practice, the heat balance is very stable, as a small increase in temperature produces carbon monoxide and hydrogen at the expense of carbon dioxide and steam.

Figure 6 shows how gas composition changes with bubble residence time in the fluid bed. The reader is referred to Kunii and Levenspiel [12] for calculation of bubble residence time.

Process Design of Equipment

The major vessels must be designed to meet the requirements set by the process calculations. The reactor design has a direct bearing on yields, so the process calculations cannot be considered final until the reactor design is complete. The reactor size and elevation and the elevation of the burner for a fluid coker or heater and gasifier for a Flexicoker must meet the requirements of the pressure balance. The design of these vessels is not complete until a satisfactory pressure balance has been made.

The mechanical design of the vessels is not considered here. Vessel construction is similar to fluid catalytic cracking units. The vessels are normally carbon steel with refractory lining.

Coking Reactor Design

The reactor (Fig. 7) vessel consists of three sections:

A stripper with sheds or disks and donuts with a minimum velocity of above 0.2 ft/s.
A dense bed zone made up of a conical section and a straight side section. The conical section should be tapered to maintain a relatively constant gas velocity in a dense bed zone.
A dilute phase disengaging section containing the cyclones. The diameter may be reduced to obtain a velocity of 2 to 5 ft/s.

The stripping steam rate is about 2 to 10 lb per 1000 lb of coke circulated.
The total coke holdup is set by process calculations and the volume of the dense bed is determined by the fluidized coke density. Expansion of the bed may be calculated by the method of Davidson and Harrison [6] and Matsen [15, 17, 18]:

$$\frac{\rho_0}{\rho} = \frac{H}{H_0} = 1 + \frac{U - U_0}{0.71\sqrt{gD_b}}$$

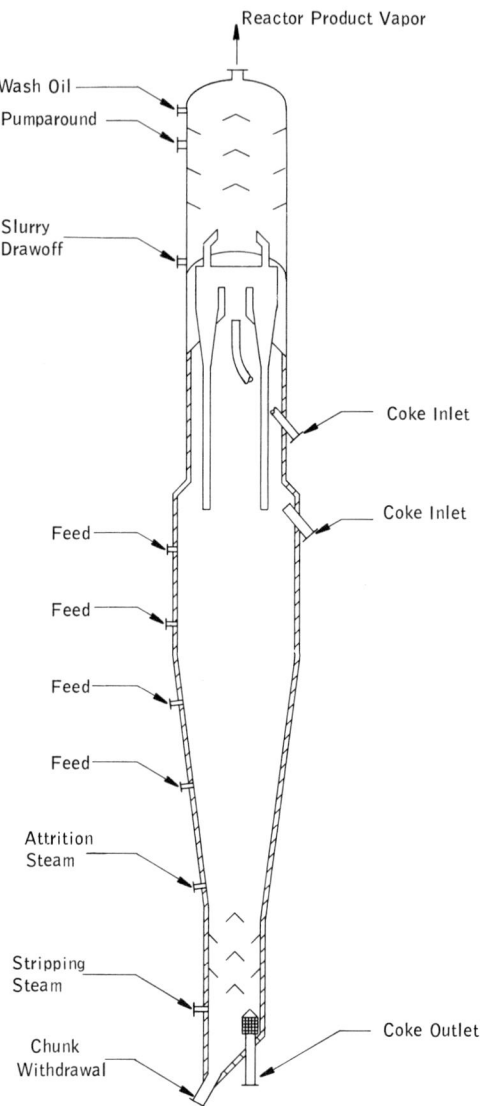

FIG. 7. Fluid coker reactor.

where ρ = average bed density
ρ_0 = bed density at minimum fluidization velocity
H = height of expanded bed
H_0 = height of bed at minimum fluidization velocity
U = superficial gas velocity
U_0 = minimum fluidization velocity
g = acceleration of gravity
D_b = bubble diameter

Bubble diameter may be estimated by the method of Kunii and Levenspiel [12] or Davidson and Harrison [6]. Some data for fluid coke are given by Matsen [15]. The equation is valid only when the vessel diameter is greater than 10 times the bubble diameter.

The design of the cyclones, the scrubber, and downstream equipment is in accordance with general industry practice.

The feed nozzles should be distributed as evenly as possible throughout the reactor bed. A simple steam atomized hose type nozzle is used [27].

Particle Size Control

Particle size control was discussed by Dunlop, Griffin, and Moser [7]. The particles grow by deposition of coke and by agglomeration in the reactor. The size is reduced by burning or gasification and by grinding. In order to maintain a constant particle size in the unit, the number of particles must be constant, as the weight of coke in the unit is more or less constant. The number of particles drawn off as product must be replaced with seed particles. The number of seed particles required is reduced by selectively withdrawing the larger coke particles from the burner by use of an elutriator.

The grinding is normally provided by steam jet attritors located in the lower part of the reactor dense bed. The steam also serves as fluidization gas. Jet velocities of 2000 to 3500 ft/s are possible with high-pressure superheated steam.

Fluid Coker Burner Design

The design of a fluid coker burner is similar to a fluid catalytic cracking unit regenerator. The burner (Fig. 8) consists of a dense bed where the burning occurs and a dilute phase, usually containing two stages of cyclones, to recover the entrained coke fines. Air is injected into the bottom of the bed through a suitable distributor. An auxiliary burner, located beneath the vessel and integral with it, provides preheated air to heat the coke to its ignition point. The burner vessel operates at a superficial velocity of 2 to 3 ft/s. The expansion of the burner bed and bed holdup can be calculated by the method used for the reactor bed.

Quench Elutriator Design

It is convenient to combine the elutriation function with product coke cooling in a single vessel before it is transported to storage. The cooling is normally obtained by spraying water into a fluid bed of coke. Any number of types of elutriators could be used. In one possible design shown in Fig. 9, the vessel consists of a lower quench section, operating at a fluidized velocity of about 1.5 ft/s, and an upper elutriation section with a velocity of about 7 ft/s. Coke is withdrawn from the burner vessel and fed into both sections. The amount of

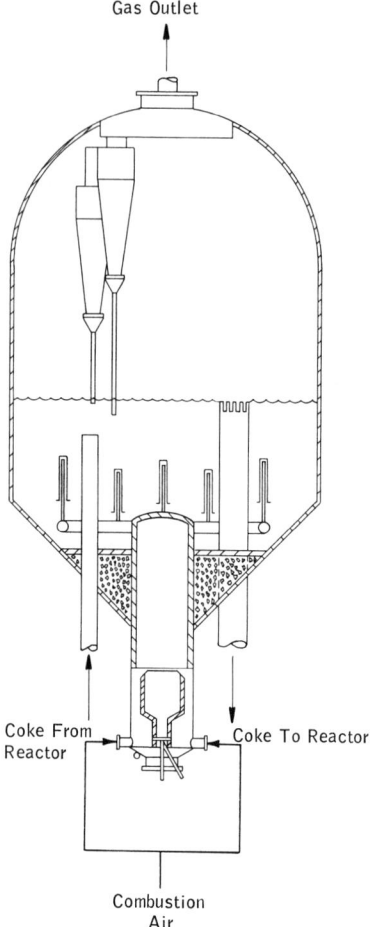

FIG. 8. Fluid coker burner.

elutriation can be controlled by the quantity fed into each section. Under the conditions given, about 33% of the feed coke is returned overhead to the burner.

Flexicoker Heater

The heater design is similar to the design of a fluid coker burner except that heat is supplied by gas and coke circulation from the gasifier rather than by burning coke. The superficial velocity is approximately the same as in the burner. The dilute phase and dust recovery sections are similar. The dense bed volume should be sized to accommodate inventory shifts due to changes in the fluidized density of the coke in the vessels when operating conditions are changed.

One critical design feature is the distribution of the very hot corrosive gas

Coking, Petroleum (Fluid)

from the gasifier into the heater. The method used to accomplish this is a proprietary feature of Flexicoking.

Gasifier Design

The critical design features of the Flexicoker gasifier are proprietary. Some general information has been published. Kett, Lahn, and Schutte [11] explain "The gasifier vessel itself contains no internal structures such as cyclones and employs similar vessel construction to that used in catalytic cracking regenerators and fluid coker burners, i.e., a carbon steel shell internally lined with commercially available refractories." The absence of cyclones sets the maximum superficial velocity in the gasifier. In order to maintain a fluid bed in the gasifier, the entrainment of coke from the gasifier to the heater must be less than the feed rate of coke from the heater to the gasifier.

One well-known problem in the burning and gasification of petroleum residua or coke is the formation of very corrosive sodium-vanadium slags. Matula, Molstedt, and Ryan [19] state:

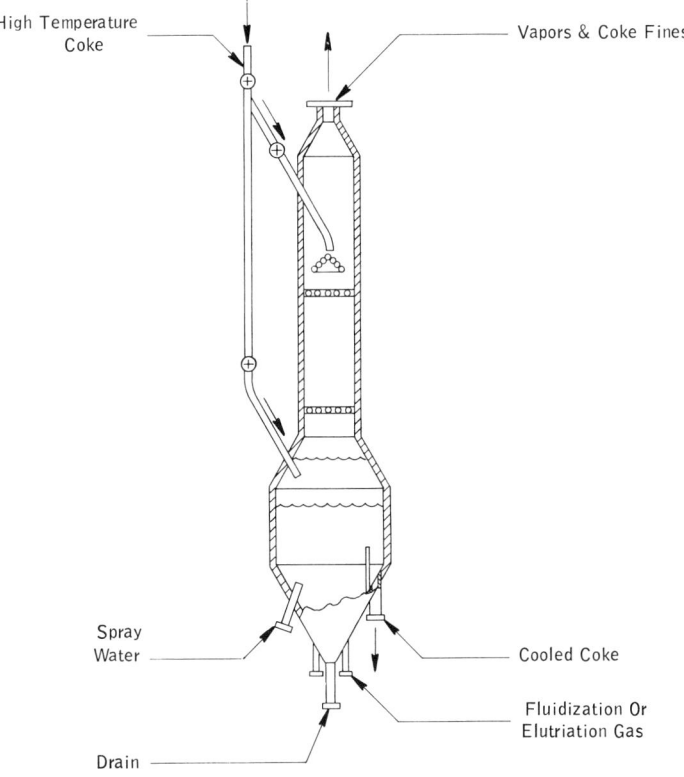

FIG. 9. Quench elutriator vessel.

In addition, because of the high levels of coke gasification achieved and the high temperature employed in the gasifier, slagging, with attendent materials problems, was another important variable studied. A specially designed gasifier air distributor has been provided in the Flexicoker to preclude slagging. Specially selected gasifier refractory and specially designed pressure and temperature instrumentation are also employed. The operation of the prototype has confirmed satisfactory slag-free operation of the gasifier even on the exceptionally high metals feedstock.

Heater Overhead Cooling and Dust Recovery

The design of the heater overhead system is fixed by the cleanup requirements of the coke gas. The gas contains coke fines, a large amount of hydrogen sulfide, and a small quantity of carbonyl sulfide. In order to remove the hydrogen sulfide, the stream must be cooled to about 100 to 160°F.

A typical overhead system is shown in Fig. 2. Sensible heat from the gas is recovered by steam generation. Most of the coke fines are recovered in tertiary cyclones. Additional fines may be recovered in a conventional venturi scrubber, preceded by a cooler. If an extremely low sulfur level is required, the carbonyl sulfide can be further converted to hydrogen sulfide in a fixed-bed catalytic reactor. The gas is further cooled and steam is condensed before it passes to sulfur recovery facilities, Stretford in the example. Design of the equipment, excepting the COS conversion step, is conventional and need not be discussed here.

Transfer Line Design

The fluid solids transfer lines used to transport fluid coke are referred to as J-Bends. A typical J-Bend is shown on Fig. 10. The main features of the J-Bends are:

A vertical standpipe section which provides a maximum pressure buildup
A sharp radius bend at the bottom of the standpipe
A slanted riser which is sloped from the horizontal
A large radius bend connecting the slanted riser with a vertical riser
A vertical riser leading directly into a receiving vessel

A slide valve in the lower portion of the standpipe is used to control the flow of the solids.

The calculation of the pressure balance for circulating the fluid coke is similar to the calculation for any other fluid solids unit where this type of solids transport is used. Whiteley and Molstedt [30] give an example of a pressure balance for a single line.

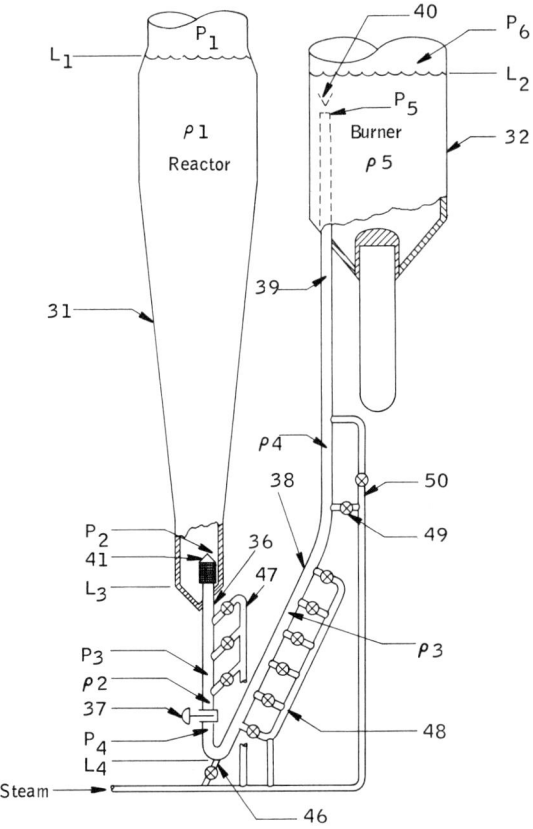

FIG. 10. Coke transfer line [30].

As a specific example of the pressures and densities applicable to this invention, a hydrocarbon oil fluid coking system as is shown in Fig. 10 containing coke particles of 40 to 800 μm, with 175 μm being the average, having a true density of 100 lb/ft^3 may have a fluid bed level L_1 in the coking vessel 80 feet above the level L_4 of the bend and 67.2 ft above the level L_3 of the entrance of the solids to the standpipe. The level L_2 of the fluid bed in the heater vessel may be about 80 ft above the level of the bend and the conveying conduit may terminate about 5 ft below this level L_2. The density ρ_1 of the fluid bed in the coker may be 40.5 lb/ft^3 and the density ρ_5 of the bed in the heater 30 lb/ft^3.

Under these conditions the coker may have a pressure P_1 of 11 lb/in.^{2} gauge and the heater a pressure P_6 of 12 lb/in.^{2} gauge. There is then a pressure P_2 at the entrance of the conduit of 30 lb/in.^{2} gauge and at the exit P_5 of 13 lb/in.^{2} gauge. The density ρ_2 of the solids suspension in standpipe 36 is 42 lb/ft^3, sufficient to increase the pressure about 1 lb/in.^{2} every 3.5 ft. If greater pressure is needed, the standpipe section can be suitably lengthened.

The pressure P_3 above the valve may be about 33 lb/in.^{2} gauge and there may be a pressure drop over the valve of about 5.3 lb/in.^{2} such that the pressure at the bend is 27.7 lb/in.^{2} gauge. A pressure loss of 1 to 3 lb/in.^{2} may be encountered at the bend.

The density ρ_3 of the solids suspension in the inclined riser 38 is about 28 lb/ft^3 and the density ρ_4 in the vertical riser 20 lb/ft after injection of about 0.0753 ft^3 of aerating gas per pound of solid. 0.0552 ft^3 of this aeration gas is added through lines 46 and 48 and 0.0201 ft^3 is added through line 49.

Using a 14-in. conduit, the rate of solids flow for the above conditions may be 12,000 lb/(min)(ft^2) with a velocity of 4.8 ft/s in the standpipe and 10 ft/s in the vertical riser.

Thus there is a pressure drop of 12.6 lb/in.^{2} over the inclined and vertical risers of which 0.7 lb/in.^{2} is accounted for by friction, leaving 13.1 lb/in.^{2} gauge as the static pressure or driving force.

A similar balance is made for each transfer line in a fluid coker or Flexicoker. The overall pressure balance sets vessel elevation. The quantity of aeration gas, which is usually steam, is set to obtain the required densities in the standpipes and risers.

Aeration requirements for standpipes and risers may be calculated by the method proposed by Masten [16]:

$$\frac{\rho_0}{\rho} = 1 + \left(\frac{U_t - U_0 - \varepsilon_0 W/\rho_0}{U_b + W/\rho_0}\right)$$

where ρ = average density of gas–solids mixture
ρ_0 = density at minimum fluidization velocity
ε_0 = void fraction at minimum fluidization velocity
W = solids mass velocity
U_t = total superficial gas velocity
U_0 = minimum fluidization velocity
U_b = bubble rise velocity = $0.35\sqrt{gD}$ [15]
g = acceleration of gravity
D = transfer line diameter

Operation

The operation of a fluid coker or Flexicoker is similar in many ways to a fluid catalytic cracking unit and requires about the same level of operating skill.

The basic startup steps are:

1. Fire auxiliary burner
2. Load coke to burner and reactor
3. Establish continuous coke circulation
4. Heat coke to ignition temperature and shutdown auxiliary burner. This may also be done prior to loading coke to the reactor

5. Bring reactor to operating temperature and establish feed to the unit
6. Line out the fractionator and start coke to storage

The shutdown procedure is simple. Feed is removed from the reactor, and the heavy oil circuits are washed with a light oil while coke is being unloaded.

There are some notable differences between the operation of a fluid coker and a fluid catalytic cracking unit (FCCU). Some of these differences tend to make the fluid coker easier to operate. The fluid coker heat balance is very easy to maintain, as there is always an excess of carbon to burn. A FCCU has a sensitive interaction between heat balance and intensity balance, and therefore between carbon burned and carbon produced, which complicates control, especially during operating changes, startup, and shutdown.

Recovery from upsets caused by loss of utilities such as steam and air is normally easier and faster with a fluid coker than with a FCCU. The fluid coker normally operates well at low feed rates. Turndown to low rate is normally limited by the ability of the tower to maintain fractionation of the products. The fluid coker proper can operate at any feed rate which will provide enough coke to heat balance.

The fluid coker has some inherent features which can create problems if proper precautions are not followed. The heavy residuum can set up if the lines are not properly heat traced and insulated. Low reactor temperature results in reactor bogging. If the particle size of the circulating coke is not properly controlled, the size can grow to the point that coke circulation problems are encountered. The feed nozzles must be maintained and occasionally cleaned to prevent poor feed distribution followed by excessive agglomerate formation. Control of the reactor bed level is critical. An excessively high bed level will flood the reactor cyclone and allow coke to be carried to the scrubber where it will plug the heavy oil circuits. Emergency instrumentation is provided to cut out reactor feed on low reactor temperature, and to close the hot coke slide valve on high reactor level.

The average service factor for all fluid cokers is currently 90%. The maximum average service factor for a single unit is 95%.

References

1. Americ, B. K., et al., *Processes for Continuous Thermocontact Treatment of Oil Stocks on Coke*, Presented at the 1959 Fifth World Petroleum Congress.
2. Anonymous, "Flexicoking Passes Major Test," *Oil Gas J.*, 73(10), 53–56 (March 10, 1975).
3. Blaser, D. E., Rionda, J. A., and Saxton, A. L., "Combine Desulfurizing and Coking," *Hydrocarbon Process.*, 50(9), 137–141 (1971).
4. Brown, J. W., Schuette, W. L., and Sherman, L. G., *Latest Developments in Flexicoking*, Presented at the AIChE Meeting, March 19, 1975.

5. Busch, R. A., "Fluid Coking: Seasoned Process Takes on New Jobs," *Oil Gas J.*, 68(14), 102–111 (April 6, 1970).
6. Davidson, J. F., and Harrison, D., *Fluidized Particles*, University Press, Cambridge, 1963.
7. Dunlop, D. P., Griffin, L. I., Jr., and Moser, J. F., Jr., "Particle Size Control in Fluid Coking," *Chem. Eng. Prog.*, 54(8), 39–43 (1958).
8. Edelman, A. M., et al., *A Flexible Approach to Fuel Oil Desulfurization*, Presented at the Japanese Petroleum Institute Meeting, May 8, 1975.
9. Esso Research and Engineering Company, "Improvements in or Relating to Process of Coking Residual Oils," British Patent 759,720 (October 24, 1956).
10. Johnson, F. B., and Wood, R. E., "New Data on Fluid Coking," *Pet. Refiner*, 33(11), 157–160 (1954).
11. Kett, T. K., Lahn, G. C., and Schuette, W. L., *Flexicoking: A Versatile Residuum Conversion Process*, Presented at the AIChE 67th Annual Meeting, December 2, 1974.
12. Kunii, D., and Levenspiel, O., *Fluidization Engineering*, Wiley, New York, 1969.
13. Martin, H. Z., Barr, F. T., and Krebs, R. W., "More on the Fluid Coking Process," *Oil Gas J.*, 53(1), 166–171 (May 10, 1954).
14. Martin, R., "Reexamine Fluid Coking," *Petro/Chem. Eng.*, 11(40), (October 1968).
15. Matsen, J. M., "Evidence of Maximum Stable Bubble Size in a Fluidized Bed," *AIChE Symp. Ser.*, 69(128), 30–33 (1973).
16. Matsen, J. M., "Flow of Fluidized Solids and Bubbles in Standpipes and Risers," *Powder Technol.*, 7, 93–96 (1973).
17. Matsen, J. M., and Tarmy, B. L., "Scaleup of Laboratory Fluid-Bed Data: The Significance of Slug Flow," *Chem. Eng. Prog. Symp. Ser.*, 66(101), 1–7 (1970).
18. Matsen, J. M., et al., "Expansion of Fluidized Beds in Slug Flow," *Chem. Eng. Sci.*, 24, 1743–1754 (1969).
19. Matula, J. P., Molstedt, B. V., and Ryan, D. F., *Flexicoker Prototype Demonstrates Successful Operation*, Presented at the 40th Mid-Year Meeting of the Division of Refining, American Petroleum Institute, May 13, 1975.
20. Matula, J. P., Weinberg, H. N., and Weissman, W., *Flexicoking—An Advanced Fluid Coking Process*, Presented at the 37th Mid-Year Meeting of American Petroleum Institute's Division of Refining, May 11, 1972.
21. Mekler, V., Schute, A., and Whipple, T. T., "Continuous Coking Process Shows Ability to Handle Heavy Feedstocks," *Oil Gas J.*, 52(28), 200–203 (November 16, 1953).
22. Metrailer, W. J., Royle, R. C., and Lahn, G. C., *Properties of Cokes Produced in the Flexicoking Process*, Presented at the American Chemical Society, Division of Petroleum Chemistry Meeting, April 7, 1975.
23. Molstedt, B. V., et al., "Method of Scouring Equipment in a Fluid Coking Process," U.S. Patent 2,735,806 (February 21, 1956).
24. Pagel, J. E., Rionda, J. A., and Fuentes, F. A., *Vanadium Recovery in the Refinery via Flexicoking of Residue*, Presented at the International Symposium on Vanadium and Other Metals in Petroleum, sponsored by University of Zulia, Maracaibo, Venezuela, August 21, 1973.
25. Rionda, J. A., et al., *Recent Advances in Residua Processing*, Presented at the National Petroleum Refiners Association Annual Meeting, April 2, 1974.
26. Sappok, R. J., and Walker, P. L., Jr., "Removal of SO_2 from Flue Gases Using Carbon at Elevated Temperatures," *J. Air Pollut. Control Assoc.*, 19(11), 856–861 (1969).
27. Soskind, D. M., Botnikov, Y. A., and Voronova, D. K., "The Feed Conditions and

Mass Feed Rate of Raw Materials into Thermocontact Cracking Reactors," *Int. Chem. Eng.*, *11*(1), 50–53 (1971).
28. Thorton, D. P., Jr., "Why Carter Likes It's Coker," *Pet. Process.*, *10*, 840–845 (June 1955).
29. Voorhies, A., and Martin, H. Z., "Fluid Coking of Residua," *Oil Gas J.*, *52*(28), 204–207 (November 16, 1953).
30. Whiteley, R. S., and Molstedt, B. V., "Method and Apparatus for Handling Fluidized Solids," U.S. Patent 2,881,133 (April 7, 1959).

D. E. BLASER

Cracking, Thermal

Introduction

The decomposition (cracking) of high molecular weight hydrocarbons to lower molecular weight, normally more valuable, hydrocarbons has long been practiced in the petroleum refining industry. Although catalytic cracking has generally replaced thermal cracking, noncatalytic cracking processes using high temperature to achieve the decomposition are still in operation. In several cases, thermal cracking processes to produce specific desired products or to dispose of specific undesirable charge streams are being operated or installed. The purpose of this article is to provide basic information to assist the practicing engineer/petroleum refiner in:

1. Determining if a particular thermal cracking process would be suitable for a specific application and could fit into the overall operation
2. Developing a basic design for a thermal cracking process
3. Operating an existing or proposed process

This article will cover in some detail three of the more commonly used thermal cracking processes:

1. Thermal cracking of gas oils and residual stocks to produce primarily gas, gasoline, and residual fuel oil
2. Visbreaking (mild thermal cracking) to reduce the viscosity of a heavy viscous stock to a more fluid stock for easier handling and for use primarily in fuel oil
3. Thermal reforming of low octane gasoline stocks to upgrade the octane number

Coking, the thermal decomposition of heavy hydrocarbons to produce gas, gasoline, and coke as the residual product rather than residual fuel oil, is covered in a separate article in this *Encyclopedia*. The thermal conversion of petroleum stocks for the production of ethylene for use in petrochemical processing is also covered in a separate article. Other processes which would fall in the catagory of thermal cracking (e.g., gas reversion, thermal polymerization, thermal alkylation, vapor-phase thermal cracking) but have never achieved widespread commercial usage will not be covered in this article.

Typical schematic flow diagrams will be presented showing the pertinent equipment and process streams of the three processes to be discussed. Correlations (along with example problems) which will permit the reader to estimate the product yields and qualities from a variety of widely different charge stocks will also be presented. Pertinent equipment requirements and design features will be discussed. Some current costs of construction and operation will be covered. A Bibliography of related published literature which were used as sources of information is presented at the end of this article for further reference by the reader.

Definition of Thermal Processes

Thermal processes are those processes which decompose, rearrange, or combine hydrocarbon molecules by the application of heat without the aid of a catalyst. More detailed definitions of the three specific thermal processes covered in this article are presented in the following paragraphs.

Conventional thermal cracking is defined as the thermal decomposition, under pressure, of large hydrocarbon molecules (higher boiling than gasoline) to form smaller molecules. The thermal cracking process is designed to produce gasoline from higher boiling charge stocks, and any unconverted or mildly cracked charge components (compounds which have been partially decomposed but are still higher boiling than gasoline) are usually recycled to extinction to maximize gasoline production. A moderate quantity of light hydrocarbon gases is also formed. As thermal cracking proceeds, reactive unsaturated molecules are formed which polymerize and ultimately create large asphaltlike molecules. These asphaltlike molecules are hydrogen deficient and tend to form coke readily. Thus they cannot be recycled without excessive coke formation and are therefore removed from the system as cycle fuel oil.

Visbreaking, an abbreviated term for viscosity-breaking or viscosity lowering, is a mild, liquid-phase thermal cracking process used to convert heavy, high viscosity petroleum stocks to lower viscosity fractions suitable for use in heavy fuel oil. This ultimately results in less production of fuel oil since less cutter stock (low viscosity diluent) is required for blending to meet fuel oil viscosity specifications. The cutter stock no longer required in fuel oil may then be used in more valuable products. A secondary benefit from the visbreaking operation is the production of gas oil and gasoline streams which usually have higher product values than the visbreaker charge. Visbreaking produces a small quantity of light hydrocarbon gases and a larger amount of gasoline. Visbreaking, unlike conventional thermal cracking, normally does not employ a recycle stream. Conditions are too mild to crack a gas oil recycle stream and the unconverted residual stream, if recycled, would cause excessive heater coking. The boiling range of the product residual stream is extended by visbreaking so that light and heavy gas oils can be fractionated from the product residual stream, if desired. In some present applications the heavy gas oil stream is recycled and cracked to extinction in a separate higher temperature heater.

Thermal reforming of gasoline is a severe, vapor-phase thermal conversion process conducted under pressure. The purpose of thermal reforming is to increase the octane number of the gasoline charge stock, but this octane number increase is not obtained without the loss of some gasoline yield. A recycle stream is not required nor normally used. Significant quantities of light hydrocarbon gases, pentanes and lighter compounds, are produced. These gases contain sizable quantities of light olefins which are useful as alkylation, polymerization, or petrochemical feed stocks. In thermal reforming the conditions are sufficiently severe to cause some polymerization of the light olefins and condensation to polynuclear compounds. These compounds form a stream with a higher boiling range than the gasoline charge. This heavy stream is removed and can be used as a fuel oil component.

History and Development of Thermal Cracking

A popular story is that thermal cracking of petroleum was discovered by accident in 1861 in a small New Jersey refinery. Petroleum was being batch distilled to produce kerosine for use in illumination and residuum for use as a lubricant. The gasoline that was present in the whole crude oil was considered an undesirable by-product and frequently it was discarded. Product quality control was not considered a fine art and the still did not require close watching. Thus the stillman left the still unattended. The residuum became exposed to cracking conditions. When the stillman returned, the stream being produced from the still was not kerosine but an odorous material which nobody would buy. This material is now known as thermal cracked gasoline.

Professor B. Silliman of Yale, in his report dated April 15, 1855, on the economic value of "rock-oil" of Pennsylvania, suggested that thermal decomposition of the oil occurred during distillation. He speculated that this probably could be used as a method for the production of valuable products.

Patents concerned with "cracking-distillation" began to appear about this time. Some of the more notable patents related to thermal cracking were:

U.S. Patent 28,246 (1860)—increase the production of illuminating oils by distilling, condensing, and returning the heavy oil vaporized to the still for cracking.

British Patent 3345 (1865)—use of pressure in distillation to keep material in the liquid state and thus raise the distillation temperature. Pressure was released prior to condensing.

U.S. Patents 419,931 and 426,173 (1890)—use of pressure in both the distillation and condensation steps.

In the latter part of the nineteenth century, numerous patents were issued in this field with the majority covering processes for the manufacture of illuminating oils. Few, if any, had much commercial value. It was not until after the turn of the century and the oncoming of the automobile with the resulting increase in gasoline demand that thermal cracking for gasoline manufacture became a commercial reality. The first commercially successful process was the Burton process which first operated in 1913. The original patent for the Burton process was U.S. Patent 1,049,667, issued in 1912.

Originally, the Burton process was a batch thermal operation producing about 30 to 35 vol.% each of gasoline, kerosine, and residual fuel oil from a gas oil charge. About 250 to 300 bbl of gas oil were charged to a horizontal still about 10 ft in diameter and 30 ft long. The oil was slowly heated by direct fire through the shell of the still until a still pressure of 75 to 100 $lb/in.^2$ gauge was obtained. Cracking normally began at about 740 to 750°F and the liquid temperature normally did not exceed 800°F. The vapors were released from the still and condensed. Heating of the still was continued while maintaining a back pressure on the still. The unit was operated on a 48-h cycle, as limited by coke deposits, with shutdown and removal of the decomposition coke after each run. Coke was removed by various means, including the installation of chains or cables which could be pulled through the upper manhead on the front of the still to remove the coke from the vessel.

An improved Burton design, called a Burton-Clark still, used the same size horizontal still but with heating tubes similar to a water tube boiler. The tubes were exposed directly to the hot furnace gases while the shell of the still was protected from the hot gases. This was a semicontinuous operation in which an additional 250 bbl of charge were added through the vapor line to the still to replace the portion of the charge distilled overhead. Again the run length was limited by coke deposits to a range of from 48 to 60 h. The entire cycle to charge, heat, cool, and clean required about 3 d. The Burton process, with improvements, dominated the thermal cracking field from 1913 to 1925 with over 1200 units being built. The decline of the Burton process was just as rapid since there were only 191 units in operation by 1931.

The Coast, Snodgrass, and Fleming processes were similar to the Burton process in that heat was applied directly to the still. A primary disadvantage of these processes was the short cycle time due to excessive coke formation in the still. There was also the inherent danger of heating large volumes of oil under pressure in a direct fired still. The development of furnace-type units caused a rapid replacement of the direct fired still processes with the gain of significantly longer run times.

The furnace-type units were generally similar in pertinent features. The charge oil, usually combined with an uncoverted gas oil recycle stream, was heated to a high temperature in the furnace tubes where cracking was initiated. The hot oil flowed from the furnace tubes into a nonfired soaking drum which allowed additional reaction time at the high temperature. The effluent from the soaking drum was quenched, usually with either raw charge or a gas oil recycle stream, immediately after leaving the soaking drum. This quenching terminated the cracking reaction, thus minimizing downstream coke formation. Pressure was broken in a flash vaporizer where everything but a heavy residual stream was flashed overhead. The overhead stream was further fractionated into a gasoline stream and a gas oil recycle stream. The residual stream was drawn off the bottoms of the flash vaporizer for use in fuel oil or other processing.

Many furnace-type thermal cracking units similar to that described in the preceding paragraph were developed. Most were developed by petroleum refining companies for their own use. Some of the more important processes were

Holmes-Manley (Texaco)
Dubbs (Universal Oil Co.)
Tube-and-Tank (Standard Oil of New Jersey)
Cross (Gasoline Products Co.)
Isom (Sinclair)

A subsequent improvement in thermal cracking units was the elimination of the high-pressure soaking drum so that all of the cracking took place in the furnace tubes. Higher temperatures were applied and contact times were significantly shorter. This required closer control of the operation which could only be achieved by the more sophisticated instruments that became available after the early development of the thermal cracking processes. Since practically all of the cracking occurs in the tubes of a pipestill, the general process was called the tubestill process. The Donnelly, the Winkler-Koch, and the later Dubbs processes were examples of tubestill processes.

In the early stages of thermal cracking process development, processes were generally classified as either liquid-phase, high pressure (350 to 1500 lb/in.^{2} gauge)–low temperature (750 to 950°F) or vapor-phase, low pressure (less than 200 lb/in.^{2} gauge)–high temperature (1000 to 1200°F). Actually, the processes were mixed phase with no process really being entirely liquid or vapor phase. The classification was still used as a matter of convenience. The processes described in the preceding paragraphs were classified as liquid-phase processes. These liquid-phase processes had the following advantages over vapor-phase processes:

1. Large yields of gasoline of moderate octane number
2. Low gas yields
3. Ability to use a wide variety of charge stocks
4. Long cycle time due to low coke formation
5. Flexibility and ease of control

The vapor-phase processes had the advantages of operation at lower pressures and the production of a higher octane gasoline due to the increased production of olefins and light aromatics. However, there were many disadvantages which curtailed the development of vapor-phase processes. High temperatures were required which the steel alloys available at the time could not tolerate. Very close control of operations was required. There were high gas yields and resulting losses since the gases were normally not recovered. There was a high production of olefinic compounds which created a gasoline with poor stability (increased tendency to form undesirable gum). This required subsequent treating of the gasoline to stabilize it against gum formation. The vapor-phase processes were not considered suitable for the production of large quantities of gasoline, but did find application in petrochemical manufacture due to the high concentration of olefins produced. Some of the better known vapor phase processes were

Gyro
DeFlorez
True Vapor Phase (Knox)

Perhaps the ultimate in the thermal cracking processes was the development of the combination unit. This unit essentially combines one or more cracking units with the whole crude fractionating system. This combination of units has the advantages of

1. Heat economy—the excess heat from the cracking process can be used to heat the crude charged to the fractionating section
2. Minimum duplication of equipment—common equipment such as fractionating towers, furnaces, and separators may be used for more than one process
3. Selective thermal processing—various boiling range stocks can be cracked individually under the optimum cracking conditions for the particular stock

Commercial Applications

Whole crude and practically any crude oil fraction can be charged to thermal cracking-type processes. Processes are available for cracking everything from hydrocarbon gases through vacuum residuum or asphalt. In the early days of thermal cracking, the gas oil fractions, kerosine through the heavy vacuum gas oils, were the primary charge stocks. Gasoline was the prime product with heavy fuel oil and light gases as by-products. With the advent of catalytic cracking, which produced increased yields of higher octane gasoline with less heavy fuel oil and no coke compared to thermal cracking, the availability of clean virgin thermal cracking charge stocks decreased. Today, catalytic cracked cycle gas oils make up most of the charge to thermal cracking units with residual stocks making up the remainder.

The utilization of improved vacuum distillation equipment resulted in the production of heavier, cleaner gas oils for catalytic cracking charge. This heavier gas oil production, of course, resulted in a higher boiling, more viscous, vacuum residuum. Although there was less vacuum residuum produced, it required more cutter stock to make the residuum stock salable as heavy fuel oil. The viscous vacuum residuum was a primary charge stock for visbreaking operations. A lower viscosity fuel oil was the primary product from a visbreaker with a small amount of gas and gasoline as by-products. Longer residuum fractions including topped crudes and even whole heavy crudes can be used as visbreaker charge stocks. In some cases it has been cited that heavy gas oils have also been used as charge to visbreaking operations.

Thermal reforming was initially designed to upgrade the octane of low octane straight run gasolines, but the process can be extended to a variety of gasoline stocks, e.g., catalytic cracked, thermal cracked, catalytic reformed, and paraffinic raffinates. The primary product, of course, was a higher octane gasoline. Light olefinic gases and a fuel oil polymer were by-products.

Presently, the combined commercial capacity of gas oil thermal cracking, visbreaking, and thermal reforming units in the United States is about 400,000 bbl/d. There has been a steady decline in capacity of these processes over the past 15 to 30 yr to a present level of about 3% of the total crude capacity for all United States refineries. Figure 1 shows the decline in capacity for each of the processes as percent of the total United States crude capacity.

At its peak in 1930, gas oil thermal cracking represented about 55% of the total crude capacity or over 2,000,000 bbl/d. Thereafter, a rapid decline of gas oil thermal cracking capacity occurred due to the construction and operation of catalytic cracking units. Today, gas oil thermal cracking capacity is about 200,000 bbl/d or less than 1.5% of the total crude capacity. Many of the thermal cracking units were converted to other operations, such as visbreaking, or were completely dismantled.

Visbreaking capacity reached a peak in the late 1950s of about 500,000 bbl/d and has declined to a present level of about 200,000 bbl/d.

Thermal reforming capacity has declined rapidly from about 500,000 bbl/d in the early 1950s to less than 25,000 bbl/d today. This rapid reduction in capacity can be attributed entirely to the development of catalytic reforming.

It may not be fair to conclude that these processes are on their way to extinction as there are still ways in which the refiner may make use of these processes in special applications. The need to refine less desirable crudes and other hydrocarbon materials such as shale oil or tar sand bitumens is becoming increasingly more important. The governmental restrictions on sulfur oxides emission from the burning of fuels may force the increased use of waxy low sulfur crudes. Thermal cracking and visbreaking are suitable means to convert these materials into easy-to-handle products. Visbreaking in combination with hydrodesulfurization is a potential means to produce low sulfur fuel oil. The gas oils produced are fractionated from the visbroken residual product, hydrodesulfurized, and recombined with the residual product. This will give a lower sulfur content fuel oil at a lower hydrodesulfurization cost than would be

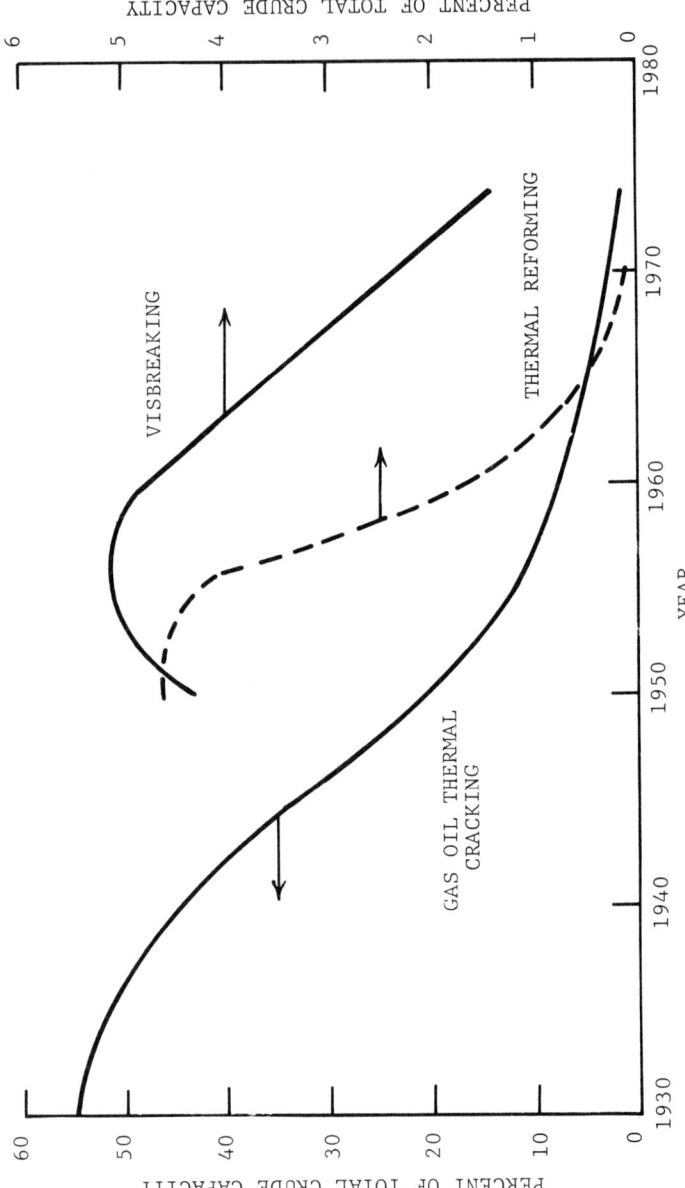

FIG. 1. Thermal cracking, visbreaking, and thermal reforming capacity decline in the United States. Basis: Total crude capacity.

incurred by hydrodesulfurizing the total visbreaker charge stock. The removal of lead from gasoline will increase the load on catalytic reformers. Thermal reforming may be the means to increase the octane of marginal stocks. The growth of the petrochemical industry will increase the demand for raw materials. Any of the three processes covered in this article has the potential to produce petrochemical stocks, particularly olefins, from heavier stocks.

Chemistry of Thermal Cracking

In thermal cracking processes many different chemical reactions occur simultaneously. Thus an accurate explanation of the mechanism of the thermal cracking reactions is difficult. The primary reactions are the decomposition of large molecules into smaller molecules. These reactions proceed by a free radical mechanism. Experimental yields from the thermal decomposition of the lighter hydrocarbons, propane through hexanes, can be predicted from mathematical relations based on a free radical theory. The heavier gas oils are too complex for precise mathematical models.

Under thermal cracking conditions, large molecules may decompose into free radicals:

$$C_{10}H_{22} \rightarrow C_8H_{17}\cdot + C_2H_5\cdot$$

Free radicals are short lived, highly reactive species that do not appear in the cracked products. Depending upon the free radical size and processing conditions, the free radical will:

1. React with other hydrocarbons. The smaller radicals, hydrogen, methyl, and ethyl, are more stable than the larger radicals. They will tend to capture a hydrogen atom from another hydrocarbon, thereby forming a saturated hydrocarbon and a new radical:

$$C_2H_5\cdot + C_6H_{14} \rightarrow C_2H_6 + C_6H_{13}\cdot$$

2. Decompose further to olefins and smaller radicals:

$$C_6H_{13}\cdot \rightarrow C_5H_{10} + CH_3\cdot$$
$$C_8H_{17}\cdot \rightarrow C_4H_8 + C_4H_9\cdot$$
$$C_4H_9\cdot \rightarrow C_4H_8 + H\cdot$$

3. Combine with other free radicals to terminate the chain reaction:

$$CH_3\cdot + H\cdot \rightarrow CH_4$$

4. React with poisons or metal surfaces which will also terminate the chain reaction.

The relative ease for cracking of various types of hydrocarbons of the same molecular weight is given in the following descending order:

1. Paraffins
2. Olefins
3. Naphthenes
4. Aromatics

Within any type of hydrocarbon, the heavier hydrocarbons tend to crack easier than the lighter ones. Paraffins are by far the easiest hydrocarbons to thermally crack with the rupture most likely to occur between the first and second carbon bonds in the lighter paraffins. However, as the molecular weight of the paraffin molecule increases, rupture tends to occur nearer the middle of the molecule. Thermal cracking of heavy stocks does not proceed directly from charge stock to end products, but stepwise through a series of progressively lower molecular weight products, e.g., heavy gas oil to light gas oil to gasoline to gas, with the reactions occurring simultaneously in a mixture of various molecular weight hydrocarbons.

The main secondary reactions occurring in thermal cracking are polymerization and condensation. These reactions are not satisfactorily explained by a free radical mechanism. Other reactions, such as cyclization, alkylation, and isomerization, also occur to some extent.

On the basis of thermodynamic relationships, the following are estimated to be the most probable thermal reactions for various types of compounds:

1. *Paraffins.* Splitting decomposition (forming a paraffin and an olefin) and dehydrogenation (forming an olefin of the same chain length). The probability of dehydrogenation decreases with increasing molecular weight. Alkylation of paraffins with olefins may occur at high pressures and temperatures. Paraffin isomerization is unlikely.
2. *Olefins.* Polymerization. Under severe thermal conditions, dienes can be formed by dehydrogenation or decomposition of olefins. A secondary reaction between dienes and olefins may produce cyclic olefins which in turn make naphthenes.
3. *Naphthenes.* Dealkylation (splitting off paraffinic side chains) and dehydrogenation to aromatics. Rupture of the naphthenic ring may also occur under severe cracking.
4. *Aromatics.* Dealkylation and condensation. Dealkylation produces paraffins, olefins, and aromatics with short side chains. Condensation occurs between aromatics, or between aromatics and unsaturates, to form polynuclear aromatic hydrocarbons which may further condense to asphaltlike compounds.

Yield and Quality Correlations

Thermal Cracking

Process Flow

Although the thermal cracking process was originally designed primarily for gas oil charge, most of the existing thermal cracking units now operate on a topped or reduced crude charge. To prevent excessive coking in the heater due to the residuum portion of the charge, fractionation of the charge into light and heavy oil fractions and charging these fractions through separate heaters is customary. This permits operating the heavy oil (contains residuum portion of the charge) heater at a lower temperature than the light oil heater to minimize coke formation. A typical flow diagram for this operation showing the major vessels and process streams is presented in Fig. 2.

The reduced crude charge is combined with a recycle gas oil stream and charged to the fractionator. The heaviest oil stream is drawn off the bottom of the fractionator and charged to the heavy oil heater. This heavy oil is heated to a nominal temperature range of 750 to 900°F, depending upon the nature of the charge, under a pressure range of 350 to 700 lb/in.^{2} gauge. The clean, fractionated light gas oil is taken as a side cut from the fractionator and routed to a separate heater where it is heated to a temperature range of 1000 to 1100°F, also under a pressure range of 350 to 700 lb/in.^{2} gauge. The light gas oil has a lower carbon residue content than the heavy oil and is less susceptible to coking and thus can be processed at a higher temperature. Although at these conditions the stream would be a mixed phase, thermal cracking is usually considered a "liquid" phase operation.

Both oil streams are partially cracked and upon leaving the heaters are recombined in the top of a reaction chamber (soaking drum) where the hotter light gas oil supplies additional heat to further crack the heavy oil stream. The total products are then removed from the bottom of the reaction chamber through a pressure-reducing valve into a flash tower.

All of the lighter cracked products and most of the remaining light gas oil are flashed overhead in the flash tower. To prevent further cracking, some of the light gas oil heater charge is injected into the top of the flash tower to quench the overhead vapors prior to routing the overhead stream to the fractionator. The residuum stream is drawn off the bottom of the flash tower through another pressure-reducing valve to a residuum flash tower where a heavier gas oil is flashed overhead and routed to the fractionator. The heavy residuum is drawn off the residuum flash tower as a fuel oil.

All of the cracked products, unconverted gas oil, and reduced crude charge are charged to the fractionator. The thermal cracked gasoline and gases are fractionated overhead and are routed to a gas–liquid separator. A portion of the liquid gasoline is returned to the fractionator as a reflux stream. The remaining gasoline is routed to a stabilizer (not shown) where the propane and lighter gases are removed from the gasoline. The final gasoline product is defined as a total butane retention gasoline.

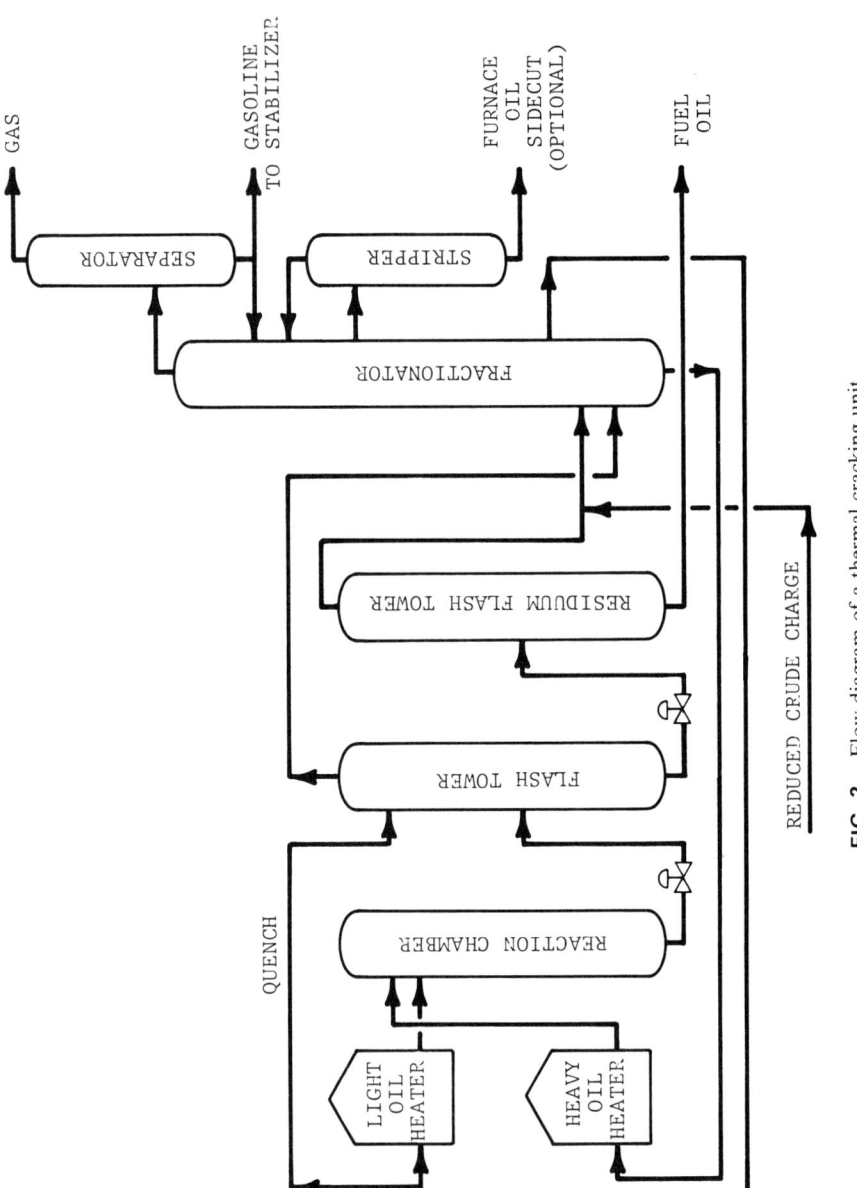

FIG. 2. Flow diagram of a thermal cracking unit.

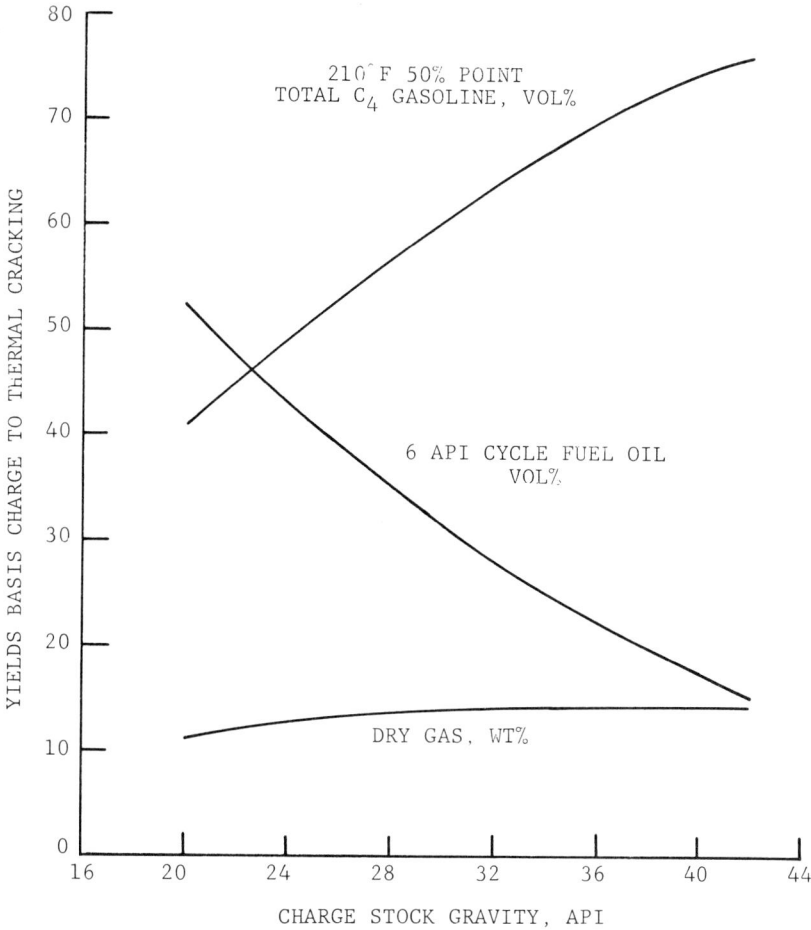

FIG. 3. Dry gas, gasoline, and cycle fuel oil yields from thermal cracking virgin stocks (gas oils, topped crudes, etc.).

The gases from the gas–liquid separator may be routed to an absorber (not shown) where a light gas oil stream (from the fractionator) is used to absorb butanes and pentanes that are carried over from the separator. The absorption gas oil containing these butanes and pentanes can be returned either to the fractionator or used as additional quench to the top of the flash tower. The butanes and pentanes are eventually removed in the total butane retention gasoline. The unabsorbed gases (propane and lighter) are combined with the stabilizer off gases and routed to subsequent gas-handling facilities.

Figure 2 also shows an optional furnace oil side cut which can be produced, if required, during times of peak demand for heating oil. This side cut effectively comes from the light gas oil stream which would have been converted to gas, gasoline, and heavy fuel oil in the thermal cracking unit.

Yield Correlations

Figures 3 through 14 present a set of simplified correlations for estimating the product yields and qualities from thermal cracking various charge stocks. These generalized correlations group the thermal cracking charge stocks into three types: virgin stocks, catalytic cracked gas oils, and thermal cracked gas oils.

Figures 3, 4, and 5 are the bases of the correlations which allow the estimation of the yields of cycle fuel oil (6° API gravity), total butane retention gasoline (210°F ASTM distillation 50% evaporated point), and dry gas (propane and lighter gases) knowing only the type and API gravity of the charge. Figure 3 is for virgin charge stocks (gas oils, topped crudes, full range crudes, kerosines, and residua). Figure 4 is for catalytic cracked cycle gas oil

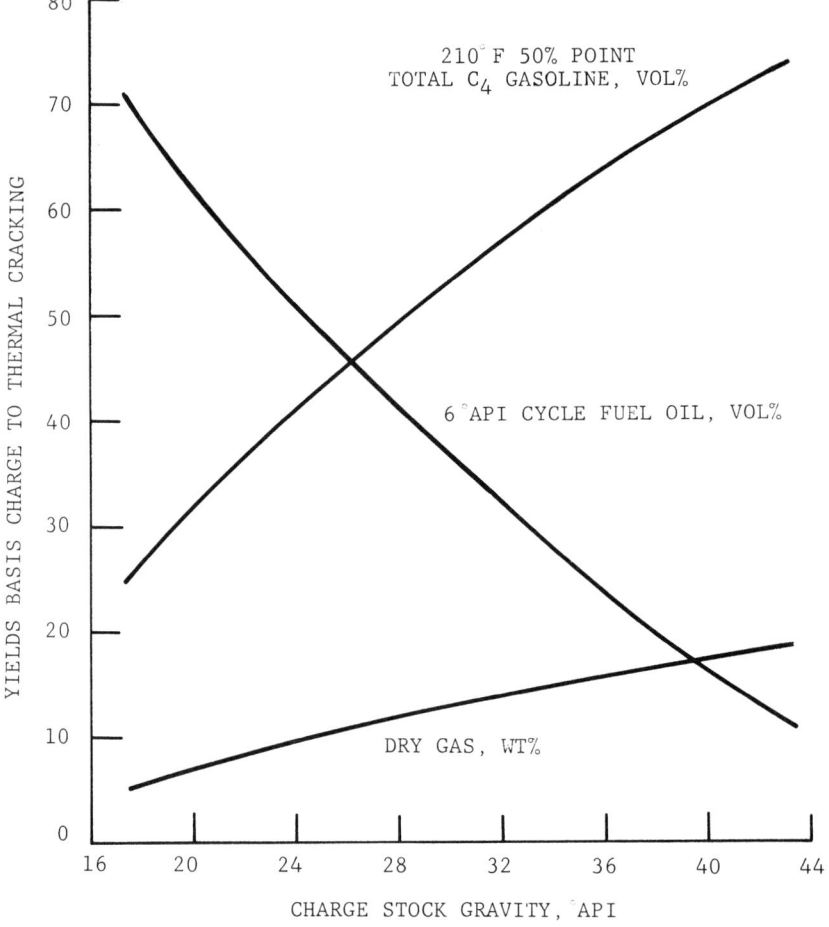

FIG. 4. Dry gas, gasoline, and cycle fuel oil yields from thermal cracking catalytic cracked cycle gas oils.

charge stocks, and Fig. 5 is for thermal cracked gas oil charge stocks which may be produced from a single-pass thermal cracking operation or one in which a furnace oil side cut is being made. These figures cannot be used to predict yields for single-pass thermal cracking operations as there is no net production of a cycle gas oil stream in the operation covered by the correlations. All gas oil is recycled to extinction, producing only gasoline, gas, and cycle fuel oil.

Since the cycle fuel oil and gasoline to be produced are not always a 6° API gravity cycle fuel oil and a 210°F 50% point gasoline, Figs. 6. through 10 are used to adjust the yields obtained from Figs. 3, 4, or 5 to the desired cycle fuel oil gravity and gasoline 50% point.

Knowing the desired ASTM distillation 50% point of the product gasoline, Fig. 6 may be used to determine the API gravity of the gasoline. Two lines are presented—one for thermal cracking of virgin charge stocks and one for thermal cracking of cracked (catalytic or thermal) charge stocks.

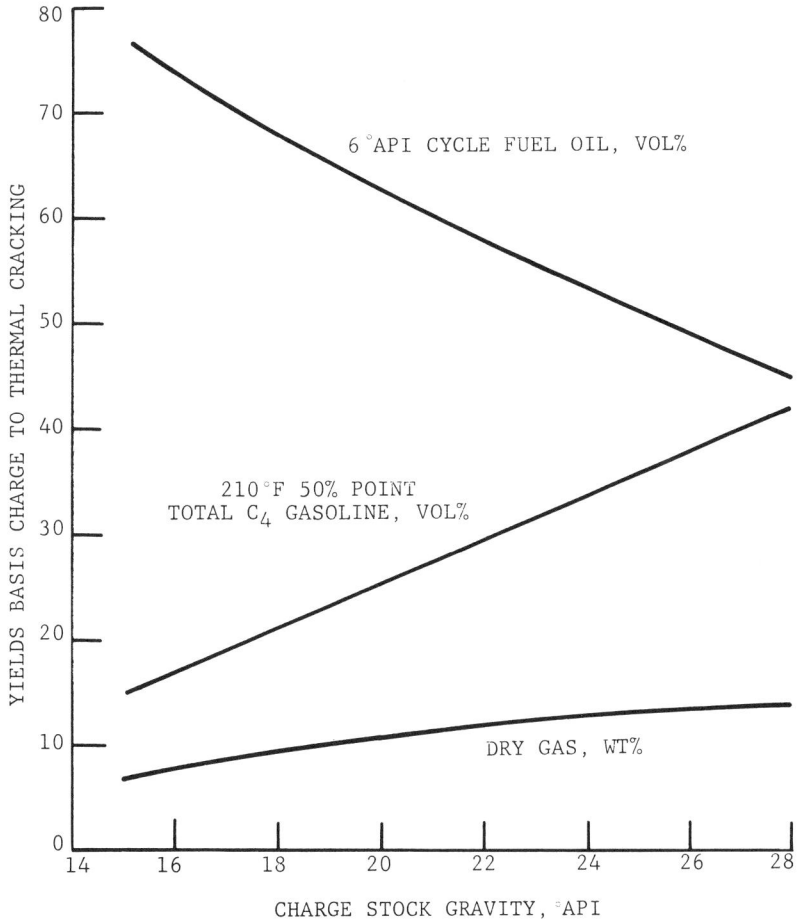

FIG. 5. Dry gas, gasoline, and cycle fuel oil yields from thermal cracking thermal cracked gas oils.

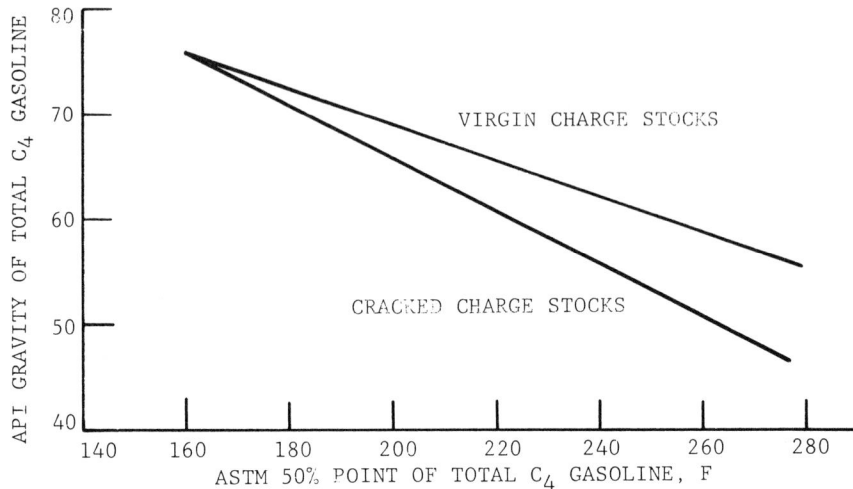

FIG. 6. API gravity vs ASTM 50% point total butane retention gasoline.

Figure 7 presents a plot for correcting the base gasoline yields (Figs. 3, 4, or 5) from the base 210°F ASTM 50% point to the desired 50% point of the total butane retention gasoline. The ratio of the gasoline yield with the desired 50% point to the gasoline yield with a 210°F 50% point is plotted against the desired gasoline 50% point. Figure 7 also presents individual correction curves for dry gas and cycle fuel oil yields as a function of the desired 50% point of the gasoline.

The gasoline 50% point is normally varied in actual operations by undercutting and changing the severity of the cracking operation. Increasing the gasoline 50% point increases the gasoline yield and results in a corresponding overall decrease in dry gas and cycle fuel oil yields. By increasing the gasoline 50% point and also maintaining cycle fuel oil viscosity, there is less recycle gas oil to be cracked to dry gas and less condensation to heavy fuel oil. Since there is less gas oil to be recycled, the throughput to the heater is reduced and the resulting increased residence time would normally produce more cracking (more dry gas make and increased cycle fuel oil viscosity) unless the operating severity is lowered. This can be accomplished by various means such as lowering the heater temperature or injecting steam into the heater coils to reduce residence time.

The viscosity of marketable cycle fuel oil may vary through a considerable range depending upon the ultimate use. For consistency in the application of the correlations being presented, a cycle fuel oil viscosity of 100 Furol at 122°F has been chosen as being representative. The minimum cycle fuel oil API gravity when producing a 100 Furol viscosity at 122°F cycle fuel oil can be estimated from the charge stock characterization factor. The characterization factor, an index used to classify petroleum stocks, can be defined as

$$K = \sqrt[3]{T_B}/S$$

Cracking, Thermal

where T_B is the molal average boiling point in degrees Rankine and S is the specific gravity at 60°F. The characterization factor can be related to viscosity, aniline point, molecular weight, critical temperature, and other tests. A method using API gravity and Saybolt Universal viscosity at 100°F to obtain the thermal cracking charge stock characterization factor is presented in Fig. 8.

An estimation of the minimum cycle fuel oil API gravity is presented in Fig. 9 for a marketable cycle fuel oil having a 100 Furol viscosity at 122°F when thermal cracking a charge stock of known API gravity and characterization factor. Although production of even lower API gravity cycle fuel oil is possible, the Furol viscosity will be greater than 100 at 122°F. This higher viscosity may not be desirable since additional low viscosity cutter stock would be required to make a marketable fuel oil. The lowest cycle fuel oil API gravity (assuming no viscosity requirements) which can be produced from a given charge stock is dependent upon equipment design and is usually limited by excessive coke formation in the furnace, reaction vessels, or fractionating towers.

After the limiting cycle fuel oil API gravity has been established from Fig. 9, the yields of gasoline, dry gas, and cycle fuel oil must be corrected for the effect

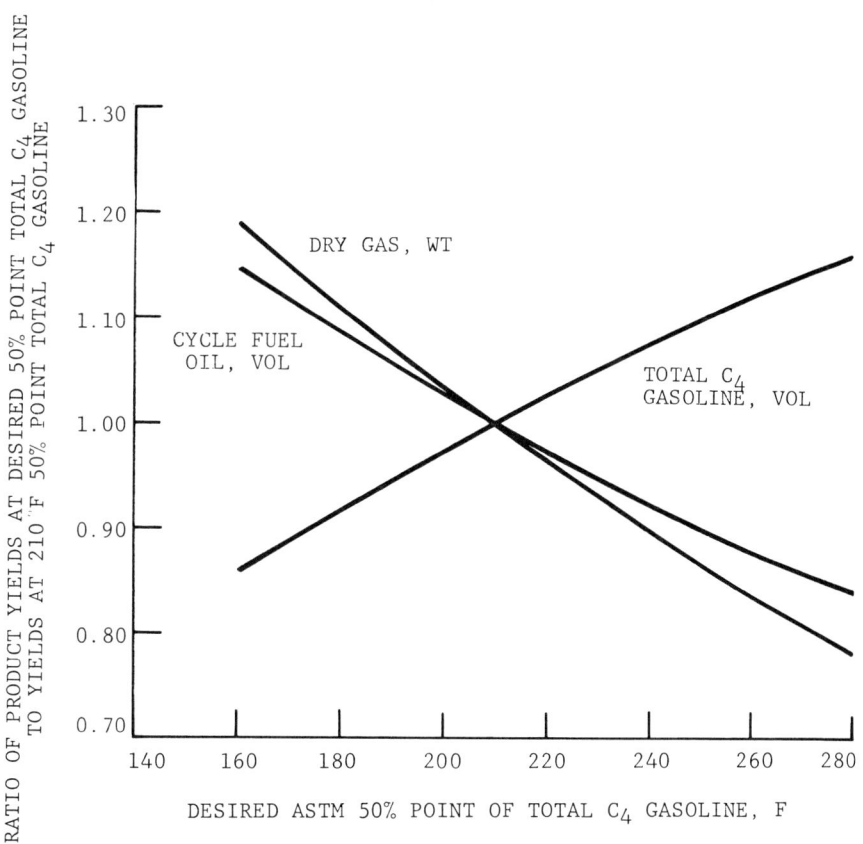

FIG. 7. Correction of yields to desired 50% point of total butane retention gasoline.

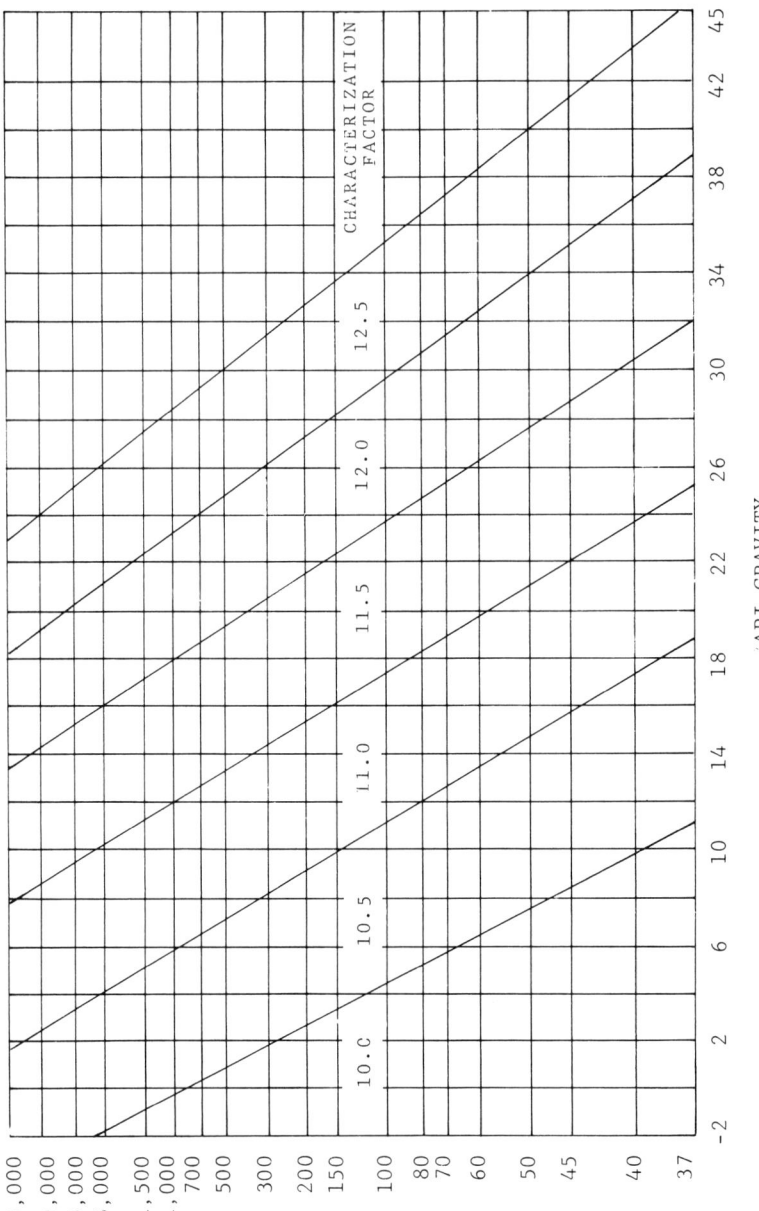

FIG. 8. Estimation of charge stock characterization factor from API gravity and Saybolt Universal viscosity at 100°F.

of changing the cycle fuel oil gravity from 6° API to the limiting gravity. Figure 10 presents the ratios of gasoline, dry gas, and cycle fuel oil yields at the desired cycle fuel oil gravity to the corresponding yields at 6° API cycle fuel oil gravity plotted against the desired fuel oil gravity. Individual correction curves are presented for gasoline, dry gas, and cycle fuel oil.

The following example problem shows how to use Figs. 3 through 10.

Example Problem 1. *Basis*: A 28° API Mid-Continent virgin thermal cracking charge stock has a viscosity of 100 SUS at 100°F. Predict yields of gasoline, dry gas, and cycle fuel oil when producing a 223°F 50% point gasoline at minimum marketable cycle fuel oil API gravity.

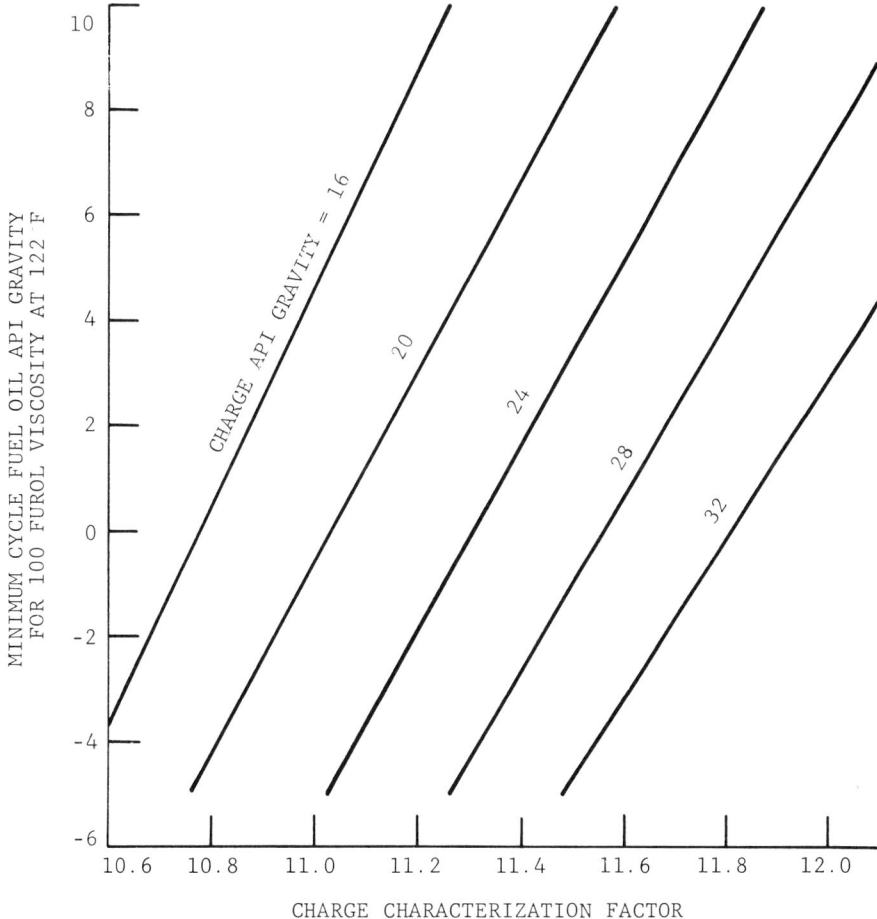

FIG. 9. Estimation of minimum cycle fuel oil API gravity when producing a cycle fuel oil of 100 Furol viscosity at 122°F.

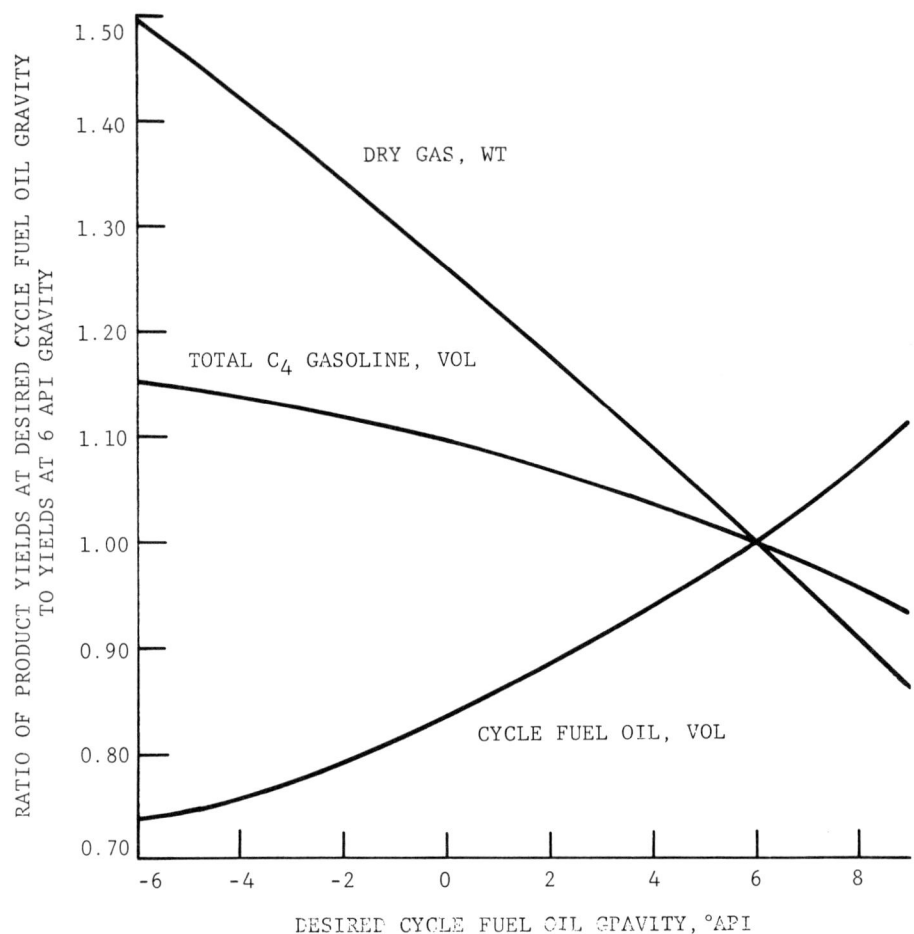

FIG. 10. Correction of yields to desired cycle fuel oil API gravity.

Solution: Since the charge stock is a virgin material, use Fig. 3 to predict thermal cracking yields.*

Dry gas (propane and lighter), wt.%	13.5
Gasoline (210°F ASTM 50% point), vol.%	56.7
Cycle fuel oil (6° API gravity), vol.%	35.1

From Fig. 6, a 223°F 50% point, total butane retention gasoline will have a 65° API gravity.

Use Fig. 7 to adjust yields for change in gasoline 50% point from 210 to 223°F.

*Note that dry gas yield is given in weight percent and gasoline and cycle fuel oil yields are in volume percent.

Cracking, Thermal

Dry gas	$= 0.955(13.5) = 12.9$ wt.%
Gasoline (223°F 50% point)	$= 1.033(56.7) = 58.6$ vol.%
Cycle fuel oil (6°API)	$= 0.966(35.1) = 33.9$ vol.%

From Fig. 8, the characterization factor for a 28°API gravity, 100 SUS viscosity (100°F) charge stock is 11.86.

Use Fig. 9 to determine the minimum cycle fuel oil API gravity when producing a salable cycle fuel oil of 100 Furol viscosity (122°F).

Minimum gravity of cycle fuel oil at 100 Furol viscosity (122°F) = 5.1°API.

Use Fig. 10 to correct all yields for the effect of varying the cycle fuel oil gravity from 6 to 5.1°API.

Dry gas	$= (1.040)(12.9) = 13.4$ wt.%
Gasoline (223°F 50% point)	$= (1.016)(58.6) = 59.5$ vol.%
Cycle fuel oil (5.1°API)	$= (0.972)(33.9) = 32.9$ vol.%

It will usually be found that after a series of adjustments for cycle fuel oil gravity and gasoline 50% point, the weight recovery will drift off from 100%. It is assumed that any error can be divided proportionally among the various yields, and the deviation from 100% weight recovery is prorated accordingly among the yields. An example proration is shown in Table 1.

The uncorrected weight percent yields of gasoline and cycle fuel oil were calculated by multiplying the volume percent uncorrected yield by the ratio of product specific gravity to charge specific gravity. The corrected volume percent yields were obtained by the reverse procedure after adjusting the weight percent yields to 100%.

After the corrected yields of dry gas, total butane retention gasoline, and cycle fuel oil have been determined, the light hydrocarbon breakdown is predicted using the values shown in Table 2.

The following example problem illustrates the use of Table 2.

Example Problem 2. *Basis*: Using yields calculated in Example Problem 1, determine the light hydrocarbon composition up to and including the total pentanes fraction. Also determine the yield and API gravity of the depentanized gasoline.

Solution: The values given in Table 3 were carried out to two decimal places to aid in following the calculations. Weight percent yields were converted to volume percent by multiplying by the ratio of charge stock specific gravity (0.887) to the individual product specific gravities.

Depentanized (DP) gasoline yields are calculated by subtracting the butanes and pentanes yields from the yield of total butane retention gasoline.

DP gasoline, wt.%	$= 48.2 - 7.57 - 7.64$	$= 33.0$
DP gasoline, vol.%	$= 59.4 - 11.43 - 10.60$	$= 37.4$

The specific gravity of the DP gasoline is calculated by multiplying the charge stock specific gravity by the ratio of weight percent DP gasoline yield to volume percent DP gasoline yield.

TABLE 1

	API Gravity	Specific Gravity	Uncorrected		Corrected	
			Vol.%	Wt.%	Vol.%	Wt.%
Dry gas (propane and lighter)	—	—	—	13.4	—	13.4
TC$_4$ gasoline (223°F 50% point)	65.0	0.720	59.5	48.3	59.4	48.2
Cycle fuel oil (100 Furol viscosity at 122°F)	5.1	1.036	32.9	38.4	32.9	38.4
Charge stock	28.0	0.887	—	100.1	—	100.0

Specific gravity, DP gasoline = $0.887 \times 33.0/37.4 = 0.783$, which corresponds to 49.2° API.

Product Octane Numbers and Sulfur Correlations

Correlations are presented in Figs. 11 through 14 relating ASTM Research and Motor octane numbers of the total butane retention gasoline with type of

TABLE 2 Distribution of Light Hydrocarbon Yields from Thermal Cracking

Component	Yield (wt.%)		Specific Gravity
	Basis Charge	Basis Fraction	
Dry gas fraction	From correlations	100.0	—
Hydrogen	—	0.4	—
Methane	—	23.2	—
Ethylene	—	1.9	—
Ethane	—	27.5	—
Propylene	—	12.2	0.522
Propane	—	34.8	0.508
Total C$_4$ fraction	0.565[a] or 0.50[b] × wt.% dry gas	100.0	0.588
Isobutane	—	15.0	0.563
n-Butane	—	45.0	0.584
C$_4$ olefins	—	40.0	0.601
Total C$_5$ fraction	0.57[a] or 0.45[b] × wt.% dry gas	100.0	0.639
Isopentane	—	20.0	0.625
n-Pentane	—	40.0	0.631
C$_5$ olefins	—	40.0	0.655

[a]Thermal cracking of virgin stocks.
[b]Thermal cracking of cracked stocks.

TABLE 3

						Wt.%	Vol.%
Dry gas						13.4	—
Hydrogen	=	13.4	×	0.4/100	=	0.05	—
Methane	=	13.4	×	23.2/100	=	3.11	—
Ethylene	=	13.4	×	1.9/100	=	0.26	—
Ethane	=	13.4	×	27.5/100	=	3.69	—
Propylene	=	13.4	×	12.2/100	=	1.63	2.77
Propane	=	13.4	×	34.8/100	=	4.66	8.14
Total C_4 fraction	=	0.565	×	13.4	=	7.57	11.43
Isobutane	=	7.57	×	15/100	=	1.13	1.78
n-Butane	=	7.57	×	45/100	=	3.41	5.18
C_4 olefins	=	7.57	×	40/100	=	3.03	4.47
Total C_5 fraction	=	0.57	×	13.4	=	7.64	10.60
Isopentane	=	7.64	×	20/100	=	1.53	2.17
n-Pentane	=	7.64	×	40/100	=	3.06	4.30
C_5 olefins	=	7.64	×	40/100	=	3.06	4.13

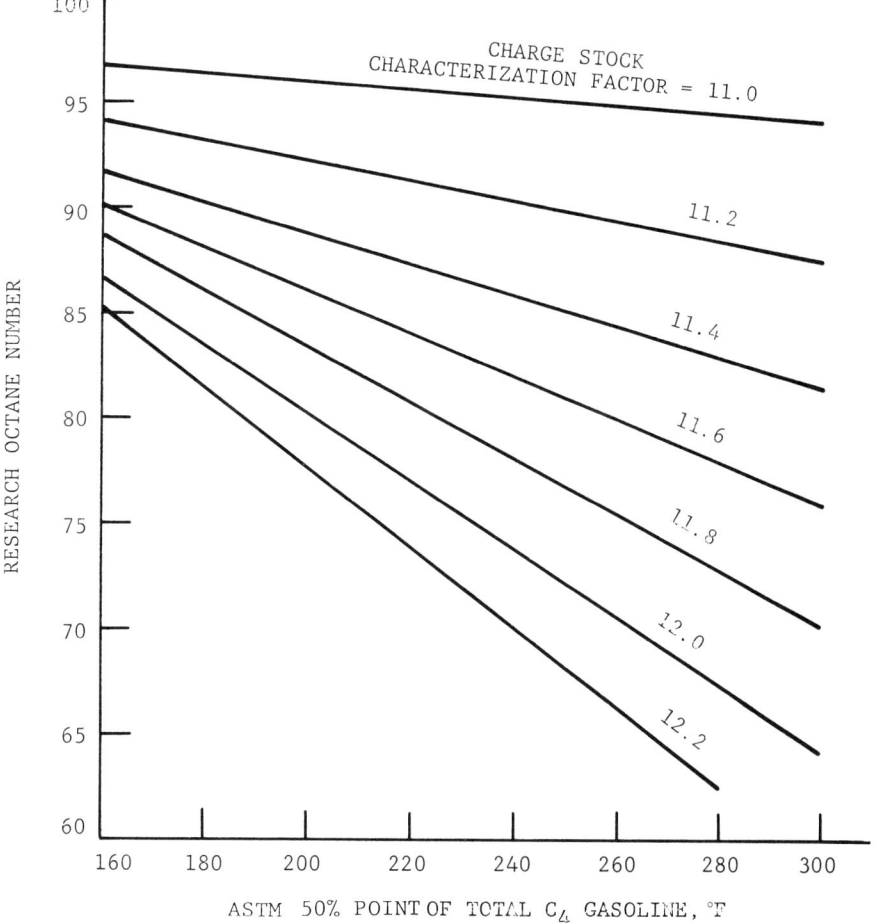

FIG. 11. Prediction of Research clear octane number of total butane retention gasoline produced from thermal cracking of virgin charge stocks.

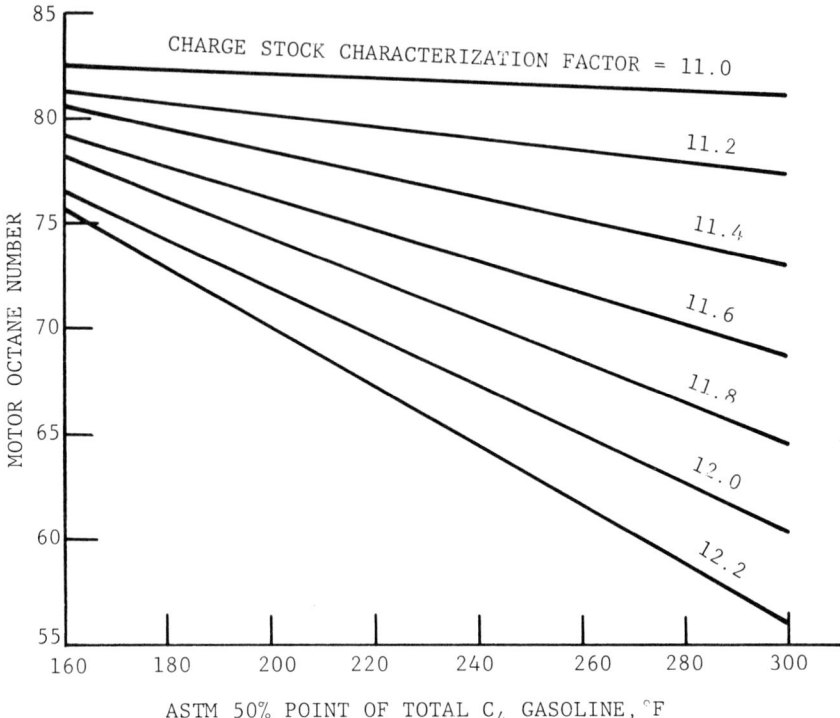

FIG. 12. Prediction of Motor clear octane number of total butane retention gasoline produced from thermal cracking of virgin charge stocks.

charge stock, charge stock characterization factor, and ASTM 50% point of the gasoline. Figures 11 and 12 are used when thermal cracking virgin charge stocks, and Figs. 13 and 14 are for cracked (catalytic or thermal) charge stocks. Use of Figs. 11 through 14 is illustrated in the following example problem.

Example Problem 3. *Basis*: Compare the Research and Motor octane numbers of the total butane retention gasoline (223°F 50% point) produced by thermal cracking each of the following stocks:

1. A virgin gas oil of 11.86 characterization factor
2. A virgin gas oil of 11.20 characterization factor
3. A catalytic cracked cycle gas oil of 11.20 characterization factor

Solution:

1. Virgin gas oil, $K = 11.86$
   Research octane number = 79.0 (Fig. 11)
   Motor octane number = 71.0 (Fig. 12)

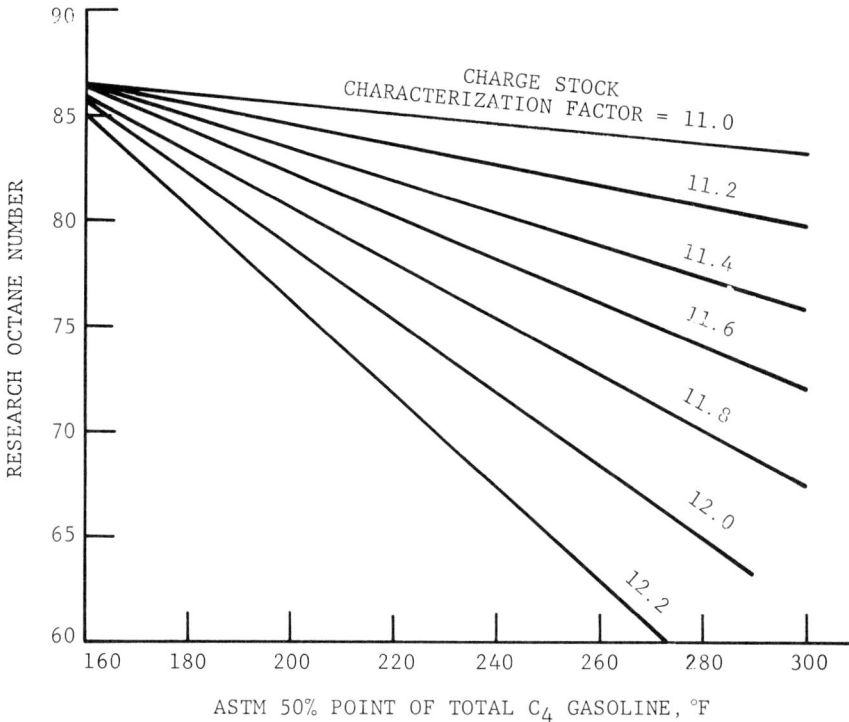

FIG. 13. Prediction of Research clear octane number of total butane retention gasoline produced from thermal cracking of cracked (catalytic and thermal) charge stocks.

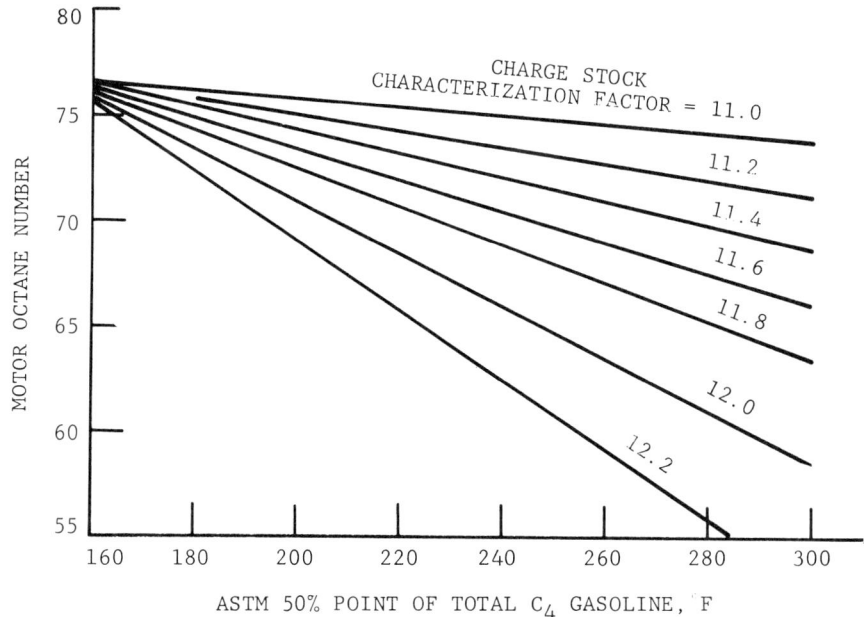

FIG. 14. Prediction of Motor clear octane number of total butane retention gasoline produced from thermal cracking of cracked (catalytic and thermal) charge stocks.

2. Virgin gas oil, $K = 11.20$
      Research octane number $= 91.5$ (Fig. 11)
      Motor octane number $= 79.5$ (Fig. 12)
3. Catalytic cracked cycle gas oil, $K = 11.20$
      Research octane number $= 83.5$ (Fig. 13)
      Motor octane number $= 74.0$ (Fig. 14)

Table 4 can be used to estimate the distribution of the sulfur in the cracked products. Note that the distribution of the sulfur in the products from thermal cracking virgin charge stocks is different than that from charging cracked charge stocks. When thermal cracking cracked stocks, more of the sulfur appears in the cycle fuel oil than when thermal cracking virgin charge stocks. This is because the easily decomposed sulfur compounds have already been removed from the cracked stocks in the prior cracking operation. The more refractory sulfur compounds are more difficult to crack and therefore appear in the cycle fuel oil stream.

Sulfur compounds crack similarly to the hydrocarbons of the charge stock, forming hydrogen sulfide, mercaptans, sulfides, disulfides, thiophenes, and asphaltic-type compounds. Hydrogen sulfide appears in the dry gas stream. Mercaptans, sulfides, disulfides, and thiophenes are found in the thermal cracked gasoline. The heavy, asphaltic sulfur compounds are in the cycle fuel oil.

Example Problem 4. *Basis*: A 28°API virgin gas oil containing 1.3 wt.% sulfur is thermal cracked to the following yields: total C_4 gasoline, 48.2 wt.%; cycle fuel oil, 38.4 wt.%; and dry gas, 13.4 wt.%. Determine the sulfur content of each product stream.

Solution: Set the problem up on the basis of 100 lb of thermal cracking charge stock or 1.3 lb of sulfur.

Weight sulfur in total C_4 gasoline	$= 1.3 \times 9/100$	$= 0.117$ lb
Weight sulfur in cycle fuel oil	$= 1.3 \times 55/100$	$= 0.715$ lb
Weight sulfur in dry gas	$= 1.3 \times 36/100$	$= 0.468$ lb
% sulfur in total C_4 gasoline	$= 0.117/48.2 \times 100$	$= 0.24$ wt.%
% sulfur in cycle fuel oil	$= 0.715/38.4 \times 100$	$= 1.86$ wt.%
% sulfur in dry gas	$= 0.468/13.4 \times 100$	$= 3.49$ wt.%

TABLE 4 Distribution of Charge Sulfur among Thermal Cracked Product Streams

	Wt.% of Charge Sulfur in Product Stream	
Charge	Virgin Stocks	Cracked Stocks
Dry gas	36	6
Total C_4 gasoline	9	5
Cycle fuel oil	55	89

Reaction Kinetics

During the initial and intermediate stages of thermal cracking, the cracking reaction can be described by a first-order rate equation:

$$K = \frac{1}{t} \ln \frac{100}{100 - X}$$

where K = first-order reaction velocity constant, $1/s$
t = residence time at thermal cracking conditions, s (based on charge liquid volume and cracking reaction section volume)
X = total butane retention gasoline production, vol.%

Figure 15 presents rate curves applicable to commercial thermal cracking operations for three types of charge stocks: heavy virgin gas oil (22° API), light virgin gas oil (36° API), and cracked charge stocks. Generally the heavier the virgin stock, the easier it is to crack and the higher the gasoline yield at a given period of time. In commercial operations a higher temperature is usually used when charging lighter stocks to reduce the longer time required.

The once-through conversion to gasoline at a fixed temperature increases with reaction time up to a certain point where competing reactions, polymerization and condensation, result in a decreasing gasoline yield. These reactions can be minimized and gasoline yield maximized in commercial operation by recycling the unconverted gas oil so that the conversion per pass is held below the point where the undesirable reactions occur. In commercial operations the gasoline yield per pass should be under 40% to minimize the above competing reactions. The fresh feed is cracked at conditions below those which would give maximum production in a single-pass process. However, gasoline yields are not lost since the maximum gasoline yields are higher for recycle operations compared to single-pass cracking. The effects of increasing conversion per pass are:

1. Lower ultimate gasoline yield
2. Higher gasoline octane number
3. Increased tendency to produce coke
4. Increased production of gas
5. Increased volatility of the gasoline

The undesirable condensation and polymerization reactions are also minimized by the withdrawal of the cycle fuel oil stream after each recycle pass. The asphaltlike, hydrogen deficient compounds are thus removed to minimize the coke formation in the cracking equipment.

The following example problem illustrates the use of Fig. 15

Example Problem 5. *Basis*: At 940°F reaction temperature, determine the reaction time required for cracking a 28° API virgin gas oil (Example Problem 1).

Solution:

$$K = \frac{1}{t} \ln \frac{100}{100 - X}$$

The total butane retention gasoline yield in Example Problem 1 was 59.4 vol.%. In a single pass normally 40% of this gasoline will be produced.

$$X = 59.4 \times 40/100 = 23.8 \text{ vol.}\%$$

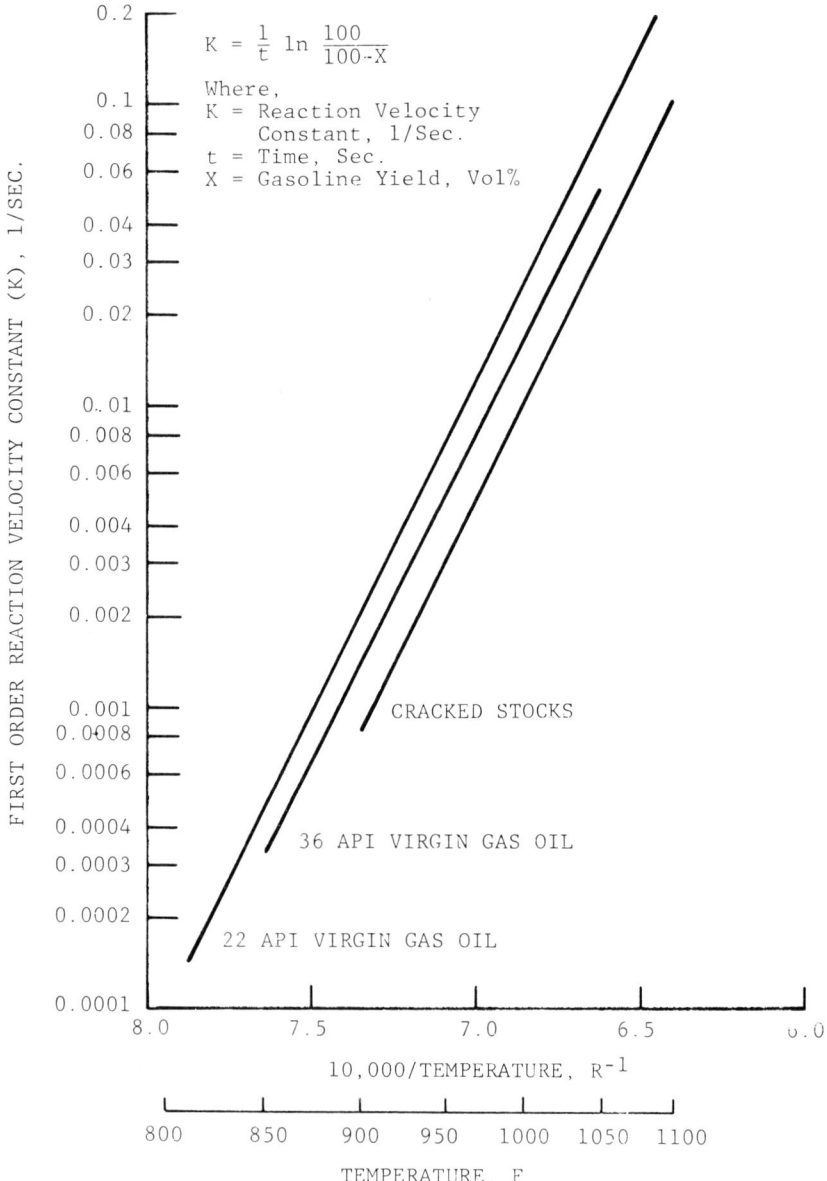

FIG. 15. First-order reaction velocity constants for thermal cracking.

From Fig. 15, at 940°F and interpolating between virgin gas oil lines at 28°API,

$$K = 0.0050$$
$$t = \frac{1}{K} \ln \frac{100}{100 - X}$$

$$= \frac{1}{0.0050} \ln \frac{100}{100 - 23.8}$$

$$= 54 \text{ s reaction time}$$

This calculated reaction time can then be used in designing a new unit to determine the volume of a reaction section required at a given charge rate and temperature to achieve the desired conversion. On an existing unit with a given reaction section volume, the reaction time can be used to calculate the maximum charge rate that can be used and still achieve the desired conversion. Later in this article the Soaking Volume Factor, which can also be used to determine reaction section volumes, will be discussed.

Visbreaking

In this section the primary purpose of visbreaking is assumed to be the maximum reduction of fuel oil production by lowering the viscosity of the residuum stream as much as possible (consistent with a stable final product) and thereby requiring less cutter stock than would be used with the nonvisbroken charge. The following discussion and correlations are based on this assumption of minimizing fuel oil production. Vacuum residuum is used as the visbreaking unit charge. Although longer residua (e.g., atmospheric residuum) can be charged, use of these residua would not result in minimum fuel oil production.

Process Flow

Figure 16 presents a simplified flow diagram of a typical visbreaking unit. The vacuum residuum from a crude still vacuum tower is charged to the visbreaking heater at relatively mild conditions of about 875°F reaction temperature (850 to 900°F range) and about 250 lb/in.² gauge operating pressure (200 to 500 lb/in.² gauge range). The visbroken products from the heater are immediately quenched to prevent undesirable cracking. The quench stream may be either an extraneous make-up gas oil or gas oils produced and recycled from the visbreaking operation. After quenching, the pressure of the product stream is reduced and the entire stream is charged to a fractionator. Dry gas, gasoline, and distillate gas oils are separated. The fractionator bottom stream is sent to a vacuum tower where additional heavy gas oils are removed from the visbroken fuel oil stream.

Recycling of the visbroken fuel oil stream is generally not practiced in commercial visbreaking operations. This stream has already experienced cracking and is therefore very susceptible to coking if high temperatures are

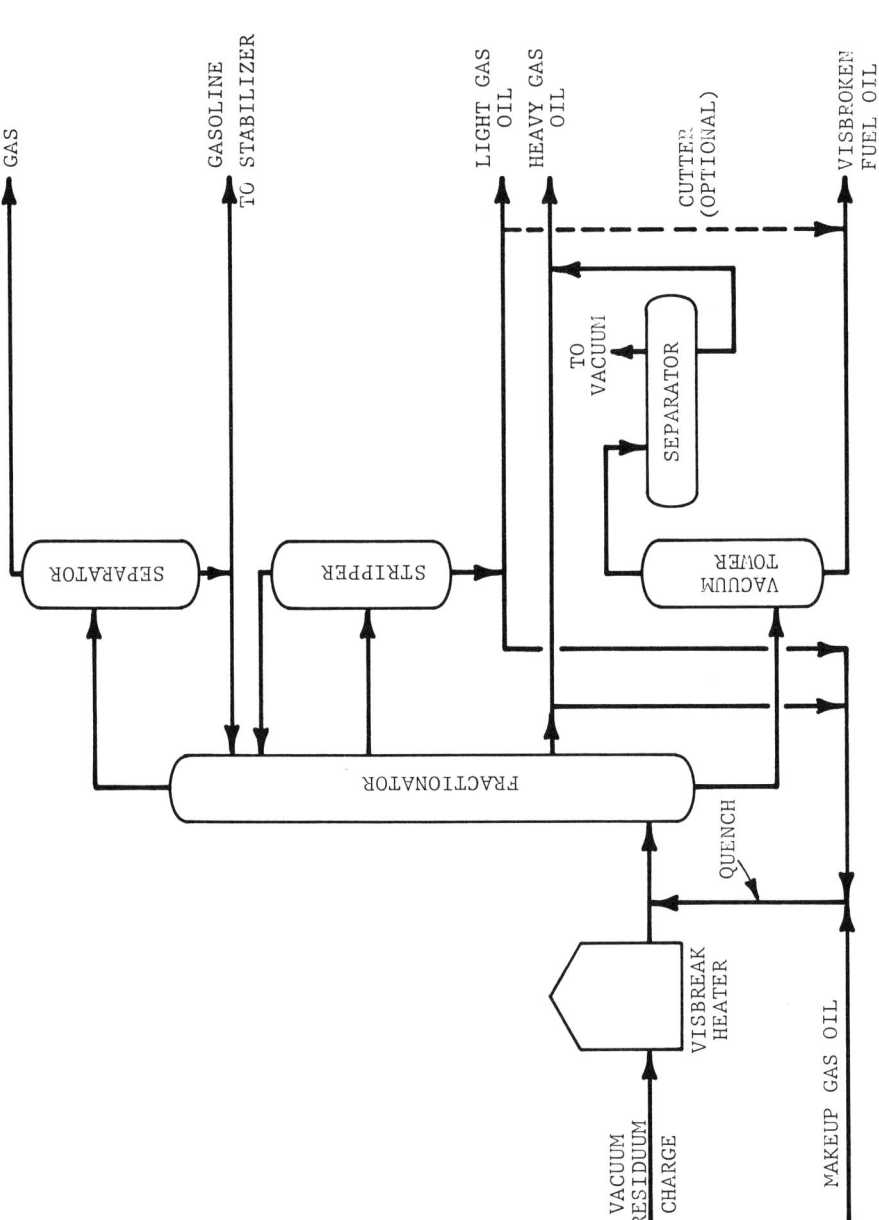

FIG. 16. Flow diagram of a visbreaking unit.

applied. The charge vacuum residuum should be free of any cracked materials. In the operation of vacuum crude towers some thermal cracking due to high skin temperatures in the reboiler is frequently possible. Elimination of the mild thermal cracking that could occur in a vacuum crude tower has been reported to increase the visbreaking unit run length from a 3-d run length to a 40-d run length.

The production of distillate gas oil streams is optional and would depend upon the economics of the individual refinery. In many instances the gas oils are left in the fuel oil as cutter, thus minimizing the requirement of cutter stocks from other sources. In some of the modern installations the heavy gas oil is recycled to extinction through a separate heater operating at more severe conditions than the visbreaking heater.

Visbreaking Severity Limits

The severity of the visbreaking operation is generally limited by the stability of the visbroken fuel oil produced. If overcracking occurs, the resulting fuel oil may form excessive deposits in storage or when used as a fuel in a furnace. One method of measuring the thermal stability of the fuel oil is the U.S. Navy special fuel oil thermal stability test [ASTM D 1661–64 (1973)]. The visbreaking correlations presented are based on operating to levels where the fuel oil quality will be limited by this test. This severity level is well within the operating limits that would be imposed by excessive coke formation in properly designed visbreaking furnaces.

The instability of the visbroken fuel oil is related to the asphaltenes present in the residuum. Asphaltenes are heavy nonvolatile compounds which can be classified according to their solubility in various solvents. The asphaltenes are difficult to thermally crack and thus are not altered in the visbreaking operations. For the purpose of visbreaking correlations, it is adequate to divide the charge and product residuum into two components: (1) asphaltenes and (2) heavy hydrocarbon oils. The asphaltenes are considered to be suspended in the hydrocarbon oil in a colloidal state. In visbreaking, some of the heavy hydrocarbons are converted to lower boiling components and removed from the residuum. The asphaltenes, being unchanged, are thus concentrated in the product residuum. If the visbreaking reaction proceeds too far, the asphaltenes tend to precipitate in the product fuel oil, creating an unstable fuel oil.

A common method of measuring the amount of asphaltenes in a petroleum stock is by extracting the hydrocarbon oil with normal pentane (ASTM D 893–69). The asphaltenes are insoluble and thus are referred to as normal pentane insolubles. Since the amount of asphaltenes in the visbreaking unit charge residuum will limit the severity of the visbreaking operations, the normal pentane insolubles content of the charge residuum is used as the correlating parameter in the following visbreaking correlations.

An estimate of the normal pentane insolubles content can be made the basis of other residuum tests. Figure 17 presents correlations with the ring-and-ball softening point test [ASTM E 28–67 (1972)] and the Conradson carbon residue

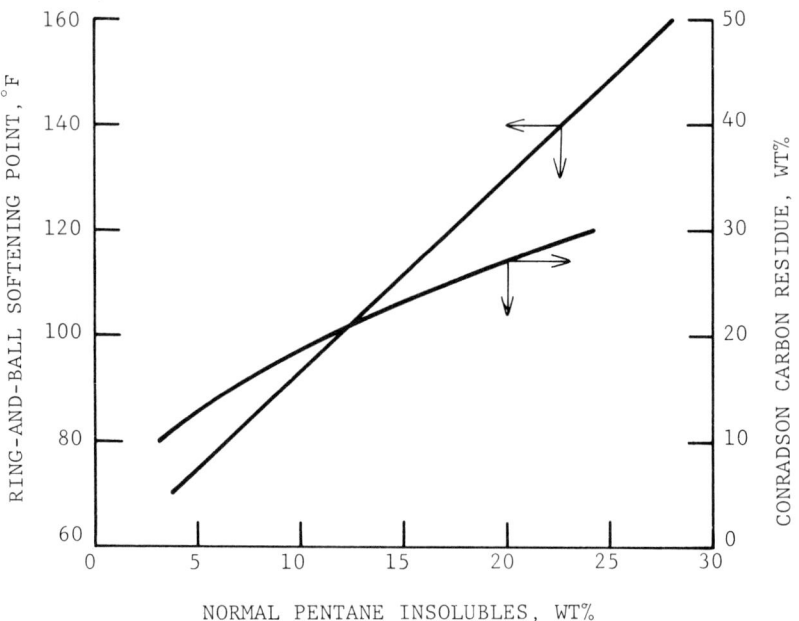

FIG. 17. Prediction of the normal pentane insolubles content of a vacuum residuum from the ring-and-ball softening point or Conradson carbon residue.

test [ASTM D 189–65 (1970)]. A correlation between the viscosity at 210°F and the normal pentane insolubles content of the vacuum residuum is shown in Fig. 18. If sufficient residuum tests are not available to permit an estimation of the normal pentane insolubles content, Table 5 can be used for estimation purposes.

Figure 18 may also be used to estimate the Saybolt Universal viscosity at 210°F of the 400°F+ visbroken product. By comparing the two curves in Fig. 18, the viscosity reduction obtained by visbreaking can be estimated.

TABLE 5 Typical Normal Pentane Insolubles Content of Vacuum Residua Prepared from Different Base Crudes

Crude Source of Vacuum Residuum	Normal Pentane Insolubles (wt.%)	
	Range	Suggested Value
Paraffinic	2–10	8
Mixed	10–20	13
Naphthenic	18–28	24

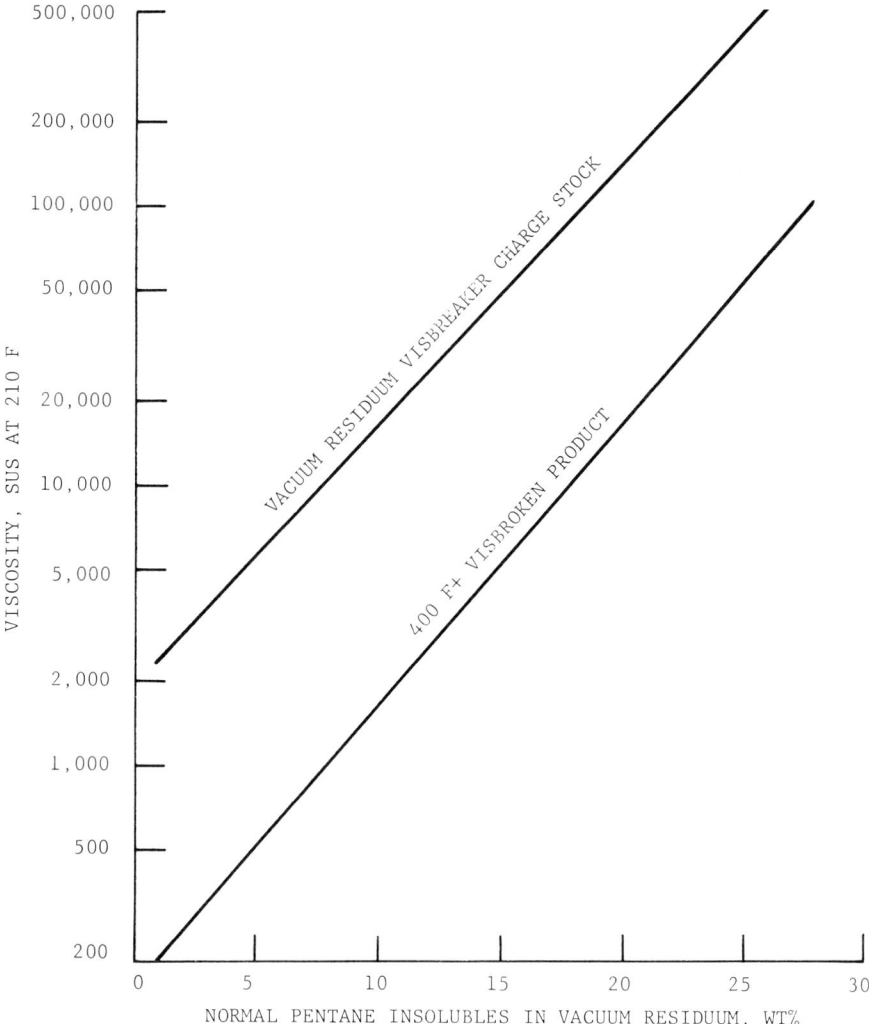

FIG. 18. Prediction of Saybolt Universal viscosity at 210°F of vacuum residuum and 400°F+ visbroken product.

Yield Correlations

By using the normal pentane insolubles content of the vacuum residuum charge, the yield of the various visbreaker products can be predicted from Figs. 19 through 27.

Figure 19 presents the volume percent yield of the 400°F+ visbroken residuum as a function of the normal pentane insolubles content. A further breakdown of the 400°F+ fraction is predicted using Figs. 20 and 21 which give the yields of 400 to 650°F light gas oil, 650 to 900°F heavy gas oil, and 900°F+ unconverted residuum. Since the visbreaker charge stock has an initial boiling

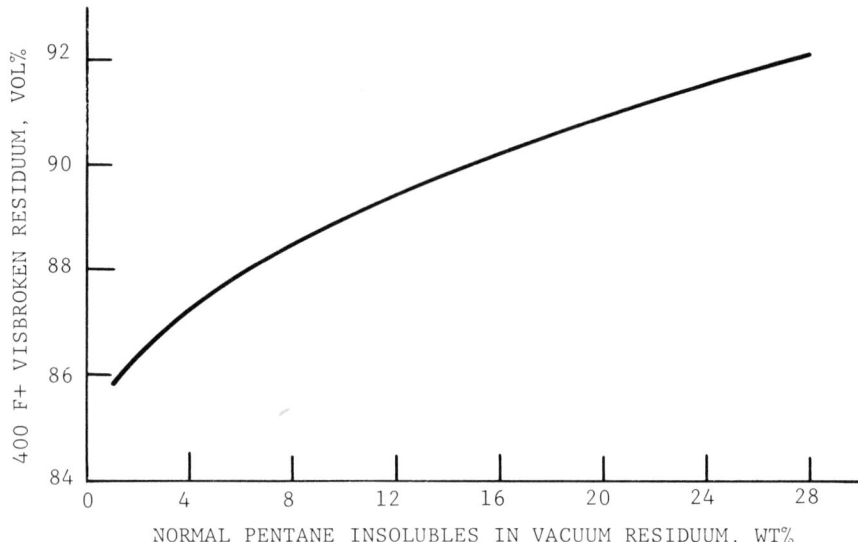

FIG. 19. Prediction of 400°F+ visbroken residuum yield from normal pentane insolubles content of visbreaker charge stock.

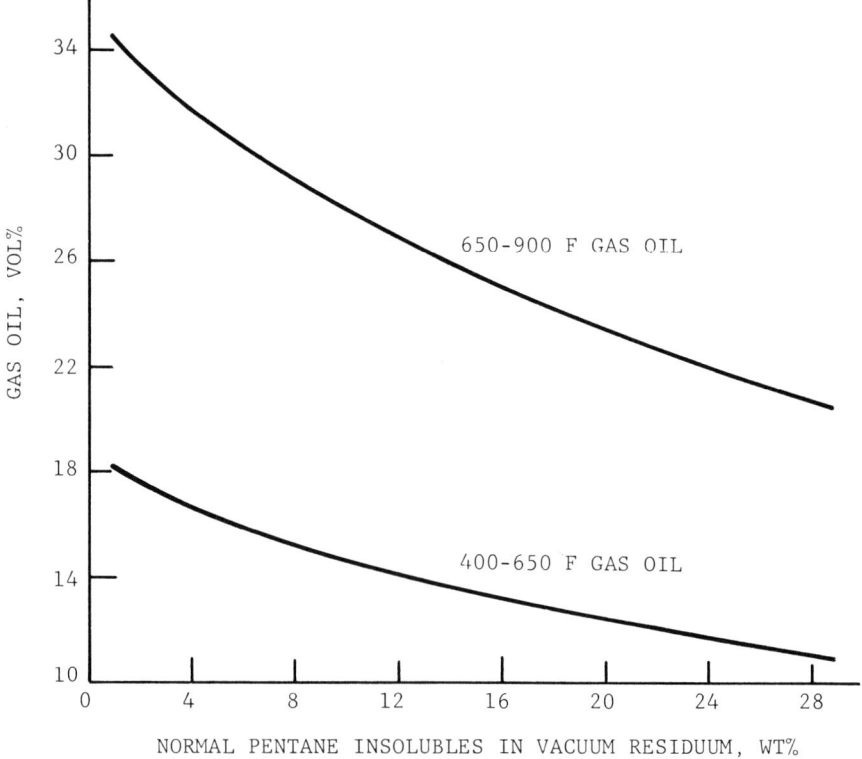

FIG. 20. Prediction of light gas oil and heavy gas oil yields from normal pentane insolubles content of visbreaker charge stock.

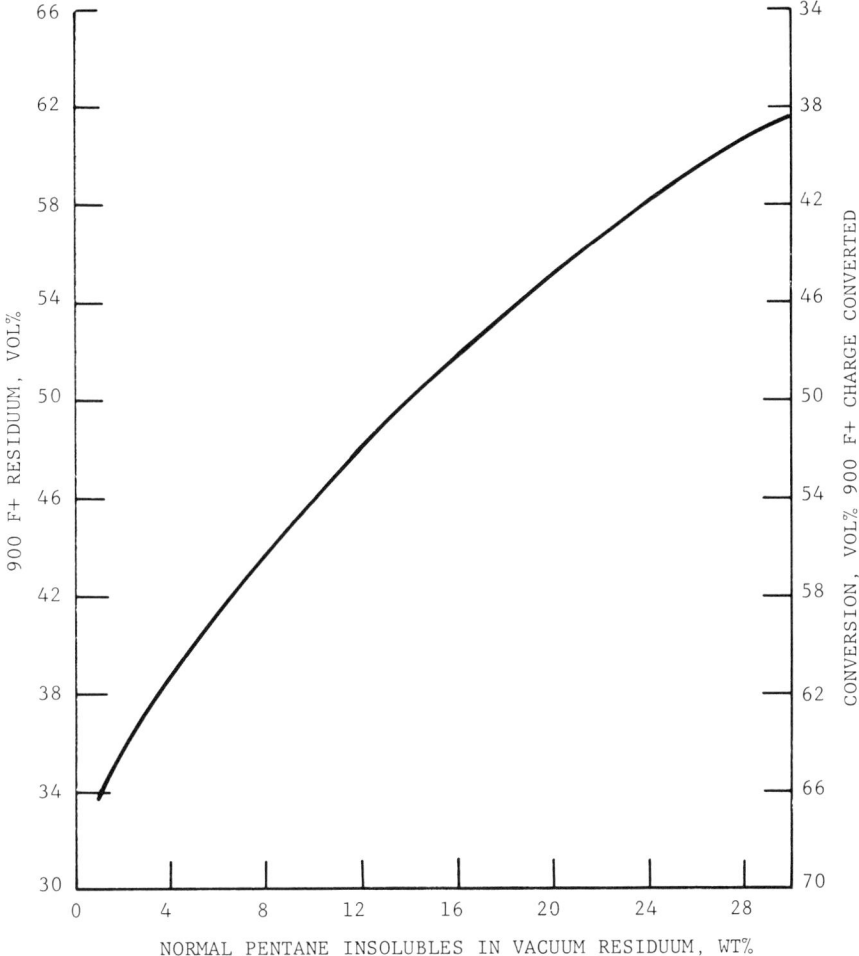

FIG. 21. Prediction of 900°F+ visbroken residuum yield from normal pentane insolubles content of visbreaker charge stock.

point of about 900°F, Fig. 21 also indicates the conversion of the original vacuum residuum into lower-boiling compounds (see right-hand scale).

Figure 22 shows the volume percent yield of total butane retention gasoline plotted against normal pentane insolubles content of the visbreaker charge. Total butane retention visbreaker gasoline octane numbers are relatively constant for most charge stocks. Suggested octane values, probably accurate within 3 octane numbers, are shown in Fig. 22. It may be desirable to cut the gasoline to a 300°F endpoint, leaving the 300°F+ visbroken products as the fuel oil stock. This results in a gain of 7 to 8 octane numbers for the total butane retention 300°F endpoint gasoline compared with the 400°F endpoint gasoline as well as reducing the viscosity of the visbroken product.

Total butanes and total pentanes yields basis visbreaker charge normal

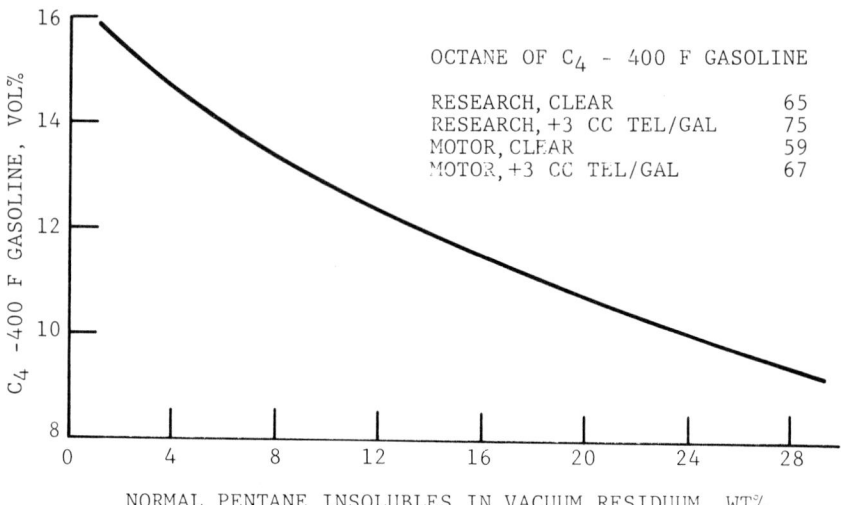

FIG. 22. Prediction of total butane retention visbreaker gasoline yield from normal pentane insolubles content of visbreaker charge stock.

pentane insolubles content can be predicted from Figs. 23 and 24, respectively. The compositions of the total butanes and total pentanes have been observed to be similar to the butanes and pentanes breakdown shown in the thermal cracking yield correlations. This is not unexpected since visbreaking is simply a mild, single-pass thermal cracking process. Likewise, after estimating the weight percent yield of dry gas (propane and lighter) from Fig. 25, the composition of the dry gas fraction can be estimated using the percentages shown in the thermal cracking correlations. For the reader's convenience, these composition percentages have been included in Figs. 23, 24, and 25.

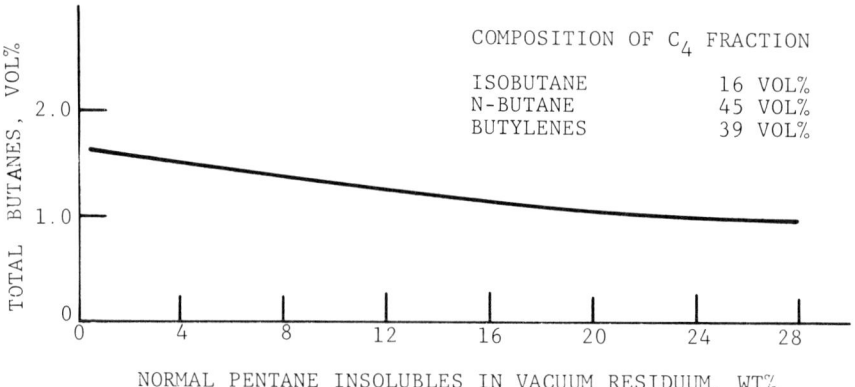

FIG. 23. Prediction of total butanes yield from normal pentane insolubles content of visbreaker charge stock.

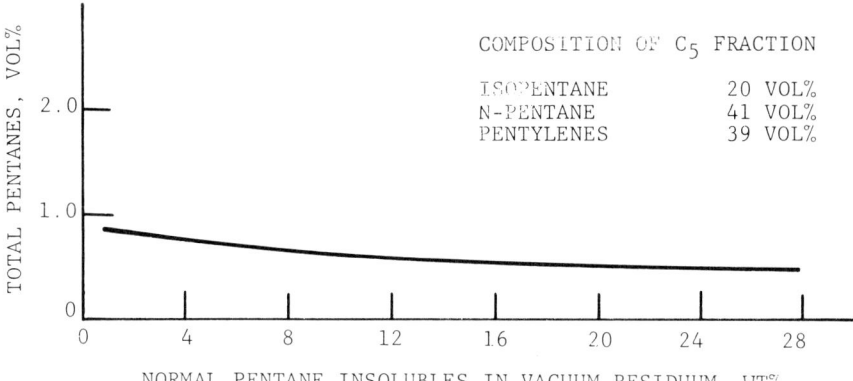

FIG. 24. Prediction of total pentanes yield from normal pentane insolubles content of visbreaker charge stock.

Figure 26 shows the quantities of finished fuel oil with a Furol viscosity of 190 s at 122°F that can be prepared from (1) 100 bbl of vacuum residuum visbreaker charge stock, (2) the 400°F+ visbroken product produced from 100 bbl of visbreaker charge stock, and (3) the 900°F+ visbroken residuum produced from 100 bbl of visbreaker charge stock. A 25 to 40 bbl reduction in finished fuel oil production can be achieved by visbreaking the residuum and blending the 400°F+ product with light cutter stock. A further reduction of 15 to 25 bbl in fuel oil production can be obtained when the gas oil boiling below 900°F is removed from the visbroken product. The finished fuel oil blends can be prepared using a light catalytic cycle gas oil cutter (36 Saybolt Universal

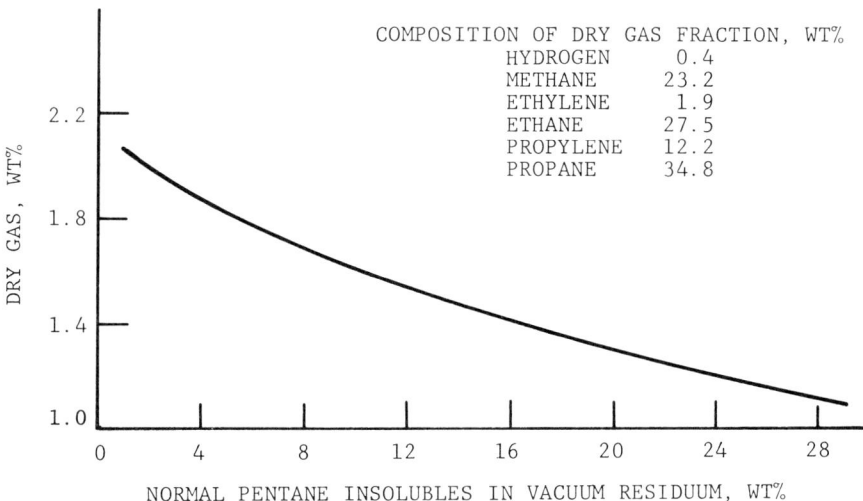

FIG. 25. Prediction of dry gas yield from normal pentane insolubles content of visbreaker charge stock.

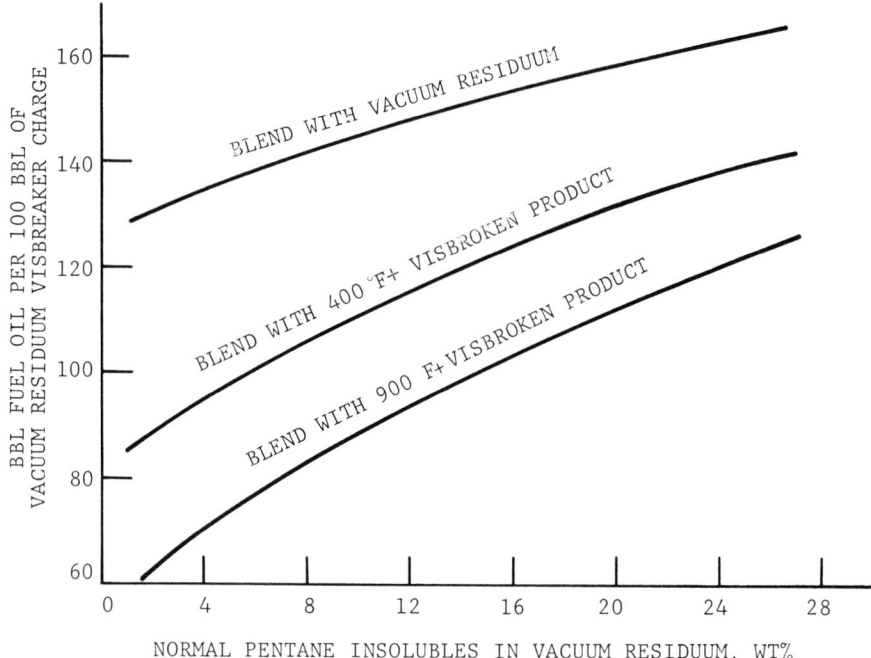

FIG. 26. Quantity of 190 Furol viscosity at 122°F fuel oil prepared from (1) vacuum residuum, (2) 400°F+ visbroken product, and (3) 900°F+ visbroken product.

viscosity at 100°F) to give a fuel oil blend with a Furol viscosity of 190 at 122°F. This fuel oil blend meets the viscosity requirement for No. 6 fuel oil.

Figure 27 presents the cutter stock requirements for (1) 100 bbl of vacuum residuum visbreaker charge, (2) the 400°F+ visbroken product from 100 bbl of visbreaker charge, and (3) the 900°F visbroken residuum from 100 bbl of visbreaker charge. Figure 27 shows that substantial savings in cutter stock are achieved when visbreaking a vacuum residuum and leaving the 400 to 900°F gas oil with the visbroken residuum, but this does not result in minimum fuel oil production. Blending the 900°F+ visbroken residuum requires about the same amount of cutter stock as the original vacuum residuum and results in a 25 to 50% overall reduction of fuel oil. This compares to an overall fuel oil reduction of 15 to 30% when leaving the 400 to 900°F gas oil in the fuel oil.

Sulfur Distribution

Sulfur distribution among the visbroken products is difficult to predict. However, for estimation purposes the following values may be used:

1. Dry gas (propane and lighter)—About 4 wt.% of the sulfur in the vacuum residuum visbreaker charge stock will appear as H_2S in the dry gas.

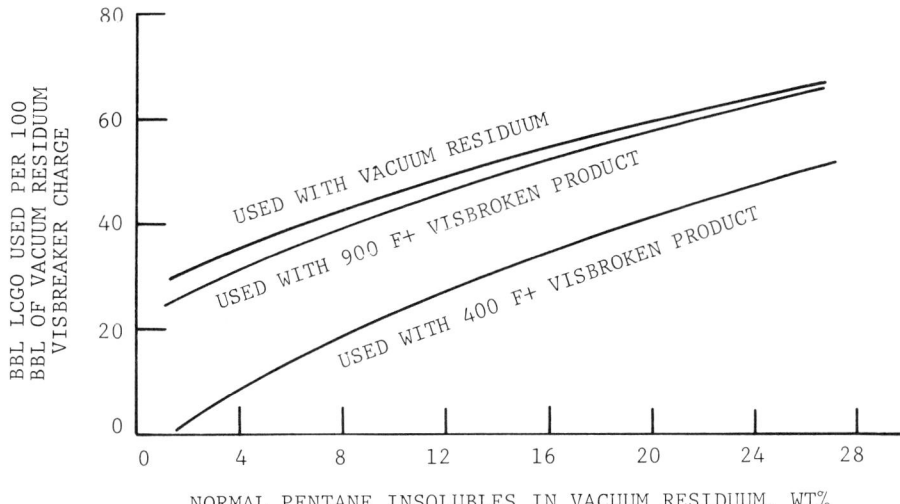

FIG. 27. Quantity of light cycle gas oil (LCGO) cutter stock required to make 190 Furol viscosity at 122°F fuel oil from (1) vacuum residuum, (2) 400°F+ visbroken product, and (3) 900°F+ visbroken product.

2. Total butane retention gasoline—Sulfur content is equal to 0.17 times visbreaker charge stock sulfur content.
3. 400°F+ visbroken product—Sulfur content is equal to visbreaker charge stock sulfur content.
4. 400 to 650°F gas oil—Sulfur content is equal to 0.5 times visbreaker charge stock sulfur content.
5. 650 to 900°F gas oil—Sulfur content is equal to 0.8 times visbreaker charge stock sulfur content.
6. 900°F+ visbroken residuum—Sulfur content is equal to 1.25 times visbreaker charge stock sulfur content.

The above sulfur distributions are approximate and the final estimated sulfur contents of the products should be adjusted to give a consistent sulfur balance with the charge stock.

The following example problem illustrates the use of the foregoing visbreaking correlations.

Example Problem 6. *Basis*: An Arabian vacuum residuum has a ring-and-ball softening point of 99°F and a sulfur content of 3.92 wt.%. Determine yields and product qualities when visbreaking at maximum severity as limited by residual fuel oil stability.

Solution: From Fig. 17, a vacuum residuum with a 99°F ring-and-ball softening point has a NC_5 insolubles content of 11.7 wt.%.

From Fig. 18, at NC_5 insolubles of 11.7 wt.% the viscosities of the Arabian vacuum residuum and the 400°F+ visbroken product are as follows: vacuum residuum, 24,000 SUS at 210°F, 400°F+ visbroken product, 2450 SUS at 210°F.

Visbreaking yields at maximum allowable visbreaking severity are determined from Figs. 19 through 27 (Table 6).

Sulfur Content: Dry gas. About 4 wt.% of sulfur in charge will appear as H_2S in the dry gas:

$$H_2S = \frac{4 \text{ wt.\%(sulfur in charge)(molecular weight } H_2S)}{(\text{dry gas yield})(\text{molecular weight sulfur})}$$

$$= \frac{0.04(3.92)(34)(100)}{1.56(32)} = 10.7 \text{ wt.\% } H_2S$$

C_4–400°F gasoline:

$$S = 0.17 \text{ (charge sulfur content)} = 0.17(3.92) = 0.67 \text{ wt.\%}$$

400°F+ visbroken products:

$$S = \text{charge sulfur content} = 3.9 \text{ wt.\%}$$

400–650°F light gas oil:

$$S = 0.5 \text{ (charge sulfur content)} = 0.5(3.92) = 1.9 \text{ wt.\%}$$

650–900°F heavy gas oil:

$$S = 0.8 \text{ (charge sulfur content)} = 0.8(3.92) = 3.1 \text{ wt.\%}$$

900°F+ visbroken residuum:

$$S = 1.25 \text{ (charge sulfur content)} = 1.25(3.92) = 4.9 \text{ wt.\%}$$

Reaction Kinetics

Visbreaking, like thermal cracking, is a first-order reaction. However, due to the visbreaking severity limits imposed by fuel oil instability, operating conditions do not approach the level where secondary reactions, polymerization and condensation, occur to any significant extent. The first-order reaction rate equation altered to fit the visbreaking reaction is

$$K = \frac{1}{t} \ln \frac{100}{X_1}$$

where K = first-order reaction velocity constant, 1/s
t = time at thermal conversion conditions, s
X_1 = 900°F+ visbroken residuum yield, vol.%

Reaction velocity constant data as a function of visbreaking furnace outlet temperature are presented in Fig. 28. The thermal conversion reactions are generally assumed to start at 800°F although some visbreaking occurs below this temperature.

Cracking, Thermal

TABLE 6

Product		Yield	Source
Dry gas		1.56 wt.%	Fig. 25
Hydrogen	1.56 × 0.4/100	= 0.01 wt.%	Table from Fig. 25
Methane	1.56 × 23.2/100	= 0.36 wt.%	Table from Fig. 25
Ethylene	1.56 × 1.9/100	= 0.03 wt.%	Table from Fig. 25
Ethane	1.56 × 27.5/100	= 0.43 wt.%	Table from Fig. 25
Propylene	1.56 × 12.2/100	= 0.19 wt.%	Table from Fig. 25
Propane	1.56 × 34.8/100	= 0.54 wt.%	Table from Fig. 25
Total butanes		1.29 vol.%	Fig. 23
Isobutane	1.29 × 16/100	= 0.21 vol.%	Table from Fig. 23
n-Butane	1.29 × 45/100	= 0.58 vol.%	Table from Fig. 23
Butylenes	1.29 × 39/100	= 0.50 vol.%	Table from Fig. 23
Total pentanes		0.63 vol.%	Fig. 24
Isopentane	0.63 × 20/100	= 0.13 vol.%	Table from Fig. 24
n-Pentane	0.63 × 41/100	= 0.25 vol.%	Table from Fig. 24
Pentylenes	0.63 × 39/100	= 0.25 vol.%	Table from Fig. 24
C_4–400°F gasoline		12.4 vol.%	Fig. 22
C_6–400°F gasoline		10.5 vol.%	By difference
400°F+ visbroken products		89.3 vol.%	Fig. 19
400–650°F light gas oil		14.3 vol.%	Fig. 20
650–900°F heavy gas oil		27.1 vol.%	Fig. 20
900°F+ visbroken residuum		47.9 vol.%	Fig. 21
Conversion into streams lower boiling than 900°F		52.1 vol.%	Fig. 21
Possible 190 SFS at 122°F fuel oil blends			
Blend 1			
Arabian vacuum residuum		100.0 vol.%	Given
Cutter (LCGO)		47.4 vol.%	Fig. 27
Fuel oil		147.4 vol.%	
Blend 2			
400°F+ visbroken product		89.3 vol.%	Fig. 19
Cutter (LCGO)		26.2 vol.%	Fig. 27
Fuel oil		115.5 vol.%	
Blend 3			
900°F+ visbroken residuum		47.9 vol.%	Fig. 21
Cutter (LCGO)		45.7 vol.%	Fig. 27
Fuel oil		93.6 vol.%	

C_4–400°F Gasoline Octane Numbers (from Fig. 22)

RON, clear	65
RON, + 3 cc TEL/gal	75
MON, clear	59
MON, + 3 cc TEL/gal	67

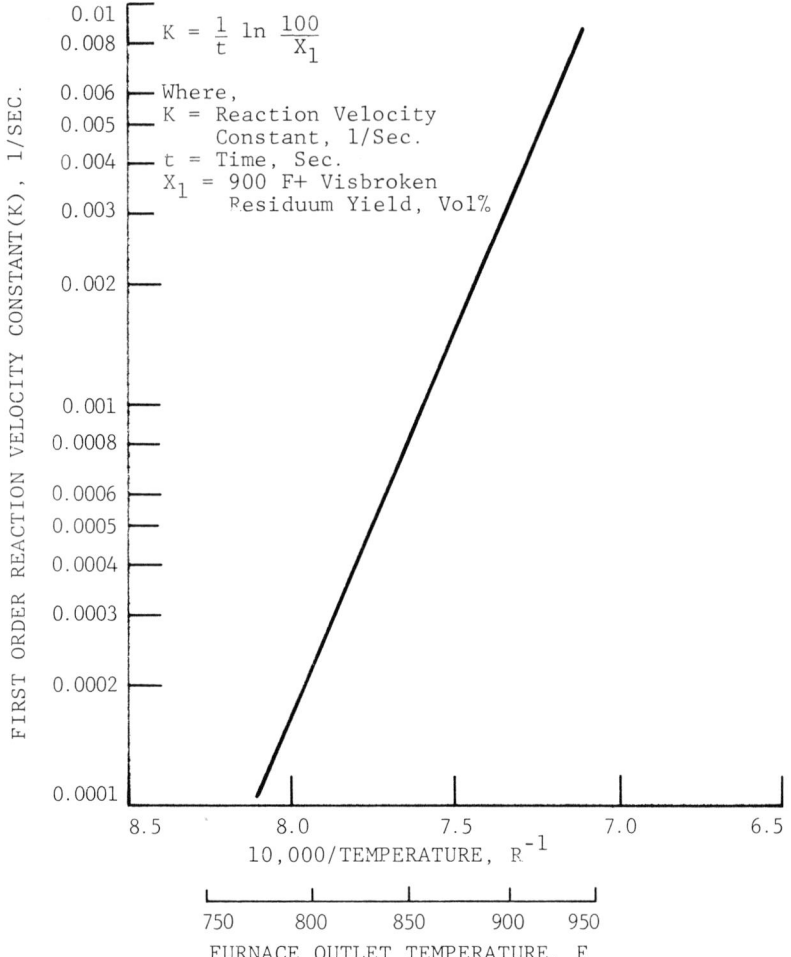

FIG. 28. First-order reaction velocity constants for visbreaking.

Example Problem 7. *Basis*: Determine reaction time required to visbreak Arabian vacuum residuum of Example Problem 6 at a furnace outlet temperature of 865°F.
Solution:

$$t = \frac{1}{K} \ln \frac{100}{X_1}$$

where $K = 0.0012$ from Fig. 28
$X_1 = 47.9$ vol.% 900°F+ visbroken residuum from Example Problem 6

$$t = \frac{1}{0.0012} \ln \frac{100}{47.9} = 613 \text{ s}$$

Cracking, Thermal

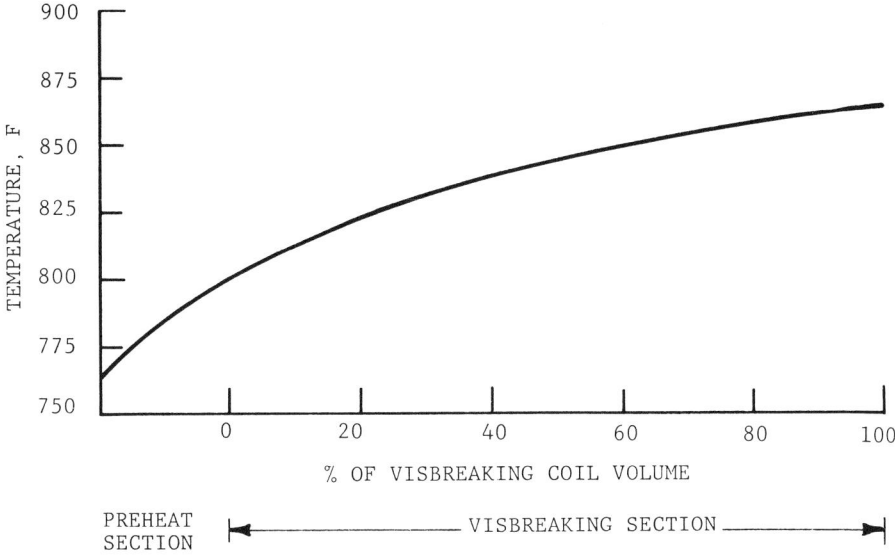

FIG. 29. Heating curve for a typical visbreaking furnace.

Therefore, the residence time in the 800 to 865°F reaction zone should be 613 s.

The optimum visbreaking furnace operations have been described by Allen, Little, and Waddell in the June 1951 issue of *Petroleum Processing* as follows:

> The heating curve of the visbreaker furnace coil which has given the best results is as nearly as possible a straight line. The last four tubes before the outlet are fired hard and have a very high heat transfer rate to hold the heating curve to a straight line when the heat of cracking tends to flatten it.

These conditions are somewhat optimistic for commercial visbreaking units. A more typical heating curve is shown in Fig. 29.

Thermal Reforming

Thermal reforming of gasoline stocks requires more severe operating conditions than thermal cracking of gas oils since the gasoline boiling range hydrocarbons are more difficult to thermally crack. Furnace outlet temperatures of 975 to 1100°F are employed with operating pressures ranging from 200 to 1000 lb/in.² gauge. The desired conversion can be achieved in single-pass operation without significant heater coking since gasoline stocks have low coking tendencies. The high conversions obtained at the high temperatures also results in production of a high octane product.

Thermal reforming can be applied to many types of gasoline stocks to increase their octane rating, but the degree of octane improvement is dependent upon the octane number of the charge stock. Generally, the lower the octane number of the charge stock, the greater the octane number increase that can be achieved.

The primary octane improving reactions in thermal reforming are (1) the cracking of paraffins into lower molecular weight paraffins and olefin chains, and (2) the dehydrogenation of naphthenes to aromatics. The elimination and conversion of the low octane normal paraffins and the production of gasoline boiling range high octane aromatics greatly improve the octane number of the gasoline stock being processed. The polymerization of the light olefins and subsequent condensation to polynuclear compounds to form a stream heavier than the gasoline charge is also significant at the high temperatures and moderate pressures used in thermal reforming.

Process Flow

Figure 30 presents a simplified flow diagram for a thermal reforming unit. The low octane gasoline is charged to the reforming heater at about 1050°F (975 to 1100°F range). After leaving the heater at about 600 lb/in.^{2} gauge (200 to 1000 lb/in.^{2} gauge range), the total product stream is quenched to about 675°F with a recycle product gasoline stream to prevent coking, and the pressure is reduced to a suitable fractionator operating pressure. The total product stream plus quench is then charged to the fractionator where most of the gasoline and lighter boiling components are distilled overhead. The overhead streams are usually separated into a dry gas stream (propane and lighter) and a total butane retention gasoline stream. The fractionator bottoms go to a stripper where the

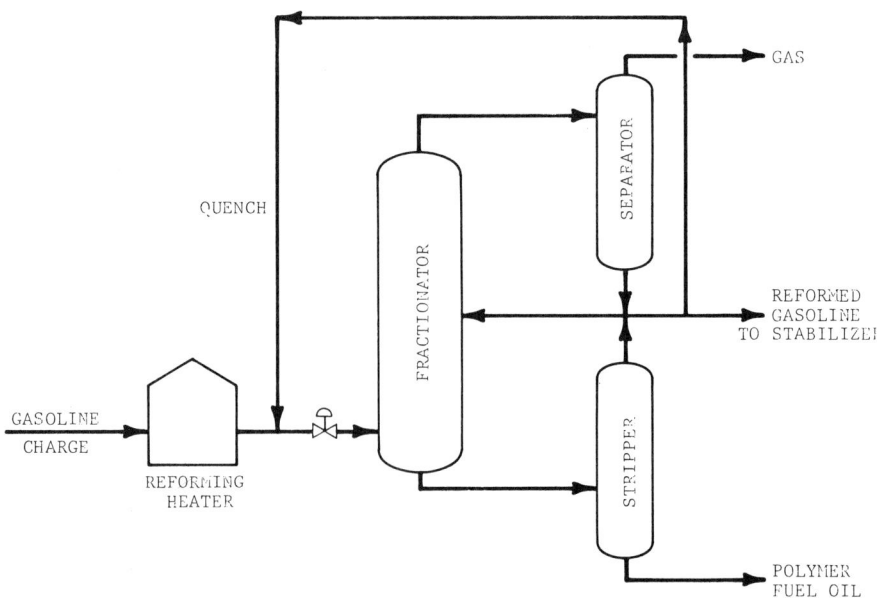

FIG. 30. Flow diagram of a thermal reforming unit.

remaining gasoline is flashed off and combined with the overhead product gasoline. The remaining polymer fuel oil stream, composed of the polymerization and condensation products formed in the thermal reforming operations, has an initial boiling point of about 400°F and a high Btu content and is suitable for use in fuel oil manufacture.

Process Correlations

Perhaps the most widely used thermal reforming yield and quality correlations were developed during World War II by the Technical Advisory Committee of the Petroleum Industry War Council (PIWC). These correlations appear in PIWC reports TML-3, TML-8, and TML-16 and use the Motor octane method as a measure of octane improvement. Since the PIWC correlations were developed, the Research octane number method has also been used as a measure of commercial gasoline quality. In general, the following thermal reforming correlations are more extensive, easier to use, and cover a wider range of charge stocks than the PIWC correlations. The prime exception is that the effect of pressure is not covered in the following correlations due to the relatively minor effect of pressure at the operating conditions normally used for commercial units. When operating at constant dry gas yield (measure of conversion), there would not be significant changes in product distribution or quality caused by a pressure difference of 100 lb/in.² gauge between two commercial units. The effect of wider differences in pressure may be determined from the PIWC reports if required.

Figures 31 through 43 present the thermal reforming correlations. Dry gas is the main correlating variable and is indicative of the severity of the thermal reforming process. The dry gas yield is used because (1) it is relatively easy and accurate to measure and (2) it undergoes a large percentage change over the range of commercial operating conditions. Correlations are presented for five different types of charge stocks: straight run gasolines, catalytic reformed gasolines, catalytic cracked gasolines, thermal cracked gasolines, and raffinate gasolines. For the correlations to be valid, the charge stock should be a depentanized full-range gasoline (C_6–400°F end point). If pentanes or butanes are present in the charge stock, they may be considered, for the purpose of calculation, to be unaffected by the thermal reforming operation.

Product Octane Numbers. Figure 31 shows how the unleaded Research octane number of the debutanized (C_5–400°F) thermal reformate varies with the weight percent dry gas yield. Dry gas is considered to include hydrogen, methane, ethylene, ethane, propylene, and propane. Figure 32 shows a similar plot for the unleaded Motor octane number. In the figures, three families of curves are shown, one for catalytic reformed charge stocks; one for catalytic cracked, thermal cracked, and straight run gasolines; and one for raffinate gasolines. The parameter in both figures is the unleaded octane number of the charge stock.

Product Yields. Figure 33 shows the yield of debutanized thermal reformate, expressed as volume percent of the gasoline charged, as a function of the weight

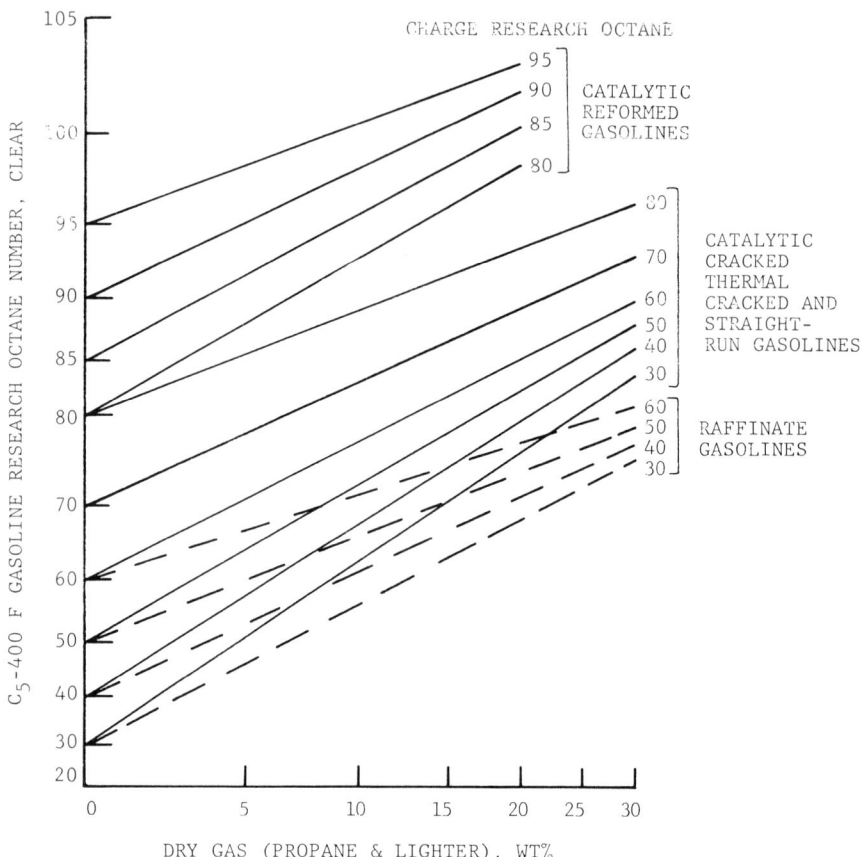

FIG. 31. Prediction of Research clear octane number of debutanized thermal reformed gasoline.

percent dry gas yield for each type of gasoline charge stock. The code letters identifying each charge stock will also be used in the remaining figures. For a given dry gas yield, raffinate gasolines give the highest yield of C_5–400°F reformate, and catalytic cracked gasolines give the lowest reformate yields. This difference in reformate yields at constant dry gas yields shows up in the polymer fuel oil yields. The raffinate gasoline, being paraffinic, cracks easier and a lower reaction temperature can be used to obtain equivalent dry gas yields. This lower reaction temperature results in less polymerization and condensation of the gasoline compounds to polymer fuel oil. In addition, the cracked gasoline contains more olefinic and aromatic compounds which would tend to polymerize and condense to the heavier fuel oil compounds.

The total butanes yield, expressed in weight percent of the gasoline charged, is shown in Fig. 34. The curve for the raffinate gasolines shows a lower butane yield for a given total dry gas yield. This indicates that a greater fragmentation of the hydrocarbon occurs in thermal reforming raffinate gasoline than for the other charge stocks. The curve for the catalytic reformed gasolines tends to be higher at high severities than the curves for the other stocks. However, valuable

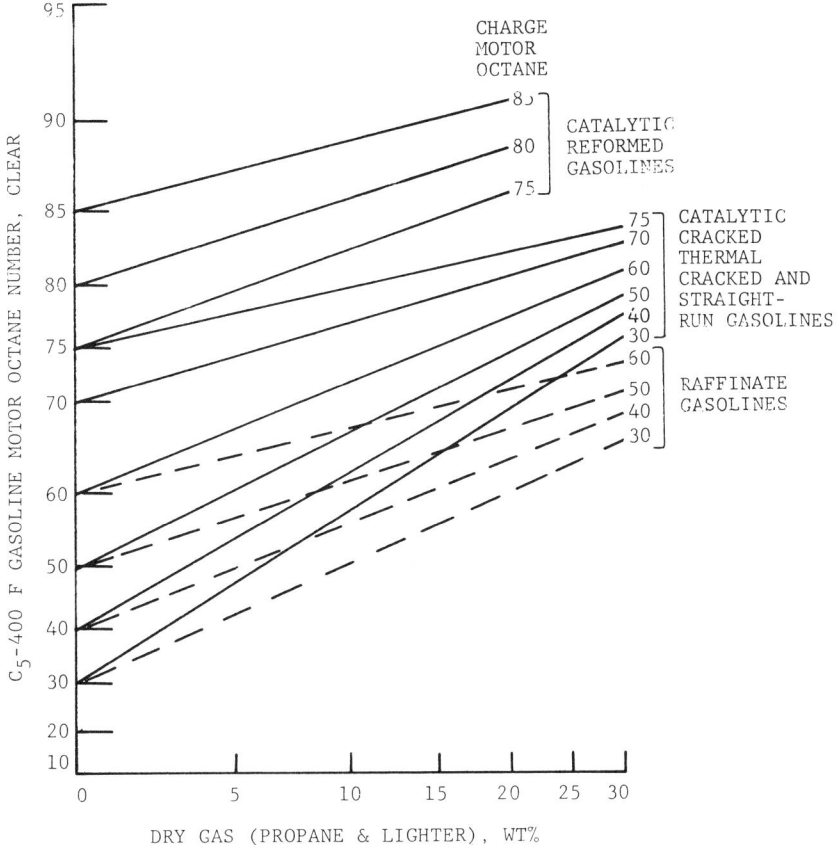

FIG. 32. Prediction of Motor clear octane number of debutanized thermal reformed gasoline.

catalytic reformed gasolines would not normally be thermally reformed to such high gas yields.

The volume percent yields of 400°F + polymer fuel oil can be obtained from Fig. 35. This stream represents polymerization and condensation compounds formed during thermal reforming which are heavier than the charge stock. This stream normally has a very high octane number but gasoline end-point specifications preclude its use in motor fuels. As discussed previously, at any given dry gas yield, raffinate gasolines make the least quantities of heavy compounds and catalytic cracked gasolines make the most.

Figure 36 shows the composition of the dry gas stream as a function of dry gas yield. With increasing severity (increasing dry gas yield), the concentration of methane and ethane in the dry gas increases and the concentration of ethylene, propylene, and propane decreases. Hydrogen concentration remains constant throughout the severity range. Note that the olefin concentrations are higher (with corresponding lower saturates concentrations) in the dry gas from the raffinate gasoline charge compared to the dry gas from the other charge stocks.

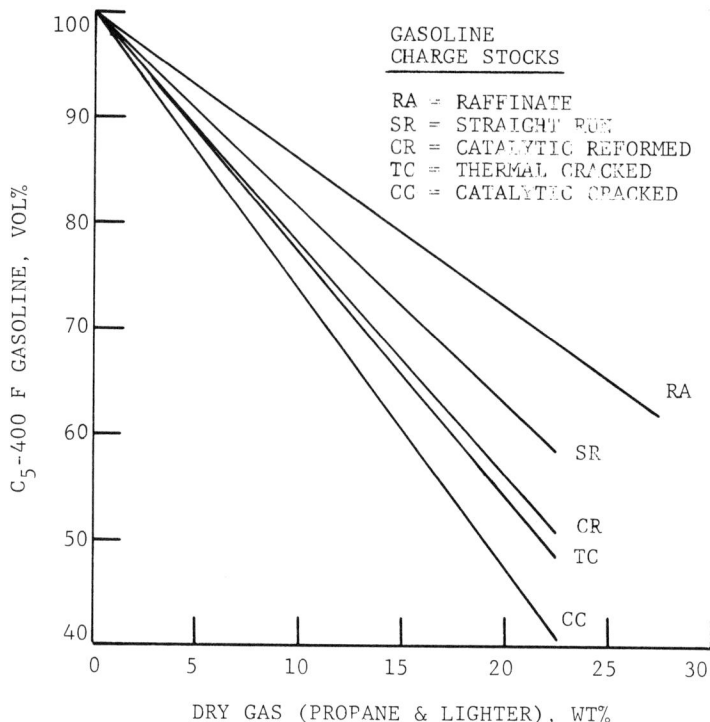

FIG. 33. Prediction of debutanized thermal reformed gasoline yield.

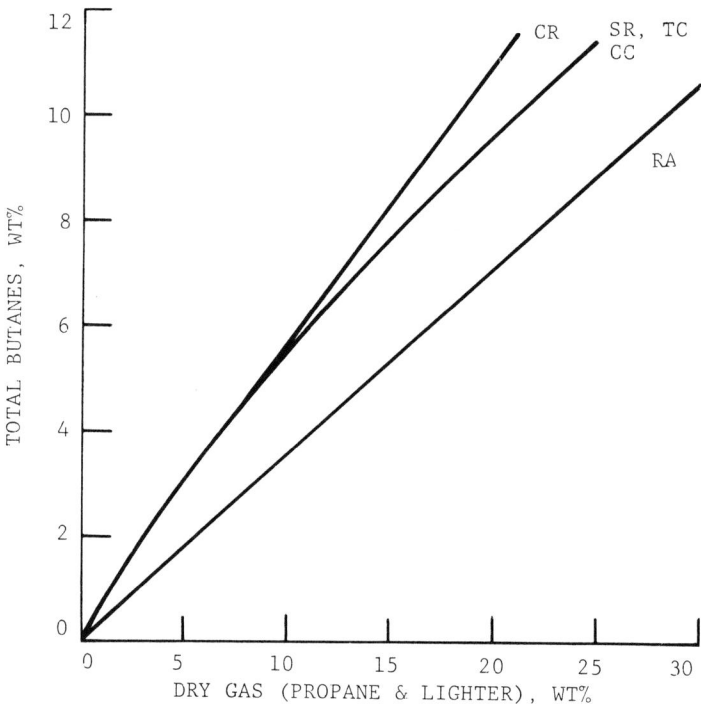

FIG. 34. Prediction of total butanes yield from thermal reforming.

Cracking, Thermal

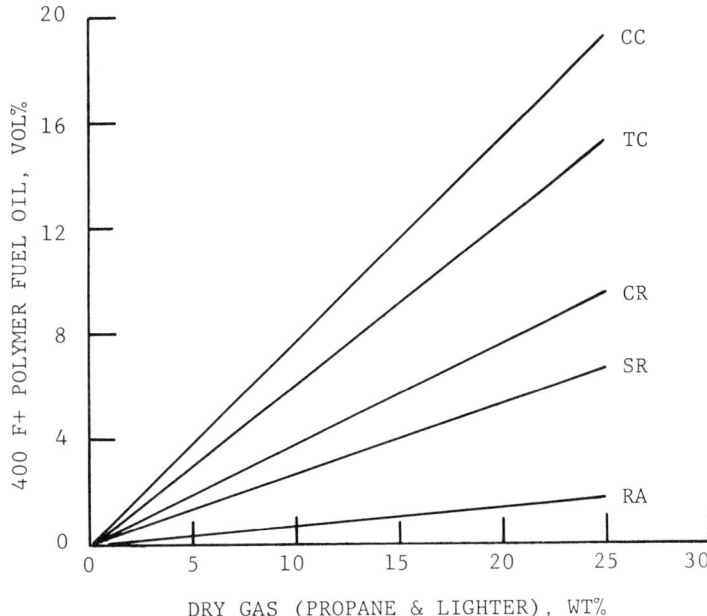

FIG. 35. Prediction of 400°F+ polymer fuel oil yield from thermal reforming.

Figure 37 shows the composition of the total butanes stream as a function of the dry gas yield. With increasing severity, the concentration of butylenes increases primarily at the expense of normal butane. It is again noted that the olefin (butylene) concentration is higher from the raffinate gasoline charge than from the other charge stocks.

Figure 38 is divided into three parts. The upper curve presents an estimation of the volume percent total pentanes yield as a function of dry gas yield. The total pentanes yield is the same for all charge stocks except raffinate gasoline which gives a lower total pentanes yield at a given dry gas yield. The composition of the total pentanes stream can be estimated from the middle portion of the figure for all charge stocks except raffinate gasoline. The lower portion of the figure may be used to estimate the total pentanes stream composition when charging raffinate gasolines. As discussed previously, any pentanes or butanes in the charge stock can be considered unaffected by thermal cracking and will appear in the final products.

Product Quality Correlations—Lead Response, Sulfur, and Gravity. Figures 39 through 43 present product quality correlations for the liquid thermal reforming products. Figure 39 shows the leaded (3 cc TEL/gal) Research octane number of the C_5–400°F thermal reformate as a function of the clear Research octane number of the reformate. The parameter is the weight percent sulfur content of the charge stock. The curves are applicable to all charge stocks. Figure 40 is similar to Fig. 39 except that it is applicable for the prediction of leaded Motor octane numbers of the C_5–400°F thermal reformate. If octane numbers at TEL concentrations other than 3 cc TEL/gal are desired, these octane numbers may be estimated from lead susceptibility charts prepared by Ethyl Corp. (Chart ECD-850) or E. I. du Pont de Nemours and Co.

The weight percent sulfur content of the C_5–400°F thermal reformate can be estimated using Fig. 41. In this figure the ratio of thermal reformate sulfur content to charge sulfur content is plotted against dry gas yield with the parameter being weight percent sulfur content of the charge stock. The figure

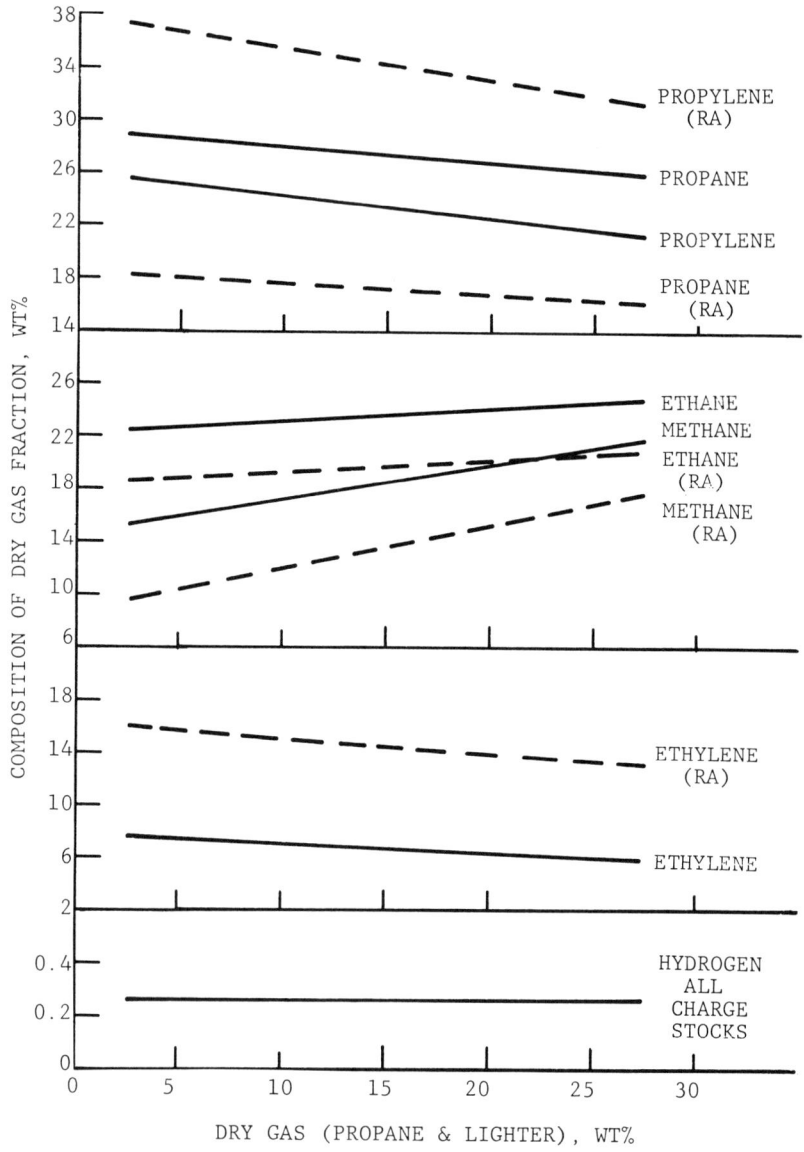

FIG. 36. Composition of dry gas fraction from thermal reforming.

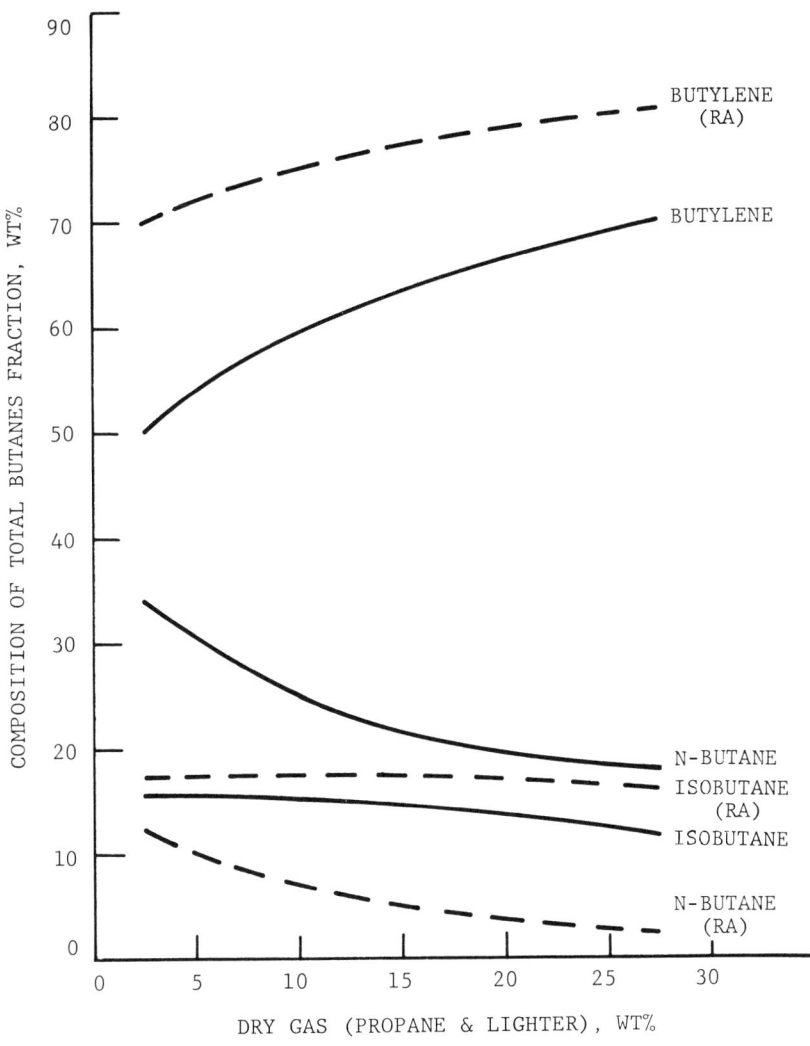

FIG. 37. Composition of total butanes fraction from thermal reforming.

was developed for straight-run gasoline charge stocks, but it is generally valid for other types of charge stocks. In cases where the charge stock sulfur is below 0.02 wt.%, the reader may use the 0.02 wt.% sulfur curve for estimation purposes.

Figure 42 shows the change in API gravity of the C_5–400°F thermal reformate as compared with the C_6–400°F charge stock. The catalytic reformed gasolines give a greater API gravity change after thermal reforming when compared to other types of gasolines. This is probably due to the high concentration of aromatic compounds (low API gravity) present in the catalytic reformed gasolines.

Figure 43 presents the API gravity of the 400°F+ polymer fuel oil as function of the dry gas yield. The catalytic reformed charge stocks produce a lower API gravity 400°F+ polymer fuel oil at the same dry gas yield compared to the straight-run, thermal cracked, catalytic cracked, and raffinate charge stocks.

The following example problem shows the use of the thermal reforming correlations.

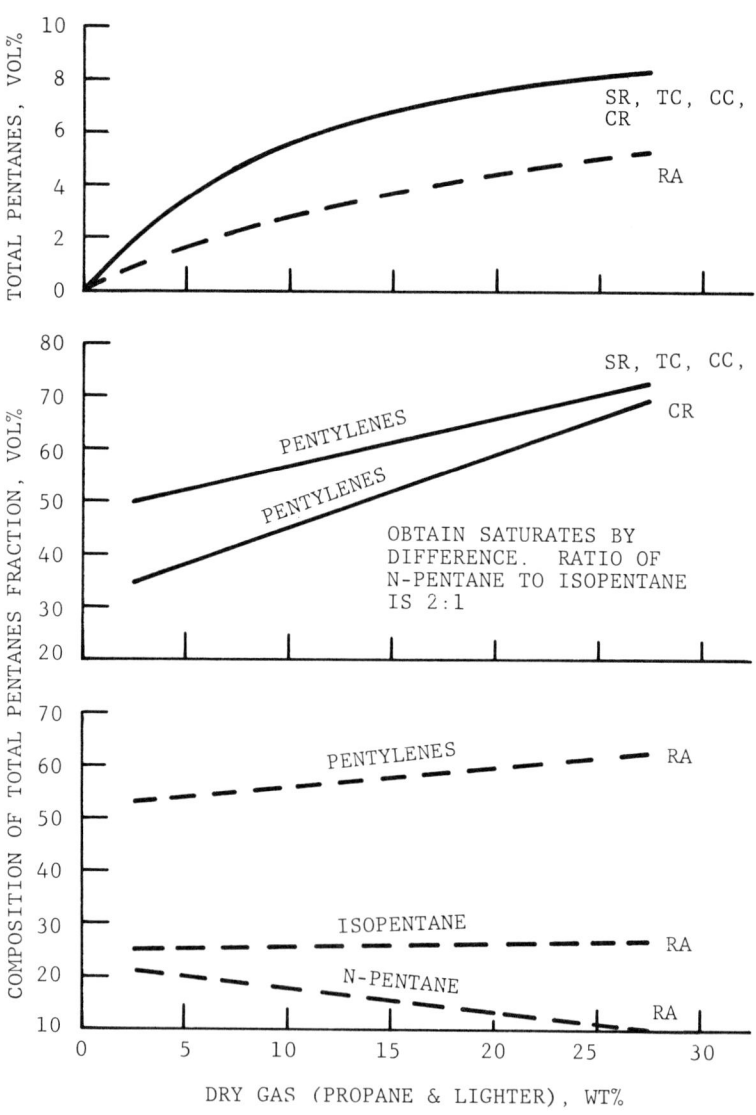

FIG. 38. Prediction of total pentanes yield and composition of total pentanes fraction from thermal reforming.

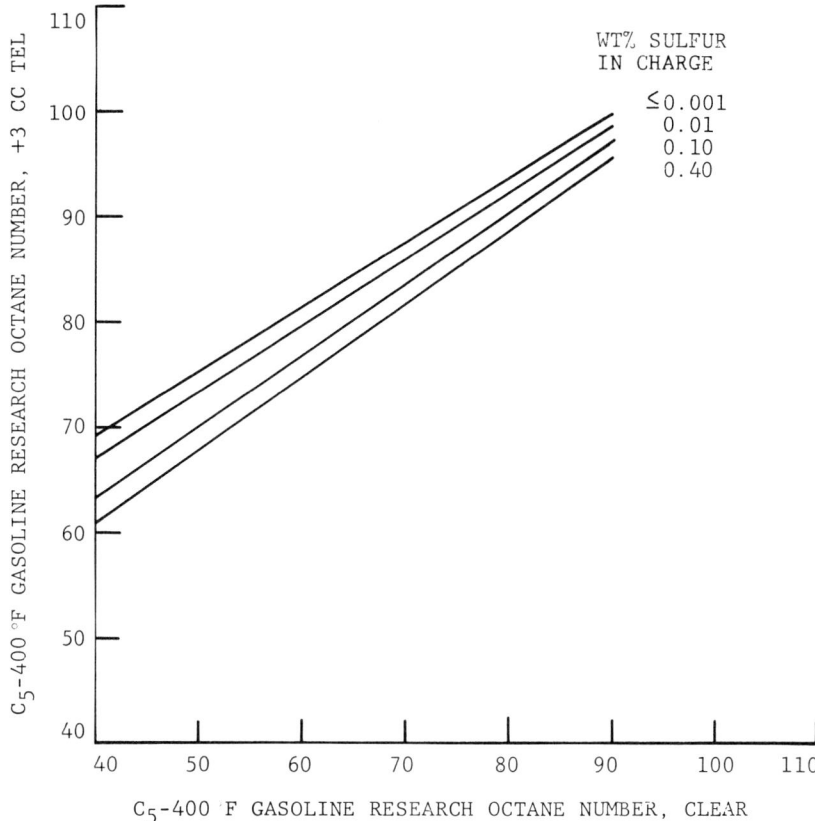

FIG. 39. Research octane number TEL response of debutanized thermal reformed gasoline.

Example Problem 8. *Basis*: A straight-run C_6–400°F gasoline has the following characteristics:

Research octane number, clear	= 60.0
Motor octane number, clear	= 50.0
Gravity, API	= 52.0
Sulfur, wt.%	= 0.2

Determine yields and product quality when thermal reforming to produce a C_5–400°F reformate having an 82.0 Research clear octane number.

Solution: From Fig. 31, the dry gas yield is 15.0 wt.% when thermal reforming the straight-run gasoline to the desired octane number. Using this dry gas yield, other product yields and qualities are determined from Figs. 32 through 43 (Table 7).

The weight recovery balance (Table 8) shows the weight percent yields add to 100.4. All yields should be prorated to a 100.0 wt.% basis as in Example Problem 1.

Reaction Kinetics

Thermal reforming is considered to be a first-order reaction. Although the cracking severity is much greater than for visbreaking or thermal cracking, the charge stock is lighter and thus more refractory, which compensates for the increased severity. Again it was necessary to slightly alter the form of the first-order equation.

$$K = \frac{1}{t} \ln \frac{100}{X_2}$$

where K = first-order reaction rate constant, 1/s
t = residence time at thermal reforming conditions, s (based on charge liquid volume)
X_2 = C_5–400°F gasoline yield, vol.%

The reaction velocity constants for thermal reforming are shown in Fig. 44 for the various types of charge stocks. Although thermal reforming is accomplished in the vapor phase, the residence time used in the first-order equation is based on the liquid volume of the charge stock. This was done to simplify

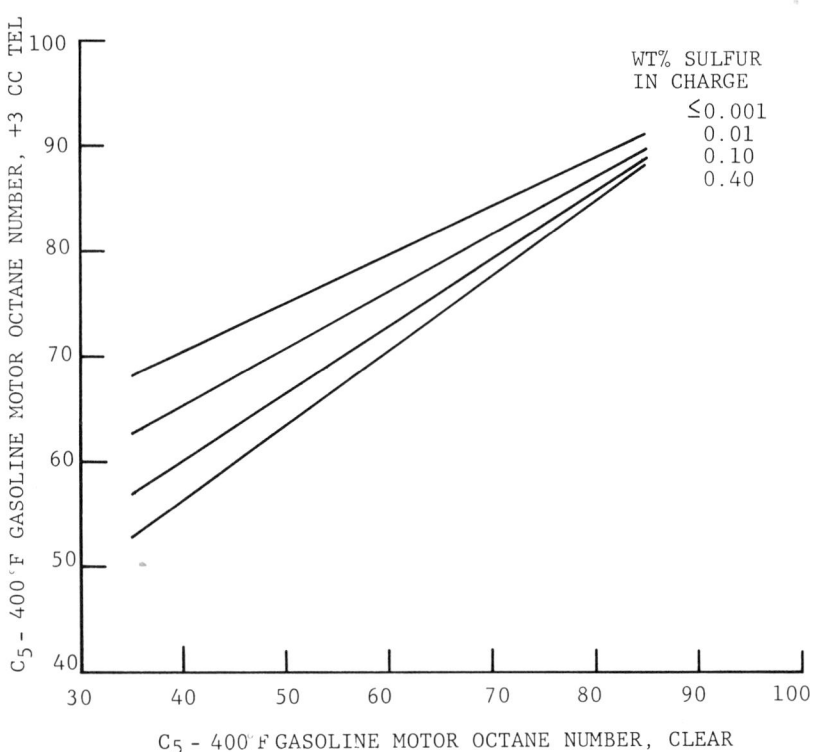

FIG. 40. Motor octane number TEL response of debutanized thermal reformed gasoline.

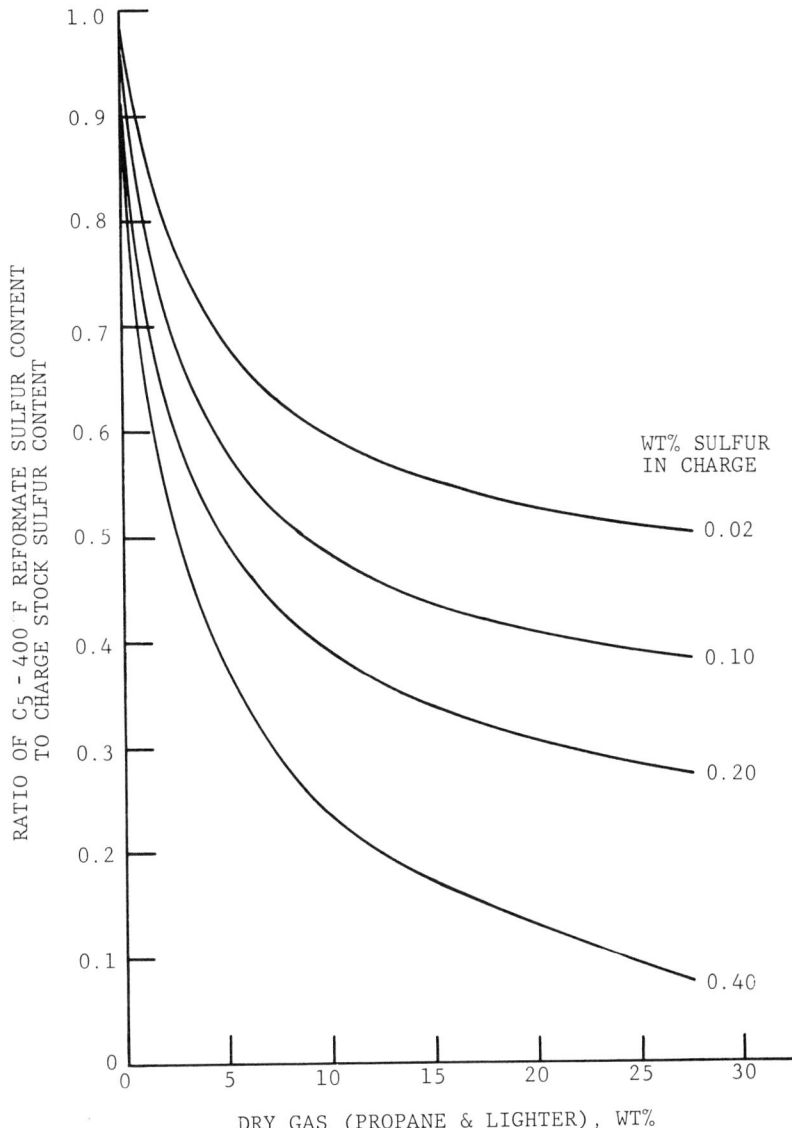

FIG. 41. Prediction of sulfur content of debutanized thermal reformed gasoline.

calculations and to maintain uniform calculation procedures among the thermal reforming, thermal cracking, and visbreaking sections. Thus the residence time can be easily calculated by dividing the reactor volume above 900°F by the charge rate. Naturally, the actual residence time of the vaporized charge stock is much shorter than the "pseudo" residence time used in the first-order equation. If required, the actual vapor residence time can be calculated by standard procedures.

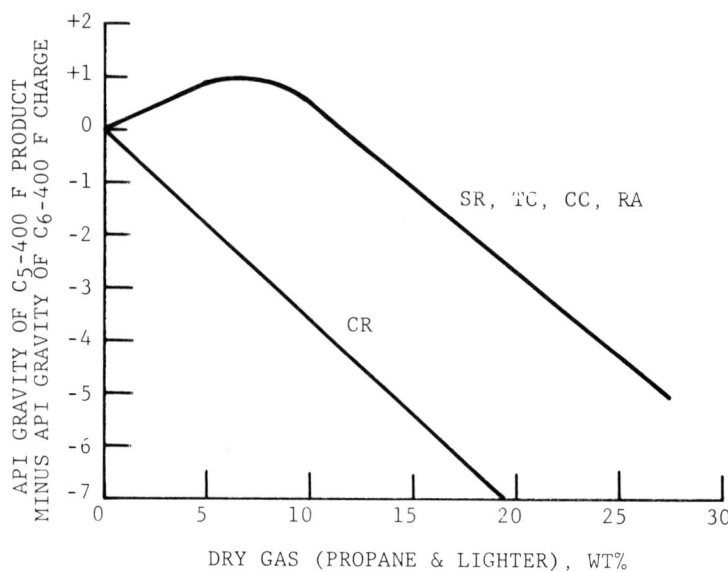

FIG. 42. Prediction of API gravity of $C_5-400°F$ thermal reformed gasoline from API gravity of $C_6-400°F$ charge gasoline.

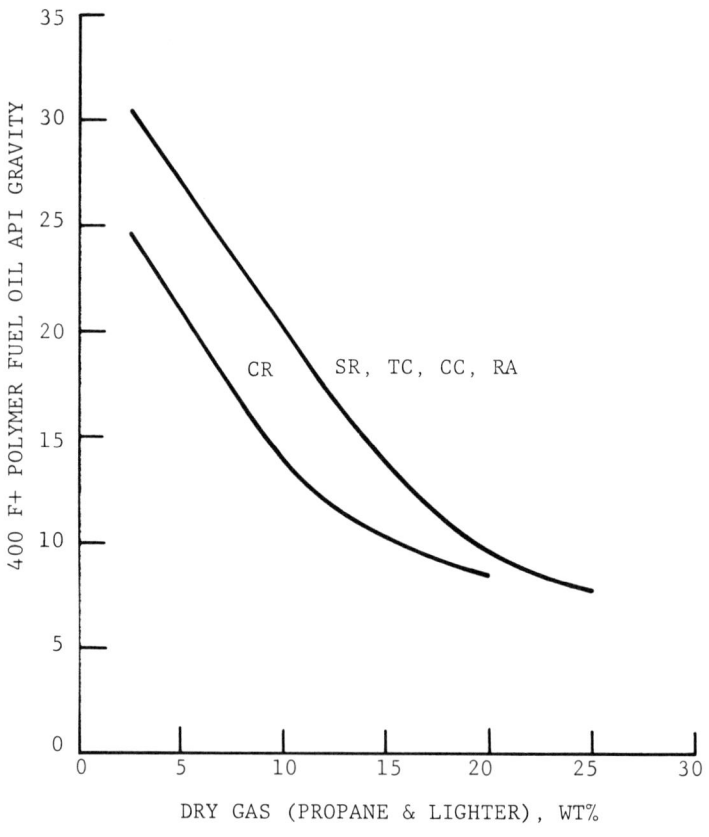

FIG. 43. Prediction of API gravity of $400°F+$ polymer fuel oil.

Cracking, Thermal

TABLE 7

Product		Yield or Quality	Source
Dry gas		15.0 wt.%	Fig. 31
Hydrogen	15.0 × 0.27/100 =	0.04 wt.%	Fig. 36
Methane	15.0 × 18.6/100 =	2.79 wt.%	Fig. 36
Ethylene	15.0 × 6.7/100 =	1.00 wt.%	Fig. 36
Ethane	15.0 × 23.7/100 =	3.56 wt.%	Fig. 36
Propylene	15.0 × 23.3/100 =	3.50 wt.%	Fig. 36
Propane	15.0 × 27.4/100 =	4.11 wt.%	Fig. 36
Total butanes		7.6 wt.%	Fig. 34
Isobutane	7.6 × 14.7/1000 =	1.12 wt.%	Fig. 37
n-Butane	7.6 × 21.7/100 =	1.65 wt.%	Fig. 37
Butylenes	7.6 × 63.6/100 =	4.83 wt.%	Fig. 37
Total pentanes		6.9 vol.%	Fig. 38
Isopentane	6.9 × 12.8/100 =	0.88 vol.%	Ratio from Fig. 38
n-Pentane	6.9 × 25.7/100 =	1.77 vol.%	Ratio from Fig. 38
Pentylenes	6.9 × 61.5/100 =	4.25 vol.%	Fig. 38
C_5–400°F gasoline		72.3 vol.%	Fig. 33
C_6–400°F gasoline		65.4 vol.%	By difference
400°F + polymer fuel oil		4.0 vol.%	Fig. 35
C_5–400°F gasoline quality			
Research octane no., clear		82.0	Given
Research octane no., + 3 cc TEL/gal		91.1	Fig. 39
Motor octane no., clear		71.2	Fig. 32
Motor octane no., + 3 cc TEL/gal		79.0	Fig. 40
Gravity, API		50.9	Fig. 42
Sulfur, wt.%		0.07	Fig. 41
400°F + polymer fuel oil quality			
Gravity, API		13.7	Fig. 43

TABLE 8 Weight Recovery Check

	API Gravity	Specific Gravity	Yields	
			Vol.%	Wt.%
Charge stock	52.0	0.771	100.0	100.0
Dry gas	—	—	—	15.0
Total butanes	—	—	—	7.6
C_5–400°F gasoline	50.9	0.776	72.3	72.7
400°F + polymer fuel oil	13.7	0.974	4.0	5.1
Total products	—	—	—	100.4

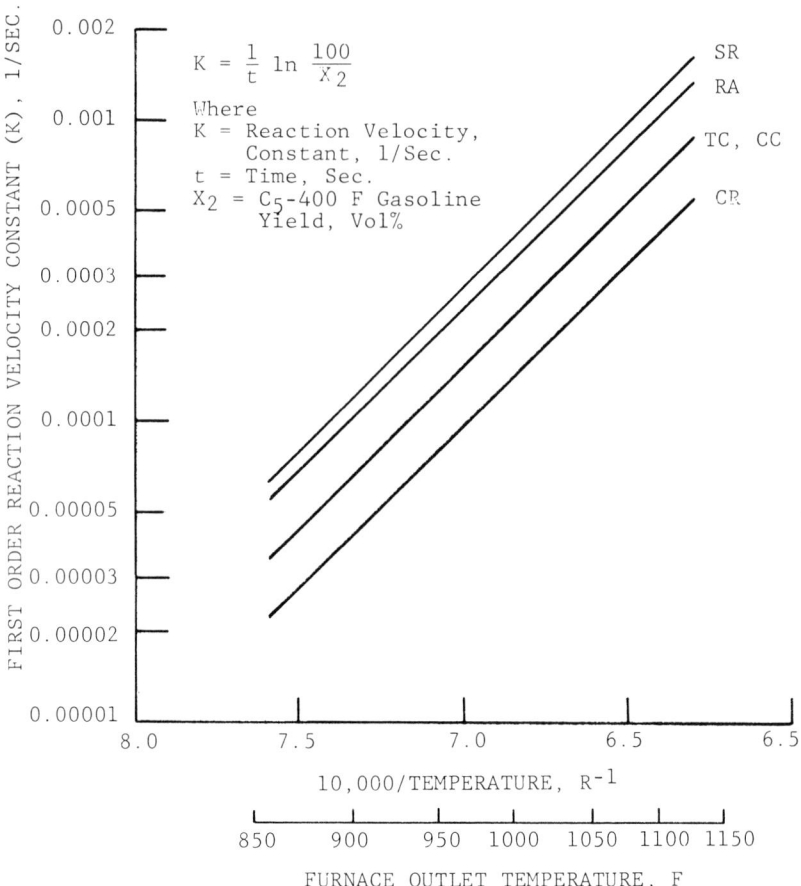

FIG. 44. First-order reaction velocity constants for thermal reforming.

Example Problem 9. *Basis:* A straight-run gasoline is charged at a rate of 8000 gal/h into a thermal reforming furnace. The furnace volume in which the charge stock temperature will be greater than 900°F is 200 ft³. Determine the desired furnace outlet temperature which will give a C_5–400°F reformate yield of 72.3 vol.%.

Solution: Calculate the "pseudo" liquid residence time.

$$t = \frac{(200 \text{ ft}^3)(7.48 \text{ gal/ft}^3)(3600 \text{ s/h})}{8000 \text{ gal/h}} = 673 \text{ s}$$

$$K = \frac{1}{t} \ln \frac{100}{X_2}$$

$$= \frac{1}{673} \ln \frac{100}{72.3}$$

$$= 0.000481$$

Using the SR gasoline line in Fig. 44, the furnace outlet temperature is determined to be 1010°F.

Equipment Design

The central piece of equipment in a thermal process is, of course, the heater. The heater must be adequate to efficiently supply the heat required to accomplish the desired degree of thermal conversion. A continually increasing temperature gradient designed to give most of the temperature increase in the front part of the heater tubes with only a slow rate of increase near the outlet is preferred. Precision control of time and temperature is usually not critical in the processes covered in this article. Usually, all that is required is to design to some target temperature range and then adjust actual operations to achieve the desired cracking. In the higher temperature processes (e.g., ethylene manufacture), temperature control does become of prime importance due to equilibrium considerations.

Design of all equipment to minimize coke formation is of importance. The excessive production of coke adversely affects the thermal cracking process in the following ways:

1. Reduces heat transfer rates
2. Increases pressure drops
3. Creates overheating
4. Reduces run time
5. Requires the expense of removing the coke from the equipment

In the following paragraphs on specific equipment, methods of reducing coke formation will be covered.

The metallurgy of the equipment, specifically the heater tubes and pumps, in the high temperature, corrosive environments must be adequate to prevent expensive destruction and replacement of equipment. In the early days of thermal cracking the metallurgy of the heater tubes was not of sufficient quality to permit extended periods of high temperatures. Modern improvements in the quality of steel has extended the durability of the thermal cracking equipment.

Heaters

The advances in heater design have reached a point where very efficient furnace and heating tube arrangements can be built which give the refiner the desired thermal cracking operation. The practice of the refiner is generally to set the specifications the heater is expected to meet for the specific application and have a heater manufacturer prepare a suitable design. Some general specifications are as follows.

Tube Size

Proper tube size selection depends upon minimizing pressure drop while obtaining good turbulence for proper heat transfer. This would depend upon the charge rate, and a general guide is shown in Table 9.

TABLE 9

Total Charge Rate (fresh feed + recycle) (bbl/d)	Internal Diameter of Tube (in.)
3,000	2–3
6,000	3–4
12,000	4–4½[a]

[a]Parallel tubes of smaller diameters would be preferable to one large tube. In this case, two 3-in. diameter parallel tubes may be preferred.

Charge Stock Velocities

The charge stock liquid velocity should be sufficient to provide enough turbulence to ensure a good rate of heat transfer and to minimize coking. A minimum linear cold (60°F) velocity of 5 ft/s basis 100% liquid charge rate should be sufficient. The maximum velocity would be limited to about 10 ft/s due to excessive pressure drop. The velocities at the higher cracking temperatures would, of course, be greater due to the partial vaporization of the charge.

Heating Rates

Most of the heat supplied to the charge stock is radiant heat. The convection section of the heater is used primarily to supply preheat to the charge prior to the main heating in the radiant section. The heat transfer rate in the convection section will range from 3,000 to 10,000 Btu/ft² of tube outside area per hour with an average rate of 5,000 Btu/ft²/h. The heating rates in the radiant section will range from 8,000 to 20,000 Btu/ft²/h depending upon the charge stock, with heavier oil generally requiring the lower heating rate.

Temperature

The heating tube outlet temperature will depend upon the charge stock being processed and the degree of thermal conversion required. The outlet temperature will vary from a minimum of 800°F for visbreaking to a maximum of 1100°F for thermal reforming. The combustion chamber temperature will range from 1200 to 1600°F at a point about 1 ft below the radiant tubes. Flue gas temperatures are usually high (800 to 1100°F). particularly since the heavy charge stock is usually entering the heater at a high temperature from a fractionating tower. An exception to the charge entering at a high temperature would be when charging gasoline to a thermal reformer. However, since thermal reforming requires high temperatures, flue gas temperatures will also be high.

Zone Temperature Control

Since it is desirable to maintain different temperature increase rates throughout the charge heating, i.e., rapid increase at the beginning of the heating coil and a lower rate near the outlet, zone temperature control within the furnace is desired. A three-zone furnace is preferred with the first zone giving the greatest rate of temperature increase and the last zone the least.

Coke Prevention and Removal

As indicated previously, coke formation limits the operation of the heater, and techniques should be employed to minimize coke formation in the heater tubes. Coking occurs on the walls of the tubes, particularly where turbulence is low and temperature is high. Maintaining sufficient turbulence assists in limiting coke formation. Baffles within the tubes are sometimes used but water injection into the charge stream is the preferred method. Water is usually injected at the inlet although water also may be injected at additional points along the heater tubes. The water, in addition to providing turbulence in the heater tubes as it is vaporized to steam, also provides a means to control temperature. The optimum initial point of water injection into the heater tubes is at the point of incipient cracking where coke would start to form. An advantage to this injection point is the elimination of the additional pressure drop which would have been created by the presence of water between the heater inlet and the point of incipient cracking.

The preferred method to decoke the heater tubes is to burn off the coke using a steam–air mixture. The heater tubes, therefore, should be capable of withstanding temperatures up to $1400°F$ (at low pressures) for limited time periods. The heater tubes along with the tube supports should be designed to handle the thermal expansion extremes that would be encountered. Mechanical means, such as drills, can also be used to remove coke, but most modern heaters use the steam–air combustion technique. Parallel heaters may be employed so that one can be decoked while permitting cracking to proceed in the other heater(s).

Metallurgy

Alloy steel tubes of 7 to 9% chromium are usually satisfactory to resist sulfur corrosion in thermal cracking heaters. If the hydrogen sulfide content of the cracked products exceeds 0.1 mol% in the cracking zone, a higher alloy steel may be required. Stabilized austenitic stainless steel, such as Type 321 or 347, would be suitable in this case. Other alloys, such as the Inconel or Incoloy alloys, could also be used. Seamless tubes with welded return bends are now normally used in heaters. Flanged return bends were used in earlier thermal cracking units to facilitate cleaning. However, use of steam–air to burn out the coke essentially eliminates the need for flanged fittings. This, in turn, reduces the possibility of dangerous leaks.

Soaking Volume Factor

A useful tool to aid in the design and operation of thermal cracking units is the Soaking Volume Factor (SVF). This factor was developed by the M. W. Kellogg Co. and combines time, temperature, and pressure of thermal cracking operations into a single numerical value. SVF is defined as the *equivalent* coil volume in cubic feet per daily barrel of charge (fresh plus recycle) if the cracking reaction had occurred at 800°F and 750 lb/in.² gauge. The equation would be:

$$\text{SVF}_{\substack{750\,\text{lb/in.}^2\text{gauge} \\ 800°\text{F}}} = \frac{1}{F} \int_0^V RK_P \, dV$$

where $\text{SVF}_{\substack{750\,\text{lb/in.}^2\text{gauge} \\ 800°\text{F}}}$ = the SVF at base reaction conditions of 750 lb/in.² gauge pressure and 800°F, cubic feet of coil volume per total charge throughput in barrels per day

F = charge (fresh plus recycle) throughput rate, barrels per day

R = ratio of reaction velocity constant at temperature T and reaction velocity constant at 800°F, K_T/K_{800}

K_P = pressure correction factor for pressures other than 750 lb/in.² gauge

dV = incremental coil volume, cubic feet

The ratio of reaction velocity constants may be obtained from Fig. 45 (based on a table given in Nelson's *Petroleum Refinery Engineering*). These ratios should not be obtained from reaction velocity constant plots such as Fig. 15 since there is a correction for the effect of temperature on the volume of the reacting material. This correction has been incorporated in Fig. 45.

The pressure correction factor, K_P, may be obtained from Fig. 46 (based on a figure presented in Nelson's *Petroleum Refinery Engineering*).

Since the temperature and pressure are not constant throughout the coil, but will vary as the charge is heated while passing through the coil, it is necessary to have or be able to calculate temperature and pressure gradients. Knowing the volume of the cracking coil and the total charge rate, the SVF may be calculated by using the following steps:

1. Plot the cracking coil temperature gradient (°F vs coil volume, ft³). Use only that portion above 800°F in the further calculations.
2. Using the temperature gradient plot from Step 1 (above 800°F) and Fig. 45, plot the ratio of reaction velocity constant, R, against coil volume, ft³.
3. Plot the cracking coil pressure gradient (lb/in.² gauge vs coil volume, ft³). Use only that portion above 800°F in the further calculations.
4. Using the pressure gradient plot from Step 3 (above 800°F) and Fig. 46, plot the pressure correction factor, K_P, against coil volume, ft³.
5. Combine the R and K_P gradient plots from Steps 2 and 4 to give a gradient of R times K_P. Plot this gradient against coil volume, ft³.

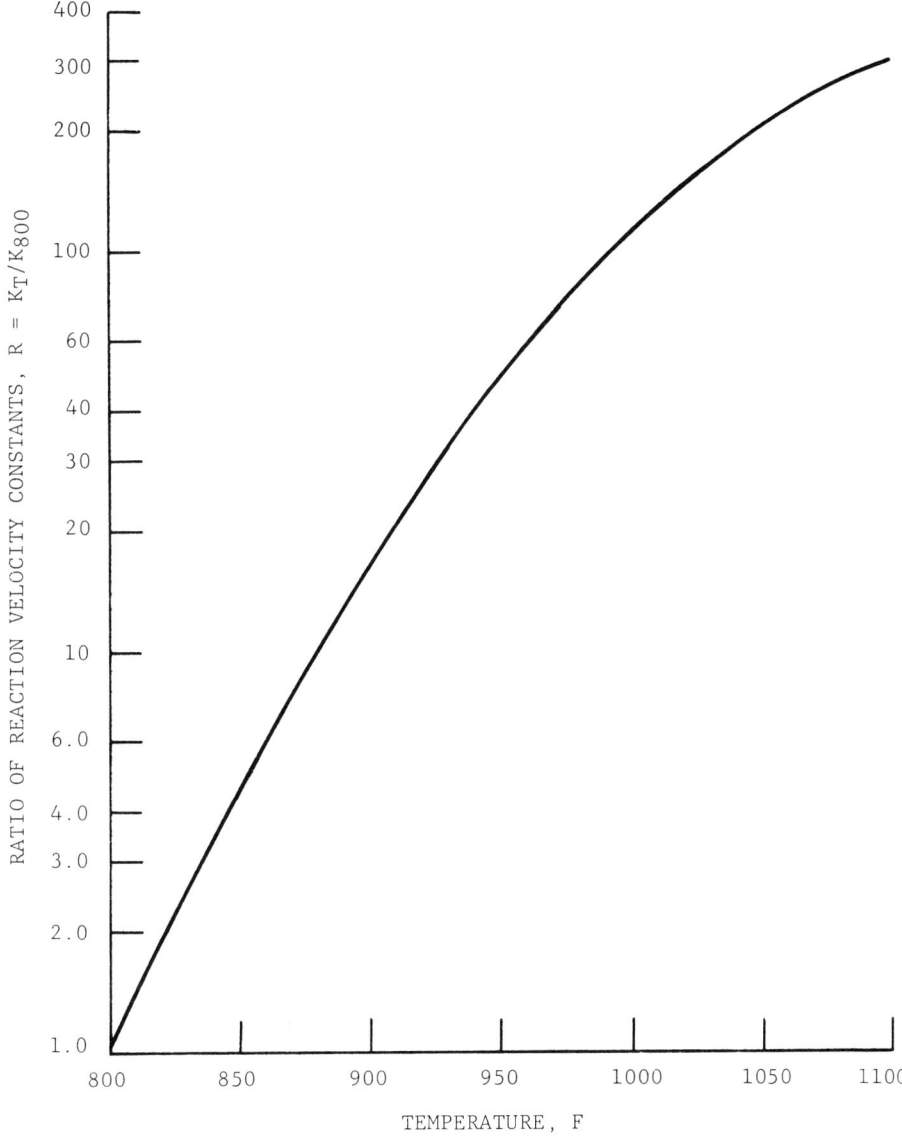

FIG. 45. Change of reaction velocity constant with temperature. Base temperature = 800°F.

6. Determine the area, A, under the curve obtained in Step 5:

$$A = \int_0^V R K_P \, dV$$

7. Divide the value obtained in Step 6 by the total charge rate in barrels per day to obtain the SVF.

When an additional soaking drum is used, then the SVF for the soaking drum should be added to the coil SVF. The SVF for the drum may be determined from:

$$SVF_D = \frac{DV}{F}(K_{TD})(K_P)$$

where SVF_D = the SVF of the drum
DV = volume of drum, ft^3
F = charge (fresh plus recycle) throughput rate, bbl/d
K_{TD} = reaction velocity constant for the mean drum temperature
K_P = pressure correction factor for the mean drum pressure

The SVF will range from 0.03 for visbreaking of heavy residual stocks to about 1.2 for light gas oil cracking.

The SVF is a numerical expression of cracking rate and thus can be correlated with product yield and quality. SVF may also be translated into cracking coils and still volumes of known dimensions under design conditions of temperature and pressure.

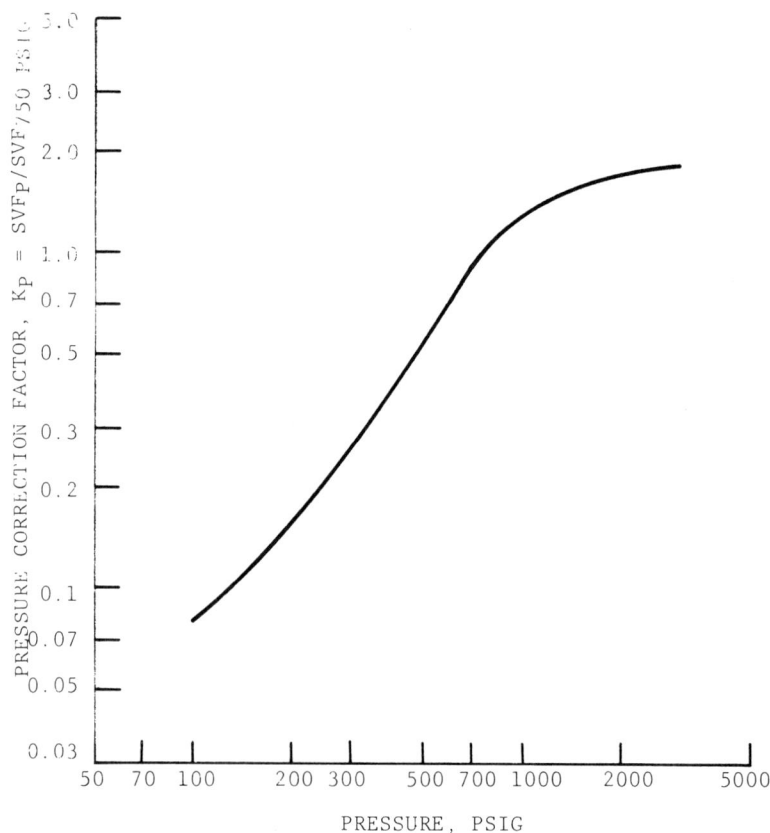

FIG. 46. Pressure correction for soaking volume factor. Base pressure = 750 lb/in.^{2} gauge.

A cracking unit seldom operates very long at design conditions. Charge stock quality changes, desired product yields and qualities change, or additional capacity is required. These changes require a SVF which is different than the design SVF. SVF may be varied by:

1. Varying pressure at constant temperature and feed rate
2. Varying temperature at constant feed rate, the pressure gradient varying with the effect upon cracking rate and fluid density in the cracking coil
3. Varying soaking volume at constant temperature and pressure by:
   a. Varying heater feed rate
   b. Varying the number of tubes in the section above 800°F

Soaking Drums

With the advent of higher firing rate, better efficiency heaters, the use of external soaking drums to provide additional reaction times is of less importance in thermal cracking operations. In modern units the coil in the heater is usually sufficient to provide the temperature–time relationships required. A possible exception would be the case where it is desirable to crack a considerable amount of heavy residual stock. The temperature required probably could not be successfully obtained in a heater coil without excessive coking. As shown and discussed relative to Fig. 2, a reaction chamber (soaking drum) is employed where the hotter, cleaner light gas oil is used to supply heat to the heavier dirty oil stream in a soaking drum. A low temperature light gas oil stream is also frequently used to wet the walls of the soaking drum to minimize coking. Parallel soakers could be used to allow one to be decoked while the other is used for the cracking operations.

Pumps

The pumps used in thermal cracking operations must be capable of operation for extended periods handling a high temperature (above 450°F and up to 650°F) corrosive liquid. In addition, since coke particles are formed in thermal cracking, the pumps must be able to withstand the potential erosion of the metal parts by the coke particles. In the early days of thermal cracking, reciprocating pumps were commonly used. In later units centrifugal pumps have been used. A preferred centrifugal pump would be of the coke-crushing type or may have open impellors with case wear plates substituted for the front rings. The metal should be 12% chromium steel alloy or a higher alloy if serious corrosion is potential.

Miscellaneous Equipment

Heat exchangers should be constructed to provide easy cleaning since the high temperatures and coke particles can create extensive fouling of the exchangers.

The downstream processing equipment (flash drums, separators, fractionating towers, etc.) would be of standard design, and no special design specifications are required other than minimizing potential coke buildup. This can be accomplished by designing the equipment so there would be no significant hold-up or dormant spots in the process equipment where coke could accumulate.

Streamline angle valves are suitable for drawing off the hot bottoms from the various vessels. Level control instrument connections within any high temperature vessels should be constructed to provide a continuous flushing of the connection with a light gas oil to prevent coke buildup.

In thermal cracking operations there is a considerable amount of excess heat which cannot be economically utilized within the cracking unit itself. When a thermal cracking unit is being considered, it is desirable to construct the unit in conjunction with some other unit, such as a crude still, which could utilize the excess heat to preheat the crude oil charge. Alternately, the excess heat could be used in steam generation facilities.

Economics

The absolute economics (investment and operating costs) of the processes covered in this article would, of course, ultimately depend upon each specific case. For a particular application, a detailed economic study would be required to determine the most profitable operation. The current economic situation of the location where the plant would be constructed and how the plant can be advantageously integrated into existing operations must be taken into consideration. The costs presented in this section should be used for general information only. The costs are based on information presented in literature by Dr. W. L. Nelson, a recognized authority on refinery costs, and have been updated to 1974 values using Nelson Cost Indexes. Specific reference sources are presented at the end of the article.

Figure 47 presents the 1974 estimated investment costs for thermal cracking, visbreaking, and thermal reforming units of various capacities. These costs include pertinent off-sites costs. Figure 47 shows that thermal cracking and thermal reforming units cost about the same with simple (no recycle) visbreaking units costing about 60% as much. If a recycle operation is employed in visbreaking, the investment cost would run 5 to 10% more than for simple visbreaking.

Ranges of 1974 unit operating costs are presented in Figs. 48 (thermal cracking and visbreaking) and 49 (thermal reforming). Pinpointing the actual operating costs is difficult because a wide variety of factors can influence the total operating cost. Thus, ranges are given. In general, the lower line of each range would be indicative of the minimum operating costs for small independent refiners while the upper line would be indicative of the maximum costs for a major refiner. A more detailed breakdown of the operating costs may be found in Nelson's *Guide to Refinery Operating Costs* (*Process Costimating*), 2nd ed.

Cracking, Thermal

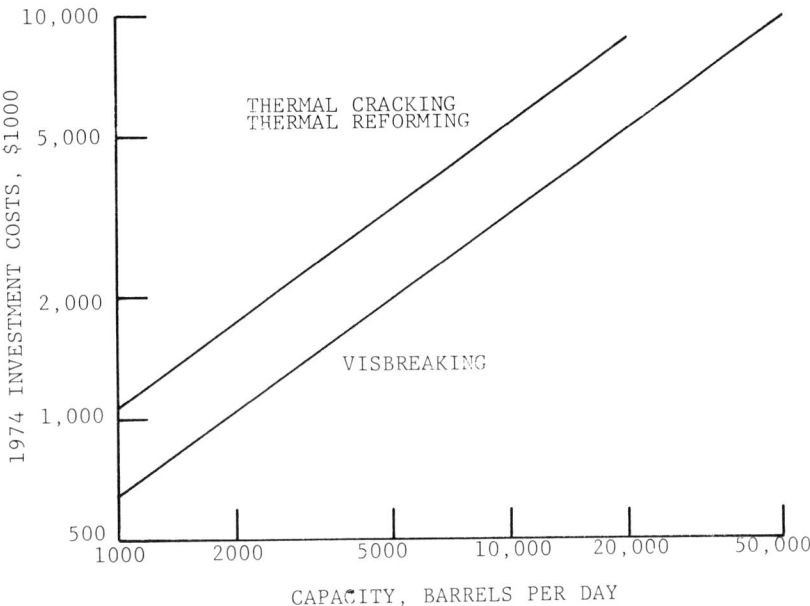

FIG. 47. Estimated investment costs for thermal cracking, visbreaking, and thermal reforming.

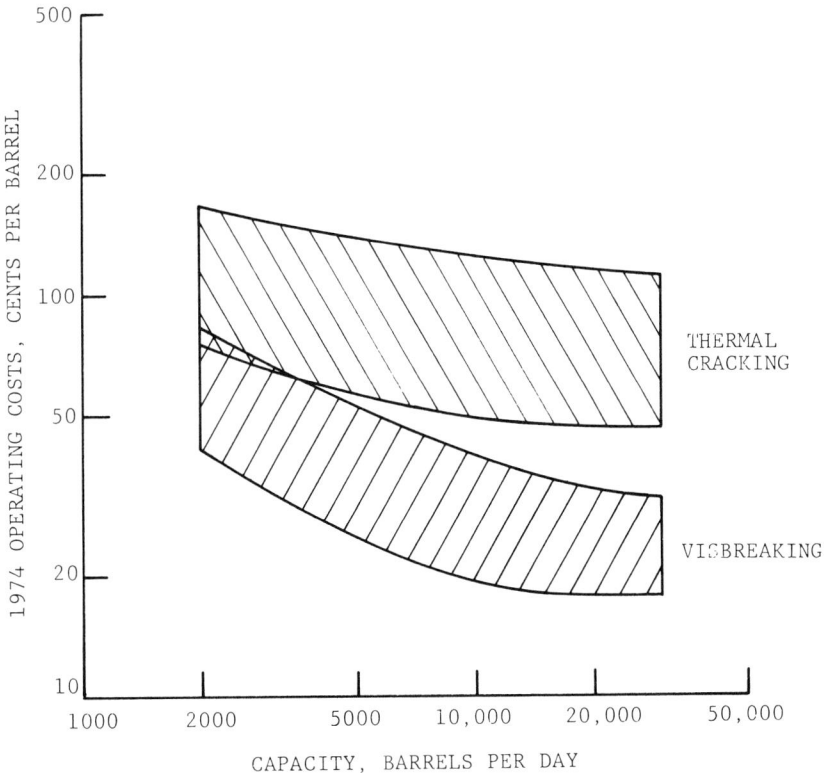

FIG. 48. Operating costs for thermal cracking and visbreaking units.

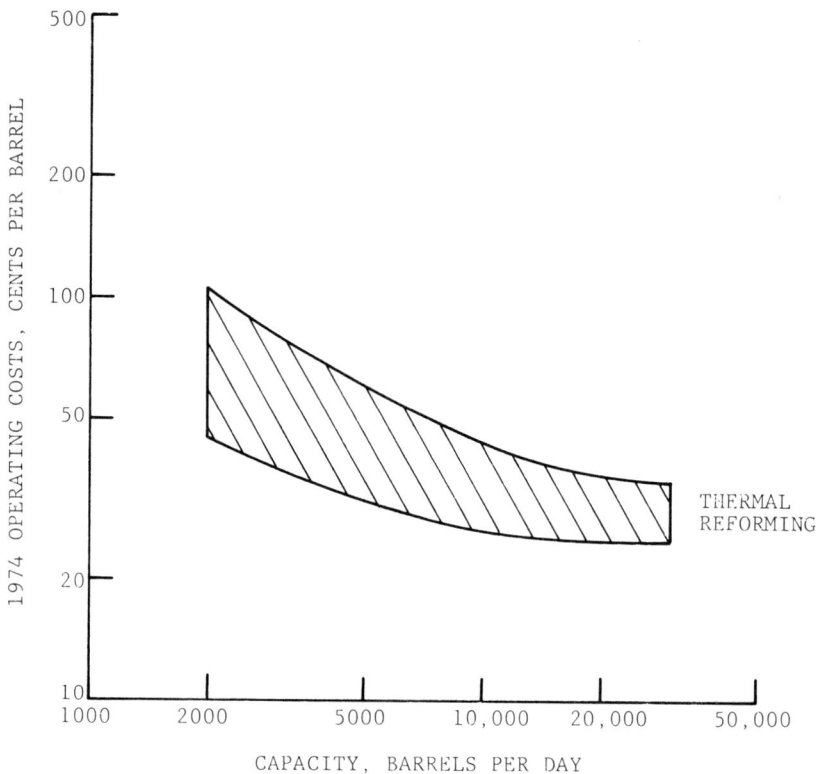

FIG. 49. Operating costs for thermal reforming units.

Bibliography

Bell, H. S., *American Petroleum Refining*, 4th ed., Van Nostrand, Princeton, New Jersey, 1959.

Bland, W. F., and Davidson, B. L., *Petroleum Processing Handbook*, McGraw-Hill, New York, 1967.

Cooper, T. A., and Ballard, W. P., "Thermal Cracking, Visbreaking and Thermal Reforming," in *Advances in Petroleum Chemistry and Refining*, Vol. 6, Wiley-Interscience, New York, 1962.

Hengstebeck, R. J., *Petroleum Processing Principles and Applications*, McGraw-Hill, New York, 1959.

Kalichevsky, V. A., *Pet. Eng.*, 27(12), C-11 (November 1955).

Kirk-Othmer Encyclopedia of Chemical Technology, 2nd ed., Vol. 15, Wiley-Interscience, New York, 1964.

Leslie, E. H., *Motor Fuels—Their Production and Technology*, Chemical Catalog Co., New York, 1923.

Nash, A. W., and Howes, D. A., *The Principles of Motor Fuel Preparation and Application*, Vol. 1, 2nd ed., Wiley, New York, 1938.

Nelson, W. L., *Guide to Refinery Operating Costs (Process Costimating)*, 2nd ed., Petroleum Publishing Co., Tulsa, Oklahoma, 1970.

Nelson, W. L., *Oil Gas J.*, *55*(15), 139 (April 15, 1957).
Nelson, W. L., *Petroleum Refinery Engineering*, 4th ed., McGraw-Hill, New York, 1958.
"Nelson Cost Indexes," *Oil Gas J.*, *73*(14), 134 (April 7, 1975).
Sachanen, A. N., *Conversion of Petroleum*, 2nd ed., Reinhold, New York, 1948.
Staff, *Hydrocarbon Process. Pet. Refiner*, *41*(9), 172 (1962).
Staff, *Hydrocarbon Process.*, *47*(9), 153 (1968).
Staff, *Hydrocarbon Process.*, *53*(9), 123 (1974).
Stephens, M. M., *Petroleum Refining*, Vol. 2, Pennsylvania State College, State College, Pennsylvania, 1941.

W. P. BALLARD
G. I. COTTINGTON
T. A. COOPER

Cracking, Catalytic

Introduction

As the uses of "benzene" developed in the early petroleum industry, the need for this product outstripped the amount available in the crude. Early distillers happened onto thermal cracking as they forced their stills, but this process, when fully developed, could not keep up with demand. In addition, some refiners, Sun Oil in particular, did not desire to extend their gasoline supply by the use of tetraethyl lead which would have to be obtained from one of their competitors, Ethyl. This decision forced them to look to other techniques to increase gasoline supplies, and they looked to catalytic cracking. Thus Socony Vacuum and Sun Oil pursued the development of the Houdry catalytic cracking process. The first commercial unit was placed into operation in 1936. This development significantly increased the yield of gasoline from the crude fed to a refinery.

In the early 1900s, two pioneers, E. W. Gray and A. M. McAfee of Gulf Oil Corp., realized the possible benefits of a cracking reaction to improve yields and experimented with aluminum chloride. However, it was not until the 1920s that A. J. Houdry invented a process which seemed feasible for commercial application. In this process a silica alumina catalyst at high temperature was used to crack gas oils to gasoline.

The main problem with the new process was the regeneration of the catalyst. One of the products of the cracking reaction is coke. This coke deposits on the catalyst and blinds the active catalytic sites. The coke can be burned off with air but a considerable quantity of heat is liberated which, if not removed, can sinter or possibly melt the catalyst. The catalyst in the Houdry process was held in fixed beds. An ingenious molten salt heat removal scheme was developed to control the temperature in the fixed beds as the catalyst was regenerated.

The first Houdry unit with a design capacity of 2000 bbl/d was placed onstream in 1936 by the Socony Vacuum Oil Co. At about the same time, Sun Oil placed a 15,000 bbl/d plant in operation. A very important development contributing to the commercial success was an automatic valve switching mechanism which provided a 10-min cracking, 5-min purging, and 10-min regeneration cycle.

The Standard Oil Development Co. (now Exxon Research and Engineering Co.) pursued the development of a simplified process which utilized powdered catalyst. Dr. W. K. Lewis of M.I.T. participated in this development, and the culmination of the effort was the fluid solid catalytic cracking process. World War II accelerated the application of this process, and, under wartime emergency agreements, many other companies participated in the development and refinement of the process. These companies included Standard Oil Development Co., Universal Oil Products, M. W. Kellogg Co., Standard Oil Co. (Indiana), Texaco Development Co., and Shell Oil Co. The first fluid unit was placed onstream in 1942 in the Baton Rouge Refinery of Standard Oil Co., Louisiana (now Exxon Co., U.S.A.). By mid-1945, 32 units were in operation in various refineries. The growth of the fluid process has continued and now there are about 350 units with a total capacity close to 10 million bbl/d. Development of cat cracking capacity not including the Soviet Union, Eastern Europe, and the Peoples Republic of China is shown in Table 1.

Fixed-Bed Catalytic Cracking

Although fixed-bed catalytic cracking units have been phased out of existence, they represented an outstanding chemical engineering commercial development [1, 2]. The unit incorporated a fully automatic instrumentation system which provided a short-time reactor/purge/regeneration cycle, a novel molten salt heat transfer system, and a flue gas expander for recovering power to drive the regeneration air compressor.

The flow in a typical unit is shown in Fig. 1. Feed was preheated with reactor effluent and then fully vaporized in a furnace and heated to about 800°F. This requirement of complete vaporization necessarily limited feeds to those with a low boiling range. Heavier liquid components were retained in a tar separator before the feed was passed into the bottom of the upflow fixed bed reactors. The catalyst consisted of a pelletized natural silica alumina catalyst and was held in

TABLE 1 Growth of Cat Cracking Capacity

Year	bbl/d
1945	1,000,000
1950	1,700,000
1960	5,200,000
1970	8,300,000
1978	9,900,000

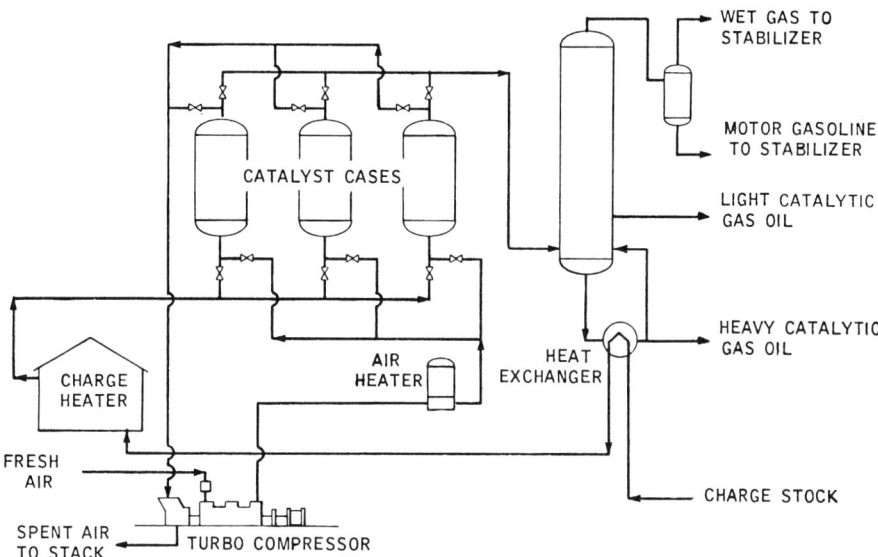

FIG. 1. Early Houdry unit.

reactors or "cases" about 11 ft in diameter and 38 ft length for a 15,000 bbl/d unit. Cracked products then passed through the preheat exchanger and were condensed and fractionated in a conventional manner. The reactors operated at about 30 lb/in.^{2} gauge and 900°F.

The heat of reaction and some of the required feed preheat were supplied by circulating a molten salt through vertical tubes distributed through the reactor beds. The reaction cycle of an individual reactor was about 10 min, after which the feed was automatically switched to a new reactor which had been regenerated. The reactor was steam purged for about 5 min and then isolated by an automatic cycle timer. Regeneration air was introduced under close control, and carbon was burned off at a rate at which the bed temperature could be controlled by the recirculating molten salt stream. This stream comprised a mixture of KNO_3 and $NaNO_2$ which melted at 284°F. This molten salt was then cooled in the reactors through which feed was being processed. The regeneration cycle lasted about 10 min. The regenerated bed was then purged of oxygen and automatically cut back into cracking service. There were three to six reactors in a unit. Gasoline yields diminished over the life of the catalyst (18 months) from 52 to 42 vol.% on fresh feed.

Equilibrium was never reached in this cyclic process. The gas oil conversion—i.e., the amount of feed converted to lighter components—was high at the start of a reaction cycle and progressively diminished as the carbon deposit accumulated on the catalyst until regeneration was required.

Multiple parallel reactors were used to approach a steady-state process. However, the resulting process flows were still far from steady state. The reaction bed temperature varied widely during reaction and regeneration periods, and the temperature differential within the bed during each cycle was considerable.

Moving-Bed Catalytic Cracking

The fixed-bed process had obvious capacity and mechanical limitations. It was improved by the Socony Vacuum Oil Co. and Houdry in parallel efforts. The results were a moving-bed process in which the hot salt heat transfer and cycle time systems were eliminated. Catalyst was lifted to the top of the structure and flowed by gravity down through the process vessels. The plants were generally limited in size to units processing up to about 30,000 bbl/d. Over the years, these units have been essentially replaced by larger fluid solids units.

In the moving-bed processes [3], the catalyst is pelletized into about $\frac{1}{8}$ in. diameter beads. These beads flow by gravity from the top of the unit through a seal zone to the reactor which operates at about 10 lb/in.^{2} gauge and 850 to 925°F. The catalyst then flows down through another sealing and countercurrent stripping zone to the regenerator or kiln which operates at essentially atmospheric pressure. In the Socony unit the regeneration air is introduced near the center of the regenerator bed and flows upward and downward. The upward flowing gas burns about 60% of the coke, heats the catalyst flowing downward, and leaves at about 850 to 900°F. The downward flowing air completes the combustion, and the catalyst reaches 1150 to 1250°F. Trim cooling is accomplished by water coils in the bed. In the Houdry unit, all the air is introduced at the bottom of the bed and flows countercurrently to the catalyst. Two or three cooling zones are provided in the bed for temperature control. The catalyst then flows into another seal zone from which it is lifted to the top of the structure to repeat the flow. Internals are carefully designed to insure uniform plug catalyst flow in the vessels even though the catalyst velocity in various parts of the system may vary by a factor of 50.

The Socony Vacuum process, the Thermofor Catalytic Cracker (TCC), is shown in Fig. 2, and the Houdry process is shown in Fig. 3 [4]. The essential equipment difference between the two units is that the Houdry unit consists of a single vessel with reaction and stripping zones separated by intermediate vessel heads. A basic process difference is that flue gas rather than air is used for catalyst lifting in the Houdry unit. This allows a higher circulation rate. The Socony Vacuum unit contains separate vessels for reaction and regeneration.

In early moving-bed units, built around 1943, bucket elevators were used to lift the catalyst to the top of the structure. In later units, about 1949, a pneumatic lift was used. This pneumatic lift permitted higher catalyst circulation rates which in turn permitted injection of all liquid feeds and feeds of higher boiling range. As shown in Fig. 4, a primary air stream is used to convey the catalyst [5]. A secondary air stream is injected through an annulus into which the catalyst can flow. Varying the secondary air rate will vary the circulation rate.

The lift pipe is tapered to a larger diameter at the top to reduce gas velocity and minimize erosion and catalyst attrition at the top. This taper is also designed so that total collapse of circulation will not occur instantaneously when the saltation concentration, or velocity, of solids is experienced (that velocity below which particles tend to drop out of the flowing gas stream). The taper can be designed so saltation is preceded by a pressure instability which can alert the operators to take corrective action.

Typically, 2 to 4% coke can be deposited on the catalyst in the reactor and

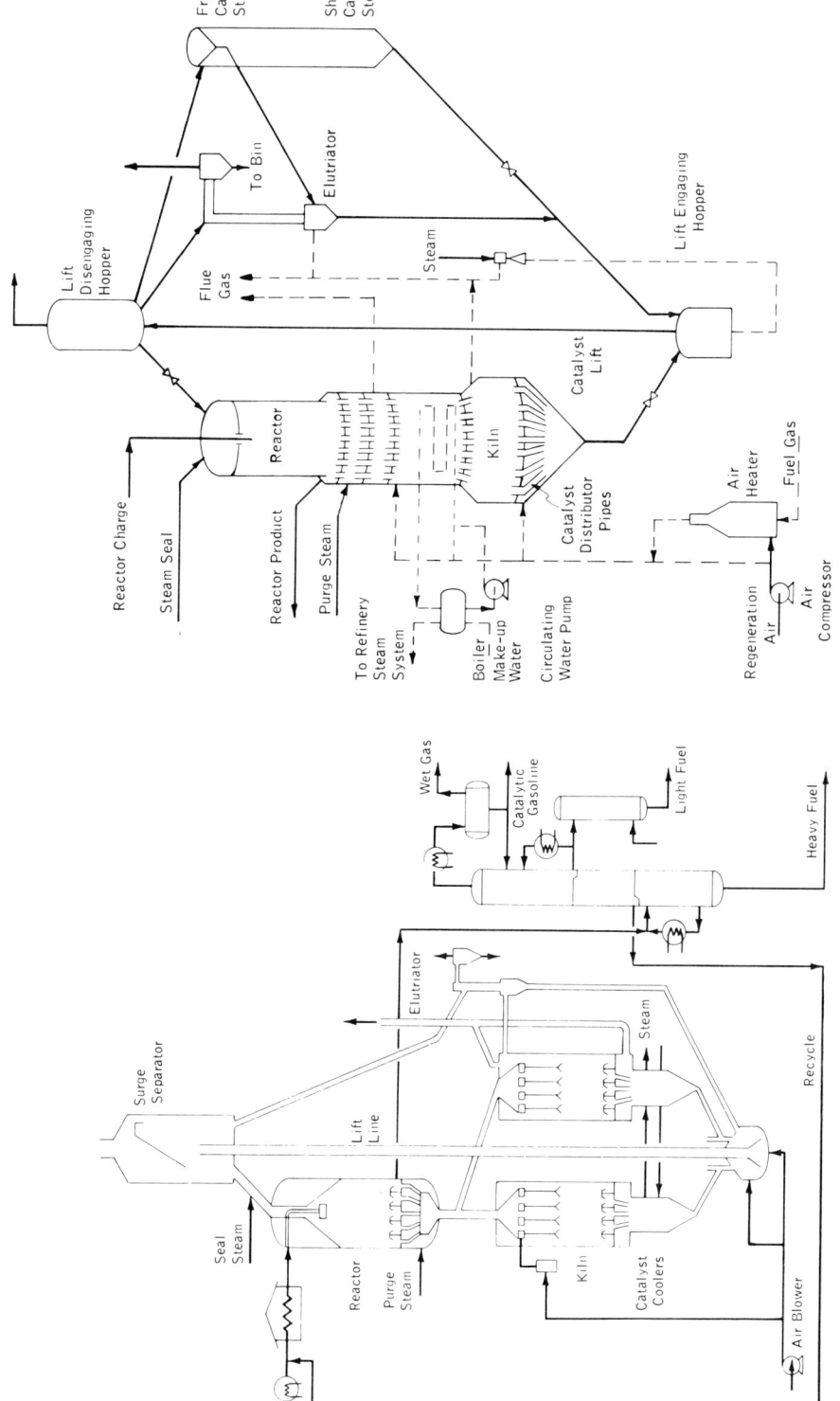

FIG. 2. Thermofor moving-bed catalytic cracking process. (Courtesy John Wiley and Sons.)

FIG. 3. Houdriflow catalytic cracking process.

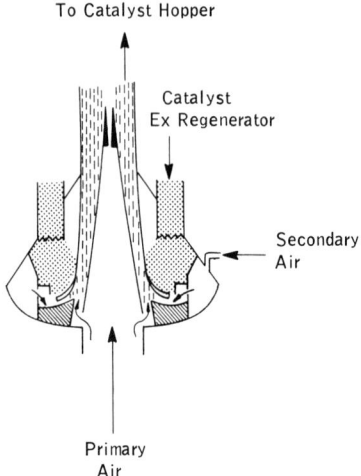

FIG. 4. Moving-bed cracker, catalyst pick-up system.

burned in the regenerator. The catalyst circulation is generally not adequate to remove all the heat of combustion, and steam or pressurized water coils are located in the regeneration zone to remove excess heat.

The oil is injected at the top of the reactor bed and flows concurrently with the beads to the bottom of the reactor zone where product vapors are collected in underflow weir channels and ducted to the fractionator. Typical operation and yields for cracking a Mid-Continent gas oil, 26° API, 750°F 50% point, 11.9 UOP K factor feed, on 3A catalyst or Durabead 5, a zeolitic catalyst, are shown in Table 2 [5].

TABLE 2 Yields from TCC Unit

Catalyst	High Alumina	D-5
Make-up, tons/d	0.75	0.35
Percent in unit	100	50
Operating conditions		
Fresh feed, bbl/d	5800	5775
Reactor temperature, °F	923	950
Throughput ratio	1.43	1.33
Conversion, vol.%	65.1	72.6
Yields		
Fuel gas, wt.%	6.0	6.4
Poly feed, vol.%	16.5	17.5
C_5+ gasoline, vol.%	46.3	57.3
Light cycle oil, vol.%	21.9	18.0
Decant oil, vol.%	7.6	7.0
Coke, wt.%	7.6	7.0
Research octane clear	—	93.6

Fluidized Solids Units

The application of fluidized solids techniques to catalytic cracking resulted in a major process breakthrough. It was possible to transfer all the regeneration heat to the reaction zone. Much larger units could be built and heavy liquid feeds could be processed.

The first fluid units (shown in Fig. 5) were Model I upflow units. They were primarily developed by Standard Oil Development Co. (now Exxon Research and Engineering Co.). In these complex multivessel units, catalyst flowed up through the reaction and regeneration zones in a riser type of flow regime. Originally, the Model I unit was designed to feed a reduced crude to a vaporizer furnace where all of the gas oil was vaporized and fed, as vapor, to the reactor. The unvaporized bottoms bypassed the cracking section. During World War II, because of the demand for C_4's for alkylation to aviation gasoline and for rubber manufacture, it was necessary to use heating oil as feed. Later, after the war when emphasis was shifted from C_4 production to gasoline production, feedstocks were made heavier again.

The design of these units was influenced by the availability of auxiliary equipment since these units were being rushed into operation at the beginning of World War II. Regenerator pressures were very low at 2 to 3 lb/in.^{2}gauge since high discharge pressure, reliable, centrifugal blowers were unavailable. As shown in Fig. 5, the cracked oil vapors and entrained catalyst left the top of the reactors which had enlarged sections so as to increase catalyst holdup. The gas velocity in these sections was in the range of 4 to 6 ft/s. All the catalyst was separated in an external 3-stage multiclone system. The catalyst dropped from the cyclones into a hopper where it was fluidized with steam to displace any entrained hydrocarbon vapors. Catalyst then flowed from the hopper down a

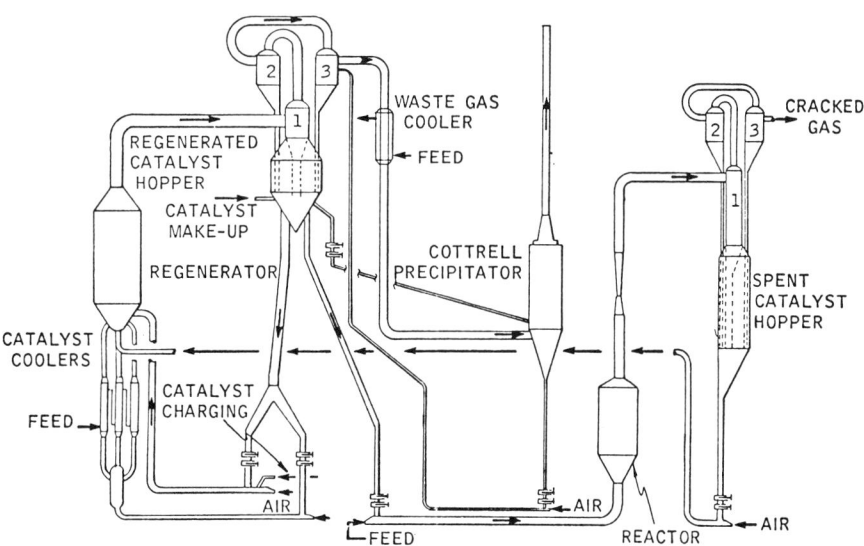

FIG. 5. Model I catalytic cracking unit.

standpipe, through a controlling slide valve located at the standpipe base, into a riser where it was carried by a stream of air into the regenerator.

The regenerator operation was similar to the reactor operation except for the use of air in place of the oil vapor. One major difference was that part of the catalyst from the regenerated catalyst hopper was returned to the regenerator through catalyst fresh feed exchangers to control regenerator temperature and to preheat the feed. In addition, there was another bypass line from the hopper to the regenerator which was used to control the dense bed level or holdup in the regenerator. The flue gas leaving the cyclones passed through heat recovery equipment into electrical precipitators where further recovery of catalyst fines was effected. The catalyst from the regenerated catalyst hopper fell into the standpipe, which had a slide valve at the base, controlling cat flow into a riser where the oil feed was injected.

Before the first Model I was placed in operation, it was realized that lowering vessel velocity to the 1.5 to 2 ft/s range would form a dense bed of catalyst in the reaction zone. This bed had a density of about 20 to 25 lb/ft^3 and was made large enough to provide sufficient catalyst holdup to complete the reaction and regeneration. A fair amount of the equipment in the Model I unit was devoted to the large cyclone installations and catalyst hoppers. All the vessels were required to be pressure vessels and were designed to withstand 20 to 30 lb/in.^{2} gauge. The lower gas velocity of the downflow unit resulted in larger diameter vessels. This considerably simplified the unit since the cyclones could be housed in the reactor and regeneration vessels. The resulting unit was the Model II or "downflow" unit. Many are still in operation.

The process operation was also modified. Carbon-burning capacity was increased to permit higher conversions. This, of course, liberated more heat in the regenerator and in turn required an increase in catalyst circulation rate to control regenerator temperature. The increased heat was sufficient to eliminate the need for a preheat furnace and allowed all the feed to be injected as a liquid. In addition, it was possible to eliminate the catalyst–fresh feed exchangers which were difficult to maintain.

With the increased catalyst circulation, a more effective means was required to strip the entrained hydrocarbon vapor from the spent catalyst leaving the reactor. An annular zone was provided around the reactor bed. Catalyst flowed into this zone countercurrent to a flow of steam which stripped out the hydrocarbon vapors. The first unit of this design was placed on steam in Baton Rouge in 1942. It was also designed by the Standard Oil Development Co. The basic configuration is shown in Fig. 6.

Regenerated catalyst flowed from the regenerator bed into the top of vertical standpipes which were approximately 100 ft long. As the catalyst flowed down the pipes, static pressure was built, much as would be built in a liquid column. The buildup was equivalent to that which would be achieved with fluid having a 35 to 40 lb/ft^3 density. An aeration gas was provided along the pipe to preserve the "fluidity" of the catalyst. A slide valve was used at the bottom to throttle the catalyst flow into the oil feed stream which was located directly below the valve. The vaporizing and reacting feed transported the catalyst to the reactor where additional holdup in the form of a fluidized bed completed the reaction. The catalyst then flowed downward to an annular

Cracking, Catalytic

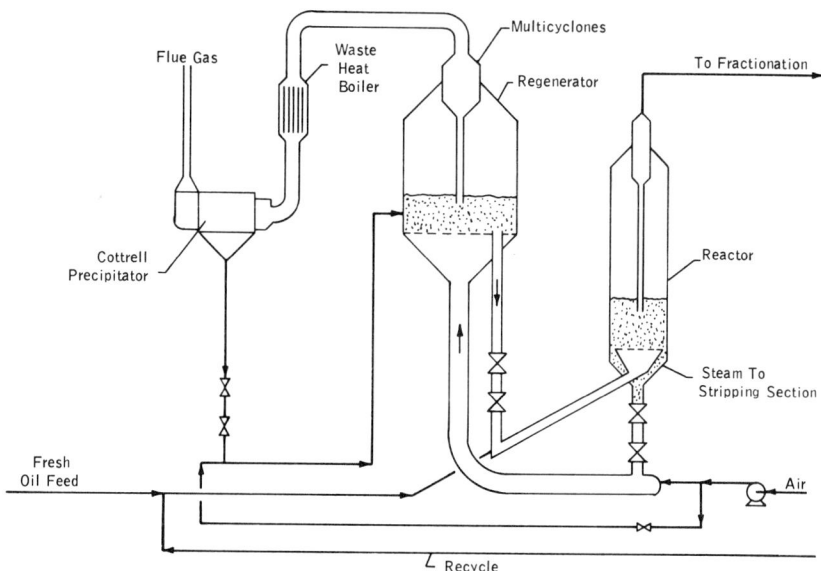

FIG. 6. Downflow Model II catalytic cracking unit.

stripping zone, a short spent catalyst standpipe, and through a controlling slide valve into the air riser which passed up into the bottom of the regenerator. The regenerator was placed high in the structure to minimize the amount of air compression required. The reactor pressure was set at about 18 to 20 lb/in.^{2} since, with the catalysts available at the time, economics indicated this to be the optimum reactor pressure.

Catalyst recovery was first accomplished by multiclones followed by electrostatic precipitation. In later units, either 2 or 3 stages of relatively large diameter cyclones were used.

The development of the catalytic cracking process proceeded rapidly in the World War II years due to wartime emergency agreements under which companies pooled technical knowledge and development. This "Recommendation 41" agreement included Standard Oil Development Co., Kellogg Co., Texaco Development Co., Universal Oil Products Co., Royal Dutch Shell Group, and Standard Oil Co. (Indiana).

The development and simplification of the process continued with the UOP side by side unit and the Model III Cat Cracker which utilized more efficient air compression equipment. The higher regenerator pressure permitted lowering the regenerator vessel to the reactor elevation, greatly simplifying the structure. The higher pressure also favored more complete regeneration.

As discussed elsewhere in this article, catalyst developments were also proceeding with equipment developments. Microspheroidal synthetic catalysts of 13% Al_2O_3 replaced ground natural clay catalyst. This change was followed by a more active 25% alumina catalyst. Zeolitic catalysts were introduced in the early 1960s [6] and caused a major swing from bed cracking to riser cracking. However, riser cracking was first practiced in the mid-1950s on 25% alumina catalyst for improved yields under special circumstances. Improved configurations of the cat cracking unit were developed in parallel with the advance in catalysts.

Subsequent to World War II, many of the Recommendation 41 participants developed modifications to the process which they either built for themselves or licensed. Others have also entered the licensing field.

Exxon Research and Engineering Co.

The catalyst circulation system of the slide valve unit was simplified in the Model IV unit, Fig. 7. In this unit, catalyst is transported between vessels by means of U-bends. Since controlling valves are not used, catalyst circulation is controlled by pressure balance or process flows. This is accomplished by varying the density of the catalyst on the upflow side of the spent U-bend by injecting a small part of the regeneration air. The density in this "control riser" varies between 10 and 20 lb/ft^3 by varying the gas velocity between 6 and 15 ft/s. An overflow well is provided in the regenerator and thus the rate of flow of catalyst leaving the regenerator automatically balances the entering flow.

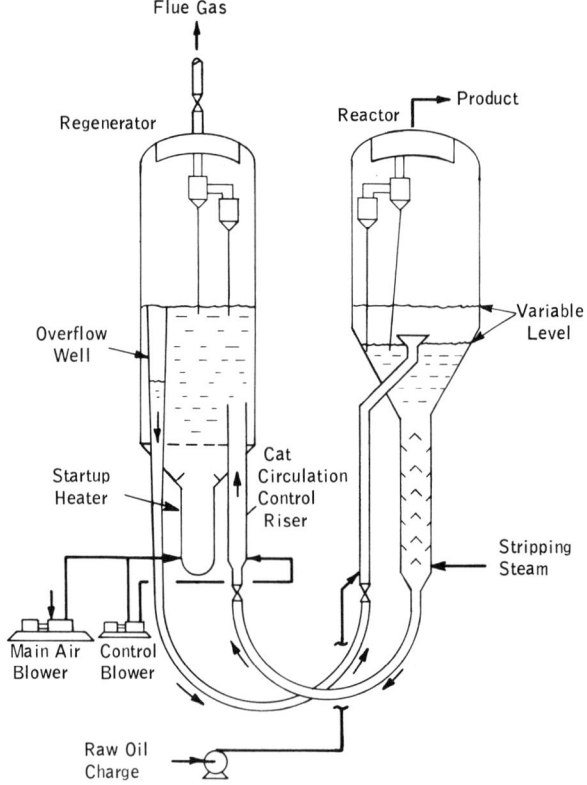

FIG. 7. Exxon Model IV.

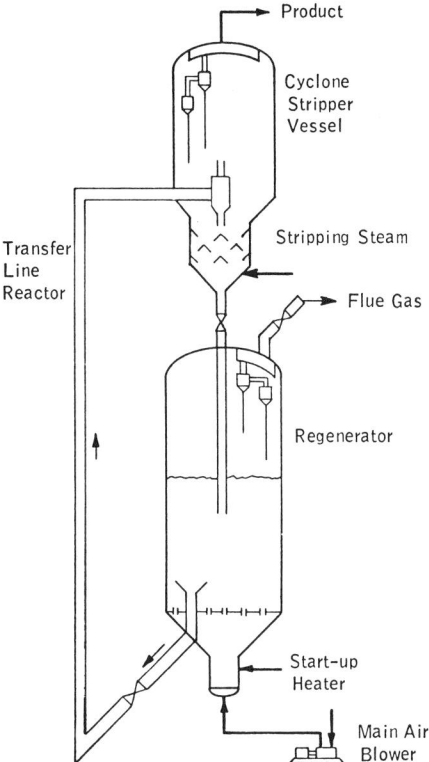

FIG. 8. Exxon transfer line configuration.

Reactor holdup is held constant because it is determined by overall unit inventory. Catalyst circulation can be varied by either shifting differential pressures between the reactor and regenerator or by varying the amount of air in the spent catalyst riser.

This basic U-bend unit was adapted to several different process schemes. A combination unit was developed in which the atmospheric residuua from a crude pipe still was fed to the top sheds in the bottom of the cat cracker fractionating tower. The hot, low molecular weight reactor products vaporized the lighter components of the atmospheric residuum to a 950°F, and sometimes as high as 1100°F, cut point. The vaporized virgin material was condensed with recycle in the fractionator, withdrawn, and then fed to the reactor. Extraneous atmospheric gas oils were also blended into this feed stream. This combination of crude distillation in the cat cracker fractionator eliminated the need for a vacuum pipe still to cut cat cracker feed.

The high velocity Model IV, developed in the late 1950s, was built with reactor and regenerator bed zones designed to run at 4 to 6 ft/s. These high velocities resulted in very low bed densities and efficient conversion and regeneration zones. Both of the vessels were tapered to a larger diameter in their upper parts to provide sufficient space to house the cyclones in the same vessel.

With the new highly active catalysts, many of the Model IV units are now run with no reactor bed. The catalyst level is held below the reactor grid in the stripper zone. Thus the catalyst in the reactor process flow stream is contacting the oil only in a dilute phase or riser type environment.

The transfer line or riser cracker was first applied in Baytown, Texas, when a Model II unit was revised to all riser cracking. This was followed by a new grass roots unit designed for riser cracking only (Fig. 8). Many of the units offered by the cat cracking licensors now feature riser or transfer line contacting. It should be noted that the advance in catalysts has made riser cracking practical in most circumstances. The catalysts are now so active that overcracking to undesirable products would occur if a bed were provided when cracking some feeds.

Universal Oil Products Co. [7, 8]

The UOP stacked unit is shown in Fig. 9. This is a very efficient unit, having a long riser reactor terminating in a reactor vessel at the top of the structure but supported on the regenerator vessel. The reactor vessel generally contains only one stage of cyclones for retaining solids in the unit. The remaining solids are scrubbed out in the lower part of the fractionator, concentrated in the slurry product, and returned to the unit with a portion of the slurry.

In more recent designs, UOP has modified the reactor configuration to an all-vertical riser design with no bed holdup. Minimum oil residence time is also provided in the dilute phase to prevent undesired secondary reactions which degrade cat cracking products (Fig. 10).

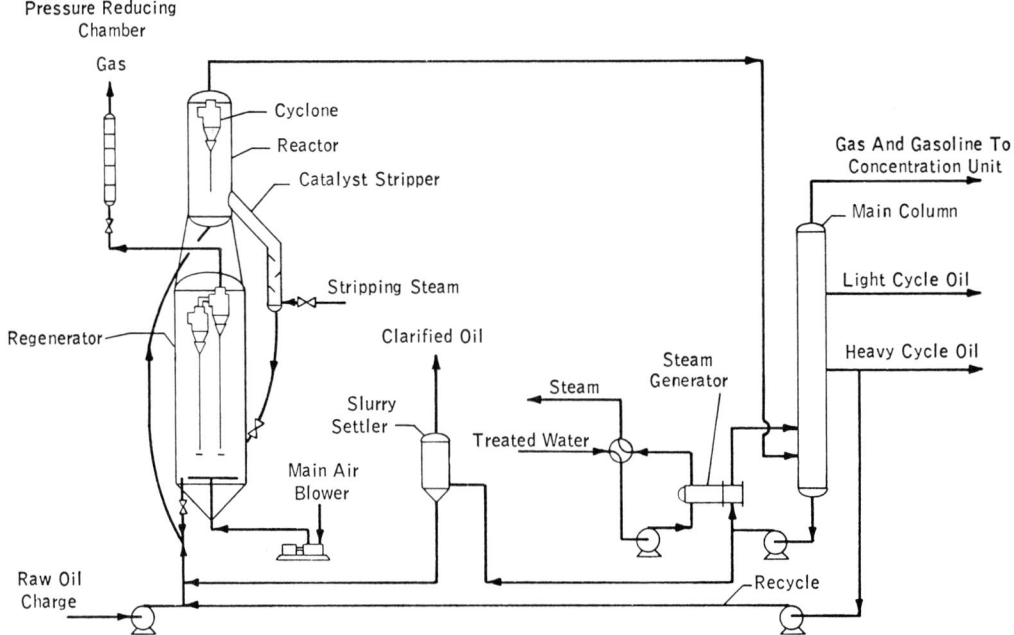

FIG. 9. UOP "stacked" fluid catalytic cracking unit.

Cracking, Catalytic

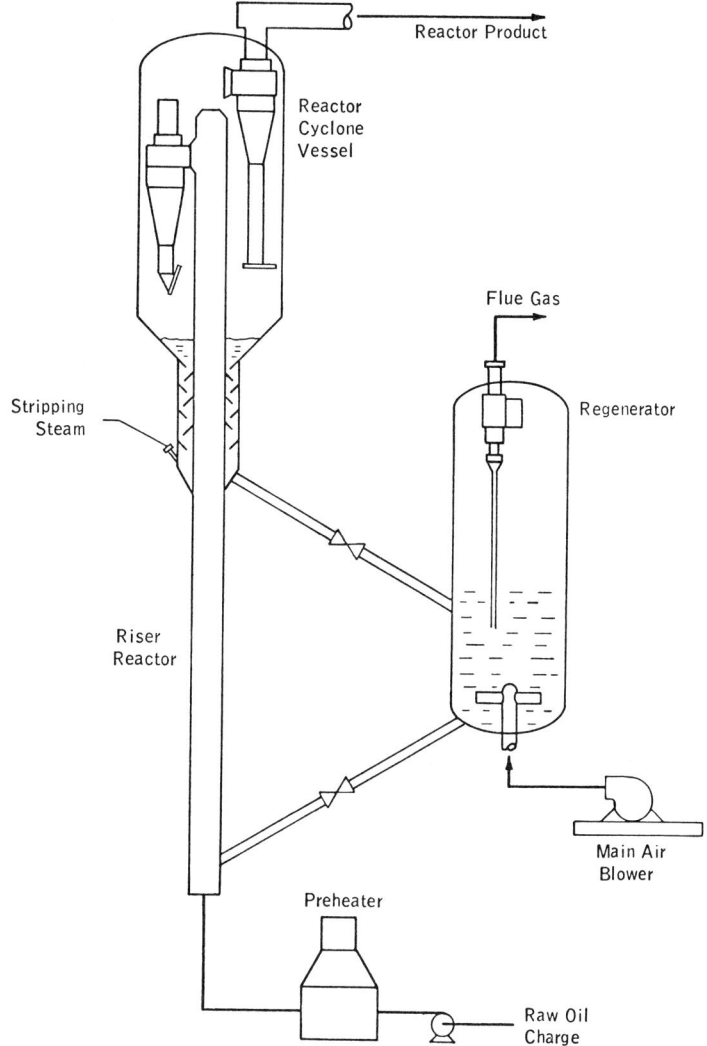

FIG. 10. UOP FCCU unit.

The latest UOP units also incorporate riser regeneration. These units are also run with a catalyst which promotes carbon monoxide conversion in the regeneration system (Fig. 11).

M. W. Kellogg Co. [9]

The M. W. Kellogg Co. was very active in the initial development of the fluid process. They have concentrated on designs (Orthoflow) which have utilized plug valves and stacked vessels (Fig. 12). They have incorporated staged

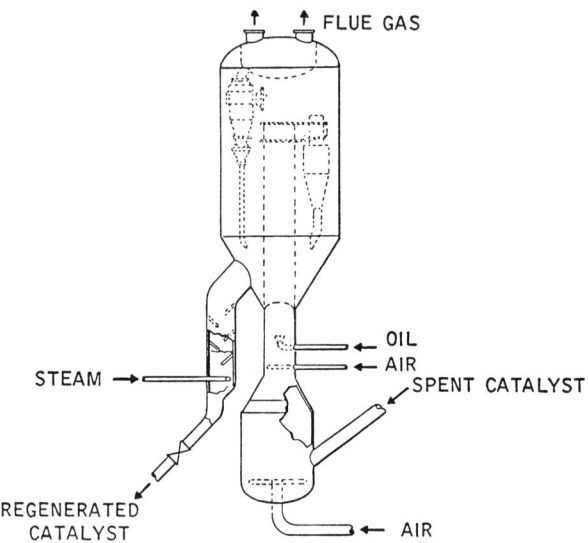

FIG. 11. UOP riser regenerator (U.S. Patent 3,919,115).

regeneration which has reduced the regeneration holdup required and/or carbon on regenerated catalyst.

Units have generally incorporated vertical feed risers for efficient riser cracking and more recently included external feed riser lines having right-angle bends which extend riser lengths and minimize erosion.

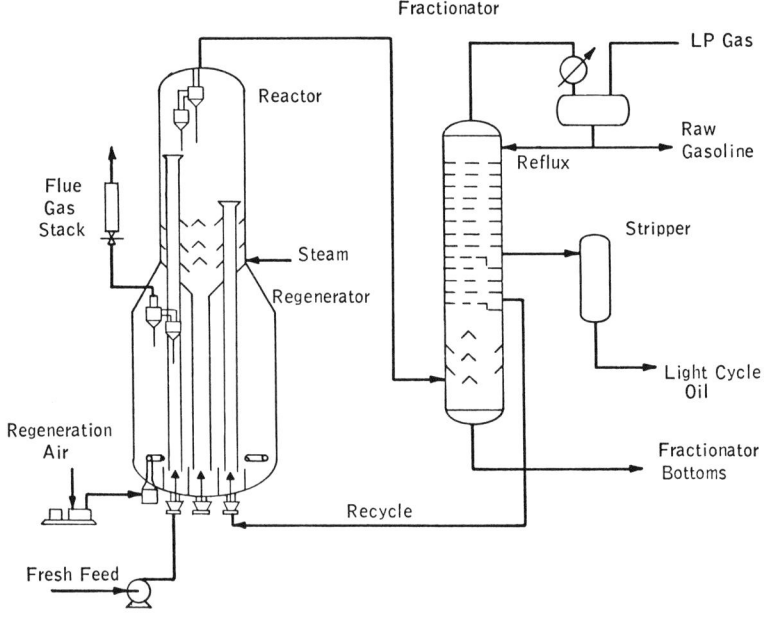

FIG. 12. Kellogg Orthoflow unit.

Cracking, Catalytic

Texaco Development Co. [10]

The Texaco process features a dual reactor system in which fresh feed and recycle can be cracked separately at optimum conditions (Fig. 13). The reactor is teardrop shaped to promote efficient contacting if a bed is retained in the vessel. The regenerator vessel is tapered to provide for a high velocity bed and a lower velocity dilute phase zone. Catalyst is introduced in the side of the regenerator vessel and directed to impart a swirl to the catalyst in the bed for more efficient regeneration. Catalyst flow is controlled by a slide valve.

Gulf Research and Development Co. [11]

The Gulf process incorporates vertical riser contacting (Fig. 14). Efficient conversion of feed is provided by injecting different components of the feed at several points in the riser to optimize the yield patterns.

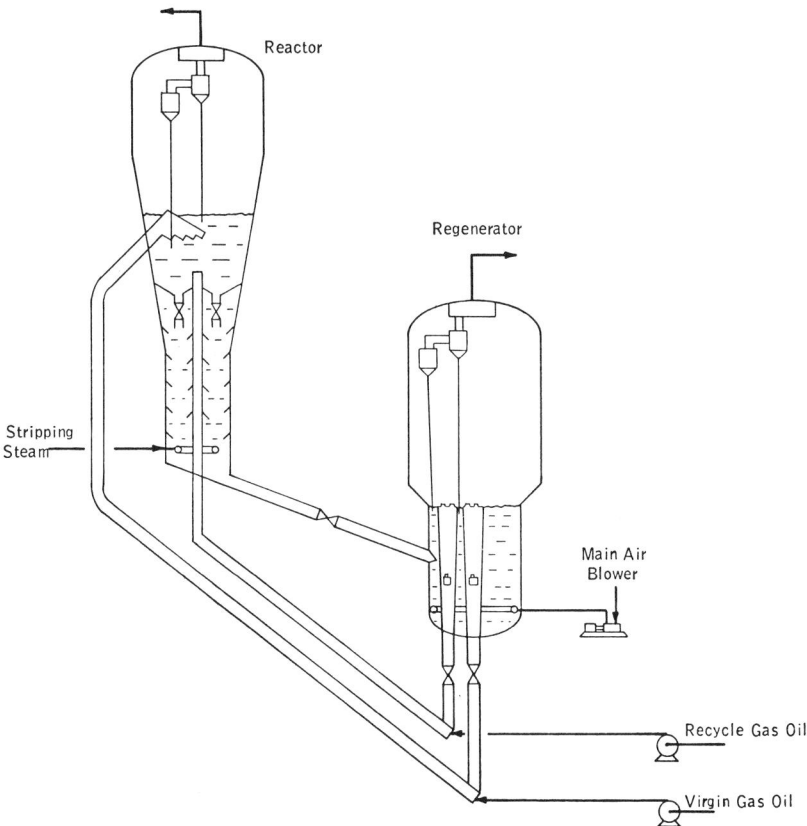

FIG. 13. Texaco fluid catalytic cracking apparatus (U.S. Patent 3,394,076).

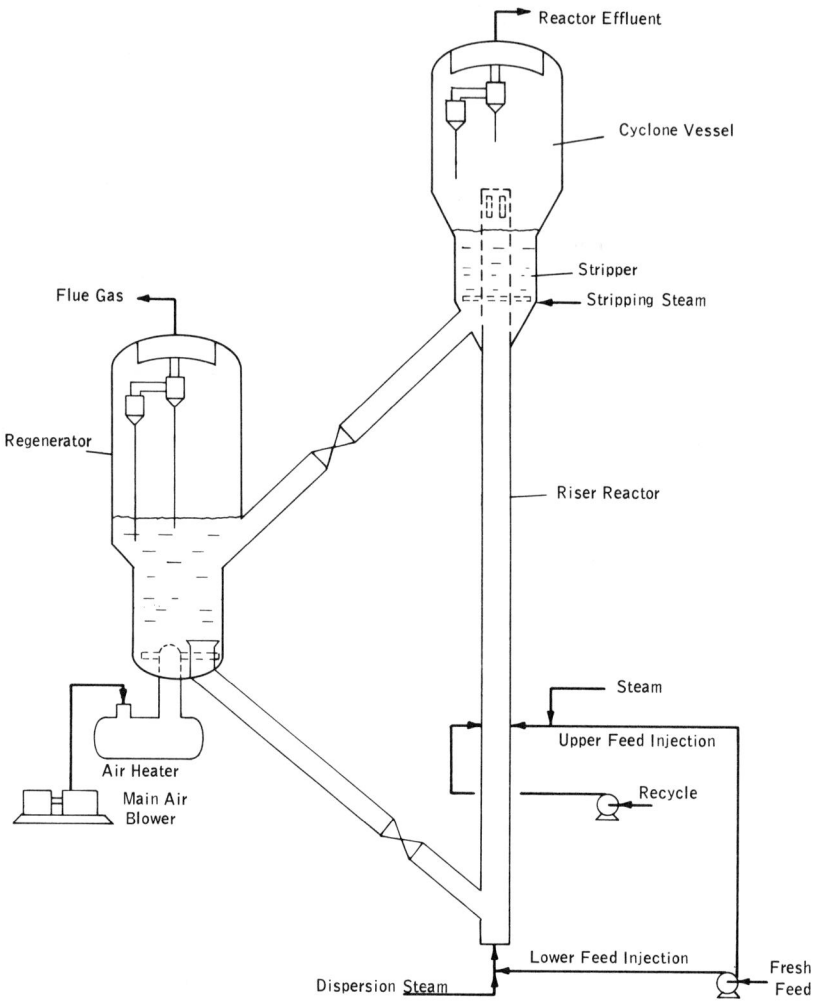

FIG. 14. Gulf catalytic cracker.

Residuum Cracking

The improvement in catalysts and unit configurations have permitted the cat cracking of poorer and poorer feedstocks. Presently, a number of licensors are offering cat cracking processes in which a residuum is fed to the unit. One such unit has been in operation in Phillip's Borger refinery for years. Companies offering the process are Kellogg, Gulf, and UOP. In some cases the residuum feed is hydrotreated in order to reduce the contaminant metals level of the feed and to improve the feed quality so that coke yields are not excessive. In the other cases a very high catalyst replacement rate is used to keep catalyst quality within the range where acceptable product yields are achieved. There is potential for applying a demetallization process to the catalyst to reduce catalyst consumption [89].

For an untreated feed, very high coke makes, in the order of 10 to 12 wt.%, might be expected for some feeds. This high coke make results in a very high heat liberation in the regenerator which cannot be fully utilized in the reactor. As a result, steam coils must be provided in the regenerator zone to recover this excess heat and control catalyst temperature in the regenerator.

The resid cracking unit which M. W. Kellogg Co. has developed in conjunction with Phillips Petroleum Co. is shown in Fig. 15. Conversions up to 85% are claimed commercially on an atmospheric residuum. The unit is similar to the Kellogg Orthoflow "C" but there are some differences which enhance performance on residuua. Catalyst flows from the regenerator through a plug valve which controls the flow to hold the reactor temperature. Steam is injected upstream of the feed point to accelerate the catalyst and disperse it so as to avoid high rates of coke formation at the feed point. Feed atomized with steam is then injected into this stream through a multiple nozzle arrangement. The flow rates are adjusted to control the contact time in the riser since the effects of metals poisoning on yields are claimed to be largely a function of the time that the catalyst and oil are in contact. The reaction is stopped by passing the mix through a rough cut cyclone.

Representative yields on an atmospheric residuum from Light Arabian

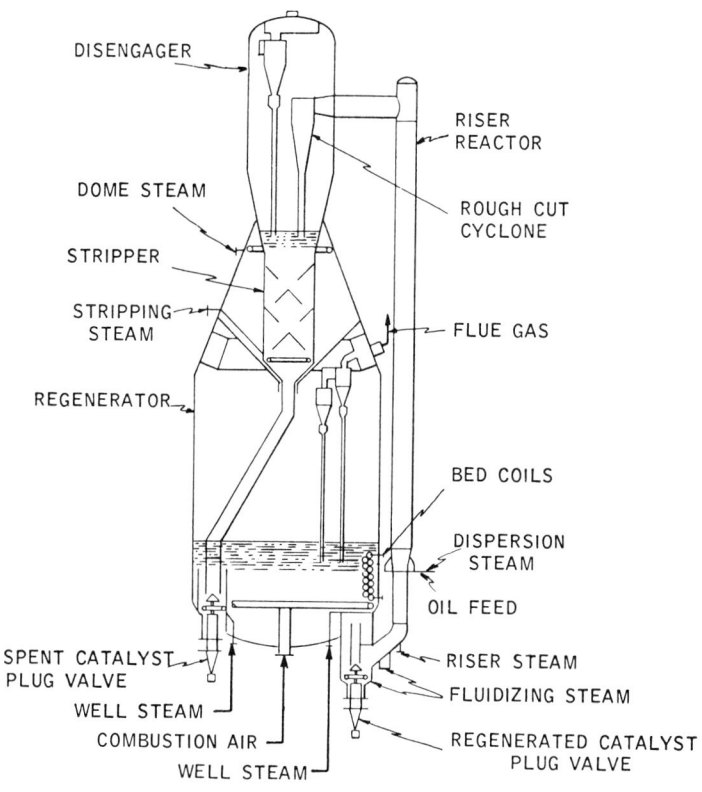

FIG. 15. Kellogg residuum cracker.

TABLE 3 Representative Yield Structure, Kellogg Resid Cracker [12]

	Vol.%		Wt.%	
Conversion, vol.% 400°F ASTM		81.7		
Yields				
C_5/400 ASTM EP gasoline	59.7		48.2	
$C_4 + C_4^{2-}$	16.1		10.1	
iC_4		5.4		3.2
nC_4		1.3		0.8
C_4^{2-}		9.4		6.1
$C_3 + C_3^{2-}$	9.5		5.2	
C_3		2.9		1.5
C_3^{2-}		6.6		3.7
Light cycle oil	13.3		13.3	
Decant oil	5.0		5.7	
H_2S				0.9
C_2 + lighter			3.3	
H_2				0.5
C_1				1.2
C_2				1.0
C_2^{2-}				0.6
Coke			13.3	
Steam, lb/h/bbl fresh feed			10,500	
Qualities				
Gasoline, °API			56	
Sulfur, wt.%			0.25	
Octane				
RON Cl			90	
+ 3 cc			97	
MON Cl			79	
+ 3 cc			86	
Light cycle oil, °API			18	
Sulfur, wt.%			2.2	
Decant oil, °API			0.0	
Sulfur, wt.%			5.5	

Crude, 18.8° API gravity, 6.0 carbon residue, with 14 ppm nickel and 54 ppm vanadium, are given in Table 3.

The Gulf residuum process consists of cracking a residuum hydrotreated to low sulfur and metals levels [13]. In this case, high conversions are obtained but coke yield and hydrogen yield are kept at conventional levels by keeping the metals on catalyst low. The hydrotreated 650+ Kuwait resid feed had properties of 23.6° API, 0.5 wt.% sulfur. It was cracked to an 80% conversion with yields as shown in Table 4.

Other licensors of residuum cracking technology are UOP and Arco. In the Arco process the cat cracking operation is a low conversion step. Metals on catalyst can be tolerated to high levels but can also be controlled to a desired level by catalyst withdrawals and correspondingly higher addition rates and/or removal of metals contaminants from the catalyst.

TABLE 4 Representative Yield Structure, Gulf Resid Cracker [13]

375°F + HDS residua (FCC charge)	0.5 wt.% S	
Conversion, vol.% 400°F ASTM	80	
Yields	Vol.%	Wt.%
$C_5/400$ ASTM EP gasoline	58.7	
$C_4 + C_4^{2-}$	17.8	
$\quad C_4^{2-}$	9.5	
$C^3 + C_3^{2-}$	12.2	
$\quad C_3$	3.6	
$\quad C_3^{2-}$	8.6	
Light cycle oil	17.2	
Decant oil	2.8	
H_2S		0.1
C_2 + lighter		3.5
$\quad H_2$		0.05
$\quad C_2^{2-}$		1.2
Coke		7.5
Qualities		
Gasoline, °API	57.3	
$\quad$ Sulfur		0.03
$\quad$ Octane, RON Cl	94	
Light cycle oil, °API	17.5	
$\quad$ Sulfur, wt.%	0.8	
Decant oil, °API	0.3	
$\quad$ Sulfur, wt.%	1.9	

Reactions and Kinetics

Catalytic Cracking Chemistry

The reactions occurring in catalytic cracking have long been recognized as distinctly different from those of thermal cracking. These differences are particularly noticeable in the distribution of light gases; C_3's and C_4's are major products of catalytic cracking, whereas C_1's and C_2's are found in thermal cracking. Numerous studies, particularly with pure compounds, have detailed the differences in the two types of cracking and have been the basis for formulation of the fundamental mechanisms. A "carbonium ion" mechanism is generally accepted as the explanation for catalytic cracking reactions, and free radical mechanisms are postulated for thermal cracking. Table 5 shows a comparison of the products of catalytic and thermal cracking [14].

Catalytic cracking reactions are frequently classified into primary and secondary reactions. The primary reaction involves the initial carbon/carbon scissions and may be represented as

$$\text{Paraffin} \rightarrow \text{paraffin} + \text{olefin}$$

Alkylnaphthene → naphthene + olefin

Alkylaromatic → aromatic + olefin

The actual reactions are, of course, not this simple, as the initial scission proceeds through a carbonium ion mechanism with several possible products other than a single olefin and a single saturated fragment. The secondary reactions of catalytic cracking are more than side reactions. They include a large number of reactions of olefins and a smaller number of independent reactions. These secondary reactions are a major factor in determining both product yields and product quality. The major secondary reactions are those of

TABLE 5   Comparative Summary of Catalytic and Thermal Cracking Characteristics of Pure Hydrocarbons [14] (temperature range, 400 to 550°C; pressure, about atmospheric)

Hydrocarbon	Catalytic Cracking	Thermal Cracking
n-Paraffins	Extensive breakdown to C_2 and larger fragments. Product largely in C_2 to C_6 range and contains many branched aliphatics. Few normal α-olefins above C_4	Extensive breakdown to C_2 fragments, with much C_1 and C_2. Prominent amounts of C_4 to C_{n-1} normal α-olefins. Aliphatics largely unbranched
Isoparaffins	Cracking rate relative to n-paraffins increased considerably by presence of tertiary carbon atoms	Cracking rate increased to a relatively small degree by presence of tertiary carbon atoms
Naphthenes	Crack at about same rate as those paraffins with similar numbers of tertiary carbon atoms. Aromatics produced, with much hydrogen transfer to unsaturates	Crack at lower rate than normal paraffins. Aromatics produced with little hydrogen transfer to unsaturates
Unsubstituted aromatics	Little reaction; some condensation to biaryls	Little reaction; some condensation to biaryls
Alkyl aromatics (substituents C_3 or larger)	Entire alkyl group cracked next to ring and removed as olefin. Crack at much higher rate than paraffins	Alkyl group cracked to leave one or two carbon atoms attached to ring. Crack at lower rate than paraffins
n-Olefins	Product similar to that from n-paraffins but more olefinic	Product similar to that from n-paraffins but more olefinic
All olefins	Hydrogen transfer is an important reaction, especially with tertiary olefins. Crack at much higher rate than corresponding paraffins	Hydrogen transfer is a minor reaction, with little preference for tertiary olefins. Crack at about same rate as corresponding paraffins

olefins catalyzed or promoted by the same acidic properties of the catalyst that initiate the primary reactions. The major secondary reactions are:

1. Cracking of olefins
2. Double bond shift, geometrical isomerization, or skeletal isomerization
3. Hydrogen transfer to an olefin from naphthenes to produce cyclic olefins and aromatics
4. Hydrogen transfer from another olefin to produce diolefins
5. Polymerization to produce higher molecular weight olefins
6. Aromatization
7. Alkylation of aromatics

Several of the above reactions proceed further to produce carbonaceous deposits on the catalyst; these coke-forming reactions include extended polymerization of diolefins or olefins, aromatization to form polycyclic aromatics, and cyclization and/or condensation of alkylated aromatics.

The application of the carbonium ion theory to the mechanism of catalytic cracking has been developed by a number of investigators and is described in detail in a number of papers by Hansford [15, 16], Greensfelder [14, 17], Haensel [18], and others. A carbonium ion is a hydrocarbon ion with a positive charge on a carbon atom, i.e., a carbon cation. Carbonium ions are formed by three routes in catalytic cracking:

1. The reversible addition of a proton from the acidic catalyst to an olefin produces a carbonium ion

$$R-CH_2-CH=CH-CH_2-R + H^+ \rightleftarrows R-CH_2-\overset{+}{CH}-CH_2-CH_2-R$$

   The small amount of olefin needed to initiate the reaction chain is thought to arise from thermal cracking or from trace olefins in the feed.

2. Carbonium ions may be formed by the abstraction of a hydride ion (H^-) from a saturated molecule by the catalyst acidic site

$$\begin{array}{c} CH_3 \\ | \\ CH_3-CH \\ | \\ CH_3 \end{array} \rightleftarrows \begin{array}{c} CH_3 \\ | \\ CH_3-C^+ \\ | \\ CH_3 \end{array} + \text{acid site}:H^-$$

3. Similarly, carbonium ions may be formed by the abstraction of a hydride ion from a saturated molecule by another carbonium ion.

$$\begin{array}{c} CH_3 \\ | \\ CH_3-CH \\ | \\ CH_3 \end{array} + R-CH^+-R \rightleftarrows \begin{array}{c} CH_3 \\ | \\ CH_3-C^+ \\ | \\ CH_3 \end{array} + R-CH_2-R$$

Carbonium ions are extremely reactive and undergo a wide variety of reactions according to specific rules regarding formation, rearrangement, and reaction. The principal rules have been described by Voge [19] and are:

1. A carbonium ion readily rearranges by methyl or hydrogen shifts to form isomerized ions:

$$\underset{+}{R-\overset{\overset{H}{|}}{C}-CH_2-R} \rightarrow R-\underset{\underset{R}{|}}{CH}-CH_2^+ \rightarrow R-\underset{\underset{CH_3}{|}}{C^+}-R$$

The order of carbonium ion stability is tertiary > secondary > primary.

2. A carbonium ion can extract a hydride ion from a saturated molecule to form a new carbonium ion as above in "3."
3. A carbonium ion can transfer a proton to an olefin and thus form a new carbonium ion and a new olefin:

$$R-CH^+-CH_3 + C_4H_8 \rightarrow C_4H_9^+ + R-CH=CH_2$$

4. If a carbonium ion is sufficiently large, it can split at the bond beta to the positive charge and form an olefin and a new carbonium ion:

$$CH_3-CH^+-CH_2-R \rightarrow CH_3-CH=CH_2 + R^+$$

5. The propagation of carbonium ions is terminated by donation of a proton to the catalyst surface and production of an olefin from the carbonium ion.

The catalytic cracking of cetane proceeds according to these rules by the following steps:

1. Cetane molecules react with either a proton or small carbonium ion on the catalyst surface to form cetyl carbonium ions; secondary cetyl carbonium ions predominate because the primary ions are slower to form and rearrange to secondary ions. For example:

$$C_{16}H_{34} + C_3H_7^+ \rightarrow C_5H_{11}-CH^+-C_{10}H_{21} + C_3H_8$$

2. Secondary carbonium ions are rearranged to tertiary ions.
3. The carbonium ions split at the carbon-carbon bond beta to the positive carbon atom, producing an α-olefin and a primary carbonium ion. For example, with a secondary cetyl carbonium ion:

$$C_5H_{11}-CH^+-CH_2-C_9H_{19} \rightarrow C_5H_{11}-\overset{\overset{H}{|}}{C}=CH_2 + C_8H_{17}-CH_2^+$$

4. The primary carbonium ion will undergo rearrangement into a secondary or tertiary carbonium ion as in Steps 1 and 2 above and then undergo beta scission. Thus

$$C_8H_{17}-CH_2^+ \rightarrow C_7H_{15}-CH^+-CH_3 \rightarrow$$
$$C_5H_{11}-CH_2^+ + CH_2=CH-CH_3$$

5. Rearrangement and beta scission continue until the carbonium ion can no longer yield fragments of three or more carbon atoms.
6. The final small carbonium ion reacts with a cetane molecule to form by hydride ion extraction a cetyl carbonium ion and a small paraffin.

The success of the carbonium ion theory in quantitatively accounting for catalytic cracking product distributions is illustrated in Fig. 16 [17] which compares the predicted (carbonium ion mechanism) and observed carbon number distribution in products from cetane. Similar agreements between experimental and theoretical results have been obtained with other hydrocarbons [17].

Isoparaffins are cracked by mechanisms similar to those for paraffins. The rate of cracking of isoparaffins is generally higher than that of paraffins due to the relatively high rates of removal of hydride ions from tertiary and secondary carbon atoms relative to primary carbons—approximately 20 and 2, respectively—compared to unity for primary carbon.

The cracking reactions of naphthenes are believed to be similar to those of paraffins; the rates of cracking of naphthenes are thought to be determined principally by the types of carbon atoms present, such as primary, secondary, or tertiary carbon atoms, rather than by the ring structure.

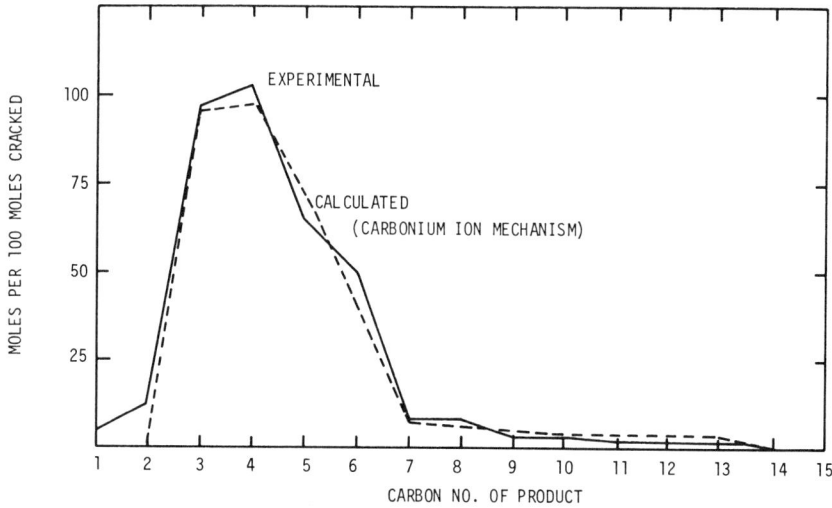

FIG. 16. Catalytic cracking of cetane at 500°C [17]. Experimental: 25% conversion over UOP-B catalyst.

The rapid reaction of olefins with protons or small carbonium ions to yield a large carbonium ion accounts for the high rate of conversion of these unsaturates. The product distribution from olefins is similar to that from paraffins of the same skeletal configuration.

Alkyl aromatics with side chains of three or more carbon atoms also react rapidly in catalytic cracking systems. This high reactivity is attributed to proton attack at the ring carbon atom which carried the propyl (or higher) substituent to form a denuded aromatic and a carbonium ion fragment which further reacts in the usual fashion.

Extensive studies of the hydrocracking of pure compounds have led to generalizations on the cracking performance of hydrocarbon types. Most of these investigations have used pure compounds with molecular weights below that of practical feeds. The same effects are generally observed with high boiling feeds; however, the rates of reaction of all hydrocarbon classes are higher as molecular weight increases, and the effects are sometimes obscured by the presence of several types of structures within a given fraction—for example, high boiling aromatic fractions usually contain large proportions of paraffinic and naphthenic carbon in addition to aromatic rings.

n-Paraffin cracking is characterized by the production of relatively large amounts of gas, predominantly C_3 and C_4, and low coke yields. The small production of coke and aromatics is attributed to secondary reactions, possibly polymerization and cyclization of olefins, and condensation of olefins with aromatics. The cracking characteristics of isoparaffins are similar to those of n-paraffins, although the yields of C_3 and C_4 are generally somewhat smaller. Naphthenes are cracked faster and usually produce more liquid products and less coke and gas than paraffins. In addition, naphthenes undergo extensive dehydrogenation. With monocyclic naphthenes this is a desired reaction leading to monocyclic aromatics and high quality gasoline rich in aromatics. With polynuclear naphthenes and high severity conditions, dehydrogenation to the less desirable polycyclic aromatics occurs. These dehydrogenation reactions of naphthenes are particularly evident with cracking catalysts contaminated with nickel, vanadium, and iron.

The cracking characteristics of aromatics depend upon the number of aromatic rings present; however, all types of aromatics exhibit the common reaction of rapid scission of the large side chains. With monocyclic aromatics this scission of side chains leads to the relatively stable aromatics in the gasoline boiling range. With diaromatics, scission of the larger side chains produces methyl or ethyl substituted naphthalenes, a major constituent of the light cycle oils produced from aromatic or naphthenic feeds. Similar scission reactions of triaromatics and higher polycyclics lead to the corresponding aromatics with only short side chains remaining. These denuded polycyclic aromatics concentrate in the heavier cycle oils; however, a substantial portion of the aromatic carbon in these structures is deposited as coke on the catalyst.

Carbon or coke formation in catalytic cracking is of exceptional importance. Not only does the buildup of the carbonaceous deposit reduce catalyst activity (Table 6), but the coke yield is a basic input into the heat balance and thus influences not only the regenerator design but also the need for feed preheat or heat recovery facilities. In addition to representing a loss of valuable liquid or gaseous products, the formation of coke or carbon alters the selectivity of the catalyst. Thus catalytic cracking over a low carbon content catalyst produces more gasoline and less coke than a similar conversion over a highly coked or spent catalyst (Table 6 [20]).

Coke formation in catalytic cracking is a complex phenomenon resulting from a number of reactions and conditions. Several principal types of coke have been recognized [21]. "Catalytic coke" is the coke which is inevitably found when any hydrocarbon is cracked over an acidic catalyst. It arises primarily from a variety of secondary reactions (polymerization of olefins and diolefins, aromatization to form polynuclear aromatics, cyclization and/or condensation of alkylated aromatics and dehydrogenation to form polycyclic aromatics). The exact mechanism of catalytic coke formation is not fully defined but size and basicity of the polynuclear aromatics are important factors. The major steps are thought to be a strong adsorption of polynuclear aromatic structures followed by condensations and reactions with olefins to produce intermediate structures of increasing size and complexity. Polynuclear aromatic structures in the feedstock are the principal contributors to catalytic coke. Coke yield of practical feedstocks is often related to the aromatic carbon content of the feed present in triaromatic or higher polynuclear aromatic structures.

A second type of coke is designated "cat-to-oil" coke. This portion of the total carbonaceous deposit arises from portions of the feed being trapped in the pores of the catalyst as it leaves the reactor and being transported to the regenerator. "Cat-to-oil" coke is frequently related to catalyst/oil ratio, porosity of the catalyst, and efficiency of the steam stripping operation as the catalyst leaves the reactor.

Some feedstocks contain small quantities of high boiling or essentially nonvolatile (at reactor conditions) hydrocarbons arising from inefficient fractionation or contamination of the feed. The fractions are deposited on the catalyst and lead to a type of coke referred to as "Conradson carbon" coke, as the coke yield from these nonvolatile fractions has been correlated with the Conradson carbon content of the feed.

Cracking catalysts contaminated with metals, principally Ni, V, and Fe, produce larger amounts of coke than "clean" catalysts operated under the same conditions. This increase in coke with contaminated catalyst is attributed to dehydrogenation of naphthenic structures and polynuclear aromatics which form additional coke by the usual mechanisms. This "contaminant" coke is

TABLE 6 Catalyst Activity and Yields as a Function of Carbon on Regenerated Catalyst [20]

% carbon on regenerated catalyst	0.1	0.2	0.4
Space velocity required for 75% conversion	6.1	5.0	3.55
Yields at 75% Conversion:			
Coke, wt.%	5.5	5.9	6.4
C_1–C_3, wt.%	8.1	8.6	9.0
C_4, vol.%	16.6	17.3	18.2
C_5/430°F, vol.%	58.8	57.7	54.9

TABLE 7 Typical Distribution of Types of Coke—Fluid Catalytic Cracking [21]

Type of Coke	% of Total Coke
Catalytic	45
Cat-to-oil	20
Conradson carbon	5
Contaminant	30
	100

determined not only by the cleanliness of the catalyst but also by the potential polynuclear aromatics in the feed. Thus coke yield in catalytic cracking reflects feedstock properties, reactor conditions including catalyst/oil ratio, physical and chemical properties of the catalyst, and stripper variables. A typical breakdown of coke from a commercial unit is shown in Table 7 [21].

Catalytic Cracking Kinetics

A theoretical or reasonably complete description of catalytic cracking is difficult due to the extreme complexity of the reacting system; many types of hydrocarbons are involved in numerous primary and secondary reactions over a catalyst whose activity and selectivity characteristics are rapidly changing. Hence kinetic treatments of catalytic cracking usually cover only one phase of: (1) the effects of feedstocks composition, process conditions, reactor configuration and catalyst types on (2) conversion of the feedstock, coke formation, gasoline production, and the formation of light gases and other products. These segregated studies have usually been brought together as an integrated model only in the proprietary correlations and process schemes of the major licensors or users of catalytic cracking.

The rate of conversion, the cracking of the feedstock to products boiling below the initial boiling point of the feed and to coke, is influenced in the following manner.

Pure compounds crack by a first-order mechanism but most commercial feeds more closely follow a second-order mechanism. This is attributed to diminution in rate of cracking of commercial feeds as the most easily converted molecules are cracked, leaving behind the more refractory molecular types.

The rate of cracking is proportional to a fractional power of oil partial pressure, and the exponent may approach unity.

Temperature increases the rate of cracking in a fashion corresponding to an activation energy of 10 to 14 kcal/gmol.

The influence of catalyst concentration (catalyst/oil ratio, C/O) on rate of cracking is somewhat less than the first power of C/O.

The rate of cracking decreases with time of contact as carbonaceous deposits build up on the catalyst and decrease instantaneous activity. The decay in activity can be represented either by an exponential, power, or other function of time-on-stream of the catalyst.

A kinetic model for conversion which incorporates most of the above effects has been developed [30] for fixed, moving, and fluid beds and both exponential and power functions of time-on-stream. Subsequently, the development was extended to include selectivity effects according to the following simplified scheme:

$$Y_1 \xrightarrow{k_0} a_1 Y_2 + a_2 Y_3 \qquad (1a)$$

$$Y_2 \xrightarrow{k_2} Y_3 \qquad (1b)$$

where Y_1 represents the gas oil charged, Y_2 represents the gasoline fraction, and Y_3 the butanes, dry gas, and coke. The a_1 and a_2 coefficients represent the masses of gasoline and gases plus coke, respectively, produced by primary reactions per mass of gas oil converted.

Several other analytical models [22–27] have been developed and applied to experimental catalytic cracking data for conversion and selectivity effects in fixed-, moving-, fluid-bed reactors using various reaction rate and catalyst activity decay functions.

Basic rate equations and the corresponding equations for conversion and gasoline yield are presented in Table 8 for the exponential catalyst activity decay and two fluidized catalyst models: the dense, fluid-bed reactor consisting of piston gas flow up through perfectly mixed solids (catalyst), and the moving-bed, riser, or transfer-line reactor consisting of cocurrent piston flows of gas and solids. This kinetic model has been tested against pilot plant data for conversion and selectivity in moving- [28, 29], fixed- [29] and fluid-bed pilot plants [30] and found to be satisfactory; it has also been used to relate nitrogen poisoning, feed composition, and recycle effects [31–33] to cracking rate and catalyst decay rate constants [31].

An important aspect of this kinetic model is the relation of gasoline yield, particularly the maximum gasoline yield, to process conditions (Table 8). The effect of process reactor temperature on the position and magnitude of the maximum gasoline yield is illustrated in Fig. 17; the higher maximum gasoline yield obtained at the lower temperature would in practice be weighed against the higher severity required, increased coke yield, and possible gasoline octane number and other product quality debits. A comparison of model predictions and moving-bed pilot plant data with a Mid-Continent gas oil and Durabead 5 catalyst at 900°F is shown in Fig. 18. This model, which includes equivalent decay functions for both the gas oil and gasoline conversion activities, predicts that the maximum gasoline yield is the same for dense-bed and riser reactions. In practice, riser crackers are generally somewhat more selective than dense beds, even at the slightly higher typical operating temperature of risers. This discrepancy is attributed to the oversimplification of the model with respect to secondary reactions and C/O ratio effects.

TABLE 8 Conversion and Selectivity in Moving and Fluid Catalytic Cracking Reactors [30]

Basic equations:
- Gas oil conversion: $dy_1/du = -K_0 y_1^2$
- Gasoline conversion: $dy_2/du = K_1 y_1^2 - K_2 y_2$
- Catalyst activity decay: $\phi = e^{-\alpha t_c}$

Model type:

Fluid bed (piston flow gas, perfectly mixed solids)	Moving bed (riser or transfer line with cocurrent piston flow of gas and solids)

Model equations:

Conversion:

$$\varepsilon = \frac{A_0}{1 + \lambda + A_0} \qquad \varepsilon = \frac{A_0(1 - e^{-\lambda})}{\lambda + A_0(1 - e^{-\lambda})}$$

Gasoline yield:

$$y_2 = r_1 r_2\, e^{-r_2 y_1} \left[\frac{1}{r_2} e^{r_2} - \frac{y_1}{r_1} e^{r_2 y_1} - \text{Ein}\,(r_2) + \text{Ein}\left(\frac{r_2}{y_1}\right) \right]$$

Maximum gasoline/conversion relation:

$$y_2^c = \frac{K_1}{K_2}(1 - \varepsilon^c)^2$$

Process variable conditions for maximum gasoline yield:

$$S^c = \frac{K_0}{(1 + \alpha t_c)\left[\sqrt{\dfrac{K_0}{K_2 y_2^c}} - 1\right]}$$

$$= \frac{K_0(1 - e^{-\alpha t_c})}{\alpha t_c \left[\sqrt{\dfrac{K_1}{K_2 y_2^c}} - 1\right]}$$

Symbols:
- A_0 = extent of reaction group for gas oil, ratio of reaction rate to space velocity, $\rho_0 k_0 / \rho_1 S = K_0/S$
- a_1 = K_1/K_0
- Ein (x) = exponential integral function, $\int_{-\infty}^{z} \dfrac{e^x}{x} dx$
- k_i = reaction velocity constant of i th reaction at $\theta = 0$, h-1 (includes initial feed concentration because of second-order reaction)
- K_0 = gas oil cracking rate
- K_2 = gasoline cracking rate
- K_1 = $a_1 K_0$
- K_i = $\rho_i k_i / \rho_i$
- Q_i = activation energy of reaction i, cal/(g)(mole)
- r_1 = K_1/K_0
- r_2 = K_2/K_0

(continued)

TABLE 8 (*continued*)

S	= liquid hourly space velocity, vol/(vol) (h)
t_c	= catalyst residence time (under oil exposure conditions), h
u	= stretched time variable; $-e^{-\lambda x}/\lambda S$ for moving beds or plug catalyst flow and $x/(1 + \lambda)S$ for fluid beds with perfectly mixed catalyst flow
V_r	= reactor volume, ft^3
x	= normalized axial distance, z/z_0
y_1	= instantaneous weight fraction of gas oil
y_2	= instantaneous weight fraction of gasoline
y_2°	= maximum gasoline weight fraction
z	= axial distance in reactor, ft
z_0	= total reactor length, ft

Greek Letters

α	= decay velocity constant, h^{-1}
β	= catalyst-to-oil ratio, vol cat/vol total oil for fixed bed; (vol cat/h)/(vol oil/h) for moving and fluid beds
ε	= instantaneous weight fraction converted
θ	= normalized time-on-stream, t/t_c
λ	= extent of catalyst decay group, $\alpha/\beta S$ or αt_c
ρ_0	= initial charge density at reactor conditions, lb/ft^3
ρl	= density of liquid charge at room temperature, lb/ft^3
ϕ	= catalyst decay function = $e^{-\alpha t_c}$

Superscript

$\circ$	= maximum condition

Application of the kinetic model to a variety of straight run gas oil feeds has allowed the development of correlations to predict the feedstock parameters needed in the model [31, 32]. The derived relations between aromatic/naphthene ratio in the feed and gas oil cracking rate and catalyst activity decay constants are shown in Figs. 19 and 20. The higher the aromaticity of the feed, the more rapid the catalyst activity decline and the more refractory nature of the feed.

Carbon formation in catalytic cracking has long been known to follow a relatively simple power function with respect to residence time [34]:

$$C = A\theta^n$$

where C = coke yield, wt.% on catalyst
 θ = catalyst residence time, min
 n = constant depending only slightly on feedstock, catalyst, and process conditions
 A = a function of catalyst, feedstock, and process conditions

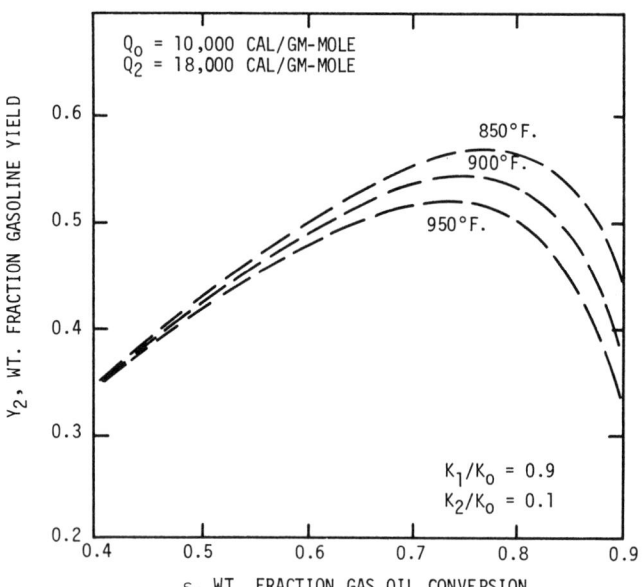

FIG. 17. Effect of temperature on gasoline selectivity [30].

The coefficient n has been found to be approximately 0.5 in most circumstances. The function A is a strong function of feedstock properties and has been related to the catalyst decay constant α and to the aromatic/naphthene ratio of the feedstock. The effect of temperature on rate of coke formation is small and

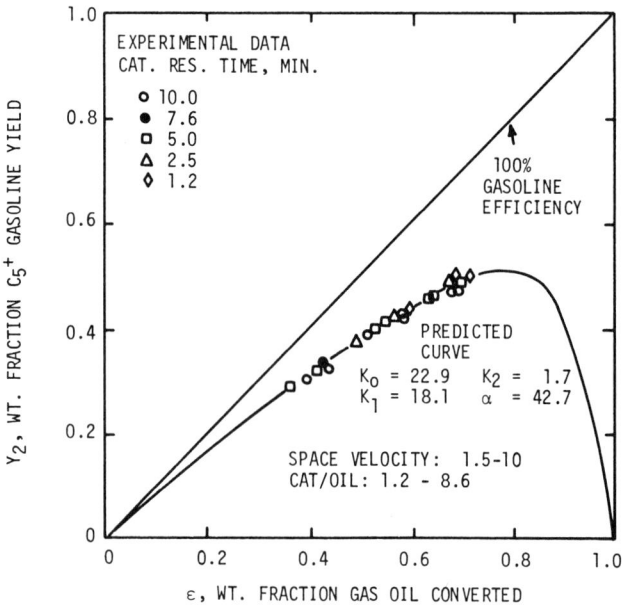

FIG. 18. Comparison of kinetic model with moving-bed data [30].

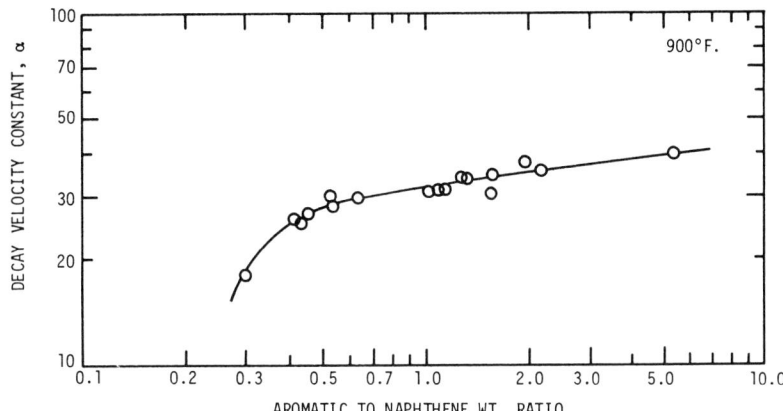

FIG. 19. Relationship between catalyst decay constant and aromatic-to-naphthene ratio [32].

corresponds to doubling the coke yield for a 190 to 200°F increase in temperature.

The above power function satisfactorily relates coke yield to residence time when the coke is primarily "catalytic coke"—the coke derived from distillate or low Conradson carbon content feeds over clean or noncontaminated catalysts with efficient stripping conditions. When these conditions are not obtained, a more complex model is required to account for the several sources or types of coke.

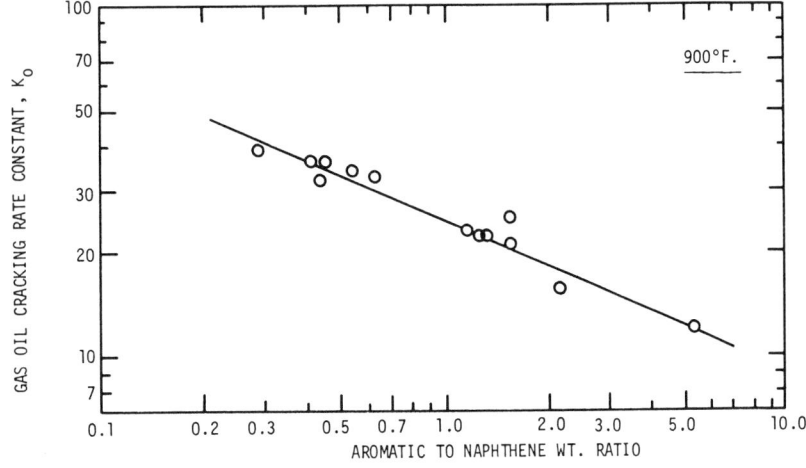

FIG. 20. Relationship between gas oil cracking rate constant and aromatic-to-naphthene ratio [32].

Operating Variable Effects

In addition to the composition of the feedstock and the condition and type of catalyst, catalytic cracking yields and product qualities are determined by a large number of operating variables and operating conditions. Not only are several of the operating variables interrelated but, in some cases, a so-called variable may be limited or even fixed by unit restrictions or heat balance requirements. Independent operating variables for an existing unit are: (1) reactor temperature, (2) recycle of unconverted feed fractions, (3) feed preheat temperature, and (4) space velocity or contact time. In addition, catalyst activity is an independent variable provided that the catalyst withdrawal and addition rates can be changed or that a catalyst of differing activity can be used. Pressure and mole fraction of hydrocarbon feed are also independent variables but are generally subject to only small variations on an existing unit or after the unit design has been established. Dependent variables are: (1) catalyst/oil ratio, (2) regeneration temperature, (3) regenerator air rate, and (4) conversion.

Process variable effects are usually illustrated with pilot plant results or laboratory data where all but one variable can be held constant or restricted to minor variations [41]. A common procedure is to develop the operating condition effects on conversion and then relate yields and product quality to conversion, with operating condition modifiers if needed. Thus the effects of operating variables can be described in an independent and straightforward manner. The effects of operating variables on a commercial unit are much more complex and generally require that the net effect of several changes be calculated or estimated. In the following sections the basic process variable effect will first be described, generally with pilot plant data, and then the corresponding and more complex commercial unit changes will be discussed.

Although conversion is a dependent variable, and is discussed below as such, a description of general conversion effects is needed to understand the role of independent variables. Conversion (that is, the amount of feed converted to products boiling below the 430°F cut point) is influenced by space velocity, catalyst/oil ratio, temperature, and catalyst activity—increasing as the latter three increase in magnitude. Generally, as these factors are increased, it is said that the severity of the reaction is increased. The quantitative effect of each of these factors is different so that the detailed yields and product qualities at a given conversion depend on the particular combination of variables which produce the observed conversion. In general, as conversion increases, the yields of all C_4 and lighter gaseous components, C_5+ gasoline, and coke increase. At high conversion levels, secondary reactions of olefins become important and the yields of all olefins decrease; the extent of these secondary reactions is determined by the manner in which the higher conversion is obtained; high activity is particularly effective in allowing high conversion levels while minimizing secondary reactions. An equally important consequence of high severity is the decrease in gasoline yield which generally occurs when conversion is increased to extreme levels. Although gasoline octane numbers increase with conversion up to and past the maximum in yield, the quality credits are generally not sufficient to offset the debits due to lower olefin yields, lower gasoline yield, and increased coke production.

Reactor Temperature

The principal effects of increasing reactor temperature at constant conversion are to decrease C_5+ gasoline yield and coke yield and to increase dry gas and total butanes yields; all olefin yields are increased; C_1 and C_2 yields increase; C_3 and C_4 paraffin yields change only slightly; and C_5+ paraffin yields decrease. Both Research and Motor octane numbers of the C_5+ gasoline increase with increasing temperature. The effects are illustrated in Table 9 [35] with pilot plant data from cracking a heavy East Texas gas oil over a silica/alumina catalyst, and in Figs. 21 and 22 [36, 37] with yield and gasoline composition data from cracking of a Mid-Continent gas oil. The increase in octane number with increasing temperature is due to the increased olefin and aromatic contents and lower paraffin content.

The effect of reactor temperature on a commercial unit is, of course, considerably more complicated as variables other than temperature must be changed to maintain heat balance. In a commercial study of increasing reactor temperature at constant fresh feed rate, recycle rate, space velocity, and feed preheat, the following changes were observed [37]:

TABLE 9 Effect of Temperature on Once-Through Moving-Bed Cracking [34][a]

	Average Reactor Temperature (°F)		
	850	900	950
Space velocity	0.8	1.3	2.0
Conversion, wt.%	55.1	55.1	55.1
H_2, wt.%	0.04	0.05	0.06
CH_4, wt.%	0.71	0.85	1.20
C_2H_4, wt.%	0.4	0.55	0.75
C_2H_6, wt.%	0.6	0.75	1.05
C_3H_6, wt.%	2.4	3.35	4.4
C_3H_8, wt.%	2.1	2.15	2.15
i-C_4H_{10}, wt.%	5.1	4.2	3.35
C_4H_8, wt.%	2.9	4.0	5.0
n-C_4H_{10}, wt.%	1.4	1.3	1.25
C_5+ gasoline, wt.%	34.6	33.5	32.2
Light fuel, wt.%	15.8	13.8	12.4
Heavy fuel, wt.%	29.1	31.1	32.5
Coke, wt.%	4.85	4.2	3.7
C_5+ gasoline gravity, °API	56.9	55.2	53.5
10 # RVP gasoline, F-1 octane no. clear	91.2	94.0	95.0
+ 3 cc TEL	97.6	98.6	99.0
RVP	7.2	7.3	7.4

[a]Charge: 28.9 °API East-Texas heavy gas-oil, 56 to 77% crude. Catalyst: 33 A.I. silica–alumina bead. Catalyst/oil ratio: 2.0.

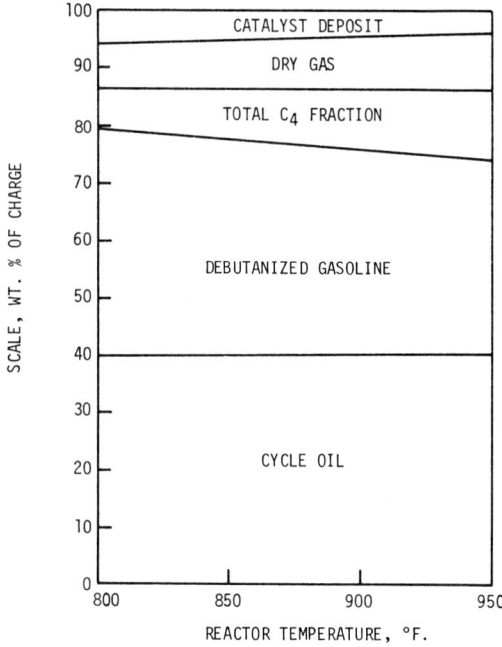

FIG. 21. The effect of temperature on product distribution at 60 vol.% conversion [36].

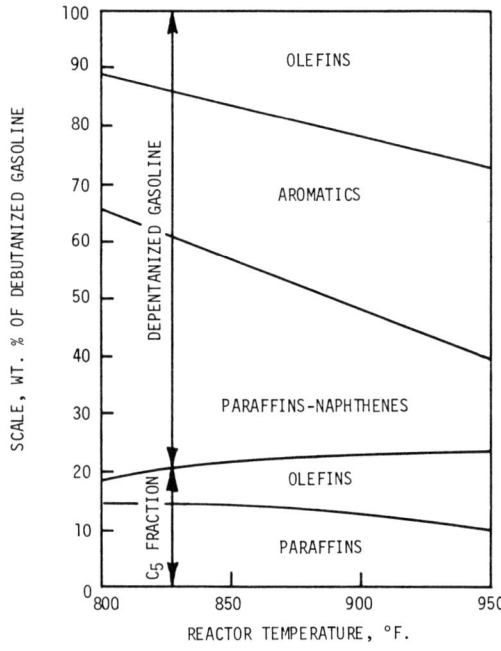

FIG. 22. The effect of temperature on debutanized gasoline composition at 60% conversion [36].

1. Catalyst/oil ratio (C/O) increased
2. Conversion increased
3. Regenerator temperature increased
4. Coke yield increased

These interrelated changes were required to maintain heat balance on the unit by increasing circulation rate and coke yield. The combined effects of higher reactor temperature and higher conversion resulted in the following additional changes:

5. C_4's, C_3's, and lighter increased
6. C_5+ gasoline yield increased
7. Light catalytic cycle oil yield decreased

Thus the effects of an increase in reactor temperature on an operating unit reflect not only the effects of temperature per se but also the effects of several concomitant changes such as increased conversion and increased C/O ratio.

Recycle Rate

With most feedstocks and catalyst, gasoline yield increases with increasing conversion up to a point and then passes through a maximum and decreases. This "overcracking" is due to the increased refractoriness of the unconverted feed as conversion increases and the destruction of gasoline through secondary reactions, primarily recracking of olefins. The onset of secondary reactions and the subsequent leveling off or decrease in gasoline yield can be avoided by recycling a portion of the reactor product—usually a fractionator heart cut with boiling points in the range of 650 to 850°F. The effect of increasing recycle rate or combined feed ratio (total feed rate, including recycle, divided by fresh feed rate) is illustrated in Fig. 23 [38] with pilot plant data obtained at a constant reactor temperature with a synthetic silica–alumina catalyst. With no recycle, the maximum gasoline yield is 42 vol.% at 60% conversion; with the volume of recycle equal to the volume of fresh feed, the gasoline yield is increased only slightly at 60% conversion, but the gasoline yield can be increased to over 55 vol.% by increasing conversion to 89% or higher.

Other pilot plant tests [37] have shown the following effects of increasing recycle rate when space velocity was simultaneously adjusted to maintain conversion constant:

1. C_5+ gasoline yield increased significantly
2. Coke yield decreases appreciably
3. Dry gas components, propylene, and propane yields decreased
4. Butane yields decreased; butylenes yields increased slightly
5. Light catalytic cycle oil and clarified oil yields increased, but heavy catalytic cycle oil yield decreased

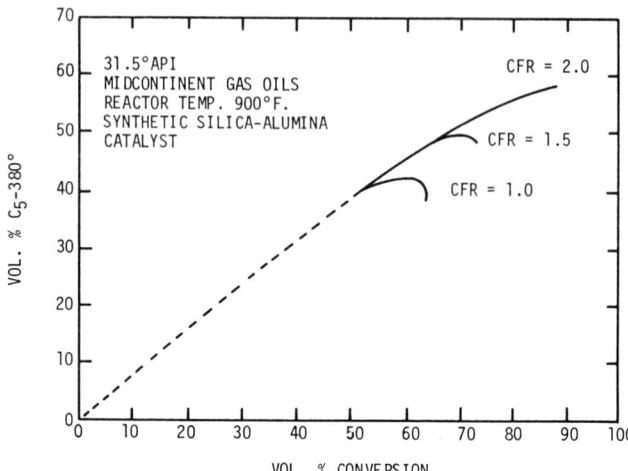

FIG. 23. The relation between combined feed ratio and gasoline yield when operating at a given conversion [38].

With the introduction of high activity zeolite catalysts, it was found that in once-through cracking operations with no recycle the maximum in gasoline yield was located at much higher conversions. In effect, the higher activity catalysts were allowing higher conversions to be obtained at severity levels which significantly reduced the extent of secondary reactions or overcracking. Thus, on many units now employing zeolitic catalysts, recycle has been completely eliminated or reduced to less than 15% of the fresh feed rate.

Feed Preheat

A more detailed description of the heat balance as affecting operation of the unit is presented later in this article. However, since the heat balance also affects the conditions of the reaction, a brief summary of this aspect is covered here. In a heat-balanced commercial operation, increasing the temperature of the feed to a cracking reactor reduces the heat that must be supplied by combustion of the coked catalyst in the regenerator. Feed preheat is usually supplied by heat exchange with hot product streams, a fired preheater, or both. When feed rate, recycle rate, and reactor temperature are held constant as feed preheat is increased, the following changes in operation result:

1. The C/O ratio (catalyst circulation rate) is decreased to hold reactor temperature constant:
2. Conversion and all conversion-related yields, including coke, decline due to the decrease in C/O ratio and severity.
3. The regenerator temperature will usually increase. Although the total heat released in the regenerator and the air required by the regenerator are

Cracking, Catalytic

reduced by the lower coke yield, the lower catalyst circulation usually overrides this effect and results in an increase in regenerator temperature.

4. As a result of the lower catalyst circulation rate, residence time in the stripper and overall stripper efficiency is increased; liquid recovery is increased and a corresponding decrease in coke usually results.

Advantage is usually taken of these feed preheat effects, including the reduced air requirement, by increasing the total feed rate until coke production again requires all of the available air. A commercial plant example of this type of "constant coke burning capacity" and the effect feed preheat can have on fresh feed rate is shown in Fig. 24 [39].

Space Velocity

The role of space velocity as an independent variable arises from its relation to "catalyst contact time" or "catalyst residence time." Thus

$$\theta = \frac{60}{\text{WHSV} \times \text{C/O}}$$

where θ = catalyst residence time, min
   WHSV = weight hourly space velocity, total feed basis
   C/O = catalyst/oil weight ratio

Catalyst/oil ratio is a dependent variable so that catalyst time becomes directly related to WHSV. When catalyst contact time is low, secondary reactions are minimized; thus gasoline yield is improved and light gas and coke yields are decreased. In dense bed units the holdup of catalyst in the reactor can be

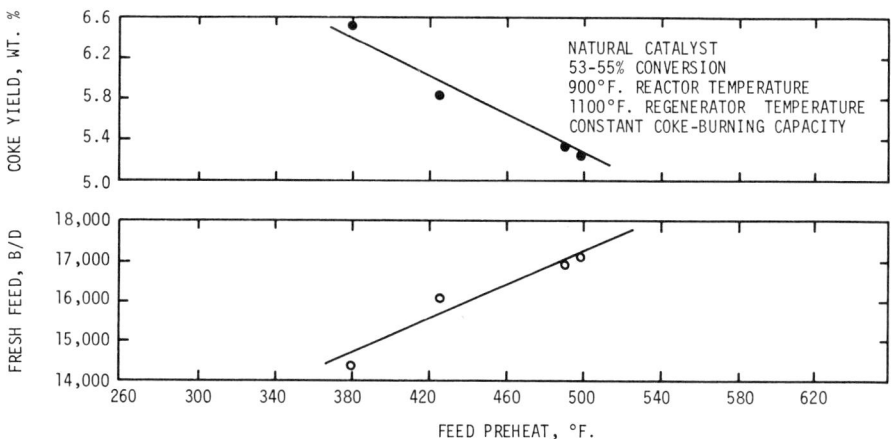

FIG. 24. The effect of feed preheat on coke yield and unit capacity [39].

controlled within limits, usually by a slide valve in the spent catalyst standpipe, and feed rate can also be varied within limits. Thus there is usually some freedom to increase space velocity and reduce catalyst residence time. In a riser type reactor, holdup and feed rate are not independent and space velocity is not a meaningful term. Nevertheless, with both dense-bed and riser-type reactors, contact times are usually minimized to improve selectivity. An important step in this direction was the introduction of high activity zeolite catalyst. These catalysts require short contact times for optimum performance and have generally moved cracking operations in the direction of minimum holdup in dense-bed reactors or replacement of dense-bed reactors with short contact time riser reactors.

Strict comparisons of short contact time riser cracking versus the longer contact time dense-bed mode of operations are generally not available due to differences in cat activity, carbon content of the regenerated catalyst, or factors other than contact time but inherent in the two modes of operation. However, in general, improvements in catalyst activity have resulted in the need for less catalyst in the reaction zone. Many modern units are now designed with only riser cracking. That is, no dense-bed catalyst cracking zone is provided and all cracking is done in the catalyst/oil transfer lines leading into the reactor cyclone vessel. However, in some of these instances the reactor temperature must be increased to the 1020 to 1050°F range in order to increase the intensity of cracking conditions to achieve the desired conversion level. This is because not enough catalyst can be held in the riser zone, since the length of the riser is determined by the configuration and elevation of the major vessels in the unit. Alternatively, superactive but more costly catalysts can be used to achieve desired conversion in the limited riser. Significant selectivity disadvantages have not been shown if a dispersed catalyst phase or even a very small dense bed is provided downstream of the transfer line riser cracking zone. In this case the cracking reaction can be run at a lower temperature, say 950°F, which will reduce light gas make and increase gasoline yield when compared to the 1020 to 1050°F operation.

Pressure and Partial Pressure of Hydrocarbon Feed

Catalytic cracker pressures are generally set slightly above atmospheric by balancing the yield and quality debits of high pressure plus increased regeneration air compression costs against improved cracking and regeneration kinetics, the lower cost of smaller vessels, plus, in some cases, power recovery from the regenerator stack gases. Representative yield and product effects are shown in Table 10. They show, at the same conversion level, that coke and gasoline yields are increased marginally and light gas yields are reduced at the higher pressure. The Research clear octane number of the gasoline is essentially unchanged, but the ASTM octane can be 2 points lower for this difference in pressure. The sulfur content of the gasoline is reduced. Some pilot data have shown there is no increase in carbon production.

Pressure levels in commercial units are generally in the 15 to 35 lb/in.^{2} gauge range. Lowering the partial pressure of the reacting gases with steam will

TABLE 10 Comparison Product Distributions at 12 and 50 lb/in.^{2} gauge Total Pressure

Total pressure, lb/in.^{2} gauge	12	50
Conversion, vol.%	65–67	65–66
Yields:		
H_2, wt.%	0.2	0.1
C_1, wt.%	1.9	1.2
C_2^{2-}, wt.%	1.4	0.9
C_2, wt.%	1.2	1.5
C_3^{2-}, wt.%	5.3	3.9
C_3, wt.%	3.9	4.4
iC_4^{2-}, vol.%	11.8	10.7
nC_4^{2-}, vol.%	2.0	1.2
iC_4, vol.%	4.4	3.5
nC_4, vol.%	2.4	3.2
iC_5, vol.%	6.3	5.6
C_5^{2-}, vol.%	3.1	3.0
nC_5, vol.%	1.4	1.6
C_6–284°F 90% pt, vol.%	16.7	19.6
Heavy naphtha, vol.%	8.2	7.2
C_6–410°F EP gasoline, vol.%	24.9	26.8
Cycle oil, vol.%	32.9	33.3
Carbon, wt.%	6.1	6.8
Quality, C_6–410°F gasoline		
API	39	44
Sulfur, wt.%	0.14	0.09
RON clear	98	98
ASTM clear	86	84

improve yields somewhat, but the major beneficial effect of feed injection steam is that it atomizes the feed to small droplets which will vaporize and react quickly. If feed is not atomized, it will soak into the catalyst and possibly crack to a higher coke make.

Both pressure and partial pressure of the hydrocarbon feed, or steam/feed ratio, are generally established in the design of a commercial unit and thus are usually not available as independent variables over any significant range. However, in some units, injector steam is varied over a narrow range to balance carbon make with regeneration carbon burnoff.

Catalyst Activity

Catalyst activity as an independent variable is governed by the capability of the unit to control the carbon content of the spent catalyst and the quantity and quality of fresh catalyst that can be continuously added to the unit. The carbon content of the regenerated catalyst is generally maintained at the lowest practical level to obtain the selectivity benefits of "low CRC"; thus catalyst addition is, in effect, the principal determinant of catalyst activity. The deliberate withdrawal of catalyst over and above the inherent loss rate through regenerator stack losses and decant or clarified oil, if any, and a corresponding increase in fresh catalyst addition rate is generally not practiced as a means of increasing the activity level of the circulating catalyst. If a higher activity is needed, the addition of a higher activity fresh catalyst at the minimum makeup rate to maintain inventory is usually the more economical route.

The general effects of increasing activity are to permit a reduction in severity and thus reduce the extent of secondary cracking reactions. Higher activity typically results in more gasoline and less coke. In other cases, higher activity catalysts are employed to increase the feed rate at essentially constant conversion and constant coke production so that the coke burning or regenerator air compression capacities are fully utilized.

Catalyst/Oil Ratio

The dependent variable catalyst/oil ratio (C/O) is established by the unit heat balance and coke make which in turn are influenced by almost every independent variable. Since C/O changes are accompanied by one or more shifts in other variables, the effects of C/O are generally associated with other effects. A basic relation, however, in all C/O shifts is the effect on conversion and carbon yield. At constant space velocity and temperature, increasing C/O increases conversion; Fig. 25 [42] illustrates this effect with typical pilot plant data. In addition to increasing conversion, higher C/O ratios generally increase coke yield at a constant conversion. This increase in coke is related to the hydrocarbons entrapped in the pores of the catalyst and carried through the stripper to the regenerator. Thus this portion of the C/O effect is highly variable and depends not only on the C/O ratio change but also catalyst porosity and stripper conditions. Typical data from experimental and commercial units are shown in Fig. 26 [43].

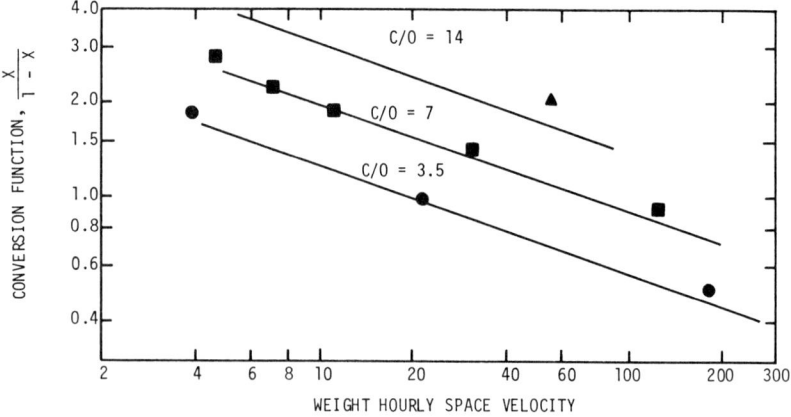

FIG. 25. Effect of C/O and WHSV on conversion [42].

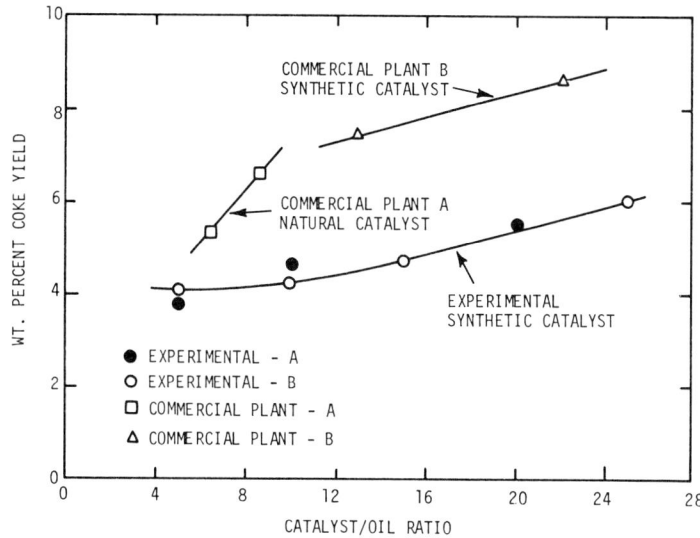

FIG. 26. Effect of catalyst-to-oil ratio on coke yield for fluid catalytic cracking at constant conversion [43].

Pilot plant and commercial unit studies have shown that the following changes, in addition to increased coke yield, accompany an increase in C/O ratio in the range of 5 to 20 at constant conversion, reactor temperature, and catalyst activity [43]:

1. Decreased H_2 yield
2. Decreased C_1-C_4 yields
3. Little effect in C_5+ gasoline yield or octane number

Regenerator Temperature

Catalyst circulation, coke make, and feed preheat are the principal determinants of regenerator temperature which is generally allowed to respond as a dependent variable within limits. Mechanical or structural specifications in the regenerator section generally limit regenerator temperature to a maximum value specific to each unit; however, in some cases the maximum temperature may be set by catalyst stability. In either case, if regenerator temperature is too high, it can be reduced by decreasing feed preheat; catalyst circulation is then increased to hold a constant reactor temperature and this increased catalyst circulation will carry more heat from the regenerator and lower the regenerator temperature. The sequence of events is actually more complicated as the shift in C/O ratio and, to a lesser extent, the shift in carbon content of the regenerated catalyst will change coke make and the heat release in the regenerator.

Regenerator Air Rate

The amount of air required for regeneration depends primarily on coke production. In the past, regenerators have been operated with only a slight excess of air leaving the dense phase; with less air, carbon content of the spent catalyst increases and a reduction in coke yield is required; with large excesses of air, burning of CO to CO_2 above the dense bed will occur. This "after burning" must be controlled as extremely high temperatures are produced in the absence of the heat sink provided by the catalyst in the dense phase [44].

Recently two new developments have resulted in different modes of regeneration. First, higher stability catalysts are available and regenerator temperatures can be increased by 100 to 150°F up to the 1325 to 1370°F range without significant thermal damage to the catalyst. At these higher temperatures the oxidation of CO to CO_2 is greatly accelerated, and the regenerator can be designed to absorb the heat of combustion in the catalyst under controlled conditions. Carbon burning rate is improved at the higher temperatures, and there usually are selectivity benefits associated with the lower carbon on regenerated catalyst; this high temperature technique usually results in carbon on regenerated catalyst levels of 0.05 wt.% or less.

A related development is the use of catalysts containing promoters for the oxidation of CO to CO_2. These catalysts produce a major effect of high temperature regeneration, i.e., low CO content regenerator stack gases and the resultant regenerator conditions may result in a lower carbon content regenerated catalyst [44].

Conversion

All of the independent variables in catalytic cracking have a significant effect on conversion which is truly a dependent variable. However, since so many yields and product quality measures and economic factors have been related to conversion in both pilot plant and commercial plant studies, it is often thought of as the major independent variable. In many operating circumstances, conversion targets are established from economic or marketing studies and then the actual independent variables are adjusted to meet this goal.

The detailed effects of changing conversion depend upon the manner in which the conversion is changed, i.e., by temperature, space velocity or catalyst oil ratio, catalyst activity, etc.—however, the general trends are similar. Increasing conversion increases yields of gasoline and all light products up to a conversion level of 60 to 80 vol.% in most cases. At this high conversion level, secondary reactions become sufficient to cause a decrease in olefin yields and C_5+ gasoline; however, the point at which this occurs is sharply dependent on the feedstock, operating conditions, catalyst activity, etc.

Feedstock Quality Effects

Feedstocks charged to cat cracking units can have extensive ranges in composition and boiling range, reflecting the flexibility of the process and, to a lesser extent, the ability of any single unit to accommodate wide variations in feed crackability, coke forming technology, and volatility. Feedstock initial boiling points are generally in the range of 600 to 650°F to exclude heating oil and diesel fuel components, but the 400 to 650°F boiling range or portions thereof are included in some cases to maximize gasoline production from crude. The usual range for the final boiling point is 950 to 1050°F as produced by vacuum flashing or vacuum distillation. In most cases the upper boiling point is limited by the Conradson carbon content or metals content of the heavy distillate; both components have adverse effects on cracking characteristics. The high boiling 1050°F+ fractions or residuum are occasionally included in cat cracker feeds; these units are referred to as "residuum cat crackers." In such cases the 1050°F+ fraction is either relatively low in Conradson carbon and metals (for example, from a waxy crude) so that these debits are relatively small or the 1050°F+ has little or low value to alternate depositions. Thus virgin distillate fractions in the boiling range 650 to 1050°F are the predominant components of most fresh feeds; many units also recycle a slurry oil (850°F+) and a heavy cycle oil stream. Gas oils from thermal cracking or coking operations, hydrotreated gas oils, lube oil extracts, and deasphalted oils are often included in feeds.

General feedstock quality effects can be indicated by characterization factor (characterization factor = $(MABP)^{1/3}$/specific gravity 60°F/60°F, where MABP is the mean average boiling point expressed in °R). However, it is obvious that a single parameter such as this can only reflect general trends. Tables 11 and 12 indicate the effect of the characterization factor of straight run and cracked stocks on gasoline yield and coke yield at a fixed conversion of 60% [45]. With both types of feeds, coke yield increases as K decreases or as the feed becomes less paraffinic. With straight run gas oils, gasoline yield increases as K (paraffinicity) decreases, but just the opposite effect is obtained with cracked stocks or cycle oils. Both of these tables refer to yields at a constant conversion—a normal basis for presenting cat cracking yield data, particularly pilot plant results. Of more interest to the refinery is a constant coke basis because this approaches the "heat-balanced" condition of commercial units. Such results are illustrated in Table 13 for a fixed coke yield of 5.3%. Under

TABLE 11 Straight-Run Feeds—Effect of Characterization Factor at a Fixed (60%) Conversion of Catalytic Yields [45]

Characterization Factor	Vol.% Gasoline Yield	% Coke Yield
11.2	49.5[a]	12.5[a]
11.4	47.0	9.1
11.6	45.0	7.1
11.8	43.0	5.3
12.0	41.5	4.0
12.2	40.0[a]	3.0[a]

[a]Not truly comparable because plants are not operated at such conditions.

TABLE 12 Cycle Oil or Cracked Feeds—Effect of Characterization Factor at a Fixed (60%) Conversion on Catalytic Yields [45]

Characterization Factor	Vol.% Gasoline Yield	% Coke Yield
11.0	35.0[a]	
11.2	37	11.5[a]
11.4	39	9.0
11.6	40	7.2
11.8	41	6.0
12.0	41.5[a]	5.3[a]

[a]Extrapolations useful only in showing general trend.

these circumstances both straight run and cracked stocks exhibit an increase in conversion and gasoline as the feedstock paraffinicity and K increase.

Molecular weight, average boiling point, or feed boiling range is an important feedstock characteristic in determining cat cracking yields and product quality. In general, for straight run fractions, crackability increases as molecular weight increases; coke and gasoline production (at constant processing conditions) also increase with the heavier feeds. These effects at constant overall composition are illustrated in Fig. 27 [46] with predicted results from correlations based on once-through cracking at constant reactor conditions and a 900°F cracking temperature. Molecular weight or boiling point trends are not always consistent with Fig. 12. For many cases the shifts in hydrocarbon-type composition with increasing molecular weight result in changes in cracking characteristics which are comparable to or greater than the changes due to molecular weight alone. This situation is illustrated in Fig. 28

TABLE 13 Effect of Characterization of Feedstock on Catalytic Cracking Yields at a Fixed Coke Deposition of 5.3% [45]

	Straight-run Feeds		Cracked Feeds	
Characterization Factor	Conversion (%)	Gasoline Yield (vol.%)	Conversion (%)	Gasoline Yield (vol.%)
11.0	—	—	30.0	20.0
11.2	50.0	39.0	39.0	26.5
11.4	52.0	40.0	45.0	31.5
11.6	56.0	41.5	51.0	35.5
11.8	60.0	43.5	57.0	39.0
12.0	70[a]	52.0[a]	60.0	41.0
12.2	[b]	66.0[b]	—	—

[a]Recycling probably necessary to cause 5.3% coke deposition.
[b]Ultimate yield about 70 to 80% maximum percentage conversion.

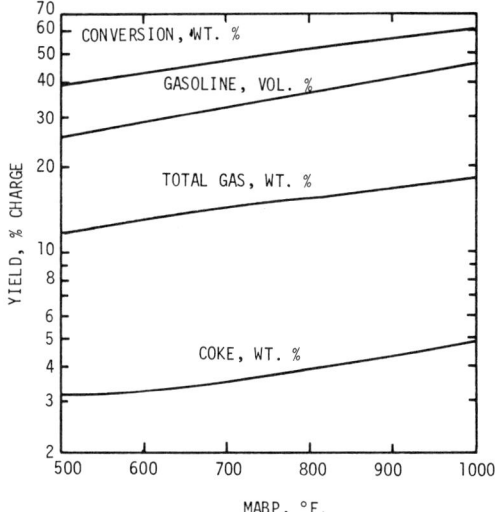

FIG. 27. Increasing the mean average boiling point results in an increase in the yield of all major cracked products [46].

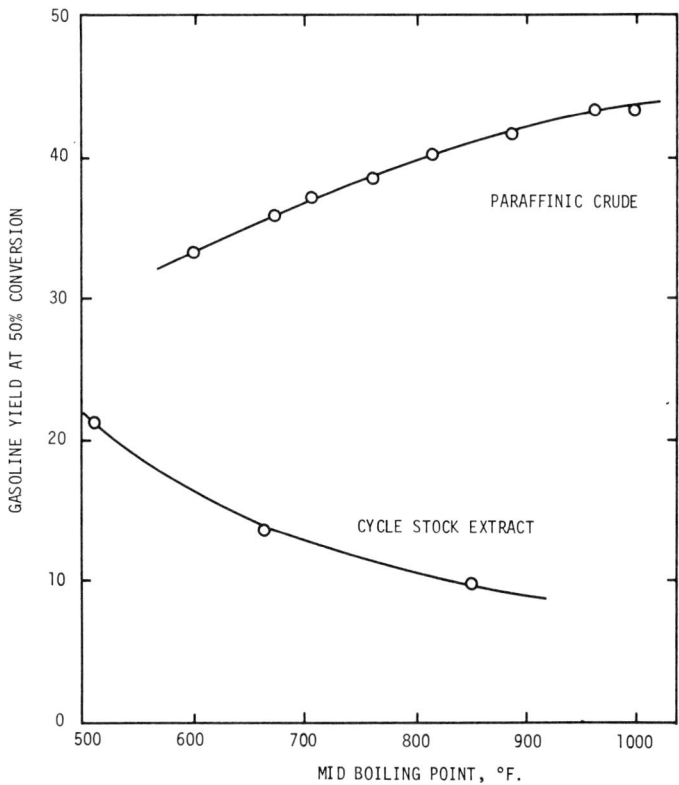

FIG. 28. Effect of mid boiling point on gasoline yield [47].

[47] wherein gasoline yield at a constant conversion of 50% shows the typical increase with increasing boiling point for a paraffinic crude, but decreases as the boiling point of highly aromatic cycle stock extracts increase. Additional feed boiling range effects are illustrated in Table 14 [48]. At approximately constant conversion, gasoline and coke yields increase as molecular weight increases, but gas yields (C_3^- and C_4's) actually are smaller with the heavier feeds. In terms of gasoline yield per unit of coke production, the medium boiling range fraction is superior to the other two. The lightest fraction results in the best quality gasoline, but the gasoline/coke ratio is lower than for the intermediate boiling range fraction.

Characterization and feed boiling range are generally insufficient to characterize a feedstock for any purpose other than approximate evaluations or comparisons. A more detailed description of the feed is needed to reflect the typical variations in feedstock composition and cracking behavior. An early correlation of this type [46] used the "n-d-M" [49] method of characterizing feedstocks. This method is dependent on refractive index (n), density (d), and molecular weight (M) of the feed. The principal cat cracking yields and naphtha

TABLE 14 Effect of Feed Boiling Range: Fluid Cracking with Silica–Alumina Catalyst [48]

	Feed Boiling Range		
	Low	Medium	High
Feedstock fraction from Texas crude	Light Gas Oil	Heavy Gas Oil	Vacuum Gas Oil
Inspections:			
Gravity, °API	33.4	24.4	20.5
20% TBP, °F	494	690	712
50% TBP, °F	536	781	900
80% TBP, °F	597	917	1200
Conradson carbon, wt.%	0.06	0.28	2.6
Operating conditions:			
Temperature, °F	975	975	981
Conversion, vol.%	60.5	63.2	60.0
Product distribution:			
C_3 and lighter, wt.%	13.4	10.3	8.8
Butanes, vol.%	9.3	4.6	2.6
Butylenes, vol.%	7.5	11.3	8.2
Gasoline, 10 lb RVP, vol.%	37.2	47.2	49.2
620°F. EP gas oil, vol.%	37.5	18.0	17.5
Heavy gas oil, vol.%	2.0	18.8	22.5
Carbon, wt.%	3.5	3.9	5.2
Octane number of 10-lb RVP gasoline			
Research, clear	99.8	97.8	94.6
Research with 1.5 cc TEL/gal	100+	99.1	96.5
Motor, clear	85.8	84.0	80.7

TABLE 15 Effect of Changing Charge Stock Properties on Cracked Product Distribution [46]

	Base	I	II	III	IV	V	VI	VII	VIII	IX	X
Property values (only changes over base case indicated):											
MABP	675	800	—	—	—	—	—	—	—	—	—
% C_P^S	44	—	—	51	54	—	54	—	—	—	—
% C_N^S	7	—	—	0	—	0	—	28	—	—	—
% C_P^A	21	—	31	—	11	—	—	—	—	—	—
% C_N^A	10	—	0	—	—	17	0	0	—	—	—
Sulfur, wt.%	2.22	—	—	—	—	—	—	—	0.20	—	—
Basic nitrogen, wt.%	0.005	—	—	—	—	—	—	—	—	0.20	—
Naphtha, vol.%	4.8	—	—	—	—	—	—	—	—	—	1.0
Predicted yields:											
Total Gas, wt.%	14.6	16.0	13.0	15.2	16.4	—	15.5	15.7	—	12.3	—
Gasoline, vol.%	31.9	36.9	—	32.1	—	32.1	—	37.4	—	24.4	—
Gasoline, wt.%	27.8	31.3	—	27.7	—	27.7	—	32.6	—	21.3	—
Octane F-1 unleaded	90.0	—	89.3	89.2	89.8	89.8	89.1	92.0	—	—	92.1
Octane F-1 + 3 cc TEL	95.4	—	95.0	95.0	95.4	95.3	95.0	96.7	96.7	—	96.2
Coke, wt.%	3.5	3.9	3.8	—	3.2	—	—	2.8	3.2	5.1	—
Conversion, wt.%	45.8	51.3	44.6	46.3	47.4	45.6	46.8	51.1	45.5	38.6	—
Gasoline/coke, wt./wt.	9.1	9.3	8.4	9.2	10.0	9.2	9.1	13.4	10.0	4.8	—
Cracking efficiency, wt.%											
100 (gasoline wt.%)/conversion wt.%	60.7	61.0	62.3	59.9	58.6	60.8	59.4	63.8	61.1	55.3	—

octane number on a once-through basis for a specific catalyst and reactor conditions were expressed as a series of equations using mean average boiling point, carbon type distribution from the n-d-M analysis, naphtha in the charge, and concentrations of total sulfur and nitrogen in the feed.

The carbon type distribution is expressed as:

C_N^A = percent of carbon atoms in naphthenic rings condensed in aromatic rings

C_P^A = percent of carbon atoms in paraffinic chains or aromatic and naphtheno-aromatic structures

C_N^S = percent of carbon atoms in naphthene rings in saturate molecules

C_P^S = percent of carbon atoms in paraffinic chains in saturate molecules

The principal effects of carbon-type distribution and other feedstock parameters are illustrated in Table 15 and Figs. 29–33 as variants on a base case of cracking straight run Middle East gas oil on a once-through basis at constant reactor conditions and 900°F cracking temperature.

Increasing the mean average boiling point (Case I) increases conversion and all yields, but the octane number is unchanged.

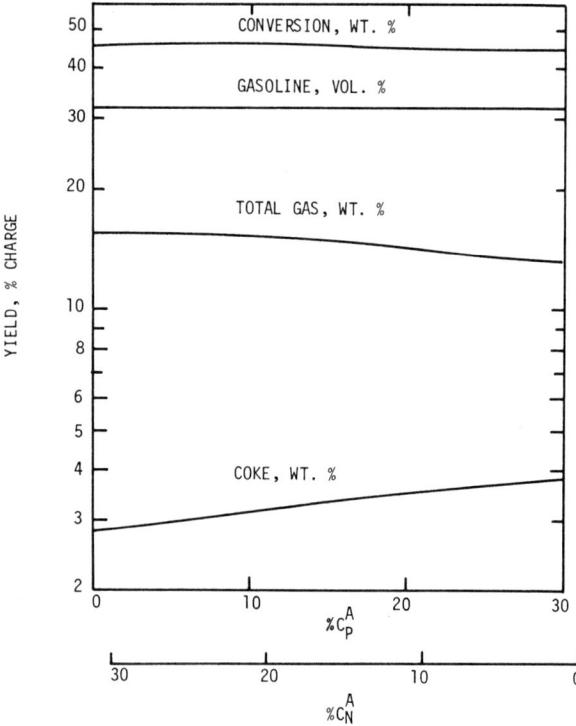

FIG. 29. Increasing the paraffin concentration of the aromatic fraction causes an increase in coke production. The double base scale shows that an increase in paraffin is accompanied by a decrease in naphthene concentration [46].

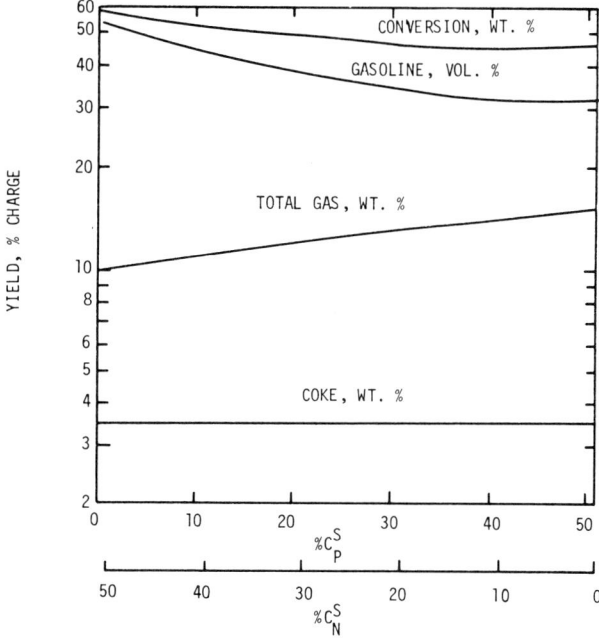

FIG. 30. Increasing the paraffin concentration of the saturate fraction causes a decrease in gasoline production [46].

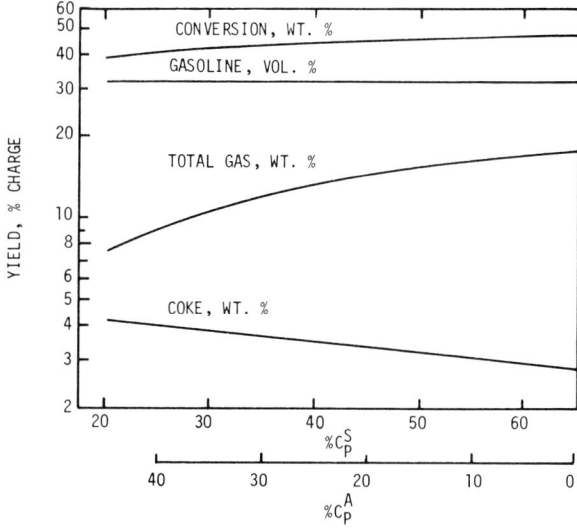

FIG. 31. Increasing the paraffin concentration of the saturate fraction at the expense of the paraffin in the aromatic fraction gives a decrease in coke production [46].

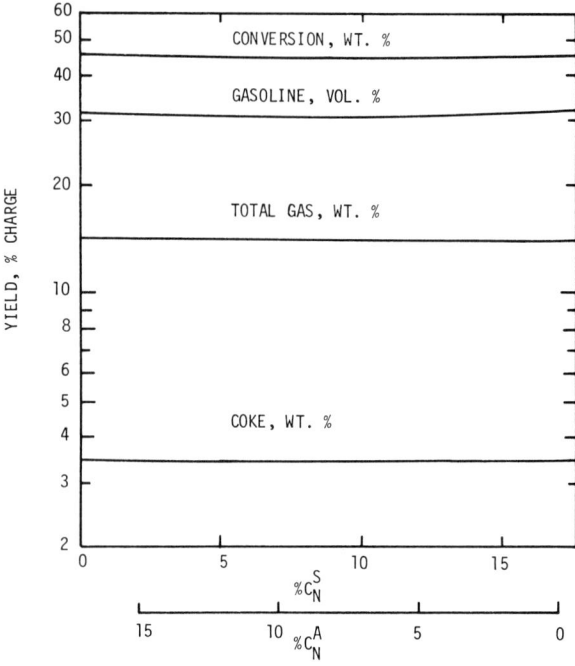

FIG. 32. Exchanging naphthene carbon atoms between the aromatic and saturate fractions has essentially no effect on product distribution [46].

Increasing the paraffinic carbon content of the aromatics fraction at the expense of the naphthenic carbon content of the aromatics fraction (Case II, Fig. 29) has only a minor effect on conversion and gasoline yield, but increases coke and decreases gas. This same type of switch of paraffinic carbon for

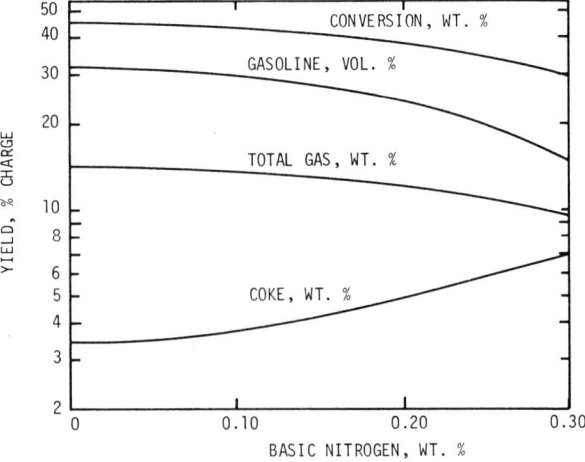

FIG. 33. Increasing the basic nitrogen lowers total gas and gasoline yields but increases the coke yield [46].

naphthenic carbon in the saturate fraction (Case III, Fig. 30) results in an increase in gas yield and a decrease in both conversion and gasoline yield; the distribution of aromatic type in the saturate portion of a feed has a greater effect on cracking yields than the paraffinic/naphthenic carbon split in the aromatics fraction.

As paraffinic carbon in the saturate fraction is increased at the expense of paraffinic carbon in the aromatics (Case IV, Fig. 31), gas yield increases and coke yield decreases. A similar switch of naphthenic carbon from the aromatics fraction to the saturate fraction has no effect on yields (Case V, Fig. 32).

Lowering the total sulfur content of a feed (Case VI) has a negligible effect on yields except for a slight decrease in coke; unleaded octane numbers are unchanged but leaded octane numbers are increased. Increasing the basic nitrogen content has a pronounced effect on all yields; conversion is decreased and coke yield is increased at the expense of naphtha and gas (Case VII, Fig. 33); octane number is not affected. A reduction in the naphtha content of the feed has no significant effect on yields, but octane number of the cat naphtha is improved due to reduction in the content of the low octane straight run naphtha in the feed (Case VIII).

Cat cracking feedstock characterization by the n-d-M or similar methods is not adequate in some cases to distinguish between feeds whose yields are significantly different. This is particularly true of the various types of aromatic carbon which are not distinguished from one another by n-d-M and related analyses; for example, carbon in monocyclic aromatic rings contributes very little to coke yield, whereas carbon in tricyclic and tetracyclic aromatic rings is largely converted to coke. It appears that a minimum number of hydrocarbon-type structures, including their average molecular weight, are necessary for an adequate determination of cracking yields. One technique [50] for a more complete feed characterization employs physical separations and mass and ultraviolet spectrometry to break down feeds into nine compound types (Table 16), including molecular weight, which are then used to predict the yields from the total feed. Predictions of yields from the nine compound types are presented in Figs. 34–37 for a 700°F mean average boiling point and once-through operations as a function of severity with an amorphous silica–alumina catalyst. The severity variable is defined as (catalyst surface area, m^2/g)/(100 × weight

TABLE 16 Compound Types Determined for Characterization of Cat Cracker Feeds [50]

Normal paraffins
Isoparaffins
Monocycloparaffins
Polycycloparaffins
Monocyclic aromatics
Dicyclic aromatics
Tricyclic aromatics
Benzanthracene and pentacyclic aromatics
Other tetracyclic aromatics

TABLE 17 Relative Crackability of Compound Types (700°F MABP) [50]

Compound Types	Conversion, wt.% to 430°F Products Plus Coke, at Severity = 0.5
Normal paraffins	24
Isoparaffins	37
Monocycloparaffins	61
Polycycloparaffins	65
Monocyclic aromatics	56
Dicyclic aromatics	13
Tricyclic aromatics	~13
Benzanthracene and pentacyclic aromatics	~56
Other tetracyclic aromatics	~16

hourly space velocity). At a constant severity the conversion levels for each component will be different and can be estimated by summing the gas, butane, gasoline, and coke yields for each component. Table 17 illustrates the conversion levels obtained at a severity of 0.5. Naphthenes, both mono- and polycyclic, are the most easily cracked type followed by monocyclic aromatics. Isoparaffins are more readily cracked than n-paraffins. The overall conversion of polycyclic aromatics, except for benzanthracene and pentacyclic aromatics, is low and most of this conversion is to coke.

Cycloparaffins and monocyclic aromatics produce the most gasoline at a given severity (Fig. 34). The yield from diaromatics is lower than that of the paraffins and, for the 700°F MBP feeds illustrated in Fig. 34, the yield of gasoline from tricyclic and higher cyclic polynuclear aromatics is negligible. Of course, for higher boiling range polycyclic aromatics containing a higher proportion of nonaromatic carbon, the yields of gasoline are correspondingly higher.

Naphthenes and isoparaffins produce the greatest yields of butanes and also relatively high yields of C_3 and lighter (Figs. 35 and 36). In these examples at constant severity and with a relatively light feed, the yields of C_4 and lighter are lowest for polycyclic aromatics; monocyclic aromatics are exceptional in producing the highest yields of C_3 and lighter.

Coke yield from saturates and monocyclic aromatics is negligible in many cases. Thus most of the coke produced arises from the polycyclic aromatics in the feed—more specifically, from the aromatic carbon in the polycyclic aromatics fraction. Typical relations among the several polycyclics are illustrated in Fig. 37.

Sophisticated cat cracking models [42, 51] are based on feedstock correlations of the preceding types or even more detailed breakdowns of hydrocarbon species. In one case [51], high resolution mass spectrometry is used to identify 50 compound types and their carbon number distributions. This type of data on over 500 cat cracker feed and product samples is used to generate predictions of feedstock effects for the model. Predictions for the basic

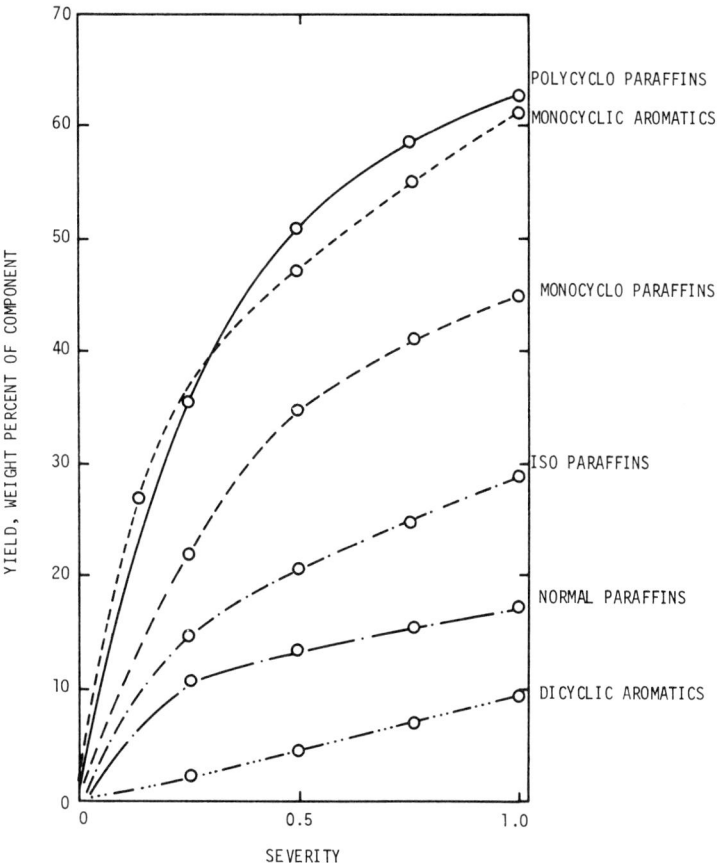

FIG. 34. Production of $C_5 - 430°F$ gasoline [50].

compound types with an average boiling point of 855°F are summarized in Table 18. These results are for a constant, once-through conversion of 70 vol.% to 430°F and lighter products plus coke, with a zeolite catalyst. Also, these predictions are for cracking of the hydrocarbon type in question when admixed with a typical feedstock of the same boiling range. Thus interaction effects between the various hydrocarbon types are reflected in these results. Crackability (proportional to rate constant for conversion) is uniformly high for saturate types and monocyclic aromatics. The negative crackability for polycyclic aromatics reflects the inhibiting effect of these components on the overall reaction rate. Similarly, the negative coke makes associated with paraffins and naphthenes reflect the interaction of saturates with polycyclic aromatics to reduce the coke make below that of the polycyclics alone.

Naphthenes are readily cracked with the highest yield of gasoline of about average octane number. The monocyclic aromatics fraction produces the highest octane gasoline and a negligible coke make, but gasoline yield is only moderate. Both 2-ring and 3 + -ring aromatics result in increased gas yields and the lowest yields of gasoline with below average octane quality.

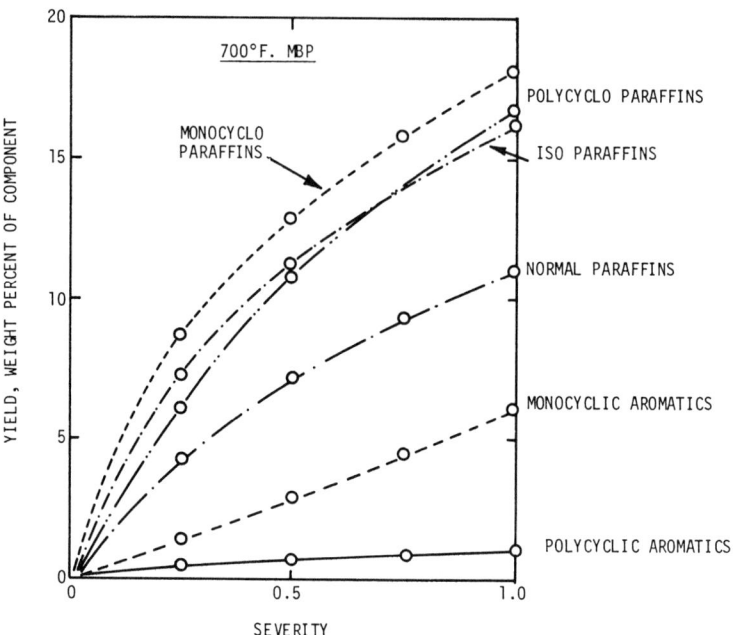

FIG. 35. Production of C_4's [50].

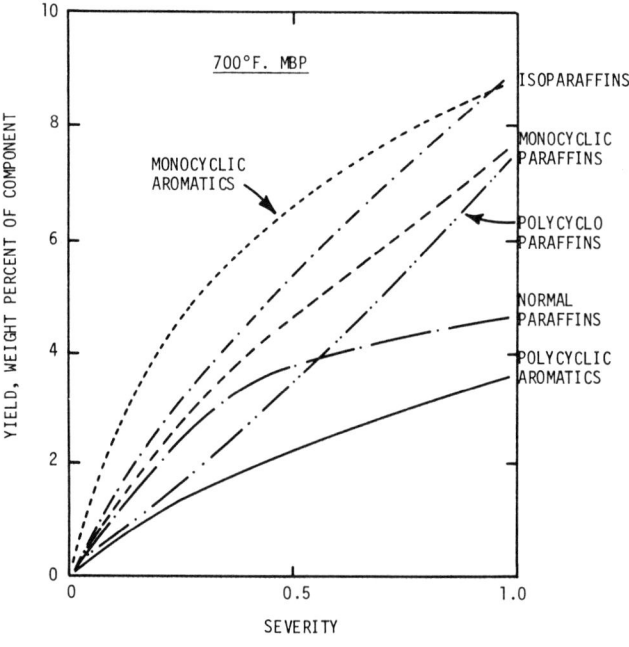

FIG. 36. Production of C_3 and lighter [50].

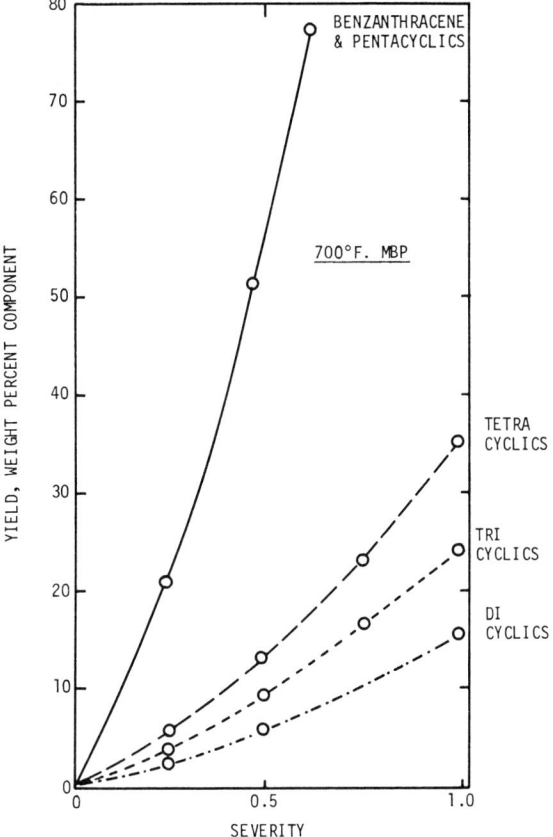

FIG. 37. Production of coke from polynuclear aromatics [50].

Other than hydrocarbons and heterocompounds of nitrogen, sulfur, and oxygen, most cat cracker feeds contain traces of nickel, vanadium, copper, and iron. These elements are present as volatile organometallic compounds and normally are not present in concentrations above 1 ppm unless the feedstock final boiling point exceeds $\simeq 1050°F$. When higher concentrations are encountered, it is due to high metals content feed components such as deasphalted oil, carry-over of residuum from the feed vacuum distillation facilities, or contamination by residuum from leaking exchangers. These heavy metal elements are deposited quantitatively on the cracking catalyst and promote an abnormally high production of coke and hydrogen; these effects are attributed to the dehydrogenation activity of the metal or metal sulfide in the cracking reactor. The relative potency of the various elements for hydrogen and coke production varies considerably; on a weight basis and with nickel as the base (unity), the relative effects of the other elements are approximately 0.2 V, 0.8 Cu, and 0.1 Fe. The relative effects of the metals or any one metallic component are influenced by the feedstock composition, operating conditions, the type of catalyst, and the "age" of the metals on the catalyst. Freshly deposited metals

TABLE 18 Catalytic Cracking of Hydrocarbon Types (70 vol.% conversion to 430°F plus coke, once-through processing with zeolitic catalyst)

Component	Crackability (conversion rate constant)	Yields						Clear Research Octane Number (C_5–430°F)
		C_1–C_3 (wt.%)	C_4 (vol.%)	C_5 (vol.%)	C_5–430°F Naphtha (vol.%)	Coke (wt.%)		
Paraffins	1.7	10.1	22.1	18.1	46.7	−0.5		89.4
Naphthenes	1.8	3.9	9.5	8.4	70.8	−0.5		92.5
1-ring aromatics	15	8.6	13.5	10.6	59.1	0.0		101.5
2-ring aromatics	0.4	12.7	16.1	10.3	51.0	3.4		90.3
3+-ring aromatics	−0.5	14.2	15.2	8.6	39.0	13.0		87.0

are most active for carbon and hydrogen production, and this activity diminishes rapidly as the catalyst remains in the unit and the metals are "passivated" or "aged." In many instances the effects of a slug of contaminated feed have disappeared after 1 or 2 weeks operation on contaminant-free feed. In addition to promoting dehydrogenation reactions in the reactor, these metallic contaminants promote the oxidation of CO to CO_2 on the regenerator side. These effects of soluble metal compounds in the feed are to be distinguished from corrosion products, scale, tramp iron, etc. The effect of these particulate solids on cracking reactions is small and usually negligible.

Sodium is not normally found in cat cracking feeds, but it is occasionally deposited on the catalyst during use. The usual sources are additives in boiler feed water or steam used on the cracking unit or contamination of the fresh feed or recycle with water or caustic during prior processing. The immediate effect of sodium contamination is to lower activity of the catalyst by neutralization of the catalytically active acid sites. Furthermore, sodium reduces the stability of the catalyst and, depending on the level of contamination, catalyst composition, and the unit operating conditions, can react with the catalyst to collapse the surface area and pore volume and increase particle density.

The extent to which variations in hydrocarbon types can actually change the cracking characteristics of typical feedstocks is best illustrated by comparing cracking results from "aromatic," "naphthenic," and "paraffinic" gas oils. Data of this type are presented in Tables 19, 20, and 21 [52]. Feedstock characteristics and the predicted operating conditions for a heat-balanced single-pass, riser operation are reported in Table 19. With the lower coke-

TABLE 19 Effects of Feedstock Type on Heat Requirements and Operating Conditions [52]

Feedstock Description	Aromatic	Naphthenic	Paraffinic
Properties:			
Gravity, °API	20	27	34
Aniline point, °F	145	188	230
Sulfur, %	0.9	0.4	0.13
Nitrogen, %	0.35	0.1	0.04
ASTM distillation:			
10% point	635	560	650
50% point	805	790	810
90% point	930	980	980
Characterization factor	11.5	12.0	12.7
Temperature, °F:			
Feed preheat	550	675	775
Riser outlet	1020	1000	1000
Regenerator	1225	1200	1200
Coke burnoff, % of fresh feed	6.3	5.4	4.8
Air rate, lb/lb of fresh feed	0.66	0.57	0.50
Steam rate, lb/lb of fresh feed	0.03	0.025	0.022
Catalyst/oil ratio, wt./wt.	7.5	6.4	5.7

TABLE 20 Effects of Feedstock Type on Product Distributions [52]

Feedstock Description	Aromatic	Naphthenic	Paraffinic
Conversion, vol.%	70	85	93
Yields, vol.%:			
Gasoline, C_5–430°F TBP end point	54.2	70.0	73.0
Butane–butene	16.8	19.0	22.5
Isobutane	5.9	7.3	8.0
n-Butane	1.4	1.9	2.5
Butenes	9.5	9.8	12.0
Propane–propylene	7.5	8.5	12.0
Propane	2.0	2.4	3.3
Propylene	5.5	6.1	8.7
Light catalytic gas oil	20.0	10.0	5.0
Decanted oil	10.0	5.0	2.0
Total	108.5	112.5	114.5
Coke, wt.%	6.3	5.4	4.8
C_2 and lighter, wt.%	3.0	2.9	2.5
H_2S, wt.%	0.4	0.2	0.1

forming tendencies of the paraffin feed, a higher preheat temperature is required to heat balance the unit at the desired high regenerator temperature. The other major shift in operating conditions due to feedstock quality is the lower cat/oil ratio required to maintain reactor temperature. Even in the heat-balanced

TABLE 21 Effects of Feedstock Type on Product Qualities [52]

Feedstock Description	Aromatic	Naphthenic	Paraffinic
Product qualities:			
Gasoline, C_5–430°F TBP			
Gravity, °API	56	58	60
Sulfur, %	0.10	0.04	0.01
Octane ratings:			
Research, clear	97	95	94
Motor, clear	85	84	83
Hydrocarbon type, vol.%:			
Aromatic	46	40	35
Olefins	16	15	20
Saturates	38	45	35
Light catalytic gas oil:			
Gravity, °API	15	17	20
Sulfur, %	1.0	0.50	0.15
Decanted oil:			
Gravity, °API	−3	0	10
Sulfur, %	2.5	2.0	0.3

operation, conversions with the paraffinic feed are higher than that of the naphthenic feed which in turn is more extensively converted than the aromatic feed. Light product yields, C_3 through the naphtha boiling range, are uniformly higher with the more saturated feeds (Table 20). Unsaturation of the C_3 and C_4 fractions is only slightly affected by feed type in these operations. Naphtha quality is highest from the aromatic feed. Octane numbers are lower from the paraffinic feed, but only by 2 to 3 units. The relative high quality of the gasoline from paraffinic feed is due both to the increased concentration of high octane number olefins and to the substantial yield of high octane aromatics from paraffinic feeds. The cycle oils from the aromatic feed are highly aromatic and thus less desirable for recycle than the other cycle stocks unless hydrogenation is used to upgrade the stocks.

Process Models

Complete process models for fluid catalytic cracking are extremely complex and generally require a computer model if anything other than a selected feature or section of the process is to be examined. Models for conversion, selectivity, coke function, carbon burning, etc., are available, but the utility of these models for final design studies or operational analysis is frequently limited by their isolation from other parts of the system. The complexity of the integrated cat cracking models is due in part to the complex interaction between pressure, heat, and chemical balance in both the reactor and regenerator systems. Also, in addition to recycle streams, the model sometimes incorporates product fractionation, flue gas power recovery, and light ends handling effects. The key role that the cat cracker plays in most refinery operations requires that the model accurately reflect all significant input changes and generate a detailed description of all products and outputs. These integrated process models are proprietary information of the petroleum process contractors and major refiners; however, several such models [42, 51] have been described in general terms and portions thereof or abbreviated models are available (see the section entitled "Catalytic Cracking Kinetics").

Amoco Oil Co. has proposed a process model [42], based on more than 400 individual pilot plant tests and validated with 180 tests from refinery fluid cat cracking units, using the following four basic relations:

$$X/(1-X) = f(z_1, \ldots, z_m)(C/O)^n (WHSV)^{n-1} \exp(\Delta E/RT_{rx}) \quad (2)$$

where C/O = catalyst to oil ratio, lb/h catalyst circulation per lb/h of oil
$WHSV$ = weight hourly space velocity, weight of total feed per hour per weight of catalyst inventory in the reaction zone, h^{-1}
ΔE = activation energy for cracking reactions, 25,000 Btu/lb mol
R = gas constant, 1.987 Btu/lb mol, °R
T_{rx} = kinetic average reaction temperature, °R
n = decay exponent, 0.65
X = volume fraction conversion

function f = a proprietary function of feed quality, hydrocarbon partial pressure, intrinsic catalyst activity, and S_{C_R}—a severity factor reflecting the effective activity at various levels of carbon-on-regenerated catalyst, C_R

$$\text{Coke, wt.\%} = g(z_1, \ldots, z_m)(\text{C/O})^n (\text{WHSV})^{n-1} \exp(\Delta E_c / RT_{rx}) \quad (3)$$

where ΔE_c = activation energy for coke yield, $\simeq 2500$ Btu/lb mol
function g = a proprietary function involving feed quality, hydrocarbon partial pressure, intrinsic catalyst activity, and the initial catalyst variable, C_R

$$\text{CCR} = F \text{ (feed enthalpy, heat of cracking, regenerator temperature, reactor temperature)} \quad (4)$$

and

$$\text{CCR} = G \text{ (reactor coke yield, heat of combustion of coke, regenerator temperature, reactor temperature)} \quad (5)$$

where CCR = catalyst circulation rate, lb/h, and F and G are the heat balance relations for the reactor and regenerator, respectively.

Simultaneous solution of the above relations defines consistent values for conversion, coke yield, and process variables. Additional correlations are then used to incorporate recycle effects and predict light ends and gasoline yields, and product quality. The yields are predicted by carbon number grouping with compound classes in each grouping. Finally, the model reports product values, utilities consumptions, and investments. This model has been employed in a wide variety of studies including an evaluation of the hydrotreating of cat cracker feed and the optimized design of a new unit.

The integrated cat cracking process model [51] used by Exxon Research and Engineering Co. in the Flexicracking process has also been described in general terms. Input/output features are shown in Fig. 38. A detailed fresh feed characterization, catalyst properties, reactor configuration, and selected independent variables are combined with recycle rate, quality of fractionation, and other specifications or limitations outside the reactor as input variables. The computer model develops a set of operating conditions consistent with the input variable restrictions and pressure, heat, and chemical balances. Extensive correlations are then used to establish detailed yields and product qualities, including compound-type distributions, and design parameters for the reactor, regenerator fractionator, and other equipment items.

Less sophisticated procedures have been described for determining product yields and qualities from straight-run gas oils [53]. These correlations are useful for preliminary design and cost studies or to estimate the effects of conversion level. Figures 39–45 are for zeolitic or molecular sieve catalysts; similar plots are available for amorphous silica–alumina catalysts. Figure 46 can be used to

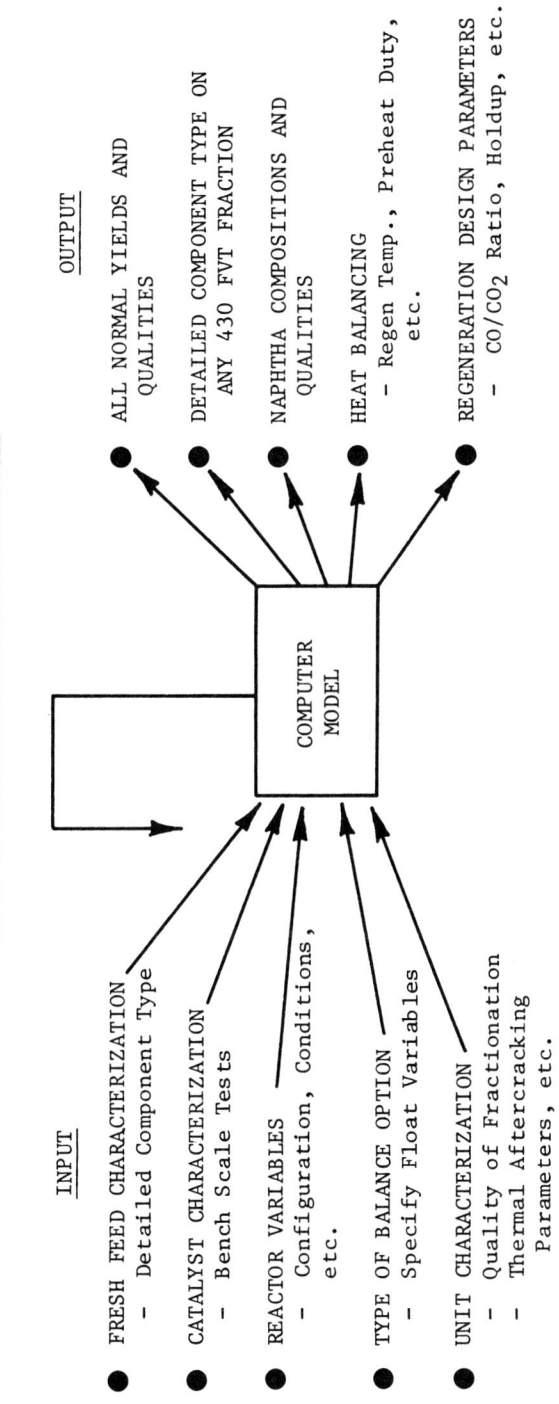

FIG. 38. Input/output feature of the Flexicracking cat cracking model [51].

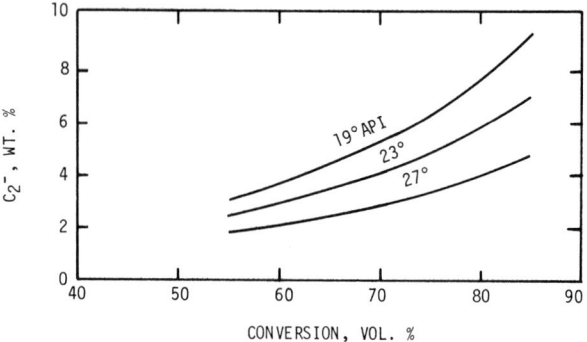

FIG. 39. Ethane and lighter yields, zeolite catalyst [53].

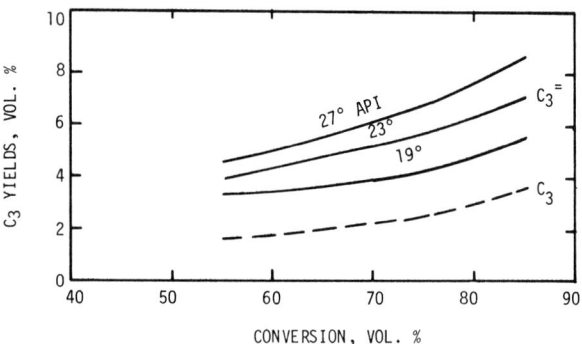

FIG. 40. Propane and propylene yields, zeolite catalyst [53].

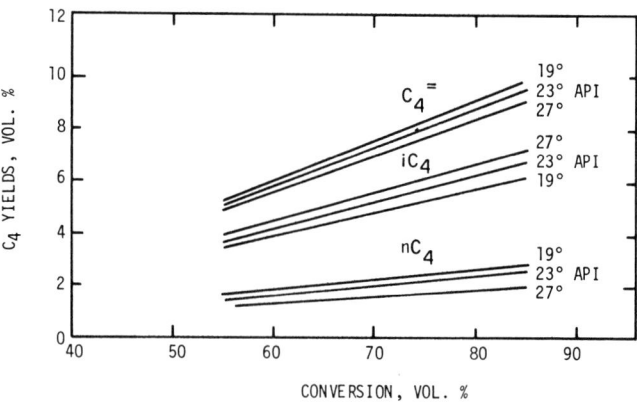

FIG. 41. Butanes and butylenes yields, zeolite catalyst [53].

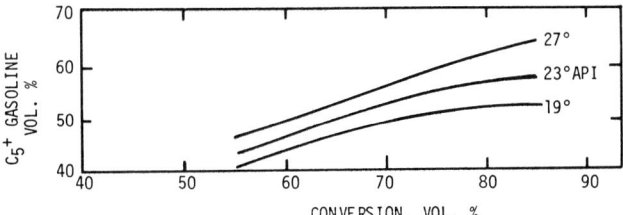

FIG. 42. C_5+ gasoline yields, zeolite catalyst [53].

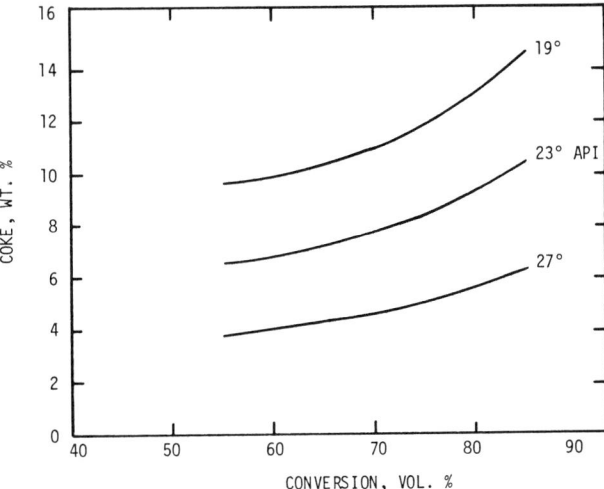

FIG. 43. Coke yields, zeolite catalyst [53].

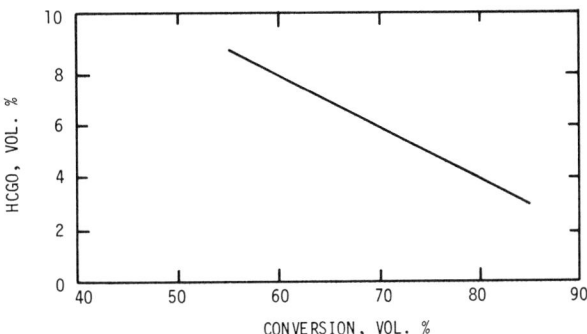

FIG. 44. Heavy cycle gas oil yields, zeolite catalyst [53].

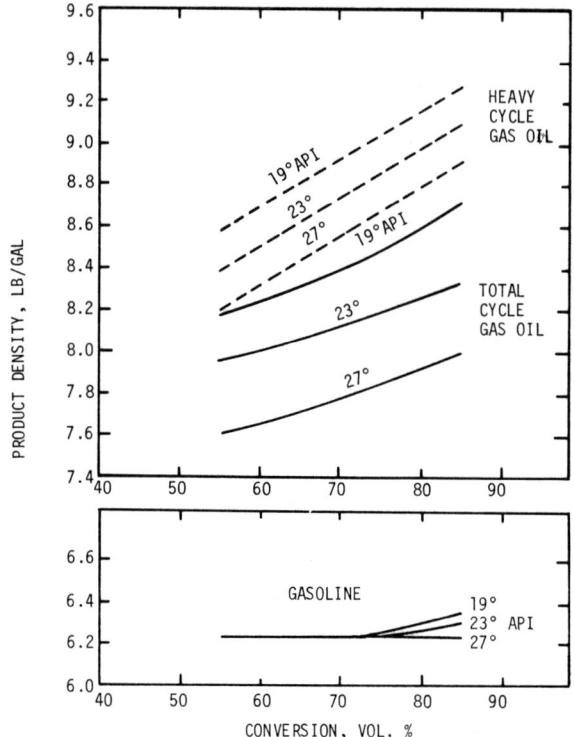

FIG. 45. FCC product gravity, zeolite catalyst [53].

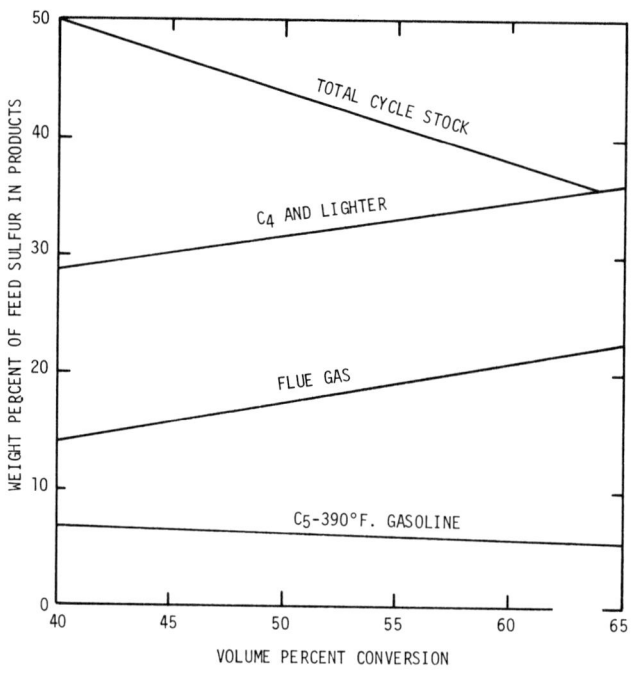

FIG. 46. Distribution of sulfur in catalytic cracking products [53].

estimate sulfur distribution. Specific gravity, or API gravity, is the only required feed parameter. The following procedure is used for a fixed conversion:

1. Fix the weight of the feed and determine the yields and weights of C_2^-, C_3^{2-}, C_3, C_4^{2-}, iC_4, C_5+ gasoline, and coke (Figs. 39–43).
2. Obtain the weight of the cycle gas oils by difference; using Fig. 45 to determine the gravity of the total cycle oil, calculate the volume of total cycle oil.
3. Calculate the volume and weights of heavy cat cycle oil from Fig. 44.
4. Obtain the volume and weight of light cat cycle oil by subtracting the yields of heavy cat cycle oil from the total cycle gas oil.

Piloting

Piloting of cat cracking operations to establish design criteria is either carried out as an integrated demonstration including reactor, regenerator, and stripper with fractionation of products and recycle, or as an isolated operation on some phase of cat cracking—catalyst characterization, feedstock evaluation, or process design, etc. The development of complete process models has allowed the process design to be established in many cases with piloting of only one or two portions of the total cat cracking unit. Frequently, only the yields and product quality from the design feed at a particular set of conditions need to be determined in pilot plant operations; the remainder of the design data is already available or can be generated from process models. Heat effects or heat balance data are seldom needed from pilot plant operations. In fact, the scale of most pilot plants is such that external heating is required to maintain the desired reactor, regenerator, and stripper temperatures. These pilot plants do not naturally seek a heat balanced condition, and it is generally necessary to specify a set of process conditions consistent with a heat balance to obtain meaningful process design data. Of course, for purposes of developing correlations or basic process information, it is not necessary that the process conditions (reactor and regenerator temperatures, catalyst circulation rate, conversion, etc.) represent a consistent set of heat-balanced conditions.

Catalyst characterization is usually limited to definition of an activity level, a carbon formation factor, and a gas formation factor. Other important catalyst properties—primarily regenerability and yield and product quality factors—are determined separately. If the catalyst in question is already in commercial use, samples of the "equilibrium catalyst" can be obtained and used in the piloting operations. However, catalyst contamination and deactivation conditions, particularly regenerator temperature and steam partial pressure, may be different for the unit being designed, and these effects must be recognized in using equilibrium catalyst from any existing unit. If the design or unit expansion or modification is to be based on a new catalyst, it is necessary to synthesize an "equilibrium catalyst." It is generally not practical to do this in pilot plant operations by starting with a charge of fresh catalyst and operating the unit until equilibrium activity is reached, due to the equilibration time of several months and the differences between pilot plant and commercial catalyst losses and deactivation conditions. New or fresh catalysts can be steamed in the laboratory to simulate the activity level of equilibrium catalysts [53]. Each catalyst requires a different and unknown set of steaming conditions to approach a valid equilibrium condition; even so, laboratory steaming does not duplicate the contamination effects of commercial units and the selective loss of fines in commercial units, but this steaming procedure to simulate equilibration in a commercial unit is used when a commercially equilibrated sample is not available.

Characterization of catalysts is carried out in small reactors which are fixed beds of pelletized or fluidized catalyst. A typical feedstock is usually run in a once-through mode of operation at typical cat cracking conditions [54–56]. The activity, coke, and gas yield results of such tests are expressed as "factors" relative to a standard catalyst. Process models are generally set up to use catalyst characterization data on this type of relative basis.

Regeneration characteristics for design purposes are determined in similar fixed or fixed-fluidized bed reactors using a charge of coked catalyst and an oxygen-containing regeneration gas. Kinetic parameters for the rates of carbon removal and CO and CO_2 formation are determined from a small number of tests at typical regenerator conditions. These rates are usually expressed relative to a standard catalyst to facilitate use with a process model.

As the cat cracking process has matured with the accumulation of vast amounts of commercial and pilot unit data and the building of complex process models, pilot plants for feedstock and process variable studies and for obtaining process design data have become smaller and simpler. Pilot plants with capacities of 1 to 10 bbl/d are now usually operated only for developmental studies such as the evaluation of new reactor designs or to produce large quantities of a product not available from commercial units. More common are laboratory-type units incorporating a reactor, stripper, and regenerator within a circulating fluidized catalyst system. A diagram of this type of unit with a charge rate of 0.01 to 0.1 bbl/d is shown in Fig. 47 [20]. A somewhat larger pilot plant feeding 0.2 to 0.4 bbl/d and using mechanized feeders for control of catalyst circulation has been used to study both dense-bed and transfer line reactors [40]. These small pilot plants may not exactly duplicate commercial unit results, but their overall performance is similar to that of the commercial units and the variations are due to recognized differences in reactor or regenerator gas/solids mixing or contacting, adiabaticity, thermal reactions, etc. With the aid of process models, small-scale pilot plant results can be compared to commercial unit data on the same or similar feeds and catalysts. Such comparisons provide the efficiency factors or other terms needed to translate the pilot plant results to a commercial unit scale. Thus laboratory cat cracking pilot plants can be used satisfactorily for feedstock evaluations, catalyst comparisons, process variable studies, and overall demonstrations of specific design cases.

Cracking Catalysts

Cracking catalysts can differ markedly in both activity to promote the cracking reaction and in the quality of the products obtained from cracking the gas oil

Cracking, Catalytic

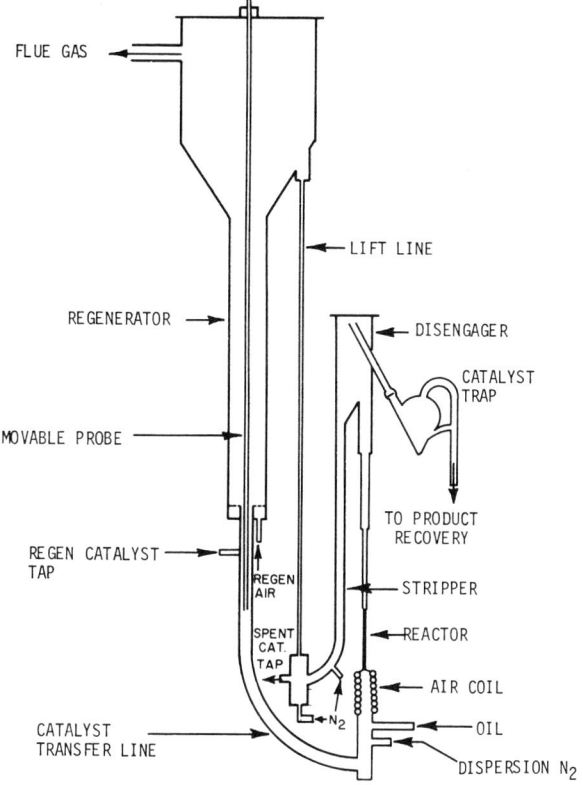

FIG. 47. Laboratory circulating fluid-bed catalytic cracking unit [20].

feeds (selectivity). Activity can be related directly to the total number of active (acid) sites per unit weight of catalyst and also to the acidic strength of these sites. Differences in activity and acidity regulate the extent of various secondary reactions occurring and thus the product quality differences. The acidic sites are considered to be Lewis- or Bronsted-type acid sites, but there is much controversy as to which type of site predominates.

Cracking catalysts generally comprise mixed metal oxides such as $SiO_2-Al_2O_3$, SiO_2-MgO, and more recently, crystalline aluminosilicate zeolites and many others.

Clay Catalysts

The first cracking catalysts were acid leached montmorillonite clays. The acid leach was to remove various metal impurities, principally iron, copper, and nickel, which could exert adverse effects on the cracking performance of the catalyst. The catalysts were first made by Houdry and later by the Filtrol Corp. and used in fixed-bed and moving-bed reactor systems in the form of shaped pellets. Later, with the development of the fluid catalytic cracking process, the clay catalysts were made in the form of a ground, sized powder. Clay catalysts are relatively inexpensive and have been used extensively for many years.

Synthetic Catalysts

The desire to have a catalyst uniform in composition and catalytic performance led to the development of synthetic catalysts. The first synthetic cracking catalyst, consisting of 87% SiO_2 and 13% Al_2O_3, was made by Houdry Corp. in pellet form and used in their fixed-bed units in 1940. Catalysts of this composition were ground and sized for use in fluid catalytic cracking units. In 1944 catalysts in the form of beads about 2.5 to 5.0 mm diameter were introduced to the industry by Socony-Mobil Co. These catalysts comprised about 90% SiO_2 and 10% Al_2O_3 and were extremely durable. These were used in the TCC process. One version of these catalysts contained a minor amount of chromia to act as an oxidation promoter.

Mixed Oxide Catalysts

In addition to synthetic catalysts comprising silica–alumina, other combinations of mixed oxides were found to be catalytically active and were developed during the 1940s. Each showed some interesting property. These systems included silica–magnesia, silica–zirconia, silica–alumina–magnesia, silica–alumina–zirconia, and alumina–boria. Of these, only silica–magnesia was used in commercial units but operating difficulties developed with the regeneration of the catalyst which at the time demanded a switch to another catalyst. Further improvements in silica–magnesia catalysts have since been made [57].

High Alumina Catalysts

During the 1940–1962 period, the cracking catalysts used most widely commercially were the aforementioned acid-leached clays and silica–alumina. The latter was made in two versions; low alumina (about 13% Al_2O_3) and high alumina (about 25% Al_2O_3) contents. High alumina content catalyst showed a higher equilibrium activity level and surface area. Principal manufacturers of the synthetic silica–alumina catalysts during this period were Davison Chemical Co., Nalco Chemical Co., American Cyanamid Co., Universal Oil Products, and Houdry Corp.; the clay catalyst was manufactured by Filtrol Corp. and the silica–magnesia catalyst by Davison and Nalco.

Semisynthetic Catalysts

During the 1958–1960 period, several manufacturers (Davison, Nalco, Cyanamid) marketed a semisynthetic (SS) grade of silica–alumina catalyst in which about 25 to 35% kaolin was dispersed throughout the silica–alumina gel. These catalysts could be offered at a lower price, but they were marked by a lower catalytic activity and greater stack losses because of increased attrition rates. One virtue of SS grade catalysts was that a lesser amount of adsorbed, unconverted, heavy products on the catalyst were carried over to the stripper zone and regenerator. This resulted in a higher yield of more valuable products and also smoother operation of the regenerator as local hot spots were minimized. Davison, which had made substantial improvements in the stability and regenerability of silica–magnesia catalyst, offered a version which contained kaolin (SM-30-S/S).

Zeolite Catalysts

Crystalline zeolites having molecular sieve properties were introduced as selective adsorbents in the 1955–1959 period. The catalytic cracking properties of several metal-exchanged versions of these materials were first disclosed in a patent [59] issued to Esso Research and Engineering Co. in 1961. In 1962 Socony-Mobil introduced the first commercial cracking catalyst which contained crystalline zeolites. This catalyst, designated Durabead 5, was made in bead form for use in TCC units. It was followed later by a spray-dried version for use in FCC units. In a relatively short time period all of the cracking catalyst manufacturers were offering their versions of zeolite catalysts to refiners. The intrinsically higher activity of the crystalline zeolites vis-à-vis conventional amorphous silica–alumina catalysts coupled with the much higher yields of gasoline and decreased coke and light ends yields served to revitalize research and development in the mature refinery process of catalytic cracking.

A number of crystalline aluminosilicates have been mentioned as having catalytic cracking properties, such as synthetic faujasite (X- and Y-types), offretite, mordenite, and erionite. Of these, the faujasites have been most widely used commercially. While faujasite is synthesized in the sodium form, the sodium is removed by base exchange with other metal ions which, for cracking catalysts, include magnesium, calcium, rare earths (mixed or individual), and ammonium. In particular, mixed rare earths alone or in combination with ammonium ions have been the most commonly used forms of faujasite in cracking catalyst formulations. Empirically, X-type faujasite has a stoichiometric formula of $Na_2O \cdot Al_2O_3 \cdot 2.5SiO_2$ and Y-type faujasite $Na_2O \cdot Al_2O_3 \cdot 4.8SiO_2$. Slight variations in the SiO_2/Al_2O_3 ratio exist for each of the types. Rare earth exchanged Y-type faujasite retains much of its crystallinity after steaming at $1520°F$ with 20% steam for 12 h; rare earth form X-faujasite, while thermally stable in dry air, will lose its crystallinity at these temperatures in the presence of steam.

Catalyst Manufacture

While each manufacturer has developed proprietary procedures for making silica–alumina catalyst, the general procedure [58] consists of the following steps: (1) the gelling of dilute sodium silicate solution ($Na_2O \cdot 3.25SiO_2 \cdot xH_2O$) by the addition of an acid (H_2SO_4, CO_2) or an acid salt such as aluminum sulfate, (2) aging the hydrogel under controlled conditions, (3) adding the prescribed amount of alumina as aluminum sulfate and/or sodium aluminate, (4) adjusting the pH of the mixture, and (5) filtering the composite mixture. After filtering, the filter cake can either be (1) washed free of extraneous soluble salts by a succession of reslurrying and filtration steps and spray dried or (2) spray dried and then washed free of extraneous soluble salts before flash drying the finished catalyst. There are a number of critical areas in the preparative processes which affect the physical and catalytic properties of the finished catalyst. Principal among them are the concentration and temperature of the initial sodium silicate solution, the amount of acid added to effect gelation, the length of time of aging the gel, the method and conditions of adding the aluminum salt to the gel, and its incorporation therein. Under a given set of conditions the product catalyst is quite reproducible in both physical properties and catalytic performance.

During the 1940-1962 period, a number of improvements were made in silica-alumina catalyst manufacture. These included continuous production lines vs batch-type operation, introduction of spray drying to eliminate grinding and sizing of the catalyst while reducing catalyst losses as fines, improving catalyst stability by controlling pore volume, and improved wash procedures to remove extraneous salts from high alumina content catalysts to improve equilibrium catalyst performance.

Zeolite cracking catalysts are made by dispersing or imbedding the crystals in a matrix. The matrix is generally amorphous silica-alumina gel and may also contain finely divided clay. The zeolite content of the composite catalyst is generally in the range of 5 to 16 wt.%. If clay (e.g., kaolin) is used in the matrix, it is present in an amount of 25 to 45 wt.%, the remainder being the silica-alumina hydrogel "glue" which binds the composite together. The zeolite may be preexchanged to the desired metal form and calcined to lock the exchangeable metal ions into position before compositing with the other ingredients. In an alternate scheme, sodium-form zeolite is composited with the other components, washed, and then treated with a dilute salt solution of the desired metal ions before the final drying step. As stated above, the matrix generally consists of silica-alumina, but several catalysts have been commercialized which contain (1) silica-magnesia/kaolin and (2) synthetic montmorillonite-mica and/or kaolin as the matrix for faujasite.

Each manufacturer employs proprietary formulations and methods of compositing the final grade of catalyst. Most manufacturers have at least several grades of zeolite catalysts and new formulations are continuing to come on the market. A representative technique for the manufacture would incorporate the following steps: (1) dilute sodium silicate ($Na_2O \cdot 3.25SiO_2$) solution is blended with kaolin, (2) the blended slurry is treated with alum solution to lower the pH to around 10 and effect gelation, (3) mixed gel is aged under controlled conditions, (4) alum solution is added to provide the necessary amount of alumina in the silica-alumina hydrogel, (5) a slurry of sodium faujasite (Y-type) is added, (6) the pH is adjusted to the desired level, (7) the composite slurry is filtered, (8) the mix is spray dried, (9) the catalyst is washed free of sodium with a dilute ammonium sulfate solution, and (10) the catalyst is reslurryed in a dilute solution of a suitable metal cation to exchange with the zeolite. This is followed by (11) filtering, rinsing, and flash drying of the finished catalyst. If precation exchange faujasite is used in Step 5, then the cation exchange treat in Step 10 is unnecessary.

Catalytic Selectivity Characteristics

In the catalytic cracking process, the most abundant products are those having 3, 4, and 5 carbon atoms. On a weight basis the 4-carbon-atom fraction is the largest. The differences between the catalysts of the mixed oxide type lie in the relative action toward promoting the individual reaction types included in the overall cracking operation. For example, silica–magnesia catalyst under a given set of cracking conditions will give a higher conversion to cracked products than silica–alumina catalyst. However, the products from SiO_2–MgO catalyst have a higher average molecular weight, hence a lower volatility, lesser amounts of highly branched/acyclic isomers, but more olefins among the gasoline boiling range products (C_4–430°F) than the products from SiO_2–Al_2O_3 catalyst. With these changes in composition, the gasoline from cracking with SiO_2–MgO catalyst is of lower octane number. These differences between catalysts may also be described as differences in the intensity of the action at the individual active catalytic centers. That is, a catalyst such as SiO_2–Al_2O_3 would give greater intensity of reaction than SiO_2–MgO as observed from the nature and yields of the individual cracked products and the Motor gasoline octane number. Titrations of these two catalysts show SiO_2–Al_2O_3 to have a lower acid titre than SiO_2–MgO, but the acid strength of the sites is higher.

While each of the individual component parts in these catalysts is essentially nonacidic, when mixed together properly they give rise to a titratable acidity as described above. Many of the secondary reactions occurring in the cracking process may also be promoted with strong mineral acids, such as concentrated sulfuric and phosphoric acids, aluminum halides, hydrogen fluoride, and hydrogen fluoride–boron trifluoride mixtures. This parallelism lends support to the concept of the active catalytic site as being acidic. Zeolites have a much higher active site density (titre) than the amorphous mixed oxides, which may account in large part for their extremely high cracking propensity. In addition, these materials strongly promote complex hydrogen transfer reactions among the primary products so that the recovered cracked products have a much lower olefin and higher paraffin content than are obtained with the amorphous mixed oxide catalysts. This hydrogen transfer propensity of zeolites to saturate primary cracked product olefins to paraffins minimizes the reaction of polymerizing the olefins to form a coke deposit, thus accounting in part for the much lower coke yields with zeolite catalysts than with amorphous catalysts.

Activity of the catalyst varies with faujasite content. Selectivity of the catalyst to coke and naphtha also varies with faujasite content; as the faujasite content drops below 5% the catalyst starts to show some of the cracking properties of the matrix, while for zeolite contents of 10% or higher very little change in selectivity patterns is noted. The various ion exchanged forms of the faujasite can result in slightly different cracking properties; e.g., using high cerium content mixed rare earths improves carbon burning rates in the regenerator, use of H-form faujasite or H^+/RE^{3+} form faujasite improves selectivity to C_3–C_5 fractions, use of a minor amount of copper form faujasite increases light olefin yield and naphtha octanes, etc.

Activity Tests

The first testing units described in the 1940s employed a fixed bed of pelleted catalyst. One standard test [61] used widely during this period used 200 cc of pelleted (3/16 in. × 3/16 in.) catalyst in a reactor at 850°F feeding a light virgin East Texas gas oil (bp 500 to 675°F, 33.3° API) at a space velocity of 0.6 V/H/V (volume of feed per hour per volume of catalyst) over a 2-h cycle period. Fresh catalyst pellets were presteamed at 1050°F for 24 h under 60 lb/in.² gauge pressure; equilibrium catalysts were regenerated coke-free before testing. Product work-up consisted of (1) a standard ASTM one-plate distillation of 100 cc of the total liquid product to determine the amount of 430°F + material remaining, thereby establishing the vol.% conversion to 430°F − product, with the gasoline yield expressed as vol.% D&L (distillate + loss); (2) the volume and density of the gas produced permitted a gas factor value; and (3) the carbon level on the spent catalyst yielded a carbon factor value.

Another fixed-bed test used extensively was the CAT-A test [62]. This test employs 200 cc of granular catalyst, light East Texas gas oil (bp 430 to 720°F, 37.0° API) feed at a rate of 1.5 V/H/V over a 10-min process period. Work up of the unit gas, coke deposit, and total liquid product provided data for (1) wt.% coke, (2) wt.% gas, and (3) vol.% i-300°F and i-410°F naphtha yields.

A variety of fixed fluid solids laboratory testing units were developed, some of which were large enough to process enough gas oil feed to provide octane number determinations on the cracked naphtha product. The Standard Fluid Testing Unit (SFTU) developed by Standard Oil Development Co. (now Exxon Research and Engineering Co.) employed a 600-g change of catalyst to a reactor heated at 950°F feeding a West Texas process gas oil (bp 600 to 750°F, 26.9° API) at a feed rate of 3.5 W/H/W (pounds of feed per hour per pound of catalyst) over a 10-min process period. The coke deposited on the catalyst was determined by ignition in air, the unit gas was analyzed by mass spectrometry, and the total liquid product distilled (15/5) to give a 430°F end point naphtha.

A main drawback to the cracking catalyst tests described above, both fixed bed and fluid bed, lay with the long cycle times (10 min to 2 h) required to process enough feed to work up the product. Commercial dense fluid bed reactor systems have a catalyst residence time of $\simeq 1$ min in the reactor. As the catalyst is continually deactivating during a cycle as carbon is being deposited, excessive cycle periods do not simulate commercial performance. This problem became acute when zeolite catalysts were developed. While these catalysts show a lower selectivity to coke than amorphous gel catalysts, their high activity leads to coke deposits of a magnitude resulting in severe loss in activity in tests with a 10-min cycle time or longer.

As improved cracking catalysts were developed, laboratory size testing equipment was specifically designed to evaluate the new catalysts. Several new laboratory catalyst tests were developed which utilize a much shorter cycle time. Because schemes have been developed to analyze the liquid product and unit gas make using chromatographic columns, the tests are "micro" in nature, rapid, and inexpensive.

A widely used cracking test developed by Atlantic-Richfield and Davison is the Micro-Activity Test (MAT). Several updated versions of this test have been made since it was first disclosed in 1967. In one version, 5 g of powdered catalyst are placed in the reactor at 900°F and contacted with 1.00 cc of a West Texas Devonian gas oil (bp 500 to 800°F, 32.4°API) at a cat/oil ratio of 5.8. The space velocity is variable over the range of 2 to 16 W/H/W or higher by changing feed rate. Principal data from the test are (1) activity expressed as vol.% converted to 430°F−, (2) a carbon factor, and (3) a gas factor. Fresh catalysts are presteamed at a variety of conditions, most notably 8 h at 1350°F and 15 lb/in.^{2} gauge pressure. Equilibrium catalysts are regenerated carbon-free before testing.

A second testing unit designed for zeolite cracking catalysts is the Micro Catalytic Cracking (MCC) unit used by Exxon Research and Engineering Co. This unit normally operates at 950°F, a catalyst charge of 4 to 20 g, feeding 1.7 ml of East Texas light gas oil (bp 500 to 675°F, 33.3°API) over a 2-min process period. The reactor assembly is vibrated at 60 cycles/s to ensure a fluidized catalyst bed. Space velocity is varied by changing the size of catalyst charge, the latter being chosen to approximate 75% conversion (430°F−). Product analyses include carbon on spent catalyst, mass spectroscopy of the unit gas, and a gas chromatograph distillation of the total liquid product. Yields are then correlated at 75 wt.% conversion (430°F−) for coke make, C_3^- dry gas, total C_4, and C_5/430°F naphtha. Activity of the catalyst is expressed as feed rate (W/H/W) required to give 75 wt.% conversion. Fresh catalysts are steamed 16 h at 1400°F and 0 lb/in.^{2} gauge; equilibrium catalysts are regenerated carbon-free before testing. Versatility of the MCC unit extends to operating temperatures up to 1200°F, process periods as short as 20 s, and space velocities as high as 100 W/H/W.

Catalyst Deactivation

A cracking catalyst should maintain its cracking activity with little change in product selectivity as it ages in a unit. A number of factors contribute to degrade the catalyst; (1) the combination of high temperatures, steam partial pressure, and time; (2) impurities present in the fresh catalyst; and (3) impurities picked up by the catalyst from the feed while in use. Under normal operating conditions, the catalyst experiences temperatures of 900 to 960°F in the reactor and steam stripper zones and temperatures of 1150 to 1325°F and higher in the regenerator accompanied by a substantial partial pressure of steam. With mixed oxide amorphous gel catalysts, the plastic nature of the gel is such that the surface area and pore volume decrease rather sharply in the first few days of use and then at a slow inexorable rate thereafter. This plastic flow also results in a loss in the number and strength of the active catalytic sites.

Zeolite catalysts comprising both amorphous gel and crystalline zeolite degrade from instability of the gel, as stated above, and also from loss in crystallinity. The latter also results from the combined effects of time, temperature, and steam partial pressure. When crystallinity is lost, the amorphous residue is relatively low in activity, approximating that of the amorphous gel matrix. The rate of degradation of the amorphous gel component may not be the same as that of the zeolite crystals; e.g., the gel may degrade rapidly and through thermoplastic flow effectively coat the crystals and interfere with the diffusion of hydrocarbons to the catalytic sites in the zeolite. Catalyst manufacturers try to combine high stability in the matrix with high stability zeolite crystals in making zeolite catalysts.

Residual impurities in freshly manufactured catalysts are principally sodium and sulfate. These result from the use of sodium silicate and aluminum sulfate in making the silica–alumina gel matrix and subsequent washing of the composite catalyst with ammonium sulfate to remove sodium. Generally, the sodium content of the amorphous gel is <0.1 wt.% (as Na_2O), and sulfate <0.5 wt.%.

With zeolite catalysts the residual sodium may be primarily associated with the zeolite, so that sodium levels may range from about 0.2 to 0.8% for the composite catalyst. Sulfate levels in zeolite catalysts are still $<0.5\%$. An excessive amount of sodium reacts with the silica in the matrix under regenerator operating conditions and serves as a flux to increase the rate of surface area and pore volume loss. Sodium faujasite is not as hydrothermally stable as other metal-exchanged (e.g., mixed rare earths) forms of faujasite. It is most desirable to reduce the sodium content of the faujasite component to <5.0 wt.% (as Na_2O) with rare earths or with mixtures of rare earths and ammonium ions.

Finally, catalysts can degrade as a result of impurities picked up from the feed being processed. These impurities are sodium, nickel, vanadium, iron, and copper. Sodium as laid down on the catalyst not only acts to neutralize active acid sites, reducing catalyst activity, but also acts as a flux to accelerate matrix degradation. Freshly deposited metals are much more effective as "poisons" to cracking catalysts than aged metal deposits. This is believed due to the metal deposited on the active exterior surface of the matrix being "buried" with time due to the thermoplastic flow of the matrix gel. Zeolite catalysts are less responsive to metal contaminants than amorphous gel catalysts [63]. Hence equilibrium catalysts can tolerate low levels of these metals so long as they have enough time to become buried. A sudden deposition of fresh metals can cause adverse effects on unit performance. Metals levels on equilibrium catalysts reflect the metals content of the feeds being processed; typical ranges are 200 to 1200 ppm V, 150 to 500 ppm Ni, and 5 to 45 ppm Cu. Sodium levels are in the range of 0.25 to 0.8 wt.% (as Na_2O).

Physical/Chemical Testing

A number of physical and chemical properties are measured to characterize a catalyst in addition to its cracking properties. These properties are obtained on fresh activated catalysts, artificially steam deactivated fresh catalysts, and plant equilibrium catalysts. Generally, the conditions for steam deactivating fresh catalysts are chosen so as to simulate the properties projected for an equilibrium catalyst.

Principal among the physical properties measured are (1) surface area (generally the BET nitrogen absorption method), (2) pore volume, (3) attrition rate, (4) bulk density, (5) metals contaminants, and (6) zeolite crystallinity data.

The fresh and equilibrium catalysts are tested to indicate qualitative relations between stability and catalytic performance. For example, with silica–alumina gel catalyst made by the same procedure and formulation, equilibrium catalyst performance could be fairly well predicted from surface area measurement.

Typical equilibrium catalyst surface areas and pore volumes for zeolite catalysts containing clay range from 80 to 110 m^2/g and 0.35 to 0.45 cc/g, respectively. Equilibrium zeolite catalysts without clay in the matrix have surface areas in the range of 125 to 150 m^2/g and pore volumes between 0.42 and 0.52 cc/g.

Attrition rate test data provide information relating break up of the catalyst into fragments which could overload cyclones in the regenerator, lead to excessive atmospheric pollution, and require increased fresh catalyst additions with accompanying economic costs. Attrition rate took on increased importance with the development of zeolite catalysts, some of which entail three separate components in the composite; zeolite, clay, and silica–alumina gel which is used as the binding material to hold the composite catalyst together. Wide variations in attrition rates were found for the early composite zeolite catalysts. Improvements in the procedures for making the catalysts are resulting in catalysts showing improved attrition resistance.

Bulk density measurements show widely different values between fresh and equilibrium catalysts of the same type. For example, a silica–alumina gel catalyst may have a bulk density of 0.46 cc/g fresh and a value of about 0.70 cc/g when at equilibrium. Such large differences could lead to preferential loss of fresh catalyst when added to the circulating equilibrium catalyst inventory. With zeolite catalysts containing kaolin, bulk density of the equilibrium catalyst can exceed 0.80 cc/g.

The active sites of mixed-oxide-type catalysts (crystalline zeolites included) are acidic and can be titrated with n-butyl amine in a nonaqueous system [60]. Use of various Hammett indicators covering a wide pK_a range allows determination of both total acid titre and the acid strength distribution of the acid sites. Other methods of measuring acid titre (active site density) include adsorption of gases such as NH_3 and pyridine.

Engineering Aspects of Fluid Processing

Regeneration

The cracking reaction deposits heavy polyaromatic coke and carbon on the catalyst. This material is removed by burning in the regenerator to reestablish the activity of the catalyst for the reaction cycle. The small amount of light hydrocarbon gases filling the voids of the catalyst as a result of incomplete stripping in the reactor stripper are also burned in the regenerator. However, this material will generally comprise much less than 10% of the regeneration requirement.

The regeneration reaction is primarily one of carbon-burning kinetics. It is complicated by carbon monoxide to carbon dioxide combustion. Regeneration is generally carried out with very little excess oxygen, and much carbon monoxide can be produced, as much as 10% by volume. If high temperatures are held in the regenerator and sufficient oxygen is provided, carbon monoxide in the flue gas can be reduced to very low levels, 500 ppm or lower. Thus a regenerator must not only burn the carbon or coke off the catalyst but can also control the carbon monoxide content of the regeneration off gases to meet CO boiler limitations or environmental requirements.

Important parameters in regeneration are:

Temperature
Oxygen partial pressure
Pressure
Amount of catalyst in the regeneration zone
Amount of carbon on the catalyst
Air and catalyst distribution
Combustion promoters
Type of catalyst
Air rate

The regeneration air rises through a bed of catalyst in small bubbles. The actual regeneration reaction takes place primarily in the thin cloud of gas in the solid emulsion phase surrounding the bubble. Figure 48 shows the bubble cloud and emulsion phases in a fluid solids bed. Regeneration also takes place but to a lesser extent in the emulsion phase and in the wake of the bubble. The regeneration gas thus flows up through the bed in essentially plug flow. This has been proven commercially in units where a localized failure has occurred in the air distributor. A high oxygen content is noted in the flue gas in the dilute phase near the cyclone inlet directly above the failure. Normal flue gas concentrations are noted on either side above the failure.

Assuming plug flow, the oxygen-driving force for the reaction will be the log mean of the entering and exit concentrations. The absolute pressure level will, of course, have a direct effect on the oxygen partial pressure. Thus higher pressure regenerators will be able to regenerate to lower carbon on catalyst levels for a given set of conditions.

Considerable regeneration can take place in the dilute phase above the bed as a considerable amount of catalyst is thrown up into this zone by the bursting bubbles. Much fine catalyst is carried up into the cyclones. In addition, the temperature of the gas increases anywhere from 10 to 150°F as the catalyst regenerates, and the carbon monoxide leaving the bed is converted to carbon dioxide before the gas enters the cyclone inlets. The 150°F rise would correspond to total CO combustion under special circumstances.

The regeneration process has evolved from the simple bed contacting zone

Cracking, Catalytic

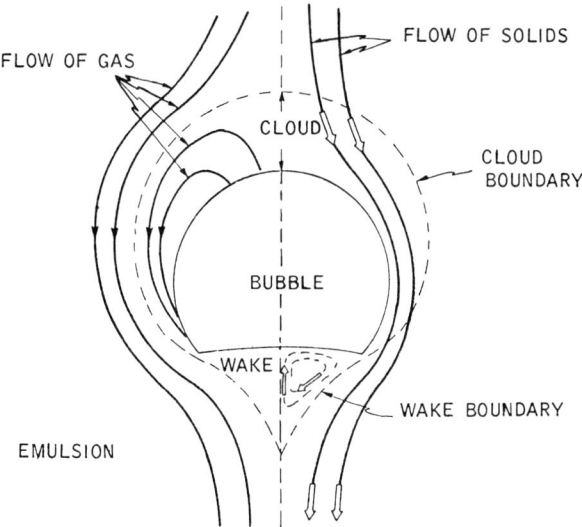

FIG. 48. Bubble model.

to staged and even transfer line contacting zones. Kellogg [9] has used staged solids flow regeneration commercially for some time. Since catalyst particles are well mixed in a regeneration bed, staging permits a significant amount of regeneration to be done at a higher average carbon level where less catalyst holdup will be required for a given amount of burning. A smaller second stage can be used to reach the final level of carbon on catalyst. Holdup required in the regeneration zone to meet a required carbon on catalyst level has been considerably reduced by staging, as shown in Fig. 49.

Other licensors have found it possible to achieve similarly low carbon levels at low catalyst holdup by increased temperature, pressure, etc. without staging the bed.

Transfer line regeneration (that is, cocurrent upflow of catalyst and air) was

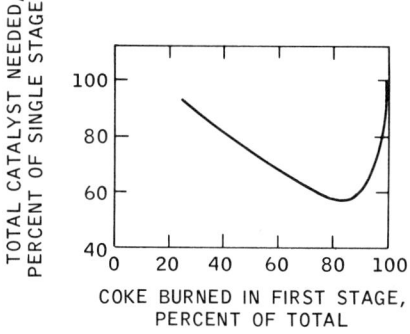

FIG. 49. Two-stage regenerator inventory vs division of burning [9].

originally applied in the Model I units. Recycle of hot solids was required to get the spent catalyst air mix to a sufficiently high temperature so that combustion could be completed in the volume provided in the regeneration zone. UOP has recently applied riser regeneration to a new unit as shown in Fig. 11 [64].

With regard to regeneration in a bed, simple relationships can be developed to characterize the carbon on catalyst level that can be obtained. Correction factors can be developed for individual units to compensate for differences in air/solids contacting since there can be inadequacies in air distribution to the bed or differences due to the location and method of introducing spent catalyst to the bed. Low CRC is desirable to improve product yields in a gasoline operation. Low CRC also results in a higher catalyst activity in the unit which will result in a greater selectivity to liquid products, gasoline, or heating oil. Low bed holdup will mean that there will be less catalyst in the unit being deactivated by the reaction and regeneration processes. Thus the fresh catalyst added to the unit will maintain a higher in-situ activity of the equilibrium catalyst.

Much information is available in the literature concerning combustion of carbon from catalyst [84–87]. The basic relationships when integrated lead to the useful relationship

$$\text{CRC} = \frac{CFT\left[-\ln\left(\frac{O_2 \text{ out}}{21}\right)\right]}{PKW}$$

where CRC = carbon on regenerated catalyst, wt.%
C = constant (derived from operating data)
F = total dry air rate, SCF/min
T = regenerator temperature, °R
O_2 out = mol% oxygen in dry flue gas
P = regenerator top pressure, lb/in.² abs
K = kinetic rate constant, s^{-1} (Fig. 50)
W = total catalyst holdup in the regenerator, tons

The combustion of carbon in the bed can proceed in the following manner:

$$C + \tfrac{1}{2}O_2 \xrightarrow{k_1} CO \tag{6}$$

$$C + O_2 \xrightarrow{k_2} CO_2 \tag{7}$$

Since these reaction rate constants are affected by temperature differently, the ratio of CO to CO_2 will shift with temperature. With an excess of oxygen, the following reaction will primarily influence the flue gas composition:

$$CO + \tfrac{1}{2}O_2 \xrightarrow{k_3} CO_2$$

Figure 51 shows the equilibrium ratio of CO_2 and CO as a function of temperature where combustion of CO to CO_2 is held to a minimum by

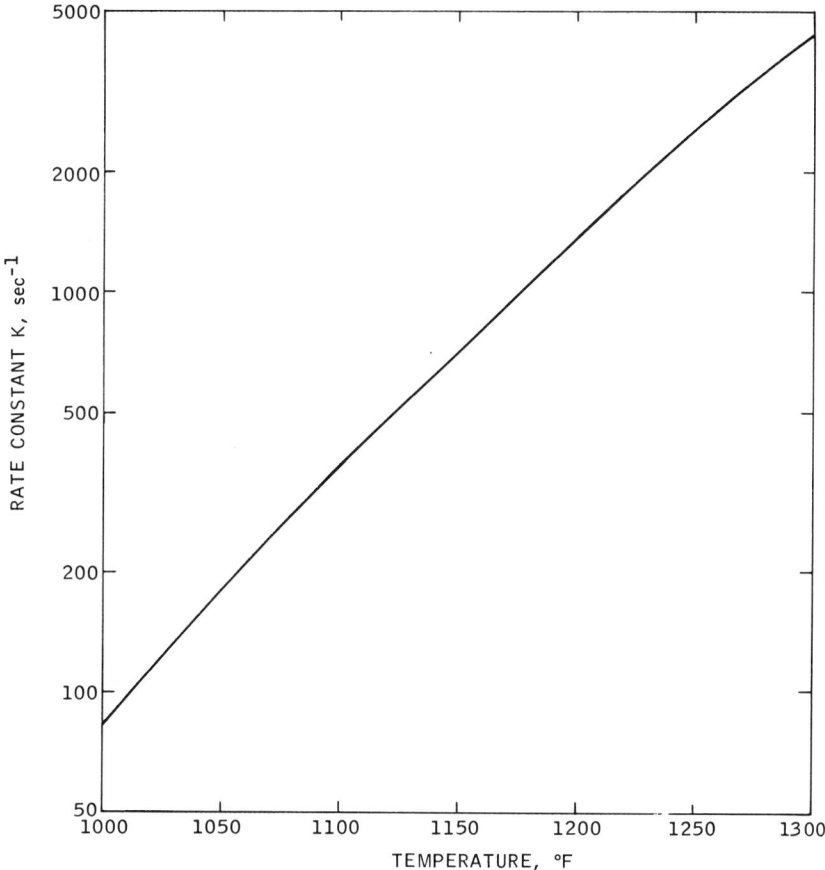

FIG. 50. Regenerator burning rate constant K.

controlling the regeneration air rate to produce a very low level of oxygen, essentially 0%, in the flue gas. This oxygen level is held low enough so that carbon does not build up on the catalyst. That is, a carbon buildup is avoided.

The proper prediction of the CO/CO_2 ratio is important in determining the heat produced in the regenerator which is available for the overall heat and process condition balance of the unit [88].

The CO combustion can become significant in the bed or above the bed in the dilute phase if oxidation promoters are added to the catalyst or if temperature and oxygen levels are sufficiently high. It is possible to reduce CO to very low levels of concentration, i.e., 20 ppm.

The kinetics of this CO combustion become very complex because of the effect of temperature and solids content changes in the dilute phase. Generally this reaction does not progress to a very great extent up to 1250°F at usual regenerator conditions. At 1250°F or above, proprietary techniques from a number of licensors are used to establish CO conversion in many commercial units [65, 66].

Cat Cracker Control

The fluid catalytic cracking unit is very complex from a control standpoint, and it can be controlled to many simultaneous limits. The exact scheme for control can vary considerably depending upon the limits which are selected and which of these constraints are met first. The use of computers in controlling cat cracking units has maximized the number of the limits which can be achieved continuously so as to maximize product values or minimize operating costs. Some of these limits are:

Feed rate
Reactor temperature
Conversion
Air rate
Carbon burning rate
Preheat
Utilities consumption
Economic yields
Cat circulation rate
Flue gas carbon monoxide level
Regenerator temperature
Flue gas oxygen level
Reactor level
Reactor pressure
Regenerator level
Regenerator pressure

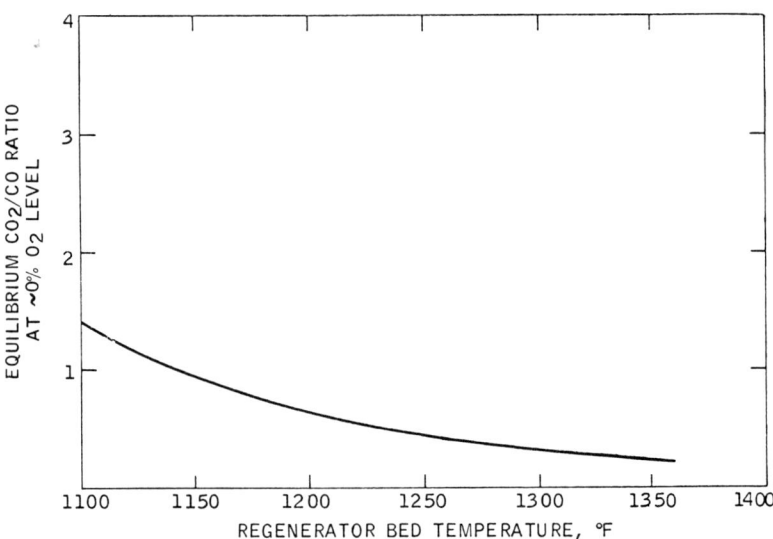

FIG. 51. Flue gas composition.

Recycle rate
Catalyst make up rate
Catalyst withdrawal rate
Pressure drop across slide valves
Gas compressor flow rate
Offgas rate

It is not possible to establish a given set of limits which will satisfy all refinery situations. For example, some refineries may desire to maximize fresh feed rate while others might desire to operate to a constant flue gas make or compressor limitation. The instrumentation provided in a cat cracker is extensive and generally will permit the unit to be driven to any of the limits desired.

Which limits are selected will depend upon the unit and how it operates. It is sometimes difficult to predict the operating response that will be ultimately obtained for a given process change in a cat cracker. This is because many of the changes not only affect the heat balance but also affect the yield pattern which in turn affects the heat balance. Some of these responses also depend upon the catalyst flow and vessel configuration. The ultimate response from an operating change will therefore depend upon the relative magnitude of the changes which are produced in a number of variables. Operating and control techniques, therefore, must be developed for specific units and situations.

In cat cracking there are two basic balances which have prime impacts on the control of the unit. One is the heat balance which stabilizes the temperature of reaction and regeneration. The other is the process yield balance. The important factor in the yield balance with regard to operating control is the coke yield. The two basic balances are interrelated since the coke burns in the regenerator and is a major factor in the heat balance. Changes made to adjust the heat balance will also affect the conditions of the cracking reaction zone and thus affect the coke make. Sometimes these factors are self-compensating and sometimes they amplify the response. When evaluating control schemes, it is important to keep unit configuration in mind. For example, a pure riser cracker will have control aspects different from a unit which has some bed holdup capability. A unit which contains an overflow well in the regenerator can be operated differently than a unit with slide valve control in the regenerated catalyst circuit.

A simple control scheme is given in Fig. 52 which depicts a slide valve control unit containing a bed of catalyst in the reactor [5].

Feed to the unit, fresh feed, and recycle are flow controlled into the feed riser. The temperature of the total feed is held constant by controlling either preheat furnace duty or a preheat exchanger bypass. In this case the reactor temperature is held constant by adjusting the slide valve in the regenerated catalyst circulating line. The regenerator operation is controlled by adjusting the carbon yield. This is done by adjusting the reactor holdup with the slide valve in the spent catalyst circulating line. If more coke is needed, the reactor level is increased. Conversion is also increased if the bed level is increased.

The indication that shows that the carbon make is correct and the unit is in carbon balance is given by the ΔT indicator between the regenerator dense bed

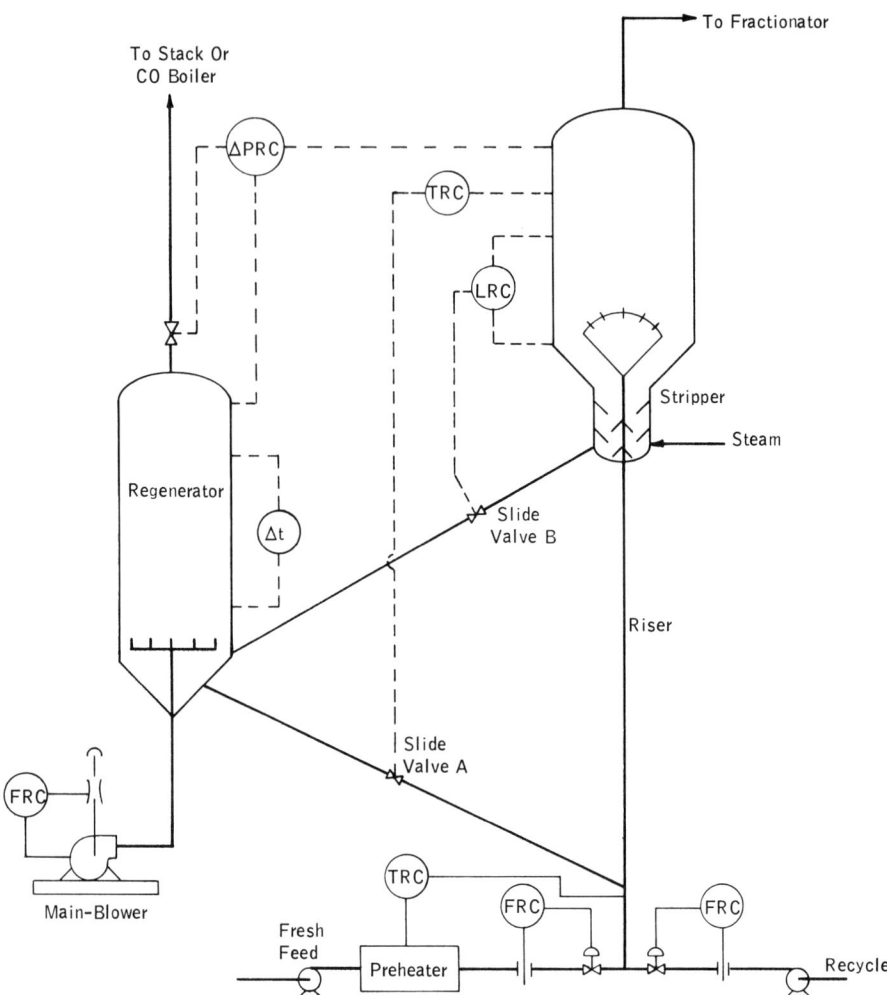

FIG. 52. Basic control system of slide valve type of unit.

and gas inlet to the cyclones. A higher ΔT shows that there is an excess of oxygen and consequently not enough carbon being made in the reactor. A lower ΔT shows that there is not enough oxygen and adjustments should be made to either decrease reactor carbon make or increase the air rate.

The regenerator pressure is controlled by valves in the flue gas line which can be driven by the differential pressure indicator between the reactor and regenerator. Reactor pressure is usually held constant by having the gas compressor hold a constant suction pressure in the overhead drum of the fractionator. The pressure balance of the unit is usually adjusted to hold about a 5-lb/in.² pressure drop across the slide valves in the catalyst circulating lines in order to minimize erosion of the valves.

One of the variations on this basic scheme is to maximize the air output of the main blower. This can be done by having the flue gas stack valves control regenerator pressure. In this case the differential pressure between the reactor and regenerator cannot be held constant and the consequent pressure fluctuations will be seen in the pressure drop across the slide valves which will render catalyst circulation somewhat more subject to minor fluctuation. However, catalyst circulation can be controlled by the normal instrumentation of the slide valves.

The development of interacting control systems by the use of a computer will permit many alternatives of the above control scheme directed to push the unit to equipment or process limits to produce the most economic performance of the unit.

The inclusion of an overflow well in the regenerator changes some of the aspects of the above scheme because one degree of operating freedom is lost. However, an important mechanical advantage is gained in eliminating a slide valve which can erode. A process advantage is that catalyst flow between the vessels will always be in balance. Thus only one operating variable change is required to change and balance catalyst circulation between vessels.

A disadvantage is that reactor holdup cannot be controlled independently. This holdup will depend on how much catalyst is loaded into the unit. Also, the amount of catalyst in the reactor will vary as the air rate or gas velocity is varied in the regenerator. This secondary factor can be very important in control. For example, if carbon make is too high as shown by too low a regenerator ΔT, adding air will fluff up the catalyst in the regenerator bed and drive more catalyst to the reactor, thus increasing the reactor bed level which in turn will make more carbon. This effect has been known to defeat the effect of the added air, and the ΔT is unaffected or driven further in the same direction. This is no problem in units where the bed level in the reactor vessel is held beneath the reactor grid because holdup will be transferred into the stripper beneath the reaction zone and the carbon make will be unaffected. In this case it is necessary to adjust the overall unit heat balance, preheat temperature, or feed rate in order to compensate for the change in heat liberated in the regenerator due to the changed air rate. Thus units with overflow wells are generally operated at fixed air rates when operated by noncomputer control.

In the riser type of reactor, the reactor product gas and catalyst are separated in a rough cut cyclone in order to provide a short contact time reaction. Thus a degree of operating freedom is lost which can have serious consequences in the regenerator side of the unit. This is because the holdup cannot be independently increased to increase coke make in the reactor and vice versa.

For this type of unit the catalyst circulation rate is an important factor in the carbon make in the reactor because catalyst holdup in the reactor is directly dependent upon the circulation rate. Thus, if the coke make in the reactor is inadequate as shown by an increasing ΔT in the regenerator, catalyst circulation must be increased to make more coke. This presumes that preheat, feed rate, and reactor temperature are at maxima. The increased circulation rate will result in a reduced regenerator temperature which could be a disadvantage.

There are alternative ways to heat balance the unit in the above situation. One is to use a more active catalyst or increase the fresh catalyst addition rates.

These are usually not adequate for good operating control of the unit. Another commonly practiced technique is to recycle slurry or fractionator bottoms and control carbon make by varying the slurry rate. This is usually an undesirable technique because the slurry contains highly aromatic components which blind the zeolitic catalyst sites and produce a poorer overall yield pattern than would be obtained without slurry recycle. A third method is to burn an extraneous fuel, torch oil, in the regenerator. This technique has been used successfully in many units. However, it is important to inject the torch oil in a number of locations in the regenerator bed so as to insure proper combustion in the bed.

Heat Balance

As noted previously, the heat balance in a catalytic cracking unit is one of the two primary balances which must be closely controlled for successful cat cracker operation. This section will review some of the details of the balance and present a representative calculation.

The circulating catalyst must remove the excess heat generated by burning the coke in the regenerator zone. The various components of the heat balance are as follows:

Heat of combustion of the coke.
Heat required to desorb the coke from the catalyst. (This is frequently overlooked in the analysis of combustion data.)
Heat required to raise the coke on the catalyst from the reactor temperature to the regenerator flue gas temperature.
Heat required to raise the incoming air to the flue gas temperature.
Make up heat required by steam purges, quenches, water sprays, and steam or water heat exchange in the regenerator area.
Heat needed to supply the radiation and convection losses.
Heat liberated from extraneous fuel injected to the bed such as torch oil.

The catalyst circulation rate is determined by dividing the net heat from the above by the difference between the reactor stripper and regenerator temperature times the specific heat of the catalyst, usually 0.26 Btu/lb/°F.

The preheat temperature of the feed is determined from the reactor heat balance. The preheat must balance the following heat loads:

Net heat from the regenerated catalyst.
Heat of cracking.
Heat liberated by absorption of coke onto the catalyst.
Heat of raising the injector stripping steam and miscellaneous steam purges to the reactor temperature.
Heat to supply radiation and convection losses.
Heat required to vaporize and heat preheated feed to the reactor outlet temperature.

Cracking, Catalytic

A representative calculation is summarized below. There are some simplifications made in this calculation to reduce the complexity of the example. These primarily involve the development of the heat of cracking and the heat content of the reactor vapor product.

Regenerator Balance

Heat of Combustion. Given:

Air rate of 22,200 SCFM at 48°F and 66% rh
Flue gas analysis at stack: 16.7% CO_2, 1.5% CO, and 0.2% O_2
Bed temperature: 1260°F
Flue gas temperature: 1370°F
Spent catalyst temperature from stripper: 925°F
Regenerator pressure: 13.5 lb/in.² gauge
Blower discharge temperature: 265°F

Regenerator Material Balance:

Flow rate equals 22,100 SCFM dry air + 100 SCFM humidity
The coke burning rate is determined from the Orsat analysis and air rate
Develop the flue gas rate:

$$\text{air in } N_2 \text{ in air, } 22{,}100 \times 0.79 = 17{,}460 \text{ SCFM}$$

$$O_2 \text{ in air, } 22{,}100 \times 0.21 = 4{,}640 \text{ SCFM}$$

The flue gas rate is developed from

Component	Orsat (dry), (%)	Flue gas rate (SCFM)		O_2 balance (SCFM)	
N_2	81.6			17,460	
CO_2	16.7	$17{,}460 \times \dfrac{16.7}{81.6} =$		3,573	3,573
CO	1.5	$17{,}460 \times \dfrac{1.5}{81.6} =$		321	161
O_2	0.2	$17{,}460 \times \dfrac{0.2}{81.6} =$		43	
	100.0			21,397	3,777

Unaccounted O_2 equals $4640 - 3777 = 863$ SCFM
Water of combustion of H_2 in coke equals $2 \times 863 = 1726$ SCFM
Flue gas rate (wet):

   Dry flue gases   21,397
   Combustion water   1,726

Humidity	100
Miscellaneous	200
	23,423 SCFM

Carbon yield:

$$CO_2 = 3573 \text{ SCFM} \times 60 \text{ min/h} \times \frac{\text{mol}}{379.4 \text{ SCF}} \times 12 \text{ lb/mol} = 6781 \text{ lb/h carbon}$$

$$CO = 321 \times 60 \times \frac{1}{379.4} \times 12 = 609 \text{ lb/h carbon}$$

Total carbon equals 6781 + 609 or 7390 lb/h

Hydrogen yield in the coke. Hydrogen in the coke is determined from the combustion water:

$$H_2 = 1726 \times 60 \times \frac{1}{379.4} \times 2.018 = 551 \text{ lb/h}$$

Thus coke burning rate = 7390 + 551 = 7941 lb/h and H/coke ratio = 6.94%.

Regenerator Heat Balance:

The heats of combustion generally used are:

14,500 Btu/lb of carbon to CO_2

4,410 Btu/lb of carbon to CO

51,500 Btu/lb of hydrogen to water

Therefore the heat liberated (in MBtu/h) is:

6,781 × 14,500 =	98.325
609 × 4,410 =	2.686
551 × 51,500 =	28.377
	129.388

Heat of desorption based on 1450 Btu/lb carbon:

$$7{,}390 \times 1{,}450 = 10.715 \text{ MBtu/h}$$

Heat coke to flue gas temperature:

$$7941 \text{ lb/h} \times (1370°F - 925°F) \times 0.4 \text{ Btu/lb/°F} = 1.413 \text{ MBtu/h}$$

Heat air to flue gas temperature:

$$22{,}200 \text{ SCFM} = 101{,}500 \text{ lb/h}$$

$$101{,}500 \text{ lb/h} \times (1370°F - 265°F) \times 0.26 \text{ Btu/lb/°F} = 29.16 \text{ MBtu/h}$$

Cracking, Catalytic

Heat to miscellaneous purges:

$$570 \text{ lb/h} \times (1370°F - 360°F) \, 0.5 \text{ Btu/lb/°F} = 2.879 \text{ MBtu/h}$$

Heat to sprays: None.
Heat losses: This will depend upon vessel size, insulation wind velocity, and ambient temperature. Loss equals 2.260 MBtu/h.
Summary

	MBtu/h
Desorption	10.715
Heat coke	1.413
Heat air	29.16
Heat purges	2.879
Heat spray	—
Heat losses	2.260
	46.427
Heat of combustion	129.388
Net heat removed by catalyst	82.961

Catalyst Circulation Rate:

Specific heat of catalyst is generally 0.262 Btu/lb.
Rate equals:

$$\frac{82.961 \text{ MBtu/h} \times 10^6}{(1260°F - 925°F) \times 0.262 \text{ Btu/lb/°F} \times 2000 \times 60} = 7.88 \text{ tons/min}$$

Reactor Heat Balance: The following is a representative reactor operation. Feed rate of 20,000 bbl/std. d of 25.6° API gravity and a molecular weight of 413 cracked to produce 2286 mols of hydrocarbon vapors (excluding coke) with an "averaged" overall product gravity of 55°. There is no recycle.

	Heat out (MBtu/h)	Heat in (MBtu/h)
Heat of cracking based on assumption of 15,000 Btu/mol generated: 2286 mol/h × −15,000 Btu/mol	−34.29	
Heat of coke absorption (liberated heat)		+10.715
Heat to steam (stripping, injector, purges): 4,800 lb/h × (360°F − 925°F) × 0.5 Btu/lb	−1.356	
Radiant and convection heat	−1.1	
Heat of coke out above −200°F, assume 0.4 Btu/lb/°F: 7,941 lb/h × (−200°F − 925°F) × 0.4 Btu/lb	−3.573	
Heat of hydrocarbon product vapor (above −200°F) leaving reactor: (262,500 − 7,941) lb/h × (768 Btu/lb)	−195.501	
Heat entering with catalyst		+82.961
Total (excluding feed in)	−235.82	+93.676
Net heat (above −200°F) entering with feed	−142.144	

$$\text{Heat content of feed} = \frac{142{,}144{,}000}{262{,}500} = 541.5 \text{ Btu/lb}$$

which is equivalent to $787°F$ (liquid) preheat temperature.

Catalyst Stripping

Catalyst leaving the reaction zone is fluidized with reactor product vapors which must be removed and recovered with the reactor product. In order to accomplish this, the catalyst is passed into a stripping zone where most of the hydrocarbon is displaced with steam.

Stripping is generally done in a countercurrent contact zone where shed baffles or contactors are provided to insure equal vapor flow up through the stripper and efficient contacting. Stripping can be accomplished in a dilute catalyst phase. Generally, a dense phase is used, but with lighter feeds or higher reactor temperature and high conversion operations a significant portion of the contacting can be done in a dilute phase.

The amount of hydrocarbon carried to the regenerator is dependent upon the amount of stripping steam used per pound of catalyst and the pressure and temperature at which the stripper operates.

Probes at the stripper outlet have been used to measure the composition of the hydrocarbon vapors leaving the stripper. Pseudohydrocarbon molecular weights of 2 to 10 have been measured. When expressed as percent of coke burned in the regenerator, the strippable hydrocarbon is only 2 to 5%. Very poor stripping is shown when the hydrogen content of the regenerator coke is 10 wt.% or higher. Good stripping is shown by 6 to 9 wt.% hydrogen levels.

The proper level of stripping is found in many operating units by reducing stripping steam until there is a noticeable effect or rise in regenerator temperature. Steam is then marginally increased above this rate. In some units, stripping steam is used as a control variable to control the carbon burning rate or differential temperature between the regenerator bed and cyclone inlets.

Troubleshooting

The catalytic cracking unit is an extremely dynamic unit, primarily because there are three major process flow streams (the catalyst, hydrocarbon, and regeneration air), all of which interact with each other. Problems can arise in the equipment and flowing streams which are sometimes difficult to diagnose because of the complex affects they can create. Table 22 summarizes some of the problems which could be encountered, symptoms, probable causes, data required to identify problems, and corrective actions.

Application of Fluidized Solids Principles to Catalytic Cracking

As noted previously, the catalytic cracking process as first commercially applied constituted a number of fixed-bed reactors which were cyclically reacted,

Cracking, Catalytic

TABLE 22 Troubleshooting Guide

	Problem	Symptoms	Causes	Data Required	Corrective Action
			Fresh Feed Quality Problems		
(a)	Metals in feed	(1) High H_2 make (2) General increase in light ends (3) Excessive coke production (4) Higher reactor velocities (5) Overloaded gas compressor	(1) Pitch entrainment at atmosphere and vacuum feed preparation unit (A&V) (2) Feed type change (3) Abnormal A&V operation	(1) Feed nickel equivalent (2) Feed CCR (3) H_2 make SCF/bbl fresh feed (4) Actual vs predicted yields (5) A&V operating conditions (6) Fresh feed color	(1) Lower metals in feed (2) Feed segregation (3) Catalyst replacement program
(b)	Feed contamination with heavy hydrocarbons	(1) Excessive coke make (2) Unexplained increase in air requirements at same conversion (3) Poor wetght balance	(1) Leak in exchanger train (2) Partly open valves (3) Fractionator bottoms entrainment in heart cut recycle	(1) Feed CCR (2) Feed RI and API (3) Recycle stream inspections	(1) Isolate leaking exchangers (2) Minimize possibility of leaks by pressure balance
(c)	Feed contamination with light hydrocarbons	(1) Unsteady preheat header pressures and flows (2) Poor weight balance (3) Shift in yields distribution	(1) Leak in exchanger train (2) Partly open valves	(1) Feed API (2) Feed front end distillation	(1) As above
(d)	Water in feed	(1) Vibration in preheat system (2) Unsteady flows and temperatures (3) Severe upset if large amounts of water in feed	(1) Water in feed tankage (2) Leaks from steam-out connections (3) Trapped water from idle equipment	(1) Water in feed	(1) Check for steam leaks on tank heaters (2) Isolate and swing to high suction on catalyst feed tank (3) Ensure that idle equipment is well drained and hot before being brought on-stream

(continued)

TABLE 22 (continued)

Problem	Symptoms	Causes	Data Required	Corrective Action
		Catalyst Problems		
(a) Catalyst contamination	(1) See metals in feed	(1) Low catalyst replacement rate (2) See metals in feed	(1) Metals on catalyst (2) H_2 production (3) Product yield distribution	(1) Consider lowering metals in feed; A&V operation (2) Catalyst replacement program (3) Feed segregation
(b) Sodium on catalyst	(1) See sintering	(1) Salt in feed (2) Treated boiler feedwater used in regenerator sprays	(1) Historical Na content of catalyst	(1) Minimize sources of Na input to system
(c) Sintering of catalyst	(1) Apparent decrease in catalyst activity (2) Increase in carbon on regenerated catalyst (3) Change in product yield distribution	(1) High Na and V on catalyst (2) Excessive regenerator temperatures (3) Low bed stability (4) Excessive or prolonged use of torch oil (5) Localized high temperatures	(1) Sintering index (2) EASC total and burnable carbon (3) Metals on catalyst (4) Yield distribution (5) Predicted and measured activity (5) Regenerator operating conditions	(1) Catalyst replacement (2) Minimize Na input with seawater or salt in feed, or with regenerator sprays (3) Consider lower V content in feed (4) Use high stability catalysts (5) Review regenerator operations
(d) Coarse catalyst	(1) Poor circulation (2) Poor regeneration (3) Change in yield distribution (4) Poorer stripping	(1) Loss of fines	(1) Roller analysis (2) H/C ratio (3) Yield distribution (4) Operating conditions in general	(1) Minimize catalyst losses by lowering regenerator velocity (2) Change to finer catalyst (3) Consider use of attriter
(e) Attrition	(1) Higher catalyst losses with increasing fines content	(1) High velocity stream into dense phase (2) Fragile catalyst	(1) Check unit for high velocity streams exceeding 200 ft/s into catalyst	(1) Eliminate or reduce high velocity fluid injection (2) Follow up fresh catalyst supplies

Cracking, Catalytic

(a)	Reactor cyclone failure	(1) High catalyst in fractionator bottoms (2) Frequent fractionator bottoms pump plugging (3) Loss of cat fines (4) Cat losses become progressively higher	(1) Minimize reactor velocity (2) Review new methods in cyclone design and repair (3) Review operating procedures and history of past occurrences
			(2) Missing ROs, blast steam, partly open valves, etc. (3) Check fresh catalyst properties (4) Study catalyst loss pattern

Reactor Problems

		(1) Erosion and/or corrosion (2) Pressure surges	(1) Catalyst content in fractionator bottoms (2) Cyclone pressure drop
(b)	Eroded or plugged grid holes	(1) Change in grid ΔP (2) Decline in reactor efficiency (3) Change in yield distribution (4) Low overflow well level if grid eroded, high if plugged	(1) Lumps of coke or refractory in catalyst (2) Failure of grid hole inserts (3) Hole velocity too high (4) Feed injector velocity too low
			(1) Grid pressure drop (2) Product yield pattern (3) Declining cat activity as determined by unit tracking despite adequate catalyst replacement rate
			(1) Review methods in grid design and repair (2) Change reactor-regenerator differential pressure controller (DPRC) setting and/or control air to maintain circulation and normal overflow well level (3) Check feed injector operation
(c)	Unsteady bed temperature and reactor pressure	See Water in feed: Failure of overflow well seals; Reactor and regenerator grid hole erosion; Surging regenerator holdup; Unsteady DPRC operation, Rough circulation	

(*continued*)

TABLE 22 (*continued*)

Problem	Symptoms	Causes	Data Required	Corrective Action
(d) Coking in overhead line	(1) Rise in ΔP between reactor outlet and fractionator inlet	(1) Condensation of reactor products in overhead line	(1) Pressure survey	(1) Ensure that overhead line insulation is in good condition
(e) Poor catalyst stripping	(1) Unexplained increase in coke (2) Higher H/C ratio	(1) Insufficient stripping steam (2) Poor catalyst/steam contacting (3) Catalyst properties (4) Low reactor temp. (5) Inaccurate steam flow controller (FRC)	(1) H/C ratios (2) Stripper tests	(1) Check stripping steam rate (2) Consider higher reactor temperature (3) Consider lower circulation rate (4) Review stripper design

Regenerator Problems

Problem	Symptoms	Causes	Data Required	Corrective Action
(a) Cyclone failure	(1) Increase in cat losses (2) Cat losses become progressively higher	(1) Excessive temperatures and allowable stresses exceeded (2) Erosion	(1) Pressure differential between dilute phase and plenum (2) Measure of cat losses (3) Regenerator velocity	(1) Minimize cat feed (2) Minimize regenerator velocity (3) Review new methods in cyclone design and repair (4) Review operating procedures and history
(b) Plenum chamber failure	(1) Increase in cat losses above normal level	(1) Excessive temperatures and stress cracking (2) Impingement of plenum sprays	(1) As above (2) Check condition of spray nozzles	(1) As above
(c) Failure of internal seals	(1) Uneven bed and cyclone inlet temperatures (2) Uneven O_2 distribution in dilute phase (3) Salt and pepper catalyst	(1) Erosion (2) Pressure bump (3) Stresses too high (4) Abnormal conditions with auxiliary burner on start-up	(1) Grid Δp (2) Catalyst appearance (3) O_2 analysis of gas at dilute phase sprays (4) Historical operating data	(1) Avoid pressure surges on startup (2) Maximize air to grid (minimize control air) (3) Lower regenerator pressure to increase velocity through grid

Cracking, Catalytic

		(4) Surging of catalyst bed		(4) Review operating history and seal design
		(5) Drop in grid Δp		
		(6) Increase in cat losses		
		(7) Afterburning		
(d)	Hole in overflow well	(1) Unstable overflow well level and cat circulation	(1) Abnormal stresses	(1) Alter standpipe aeration
		(2) Unsteady overflow well density	(2) Erosion	(2) Adjust DPRC and circulation
		(3) High regenerator holdup		(3) Maximize control air (minimize air to grid)
		(4) Uneven temperatures		(4) Review design
		(5) Uneven O_2 distribution		
(e)	Grid hole plugging and or erosion	(1) Uneven bed and cyclone temperatures	(1) Lumps of catalyst or refractory in cat bed	(1) Maintain grid pressure drop at 30% bed pressure drop
		(2) Uneven O_2 distribution	(2) Low bed stability	(2) Maximize air through grid
		(3) Increased cat losses		(3) Review grid design
		(4) Surging of catalyst bed		(4) Check condition of cat hopper to ensure that debris does not enter regenerator
		(5) Increased singering		
(f)	Stuck or failed trickle valves	(1) High catalyst losses but no increase with time	(1) Binding of hinge rings by:	(1) Careful examination of trickle valves on shutdown
			(a) Clearances not large enough	
		(2) Loss of catalyst fines. Increase in coarseness index	(b) Oxidation scale or pieces of lining	(1) Revise or replace as required to meet specifications
			(c) Nonuniform wear of moving parts	

(*continued*)

TABLE 22 (continued)

Problem	Symptoms	Causes	Data Required	Corrective Action
(g) Plugged diplegs	(1) As above	(2) Installation angles not correct (3) Improper material (1) Spalled refractory forming partial and eventually final plug (2) Air out periods with a lot of water/steam in vessel	(1) Careful examination of diplegs or use of probes, lights, balls, etc. to ensure diplegs are free before startup	(1) Lower bed level to allow catalyst to flow out of dipleg
(h) Excessive input of steam or air	(1) Higher level of catalyst losses (2) High cyclone pressure drop (3) Indication of attrition (higher catalyst fines content despite higher losses)	(1) Missing restriction orifices (2) Large restriction orifices (3) Partially open valves on steam, water, or air lines (4) Malfunctioning steam traps (5) Metering errors	(1) Equipment survey to make sure all restriction orifices (ROs) are in place, etc.	(1) Careful detailed check of unit for and correction of equipment conditions listed under causes
(i) Bed stability too low	(1) Increase in temperature below the grid	(1) Insufficient air through grid to support catalyst bed (2) Eroded grid holes	(1) Grid ΔP (2) Historical records of temperature below grid	(1) Maximize air through grid (2) If problem expected to persist, redesign grid for lower air rates
(j) Surging of catalyst bed	(1) Erratic or cycling instrument records on holdup, density, and overflow well (2) Unsteady circulation and heat balance (3) Catalyst sintering (4) Uneven catalyst regeneration	(1) Seal failures (2) Grid hole erosion (3) Hole in overflow well (4) Poor bed stability (5) Poor DPRC control	(1) Pressure survey (2) Operating condition survey (3) Bed stability calculation	(1) See action required for seal failures, grid hole erosion, poor bed stability, poor DPRC control

Cracking, Catalytic

(m)	Unsteady pressure differential control	(1) Fluctuating regenerator pressure
		(2) Unsteady circulation and overflow well level
		(3) Unsteady reactor temperature
		(4) Catalyst shifts between reactor and regenerator
			(1) Poor stack slide valve performance
			  (a) Sticky slide valves
			  (b) Poor slide valve instrument performance
			(2) Unsteady circulation
				(1) Unit performance records	(1) Check out slide valves and associated instrumentation
					(2) Check adequacy of U-bend or standpipe aeration

(n)	Regenerator hot spots	(1) High temperatures on vessel shells or U-bends	(1) Damaged refractory	(1) Change in paint color (2) Glow at night (3) Surface temperature measurements	(1) Cool the spot with steam or water, avoiding abrupt changes in temperature due to interruption of cooling
(o)	Rough catalyst circulation	(1) Unsteady regenerator and reactor temperatures and holdups (2) Unsteady control air (3) Fluctuating overflow well level and U-bend densities (4) U-bend vibration	(1) Improper aeration (2) Coarse catalyst (3) Fluctuating DPRC (4) Water in aeration medium (5) Poor performance of control air system	(1) U-bend aeration pattern and pressure survey (2) Standpipe and U-bend densities (3) Circulation rate (4) Catalyst roller analysis	(1) Aeration changes (2) Put control air on manual to determine if it is the cause (3) Check DPRC system (4) Make sure aeration medium is water free

(continued)

TABLE 22 (continued)

	Problem	Symptoms	Causes	Data Required	Corrective Action
(k)	High carbon on catalyst. Carbon buildup	(1) Dark catalyst (2) Dilute phase temperature decreases relative to dense bed temperature (3) Low excess O_2 (4) Apparent loss in cat activity	(1) Excessive coke from: 　(a) Operating intensity increase 　(b) Poorer feedstock 　(c) Poor catalyst stripping 　(d) Heavier recycle 　(e) Leakage of fraction bottoms into feed (2) Low excess O_2 (3) Poor air distribution (4) False O_2 recorder readings (5) Feed and recycle meter errors	(1) Carbon on catalyst analysis (2) Feed quality (3) Recycle boiling range and CCR (4) Check O_2 levels in regenerator (5) Check O_2 recorder for accuracy (6) Check stripper performance	(1) Lower intensity of operation (2) Increase air to regenerator (3) Inject lighter feed (4) Improve normal feed and/or recycle if possible (5) Improve catalyst stripping (6) Check meter accuracy
(l)	Afterburning	(1) Excessive dilute phase and cyclone temperatures (2) Trend to lighter catalyst (3) Increase in dilute phase temperature relative to dense bed (4) High excess O_2 (5) Higher CO_2/CO ratios	(1) Insufficient coke production: 　(a) Decrease of intensity 　(b) Swing to better feed 　(c) Lighter recycle (2) Too much excess air (3) False O_2 recorder readings (4) Feed and recycle meter errors (5) Cyclone steam meter errors	(1) Same as in (k) above except (6)	(1) Decrease air (2) Marginal use of torch oil (3) Increase operating intensity (4) Check cyclone stream rate

Cracking, Catalytic

	Problem	Causes	Solutions
(p)	Low catalyst circulation rate	(1) Inability to lower regenerator temperature (2) Excessive feed preheat requirements	(1) Partial blockage of U-bends (2) Too much stripping steam (3) Improper aeration (4) Control air too low (5) DPRC setting inadequate	(1) Change DPRC setting (2) Check control air system and increase control air rate (3) Ensure adequate aeration (4) Check unit design

Fractionator Problems

	Problem	Causes	Solutions
(a)	Poor split between LCGO and recycle	(1) High overlaps between heating oil and recycle stream distillation (2) Shift in yield pattern (3) Increase in recycle volume at constant operating conditions	(1) Inadequate steam to recycle stripper (2) Improper tray loading in tower (3) Stripper malfunction	(1) Ensure adequate stripping steam rates (2) Adjust pumparound heat duties (3) Consider equipment changes
				(1) Heating oil and recycle distillations (2) Fractionator pressure survey (3) Pumparound heat removal (4) Detailed fractionator and stripper analysis (5) Consider equipment X rays
(b)	Fractionator bottoms too light	(1) High bottoms API (2) Excessive bottoms yield	(1) Too much heat removal in bottoms pumparound	(1) Lower heat removal from tower bottoms consistent with
			(1) Fractionator bottoms inspections (2) Tower operating	

(continued)

TABLE 22 (continued)

Problem	Symptoms	Causes	Data Required	Corrective Action
	(3) Dark color of recycle stream	(2) Tower bottom liquid too cold (3) Poor liquid/vapor contacting in shed section (4) Lower operating intensity (5) Heart cut recycle rate too low (6) Leaks into bottoms system	conditions in the lower half (3) Check on cracking intensity (4) Check recycle flow rate; meter accuracy	coking considerations (2) Maximize heart cut recycle (3) Consider equipment changes
(c) Coking	(1) Frequent plugging of exchangers and pumps with coke (2) Poor heat transfer in exchangers (3) High insolubles in fractionator bottoms	(1) Excessive temperatures in tower bottom (2) Low bottom API (3) High liquid residence time in tower bottom	(1) History of bottoms API and bottoms sediment and water content (2) Tower bottom operating conditions (3) Calculations of bottoms liquid residence time (4) History of equipment fouling	(1) Review recommended limitations of tower bottoms operating conditions and adjust operation accordingly (2) Consider changing pump screen size
(d) Salt fouling of fractionator top	(1) Higher pressure drop in top trays (2) Flooding in the top section (3) Poor split between overhead and first sidestream (4) Salt in TPA pumps (5) Chlorides in overhead water	(1) Salts in fresh feed (2) TPA return too cold	(1) Pressure and temperature survey of tower top (2) TPA operating conditions (3) Chlorides in fresh feed (4) Chlorides in distillate drum water (5) Split between overhead product and first sidestream	(1) Cautious water injection to TPA (2) Adjust tower temperatures (3) Adjust TPA return temperature (4) Consider installing screens in TPA pump suctions

Cracking, Catalytic

Miscellaneous Problems

	Problem	Possible Causes	Check	Remedy
(a)	Air blower low turbine efficiency	(1) Excessive steam usage (2) Unusual exhaust steam conditions (3) Decrease in rev/min at maximum steam	(1) Turbine fouling (2) Turbine blade wear (3) Quality of steam supply	(1) Steam supply and exhaust conditions (2) Calculation of turbine efficiency (3) Blower operating conditions	(1) Improve steam quality (2) Consider turbine wash or shut down for cleaning
(b)	Surplus heat in unit	(1) High regeneraton temperature (2) Low control air blower rate (see also low circulation rate)	(1) Too much feed preheat (2) Reactor temperature set too low	(1) Unit temperatures and heat balances (2) Circulation rate (3) Coke make	(1) Back off on feed injection temperature (2) Increase circulation rate (3) Increase reactor temperature and lower holdup (4) Adjust DPRC (5) Use dilute phase sprays

(c)	Insufficient heat in unit	(1) Low regenerator temperature
		(2) High control air if on automatic setting
			(1) Insufficient feed preheat
			(2) Reactor temperature set too high
			(3) Not enough carbon produced:
			  (a) Cat activity low
			  (b) Lighter recycle
			  (c) Lighter feed
			  (d) Lower reactor holdup
				(1) Unit temperatures and heat balances
				(2) Operating intensity level
				(3) Circulation rate
				(4) Coke production
				(5) Feed and recycle inspections
					(1) Increase preheat if possible
					(2) Increase fresh cat additions
					(3) Take out sprays if any
					(4) Use torch oil

purged, regenerated, purged, etc. These units handled rather small capacities and were quite complex. A basic difficulty arose in the fact that a fair amount of carbon was deposited on the catalyst during the reacting cycle. Thus the regeneration cycle consumed a significant period of time and large amounts of heat had to be transferred out of the catalyst bed.

Very clearly there was a need for a moving catalyst system where hot catalyst could enter the reactor, provide heat to the reactor feed, contact and convert the feed, and be removed to a regeneration zone where the carbon could be burned off and the catalyst heated to provide hot catalyst to the reactor.

There is a significant problem in the transfer of the catalyst between the reactor and regenerator. Positive seals must be provided between the two environments not only to recover the reactor products but to prevent the formation of combustible mixtures in the contact area.

Reducing the size of the catalyst particles so that the average size is in the 50 to 80 μm diameter size range permits the solids to be fluidized. That is, when a small amount of aerating gas is passed up through the solids, the particles are separated to a sufficient degree that they move about freely in the bed. The bed then behaves much like a fluid. Thus fluidization provides a technique whereby hydraulic seals can be provided between contact zones. Since the fluidity of the gas–solids mix is excellent, high capacity solids flows are possible. This in turn permits the easy transfer of the large amount of regeneration heat to the reactor zone. Steam or water cooling coils can be eliminated from the regenerator zone. The advantage of the fluid system can be attributed to the intense agitation and mixing caused by the rapid movement of the small catalyst particles. This movement promotes reaction rates by providing high mass transfer rates between the catalyst surface and the gas. In addition, there is virtual elimination of hot spots in the fluidized zone, thus making the regenerator and reactor beds essentially isothermal zones. A disadvantage is that, in fluid beds, the majority of the reactant vapors pass through the beds in bubbles, with only the gas in a cloud around the bubble actually contacting the solids. However, the move to dilute phase or riser contacting has virtually eliminated the dense bed or bubble contact zone. The success, of course, of a dilute phase system is dependent upon having sufficient catalyst, temperature, or catalyst activity present to achieve the desired conversion.

The first clear demonstration of the possible use of the fluid solids principle in this process occurred in the laboratories of the Standard Oil Development Co. (now Exxon Research and Engineering Co.) in 1939. An 8-in. diameter 40 ft long vertical pipe was filled with catalyst and aerated. A significant pressure buildup was observed, demonstrating that a hydraulic seal was possible. When the bottom flange was removed, the catalyst flowed out at a high rate, showing that a commercial cat transfer system could be developed.

The following paragraphs discuss how this pressure buildup and solids transfer can be engineered into present commercial applications.

Fluid Catalytic Cracking Catalyst Physical Properties

Fresh fluid catalytic cracking catalyst is available in a number of size ranges usually designated as fine, medium, and coarse particle size. The grade chosen

TABLE 23 Fresh Catalyst Properties

	Fine	Medium	Coarse
% off at 20 μm	3–5	2–3	2
40 μm	25	19	17
80 μm	13	23	32
Average particle size, μm	51	59	65

for a unit will depend upon how efficiently the catalyst collection system is operating and the sensitivity of the catalyst circulation system to coarse catalyst particles. It is generally desirable to operate with as coarse a particle as can be circulated stably, since there are fewer small particles in the catalyst which will be directly lost on addition. The particle size distribution of the grades is summarized in Table 23.

The pore volume of fresh catalysts can vary from 0.4 to 0.88 cc/g and apparent bulk densities from 0.37 to 0.8 g/cc. The higher pore volume and low bulk density catalysts will be entrained more from fluid beds and result in greater catalyst losses. Newer catalysts generally are trending to pore volumes of 0.4 cc/g and bulk densities of 0.75 to 0.8 g/cc. Surface area of fresh catalysts can vary anywhere from 600 to 250 m^2/g.

Catalyst properties equilibrate in commercial units. There is generally a large drop in surface area from the fresh catalyst range and some reduction in pore volume and increase in bulk density. The particle size is largely determined by the performance of the catalyst recovery equipment. Representative physical properties of equilibrium catalyst are presented in Table 24. The catalyst recovery equipment largely determines the particle size of the equilibrium catalyst in the unit and strongly affects the amount of 0 to 40 μm fines in the catalyst. The fresh catalyst composition and catalyst loss rate from the recovery equipment largely influences the amount of larger particles in the catalyst above 80 μm.

It is important to control the particle size of the catalyst, as the size and size distribution affects the fluidization characteristics of the catalyst. This in turn affects operating stability, processing capacity and, of course, design criteria.

There are several important factors that determine how much 0 to 40 μm catalyst is held in the circulating catalyst. Providing two stages of cyclones in

TABLE 24 Equilibrium Catalyst Properties

	Mean	Range
Particle density, g/cc	0.74	0.62–0.86
Pore volume, cc/g	0.43	0.33–0.53
Surface area, m^2/g	112	70–156

each process vessel, reactor, and regenerator will generally result in 3 to 15% 0 to 40 μm size particles (fines) in the equilibrium catalyst in the unit. The variation will primarily depend upon vessel superficial velocity and the efficiency of the cyclone system.

If solids that are collected in external cyclones are returned to the unit, or if precipitator fines are returned, the 0 to 40 μm size might increase to as much as 25%. It is not necessary to hold this much fines in the unit from the standpoint of the catalyst fluidity required for circulation and contacting, but catalyst losses are much reduced as considerably more attrition must take place before the catalyst is lost from the unit.

Dense Phase Flow in Lines

When gas is passed into a bed of fluid solids, pressure drop across the bed increases as the gas rate is increased as shown by the basic phase diagram in Fig. 53. The velocity and pressure drop reach the point where the particles of the catalyst are separated and supported by the gas flowing through them. This is the minimum fluidization velocity.

The bed then expands and gas starts to pass through the bed in bubbles. The density of the bed drops as the velocity and the amount of gas passing through the bed in bubbles increases. A point is reached where the particles are all entrained or transported from the vessel. Further pressure drop is primarily frictional pressure drop due to the higher velocity. It is not the purpose of this discussion to present a detailed review of fluidization principles, but only to present them in sufficient detail to orient the reader to this area of importance. More detailed discussions of fluidization can be found in Refs. 69–72.

The fluid solids flow in standpipes is depicted in the area of the figure slightly to the right of the minimum fluidizing velocity. It is desired to maximize density but also preserve fluidity. Therefore, there is a fairly narrow range in which the gas velocity relative to the solids must be controlled. Generally, the desirable range is from 0.05 to 0.25 ft/s. This range depends upon fluidization characteristics of the solid.

As the solids flow down the standpipe, the pressure increases, which reduces the volume of the gas aerating the solids. Thus the bubbles become smaller and

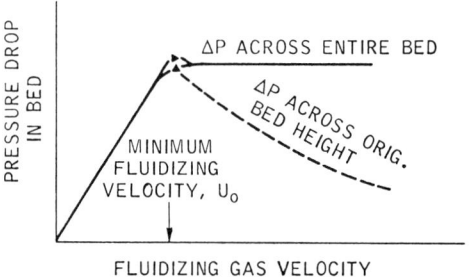

FIG. 53. Pressure drop in fluid beds.

Cracking, Catalytic

eventually the solids defluidize, which causes instability in catalyst flow and then bridging, which stops the catalyst flow. This problem is eliminated by providing aeration gas at various points in the pipe to maintain the differential velocity between the solids and gas in the range required for good fluidization. This is shown in Fig. 54.

A useful equation to determine the quantity of aeration required in a standpipe is

$$W_A = \frac{0.70 e\, W_c \rho_A}{T}$$

where W_A = aeration steam, lb/h/ft of standpipe between aeration levels
  W_c = catalyst circulation rate, lb/min
  ρ_A = standpipe pressure buildup. lb/ft²/ft of length (apparent density in standpipe, lb/ft³)
  T = standpipe temperature, °R
  e = void (gas phase) volume in standpipe, ft³/lb cat:

$$e = \frac{1}{\rho_F} - \frac{1}{\rho_S}$$

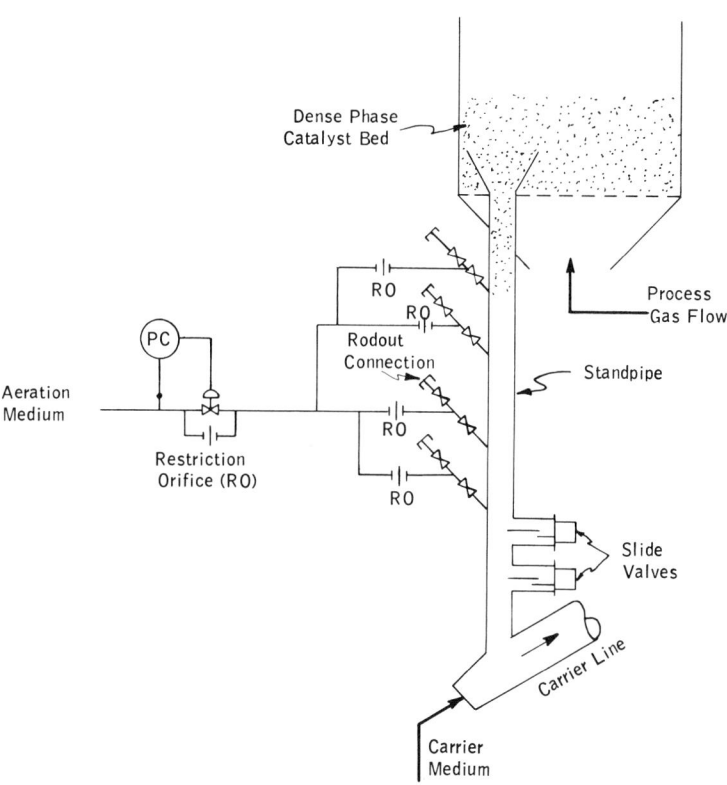

FIG. 54. Catalyst standpipe aeration.

where ρ_F = actual fluidized density of catalyst, lb/ft³
        ρ_S = skeletal density of catalyst, lb/ft³

Generally, as catalyst becomes coarser or as fines are lost from the catalyst, more aeration must be provided and it is more difficult to avoid fluidization problems.

There are interactions of pipe size, flowing velocity, and bubble size which may result in the formation of stable bubbles in catalyst circulating lines [67]. The stable bubble usually forms at junctions between vertical and sloped portions or at constrictions. This can happen when the emulsion superficial velocity, $W\rho_0$, is approximately equal to the bubble rise velocity, defined as $0.35\sqrt{gD}$ where the bubble diameter, D, has grown to pipe size. The bubble will flow up the standpipe if the circulation rate is significantly reduced or will be carried along with the solids if the rate can be appreciably increased. This is, of course, an unsatisfactory condition for plant operation.

Sloping standpipes create special problems as gas is more apt to deaerate to the top of the sloped pipe and migrate to the vertical section of the line where a bubble can form and interfere with flow. This problem has been solved with vent pipes, close control of aeration, or providing solids with better fluidizing properties.

Slide Valves in Catalyst Lines

The flow of catalyst in lines between the reactor and regenerator vessel is usually controlled by variable port slide valves or plug valves.

The port area of a slide valve is dependent primarily upon flow rate and can be determined by [68]

$$\text{lb/min} = 31.4 C_S \rho_B^{1/2} (P_1 - P_2)^{1/2} (D_0 - 1.5 D_p)^2$$

where $C_S = 0.45$
        ρ_B = bulk density of solids, lb/ft³
        D_0 = hole area, in.
        D_p = particle diameter, in.
        P_1 = upstream pressure, lb/in.²
        P_2 = downstream pressure, lb/in.²

Slide valve areas for commercial flows can be approximated by

$$A = 50 Q / C \sqrt{\rho \Delta P}$$

where A = area of port, in²
        Q = circulation rate tons/min
        C = flow coefficient (0.85 to 0.95)
        ρ = flowing density, lb/ft³
        ΔP = pressure drop, lb/in.²

Cracking, Catalytic

Bed Density

In fluid units, dense beds are usually provided in the regenerator and reactor stripper. Also, in some units, the reaction is completed in a dense bed. The density of catalyst in the bed is dependent upon the vapor velocity of gas passing up through the bed and the density of the particles in the bed. Figures 55 and 56 show a general relationship of density and velocity for cat cracking catalysts in reactors and regenerators.

Catalyst Entrainment

Gas leaving a dense bed will entrain a considerable amount of catalyst into the dilute phase as the gas bubbles break at the surface. Most of this catalyst will fall back into the bed, but at a certain distance above the bed the entrainment

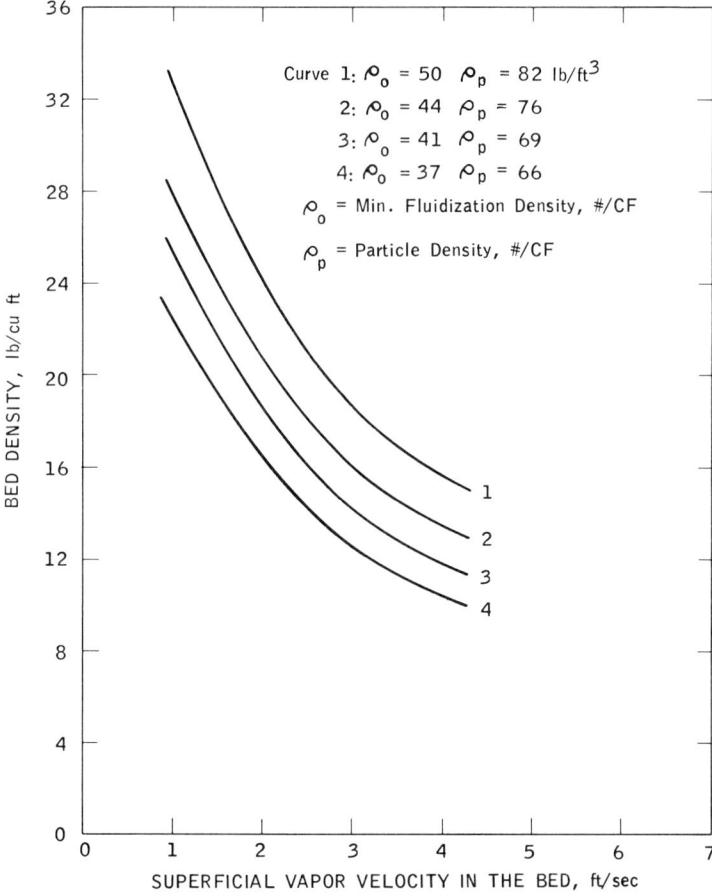

FIG. 55. Reactor-bed densities for microspheroidal catalyst.

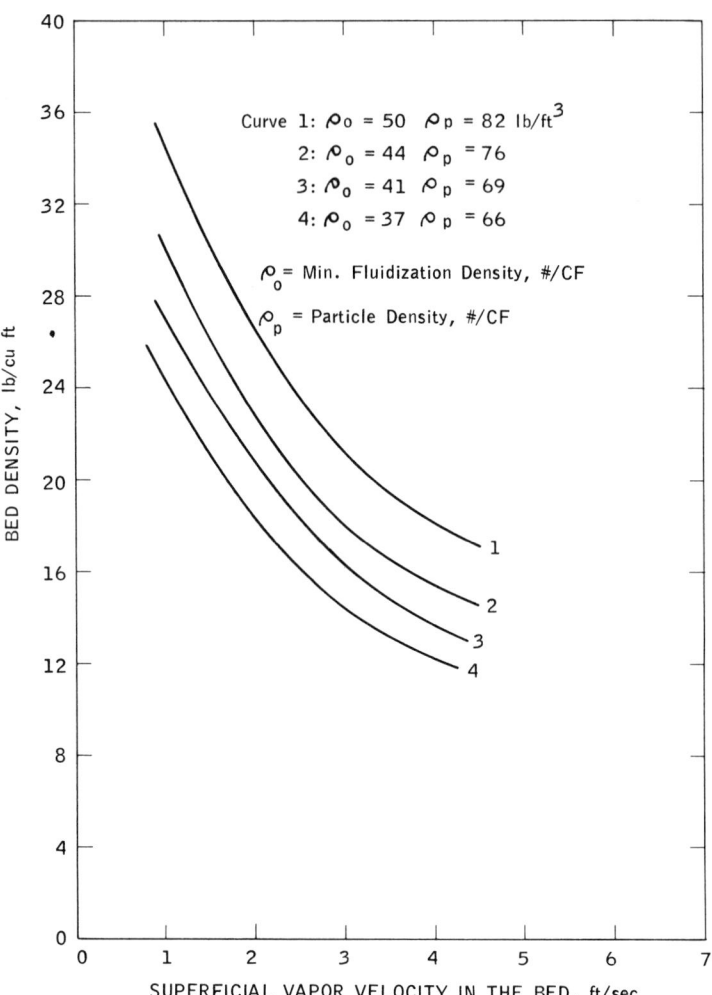

FIG. 56. Regenerator-bed densities for microspheroidal catalyst.

becomes constant. This height is called the transport disengaging height (TDH). Numerous methods are available to develop this entrainment rate. A simplified method is given in Figs. 57–59. Figure 59 is based primarily on regenerator gas characteristics. The rate should be halved for reactor entrainment because the gas viscosity is about half.

It is important to know the entrainment rate in order to design cyclones and cyclone diplegs properly.

Entrainment can be affected considerably by poor distribution of gas in the bed. If bed densities are significantly different from the density that would be observed in a well-fluidized bed, as shown in Figs. 55 and 56, gas is not being equally distributed through the bed. This condition can result in excessive localized entrainment which will result in shorter cyclone life, increased losses, and possibly process difficulties, particularly in the regenerator.

Cracking, Catalytic

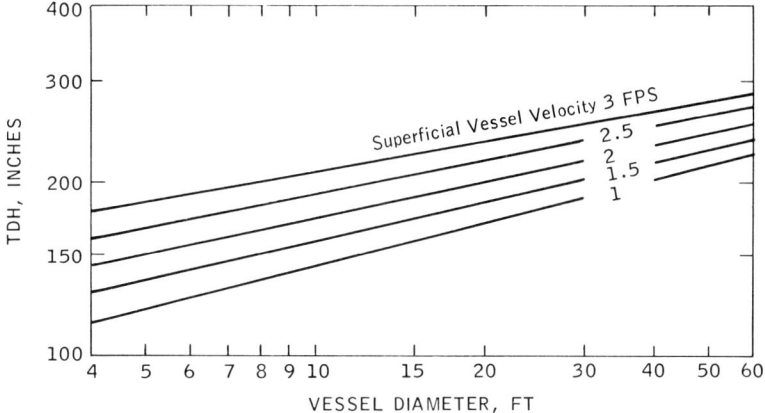

FIG. 57. Approximate transport disengaging height. (Courtesy Emtrol Corp.)

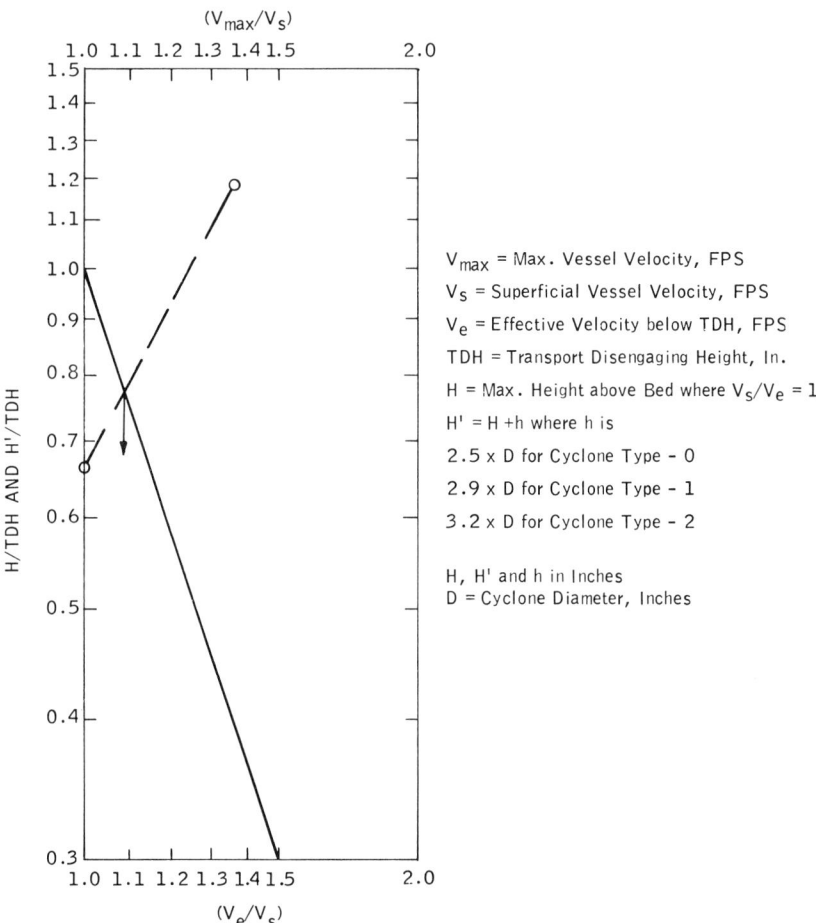

V_{max} = Max. Vessel Velocity, FPS
V_s = Superficial Vessel Velocity, FPS
V_e = Effective Velocity below TDH, FPS
TDH = Transport Disengaging Height, In.
H = Max. Height above Bed where V_s/V_e = 1
H' = H + h where h is

2.5 × D for Cyclone Type - 0
2.9 × D for Cyclone Type - 1
3.2 × D for Cyclone Type - 2

H, H' and h in Inches
D = Cyclone Diameter, Inches

FIG. 58. Approximate solution for determining effective velocity V_e. (Courtesy Emtrol Corp.)

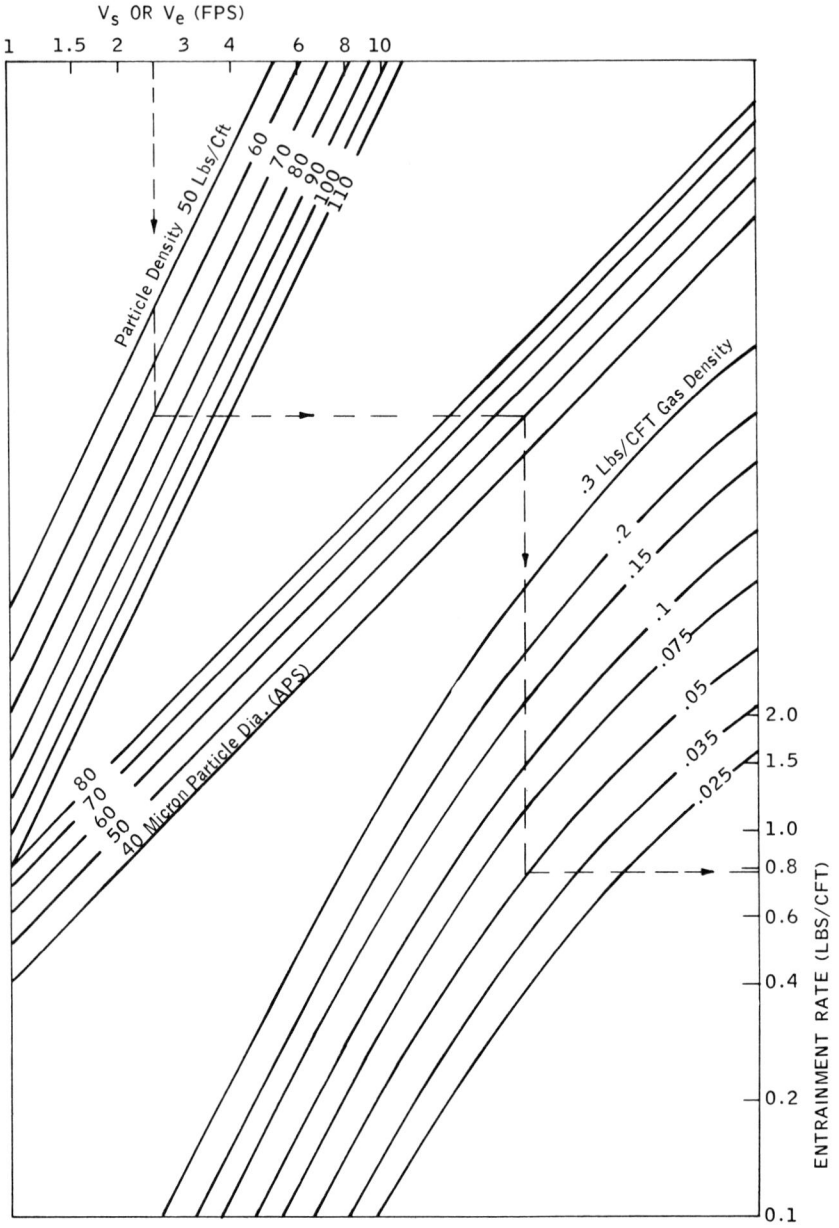

FIG. 59. Approximate entrainment curve. (Courtesy Emtrol Corp.)

Catalyst Holdup in Circulating Lines

The catalytic cracking catalysts have been developed to the extent where most of the cracking reaction can be accomplished in the line conducting the catalyst to the reactor vessel. Thus it is important to know how much catalyst exists in this transfer or riser line. There is a rapid increase in mols as the feed cracks, so the gas velocity in the line is rapidly changing. However, it can be shown that the holdup in any section of line will be equal to

$$dH = \frac{(CCR)(SF)\,dl}{60v}$$

where H = catalyst holdup, tons
CCR = cat circulation rate, tons/min
SF = slip factor, vapor velocity/cat velocity
l = length, ft
v = vapor velocity, ft/s

The slip factor over the entire transfer line is generally about 1.5. The slip factor can also be calculated at the entry point by ratioing the fully vaporized feed and gas velocity to the entering catalyst velocity. The slip factor will approach 1.1 at the outlet section of the line. The distance in which catalyst is accelerated to final velocity will depend upon the amount of catalyst present and the final velocity. However, this vertical distance can be generally assumed to be 20 to 40 ft. Some laboratory data on accelerating zones of catalyst in gas have shown the log of the slip factor to vary linearly with height. Data have also

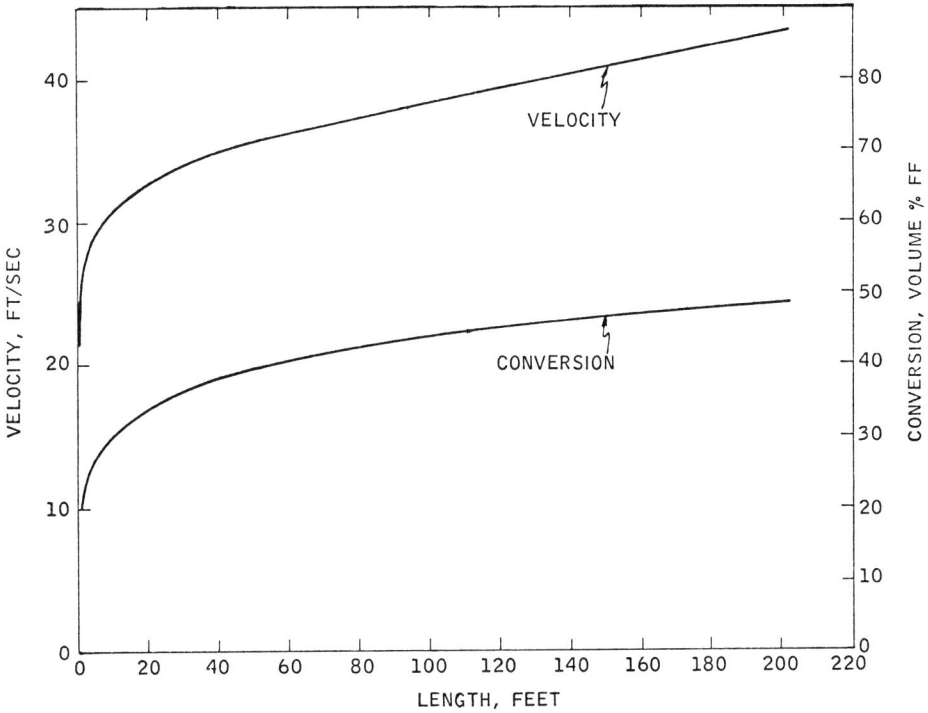

FIG. 60. Velocity profile in a transfer line reactor.

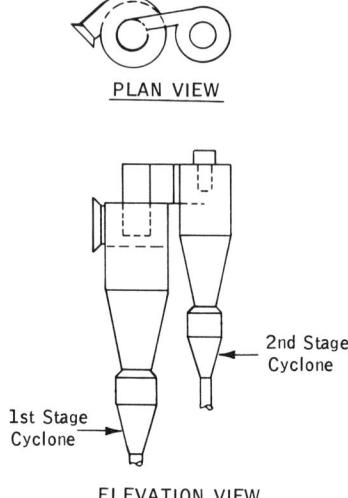

FIG. 61. Two-stage cyclone system.

shown that the generation of mols of cat cracker products varies linearly with conversion. Thus a rigorous calculation method can be developed for transfer line holdup. The procedure can be simplified by the use of average slip factors and velocity profiles, which can be used in conjunction with the general equation presented above to calculate the holdup for a riser reactor.

A representative profile of velocity and conversion is shown in Fig. 60.

Cyclone Design

The entrained catalyst is retained in the fluid solids vessels by cyclones located in the upper part of the vessel. Generally, two stages of cyclones are used although three stages are provided in some units. The cyclone, Fig. 61, is a simple inertial collection device in which the process gases are swirled and particles removed by centrifugal force. The degree of separation in the cyclones is affected by the cyclone design and conditions of operation. These are highlighted below. A more complete exposition can be found in a recent API publication, *Manual for Disposal of Refinery Wastes—Volume on Atmospheric Emissions*, Chapter 11, "Cyclone Design," API Publication No. 931, May 1975.

There are two important factors in the design of a cyclone system. The first is the pressure balance and the second is the efficiency. It is mandatory that the pressure balance be satisfied. If it is not, the cyclone diplegs will flood with catalyst, and catalyst losses will be prohibitive. The unit would be considered inoperable at the velocities which caused the flooding and it would have to be run at lower gas rates where the cyclones would not flood. However, the unit would be generally considered operable at desired feed rates with cyclones of lower efficiency, although the catalyst losses might be 20 to 50% higher than

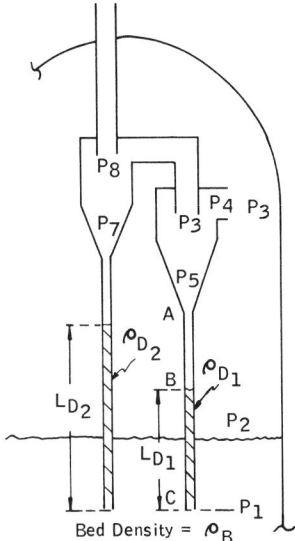

FIG. 62. Cyclone pressure balance relationships.

desired and the circulating catalyst may be coarse and sensitive to circulation problems. The high cost of catalyst, and increased emphasis on smooth operations and low maintenance, usually result in justifying high efficiency, large cyclones for a commercial installation.

The pressure of process gas is reduced as it passes through the cyclone system. Since the separated solids must be returned to the bed by means of diplegs, the level of the catalyst in the diplegs must be high enough so that the static head of the catalyst in the dipleg will balance the pressure drop that has been taken in the gas stream at the top of the diplegs. Sufficient outage between this catalyst level in the dipleg and the bottom of the cyclone must exist for expected operating fluctuations.

The pressure balance through the cyclone system is illustrated in Fig. 62. The balanced pressure point is at the bottom of the dipleg, P_1.

The pressure drop through the bed is equivalent to

$$P_1 - P_2 = \frac{\text{dipleg submergence, ft} \times \text{bed density, lb/ft}^3}{144}$$

The pressure drop to the cyclone inlet is equal to

$P_2 - P_3 = $ Height of inlet above the bed (ft) × average density of catalyst in the dilute phase (lb/ft^3) ÷ 144

The pressure drop across the cyclone inlet is primarily an acceleration pressure drop:

$$P_3 - P_4 = \frac{1.1(\text{inlet velocity, ft/s})^2(\text{gas density, lb/ft}^3 + \text{entrainment, lb/ft}^3)}{144 \times 2g_c}$$

The pressure drop through the cyclone system will depend upon the cyclone design and amount of solids present. The important factors in cyclone design are velocity and outlet area/inlet area ratio. The effects of these factors are shown in Figs. 63 and 64. Cyclone model types will affect pressure drop somewhat. As shown, models designated as A, B, X, and Y have slightly different velocity head relationships. Other models could be expected to follow similar curves.

In a primary cyclone the entrained solids slow down the spinning of gas in the cyclone such that at, say, 1 lb/ft^3 entrainment rate the velocity head pressure drop of the cyclone is only 35% of what it would be if no solids were present.

Pressure drop to the top of the primary cyclone dipleg is equal to

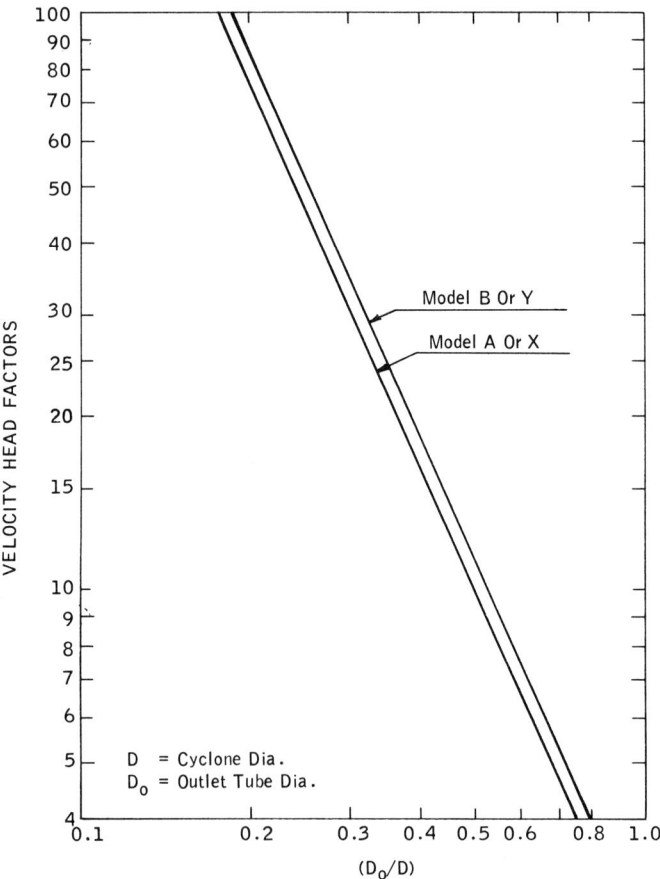

FIG. 63. Velocity head factors for cyclone pressure drops (Models A, B and X, Y). (Courtesy Emtrol Corp.)

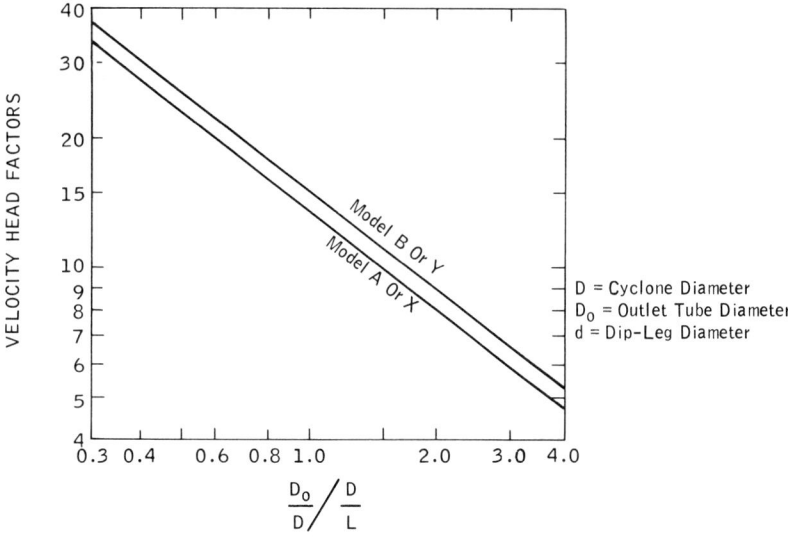

FIG. 64. Dipleg velocity head factors for Models A, B and X, Y. (Courtesy Emtrol Corp.)

$$P_4 - P_5 = \text{velocity head drop} \times \text{entrainment correction factor}$$
$$\times \frac{(\text{inlet velocity, ft/s})^2}{144 \times 2g_c} \times \text{gas density, lb/ft}^3$$

The height of the catalyst in the cyclone dipleg is determined by the difference in pressure between P_5 and P_1 and can be determined by

$$\text{Ht}_{\text{dipleg}} = \frac{(P_1 - P_5)144}{\text{catalyst density in dipleg, lb/ft}^3}$$

The pressure to the secondary dipleg top and to the plenum chamber can be determined in a similar manner.

Many different models of cyclones are available from a number of vendors. Particle collection efficiency can range considerably depending upon the operating conditions, basic cyclone proportions, and particle characteristics.

For a given area of the inlet of a cyclone, the general effect of changing factors in cyclone design on efficiency are listed below.

Increasing	*Efficiency*
Quantity of gas (velocity)	Increases
Cyclone diameter (same outlet pipe)	Increases
Outlet pipe diameter	Decreases
Barrel length	Increases
Cone length	Increases

It is generally not feasible to install the most efficient cyclone available, as the size of the vessel to house such a cyclone would become prohibitive. Therefore, cyclone design involves a compromise between mechanical and process features. Since the dimensions of cyclones can vary considerably and larger cyclones for a given inlet area are more efficient but more costly, it is essential in the design and evaluation of competitive cyclone systems that the systems have equivalent collection efficiencies.

A number of companies manufacture cyclone systems for fluid catalytic cracking units. Some of them are the Ducon Co., Envirotech Corp., Emtrol Corp., and Vantongeren Corp.

Energy Recovery

A considerable quantity of recoverable heat leaves the regenerator in the flue gas stream. The gases are hot, 1150 to 1400°F, at a pressure of 10 to 40 lb/in.² gauge, and may contain as much as 10% carbon monoxide.

To recover some of this available heat, flue gas coolers or steam generators were installed in the early cat units upstream of the pressure relief valves. Only the specific heat of the gas was recovered. These boilers required much maintenance since they were subject to erosion of the fine catalyst escaping the cyclone recovery systems. Moreover, the gas velocity in the tubes must be high enough to prevent fine catalyst from depositing on the surface and reducing heat transfer. Heat transfer coefficients of 10 to 15 Btu/h/ft²/°F are generally realized at gas velocities of 40 ft/s.

Carbon monoxide boilers have generally replaced the coolers and are installed in almost all new units. These are merely large boilers which are designed to burn low Btu combustion gas. They can generally be fired with extraneous fuels. These boilers are located downstream of the pressure control valves in the flue gas system and operate at only a few inches of water pressure. Since the air required for the combustion of the flue gas must be compressed into the system, low pressure fans are provided in this service.

The pressure head of the regeneration gases can be recovered in expanders. The very first fixed bed units utilized CO boilers and expanders. However, more recent power recovery systems in cat crackers have become more complex as shown in Fig. 65. Complex control and flue gas diversion systems have been added to insure a safe and reliable expander system.

The amount of energy recoverable in the regenerator off gases [73] will, of course, depend upon the temperature, pressure and quantity of gas. The following relationship can be used to determine expander horsepower:

$$\text{Shaft horsepower (shp)} = E\left(\frac{GRT_I}{500}\right)\left(\frac{K}{K-1}\right)\left(1 - \frac{P_E}{P_I}\right)^{(K-1)/K}$$

where E = overall turbine efficiency ($\sim 80\%$)
       G = mass flow of flue gas, lb/s
       R = gas constant (55.3 for flue gas)
       T_I = expander inlet temperature, °R

recovery possible in an expander is shown in Fig. 66. The energy which is recovered in the expander can be used to drive the main blower and/or be converted to electricity.

Mechanical Considerations

Operation of a catalytic cracking unit provides the engineer with complex problems in erosion, high temperature exposure, fluid flow, cyclone catalyst separation, refractory usage, and metallurgy. Good knowledge in these fields, i.e., materials selection coupled with specialized design concepts, fabrication techniques, and maintenance procedures, is the key to profitable operation. A measure of unit performance and maintenance upkeep is the unit run length (time in operation between shutdowns). Unit run lengths in excess of 6 years have been reported but run lengths of 2 to 4 yr are more common, especially if unit throughput greatly exceeds the original design.

The following paragraphs discuss some of these conditions and offer some guidelines for control or handling of specific problems. The major portion of the information presented is based on empirical development from "trial and error" application. Included are sections covering material selection, special design factors, operation precautions, equipment inspection and maintenance suggestions, and desirable design features. Particular reference is made to the section entitled "Troubleshooting" since many of the problem areas encountered in catalytic cracking operation are covered.

Material Selection and Design Aspects

Important factors affecting selection of construction materials for fluid catalytic cracking units are:

Erosion
Allowable stress at elevated temperature
Facility for field repair of equipment
Afterburning in regenerator section
Corrosion
High temperature regeneration

The selection of materials must consider strength, erosion/corrosion resistance, and ease of construction, i.e., design. Also, refiners prefer materials which will insure low maintenance since substantial production losses are incurred each day the units are shut down. For this reason equipment could be designed for specific performance and not necessarily for specific design life.

A discussion on selecting materials of construction is given below. It is essential whenever selecting materials that they are practical and conform with good mechanical practice, i.e., trade skills and equipment readily available.

recovery possible in an expander is shown in Fig. 66. The energy which is recovered in the expander can be used to drive the main blower and/or be converted to electricity.

Mechanical Considerations

Operation of a catalytic cracking unit provides the engineer with complex problems in erosion, high temperature exposure, fluid flow, cyclone catalyst separation, refractory usage, and metallurgy. Good knowledge in these fields, i.e., materials selection coupled with specialized design concepts, fabrication techniques, and maintenance procedures, is the key to profitable operation. A measure of unit performance and maintenance upkeep is the unit run length (time in operation between shutdowns). Unit run lengths in excess of 6 years have been reported but run lengths of 2 to 4 yr are more common, especially if unit throughput greatly exceeds the original design.

The following paragraphs discuss some of these conditions and offer some guidelines for control or handling of specific problems. The major portion of the information presented is based on empirical development from "trial and error" application. Included are sections covering material selection, special design factors, operation precautions, equipment inspection and maintenance suggestions, and desirable design features. Particular reference is made to the section entitled "Troubleshooting" since many of the problem areas encountered in catalytic cracking operation are covered.

Material Selection and Design Aspects

Important factors affecting selection of construction materials for fluid catalytic cracking units are:

Erosion
Allowable stress at elevated temperature
Facility for field repair of equipment
Afterburning in regenerator section
Corrosion
High temperature regeneration

The selection of materials must consider strength, erosion/corrosion resistance, and ease of construction, i.e., design. Also, refiners prefer materials which will insure low maintenance since substantial production losses are incurred each day the units are shut down. For this reason equipment could be designed for specific performance and not necessarily for specific design life.

A discussion on selecting materials of construction is given below. It is essential whenever selecting materials that they are practical and conform with good mechanical practice, i.e., trade skills and equipment readily available.

Erosion [76]. In fluid catalytic cracking, the process is carried out by fluidized catalyst which flows between the reactor and regenerator. The fluidized catalyst consists of small, hard particles in a gaseous medium. The particles are in constant contact with each other and with the walls of the containing equipment. The particles usually are harder than the surfaces they contact and as a result erosion of the surfaces occurs.

The factors affecting catalyst erosion are:

Velocity
Density
Change in direction of flow (angle of impingement)
Turbulent flow (orifices, grid inserts)

In general, the erosion rate can be assumed proportional to the density of the fluidized catalyst and proportional to an exponential function of flow velocity, i.e., V^x where $x > 2$.

The common construction materials, such as carbon steel and alloy steel, do not have adequate erosion resistance for all types of exposure and it is necessary to use special erosion barriers. Refractory linings [77–79] over the steel to protect it from erosion have found widespread use. When it is not practical to use refractory linings, the following other methods can be employed:

Add heavy corrosion/erosion allowances to the metal
Provide erosion-resistant hard-surfacing deposits
Install replaceable steel/alloy wear plates

The application of hard-surfacing deposits directly to the wall of pressure-containing components is not favored since the deposits form numerous cracks on cooling. On vibrating equipment the cracks can permeate into the base material (wall) and cause cracking failures. Hard-surfacing applied to special liners would resist erosion without causing damage to the equipment.

Specific locations where erosion is encountered are cyclone systems, catalyst transfer lines, flue gas systems, and reactor overhead lines. In the regenerator and reactor vessels erosion is usually not a problem since the flow velocity is low, i.e., <1.5 m/s (5 ft/s).

Erosion in cyclone systems is particularly severe in regenerator cyclones. The erosion in the primary cyclones is concentrated in the inlet horn and main barrel sections. In the secondary cyclones the erosion is concentrated in the top of the diplegs, dustpots, and lower $\frac{1}{3}$ of the main cones. To control or reduce erosion of the cyclones to a tolerable level, the gas inlet velocities m/s (ft/s) may be limited to the approximate values shown below. It is important that proper cyclone velocities be used since the cyclone separation efficiency is a function of cyclone inlet velocity; i.e., low velocity would insure a low erosion rate but more importantly would result in low efficiency. The cyclone size and number of cyclones (i.e., velocity) are functions of the vessel size. Cyclone access, i.e., size, is also a factor when considering maintainability.

Reactor	Regenerator	
Primary and secondary	Primary	Secondary and tertiary
18 m/s	18.5 m/s	23 m/s
(60 ft/s)	(60 ft/s)	(75 ft/s)

Dipleg erosion, particularly in the primary diplegs where the mass flow rate is higher than in secondary diplegs, has been found to be concentrated at internal weld projections. It is desirable that welding and fabrication procedures for diplegs insure smooth internal surfaces in the diplegs.

On the reactor side, the grid, grid supports, grid holes, and catalyst deflector are susceptible to severe erosion. The grid plate itself (underside), grid supports, and deflector are exposed to a combination of erosion/corrosion that produces pitting and surface irregularities. The increased surface area of the exposed steel and the localized turbulence at the surface irregularities can lead to accelerated attack. For FCCU feed stocks with sulfur content less than 0.5 wt.%, the corrosion/erosion rate is usually acceptable without special precaution, i.e., carbon steel construction is satisfactory. For higher sulfur content in the feed (>0.5 wt.%), 12 Cr (Type 410) alloy steel cladding on the underside of the grid may eliminate the corrosion/erosion attack. The supports and catalyst deflector can be carbon steel for the lower sulfur content, and 5 Cr or 18/8 Cr-Ni Type 304 alloy steels for the higher sulfur content feed. The Type 304 alloy steel may be a more practical choice since fabrication by welding is simplified.

The leading edges of supports and the edge of the catalyst deflector may be hard-surfaced with Stellite 6 to reduce erosion. All edges or surfaces should be, as far as practical, positioned normal to the flow to take advantage of a right-angle approach (less erosion since the catalyst particles by collision take the brunt of the energy dissipation). When a normal approach angle is not possible, it is desirable to use an impingement angle $<20°$ to reduce erosion.

The reactor grid holes are exposed to severe erosion/corrosion since the gas/catalyst velocity through the flow restriction may be high (about 52.5 m/s (175 ft/s)). Cast Stellite Grade 6 inserts have been found effective in reducing the erosion rate of the grid holes and hence, maintain hole diameter so that the differential pressure across the grid is maintained reasonably constant for the particular flow rate. The grid hole surface of the Stellite 6 casting is smooth, and accurate hole dimensions are possible with the casting technique. This is important since surface irregularities within the hole will produce localized turbulence which can lead to accelerated erosion. The Stellite 6 alloy can be cast into a carbon or alloy shell to facilitate welding to the grid. Stellite 6 cannot be satisfactorily welded since it lacks ductility. Grid inserts produced by weld deposits of hard-surfacing alloys are susceptible to accelerated erosion due to the rough "as welded" surface in the holes.

From a sulfur corrosion standpoint, all of the hard-surfacing alloys except the Ni base alloys are suitable for reactor side exposure.

In the design of flow restriction orifices, such as slide valves [80] or grid holes (refer to the section entitled, "Application of Fluid Solids Principles to Cat Cracking"), erosion can be reduced to acceptable levels, i.e., run length, by limiting the pressure drop kPa (lb/in.^{2}) across the orifice to the approximate values listed in Table 25.

Allowable Stress at Elevated Temperature. All pressure vessels and piping are assumed to be designed in accordance with the appropriate codes. At elevated temperatures, above 480°C (900°F), the allowable stress values drop significantly for carbon steel, and alloy steels may be more economical. As a result, an alloy steel in place of carbon steel is often specified because of the superior strength of the alloy at the particular temperature.

At temperatures above 650°C (1200°F), austenitic stainless steels such as 18/8, 25/12, and 25/20 Cr-Ni steels are required. These steels have high temperature strength and good oxidation resistance for FCCU exposure.

While Cr or Cr-Ni alloy steels are often used for small diameter piping, this is not the case for large diameter piping and vessels. It is usually economical to fabricate such equipment from carbon steel and provide an internal thermal lining so that the metal temperature is reduced to a level where the carbon steel has adequate strength. This will also reduce the thermal expansion of a piping system because of the lower temperature so that a piping system with less flexibility is required. For fluid temperatures above 480°C (900°F), the wall thickness for internally lined (thermal) equipment for pressure may be based on a 345°C (650°F) metal temperature. The metal temperature for lined piping flexibility design may be the actual metal temperature (based on lining characteristics) in still air and the highest ambient temperature.

The diameter (or dimensions) of the internally lined equipment (cold wall) will be considerably larger and generally will weigh more than unlined equipment (hot wall). When combining alloy (hot) equipment and internally lined (cold) equipment, for example, hot valves and cold piping, the designer must consider the effect of the elastic behavior of the cold pipe on the hot valves which may be susceptible to creep deformation. Special guidance or supports may be required to avoid distortion of the valves since they may act as "plastic hinges" in the line.

Economic design is a vital factor in the selection of construction materials

TABLE 25 Maximum Desirable Pressure Drop from Erosion Standpoint

	kPa	lb/in.^{2}
Catalyst transfer line slide valves (throttling)[a]	41	6
Flue gas slide valves (pressure control)	103	15
Regenerator grid holes (with catalyst passing through grid)	7[b]	1[b]
Regenerator grid holes (with air passing through grid)	21[b]	3[b]
Reactor grid holes	14[b]	2[b]

[a]More than one valve in series is frequently used to insure long run lengths, i.e., >2 yr.
[b]The catalyst may be attrited at this and higher pressure drop, and grid hole erosion will be experienced.

although the trend in recent years has been to choose materials for reliability. Accordingly, most vessels can be designed to provide at least 10 years maintenance-free service where practical. In most instances the cost differences between various construction materials, i.e., carbon and alloy steels, may be obvious so that detailed estimates may not be necessary.

Internals in the reactors and regenerators are not governed by Code regulations except when they affect the pressure shell. Thus it is possible to fabricate from carbon steel even though the internal temperature is above 600°C (1100°F) which is the Code (piping) design limit for carbon steel. This facilitates designs since the internals of the reactors and regenerators may generally be subject to modification and usually require maintenance during turnarounds.

When service life requirements can be specified and/or thermal or load cycles can be established, special component designs can be based on actual time conditions and/or fatigue properties of the carbon or alloy steels as mentioned below. This design concept may lead to some distortion of internals if conditions change, but would be tolerable provided that failure does not take place. The elevated temperature makes the steel more "plastic" and more readily deformed without "locked-up stress." Temperature/stress information at elevated temperatures for nonpressure-containing internals carbon and alloy steels are usually covered in steel company publications.

Vessel internals (and some piping) may be exposed to varying conditions during a run, i.e., a long-time operating condition and short-time upset conditions. The design stress for long-time operation is generally that based on 1% creep in 100,000 h at the design temperature. Short-time design stress is generally based on stress rupture in 1000 h at the upset temperature. Other conditions may require an intermediate design condition which could be based on, say, 1% creep in 10,000 h at the intermediate design temperature.

The cyclone system in the regenerator, for example, can be designed using these criteria. For instance, stress could be based on 1% creep in 100,000 h at a design temperature of 650°C (1202°F) for long-time operation, and also for a short-time operation (afterburning) based on rupture in 1000 h at, say, 982°C (1800°F) for 18/8 Cr-Ni alloys. Carbon steel cyclone systems would use a much lower maximum temperature and such systems would rely on a responsive quench water spray system for temperature control. Grids also may be designed for normal operation, startup, slump, and blasting conditions, each of which may have different pressure/temperature conditions.

Facility for Field Repair of Equipment. It is essential to use materials which can be readily repaired (welded) in the field by refinery forces or contractors. Carbon steel is ideally suited since it is economical, readily available, and weldable. However, the strength of carbon steel decreases rapidly at elevated temperatures, and at a temperature of 540°C (1000°F) carbon steel normally is not economical for pressure-containing equipment. The Cr alloy steels which are the most economical materials at temperatures above 540°C (1000°F) require special heat treatment after welding in order to produce satisfactory welds. The heat treatment is done most conveniently in the shop where furnaces are available. There are methods available to provide field weld heat treatment, but the field work is time consuming and requires special equipment and close quality control.

Often 18/8 Cr-Ni alloy steels are used for the pressure-containing shell and for internals where it is not practical to use carbon steel. By using 18/8 Cr-Ni alloy steel, the welding and fabrication problems can be minimized. Field repair problems also may be reduced since this alloy undergoes essentially no change in properties with exposure to catalytic cracking conditions. The 18/8 Cr-Ni alloys provide increased strength and corrosion resistance over the Cr steel alternatives. These advantages often outweigh the higher initial cost of 18/8 Cr-Ni alloy steel construction.

The designer must recognize the poorer thermal conductivity (less than $\frac{1}{2}$ relative to carbon steel) of the 18/8 Cr-Ni alloys which can make the alloys susceptible to deformation when exposed to localized high temperatures. Also, this property makes these alloys susceptible to cracking if exposed to severe temperature fluctuation such as generated by water droplet impingement from spray nozzles.

Afterburning. In catalytic cracking regenerators, the carbon is burned off the fluidized catalyst in an atmosphere of excess oxygen (roughly 1%). The CO in the regenerator flue gas tends to further oxidize to CO_2, resulting in rapid temperature excursions of the flue gas. This condition, called afterburning, can produce flue gas temperatures up to 982°C (1800°F) in the dilute phase section (top) of the vessel and in the downstream flue gas system.

Generally, carbon steel cyclones and plenum chambers are satisfactory for service up to 620°C (1150°F) on a long-time basis. However, above 620°C (1150°F) alloy construction materials are usually required from both strength and oxidation standpoints. The alloy steel normally used is 18/8 Cr-Ni. The cyclone system and equipment downstream in the flue gas system are usually designed for short-time exposure to afterburning. Frequently a temperature limit of 982°C (1800°F) is used in designs to insure that adequate strength has been provided for flue gas components.

An emergency spray water system is normally used in the dilute phase of the regenerator to quench the flue gas as soon as afterburning is detected. Similarly, steam can be injected into the primary cyclone outlets to reduce the temperature of the flue gas. At this location the flue gas is even more susceptible to ignition since less catalyst is entrained in the flue gas which acts as a "heat sink" to reduce rapid temperature fluctuations. The spray system can be actuated by a temperature rise within the cyclones.

Water spray systems are frequently provided for afterburning control in the downstream flue gas system for protection of waste heat exchangers, pressure-controlling slide valves, orifice plates, muffler, etc. Since the sprays must operate in a confined space, the possibility of damaging the internals by spray water impingement (quenching) is a risk that must be considered.

Stress corrosion cracking and intergranular corrosion of 18/8 Cr-Ni alloy steel equipment is not considered to be a problem for internals operating at elevated temperature under FCCU conditions.

Corrosion. The main cause of deterioration of materials and linings in fluid catalytic cracking units, as previously mentioned, is erosion. The presence of vaporized oil with the fluidized catalyst in the reactor can significantly increase the metal loss rate (erosion/corrosion) if it contains sulfur [81]. The protective measures and the construction materials used to minimize sulfur erosion/corrosion have been discussed earlier in the section entitled "Erosion."

The presence of excess oxygen in the regenerator can cause serious oxidation of carbon steel inside the vessel. Carbon steel is subject to excessive scaling at about 635°C (1175°F). Since most regenerators are designed for temperatures at or above this temperature, alloy steel for the exposed internals may be provided. With the possible process incentives for operation above 650°C (1200°F), 18/8 Cr-Ni alloy steels are required. Below 625°C (1160°F), carbon steel and Cr alloys are satisfactory.

The dual layer type internal lining using hexmesh steel reinforcement must have adequate oxidation resistance for long-time exposure. This applies equally to the hexmesh and the submerged support studs. For regenerator service, 12 Cr (Type 410) alloy steel has been found adequate. This alloy resists oxidation up to 760°C (1400°F). For the reactor side, carbon steel and support studs are generally satisfactory for all feedstocks.

Where possible, single layer (monolithic) linings are used since they avoid exposing the support steel (studs) directly to the process fluid. They also reduce cost since alloy studs are not required and installation is reduced to a one-step operation.

A typical refractory specification for hexmesh reinforced and oneshot linings covering desirable physical property requirements, for the solids section of a catalytic cracking unit is shown in Table 26. It indicates where the specific refractories can be used and the testing suggested to insure conformance with the specification. These type linings are not considered to be susceptible to corrosion attack even with high sulfur content feeds. Experience indicates that corrosion of pressure vessel shells and piping, when protected by refractory linings, has been negligible after as much as 30 yr of service.

In unlined reactors (hot shell), erosion-resistant lining with hexmesh and refractory (Type B from Table A) can offer adequate protection to the shell where the corrosion allowance has been depleted. This may occur frequently in the dense phase of reactors when the sulfur content of the feed is increased. The lining is simple to install and maintain, and the materials are readily available.

Because of the satisfactory performance of the refractory-type linings, a reduction of corrosion allowance on new vessels is frequently considered. On older units the vessels have in some cases been uprated for higher pressure operation through the unused corrosion allowance.

Alloys which have Ni content greater than 50% are susceptible to sulfur attack at temperatures above 370°C (700°F). Use of these materials where they are directly exposed to the process in fluid catalytic crackers, especially reactors, is not desirable. This applies to Ni base hard-surfacing deposits as well as Ni alloys.

High Temperature Regeneration. In recent years, some catalytic cracking units have been converted to operate with the regenerator temperature as high as 760°C (1400°F). It can be assumed that most units will eventually be so

TABLE 26 Typical Castable Specification for FCCU Use

Castable "A": Use for "one shot" applications and backup insulation in dual layer linings.
Medium weight castable, 70 to 85 lb (maximum 90 lb gun applied)/ft^3 installed and dried for 18 h at 220°F.
Shrinkage: Maximum ±0.3% *as cast* after drying at 220°F for 18 h and heating to 1500°F in 1 h and holding at 1500°F for 5 h.
Strength: Minimum 500 lb/in.^{2} cold crushing strength *as cast* after drying at 220°F for 18 h and heating to 1500°F in 1 h and holding at 1500°F for 5 h.
Maximum use temperature: 2300°F.

Castable "B": Use for refractory in hexmesh reinforcing in dual linings.
Super erosion-resistant refractory, 125 to 170 lb/ft^3 installed and dried for 18 h at 220°F.
Shrinkage: 0.3% maximum as cast after drying at 220°F for 18 h and heating to 1500°F in 1 h and holding at 1500°F for 5 h.
Strength: Minimum 5000 lb/in.^{2} cold crushing strength *as cast* after drying at 220°F for 18 h and heating to 1500°F in 1 h and holding at 1500°F for 5 h.
Maximum use temperature: 2600°F.
Al_2O_3 *content of aggregate*: Minimum 85% by weight.

Castable "C": Use for cyclone linings in hexmesh and in hexmesh for dual layer lining at severe erosion locations.
Same as "B" but maximum use temperature to be 3300°F.

Notes:

(1) Test shall be made to determine cold crushing strength when the material arrives on-site by a reputable independent laboratory approved by owner. The test shall be in the *as-cast form*. Manufacturer's data will not be acceptable in lieu of tests.

(2) Tests shall be made when the castables are being installed to determine density as applied, i.e., the sample shall be pneumatically applied if this method is used for the installation. A test panel of at least 3 ft × 3 ft shall be used. It should be weighed prior to and after lining application.

(3) All materials must meet the specified density, shrinkage, and strength requirements or the lining materials can be rejected by owner.

(4) The number of test samples shall be determined based on the following:

One sample per 2 tons of refractory
One sample per 4 tons of castable insulation

A sample shall consist of at least three specimens. Castables from several batches, i.e., manufacturers' mixing batch size, may be combined and mixed to obtain the testing sample per the above. Specimen size shall be 2 in. × 2 in. × 2 in. cast in plastic or metal forms.

(5) Refractory and insulating castable binder cements used for compounding the castable materials must be certified to be less than 3 months old at time of usage. The supplier shall supply such certification.

(6) Castables having silicate binders shall not be used.

TABLE 27 Typical Regenerator Side Materials Suggested for High Temperature Regeneration

Cyclones and plenum chamber	18/8 Cr-Ni
Diplegs, bracing, and trickle valves	18/8 Cr-Ni
Flue gas slide valves and catalyst throttling slide valves (assuming external insulation)	18/8 Cr-Ni
Internal nozzles and piping	18/8 Cr-Ni
Grid distributor	18/8 Cr-Ni
Overflow well and spent catalyst inlet	18/8 Cr-Ni
Lining	4 to 5 in. single layer medium weight castable insulation supported by CS studs

modified because of the possible process incentives. The apparent benefits of the 760°C (1400°F) regeneration are reduced carbon make and increased yields of gasoline.

The regenerator, regenerated catalyst transfer lines and catalyst slide valve, oil injectors, and the regenerator overhead system (flue gas) may be significantly affected if 760°C (1400°F) regeneration temperature is considered. Specific material requirements for this design condition are shown in Table 27. At 760°C (1400°F) metal temperature, only austenitic stainless steels, for example, 18/8 Cr-Ni, can be used since the lower alloy steels have inadequate strength and oxidation resistance.

The temperature of the regenerator overhead system can be reduced by water sprays in the regenerator plenum chamber or flue gas line at some risk of failure as discussed earlier. This would allow the use of Cr alloy or carbon steels in the stack, stack valves, water seal tanks,* and other downstream equipment. The loss of energy due to the nonrecoverable heat of vaporization of the spray water in the flue gas may be considerable and would have to be economically justified. This scheme may be considered on existing units to defer or avoid the costly replacement of an existing flue gas system.

Special Design Factors

A checklist on some of the important factors which should be considered for all new construction or for all major modifications of existing units are:

1. Velocities in catalyst transfer and flue gas lines should be limited to minimize erosion.
2. Lined versus unlined reactor/stripper shell should be based on economics.
3. Refractory lining may be required for the fixed orifice and for the top slide

---

*Water seal tanks for switching flue gas between CO boiler and bypass stack in the regenerator overhead system would have to be eliminated since 18/8 Cr-Ni internals would be susceptible to thermal shock and failure by water exposure.

surface for throttling catalyst slide valves. Slide/guide bearing surfaces can be hard-surfaced with Stellite 6.
4. In the stack flue gas throttling slide valve, the fixed orifice and slides, and slide/guide bearing surfaces can be hard-surfaced with Stellite 6.
5. Line restrictions in flue gas lines, such as stack valves, may cause downstream turbulence. For protection of thermal lining, a bare stack erosion liner, $2\frac{1}{2}$ to 3 pipe diameters in length, may be required immediately downstream of stack valves.
6. Material selection for the catalyst transfer lines, catalyst slide valves, and flue gas slide valves should be based on the specified design temperatures including emergency conditions and considering ease of maintenance.
7. Selection of materials for reactor in the grid location depends on the sulfur content in the feed.
8. All cyclones may require an erosion-resistant refractory lining.
9. Reactor grid hole inserts may be required to minimize grid hole erosion.
10. Protective shields may be required on the regenerator cyclones and diplegs where they are in line with the dilute phase-quenching sprays.
11. An anticoking baffle can be used in the top of the reactor above the cyclone inlets to eliminate dangerous coke formation.
12. Metals or weldments with nickel contents above 50% should not be used if service temperatures are above 370°C (700°F) in the presence of sulfur.
13. When designing regenerator internals, both long- and short-term design conditions should be considered.
14. In throttling slide valves in catalyst or flue gas service, the center of the orifice opening (variable) should be located as close to the centerline of the valve as practical to minimize downstream valve body erosion.
15. Avoid plug-type manways on the reactor side since coke may form in the annulus between the plug and the nozzle and prevent removal of the plug.

Suggested Operating Precautions

Some of the operating precautions suggested to obtain maximum service life of fluid catalytic cracker equipment are:

1. Use of blast facilities in catalyst transfer lines and slide valves should be minimized since excessive use will cause localized erosion (failure) as well as catalyst attrition.
2. Steam purge should be maintained through the regenerator dilute phase sprays (and plenum chamber sprays if used) while not in use to prevent piping or spray nozzle tip overheating and subsequent failure by thermal shock when placed in operation.
3. The grid design pressure drop should not be exceeded at the grid design temperature during start-up, or grid distortion can result. This is particularly important for dished grids which may have been designed based on buckling.
4. If the unit is started up with new or extensively repaired lining, the rate of

temperature increase should be controlled to prevent damage of the new lining due to rapid vaporization of water contained in the lining.
5. The disks of stationary slide valves (Model IV cracking units) should be partially moved at least once a week to minimize slide sticking.
6. Slide valve external insulation (if specified) must be maintained since loss of body insulation will reduce internal clearances between slide and guides.
7. Regenerator bed level should be controlled to minimize cyclone erosion resulting from higher cyclone catalyst loadings caused by above normal bed levels.
8. Cyclone velocities should be kept within design limits to keep erosion within tolerable levels.
9. In double disk flue gas slide valves, the orifice opening formed by the disks should be centered in the valve to minimize downstream erosion.
10. When two or more flue gas slide valves are used in series, equal pressure drop should be taken by each valve. This will result in more uniform erosion on valve internals, insuring longer slide valve life. Noise generation also will be reduced if the pressure drop taken by the two valves is equal.
11. In high regenerator temperature, 650 to 760°C (1200 to 1400°F), units, loss of feed can expose the reactor and riser/reactor nozzle to higher than reactor design temperature. Instrumentation must be provided for quick closure of the regenerated catalyst slide valve(s) and to inject emergency steam into the riser for temperature control of the riser and reactor.
12. Interstage cyclone steam in the regenerator should not be shut off since readmittance of steam can thermally shock and damage the piping. A failure in the piping would permit flue gas to enter the cyclone and cause increased catalyst losses.

Inspection and Maintenance

Inspection/Onstream. Erosion is a constant concern in all piping which carries even small amounts of entrained catalyst. Thickness measurements using ultrasonic techniques are usually complicated in fluid catalytic cracking units by metal temperatures in excess of 370°C (700°F). Methods are available for such measurements up to 540°C (1000°F) through the use of water-cooled transducers. Piping of concern includes the main catalyst transfer lines, the flue gas line, oil riser, the oil–catalyst slurry piping, and pressure controlling slide valve bodies, orifice areas, etc. In particular, areas of flow direction changes such as elbows or points of high turbulence, i.e., injection points for aeration or blast media, may be checked on a periodic basis.

For internally insulated (refractory lined) equipment, visual observation of painted steel surfaces will provide by paint deterioration an indication of higher than normal metal temperatures which would be the result of internal lining damage.

For externally insulated equipment, visual inspection is of little or no value since the surface shows no hidden damage. Generally, experience shows that well-maintained and well-operated equipment can be expected to be trouble-free for at least 2 years provided the design limits of the unit are observed.

Inspection/Shutdown. Erosion probably accounts for the largest inspection effort during a shutdown [82, 83]. This involves the cyclone systems in regenerator and reactor, slide valves used for pressure/flow control, grids, slip-type seals, expansion joint-type seals, catalyst and flue gas transfer lines, stripping steam injection nozzles, and aeration and blast connections.

In the regenerator the cyclone system may have been exposed to temperature cycles from afterburning. These cycles can impose thermal movements in the cyclones which can cause fatigue cracking of the cyclone bodies, interconnecting ducts, and outlet pipes. All such surfaces should be examined for failures or cracks and repaired before start-up. Otherwise, high catalyst losses will take place forcing a shutdown. The equivalent of a 6.4 cm^2 (1 in.^{2}) hole in the secondary cyclone outlet of a fluid catalytic cracking unit will pass approximately 1000 lb/h of catalyst. At this rate the hole will enlarge rapidly, increasing the catalyst leakage rate.

The seals between the grid and pipes passing through the grid in the regenerator must be inspected. Any cracks in expansion joint-type seals will normally require replacement of the seal unless sound repairs can be made. A crack, if undetected, will lead to localized erosion of the seal and rapid failure. The loss of a seal can lead to oxygen breakthrough to the dilute phase with resultant severe afterburning and cyclone cracking.

Internal linings must be checked for soundness, especially looseness of lining sections which could fall from the shell or piping wall during operation and cause hotspots. Local spalling of linings, which can reduce the lining thickness below, say, 50%, also may require replacement. At locations where erosion has caused damage or lining failure, steps to repair or modify the lining must be taken. This may involve the use of special refractories and/or lining design changes.

A check on oxidation or corrosion must be made to insure that adequate metal remains on components. This will give clues to the rate of metal loss for future replacement planning. On slurry lines, wall thickness must be checked at points of turbulence such as elbows and downstream of orifices so that planned replacements can be made without emergencies. If corrosion is part of the problem, steels with higher corrosion resistance, i.e., 5 Cr versus CS, can significantly reduce the metal loss.

Maintenance. As mentioned earlier, the FCC unit should be designed for ease of maintenance since repair and/or modifications of eroded or worn equipment is certain at turnarounds. For this reason, the following are *desirable features* which are suggested to be part of the design philosophy.

Provide adequate access manways
Design scaffolds and keep them in top repair (for safety)
Locate and size main manways to facilitate cyclone removal
Use materials that are readily available, fabricated, and installed considering

factors such as weldability, hardness, needs for heat treatment, special curing, and special dryout

Use flat plate, all welded construction, where possible, i.e., grids and internals

Avoid slip-type seals in catalyst service

Maximize use of welded construction

Avoid bolted joints or bolted construction inside vessels

Use valves in which internals replacement can be made through the bonnet

Design cyclones so that access is available for localized repairs. Cyclones with less than 4 ft barrel diameter are difficult to enter for proper repairs

Provide permanent facilities for erecting scaffolds inside vessels, i.e., support clips nozzles in vessels, and nozzles in lines for bosun chair lift cables

Avoid permanent internal scaffolds since oxidation/corrosion damage during operation is difficult to ascertain

List and maintain spares for throttling valve internals, reactor grid inserts, valve operators, spray tips, and expansion joints

The following lists maintenance items that require attention at turnarounds.

Spalling, loss, and erosion of refractory and insulation
Distortion and erosion of the regenerator grid distributor
Grid seal erosion or cracking
Cyclone erosion
Cyclone body and duct cracking
Regenerator dipleg spray shield damage from water impingement
Feed injector erosion
Regenerator dilute phase spray gun and spray nozzle erosion
Erosion of slide valve bodies and internals
Expansion joint erosion and/or cracking
Inadequate provisions for thermal expansion of components
Erosion in catalyst transfer lines at blast and aeration connections
Reactor grid, grid supports, deflector plate, and grid hole erosion/corrosion
Coke removal from inside anticoking baffle
Coking at catalyst stripper entrance
Stripping steam nozzle erosion
Erosion of slurry lines at points of turbulence
Erosion inside of diplegs at weld projections
Reactor overhead line erosion in bends
Thermowell erosion

References

1. E. J. Houdry et al., "Catalytic Processing of Petroleum Hydrocarbons by the Houdry Process," *Proc. Am. Pet. Inst. Section III*, *19*, 133–148 (1938).
2. M. Sittig, *Pet. Refiner*, *31*(9), 263–316 (1952).

3. A. Danner, *Pet. Refiner*, *29*(9), 179–182 (1950).
4. J. B. Holiman, "Mechanical Aspects of Houdriflow Catalytic Cracking Process," *Adv. Pet. Chem. Refining*, *2*, 580 (1959).
5. G. D. Hobson, *Modern Petroleum Technology*, 4th ed., Wiley, New York, 1973, pp. 288–309.
6. E. J. Demmel et al., *Proc., 31st, API Div. Refining Meeting*, pp. 165–171 (May 1966).
7. C. L. Helmer and W. L. Vermilion, "Developments in Fluid Catalytic Cracking," *1973 Technology Conference, Universal Oil Products Company, Des Plains, Illinois*, 1973.
8. W. L. Vermilion, "Modern FCC Design: Evaluation and Revolution," *Belgian Petroleum Institute Meeting, Antwerp, Belgium*, November 4, 1974.
9. E. L. Whittington et al., "Catalytic Cracking—Modern Designs," *Am. Chem. Soc., Div. Pet. Chem., Prepr.*, *17*(3), B66–82 (July 1972).
10. D. P. Bunn et al., "The Development and Operation of the Texaco Fluid Catalytic Cracking Process," *AIChE 64 National Meeting, New Orleans, Louisiana*, March 16–20, 1969.
11. M. C. Bryson et al., "Gulf's FCC Process," *Proc., 37th API Div. Refining Meeting*, pp. 377–386 May 1972.
12. J. A. Finneran et al., "Application of Heavy Oil Cracking in a Fuels Refinery," *AIChE 74th National Meeting, New Orleans, Louisiana*, March 11–15, 1973.
13. R. E. Hildebrand et al., "Desulfurization and Catalytic Cracking of Residua," *Delaware Valley Section Meeting, AIChE, Drexel University, Philadelphia*, March 19, 1974.
14. B. S. Greensfelder, "Theory of Catalytic Cracking," in *Chemistry of Petroleum Hydrocarbons II*, Reinhold, New York, 1955, pp. 137–164.
15. R. C. Hansford, "A Mechanism of Catalytic Cracking," *Ind. Eng. Chem.*, *39*, 49 (1947).
16. R. C. Hansford, "Chemical Concepts of Catalytic Cracking," *Adv. Catal.*, *4*, 1–30 (1952).
17. B. S. Greensfelder et al., "Catalytic Cracking and Thermal Cracking of Pure Hydrocarbons," *Ind. Eng. Chem.*, *41*, 2573–2584 (1949).
18. V. Haensel, "Catalytic Cracking of Pure Hydrocarbons," *Adv. Cat.*, *3*, 179–197 (1951).
19. H. H. Voge, "Catalytic Cracking," in *Catalysis*, Vol. 6 P. H. Emmett, ed., Reinhold, New York, 1958, p. 407.
20. S. J. Wachtel et al., "Laboratory Circulating Fluid Bed Unit for Evaluating Carbon Effects in Catalytic Cracking Selectivity," *Am. Chem. Soc., Div. Pet. Chem. Prepr.*, *3*, A55–62 (1971).
21. H. R. Crane et al., "Poisoning Effects of Metals," *Pet. Refiner*, *40*(5), 168–172 (1972).
22. A. Sadana and L. K. Doraiswamy, "The Effect of Catalyst Fouling in Fixed-, Moving-, and Fluid-Bed Reactors," *J. Catal.*, *23*, 147–157 (1971).
23. D. R. Campbell and B. W. Wojciechowski, "Theoretical Patterns of Selectivity in Aging Catalysts with Special Reference to the Catalytic Cracking of Petroleum," *Can. J. Chem. Eng.*, *47*, 413–417 (1969).
24. D. R. Campbell and B. W. Wojciechowski, "Selectivity of Aging in Static, Moving, and Fixed Bed Reactors," *Can. J. Chem. Eng.*, *48*, 224–281 (1970).
25. R. A. Pachovsky and B. W. Wojciechowski, "Theoretical Interpretation of Gas Oil Conversion Data on an X-Sieve Catalyst," *Can. J. Chem. Eng.*, *49*, 365–369 (1971).
26. R. A. Pachovsky et al., "Theoretical Interpretation of Gas Oil Selectivity Data on an X-Sieve Catalyst," *AIChE J.*, *19*, 802–806 (1973).

27. D. R. Best et al., "Application of the Time On Stream Theory of Catalytic Decay," *Ind. Eng. Chem., Process Des. Dev.*, *12*(3), 254–261 (1973).
28. J. A. Paraskos et al., "A Kinematic Model for Catalytic Cracking in a Transfer Line Reactor," *Ind. Eng. Chem., Process Des. Dev.*, *15*(1), 165–169 (1976).
29. V. W. Weekman, Jr., "A Model of Catalytic Cracking Conversion, in Fixed, Moving, and Fluid Bed Reactors," *Ind. Eng. Chem., Process Des. Dev.*, *7*(1), 90–95 (1968).
30. V. W. Weekman, Jr., and D. M. Nace, "Kinetics of Catalytic Cracking Selectivity in Fixed, Moving, and Fluid Bed Reactors," *AIChE J.*, *16*(3), 397–404 (1970).
31. D. M. Nace et al., "Application of a Kinetic Model for Catalytic Cracking—Effect of Charge Stocks," *Ind. Eng. Chem., Process Des. Dev.*, *10*(4), 530–538 (1971).
32. S. E. Voltz et al., "Application of a Kinetic Model for Catalytic Cracking—Some Correlations of Rate Constants," *Ind. Eng. Chem., Process Des. Dev.*, *10*(4), 538–541 (1971).
33. S. E. Voltz et al., "Application of a Kinetic Model for Catalytic Cracking—III. Some Effects of Nitrogen Poisoning and Recycle," *Ind. Eng. Chem., Process Des. Dev.*, *11*(2), 261–265 (1975).
34. A. Voorhies, Jr., "Carbon Formation in Catalytic Cracking," *Ind. Eng. Chem.*, *37*(4), 318–322 (1945).
35. A. G. Oblad, T. H. Milliken, and G. A. Mills, "The Effects of the Variables in Catalytic Cracking," in *The Chemistry of Petroleum Hydrocarbons II*, Reinhold, New York, 1955, pp. 165–188.
36. C. R. Olsen and M. J. Sterba, "Effect of Reactor Temperature on Product Distribution and Product Quality in Fluid Catalytic Cracking," *Chem. Eng. Prog.*, *45*, 692–700 November (1949).
37. J. A. Montgomery, "The Effect of Operational Variables," in *The Davison Chemical Guide to Catalytic Cracking*, Petroleum Chemicals Dept., Grace and Co., Baltimore, Maryland, 1975, pp. 15–21.
38. J. B. Pholenz, "How Operational Variables Affect Fluid Cat Cracking," *Oil Gas J.*, *61*(13), 124–143 (April 1, 1963).
39. J. W. Moorman, "Operating the Cat Cracker—5. What is the Effect of Feed Preheat in Catalytic Cracking?" *Oil Gas J.* *53*(36), 68–73 (January 10, 1955).
40. W. M. Haunschild, D. O. Chessmore, and D. G. Spars, "A Pilot Plant Comparison of Riser and Dense Bed Cracking of Hydrofined Feed. Stocks," *AIChE 79th National Meeting, Houston, Texas*, March 16–20, 1975.
41. F. H. Blanding, "Reaction Rates in Catalytic Cracking of Petroleum," *Ind. Eng. Chem.*, *45*(6), 1186–1197 (1953).
42. E. G. Wollaston et al., "FCC Model Valuable Operating Tool," *Oil Gas J.*, *73*(38), 87–94 (September 22, 1975).
43. J. W. Moorman, "Space Velocity and Catalyst-to-Oil Ratio Together Provide a Measure of Cracking Severity," *Oil Gas J.* *52*(44), 108–109 (March 8, 1954).
44. J. J. Blazek, "High Temperature Regeneration," *Catalgram* (Davison Chemical Co.), *48* (1975).
45. W. L. Nelson, "Paraffinic Crudes Are Superior for Catalytic Cracking," *Oil Gas J.*, *60*(24), 161 (June 11, 1962).
46. H. E. Reif, R. F. Kress, and J. S. Smith, "How Feeds Affect Cat Cracker Yields," *Pet. Refiner*, *40*(5), 237–244 (1961).
47. P. J. White, "Effect Feed Composition on Cat Cracker Yields," *Oil Gas J.*, *66*(21), 112–116 (May 20, 1968).
48. R. V. Shankland, "Industrial Catalytic Cracking," *Adv. Cat. 6*, 392 (1954).
49. K. Van Nes, *Aspects of the Constitution of Mineral Oils*, Elsevier, Amsterdam, 1951.

50. P. J. White, "Effect Composition on Cat Cracker Yields," *Oil Gas J.*, 66(21), 112–116 (May 20, 1968).
51. W. L. Pierce et al.," Innovations in Flexicracking," *Hydrocarbon Process.*, 51(5), 92–97 (1972).
52. E. L. Whittington, J. T. Murphy, and I. H. Lutz, "Striking Advances Show Up in Modern FCC Design," *Oil Gas J.*, 70(44), 49–54 (October 30, 1972).
53. J. H. Gary and G. E. Handwerk, "Catalytic Cracking," in *Petroleum Refining Technology and Economics*, Dekker, New York, 1975, pp. 100–108.
54. W. S. Letzch et al., "Lab Can Evaluate Zeolite FCC Catalyst," *Oil Gas J.*, 74(4), 130–144 (January 26, 1975).
55. F. G. Ciapetta and D. S. Henderson, "Microactivity Test for Cracking Catalysts," *Oil Gas J.*, 65(42), 88–93 (October 16, 1967).
56. W. R. Gustafson, "A Modified Micro-Activity Test for Cracking Catalyst," *Am. Chem. Soc. Div. Pet. Chem., Prepr.*, 14(3), 56–69 (1969).
57. J. J. Blazek et al., *Proc. Am. Pet. Inst.* 42(III), 277–286 (1962).
58. K. D. Ashley and W. B. Innes, *Ind. Eng. Chem.*, 44, 2857 (1952).
59. C. N. Kimberlin, Jr., and E. M. Gladrow, U.S. Patent 2,971,903 (February 14, 1961).
60. H. A. Benesi, *J. Phys. Chem.*, 61, 970 (1957).
61. M. E. Conn and G. C. Connally, *Ind. Eng. Chem.*, 39, 1138 (1947).
62. J. Alexander and H. G. Shimp, *Natl. Pet. News*, 36, R537 (1944).
63. C. L. Thomas, in *Catalytic Processes and Proven Catalysts*, Academic, New York, 1970, Chap. 4.
64. U.S. Patent 3,844,973.
65. Anon., "Complete Combustion of CO in Cracking Process," *Chem. Eng.*, 82, 25 (November 24, 1975).
66. H. U. Hammershaimb et al., "Recent Developments in Catalytic Cracking CO Combustion—A Tool for the Refiner," *1976 NPRA Annual Meeting*, San Antonio, March 29, 1976, Paper AM-76-26.
67. J. M. Matsen, *Powder Technol.*, 7, 93–96 (1973).
68. P. U. Bulsara, "Pressure and Additive Effects on Flow of Bulk Solids," *Ind. Eng. Chem., Process Des. Dev.*, 3(4), 348–355 (1964).
69. F. A. Zenz and D. F. Othmer, *Fluidization and Fluid Particle Systems*, Reinhold, New York, 1960.
70. M. Leva, *Fluidization*, McGraw-Hill, New York, 1959.
71. J. F. Davidson and D. Harrison, *Fluidized Particles*, Oxford University Press, Cambridge, 1963.
72. J. F. Davidson and D. Harrison, *Fluidization*, Academic, New York, 1971.
73. W. C. Meyer and L. M. Stettenbeng, "The Power Recovery Gas Expander," *ASME Paper 64-PET 13*, September 1964.
74. A. P. Krueding, "Power Recovery Techniques as Applied to Fluid Catalytic Cracking Units Regenerator Flue Gas," *AIChE 79th National Meeting*, March 1–20, 1975.
75. L. M. Stettenbeng, "The Power Recovery Flue Gas Expander in the Fluid Bed Catalytic Cracking Cycle," *47th Annual Fall Meeting*, Western Gas Processors and Oil Refiners Association, October 1972.
76. J. S. Clarke, "How to Combat Erosion in Fluid Cat Crackers," *Oil Gas J.*, 51(46), 262–264, 268–269, 272, 276 (March 23, 1953).
77. J. S. Clark, "Maintenance of Catalytic Cracking Units—3. Here's How Heat-Insulating Linings are Maintained," *Oil Gas J.* 54(67), 118–122 (August 13, 1956).
78. M. S. Crowley, "Refinery Usage of Refractories," *ASME Petroleum Mechanical Engineering Conference*, New Orleans, Louisiana, September 18–21, 1967, Paper 66-PET-4.

79. M. S. Crowley, "Refractory Problems in Refineries," *NACE 1972 National Meeting*, Paper 64.
80. J. S. Clarke, "Maintenance of Catalytic Cracking Units—5. Slide Valves—Their Upkeep and Adjustment," *Oil Gas J.*, *54*(73), 112–116 (September 24, 1956).
81. K. L. Moore, "Alloy Stops Corrosion in Fluid Coker," *Pet. Refiner*, *40*(5), (1961).
82. J. S. Clarke, "Maintenance of Catalytic Cracking Units—1. When Trouble Comes... Shut It Down, or Make Hot Repairs?" *Oil Gas J.*, *54*(61), 90–92 (July 2, 1956).
83. J. S. Clarke, "Maintenance of Catalytic Cracking Unit—6. Planning and Executing a Turnaround," *Oil Gas J.*, *54*(77), 100—103 (October 22, 1956).
84. W. F. Pansing, "Regeneration of Fluidized Cracking Catalysts," *AIChE J.*, *2*(1), 71–74 (1956).
85. P. B. Weisz and R. B. Goodwin, "Combustion of Carbonaceous Deposits Within Porous Catalyst Particles—II, Intrinsic Burning Rate," *J. Catal.*, *6*, 227–236 (1966).
86. S. Tone et al., "Kinetics of Oxidation of Coke in Silica Alumina Catalysts," *Bull. Japan Pet. Inst.*, *14*(1), 76–82 (1972).
87. M. F. L. Johnson and H. C. Mayland, "Carbon Burning Rates of Cracking Catalysts in the Fluidized State, *Ind. Eng. Chem.*, *47*(1), 127–132 (1955).
88. W. D. Ford et al., "Modeling Catalytic Cracking Regenerators," *1976 NPRA Annual Meeting, San Antonio*, March 29, 1976, Paper AM-76-29.
89. R. R. Edison et al., "Catalytic Cracking of High Metal Content Feedstocks," *1976 NPRA Annual Meeting, San Antonio*, March 30, 1976, Paper AM-76-37.

E. C. LUCKENBACH
A. C. WORLEY
A. D. REICHLE
E. M. GLADROW

Heavy Oil Cracking

Background

Heavy oil cracking (HOC) is a process for catalytically cracking reduced crudes into lighter, more valuable products. The important difference between HOC and other refinery processing schemes is the elimination of residual fuel as a refinery product and the production of large amounts of steam that can provide energy for normal refinery energy applications. In many refineries this steam can be used advantageously; in others it may offer potential for cogeneration facilities.

The HOC process is really a specialized version of the FCC (fluid catalytic cracking) process adapted to the conversion of reduced crudes having a wide range of qualities. Depending on quality and product objectives, feedstocks with vanadium-plus-nickel content of 5–30 ppm and carbon residues of 5–

Heavy Oil Cracking

10% can be processed without feed pretreatment. HOC offers the refiner a flexible resid upgrading process which, when coupled with HDS (hydrodesulfurization), provides a particularly effective means for meeting demands for gasoline, furnace or diesel fuel blending stock, and light olefins while conserving available crude oil.

One of the first HOC units was installed for Phillips Petroleum at its Borger, Texas refinery. A later unit of more advanced design was put onstream for Phillips to process 50,000 BPSD of desulfurized vacuum reduced crude at the company's Sweeny refinery. A more recent HOC installation is for Valero Energy at Corpus Christi. This is a state-of-the-art refinery that can operate entirely on reduced crude rather than whole crude.

The heavy oil cracking process is very similar to the fluid catalytic cracking process in configuration, catalysts, and product handling. It differs from gas oil cracking in that the heat release, due to burning the coke produced from the asphaltenes in the charge, is considerably greater. In addition, the much higher content of feedstock metals, particularly nickel and vanadium, requires special consideration in catalyst development and in operation. With the current need to convert residual fuels to gasoline and middle distillates, the installation of new heavy oil cracking units, as well as the conversion of FCC units to reduced crude, cracking has become of great interest in the refining industry.

Feedstocks

When the price of crude oil skyrocketed in the 1970s, refiners looked for more reasonably priced crude oils or residual oils to process into gasoline, diesel fuel, naptha, and other liquid products. These lower priced feedstocks often had heavy viscosities, were high in sulfur content, contained heavy metals, and varied considerably in respect to asphaltene content. Therefore, the design of new HOC units varies depending upon the crude being fed to the refinery. These generally fall into the following four categories.

1. High quality reduced crudes: This includes reduced crudes of metal contents less than about 10 ppm and carbon residues of less than 5%. Feedstocks of this quality can be handled in a modified FCC unit. The usual problem is developing a design which will minimize coke formation so that external means to remove heat from the regenerator are not required. For borderline heat removal requirements, light oil recycle is used to remove heat which eventually increases the steam production from the slurry steam generator. Catalyst and operating variables to achieve the low coke yield include high catalyst activity, use of metal passivators, low pressures, and low contact time. An illustration of the yield that can be achieved with such feedstocks is shown in Table 1 for both high and low severity operation, wherein it is designed to maximize gasoline or middle distillate.

TABLE 1 Range of Product Distributions in Heavy Oil Cracking

Feedstock inspections	Gravity	23.3°API
	Distillation	650°F–10% point
		980°F–80% point
	Carbon residue content	3.9%
	Metals content, ppm	Ni, 2.4
		V, 3.3

Type of Operation	Maximum Gasoline	Maximum Middle Distillate
Conversion, vol.%	79.1	56.2
Gasoline, C_5–430°F TBP, vol.%	63.6	48.4
C_4–$C_4^{2=}$, vol.%	15.9	8.2
C_3–$C_3^{2=}$, vol.%	9.2	5.0
Light catalytic gas oil, vol.%	15.1	32.1
Heavy catalytic oil + decanted oil, vol.%	5.7	11.7
Total, vol.%	109.7	105.6
Coke, wt.%	8.0	7.1
C_2 and lighter, wt.%	2.2	1.5

This feedstock has a 3.9% carbon residue, 5.7 ppm metals content, and a 23.3° API. The normal coke yield in an FCC unit which easily satisfies the heat requirements is about 5%. The 3.9% carbon residue requires specialized conditions to avoid excessive regenerator temperature.

Note also that there is a good flexibility for varying conversion and yields in HOC units. Pilot-plant data show the maximum gasoline operation produces slightly less than 64% gasoline at 80% conversion. At milder severities a 32% yield of light catalytic gas oil is achieved. In both cases, liquid recoveries greater than 105% are realized.

2. Reduced crudes with metals content up to about 30 ppm and carbon residues of 5 to 10%: This category is illustrated by the operation of the original HOC unit at Phillips' Borger, Texas refinery, and in operation since 1961. Typical data are shown in Table 2. The feedstock contains 22 ppm of nickel plus vanadium and has a 5% carbon residue. The yields are poorer than for the Category 1 feed, giving about 6% less gasoline and 3% more coke. Because of the additional coke, steam coils located within the regenerator bed are needed to remove the excess heat. This shows up in the production of about 250 lb of steam per barrel of feed to the HOC unit.

3. Feedstocks with metals content from 30 to about 150 ppm and carbon residues of about 10 to 20%: Such feedstocks require feed pretreatment, such as residual desulfurization, to lower the metals and carbon residues so that they can be processed without excessive catalyst makeup rate and without excessive regenerator cost and size. The HOC units operating at

the Phillips refinery in Sweeny, Texas, and that designed for Valero fit into this category. The Phillips unit at Borger, Texas, can now run heavier, sour crudes with the installation of a resid HDS unit before the existing HOC unit.

Table 3 shows that hydrodesulfurization at about 90% desulfurization also gave a 50% reduction in carbon residue and an 80% demetallization. In addition, the bulk quality of the feedstock was improved as indicated by the increase in API gravity and the 40°F increase in aniline point. These improvements in properties show up in heavy oil cracking as a 13% increase in gasoline yield, a 13% increase in total liquid recovery, and a 4.5% decrease in coke yield. The dramatic improvement in gasoline yield and total liquid recovery significantly improves the economics of hydrodesulfurization of reduced crudes.

TABLE 2 Heavy Oil Cracking of Medium Quality Reduced Crudes (Texas Panhandle atmospheric resid)

Feedstock properties:		
Gravity, °API	21.3	
Sulfur, wt.%	1.5	
Conradson carbon, wt.%	5.0	
Metals, ppm:		
Nickel	15	
Vanadium	7	
Conversion vol.% of fresh feed	76	
Product yields:[a]		
Gasoline, vol.%	57.8	
Butane–butylene:	12.0	
Isobutane		3.1
n-Butane		0.9
Butylenes		8.0
Propane–propylene:	7.0	
Propane		1.5
Propylene		5.5
Light cycle oil	15.0	
Decanted oil	9.0	
Total		100.8
Coke, wt.%	11.0	
C_2 and lighter, wt.%	3.0	
Hydrogen, wt.%	0.2	
Gasoline octane ratings:		
Research, clear	88	
Motor, clear	77	
Export steam (600 lb/in.^{2}gauge), lb/bbl fresh feed	250	

[a]Using Phillips passivator.

TABLE 3 Effect of Hydrodesulfurization on Yields from Heavy Oil Cracking

Charge Stock	Untreated Atmospheric Tower Bottoms	Hydrodesulfurized Atmospheric Tower Bottoms
Inspections:		
Gravity, °API	16.1	23.3
Carbon residue content	8.6	4.2
Aniline point, °F	161	202
Sulfur, %	3.5	0.3
Metals, ppm:		
Vanadium	51	9
Nickel	16	6
HOC yields at constant conversion of 77%:		
Gasoline, C_5–430°F TBP, vol.%	52.6	65.0
C_4–C_4^2, vol.%	15.5	16.0
C_3–C_3^2, vol.%	11.2	8.0
Catalytic gas oil, vol.%	23.0	23.0
Total, vol.%	99.3	112.0
Coke, wt.%	12.4	8.0
C_2 and lighter, wt.%	4.2	1.4
Gasoline octane ratings:		
Research, clear	92.7	89.3
Motor, clear	78.4	79.0

TABLE 4 HOC Feedstock Categories

Category	Carbon Residue (%)	Metals (Ni + V) (ppm)
1 (good)	<5	<10
2 (medium)	5–10	10–30
3 (poor)	10–20	30–150
4 (bad)	>20	>150

4. Feedstocks with metals content about 150 ppm and carbon residues about 20%: The economics of HDS pretreatment for such feedstocks are very poor because of low catalyst life. At this time, such poor quality reduced crudes can only be charged to either delayed or fluid coking with the subsequent coker gas oils being charged to an FCC unit, preferably after hydrotreatment. The recently announced Englehard ART Process (see Appendix) is an alternative to coking for these very poor feedstocks. Subsequent coker or ART gas oils would then be hydrotreated and charged to an HOC or FCC unit. The ART process is much more selective in removing asphaltenes and metals than the normal coking processes and may receive serious consideration for processing poor quality feedstocks.

To summarize, feedstocks can vary over a wide range (see Table 4) because current HOC technology includes use of advanced catalysts, metal passivation, riser cracking, modern regenerator heat removal, various HDS technologies, and increased conversion efficiency. It provides the refiner with a flexible upgrading process which can be the answer to many bottom-of-the-barrel processor problems caused by heavy crudes containing high amounts of metal, sulfur, or asphaltenes.

HOC Unit Design

In the basic configuration of an HOC converter (Fig. 1), the disengager is directly above the regenerator with an external vertical riser. The essentially vertical flow of solids provides an extremely stable unit with minimum erosion problems. The major problem in designing heavy oil cracking units is avoiding extremely high regenerator temperatures which cause excessive catalyst deactivation and equipment problems, and lead to high amounts of thermal cracking in the reaction system.

Regenerator bed temperatures are limited to around 1300°F, and feed introduction systems are designed to provide efficient mixing of oil and catalyst and rapid quenching of catalyst temperature to the equilibrium mix temperature. Previous experience in FCC units operating at high regenerator temperature conditions has shown that excessive thermal cracking, which leads to high light hydrocarbon production, can occur if catalyst temperatures

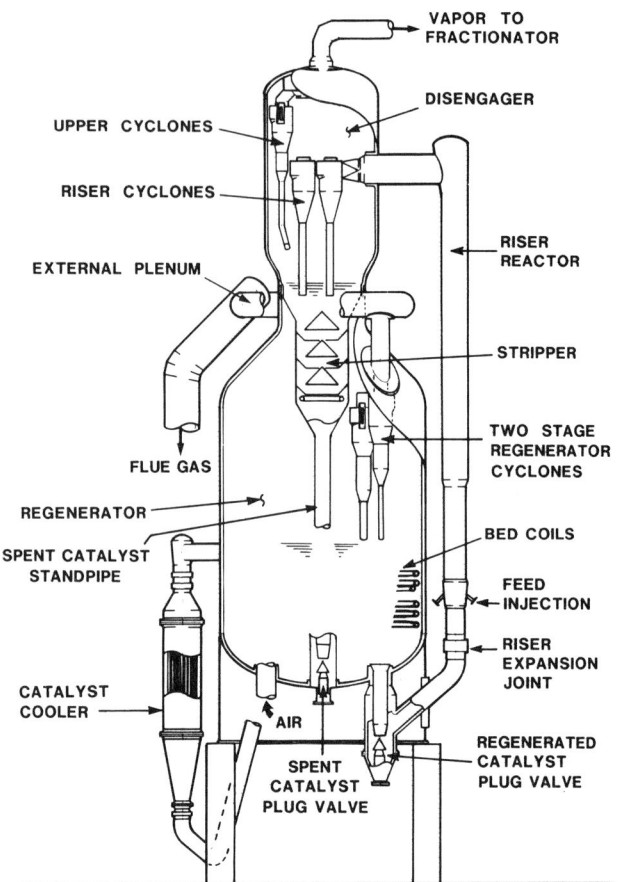

FIG. 1. Kellogg HOC converter with complete CO combustion regeneration.

are not rapidly lowered. In cracking of reduced crudes, an additional means, i.e., high dispersion steam rates, is used to further promote mixing and vaporization of the feedstock. The reaction system is an external vertical riser providing very low contact times and terminating in the riser cyclones for rapid separation of catalyst and vapors. An additional problem, somewhat more pronounced in cracking of reduced crudes, is the tendency for coking to occur on the cyclones and disengager walls. Specialized systems are provided to purge and sweep the hydrocarbons from this area rapidly.

A conventional baffle stripper is used to remove hydrocarbons occluded in the voids in and between the catalyst particles. In the regenerator the coke is burned off in countercurrent regeneration which avoids high initial burning rates and excessive particle temperatures. The high amounts of coke produced in cracking of reduced crudes can cause extreme temperatures and excessive catalyst deactivation. The Kellogg heavy oil cracker design utilizes steam coils located within the regenerator bed and/or external catalyst coolers to remove the excess heat produced by the high coke yields. In cases where a high degree of flexibility is required, the lowest heat load, as dictated by the best feed, is satisfied by the steam coils located within the bed. The additional heat removal requirements up to the maximum are satisfied by external catalyst coolers wherein the catalyst circulation rate is controlled to adjust the heat removal. The steam coils located within the bed are usually simple hairpin coils. The coil enters the regenerator at the wall and runs along the periphery of the vessel, makes one or more turns, and exits the other point at which it entered. The number of coils used depends on the excess coke and the amount of steam to be made. The steam production in these coils is generally in the order of 10–15 lb of steam/lb of excess coke.

Equipment design plays an important role in HOC operational efficiency. For example, one catalyst cooler design uses a single-pass fixed tubesheet exchanger. Special ceramic ferrules at the tube inlets and outlets provide erosion protection at these points. Using carefully designed transition sections at the cooler entrance and exit assures even catalyst distribution across the cooler. The heat removed by the cooler may be operated from zero duty up to full design heat removal.

Heat Removal System

The major problem in designing for catalytic cracking of reduced crudes is providing for the heat removal developed from processing these high coke-former stocks. Regenerator temperature is generally a limiting variable from both catalyst stability criteria as well as equipment limitations. In general, FCC units cracking clean gas oils require about a 5% coke yield to provide the heat necessary for vaporizing feed, for carrying out the cracking reaction, and for heating air and process steam. Safe regenerator temperatures can be achieved at these conditions. The 5% coke is that associated with the cracking reaction. In reduced crude cracking, an additional amount of coke essentially equal to the carbon residue is produced. Additional coke results from adverse dehydrogenation reactions caused by the metals poison. This latter effect is lowered with the use of metal passivators.

The heat removal requirement in cracking reduced crudes can be reduced by first minimizing coke yield. Above this, means such as direct heat exchange with flowing streams of indirect heat exchange through steam coils or waste heat boilers are required. Lowering coke yield at a given conversion level is achieved by adjusting operating variables in the manner shown below:

1. Decreasing oil partial pressure
   (a) By lowering total pressure
   (b) By increasing dispersion steam rate
2. Increasing activity and lowering contact time
3. Improving oil-catalyst contact
4. Increasing temperature, decreasing catalyst-to-oil ratio

5. Lowering or eliminating heavy oil recycle
6. Improving catalyst selectivity by higher makeup rates and by use of a passivator

Pressure has made a full swing in its use as a design parameter in catalytic cracking. In the early days of FCC operations, pressure was kept low (at about 10 lb/in.^{2}gauge) because of equipment limitations and the benefits of low air blower horsepower requirements. Pressure was gradually increased to about 30–35 lb/in.^{2}gauge to obtain the benefits of increased oxygen partial pressures on regenerator catalyst. Along with higher pressure, regenerator temperatures have risen considerably (about 200–300°F) because of complete CO combustion and more stable catalysts. The much higher regenerator temperatures have offset the need for high oxygen partial pressure. The current trend, particularly in reduced crude operations, is to lower pressure to obtain lower coke yields and less heat release. Lower oil partial pressure can be realized by lowering total pressure and/or increasing dispersion steam rate. A decrease of 10 lb/in.^{2} in oil partial pressures provides about 0.6–1.5% less coke at otherwise constant conditions.

Catalysts

Special catalysts are required for HOC units because of the required specifications for activity and selectivity. Most catalyst suppliers are providing customized catalysts for HOC units. Some have incorporated a greater zeolite content or they have pore structures which avoid trapping large molecules and causing coke production.

Poisons such as sodium and vanadium accelerate the deactivation rate of catalyst. High amounts of sodium are usually avoided by double desalting of the crude. Means of avoiding deactivation due to vanadium have become the major research study of the catalyst suppliers. Some studies indicate that vanadium forms eutectics and migrates to the zeolite site, causing deactivation. Increasing temperature accelerates deactivation by this means. Thus designs for low temperatures result in higher equilibrium activities and lower catalyst costs. The optimum combination of temperature and vanadium concentration is a continuing study. Many researchers feel that up to 10,000 ppm of nickel plus vanadium on catalyst and temperatures of 1350°F and less, with an equilibrium activity of about 65 to 70, is ideal for HOC units. The catalyst makeup rate to achieve this metals concentration can be readily determined by a simple metals balance. For a feedstock of 30 ppm nickel plus vanadium, a makeup rate of metals-free catalyst equal to 1.0 lb/bbl would be required. This makeup can be a mixture of fresh plus equilibrium catalyst, as long as a desired microactivity of 65 to 70 is being achieved. The nickel contaminant, although it does provide a slight deactivation, primarily affects the selectivity of the catalyst by increasing hydrogen and coke yield. Antimony passivators developed by Phillips Petroleum mitigate the effect of

TABLE 5 Ten Month Operating Comparison Before and After Passivation with Phil AD CA from Phillips Petroleum

	Percent Change of Mean Values for 10 Months Before and After Passivation
Fresh feed rate, bbl/d	+11.9
Conversion, liquid volume %	+5.4
Gasoline produced	+20.4
Gasoline octane Research Octane Number, clear	+1.3
Coke, wt.% fresh feed	−9.1
Hydrogen, SCF/bbl converted	−61.2

nickel on selectivity. The improvement is substantial, and use of passivators has received widespread acceptance. It is generally incorporated as a design parameter for both new units and revamps. Table 5 shows the improvement in operation of the Phillips Borger HOC unit due to use of the antimony (Phil AD-CA) passivator.

Process Information

While customized catalysts can improve HOC operations, optimization can also take place by closer process control. For example, high cracking activity, if used correctly, can override the adverse dehydrogenation activity of the metals' poisons. Thus, low contact time in risers, along with rapid and efficient separation of catalyst and oil vapors, is preferred in reduced crude cracking.

In recent years, with the advent of complete CO combustion and very high regenerator temperatures, the need for highly efficient distribution of feedstocks over the catalyst has been demonstrated. The need for good distribution is equally is not more important in reduced crude cracking as it is in gas oil cracking. This has resulted in development of specialized feed introduction systems that efficiently disperse the feed and allow each particle of catalyst to do its share of the work.

Recycle of heavy catalytic gas oil or slurry oil produces more coke and heat release than it removes through vaporization, and therefore should be minimized or eliminated.

In addition to reducing coke yield, lower regenerator temperatures can be realized by direct heat removal or lower heat generation. Some methods are:

1. Incomplete combustion of coke by limiting air rate to give high CO/CO_2 ratios.
2. Use of dilute phase spray water systems (not usually recommended).
3. Use of excess air at complete CO combustion conditions.
4. Use of water partially or wholly in place of dispersion steam (heat of vaporization).
5. Increasing regenerator temperature, i.e., flue gas temperature—heat is removed in flue gas.
6. Light oil recycle.
7. Low feed preheat temperature.

Indirect heat removal systems in general involve steam generation. Heat removal by steam generation can be applied in various areas of the unit. These are:

1. In slurry reflux steam generators by the use of light gas oil recycle. Light oil recycle streams are preferred since they remove more heat through vaporization than they input coke production.

2. In the flue gas waste heat boilers by operating with excess air and by increasing regenerator flue gas temperature.
3. Internal bed coils for fixed heat removal. Such systems require continuous rather than intermittent service because of the problems associated with temperature shocking of the steam coils.
4. External catalyst coolers to provide a variable heat removal system. These involve circulating a portion of regenerator catalyst downward through a standpipe and control valve in dense phase flow, and then upward with air in dilute phase flow through an exchanger generating steam. The amount of heat released is controlled by the catalyst circulation rate through the external exchanger.

Keep in mind that the high carbon residue of most residual oils results in a coke yield in excess of that required to supply the heat needed for the process. This excess heat, which is produced in the regenerator, must be removed if regenerator temperatures are to be kept within reasonable limits.

A cracker processing an average residual feed without regenerator heat removal would experience a regenerator bed temperature in excess of 1400°F. This high temperature would result in severe catalyst deactivation. This is shown in Fig. 2. This figure, drawn from actual deactivation data on catalyst containing 6000 ppm nickel plus vanadium, shows that as temperatures increase above 1300°F in a hydrothermal environment, catalyst activity decreases rapidly. Most important of all, the higher regenerated catalyst temperature will result in a low catalyst-to-oil ratio which will reduce liquid

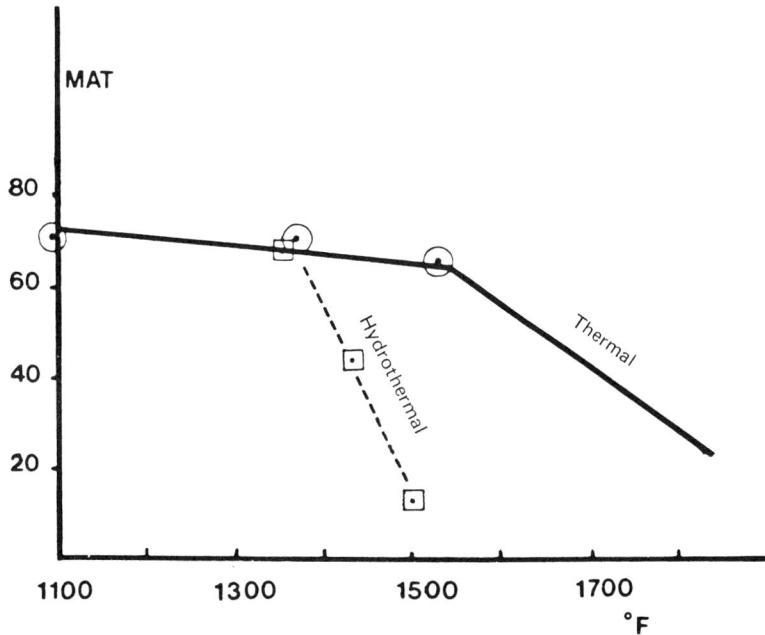

FIG. 2. Catalyst activity vs regenerator temperature.

yields and selectivity due to thermal cracking. In the meantime, the higher catalyst temperature produces high dry gas yields.

Catalyst activity can be maintained at higher temperatures in a thermal (dry) environment. A dry environment is, however, difficult to achieve in a catalytic cracking regenerator since the coke contains some hydrogen. It may be possible to reduce the quantity of steam in contact with high temperature catalyst by using two-stage regeneration. This would minimize the catalyst deactivation. Two-stage regenerators do not, however, operate in complete CO combustion, and therefore may require a CO boiler to meet environmental constraints. In addition, two-stage regenerators are mechanically complex and are difficult to control. These factors add to the overall cost of the unit as well as to the potential for serious operational upsets.

Many of the older FCC units employ two-stage catalyst regeneration. At the time these units were built, this was the most modern regeneration scheme available. Since that time, however, regenerator technology has advanced. The modern converter uses single-stage complete CO combustion. This design provides a mechanically simple system which operates in a very stable mode. Under complete CO combustion, there is little risk of excessive afterburn since all of the carbon monoxide is consumed in or near the regenerator bed.

For applications where regenerator heat removal is necessary, the regenerator is fitted with appropriate heat removal equipment. This equipment may consist of bed coils and/or external catalyst coolers.

Bed coils (Fig. 3) consist of layers of simple hairpin tubes located near the

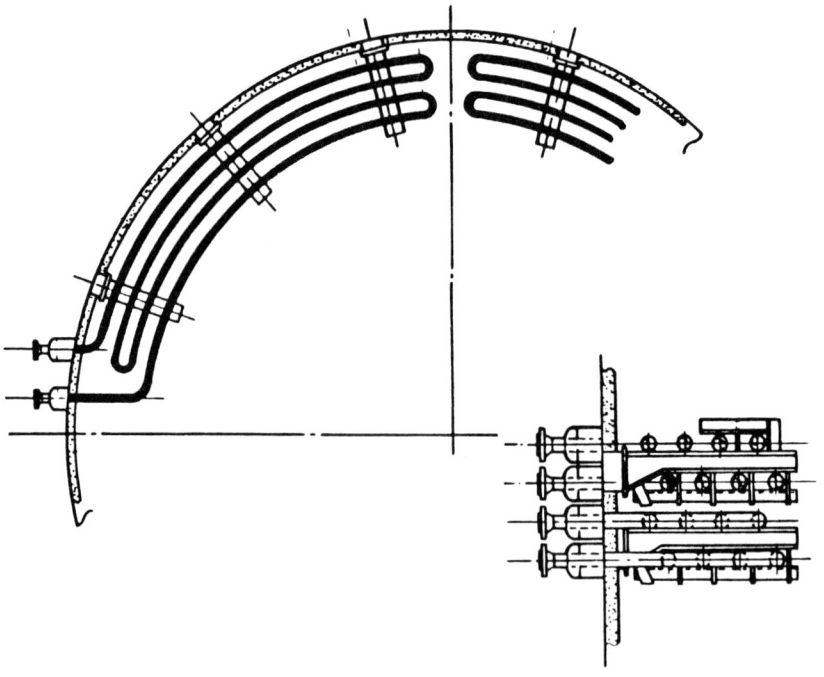

FIG. 3. Bed coils.

regenerator wall. Boiler feedwater is circulated through the tubes for production of high-pressure steam. Bed coil installations such as this have been in service for over 20 years with very good results. Erosion by catalyst is negligible and is not a factor in the service life of the coils.

The bed coils are essentially constant heat removal devices; the amount of heat removed is proportional only to the temperature driving force between the bed and the steam. Since regenerator temperature should be kept between 1275 and 1325°F to maintain high catalyst activity and achieve complete CO combustion, temperature is not a practical control variable for bed coil heat removal.

Where refiners desire high operating flexibility for changing throughput, external catalyst coolers are used in place of regenerator bed coils. A unit equipped with catalyst coolers can operate with feedstocks ranging from traditional vacuum gas oils up to the heaviest design feedstocks. The catalyst coolers provide positive control over the unit heat balance and allow the refiner to select the optimum operating conditions for any feed and operating objective.

The catalyst cooler (Fig. 4) works by withdrawing catalyst down a standpipe and across a slide valve. Air injected at the base of the cooler carries the catalyst in dilute phase through the vertical tubes of the cooler and back into the regenerator. Catalyst is cooled by generating high-pressure steam on the shell side of the exchanger. Heat removal control is obtained by throttling the flow of catalyst to the cooler.

Modern catalyst coolers have been much improved over the units installed in early cat crackers during the 1940s. These early units suffered erosion of the inlet tubesheets and tube ends as a result of uneven distribution of catalyst at the inlet tubesheet and erosion caused by eddy currents at the tube inlet. Current designs use ceramic ferrule inserts on the inlet tube ends to allow the air–catalyst flow to become fully developed before the metal tube is reached, thereby minimizing erosion due to turbulence.

Most HOC units can use catalyst coolers only. This gives the unit the highest operational flexibility. Should a refiner wish to specify a minimum heat removal case, it is possible to design the unit with both bed coils and external catalyst coolers.

Operating Characteristics

As previously mentioned, the operating characteristics of an HOC unit on the reactor–fractionator side, are very similar to those of an FCC unit. Since the carbon residue fraction passes through the riser and stripper adhering to the catalyst, only the hydrocarbon portion of the residue is cracked, just as the gas oils in an FCC unit are cracked.

Consequently, fluctuations in resid feed rate similar in magnitude to those in an FCC reactor can be handled in an HOC reactor, while variations in resid quality lead to expected lower conversion and gasoline yield as the feedstock becomes more aromatic.

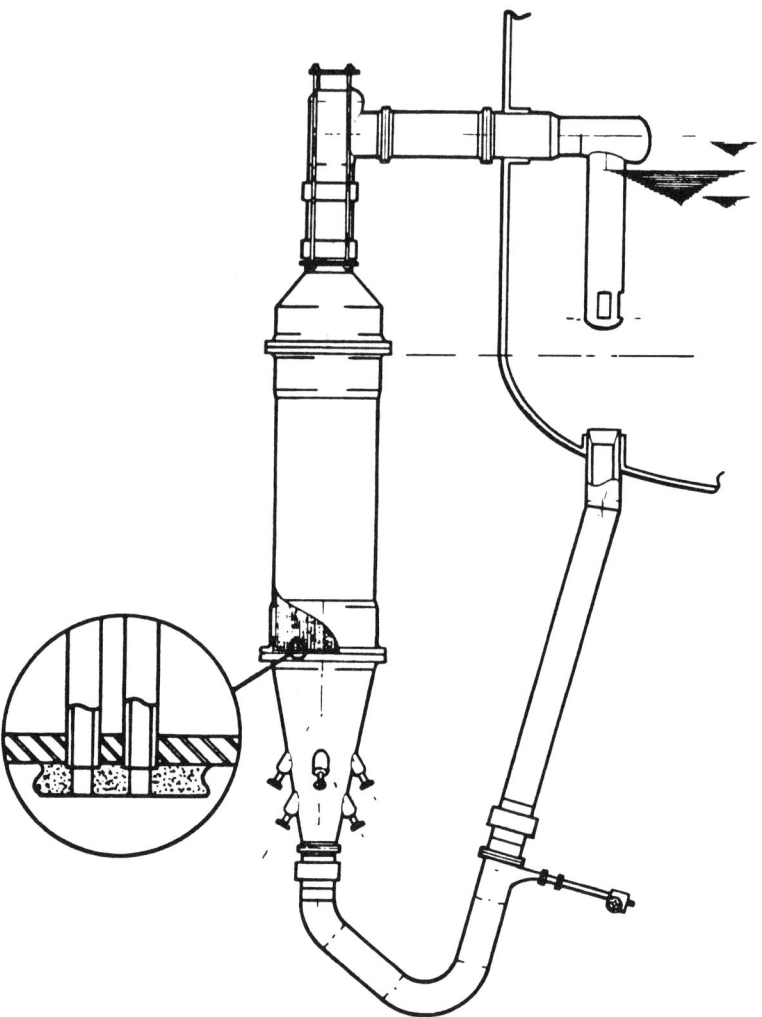

FIG. 4. Catalyst cooler.

Flexibility is required in the regenerator, to permit efficient operation with feedstocks having a broad range of carbon residue content. As the carbon residue increases, a greater degree of heat removal and coke burning capacity must be provided. Normal ranges in carbon residue content of about 2 to 3% can be handled by adjusting the air blower rate and the quantity of heat removed by the existing bed coils.

Where additional flexibility is required to handle a much greater increase in carbon residue content, an external catalyst circulating system is used. In this system the catalyst is removed from the regenerator and is passed upward with air through an exchanger where it is cooled by indirect heat exchange with water to produce saturated steam.

The run length in HOC units is upwards of 3 years, similar to that of FCC units. The only additional complexities are the steam coils in the regenerator and/or catalyst coolers. The design involves facilities to allow removal of one or more steam coils individually from the system without seriously affecting the overall heat removal capacity of the unit. Features are incorporated into the HOC design which have been proven successfully in traditional FCC units to foster long run lengths.

Flow through an HOC converter is very similar to flow in an FCC designed to crack gas oils.

The reaction system is an external vertical riser terminating in the riser cyclones for rapid separation of catalyst and vapors. Dispersion steam in amounts considerably higher than that used for gas oils are required to assist in the vaporization of the reduced crudes.

A conventional baffled stripper is used to remove hydrocarbons occluded in the voids within and between the catalyst particles. It should be noted that the carbon residue of the feedstock, since it goes essentially 100% to coke and since it is nonvaporizable, travels through the riser and stripper and into the regenerator as a film on the catalyst.

In the regenerator the carbon residue coke is burned off along with the catalytic coke. Although the carbon residue fraction must travel through the reactor and stripper as a liquid film on the catalyst, it does not affect the catalyst flow or the high quality of fluidization inherent in FCC units.

Catalyst that has been regenerated is withdrawn from the bed at a controlled rate through a plug valve which regulates flow to the riser.

Cracked hydrocarbon vapor, steam, and inert gas flow from the HOC reactor to a main fractioning tower. The net bottoms from the tower, which consist of catalyst slurry in recycle oil, can flow back to the reactor. Other fractionation products are withdrawn from the tower in the conventional manner.

Flue gas from the regenerator bed, with entrained catalyst, enters two-stage cyclones internal to the regenerator, where the major portion of the entrained catalyst is removed. Flue gas leaving the regenerator passes through an external catalyst separator for final catalyst removal.

In some instances the large volumes of flue gas developed in the HOC regenerator may justify use of an expander for energy recovery. The HOC regenerator can be designed for either partial or complete CO combustion. With partial CO combustion, a CO boiler would be used to convert carbon monoxide to carbon dioxide. The flue gas will, in some instances, require removal of sulfur dioxide before discharge to the stack to comply with effluent regulations.

HOC Equipment Features

The features of an HOC converter, starting from the regenerated catalyst standpipe, include the following.

Precise Riser Temperature Control. This is achieved by direct regulation of regenerated catalyst flow with a plug valve that is positioned by the riser outlet temperature controller.

Even Distribution of Feed on Catalyst. Catalyst is accelerated with steam at the base of the riser to impart an upward velocity before mixing with feed. The heated residuum is also mixed with dispersion steam before entering the riser. This steam serves to atomize the feed evenly over the catalyst.

All Vertical, External Riser. The even oil distribution is further enhanced by the use of a straight, vertical riser designed to minimize slippage between the catalyst and vapor phases. The riser's external location simplifies erection and provides easy access for inspection.

Right-Angle Turn. The catalyst/vapor mixture flows from vertical to horizontal in a special right-angle turn designed to minimize erosion.

Rapid Catalyst/Hydrocarbon Separation. Roughcut cyclones are directly coupled to the riser outlet to make the sharp separation of catalyst and hydrocarbon required to minimize recracking. The reaction vapors then pass through upper cyclones where all but a trace of catalyst is removed.

Steam Stripping of Catalyst. Catalyst separated from both sets of cyclones is passed through a stripping zone countercurrent to a flow of steam. The steam displaces hydrocarbon from between catalyst particles and allows diffusion of more volatile hydrocarbon from catalyst pores.

External Flue Gas Plenum. The use of an external plenum design avoids the numerous cracking problems leading to high catalyst losses encountered with internal plenums. The design also reduces areas of high local stress, permits the use of carbon steel, and makes the plenum much more accessible for inspection.

Self-Supporting Converter. The plug valves, designed and patented by Kellogg, not only provide excellent control and long wear, but their vertical action also facilitates the use of stacked regenerator and disengager vessels. This in turn allows a compact self-supporting converter design that uses a minimum of structural steel and plot area.

HOC Yields and Utilities

Previous tables gave some yield data, but this topic needs further elaboration, especially when HOC is combined with hydrocracking(HC)/hydrodesulfurization (HDS). Tables 6 through 11 give an example of resid processing using these process combinations.

TABLE 6 Properties of HDS + HOC Feedstock

Crude source, vol.% resid:	
Arabian light	6
Iranian heavy	33
Gravity,°API	17.5
Sulfur, wt.%	2.8
Nitrogen, wt.%	0.34
Conradson carbon, wt.%	7.8
Metals, wppm:	
Nickel	18
Vanadium	57
Cutpoint,°F	650

The resid feedstock, shown in Table 6, is 33 vol.% Iranian heavy atmospheric residue and 67 vol.% Arabian light residue. Table 7 illustrates the results of processing this feedstock in a HC/HDS plant to provide feed for an HOC.

The 650°F + HOC charge contains 25 ppm metals, 0.38% sulfur, and 0.197% nitrogen.

Decreased Conradson carbon and increased hydrogen content of the atmospheric resid from the desulfurizer also are beneficial to the HOC plant operation, resulting in relatively high liquid product yields and low coke yields.

The utilities for the HC/HDS unit are given in Table 8. The plant shown in the table contains two identical parallel reactor trains with common product separation. A charge rate of 60,000 bbl/d was chosen as the base.

The yields for HOC cracking of the 650°F + fraction from the HC/HDS

TABLE 7 Unicracking/HDS Product Yields, Properties, and H_2 Consumption

Hydrogen consumption, SCF/bbl feed, chemical			500
Product yields and properties:			
C_1–C_4 total, wt.% feed			0.23
H_2S, wt.% feed			2.64
NH_3, wt.% feed			0.19
	C_5–400°F	400–650 °F	650°F+
Yield, vol.% feed	0.9	14.6	85.2
Gravity,°API	52.0	30.7	20.4
Sulfur, wt.%	<0.0005	0.023	0.380
Nitrogen, wt.%	0.0019	0.124	0.197
Conradson carbon, wt.%			3.6
Pour point,°F			+80
Metals, wppm Ni + V			25

TABLE 8 Unicracking/HDS Utilities

Fuel fired, MBtu/bbl	64
Power, kWh/bbl	5.7
Steam, lb/bbl	42
Cooling water, gal/bbl (20°F rise)	77
Process water, gal/bbl	1.9

unit are given in Table 9. The yields reflect the high boiling nature and prior hydrodesulfurization of the feedstock.

The yield of gasoline of 90 Research octane, clear, is 64 vol.% at 80% conversion. The C_3–C_4 yields are low and the degree of unsaturation is relatively high due to both the high boiling material and the metals poisoning of the catalyst.

Of interest is the large amount of steam production from the steam coils. In this case 280,000 lb/h was produced at 600 lb/in.² gauge and 750°F. The

TABLE 9 Heavy Oil Cracking of 650°F + Resid from Unicracking/HDS[a]

Heavy oil cracking charge, BPSD	51,000
Conversion, vol.%	80
Product yields, vol.%:	
C_5–400°F ASTM EP gasoline	64.0
Butane–butene:	13.5
Isobutene	3.5
n-Butane	1.0
Butenes	9.0
Propane–propylene:	8.4
Propane	1.8
Propylene	6.6
Light catalytic gas oil	14.6
Heavy catalytic gas oil	2.4
Fractionator bottoms	3.0
Total	105.9
C_2 and lighter, wt.%	4.2

Steam produced in regenerator coils, lb/h = 280,000 at 600 lb/in.² gauge, 750°F

Product Properties

	Gasoline	Light Catalytic Gas Oil	Heavy Catalytic Gas Oil	Fractionated Bottoms
Gravity, °API	56.6	17.8	4.2	0.4
Sulfur, wt.%	0.05	0.40	0.94	0.94
Octane no.:				
Research, clear	90			
Motor, clear	78			
Fresh catalyst makeup rate: lb/bbl = 1.0				

[a] Of a mixture of light Arabian and heavy Iranian residua.

TABLE 10 Heavy Oil Cracking Unit Utilities[a] (Steam: 1000 lb/h)

600 lb/in.² gauge, 750°F:		
Production:	Regenerator coils	280
	Flue gas waste heat boiler	260
	Total	540
Consumption:	Air blower steam turbine	241
	Gas compressor steam turbine	32
	Let down to 150 lb/in.² gauge	8
	Other	2
	Total	283
Export		257
150 lb/in.² gauge 450°F:		
Production:	Slurry boiler	40
	Let down from 600 lb/in.² gauge	8
	Total	48
Consumption:	Converter process	6
	Turbine driven pumps	42
	Total	48
50 lb/in.² gauge 300°F:		
Production:	Turbine exhausts	42
	Blowdown	4
	Total	46
Consumption:	Process	13
	DEA stripper	32
	Surface condenser ejector	1
	Total	46
Boiler feedwater (gal/min)		1200
Condensate return (gal/min)		603
Electrical consumption (kYA)		1090
Cooling water, 25°F rise (1000 gal/min)		42
Washwater (gal/min)		106

[a]For converter, fractionator, vapor recovery unit—no power recovery, complete CO combustion, maximum use of water coolers.

catalyst makeup required to maintain a 0.5% vanadium content on the catalyst is 1.0 lb/bbl.

Table 10 shows the utilities for the HOC unit. About 540,000 lb/h of steam is produced in the regenerator coils and the waste heat boiler in the flue gas line.

In this unit, without power recovery, about half this steam is used to drive the air blower and gas compressor with the remainder being available for general refinery use. This is a significant percentage of refinery steam requirement and demonstrates the need for careful integration of the HOC steam production with overall refinery steam requirements.

Noteworthy is that this steam is produced by burning the poorest portion of the crude, i.e., the carbon residue, in the HOC regenerator. The combined yields for the Unicracking/HDS and the HOC units are shown in Table 11.

As can be seen, essentially complete conversion of the 650°F + residuum

TABLE 11 Overall Yields for Unicracking/HDS Plus HOC (basis: fresh feed of 17.5°API, 650°F + resid—67% light Arabian + 33% heavy Iranian by volume; all percentages are based on fresh resid feed to HDS)

	HDS		HOC		HDS + HOC	
	vol.%	wt.%	vol.%	wt.%	vol.%	wt.%
Feed:						
Resid	100.0	100.0	85.2	83.6	100.0	100.0
H_2 at 500 SCF/bbl		0.8				0.8
Total		100.8				100.8
Product yields:						
C_1–C_4		0.2		14.6		14.8
H_2S		2.6		0.1		2.7
NH_3		0.2				0.2
Coke				8.2		8.2
C_5–400°F ASTM	0.9	0.8	54.5	43.2	55.4	44.0
400–650°F ASTM	14.6	13.4	12.4	12.4	27.0	25.8
Heavy cycle oil			1.9	2.3	1.9	2.3
Fractional bottoms			2.4	2.8	2.4	2.8
Total		17.2		83.6		100.8

	HDS		HOC		HDS + HOC	
	°API	wt.% S	°API	wt.% S	°API	wt.% S
Product properties:						
C_5–400°F ASTM	52.0	0.0005	56.6	0.05	56.5	0.05
400–650°F ASTM	30.7	0.023	17.8	0.40	24.5	0.21
Heavy cycle oil			4.2	0.94	4.2	0.94
Fractional bottoms			0.4	0.94	0.4	0.94

has been achieved. The yield of 650°F + product is only 4.3 vol.% of the starting material. The conversion has been primarily to gasoline (55.4% of atmospheric resid) and to light catalytic gas oil (27.0% of atmospheric resid). These materials have sufficiently low sulfur contents to be used directly as blending components of marketable products.

Retrofitting a Refinery to Include HOC

Used in conjunction with atmospheric residual hydrodesulfurization or other pretreatment processes, HOC is an attractive economic alternative to other bottom-of-the-barrel processing. This is especially true when a refinery has unused FCC capacity that is suitable for and available for HOC retrofit.

The combination of unused FCC capacity, plus the high cost of new reduced crude cracking units, has placed considerable emphasis on revamping units designed for gas oil cracking to include all or part of the vacuum tower bottoms as charge. Since many FCC units run at reduced capacity, there is usually sufficient room in the fractionator and vapor recovery unit to handle the additional liquid and vapor loads. The major change required is modifying the regenerator to permit burning the higher amount of coke and to satisfy the increased heat removal requirements.

The modification of FCC units to handle portions of reduced crude can be classified as follows:

1. No major changes in equipment. Bottoms are added to gas oil fed up to a regenerator temperature or gas compressor limitation. A passivator and/or specialized catalysts are used. Increased steam rates and sometimes water are used to improve feed vaporization and lower the partial pressure.
2. Further increase in amount of bottoms so that regenerator temperature limitations are exceeded. The modification required is to insert steam coils into the regenerator and/or include external catalyst coolers. Additional air blower capacity may be included to give a regenerator superficial velocity higher than normal, i.e., up to about 3.5 ft/s. Oxygen-enriched air is used in some cases to give an increase in coke burning capacity at the same regenerator velocity.
3. Full modification of FCC units to heavy oil cracking. This can be accomplished by the installation of another vessel in series with the regenerator. Figure 5 shows this modification.

The spent catalyst coming from the stripper flows to the existing regenerator where the carbon on catalyst is partially burned off at the regenerator blower limitations to produce a flue gas with substantial quantities of CO so that it is routed to an existing CO boiler. The temperature in this initial regenerator will be on the order of 1200 to 1250°F. The partially regenerated catalyst is then routed to the second regenerator where the remaining carbon is burned off at complete CO combustion conditions. This second regenerator would also contain steam coils to control the regenerator temperature and produce steam. The completely regenerated catalyst would then flow to the riser for cracking of the reduced crude. The second regenerator is designed so that if the remaining facility is shut down, catalyst circulation can be discontinued with this vessel operating as a fluidized fixed bed combustor. Torch oil inserted through a specially designed system would be used as the heat source. The design is such to prevent excessive catalyst temperatures and deactivation.

A comparison of the modified FCC process versus delayed coking of vacuum tower bottoms plus FCC of the virgin and coker gas oils from light Arabian atmospheric tower bottoms is shown in Table 12. As shown, the heavy oil cracking scheme produces about 3.5% more gasoline and considerably more C_3–C_4 olefins for alkylation. Of prime importance is the 4.5% greater total liquid (C_3+) recovery. Noteworthy is the fact that in delayed coking, about 10.5 wt.% of the atmospheric tower bottoms is marketed as coke, whereas in the modified FCC process nearly the same amount is

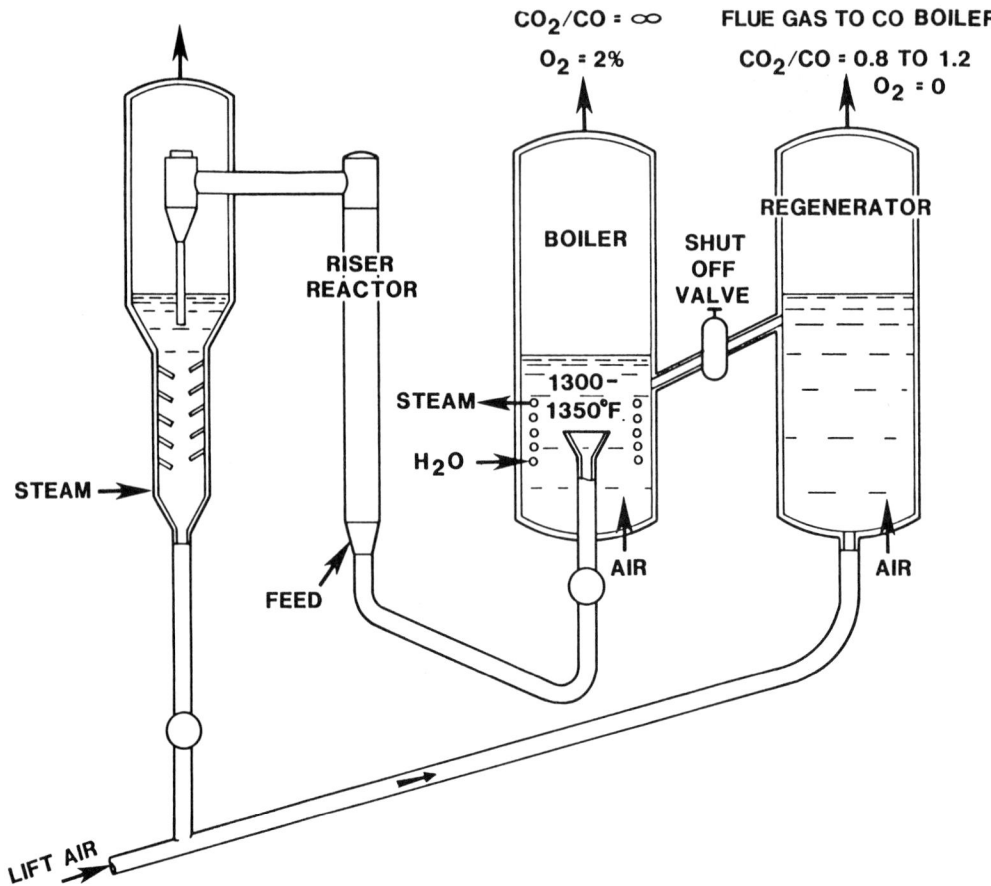

FIG. 5. Revamped FCC to HOC units.

converted to high-pressure steam for general refinery use. With the boiler modifications mentioned above, this steam will be available for operating the remainder of the refinery when the FCC unit is down. If more steam is produced than needed for normal refinery use, consideration can be given to producing electric power through steam turbines.

Orientation cost estimates have been provided for the above comparisons. For a 15,000 bbl/d delayed coker on the vacuum tower bottoms, an estimated bare cost of $30,000,000 is expected. For a converted FCC unit handling the full 50,000 bbl/d of atmospheric tower bottoms, the conversion, which would include a new riser, an air blower, the boiler vessel with steam coils, and the flue gas line with a waste heat boiler, would cost $26,000,000. Thus an initial assessment shows a higher valued product at slightly lower investment costs for the modified fluid catalytic cracking unit versus a delayed coker. Economic studies should be carried out for specific cases since the relative amounts of gas oils and vacuum tower bottoms and metals level will differ considerably and will have a significant effect on these comparisons.

Heavy Oil Cracking

TABLE 12 Comparison of Heavy Oil Cracking Versus Delayed Coking Plus FCC of Light Arabian Atmospheric Tower Bottoms

	Atmospheric Tower Bottoms	Blend of 92.8% Atmospheric + Vacuum Gas Oil and 7.2% Coker Gas Oil	Vacuum Tower Bottoms
Steam rate, BPSD	50,000	37,700	15,000
Inspections:			
Gravity °API	16.9	22.5	8.0
Sulfur, %	3.0	2.5	4.2
Carbon residue content, %	7.6	0.7	16.0
Metals, ppm:			
Vanadium	26	—	70
Nickel	8	—	24

Process	Heavy Oil Cracking	FCC	Delayed Coking	Total: FCC + Delayed Coking
Conversion, vol.%	84	75	52	72.2
Yields, vol.% of:	Charge	Charge	Charge	Atmospheric tower bottoms
Debutanized gasoline	56	62	19	52.4
C_4–C_4^{2-}	17	13	5.0	11.3
C_3–C_3^2	10	8	3	6.9
Light gas oil	13	20	30	24.1
Heavy gas oil	2	—	18	3.8
Decanted oil	5	5		
Total	103	108	75	98.5
Coke, wt.%	12	5.5	33	10.5[a]
C_2 and lighter, wt.%	3.5	4.2	3.5	4.1
Steam production (600 lb/in.² gauge), lb/h (from coils plus waste heat boiler)	540,000	100,000 from waste heat boiler		

[a] 874 tons/d of delayed coke. Total coke (FCC + delayed coke) = 15 wt.% on atmospheric tower bottoms.

Environmental and Energy Factors

The HOC process uses complete CO combustion regeneration. The SO_x in the flue gas can be reduced significantly by using an SO_x adsorption additive. This may eliminate the need for a flue gas scrubber.

SO_x adsorption additives are only effective when used in complete CO combustion regenerators. In these units, 67 to 75% if the SO_x produced in the regenerator will be transferred to the riser and reduced to hydrogen sulfide. This transfer is most effective when regenerator temperatures are below 1350°F. These conditions—complete CO combustion with regenerator temperatures below 1350°F—are exactly the conditions which exist in the HOC regenerator.

The combination of controllable regenerator heat removal and complete CO combustion results in a unit which can adjust to changing feedstock quality and operating objectives and still maintain optimum operating condition. Units without regenerator heat removal are restricted to their natural heat balances. As feedstocks change, the operating conditions (regenerator temperature, catalyst-to-oil ratio) change as dictated by the heat balance demands of the process. In residual oil cracking these uncontrolled variations in unit operating conditions can be well outside the range of optimum operation.

By using controllable regenerator heat removal, the HOC process gives the operator complete control of the unit heat balance. By adjusting the heat removed from the regenerator, he can hold the unit at optimum conditions for all feeds and operating objectives.

The use of single-stage complete CO combustion regeneration allows the HOC process to retain high power recovery efficiency even while processing light feeds or operating at reduced throughput. Power recovery expanders are constant volume machines and as such are designed for a specific flue gas rate. Should the flue gas flow decrease due to a reduction in feed rate or due to a feedstock change, the power produced by the expander will decrease. This decrease in recovered power will exceed the decrease in power required by the air blower. The resulting power shortage must be made up by either a steam or electrical driver.

The problem is solved by allowing the flue gas oxygen content to increase for light feeds or for reduced throughput. By doing this, the air flow to the regenerator can be held constant. Since the air represents over 90% of the total flue gas volume, this approach minimizes the decrease in flue gas volume. This maximizes the unit's ability to process various feeds at different throughputs without suffering an increase in utility consumption.

This technique is only possible with complete CO combustion regeneration. Any attempt to raise the flue gas oxygen level in a unit operating with partial CO combustion would result in afterburning in the cyclones and the flue gas line. This would result in severe equipment damage to the converter and possibly to the power recovery expander.

Commercial Performance of an HOC Unit

One of the latest heavy oil cracking units built in the United States is at Valero Refining Company's Corpus Christi, Texas refinery. This facility has successfully processed a hydrotreated atmospheric resid at feed rates up to 30% above design. Gasoline yields have averaged 5 to 10 vol.% above design based on

Heavy Oil Cracking

fresh feed. These yields verify the choice of HOC by Valero who compared processing alternatives on the basis of return on investment (ROI).

The Valero (formerly Saber) refinery (Fig. 6) was designed to process 46,100 BPSD, ranging in quality from light to heavy Arabian atmospheric resid. The process units include:

Resid desalter
Resid HDS
HOC
Hydrogen plant with PSA unit
HF alkylation
Dimersol unit
Citrate scrubbing
Gasoline and LPG sweetening
Sour water stripper
Sulfur plant

Properties of the design feedstocks are shown in Table 13.
Primary considerations in the design of the Valero HOC unit included:

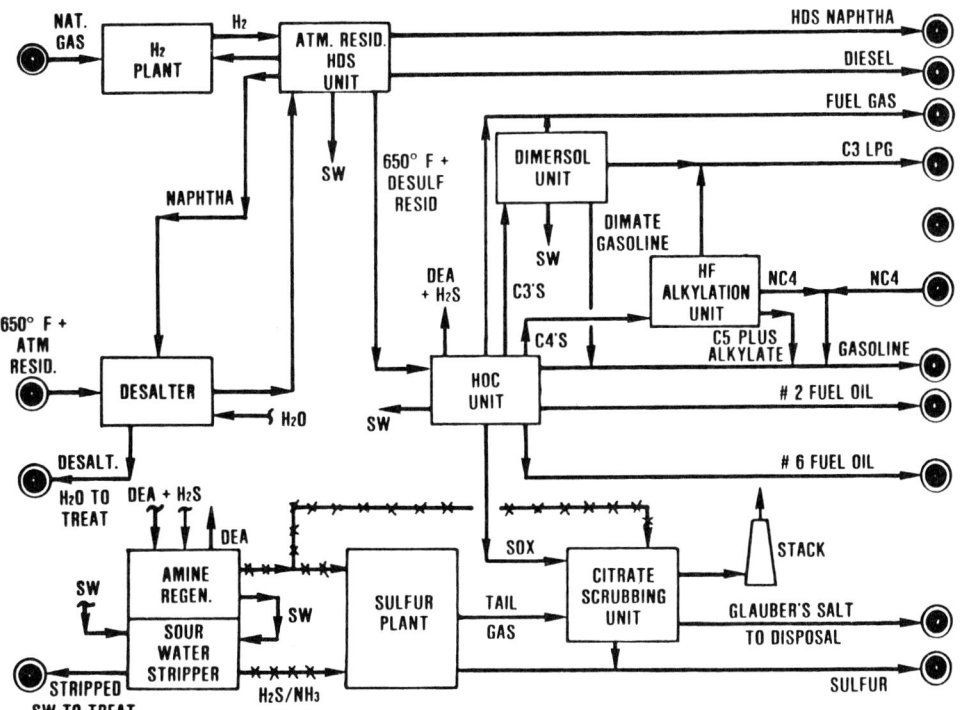

FIG. 6. Valero refinery.

TABLE 13 Valero Refining Design Feedstocks

	Light Arabian	Heavy Arabian
Feed to resid desalter:		
Feed rate, BPSD	46,100	46,100
TBP cut point, °F	650+	650+
Gravity, °API	16.2	12.6
Sulfur, wt.%	3.2	4.34
Total nitrogen, ppm	2,000	2,700
Carbon residue, wt.% (Ramsbottom)	8.3	13.3
Metals, ppm Ni + V	37.2	125
Salt, PTB	125	125
Feed to HOC:		
Feed rate, BPSD	41,500	40,000
TBP cut point, °F	650+	650+
Gravity, °API	22.9	22.3
Sulfur, wt.%	0.20	0.36
Total nitrogen, ppm	1,200	1,800
Carbon residue, wt.% (Conradson)	3.6	7.1
Metals, ppm:		
Ni	0.7	5.6
V	1.3	4.4

Complete CO combustion with promoter
Heat removal from regenerator by coils immersed in regenerator bed and external catalyst coolers
Maximum recovery of energy by use of a flue gas expander
Use of Philips antimony passivation
High C_3 recovery in vapor recovery unit.

The Valero regenerator is designed for promoted complete CO combustion in a single stage to meet environmental requirements without the need for a CO boiler. The regenerator temperature is controlled below 1300°F by the removal of heat produced by the combustion of excess coke from the heavy feedstocks charged to the unit. Too high a regenerator temperature will result in severe hydrothermal deactivation of the catalyst and poor gasoline selectivity.

An analysis of feedstock properties (Table 14) demonstrates the range of feedstock qualities experienced over the first 2 years of operation. The low API gravity feed had an intermediate carbon residue content and a higher than design sulfur, metals, and nitrogen levels. The metals and nitrogen levels in the high API gravity feed were somewhat lower than the low API gravity feed, but the carbon residue and sulfur content were considerably lower.

Tables 15, 16, and 17 contain operating and yield data for actual operation. During periods of charging the low API gravity feed, the fresh feed rate was only slightly above the design rate of 41,500 BPSD for light Arabian. Even though the feedstock nitrogen and metals were higher than design, the HOC

Heavy Oil Cracking

TABLE 14 Valero HOC Unit: Hydrotreated Feedstock Analyses

	Design Light Arabian	Low API Gravity Feed	High API Gravity Feed	Test Run
Gravity, °API	22.9	21.5	23.8	23.4
Carbon residue, wt.%	3.6	4.4	2.9	2.2
Sulfur, wt.%	0.2	0.46	0.24	0.3
Total nitrogen, wppm	1200	2000	1800	1060
Metals, wppm:				
Ni	0.7	5.4	5.9	0.6
V	1.3	9.6	5.6	2.0

TABLE 15 Valero Heavy Oil Cracking Unit: Operating Parameters

	Design Light Arabian	Low API Gravity Feed	High API Gravity Feed	Test Run
Fresh feed, BPSD	41,500	42,840	52,480	45,664
Combined feed ratio	1.05	1.07	1.07	1.05
Catalyst/oil	6.9	5.9	5.5	6.7
Catalyst circulation, short tons/min	31	30.8	34.4	36.3
Feed preheat, °F	473	488	497	510
Riser outlet, °F	980	975	975	970
Regenerator bed, °F	1,280	1,319	1,325	1,299
O_2 in flue gas, mol.%	6.9[a]	4.8	4.2	2.8
CO in flue gas, ppm	500	271	74	297
C_3 recovery, %	92			97

[a]Based on full air blower capacity to maximize power recovery when processing light Arabian resid. Design oxygen level in flue gas when processing heavy Arabian resid was 2.0 mol.%.

TABLE 16 Valero HOC Unit: Catalyst Data

	Design Heavy Arabian	Low API Gravity Feed	High API Gravity Feed	Test Run
Activity, MAT	68	65	68	73
Carbon on regenerated catalyst, wt.%	0.10	0.08	0.05	0.07
Metals on equilization catalysts, ppm:				
Ni	5000	1949	2310	1924
V		3384	2853	2435
Catalyst addition, lb/bbl	0.64	0.68	0.48	0.29
Catalyst losses, lb/bbl	0.15	—	—	0.12

TABLE 17 Valero HOC Unit: Yield Performance

	Design Light Arabian	Low API Gravity Feed	High API Gravity Feed	Test Run
Conversion to 430°F, vol.%	81.4	77.7	78.5	84.8
Conversion to 650°F, vol.%	93.7	92.6	96.3	96.7
Dry gas, SCFB	291.4	260.8	260.7	191.4
Wt.%	3.4	—	—	2.6
Propane/propylene, vol.%	2.2/9.0	2.4/7.3	2.6/8.1	2.8/8.6
Butanes/butylenes, vol.%	7.6/11.3	6.1/8.4	6.0/7.4	8.7/9.0
C_5–430°F ASTM EP gasoline, vol.%	60.6	59.1	63.2	67.1
Light cycle oil (650°F EP), vol.%	12.3	14.9	17.8	11.9
Decanted oil, vol.%	6.3	7.4	3.7	3.3
Total C_3 + liquid, vol.%	109.3	105.6	108.9	111.4
Coke, wt.%	8.6	10.9	7.5	7.6
Gasoline selectivity	0.744	0.761	0.805	0.791
Coke selectivity	0.106	0.140	0.096	0.090

yielded 59.1 vol.% gasoline at a conversion of 77.7 vol.%. During the period, while charging the high API gravity feed, the feed rate averaged 52,480 BPSD, more than 30% above the original design rate for heavy Arabian. Despite this increased throughput, the HOC gasoline yield increased.

The product quality data presented in Table 18 show that the gasoline octane numbers and the light cycle oil cetane index for all operations are within the expected ranges. Of special interest is the fact that the Cetane Index of the light cycle oil remained in the low twenties even though the yield of decanted oil fell by about 50%.

Figures 7 through 12 are plots of fresh feed rate, feed gravity, and yields for the period of April through November 1984. The open area during July 1984 was when the unit was down for its annual turnaround.

TABLE 18 Valero HOC Unit: Product Properties

	Design Light Arabian	Low API Gravity Feed	High API Gravity Feed	Test Run
Gasoline:				
Research Octane no., clear	90.3	91.9	90.4	89.9
Motor Octane no., clear	79.3	80.6	79.8	80.5
(Research Octane no. + Motor Research no.)/2	84.8	86.2	85.1	85.2
Light cycle oil:				
Cetane Index	19.0	18.5	20.1	22.3

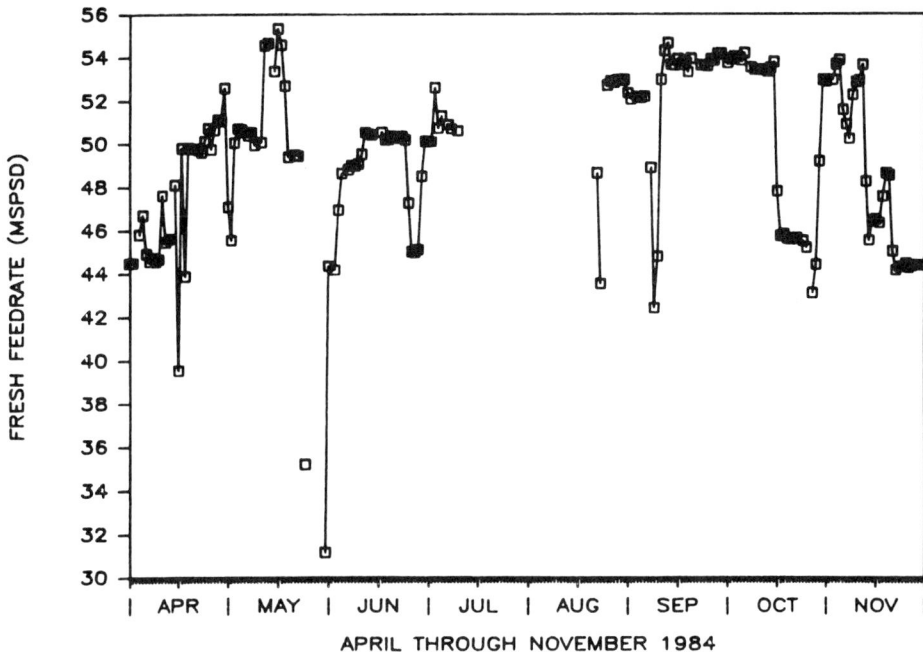

FIG. 7. Valero Refining Co.: HOC summary. Fresh feed rate.

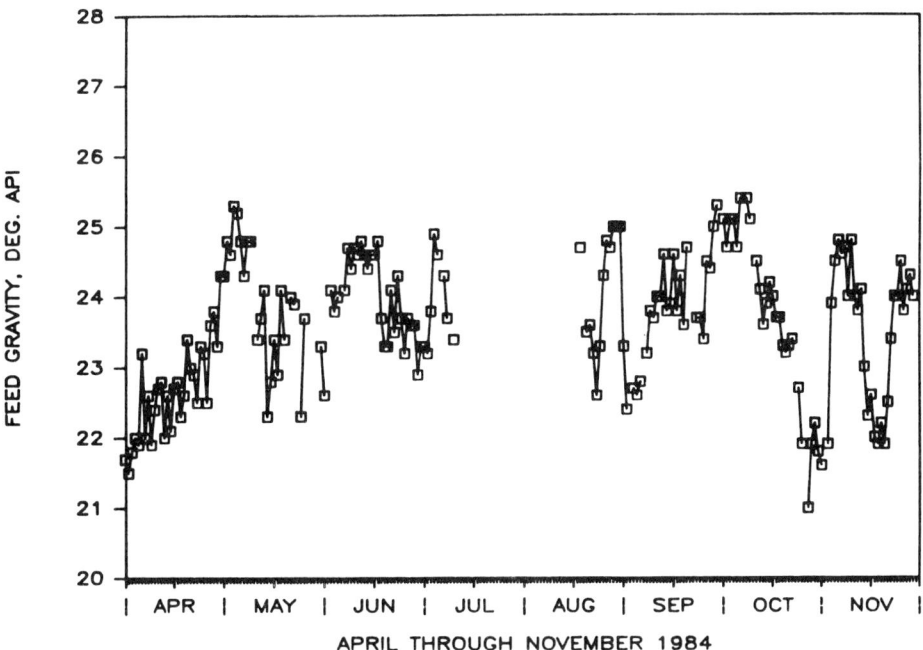

FIG. 8. Valero Refining Co.: HOC summary. Feed gravity.

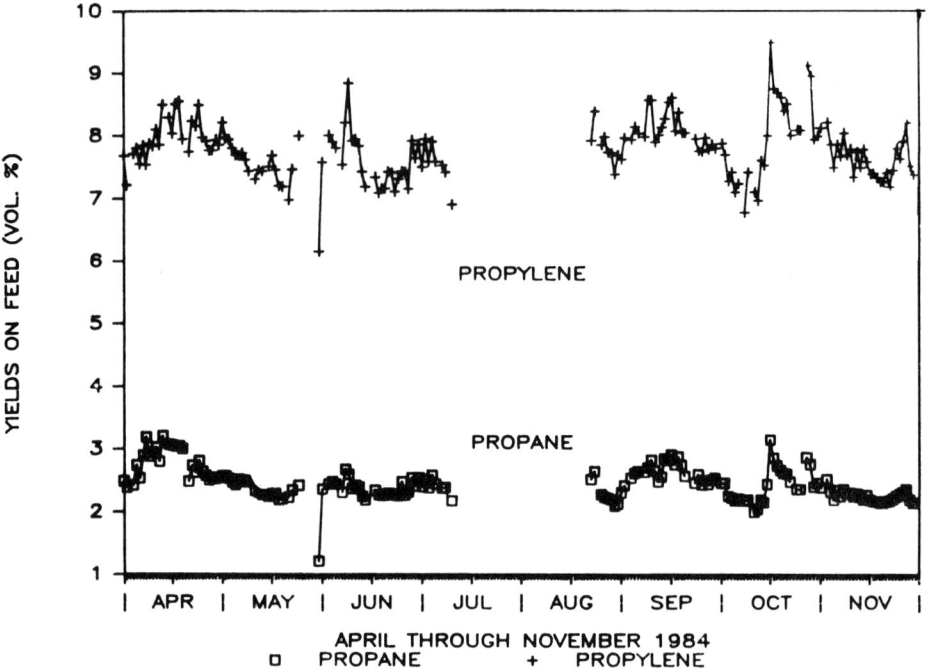

FIG. 9. Valero Refining Co.: HOC summary. Yields on feed (C_3s).

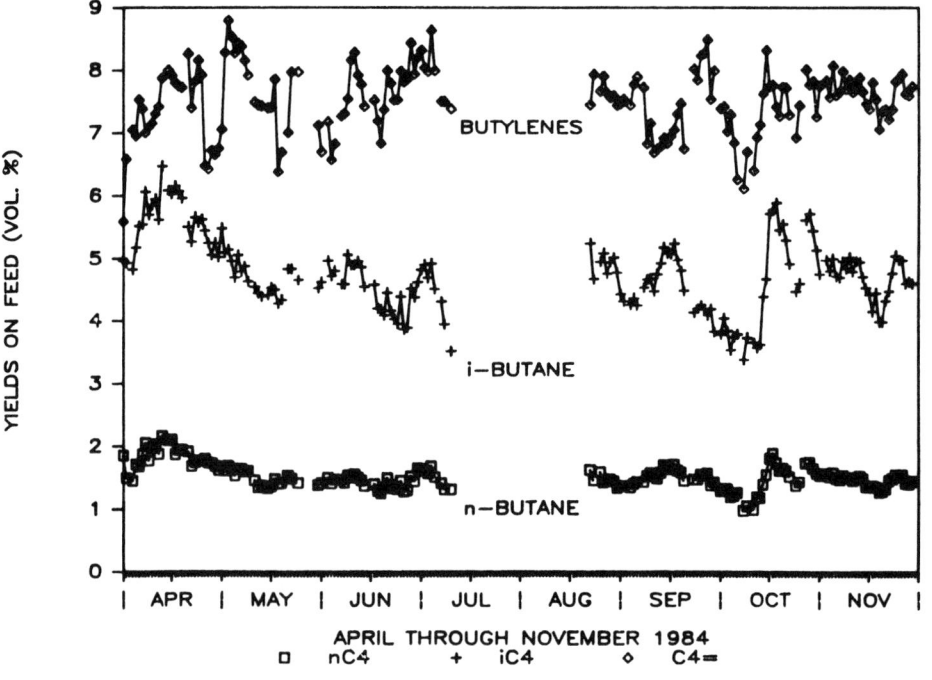

FIG. 10. Valero Refining Co.: HOC summary. Yields on feed (C_4s).

Heavy Oil Cracking

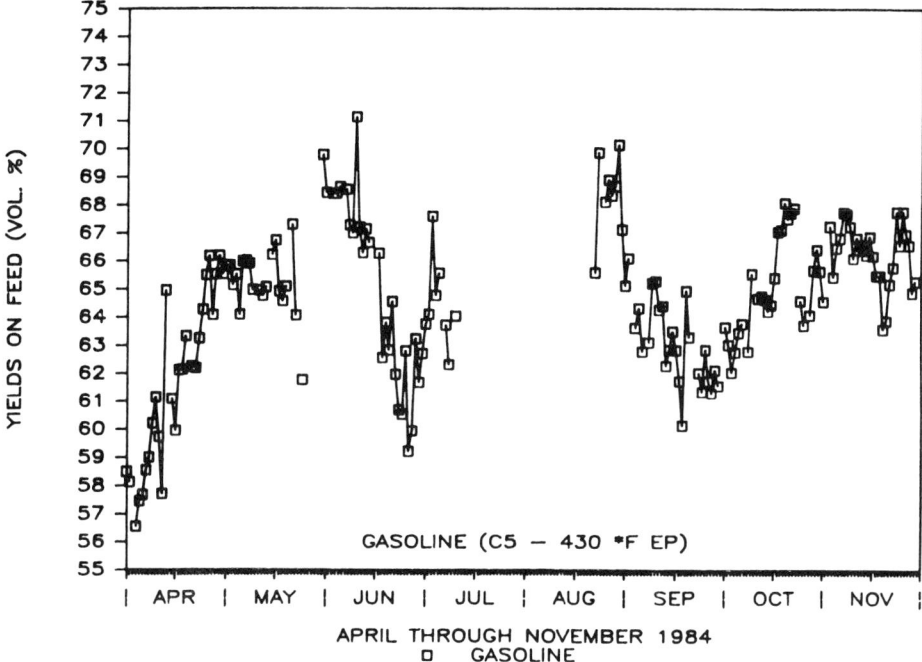

FIG. 11. Valero Refining Co.: HOC summary. Yields on feed (C_5s to 430°F).

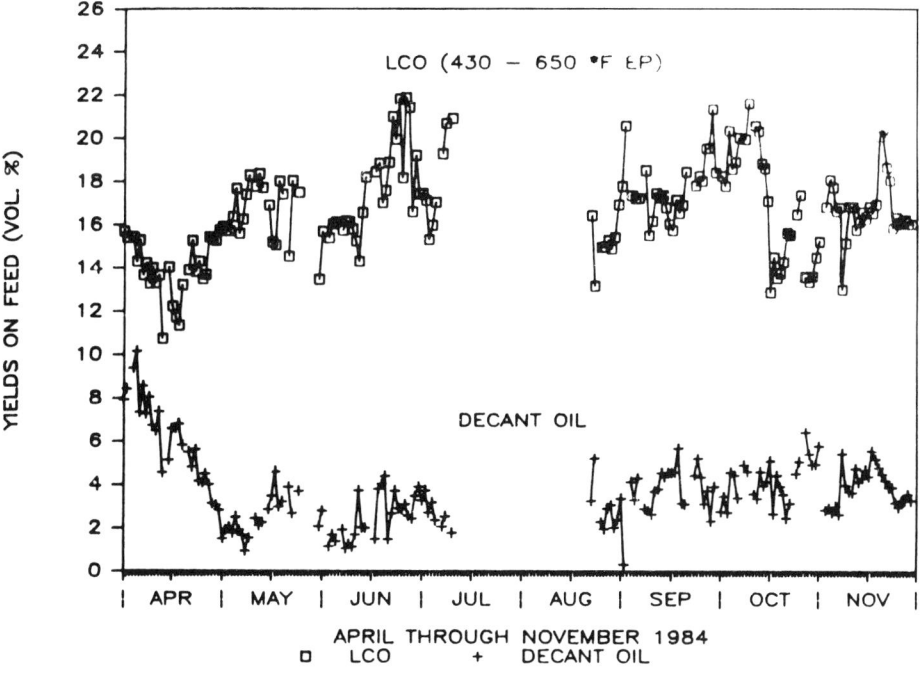

FIG. 12. Valero Refining Co.: HOC summary. LCO and decant oil yields.

Effect of Metal on Catalyst

The catalyst metal content in the Valero HOC has varied from a low of 4000 ppm to a high of 7500 ppm nickel plus vanadium. The effects of this change in catalyst metals on both the catalyst activity and on the operation of the unit are summarized in Table 19. As would be expected, increased catalyst metals resulted in a higher yield of dry gas. The catalyst activity decreased by four numbers while the vanadium level on the catalyst increased by 1567 ppm or 67%. This resistance to metals deactivation demonstrates the value of controlling the regenerator temperature. By holding the regenerator temperature in the 1300–1325°F range, Valero can minimize the harmful effects of vanadium on the catalyst.

Operations and Maintenance

The refinery was commissioned on June 9, 1983.

During June and July 1983, mechanical problems typical to start-up of a new unit were resolved. During the precommissioning period, the internal components of the plug valves were hard surfaced to improve their durability. Following this period, the on-stream factor for the HOC complex was 75% based on 390 days in operation out of a total of 520 days. Fifty-five percent of this downtime was associated with problems in the citrate unit and the flue gas cooler (FGC). The citrate unit represented new technology for SO_2 removal and was the first commercial unit in operation. Developmental problems which occurred during this period have been resolved and the citrate unit is performing very well.

Several problems external to the HOC complex contributed about 5% to the overall downtime. Much of the remaining 40% of the downtime was

TABLE 19 Valero HOC Unit: Effects of Metals on Catalyst

	Low Metals	High Metals
Metals on catalyst, ppm:		
Ni	1656	3530
V	2349	3916
Catalyst activity, MAT	70	66
Carbon on regenerated catalyst, wt.%	0.09	0.03
Conversion to 430°F, vol.%	81.7	75.1
Conversion to 650°F, vol.%	97.1	95.1
C_5–430°F ASTM EP gasoline, vol.%	68.3	62.8
Dry gas, SCFB	195.4	254.1
Hydrogen/methane mole ratio	0.90	1.19

associated with the annual turnaround in July 1984. If we calculate the on-stream factor, eliminating problems external to the HOC, the on-stream factor would be 90%.

Appendix: Asphalt Residual Treating (ART)

The asphalt residual treating (ART) process is a combination of selective vaporization and fluid decarbonization and demetallation. The configuration (Fig. 13) is similar to that of a riser FCC unit and utilizes essentially the same design principles and equipment proven in more than 40 years of fluid catalytic cracking experience. The process, developed by Engelhard Corporation, utilizes a proprietary fluidized solid called ARTCAT.

This process is a noncatalytic technological breakthrough in contaminant removal and will remove over 95% of the metals, essentially all the asphaltenes, and 30 to 50% of the sulfur and nitrogen from residual oil while preserving the hydrogen content of the feedstock. This provides greatly improved cost effectiveness by producing less unwanted by-products and consuming less energy than competing processes. The process also enables the subsequent conversion step in residual oil processing to be accomplished in conventional downstream catalytic processing units.

In the process, residual oil charge is contacted with hot ARTCAT fluidizable contact material in a short residence time contactor. In the contactor the lighter components of the feed are vaporized; asphaltenes and the high molecular weight compounds are thermally cracked to yield lighter compounds and coke. The metals present, as well as some of the sulfur and nitrogen bound in the complex unvaporized compounds, are retained on the ARTCAT contact material. At the exit of the contacting zone, the oil vapors are separated from the ARTCAT material and are rapidly quenched to minimize incipient thermal cracking of the products. ARTCAT contact material, which now holds metals, sulfur, nitrogen, and coke, is transferred to the regenerator where the combustible portion is oxidized and removed. Regenerated contact material, bearing metals but very little coke, exits the regenerator and passes to the contactor for further removal of contaminants from the charge stock.

The metals level on the ARTCAT contact material in the system is controlled by the addition of fresh ARTCAT material and the removal of spent material. A high metals level can be maintained without detrimentally affecting its performance. During commercial tests a metals level of 3.1 wt.% nickel and vanadium plus 0.6 wt.% sodium was successfully achieved. Much higher metal levels are believed possible.

Because the contact material is essentially inert, very little molecular conversion of the light gas oil and lighter fractions takes place. Therefore the hydrogen content of these streams is preserved. In other words, the lighter compounds are selectively vaporized. The molecular conversion which does take place is due to the disproportionation of the heavier, thermounstable compounds present in the residual feedstock.

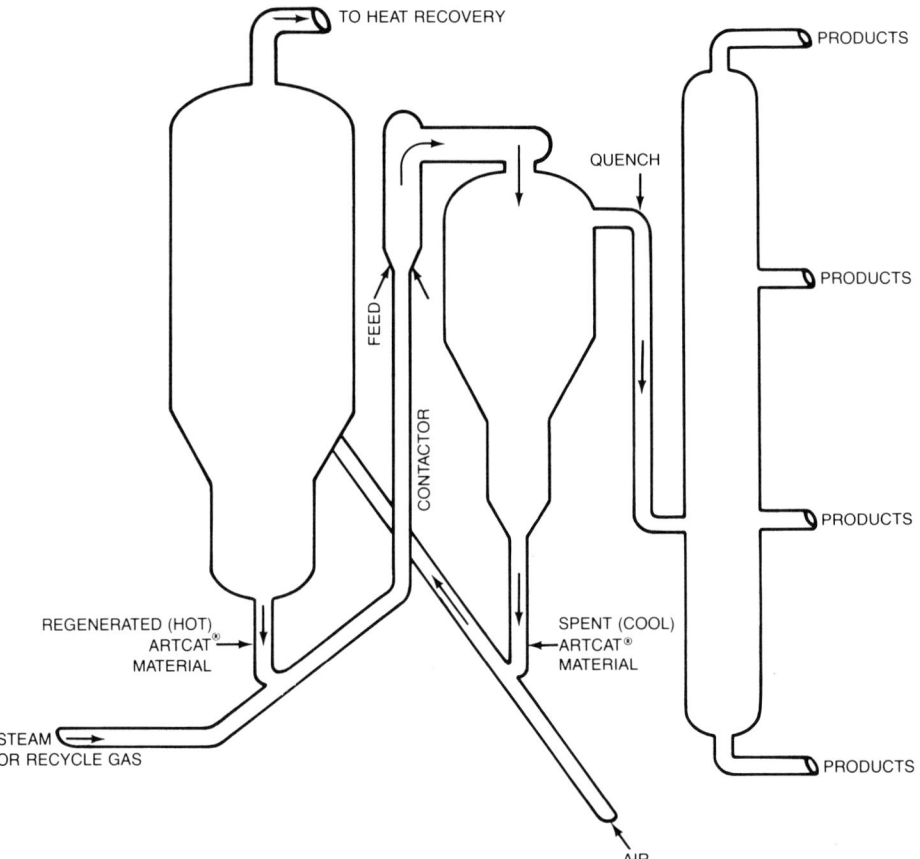

FIG. 13. The ART Process.

Coke production is typically equivalent to 80–90% of the feedstock Ramsbottom Carbon Residue content. Heat from the combustion of coke is used internally. Surplus heat is recovered as steam or electric power. No coke product is produced. In contrast, delayed and fluid cokers yield a coke product equivalent to 1.3 to 1.7 times the Ramsbottom Carbon Residue.

A small amount of gasoline and lighter compounds are produced in an ART unit. Because ARTCAT contact material is substantially inert, the cracking that takes place is primarily thermal in nature. The C_5+ gasoline has an olefin content of about 50%. The octane level is on the order of 70 Motor Octane Number and 80 Research Octane Number. The gasoline can be stabilized by mild hydrotreating to preserve olefin content; full hydrotreatment followed by reforming, or alternately, it can be charged to the base of a FCC riser prior to feed injection. ART naphtha fed to the base of a FCC riser will crack to yield approximately some LPG and a higher octane gasoline.

Finished products also contain a small amount of butadiene. However, in comparison to a refinery's total C_4 stream, the butadiene increase is very slight. Sulfur removal is moderate at 30–50%. most of the sulfur removed leaves the system with the flue gas stream in the form of SO_2 and SO_3. There are many processes to remove these sulfur oxides from flue gas.

This process is capable of treating whole crudes, reduced crudes, tar sands bitumen, or vacuum bottoms. It is a *treating* process, designed to minimize molecular conversion of the noncontaminant, hydrocarbon portion of the feedstock, thereby maximizing the hydrogen content of the ART Process' liquid products. This is accomplished while, at the same time, removing or reducing the constituents of the feedstock which would poison conventional catalysts contained in downstream conversion units. End products are therefore suitable for processing in any of the conventional downstream catalytic refinery processes such as FCC, hydrotreating, and hydrocracking. The 343°C+ ART product is no longer a resid but behaves as a vacuum gas oil. In addition, because the ART process destroys asphaltenes, the production of specialty coke from the 566°C+ product in a downstream delayed coker is a possibility.

This article was prepared from material provided by and with the cooperation of the M. W. Kellogg Company, Houston, Texas. The costs and yield reflect actual figures at the time of preparation. There have been improvements in economic and yield data since the original text was written. The author particularly acknowledges the assistance of E. Louis Whittington, Manager of Process Engineering Department; J. William Wilson, Process Manager; Richard E. Wrench, FCC Manager; Lewis Yen, Manager Technical Data; and Raymond Waters, Manager Public Relations.

GUY E. WEISMANTEL

Cracking, Catalytic, Optimization and Control

Today, because of many factors greatly affecting the petroleum industry, improved control and more efficient management of processes are mandatory. We have become accustomed in the past to normal competitive "squeezes" on profits but, while these must be served, new pressures have entered the industry and other dormant old ones have been given new life. All require greater efforts to increase productivity and maintain a profitably operative industry.

The Arab embargo on oil shipments put us face to face with the need to maximize energy recovery, minimize dependence on imports, exploit other energy forms, and generally upgrade all industrial and private conservation measures. Increased government regulation has brought about requirements to eliminate exposure to health hazards and clean up our environment. All these items require large investments of capital and considerable operating expense. Other than "trimming" up (cutting losses from leaks and spills and eliminating obvious hazards), the majority of plants has done little more to improve their operations. Further "tightening" of economic position and erosion of profits may be expected; overall operating efficiencies must be improved to minimize these effects.

This discussion is concerned with a major step toward the "ultimate" in processing and plant operating efficiency. While the improvement is developed in terms of the Fluid Catalytic Cracking (FCC) process, the philosophic general approaches and benefits can be applied to any plant or process unit. The steps outlined have proved a successful route to increased efficiency and higher profits from petroleum processes.

Catalytic cracking capacity in most refineries is second only to that of crude oil distillation units. Up to 50% of a crude oil charged is cracked in some refineries. This ranking means that greater total benefits will result from smaller improvements in the FCC process than from much larger percentage gains on less significant processes. The complexity of these units as well as capacity consideration directed earliest attention to upgrading performance of the cracking process. Each unit combines a reactor and regenerator, with fractionation and gas plant auxiliaries. Operation requires the manipulation of a large number of controlled variables having direct effects on performance. Accompanying these direct effects are interaction between variables—effects generally not understood or taken into account during manual operation.

The volumes and distribution of the wide range of products resulting from cracking are mainly functions of the controlled variables; however, external disturbances and uncontrolled variables have substantial effects. These are typically results of (1) seasonal changes in energy demands, (2) swings in feedstock quality (accentuated now by the necessity to run oils from many sources), (3) changes in catalyst quality (one result of varying crude oil supply source), and (4) abrupt changes in feedstock because of emergency conditions—either in the cracking unit or another process unit (crude still, slop tank, recycle streams, etc.).

The obvious requirement is for any process unit to produce the maximum quantity of highest valued products for minimum cost. This, for cat cracking, means that the value of the overall distribution of products must "average highest" or be the maximum. Concurrently the distribution will show generally reduced yields of lowest valued products such as hydrogen and coke.

In this article the terms improved *control* or *regulation* refer to the objective of reducing uncontrolled variation of process performance variables. *Optimizing control* is realized when a control loop drives and holds against at least one *constraint*, or limit, to operating performance. *Optimization*, or *steady-state optimization*, is used to refer to a strategy by which more than one operating (independent) variables are set to realize the "best" performance with respect to a specific operating *objective* (e.g., highest product value upgrade), simultaneously avoiding violation of any of a number of operating constraints.

The optimum operation of the cracker to maximize profits (or attain any other important objectives) requires that the unit be operated near one or more of the process equipment limits. The absolute optimum will be *on* one or more of the constraints, but the inability to completely eliminate process "noise" (i.e., the fact is that there is no "perfect" control) dictates that operation be on the safe side or somewhat away from this point. A plant operator cannot assimilate, analyze, and act on all the information necessary to hold the process nearly as close to constraint levels as desired. The operator's departure from constraint level operation is from two causes—the above noted physical inability to hold conditions closely and *his* requirement to keep the unit operating smoothly far enough from process limit to avoid hazards or upsets. He really operates within a "comfort zone" with a margin of safety—where the unit will perform without too much attention. This deviation from optima or constraint levels (limits) on the part of the operator is itself subject to considerable human variability. Each operator will find his own operating area, often resulting in three or more distinct levels of operations per day.

The approach to FCC unit control and optimization discussed here recognizes all the foregoing problems and was developed with the knowledge that an automated system which would enforce the optima would show rapid payout and enhance the total refinery profit picture. Additional side benefits accrue to other process units as well—e.g., increased alky olefins produce more alkylate, in turn adding to no-lead fuel volumes and/or decreasing overall additives required in motor fuel blends.

Development of the FCC control and optimization system is a result of over 15 years practical experience and detailed study of several commercial FCC units. Progressive improvements led to the approach to FCC control and optimization in stages—sometimes called a "three-tiered" (or hierarchical) approach. This method of achieving a total system permits progress to any desired level with each step a logical move toward total on-line optimization and control. The staged design permits liberal adaptation to any plant with justification for each succeeding stage readily available from results of work completed.

This system is diagrammed in Fig. 1. From bottom to top the relation of the three functions—computer control (I), dynamic optimization (II), and steady-state optimization (III)—are shown. The basic functions recommended for

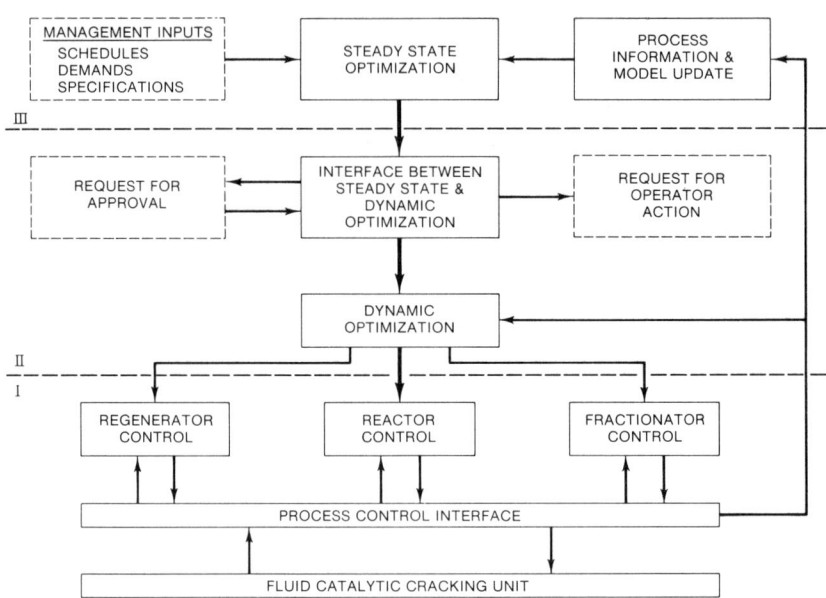

FIG. 1. Three-step approach to FCC optimization and control.

every application are computer control and dynamic optimization. On-line steady-state optimization may not be required when major changes in process operation occur no more than 2 or 3 times per week. Off-line use of the technique is recommended in this case.

Before any new or significantly different operating system can be completely successful it should be noted that the system installed must meet with operator acceptance. The old truism "nothing will work if the operator doesn't want it to" still applies. This acceptance will depend largely on the effectiveness of management preparation and, later, the surveillance program. If the operator can rely on its function to take timely, proper action and to alarm only when a serious condition exists, he will accept the system as his full operating partner. However, if the computer issues unnecessary alarms or otherwise gets him into trouble (in hot water), the operator will remove the system from computer control at the first opportunity and leave it in completely manual mode. The system prescribed has met with favorable operator reaction. Proper orientation and schooling in its operating functions provided the operator with confidence in its capabilities and the desire to make maximum use of the system.

The first indicated phase in FCC control and optimization (I in Fig. 1) is to achieve more precise regulation with correspondingly reduced process variation. The process control interface involves the reactor, regenerator, and main fractionator. Immediate attention is given to insure that all basic plant instruments are adequate. Deficiencies must be corrected and marginal items upgraded. On-stream analyzers for gas and liquid streams provide the backbone for material and heat balances as well as furnishing the individual component yields. These analyzers coupled with a sophisticated data gathering system and appropriate software provide the total system analysis "tools."

The reactor–regenerator–fractionator–vapor recovery are inseparable elements in the FCC unit system. All enter the control and optimization scheme. Regenerator and fractionator control must be adequate to compensate each reactor change. A basic premise in the present approach is that the computer must make the necessary changes on the regenerator and fractionator to compensate computer-initiated changes on the reactor. If operator intervention is required whenever the computer changes a setpoint, the fundamental purpose of the control system is defeated. Following the above premise, the result is stable, responsive regulation of the entire FCC system.

Reactor control with minimum process "noise" is of paramount importance in the total process control order. A "wallowing" reactor cannot operate efficiently and will not be amenable to the high levels of optimization we seek. Yield estimates will be highly questionable. Control loops for the reactor include riser outlet temperature and fresh feed charge rate.

Optimum FCC performance requires that carbon on catalyst be burned to low levels. This generally implies burning at the highest temperature possible without exceeding established temperature limits—commonly metallurgical limits.

Relatively long lags are involved in regenerator control; the system must compensate for these lags. Coke burning rates depend on both burning temperature and coke concentration on the catalyst. These complications introduce nonlinear relations into the regenerator air rate control systems. Feedforward prediction of coke production in the reactor is included in the control system as well as safety overrides on critical temperatures. The control loops installed on the regenerator regulate the catalyst bed temperature as well as the regenerator temperature differential.

Feedforward control equations are required in order for the fractionator to respond to changes in reactor operating conditions without operator intervention. Interactions exist in heat and material balances. Changes in heat input from the reactor and removal from the tower bottom (via slurry boilers and exchangers) affect compositions, tower loading, and draw rates. Main fractionator control loops are concerned with product specifications and these draw rates.

Experience has shown control problems assume varying importance on different FCC units. Some reactors, regenerators, or fractionators require major upgrading of control while others may be relatively stable. Throughout both regenerator and fractionator systems, control was improved by decoupling and using nonlinear predictive yield equations—coke for regenerator control and all other yields for fractionator operation. Product specifications or other operating requirements may add to the controls required. Detailed study should be given the process to insure controls are adequate for specified tasks. Controls for each unit should be evaluated by a competent control engineer experienced in FCC unit operation.

The second level (II) in the overall cat cracking control structure is labeled and named dynamic optimization. Dynamic optimization is used to refer to a specific real-time, multivariable, multiple-constrained, optimizing control strategy. This stage requires an optimization strategy as well as computer-supervised control of operations. This phase receives target values for operating variables from higher levels of optimization (steady-state) or directly from operations management. Dynamic optimization, based on a relatively simple model, continuously enforces operating policies known to be most profitable for the unit and at the same time takes maximum advantage of changes in capacities of each segment of the FCC reactor–regenerator system or the main fractionator.

The typical cat cracker frequently experiences uncontrollable changes in process conditions which in turn create unit capacity changes. Typical of these conditions are feed quality and mass flow rate, catalyst activity, equipment condition, and ambient effects. The dynamic optimization controller senses these capacity changes. If significant, a model is used to calculate the increment of change in key variables required to hold the process at limiting conditions (constraints) in the most profitable manner.

The computer adjusts key variables about every 10 min to hold the unit near optimum conditions. The rates of change are fitted to the dynamics for the specific process unit. A diagram of this function is shown in Fig. 2. Throughout this process as described, the dynamic optimizer performs all operator functions. It attends each detail of operation and enforces safe practice.

A surveillance routine is included in the dynamic optimizer which typically runs every 15 s. Its functions are to (1) detect actual or potential violations of process limits, (2) print or otherwise alert control rooms to alarm status, and (3) initiate corrective action by the dynamic optimizer. Fifteen or 20 key variables are scanned and checked for violations of limits.

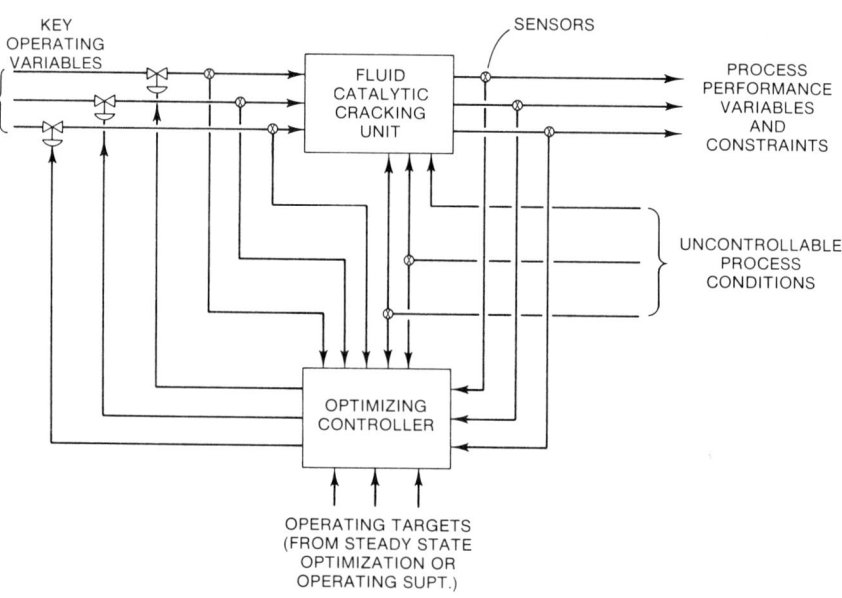

FIG. 2. Dynamic optimization functions.

Cracking, Catalytic, Optimization and Control

Three conditions are recognized by surveillance tests.

1. The process is operating safely and there is available capacity for additional feed or a higher level of conversion which can be used. A new move will be initiated to use this capacity if appropriate.
2. The process is operating safely with no room to move to a more profitable area. No action is taken.
3. The process is approaching or is operating too close to process limits, and conditions should be relaxed slightly. The dynamic optimization program is called to initiate the change to safer or more conservative operating levels. The plant operator is called by output message and/or audible alarms if action is beyond the scope of activities under computer control. A typical latter case might require starting a spare pump and switching flows as costs of automating these events may not be justified if occurrences are infrequent.

Advanced computer control of the unit coupled with dynamic optimization will gain a large part of the economic benefits available from optimizing FCC operation. The justification for implementing dynamic optimization comes from the incremental improvement brought about by both (1) continuously enforcing most profitable operating conditions and (2) taking advantage of all changes in unit capacity. Figure 3 is a plot of actual plant data showing how the value of FCC products (expressed as $/bbl fresh feed charged) varies with conversion level. Figure 4 contains plant data showing the variation in actual conversion which took place in a 36-h period. This illustrates the wide swings which occurred on this particular cracking unit with normal operator attention and conventional controls.

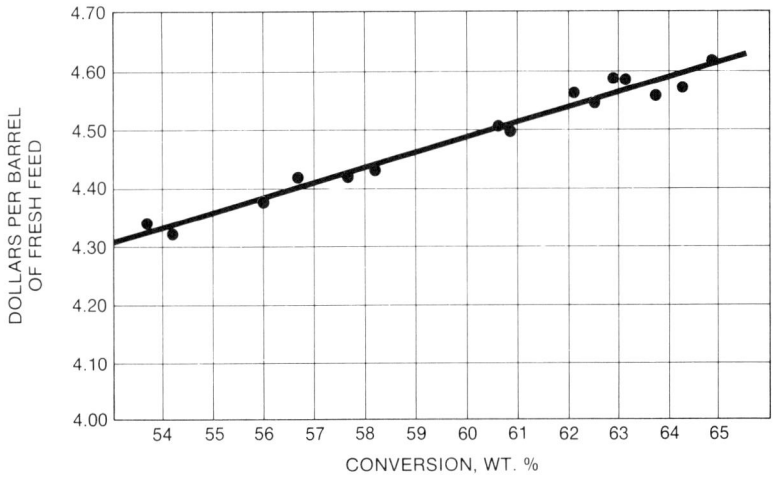

FIG. 3. Product value vs conversion, wt.% (based on 12-h calculated averages).

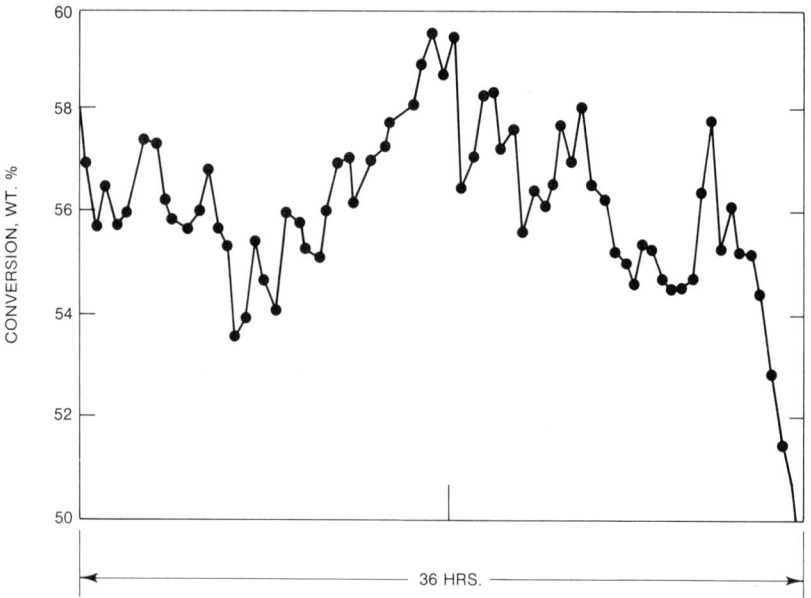

FIG. 4. Conversion, wt.% vs time.

Significant increases in profitability usually can be obtained by better regulation of conversion and maintaining its level nearer process constraints. Figure 5 demonstrates the type of benefits for improvement in operation that can be realized through computer control. Curve A is the normal distribution of conversion values from data in Fig. 4. Its mean μBC and standard deviation σBC represent the average conversion level and its variability under normal operator control of the process. Mean conversion obtained is well below the maximum possible level because of the wide variation encountered. Computer control of conversion generally reduces the variability over 40%. This resulting curve with mean μBC and standard deviation σCC is shown at B. The reduced variability permits moving mean conversion nearer the limit as shown by μCC with resultant Curve C. The new value attainable by computer control is equal to $\mu CC = \mu BC + 2\sigma BC - 2\sigma CC$. This shows how the average conversion can be increased with the computer-controlled process to provide significantly higher profitability.

The utility of this system is not limited to maintaining maximum conversion. The system is equally effective at minimum conversion levels (maximum fuel oil or minimum motor fuel). In short, no matter what constraint or limit to the process is involved, dynamic optimization and computer control move the process in a more profitable direction to the constraints while holding operations stable.

The highest level in the overall approach used to FCC unit control and optimization has been called "steady-state optimization." The function for this phase is to provide to the dynamic optimizer the operating targets and a list of constraint values presently limiting profitability for the unit. Optimum

Cracking, Catalytic, Optimization and Control

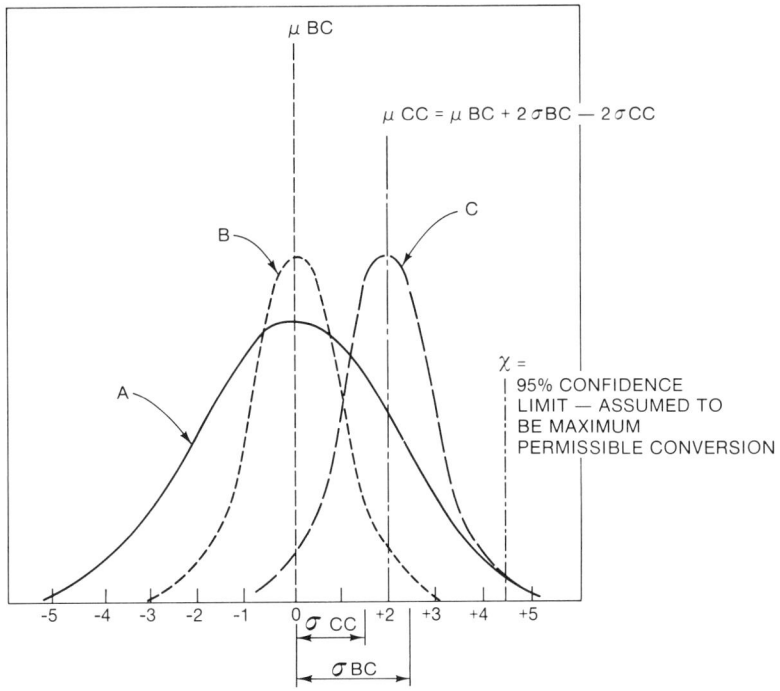

FIG. 5. Typical improvements from dynamic optimization and control of conversion.

operation of the process is always limited by constraints—air blower capacity, temperature limits, compressor capacity, fractionator overhead cooling, management requirement of minimum quantity of one or more products, product quality specifications, etc. The same constraints are not always limiting. The air blower capacity may limit because of excess coke make when processing a certain oil, while for another type of feed at maximum conversion conditions, excess gas production may exceed compressor capacity. Individual constraints usually do not change unless major processing changes take place. The steady-state optimizer is operated infrequently under these conditions—perhaps only once per day or on demand by process engineers.

This step in the procedure employs a comprehensive model of the process—one which must be "tailor-made" for a particular FCC unit and its conditions. This comprehensive model depends on material and energy balances around the reactor, regenerator, and fractionator. For reactors with multiple risers, each riser is "balanced" as well. Yields are predicted using equations containing functions of several operating variables as well as the character of feeds to the unit. These equations represent the theoretical mechanisms of cracking insofar as possible, but where theory is inadequate, empirical equations based on many years experience must be developed. The model is nonlinear and its accuracy is insured by updating from on-line yield analysis.

The most important part of development of the comprehensive FCC model and the yield equations is the data-based plant test program. The total system requires as stated earlier—a highly workable model involving many variables and process constraints. Best operating policy depends on knowing the precise relation and/or interaction the independent variables and constraints have with each other. This requires a carefully designed in-plant test program and exhaustive data analysis. Results rely heavily on precise process measurements, since small increments of difference between differences of large numbers (second derivatives) are required to evaluate the interactions. Existing measurements are generally lacking in the precision required. Data gathering other than by the on-line system is too cumbersome and filled with delays to provide the correlations needed. Large masses of data which go into the model require a sophisticated system to process the data to avoid excess man-hours in deriving a satisfactory model and yield equations. Applied Automation, Inc. has developed such highly specialized means to process data and routinely use them for analysis and correlation of all measurements and results calculated on-line. This permits much earlier and more comprehensive definition of the essential model for the optimization sectors of the system. All the "know-how" of 15 years' study experience is built into the model plus the added trimming or tailoring required to adapt it to the unique features of a customer's plant.

The test programs discussed here have always presented extra or "bonus benefits" aside from their contributions to the process model. Results of studies reveal improved modes of operation, and in some cases these immediate (and continuing) benefits have been great enough to pay out an entire system in a year or less.

The comprehensive model we are discussing includes an updating section. This uses current process data (measured on-line) to adjust the coefficients in the model. This insures that the model will reflect current conditions and not issue predictions based on obsolete data. This updating procedure by frequent adjustment also will include intermediate to longer term effects such as drifts in catalyst quality, feedstock changes, and equipment performance changes. Changes in management objectives which result in maximum or minimum charge rates, specified volumes of a certain product, or changed product quality specifications are supplied by process engineers. A nonlinear "optimizer" uses all this currently revised operating and management information as inputs to the process model to produce an output of values for the most profitable operating conditions. These conditions and proposals to change strategy are printed out for verification and review by appropriate personnel. All accepted strategy developed by the optimizer and steady-state model is transferred to the dynamic optimizer where these conditions become the target about which optimization and control of this unit are centered.

The efficiency of the entire optimizer-control system now can be seen to hinge on three prime factors—the degree of control, precision achieved in data taking, and a properly designed, executed, and analyzed experimental program.

Figure 6a illustrates how steady-state optimization affects operations in the face of constraints, feed changes, etc. Note that although the representations of constraints are shown as straight lines, many are highly nonlinear and require compensation for this factor. Constraints limiting operation in this figure are labeled 1, 2, and 3. Their bounds define the available operating space in terms of fresh feed rate and quantity of fresh feed converted. Also, note that the quality

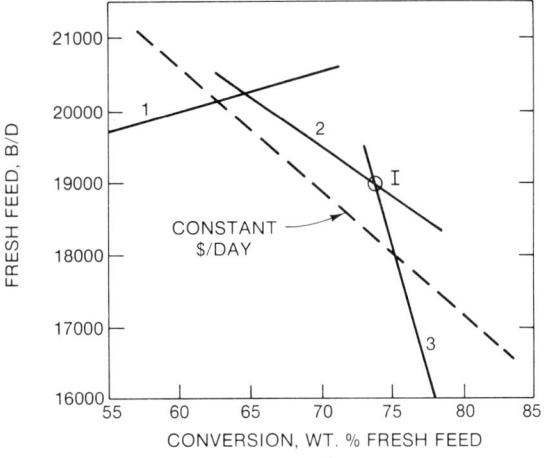

FIG. 6a. Feed–conversion operating space (for high-quality feed).

of feed converted is a function of several operating variables. The optimum is located at the intersection of two or more constraints and generally as the highest on the constraint boundaries.

Figure 6b shows a similar diagram for the results when a poorer feed is charged. Note that the optimum point is at considerably lower conversion although nearly the same throughout as for the better feed (Fig. 6a). The constraints limiting operation for these two cases may or may not be the same. The poorer feed II loads the cracker when either feed or conversion is at a lower level than for the better feed. The total product value curve (total $/d) for this poorer feed is much more "flat" than that for the better feed. A drop in feed rate requires a correspondingly greater increase in conversion to maintain constant

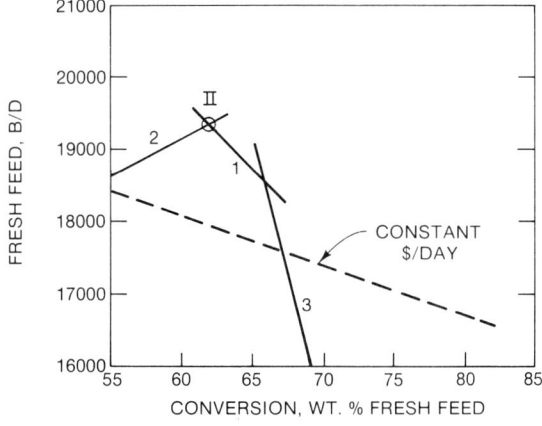

FIG. 6b. Feed–conversion operating space (for poorer-quality feed).

dollar output when compared to similar data shown in Fig. 6a. The steady-state optimizer will furnish new target operating points to dynamic optimization whenever substantial changes in feedstock such as this occur.

A fully automated and controlled FCC unit will employ the steady-state optimizer on-line. Appropriate management directives (total feed available, special product requirements, etc.) will be entered manually. The new optimum points determined by the steady-state optimization will be printed and subjected to approval by process engineers. After acceptance, the new targets are transferred to the dynamic optimizer for enforcement in the most profitable manner. A plant which has only a limited process computer control capability may alternately run the steady-state optimizer off-line. Data for updating in this case may be supplied manually or by means of a data link to the laboratory or other information source.

Some cases may require that the scope of FCC control and optimization be an absolute minimum. Here only dynamic optimization is appropriate. Dynamic optimization target values must be specified to the dynamic optimizer rather than being received from the on- or off-line steady-state optimizer. The process engineer in this case must know detailed process objectives and know or have additional/other means to determine optimum conversion levels and fresh feed rates.

Reliable computer hardware and advanced on-stream process analyzers are now available. Both are essential steps in obtaining on-line plant material balances. New analyzers required the concurrent development of auxiliary sampling systems. Some of these require considerable sophistication in design to overcome the biggest single source of off-line laboratory "error"—the failure to obtain a representative sample of material for analysis. Analyses times have been reduced phenomenally, and computer-controlled on-stream instrument systems are now available to manage 32 or more analyzers, such as chromatographs, and communicate the results to the material balance and economics programs within the framework of the dynamic optimizer.

Overall system costs have decreased sharply despite greater speed, capacity, and reliability. The incentives for computer control and optimization, though readily identifiable in the past, are now becoming essential for increased profit and more efficient use of our energy resources. The result of these factors—(1) development of a capable control system, (2) availability and lower cost of reliable hardware, (3) recently developed (new) on-line process stream analyzers, and (4) increased incentives—will be a rapid increase in the number of FCC units to be placed on computer control.

Computer controlled processes provide operational flexibility. Changes in feedstock, product slate, process economic factors, schedule requirements—all are pieces of the complex puzzle solvable by the computer and steady-state optimizer. This capability to keep the process "busy" will turn the plant to higher productivity—ergo, a state of higher profitability.

The approach to cat cracking control and optimization presented here provides a stepwise procedure helpful in developing the total system and useful in new applications if less than the "whole bit" is contemplated at one time. Each module in the series is tailored to fit the particular unit. The ultimate system can capably perform most operating functions, manage process upsets, accommodate short-term changes (such as ambient conditions), and compensate longer term changes such as revised process objectives. All this is accomplished while optimizing the profit output (or other desired objective function) for the process.

J. A. FELDMAN
B. E. LUTTER
R. L. HAIR

Deasphalting

Introduction

Processing of crude normally involves separation into various fractions which require further processing in order to produce marketable products. The initial separation process is distillation, which separates crude into increasingly higher boiling range fractions. Since petroleum fractions are subject to thermal degradation, there is a limit to the temperatures that can be used in simple separation processes. The crude cannot be subjected to temperatures much above 700°F without encountering some thermal cracking. Therefore, to separate the higher molecular weight and higher boiling fractions from crude, special processing steps must be used. Normally, the first step is to subject the bottoms fraction from an atmospheric crude tower to vacuum distillation. Vacuum distillation also has its limitations in the degree of pressure reduction feasible in large-size commercial equipment. After vacuum distillation, there are still some valuable oils left in the vacuum-reduced crude. These valuable oils are recovered by solvent extraction. The first application of solvent extraction in refining was the recovery of heavy lube oil base stocks by propane deasphalting. In order to recover more oil from vacuum-reduced crude, mainly for catalytic cracking feed, higher molecular weight solvents such as butane, and even pentane, have been employed.

Deasphalting was developed more than 40 years ago as a joint effort of Kellogg, Standard Oil Co. of New Jersey, Standard Oil Co. of Indiana, and

Union Oil Co. of California. The objective of their combined research efforts was the recovery of very heavy lubricating oils, generally known as "bright stocks" or cylinder stocks, from asphalt base crudes. Since these valuable heavy lubes could not be distilled readily without thermal degradation, they were available only from paraffin-base crudes, such as Pennsylvania, which contain little or no asphalt. From these crudes, the heaviest fractions could be economically recovered by steam or clay refining.

The separation of residual fractions from asphalt base crudes into oil and asphalt fractions was first performed on a production scale by mixing the residue of vacuum distillation with propane (or mixtures of "normally gaseous" hydrocarbons) and continuously decanting the resulting phases in a suitable vessel. Temperature was maintained within about 100°F of the critical temperature of the solvent, at a level which would regulate the yield and properties of the deasphalted oil in solution and which would reject the heavier undesirable components as asphalt.

From this original process, pilot plant development in Kellogg's laboratory progressed through multistage countercurrent designs. As a result of this work, the advantages of maintaining a temperature differential between the final stage at the extract (oil) end and the feed stage were discovered. The multistage settlers used in previous designs were replaced by the countercurrent extraction column. The design of the column was developed by Kellogg in its pilot plants and applied successfully to industrial installations (Fig. 1).

Today, deasphalting and delayed coking are used frequently for residuum conversion. The high demand for petroleum coke, mainly for use in the aluminum industry, has made delayed coking a major residuum conversion process. However, many crudes will not produce coke meeting the sulfur and metals specifications for aluminum electrodes, and coke gas oils are less desirable feedstocks for fluid catalytic cracking than virgin gas oils. In comparison, the solvent deasphalting (SDA) process can apply to most vacuum residues. The deasphalted oil is excellent feedstock for both fluid catalytic cracking and hydrocracking [1, 2]. The amount of recoverable oil from SDA is generally higher than that from delayed coking. Since it is relatively less expensive to desulfurize the deasphalted oil than the heavy vacuum residuum, the SDA process offers a more economical route for disposing of vacuum residuum from high sulfur crude [3, 4].

Although coke of high quality is worth a lot more than asphalt, coke of high sulfur and metals has about the same value as asphalt of the same quality. They both require expensive stack gas clean-up facilities when used as fuel.

Deasphalted Oil (DAO) and Solvent Asphalt

The deasphalted oils are high in viscosity, often higher than 200 SUS at 210°F. They usually are solid at room temperature. Their exceptionally paraffinic character is revealed by the presence of a distinct high melting point, a

Deasphalting

FIG. 1. This 45,000 bbl/stream d solvent deasphalting unit at Socal's Richmond refinery is the largest of its kind. Kellogg built the facility.

characterization factor of 12 or better, and a remarkably high gravity, 16 to 24°API. The carbon residue of DAO is about 1 to 6 wt.%. This low quantity represents a range of 60 to 80% rejection of carbon residue to the asphalt phase at 50 to 70 vol.% of the DAO yield. Unlike the distillate stocks, the carbon residue in the DAO has much less effect on its cracking characteristics. The DAO actually makes less coke and more gasoline during catalytic cracking operation than virgin distillates of lower carbon residue, and much less than visbreaker and coker gas oil. While the high molecular weight paraffins in the DAO go largely to tar and coke under the prolonged thermal cracking conditions of Conradson or Ramsbottom test procedures, they are thought to be broken easily into low molecular weight oils by catalytic cracking [1].

The metals content of the DAO is very low. At 50 to 60 vol.% of the DAO yield, only 2 to 3% total Ni + V in the feed remains in the DAO. They increase rather drastically at higher DAO yields. Nitrogen and sulfur contents in the DAO are also related to the DAO yield. Their reduction rates are much less than the metals. About 25% of the nitrogen and 40% of the sulfur in the feed appears in the DAO at 60 vol.% DAO yield [5, 6] (Fig. 2).

Unlike DAO, solvent asphalts from different crude sources vary considerably. They are the remainder of the vacuum residue after removal of DAO

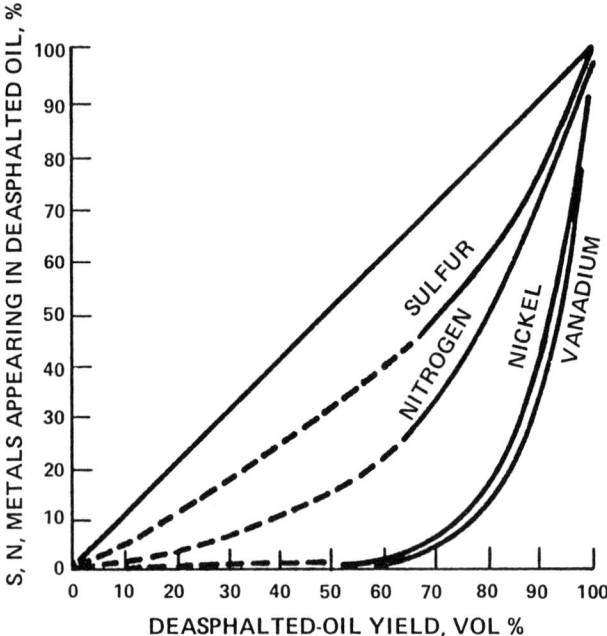

FIG. 2. Selectivity in solvent deasphalting [6].

of desired quality. As such, they normally are not produced to any particular specification and may, in fact, be limited only by handling problems. They may have ring-and-ball softening points of 300°F or higher and zero penetration at 77°F. They can be added to fluxes for making penetration asphalts. Otherwise, the flux would require blowing to increase asphaltene concentration by polymerization of lower molecular weight hydrocarbons in the flux [5].

Although solvent asphalts are free of wax, their viscosity is still too high for fuel. In some cases they can be blended with refinery cutter stocks to make No. 6 fuel oil. But when the sulfur content of the original residue is very high, even the blend fuel oil will not be able to meet the sulfur specification of fuel oil unless stack gas clean-up is available.

Typical deasphalting yields and their product properties are shown in Tables 1 and 2 [1].

Process Applications

The deasphalted oil and solvent asphalt are not finished products. They usually require further processing or blending. There are many possible routes to

Deasphalting

TABLE 1 Typical Yields and Product Properties for Lube Feedstock Preparation

Crude Source	Oklahoma	Peru	East Texas	Kuwait	Kuwait
SDA Feed					
°API	19.3	14.9	14.3	8.6	5.4
SUS at 210°F	385	740	920	950	23,000
Con carbon, wt.%	7.3	6.7	11.5	16.0	24.0
DAO					
Vol.% on feed	77.0	76.3	60.0	36.8	25.0
°API	23.3	19.4	23.0	24.3	21.2
SUS at 210°F	150	207	155	94	163
Con carbon, wt.%	1.7	1.7	1.5	0.7	1.3

TABLE 2 Typical Yields and Product Properties for Cracking Feedstock Preparation

Crude Source	Arab	West Texas	California	Canadian	Kuwait	Kuwait
SDA feed						
Vol.% crude	23.0	29.2	20.0	16.0	22.2	32.3
°API	6.8	12.0	6.3	9.6	5.6	8.1
Con carbon, wt.%	15.0	12.1	22.2	18.9	24.0	19.7
SUS at 210°F	75,000	526	9,600	1,740	14,200	3,270
Metals, wppm						
Ni	73.6	16.0	139	46.6	29.9	29.7
V	365.0	27.6	136	30.9	110.0	89
Cu + Fe	15.5	14.8	94	40.7	13.7	7.5
DAO						
Vol.% feed	49.8	66.0	52.8	67.8	45.6	54.8
°API	18.1	19.6	18.3	17.8	16.2	17.1
Con carbon, wt.%	5.9	2.2	5.3	5.4	4.5	5.4
SUS at 210°F	615	113	251	250	490	656
Metals, wppm						
Ni	3.5	1.0	8.1	3.9	0.9	0.6
V	12.4	1.3	2.3	1.4	0.7	4.0
Cu + Fe	0.2	0.8	3.5	0.2	0.8	0.8
Asphalt						
Vol.% feed	50.2	34.0	47.2	32.2	54.4	45.2
°API	−1.3	−0.9	−5.1	−5.1	−1.3	−2.0
Softening point (R&B), °F	278	171	246	224	201	193
Penetration at 77°F	1	—	7	0	0	0
% metals rejected to asphalt	96.5	96.6	98.2	97.1	98.7	95.8

integrate SDA with other facilities in a refinery. The most frequent ones are discussed below.

DAO—Lube Oil

In this case the deasphalted oil from vacuum residue is extracted further with either phenol or furfural and followed by dewaxing [7]. The finished product will have a fairly high viscosity index, 95 or better. The addition of a deasphalting unit increases the total production of lube oil by about 20%. A significantly higher yield of lube oil can be produced from the DAO by replacing conventional phenol extraction with mild hydrocracking. The finished product may have a range of viscosity indexes from 90 to 150, depending upon the cracking severity [8].

DAO—Cracking

The combined demand for more gasoline and less fuel oil per barrel of crude led to the application of deasphalting to provide more catalytic cracking and hydrocracking feeds. As much as one-third of the catalytic cracking feed in a refinery can be deasphalted oil. In fact, more than 70% of the DAO produced today is being used for cracking.

DAO—Hydrodesulfurization

Direct hydrodesulfurization of the vacuum residues from high sulfur crude is, theoretically, the best way to enhance their value. However, the concentrations of asphaltenes and metals in the residues can severely limit the performance of hydrosulfurization catalysts. Although technically possible, the process requires high pressure, low space velocity, and high hydrogen recycle ratio. Thus the direct hydrodesulfurization is economically unattractive, and the advantage of using the deasphalting process to remove the troublesome compounds becomes obvious. The deasphalted oil, with its minimum amount of asphaltenes and metals, is easier to desulfurize than the heavy vacuum residues. Published data show that desulfurization of the vacuum gas oil and deasphalted oil consumes only 65% of the hydrogen requirecd for direct desulfurization of topped crude. The total investment of the former process could be 20% less than the latter one [9].

Solvent Asphalt—Blending

If solvent asphalt is used for making fuel oil, cutter stocks, such as light cycle oils, are required for blending. Pollution control often limits the use of high sulfur solvent asphalt for fuel oil blending. "Tailor made" asphalts, such as

Deasphalting

paving asphalt, roofing asphalt, and emulsified asphalt, can be made by blending the solvent asphalt with resins [10].

Solvent Asphalt—Partial Oxidation

Solvent asphalt can be used as feed to a partial oxidation unit to make a hydrogen-rich gas. After treatment for H_2S removal, the gas can be used as raw material for making ammonia, methanol, or hydrogen.

Solvent Asphalt—Visbreaker

A visbreaker can be used to reduce the viscosity of solvent asphalt, therefore minimizing the need for cutter stock to be blended with the solvent asphalt for making fuel oil. The gas and gasoline produced are also useful by-products.

Process Description

A typical solvent deasphalting process flow diagram is shown in Fig. 3. The charge oil is mixed with dilution solvent from the solvent accumulator and then cooled to the desired temperature before entering the extraction tower. Because of its high viscosity, the charge oil can neither be cooled easily to the required temperature nor will it mix readily with solvent in the extraction tower. By adding a relatively small portion of solvent upstream of the charge cooler (insufficient to cause phase separation), the viscosity problem is avoided.

The charge oil, with a small amount of solvent, enters the extraction tower at a point about two-thirds up the column. The solvent is pumped from the accumulator, cooled, and enters near the bottom of the tower. The extraction tower is a multistage contactor, normally equipped with baffle trays. The heavy oil flows downward while the light solvent flows upward. As the extraction progresses, the desired oil goes to the solvent and the asphalt precipitates toward the bottom.

As the extracted oil and solvent rise in the tower, the temperature is increased in order to control the quality of the product by providing adequate reflux for optimum separation. Separation of oil from asphalt is controlled by maintaining a temperature gradient across the extraction tower and by varying the solvent/oil ratio. The tower top temperature is regulated by adjusting the feed inlet temperature and the steam flow to the heating coils in the top of the tower. The tower bottom temperature is maintained by the temperature of the entering solvent.

The deasphalted oil–solvent mixture flows from the top of the tower under pressure control to a kettle-type evaporator heated by low-pressure steam. The vaporized solvent flows through the condenser into the solvent accumulator.

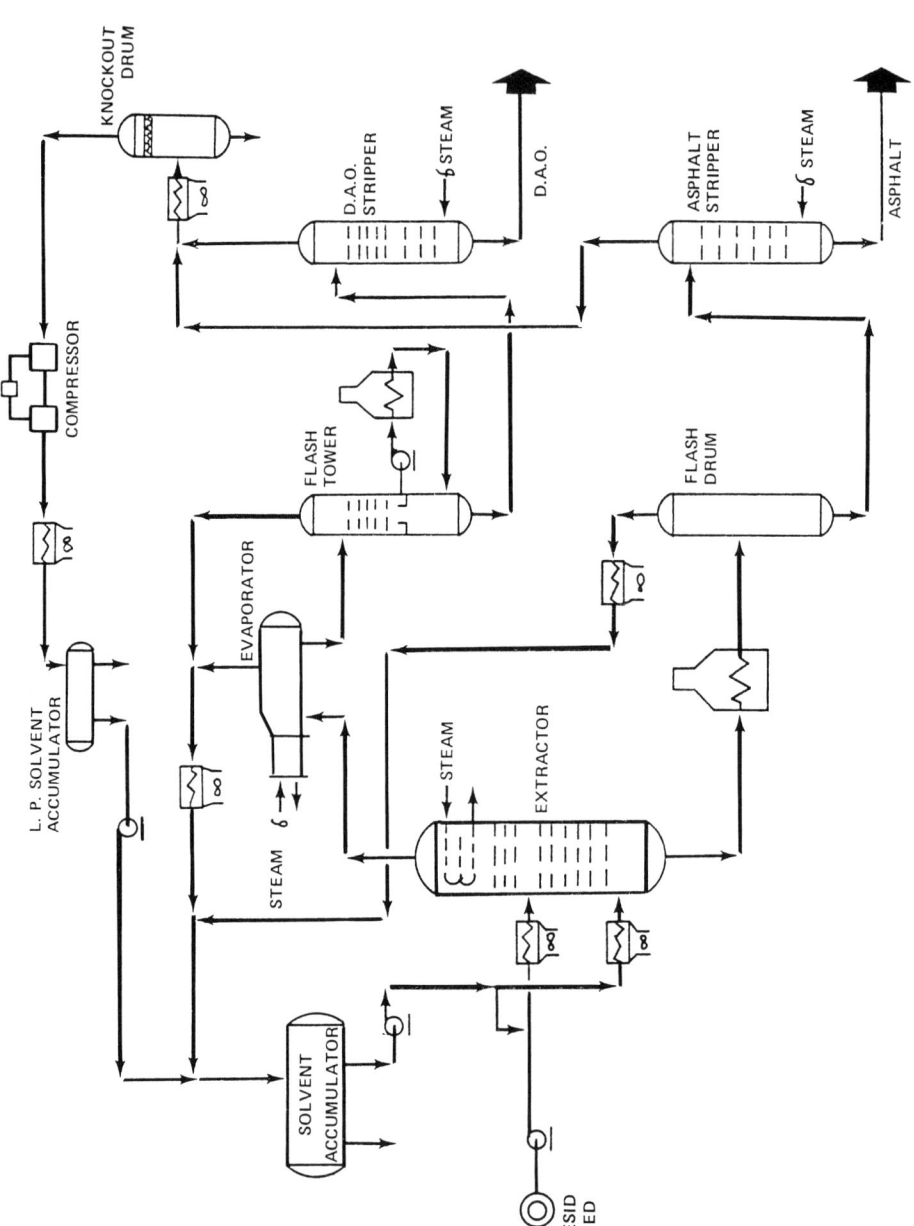

FIG. 3. Process flow of a solvent deasphalting unit.

The liquid phase flows from the bottom of the evaporator, under level control, to the deasphalted oil flash tower which is reboiled by means of a fired heater. In the flash tower, most of the remaining solvent is vaporized and flows overhead, joining the solvent from the low-pressure steam evaporator.

The deasphalted oil, with relatively minor solvent, flows from the bottom of the flash tower under level control to a steam stripper operating at essentially atmospheric pressure. Superheated steam is introduced into the lower portion of the tower. The remaining solvent is stripped out and flows overhead with the steam through a condenser into the compressor suction drum where the water drops out. The water flows from the bottom of the drum under level control to appropriate disposal.

The compressor takes suction on the drum and discharges through a condenser into a low-pressure solvent accumulator. After water settling, the solvent is pumped under level control to the main high-pressure solvent accumulator. The deasphalted oil is pumped under level control from the bottom of the stripper to other processing units, or it may be cooled and sent to storage.

The asphalt–solvent mixture is pressured from the extraction tower bottom on flow control to the asphalt heater and on to the asphalt flash drum, where the vaporized solvent is separated from the asphalt. The drum operates essentially at the solvent condensing pressure so that the overhead vapors flow directly through the condenser into the solvent accumulator.

Hot asphalt with a small quantity of solvent flows from the asphalt flash drum bottom, under level control, to the asphalt stripper. The stripper is operated at near atmospheric pressure. Superheated steam is introduced in the bottom of the stripper. The steam and solvent vapors pass overhead, join the deasphalted oil stripper overhead, and flow through the condenser into the compressor suction drum. The asphalt is pumped from the bottom of the stripper, under level control, to storage.

Solvent Extraction

Selection of Solvent

In the process design of a deasphalting unit, the selection of a solvent or solvent mixture seriously affects the economics, flexibility, and performance of the plant. The solvent must be suitable, not only for the extraction of the desired oil fraction, but also for control of the yield and/or quality of the DAO at temperatures which are within the practical operating limits. On one hand, if the temperature is too high, that is, too close to the critical temperature of the solvent, the operation becomes unstable. On the other hand, if the temperature is too low, the charge oil may be too viscous to handle, and good contacting in the extractor will not be possible.

Propane deasphalting has been used for several decades in the manufacture of lubricating oils. Propane is by far the most selective solvent among the light hydrocarbons used for deasphalting. At temperatures ranging from 100 to

150°F, paraffins are completely soluble in propane while asphaltic and resinous compounds precipitate. The rejection of these compounds drastically reduces the metals and nitrogen content in the deasphalted oil. Furthermore, consistent with high paraffinity, the lower carbon-to-hydrogen ratio in DAO indicates the rejection of condensed-ring aromatics which are undesirable for cracking processes because of their coking tendency. Although DAO from propane deasphalting has the best quality, the yield is usually less than deasphalting with a heavier solvent. A relatively large quantity of propane has to be circulated in order to produce commercial quantities of DAO. The ratios of propane–oil required vary from 6–1 to 10–1 by volume, with the ratio occasionally being as high as 13–1 [11]. Since the critical temperature of propane is only 206°F, this limits the extraction temperatue to about 180°F. Therefore, propane alone is not suitable for feedstocks of high viscosity because of the low operating temperature.

Isobutane and n-butane are more suitable for deasphalting heavier feeds. Since their critical temperatures are higher, 273.2 and 305.6°F, respectively, higher extraction temperatures can be used to reduce the viscosity of the heavy feed and to increase the transfer rate of oil to solvent. However, butanes lose their selectivity when used for treating lighter stock.

High selectivity, the ability to produce high yields of good quality oil (low carbon residue–low metals), generally is not affected by solvent composition. However, loss of selectivity becomes noticeable when the solubility of the oil in the solvent is very high. Such is the case when relatively light oil is treated with solvent containing too much butane. Thus no loss of selectivity would be expected with n-butane or isobutane when processing a heavier feed, but loss of selectivity would be expected if a lighter stock was processed.

In addition, when treating a light oil with butanes, the extraction temperature has to be raised in order to limit the oil solubility to achieve the desired quality of DAO. The required temperature may approach the critical temperature of the solvent, causing the operation to become unstable.

Although n-pentane is less selective for metals and carbon residue removal, it can increase the yield of DAO from a heavy feed by a factor of 2 to 3 over propane [9, 12]. In most cases the metals and carbon residue content in the pentane DAO is considered too high for direct cracking because of short catalyst life. However, a blend of vacuum gas oil and pentane DAO, after hydrodesulfurization, makes an excellent cracking feedstock [9, 13]. The desulfurized oil also can be blended with asphalt for making low sulfur fuel oil.

It is obvious that the use of a single solvent limits the practical range of feedstocks which can be processed. When a unit is required to handle a variety of feedstocks and/or produce various yields of DAO, only a dual solvent can provide this kind of flexibility. For instance, a mixture of propane and n-butane would suit both the heavy resid and the lighter feed. By adjusting the solvent composition, the desirable product quantity is obtainable within the range of temperature control.

It has been general experience that the useful life of an SDA unit far exceeds the time interval over which a constant crude source can be assured. It is inevitable that there will be a change in the charge to the plant which will exceed the narrow limits imposed by a single solvent. Such feed changes have occurred in almost every plant built to date, and many plants would have been virtually useless if designed for a single solvent only.

The selection of solvent for deasphalting may be summarized as follows:

1. Propane is the best choice for lube oil production due to its ability to extract only the paraffinic hydrocarbons and to reject most of the carbon residue.
2. A mixture of propane and butane should be considered for preparing feedstocks for the catalytic cracking processes. Their ability to remove the metal-bearing components is critically important to the DAO's quality.
3. Pentane deasphalting, plus hydrodesulfurization, can produce more feed for catalytic cracking or low sulfur fuel oil.

Solvent/Oil Ratio

Besides the solvent composition, the solvent/oil ratio also plays an important role in SDA operation. Solvent/oil ratios vary considerably and are governed by feedstock characteristics and desired product qualities. For each individual feedstock, there is a minimum operable solvent/oil ratio. This ratio is lowest for short residues produced from asphaltic or naphthenic base crude oils, and highest for long paraffinic base crude residues, very resinous residues, and residues from poor vacuum distillation [10].

It should be mentioned that increasing the solvent-to-oil ratio almost invariably results in improving the DAO product quality at a given yield. However, this effect also has a diminishing return. In addition, the cost of the plant (both capital investment and operating) is proportional to the solvent-to-oil ratio, and one should seek an optimum ratio for designing the plant.

Extraction Temperature and Pressure

Extraction Temperatures

The main consideration in the selection of a practical operating temperature is its effect on the yield of deasphalted oil. Although it is a basic law that lowering the temperature increases oil solubility, the solvent phase may not be completely saturated with oil as the temperature is lowered. The increase of oil yield may not occur because the viscosity effect on the mass transfer rate, from the heavy to the light phase, overrides the temperature effect on solubility. Thus, for practical applications, the lower limits of operable temperature are set by the viscosity of the oil-rich phase.

When the operating temperature is near the critical temperature of the solvent, control of the extraction tower becomes difficult. This loss of control, or instability, occurs since the rate of change of solubility with temperature becomes very large at conditions close to the critical point of the solvent. Such large changes in solubility cause large amounts of oil to transfer between the solvent-rich and the oil-rich phases. This, in turn, causes "flooding" and/or uncontrollable changes in product quality. Therefore, the upper limits of operable temperatures must lie below the critical temperature of the solvent in order to insure good control of the product quality and to maintain a stable condition in the extraction tower. When a propane and butane mixture is used as solvent, the extraction temperatures range from 150 to 250°F [11].

Temperature Gradient

The temperature gradient across the extraction tower influences the sharpness of separation of DAO and asphalt by generation of internal reflux. This phenomenon occurs when the cooler oil/solvent solution in the lower section of the tower attempts to carry a large portion of oil to the top of the tower. When the oil/solvent solution reaches the steam-heated, higher-temperature area near the top of the tower, some oil of heavier molecular weight in the solvent solution is rejected. This rejection takes place because the oil is less soluble in solvent at the higher temperature. The heavier oil dropped out of solution at the top of the tower attempts to flow downward and causes the internal reflux. It is generally true that the greater the temperature difference between the top and the bottom of the tower, the greater will be the internal reflux and the better will be the quality of the DAO. However, too much internal reflux can cause tower flooding and jeopardize the extraction process.

Operating Pressure

The extraction tower pressure normally is not considered to be an operating variable. It must be higher than the vapor pressure of the solvent mixture at the tower operating temperature. Thus the tower pressure should be changed only when there is a desire to change the solvent composition or the tower temperature. The excess pressure is necessary to prevent vaporization of light hydrocarbon contaminants, such as ethane, in the solvent as well as to prevent vaporization of the solvent mixture. When a mixture of propane and butane is used as solvent, the operating pressure ranges from 400 to 600 $lb/in.^2$ gauge [11].

Extraction Tower

As in all solvent extraction systems, the contacting device is most important to the separation of DAO and asphalt. It is difficult to achieve a uniform distribution of heavy oil and light solvent across the entire extraction tower. Proper contact and distribution of the oil and solvent in the extractor are essential to the efficient operation of any SDA unit. In the early days, mixer-settlers were used as contactors. They are less efficient than the countercurrent contacting devices. Packed towers are difficult to operate in this process because of the large differences in viscosity and density between the asphalt and solvent-rich phases. Towers with sieve trays have been used in a few installations. At present, nearly all deasphalting units use either baffled towers or rotating disk contactors (RDC).

General Design of the Extraction Tower

The extraction tower for solvent deasphalting consists of two contacting zones—a rectifying zone above the oil feed and a stripping zone below the oil feed. The rectifying zone contains some elements designed to promote contacting and to avoid "channeling." Steam-heated coils are provided to raise the temperature sufficiently to induce an oil-rich reflux in the top section of the tower. The stripping zone has disengaging spaces at the top and bottom and consists of contacting elements between the oil inlet and the solvent inlet.

Static Baffle Tower

A countercurrent tower with static baffles has proved to be a successful design at Kellogg and is widely used in solvent deasphalting service. The baffles consist of fixed elements formed of expanded metal gratings in groups of two or more to provide maximum change of direction without limiting capacity. Under actual field conditions these towers have retained good contacting efficiency over a very broad throughput range—at least to 50% of the design capacity. The designs methods for deasphalting towers are largely empirical. Most of the baffled towers were designed from pilot plant data as well as the data from existing units. Reasonable adjustments may be necessary for the effects of viscosity and densities of the phases.

It was assumed that commercial internal tower geometry consistent with the pilot plant would produce similar effects on droplet size and turbulence which, in turn, might affect limiting velocity. Therefore, a relative velocity factor to be used for sizing the baffled tower was developed from the viscosity and density data. This design method was entirely successful. The efficiency of the towers was equal to, or better than, that of the pilot-plant extracters at the design throughput [14].

Rotating Disk Contactor

Originally developed by Shell Oil Co. for furfural extraction of lubricating oils, the rotating disk contactor has been employed in several commercial deasphalting units [15]. Like the baffled tower, the RDC also consists of two contacting zones, a rectifying zone and a stripping zone. In the RDC, however, disks connected to a rotating shaft are used in place of the static baffles of the tower. The rotating element, which is driven by a variable speed drive at either the top or the bottom of the column, can be operated between 10 and 60 r/min. Although operating flexibility is provided by controlling the speed of the rotating element, and thus the amount of mixing in the contactor, the relatively complicated mechanical seal of the rotating shaft could be a disadvantage. While comparable to the baffled tower in performance, the RDC is substantially more expensive and has lower capacity in larger applications.

Solvent Recovery System

In the SDA process, the solvent is recovered for circulation. Most of the equipment costs and almost all of the energy expended go directly to the solvent recovery system. Thus the economics and operability of an SDA unit are critically dependent on the design of the solvent recovery system.

Solvent Recovery from Deasphalted Oil

Solvent may be separated from the deasphalted oil in several ways. Conventional evaporation, in which a substantial amount of solvent is first vaporized by low-pressure steam and then condensed, usually can contribute to operating cost economy. However, as the solvent in the liquid mixture is reduced, the further formation of vapor results in uncontrollable foaming. Because of the relatively high vapor pressure of the solvent, as the solvent boils off, the temperature rise of the oil is insufficient to retain the fluidity required to prevent foaming. The problem is particularly troublesome when the operation is running at relatively low temperature and pressure.

The use of "flash tower and fired heater" by Kellogg has solved the foaming problem. In the tower system, characterized by high temperature and pressure, carryover is prevented by the countercurrent flow of liquid on the trays. The same solvent-rich liquid also removes all the superheat from the vapors generated by the heater, and consequently reduces the desuperheating duty in the condenser.

In addition, the tower system can be used in combination with conventional evaporation when excess steam is available. In such a combination the critical portion of solvent vaporization takes place in the tower where the high temperature and pressure eliminates foaming. As steam consumption can be variable, the system may be used as a disposal for the excess steam.

It is most economical to recover the solvent at a temperature close to the extraction temperature. If a higher temperature of recovery is set, heat is wasted in the form of high vapor temperature. If a lower temperature is set, the solvent must be reheated; the reheating process represents wasted heat. The solvent recovery pressure should only be low enough to maintain a smooth flow under pressure from the extraction tower. If the pressure is reduced substantially below the extractor pressure, the cost increases both because the higher pressure difference increases the cost of pumping and the larger vapor volume at lower pressure increases the cost of equipment.

Solvent Recovery from Asphalt

The asphalt solution from the bottom of the extraction tower contains less than an equal volume of solvent. A fired heater is used to maintain the temperature of the asphalt solution well above the foaming level. The high temperature is also required to keep the asphalt phase in a fluid state, thus avoiding plugging problems. A specially designed flash drum is used to separate the solvent vapor from asphalt. The purpose of the special design is to prevent carryover of asphalt into the solvent outlet line and to avoid fouling the downstream solvent condenser. This fouling would require cleaning with wash oil.

Since the solvent recovery system from asphalt is not subject to variations associated with the extractor conditions to the same degree as the solvent recovery system from deasphalted oil, it is desirable to operate at constant temperature and pressure with a separate solvent condenser and accumulator. The recovered solvent is pumped to the main solvent stream for circulation.

Low-Pressure Solvent Recovery System

Solvent remaining in the products from the high-pressure flashing system is stripped with superheated steam at low pressure. The stripper overhead vapors are combined and cooled to condense the steam. The wet hydrocarbon vapor is compressed, condensed, and sent to the main solvent stream.

The cost of the low-pressure condenser system is more than offset by the lower compressor cost and utilities. Furthermore, it is better to separate water from a small amount of solvent rather than from the whole solvent system.

Solvent Make-Up

The estimated solvent make-up for a 30,000 bbl/d SDA unit is 20 bbl/d. This includes leakage, sample draw-offs, and traces of entrained solvent in water draw-off from the compressor suction drum and the solvent accumulators.

Operating Problems

In the solvent deasphalting operation, the temperature and solvent/oil ratio are the most important process variables. Since varying the extraction tower temperature affects the DAO yield, it is desirable to keep the tower temperature constant for a stable operation. In addition, the solvent/oil ratio, which has a profound effect on the DAO quality, can be adjusted to keep the quality of DAO from various feedstocks nearly constant.

Foaming may occur in the evaporator, solvent flash tower, and asphalt flash drum. The foaming is caused by operating conditions rather than by equipment design. Fouling is very slight and is usually confined to the condensers, which can be washed effectively with hot gas oil.

All asphalt lines should be heated adequately and insulated so as to provide the viscosity reduction which is necessary for good flow characteristics. Gas oil lines should be provided for start-up and flushing of all asphalt circuits. Reciprocating steam pumps are suggested for pumping asphalt.

Perhaps the only polluter in an SDA unit is foul water. The water draw-off from the solvent accumulators and the compressor knockout drum will contain traces of solvent. There may be small quantities of DAO and/or asphalt in the water during upset periods.

Solvent deasphalting is a low temperature process; therefore, there is virtually no corrosion. Many commercial units have achieved on-stream factors in the range of 95 to 98% [10]. Turnarounds may be scheduled once in 3 years.

Utilities

In an SDA unit a substantial amount of utilities is consumed in solvent recovery. Thus utilities consumption is related to solvent composition and the solvent/oil ratio. The heavier solvent needs not only more heat to vaporize and condense, but also more pumping cost. A plant designed for n-butane alone

TABLE 3 Utility Summary for a 30,000 bbl/stream d SDA[a]

Deasphalted oil yield, Vol.%	65
Solvent composition, Vol.% C_3/C_4	40/60
Steam[b,c]	
Low-pressure steam, 35 lb/in.^{2} gauge, to condensate, lb/h	116,000
Superheated low-pressure stripping steam, 500°F, to sewer, lb/h	17,000
High-pressure steam, 150 to 35 lb/in.^{2} gauge, for pump drivers, lb/h	5,000
High-pressure steam, 150 lb/in.^{2} gauge, for fuel atomization, lb/h	4,000
Electricity	
2300 V supply for pump and compressor drivers, kW	2,600
440 V supply for pump drivers and air cooler fans, kW	800
110 V supply for lights and instruments, kW	100
Fuel	
Total liberation, MM Btu/h	165
Cooling water[d]	
gal/min	60

[a]No safety factors have been included in any of the figures.
[b]Low-pressure steam requirements represent gross consumption. Net imported low-pressure steam requirements will be gross, less low-presure steam production from pump driver exhaust.
[c]Steam consumption does not include tank or building heating or steam tracing.
[d]Cooling water used for pumps and compressor which may require cooling of jackets and lube systems.

may require utilities valued at 20% more than using a mixture of 50% propane–50% *n*-butane. When using pentane as solvent, the solvent and the deasphalted oil are separated at a supercritical temperature. Approximately 90% of the solvent is recovered directly as liquid rather than as a vapor by evaporation. The saving in utilities is about 50% over the usual evaporative method [12]. However, the poor selectivity of pentane limits its applications.

Although the power demand can be met by 100% electric drivers, putting in some steam drivers may be more economical. For example, it may become expedient to use 600 lb/in.^{2}gauge steam on one or more large drivers. The 600 lb/in.^{2}gauge steam exhausts at 200 lb/in.^{2}gauge, which may be used for an equivalent number of smaller drivers. The 200 lb/in.^{2}gauge steam exhausts at 30 lb/in.^{2}gauge, which may be used to vaporize solvent. On the other hand, if some amount of exhaust steam is available from other sources, it may be preferable to use electric drivers and to use exhaust steam for vaporizing the solvent in the SDA unit. A combination of these alternatives, as well as others, should be examined in an effort to minimize the utilities in an SDA unit.

A utility summary for a 30,000 bbl/stream day SDA unit is shown in Table 3.

Economics

Although solvent deasphalting was originally considered as a substitute for vacuum distillation, it has now found an economic place in the refiner's overall processing scheme as a supplement to vacuum distillation. Without using solvent deasphalting, the vacuum residue would be blended into heavy fuel oil or charged to a thermal cracker, visbreaker, or coking unit.

Solvent deasphalting is especially well suited for crude oils which are difficult to reduce, and it can be applied to almost any crude oil when clean gas oil and specification asphalt products are desired.

Generally, the investment of an SDA unit increases with the DAO yield and the solvent/oil ratio for which the unit is designed. If a large range of DAO yields is desired for a given feed rate, the unit must be designed for maximum DAO yield as well as maximum asphalt yield. Such a unit, of course, will cost more than a unit with less flexibility. A unit designed for a higher solvent/oil ratio will require not only more investment due to larger equipment, but also more operating cost for solvent recovery.

References

1. P. T. Atteridg, "A Fresh Look at Solvent Decarbonizing," *Oil Gas J.*, pp. 72–77 (December 9, 1963).
2. D. H. Stormont and C. Hoot, "SOCAL Sets New Process Route with Richmond Hydrocracking Complex," *Oil Gas J.*, pp. 146–167 (April 25, 1966).

3. W. J. Rossi, B. S. Deighton, and A. J. MacDonald, "Get More Light Fuel from Resid," *Hydrocarbon Process.*, pp. 105–110 (May 1977).
4. J. G. Ditman, "Deasphalting Paves Way for Low Sulfur Product," *Oil Gas J.*, pp. 84–85 (February 18, 1974).
5. J. G. Ditman, "Deasphalt to Get Feed for Lubes," *Hydrocarbon Process.*, pp. 110–113 (May 1973).
6. J. G. Ditman, "Solvent-Deasphalting-Versatility Preparation for Lube-Hydrotreat Feed," *Oil Gas J.*, pp. 45–48 (January 14, 1974).
7. M. Soudek, "What Lube Oil Process to Use," *Hydrocarbon Process.* pp. 59–66 (December 1974).
8. M. C. Bryson, W. A. Horne, and H. C. Stauffer, *Gulf's Lubricating Oil Hydrotreating Process*, Presented at the 34th Mid-Year Meeting of API, Chicago, May 12, 1969.
9. A. Billon, J. P. Peries, E. Fehr, and E. Lorenz, "SDA Key to Upgrading Heavy Crudes," *Oil Gas J.*, pp. 43–48 (January 24, 1977).
10. S. H. Alexander and J. D. Shurden, "Petroleum Asphalt-Chemistry, Refining, and Products," *Adv. Pet. Chem. Refin.* 5, 250–320 (1962).
11. W. F. Bland and R. L. Davidson, "Propane Deasphalting," in *Petroleum Processing Handbook*, McGraw-Hill, New York, 1967, pp. 3–82 and 3–83.
12. J. A. Gearhart and L. Garwin, *A New Economic Approach to Residuum Processing*, Presented at the NPRA Annual Meeting, San Antonio, March 30, 1976.
13. C. W. Selvidge and F. O. Torrea, *Processing High Metal Residues by the DEMEX Process*, Presented at the NPRA Annual Meeting, San Antonio, April 1, 1973.
14. Pullman Kellogg, *SDA Design Notes*, June 19, 1964.
15. S. Marple, K. E. Train, and F. D. Foster, "Deasphalting in a Rotating Disc Contactor," *Chem. Eng. Prog.*, 57(12), 44–48 (1961).

CARL PEI-CHI CHANG
JAMES R. MURPHY

Dehydrogenation

Dehydrogenation is a class of chemical reactions by means of which less saturated and more reactive compounds can be produced. There are many important conversion processes in which hydrogen is directly or indirectly removed. In principle, any compound containing hydrogen atoms can be dehydrogenated. But in this section only dehydrogenations of carbon compounds are treated, mainly those of hydrocarbons and alcohols. The largest-scale dehydrogenations are those of hydrocarbons, and these are exemplified here in particular detail for the conversion of ethylbenzene to styrene, a very prominent current dehydrogenation process. Also prominent are conversions of paraffins to olefins, olefins to diolefins, cycloparaffins to aromatics, and alcohols to aldehydes or ketones.

Dehydrogenations of less specific character occur frequently in the refining and petrochemical industries, where many of the processes have names of their own. Some in which dehydrogenation plays a large part are pyrolysis, cracking,

Dehydrogenation

gasification by partial combusion, carbonization, and reforming. These processes are described elsewhere in this *Encyclopedia*. Earlier reviews on dehydrogenation are those of Kearby [17] and Thomas [32].

In general, dehydrogenation reactions are difficult reactions. They require high temperatures for favorable equilibria as well as for adequate reaction velocities. Pure dehydrogenations are endothermic by 15 to 35 kcal/g-mol, and hence have large heat requirements. Active catalysts are usually necessary. Furthermore, since permissible hydrogen partial pressures are inadequate to prevent coke deposition, periodic regenerations are often necessary. Because of these problems with pure dehydrogenations, many efforts have been made to use oxidative dehydrogenations in which oxygen or another oxidizing agent combines with the hydrogen removed. This expedient has been successful with some reactions where it has served to overcome thermodynamic limitations and coke-formation problems.

The endothermic heat of pure dehydrogenation may be supplied through the walls of tubes (2 to 6 in. i.d.), by preheating the feeds, by adding hot diluents, by reheaters between stages, or by heat stored in periodically regenerated fixed or fluidized solid catalyst beds. Usually, fairly large temperature gradients will have to be tolerated, either from wall to center of tube, from inlet to outlet of bed, or from start to finish of a processing cycle between regenerations. The ideal profile of a constant temperature (or even a rising temperature) is seldom achieved in practice. In oxidative dehydrogenations the complementary problem of temperature rise because of exothermicity is encountered.

Other characteristic problems met in dehydrogenations are the needs for rapid heating and quenching to prevent side-reactions, the need for low pressure drops through catalyst beds, and the selection of reactor materials that can withstand the operating conditions.

Enthalpies and Equilibria

Table 1 shows enthalpies, in cal/g-mol, for some typical dehydrogenation reactions. The data are calculated at 800 K (527°C), mainly from American Petroleum Institute Project 44 data [27], but because the enthalpies change only slightly as the temperature is varied within the operating range, these numbers may be used to calculate adiabatic temperature changes or heat transfer requirements for the reactions. Tables of gaseous heat contents, or average values for gaseous heat capacities may be used for this purpose. It is apparent from Table 1 that the reactions become strongly exothermic when oxygen is added.

Equilibria can be readily calculated from free energies of formation (ΔG_f) or from $\log_{10} K_f$ in standard tables. The possible equilibrium conversions are essential guides in process design. As an example of a dehydrogenation equilibrium, calculated values for the dehydrogenation of propane are given in Table 2. This table illustrates the dominant effect of temperature on the equilibrium, as well as the very significant roles of pressure and of added diluents. Added hydrogen diluent, with the total pressure maintained at 1 atm

TABLE 1 Enthalpies of Reaction for Vapors at 800°K (527°C) (data from Ref. 27, except alcohols from Ref. 7)

Reaction	ΔH_{800} (cal/g-mol)
Ethane → ethylene + H_2	34,300
Propane → propylene + H_2	30,900
Butane → 1-butene + H_2	31,300
Butane → t-2-butene + H_2	28,500
n-Dodecane → 1-dodecene + H_2	31,200
1-Butene → 1,3-butadiene + H_2	28,400
t-2-Butene → 1,3-butadiene + H_2	31,100
1-Butene + $\frac{1}{2}O_2$ → 1,3-butadiene + H_2O	−30,500
Methylcyclohexane → toluene + $3H_2$	51,500
Methanol → formaldehyde + H_2	21,700
Methanol + $\frac{1}{2}O_2$ → formaldehyde + H_2O	−37,200
Methanol + H_2O → $3H_2$ + CO_2	15,900
Isopropyl alcohol → acetone + H_2	14,200
Isopropyl alcohol + $\frac{1}{2}O_2$ → acetone + H_2O	−44,700
Ethylbenzene → styrene + H_2	29,700

abs, severely depresses the equilibrium conversion. Nevertheless, hydrogen is sometimes added with the feed to diminish the rate of catalyst deactivation.

Temperatures at which 50% dehydrogenation is possible at atmospheric pressure without added diluents are shown for several reactions in Table 3. Ethane requires quite a high temperature whereas isopropyl alcohol can be dehydrogenated at a rather low temperature if a sufficiently active catalyst is used. These temperature values serve as rough guides, but in design work it is necessary to make exact equilibrium calculations as illustrated by the examples that follow. In making these calculations, the tables of Noddings and Mullet [22] are very helpful. A limitation on the accuracy of such calculations, however, is imposed by uncertainties in the basic thermodynamic data, and it may well happen that experimental equilibrium results, based on careful sampling, analyses, and temperature measurements, will differ somewhat from calculated equilibria. But before such a conclusion is reached, it is essential that the equilibrium be approached from both directions, that sampling be carefully done, and that side reactions be minimal.

Equilibrium curves for the conversion of n-butane to n-butenes and

TABLE 2 Equilibrium Conversions in Percent for Propane → Propylene + H_2

Temperature (°C)	1 atm, undiluted	1 atm, 10 mol inert	1 atm, 10 mol H_2	2 atm, undiluted	10 atm, undiluted
427	6.35	19.0	0.45	4.49	2.0
527	24.6	56.5	7.0	17.7	8.0
627	59.8	87.5	37.1	46.6	23.0
727	87.0	96.5	76.5	78.0	36.2

Dehydrogenation

TABLE 3 Temperatures at Which 50% Dehydrogenation Is Possible at Atmospheric Pressure with No Diluents Added

Reaction	Temperature (°C)
Ethane → ethylene + H_2	722
Propane → propylene + H_2	600
Isobutane → isobutylene + H_2	543
n-Dodecane → 1-dodecene + H_2	620
Ethylbenzene → styrene + H_2	621
t-2-Butene → 1,3-butadiene + H_2	656
Methanol → formaldehyde + H_2	411
Isopropyl alcohol → acetone + H_2	178

butadiene at constant pressure are shown in Fig. 1. Here the butenes are an equilibrium mixture of 1-butene, *trans*-2-butene, and *cis*-2-butene, but the butadiene is solely 1,3-butadiene, since 1,2-butadiene can only form in very small amounts. An inert diluent (steam or nitrogen, for example) increases possible conversions to olefins and dienes at a given temperature. A reduced pressure could also be used for this purpose, with nearly equivalent results at the lower conversions and even more effect at higher conversions.

Equilibrium calculations for the conversion of ethylbenzene to styrene in a constant pressure flow-system will be given here in greater detail. The process is carried out with an added inert diluent, normally steam. The reation is written

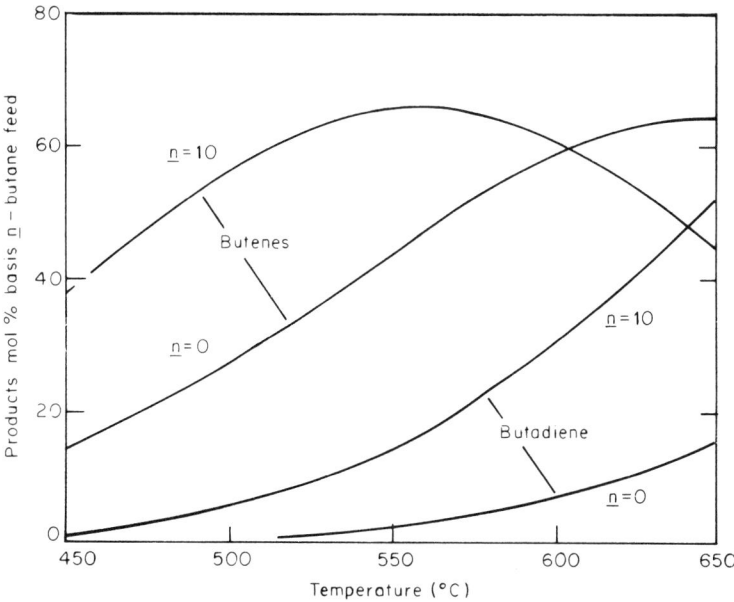

FIG. 1. Equilibrium in the system n-butane ⇌ n-butenes ⇌ 1,3-butadiene at 1 atm with n mole inert per mole total C_4.

$$\text{Ethylbenzene} \rightleftarrows \text{styrene} + H_2$$

The equilibrium constant is $K_p = (P_H)(P_S)/(P_E)$, where the P's are partial pressures in atmospheres absolute. In the general case there will be n moles of steam per mole of ethylbenzene initially present, and n can be taken as a constant since the consumption of steam by reaction to form H_2 and CO_2 is small. However, the side reactions become significant in the case of hydrogen, and somewhat more than 1 mol of H_2 is formed per mole of styrene, say a mol/mol. Thus, if the fractional conversion of ethylbenzene at equilibrium is x, the partial pressures at equilibrium will be

$$P_E = (1-x)P/(1+n+ax)$$
$$P_S = xP/(1+n+ax)$$
$$P_H = axP/(1+n+ax)$$

Then the conversion at the temperature and pressure of the reactor exit cannot exceed that given by

$$K_P = ax^2 P/(1-x)(1+n+ax)$$

where $\ln(K_p) = -\Delta G/RT$, P is in atmospheres absolute, ΔG is the standard free energy change in cal/g-mol, $R = 1.987$ cal/g-mol°K, and T is in °K. From the tables of American Petroleum Institute Project 44 [27], the following values of K_p are obtained:

Temperature, °C	427	527	627	727
K_p, atm	0.00329	0.0469	0.376	2.00

Using these numbers and the interpolated data, the curves of Fig. 2 have been calculated. A value of $a = 1.3$ at $x = 0.4$ is reasonable from the data of Wenner and Dybdal [35]. The calculated equilibria are in fairly good agreement with experimental data, though there are some data indicating that true equilibrium may be at a little lower conversion than calculated from the thermodynamic data. In practice, a high temperature, a low pressure, and dilution with an inert material (steam, etc.) are used to favor a high equilibrium conversion.

Design Principles

Selection of operating conditions for a straight dehydrogenation reaction often requires a compromise. The temperature must be high enough for a favorable equilibrium and for a good reaction rate, but not so high as to cause excessive cracking or catalyst deactivation. The rate of dehydrogenation reactions

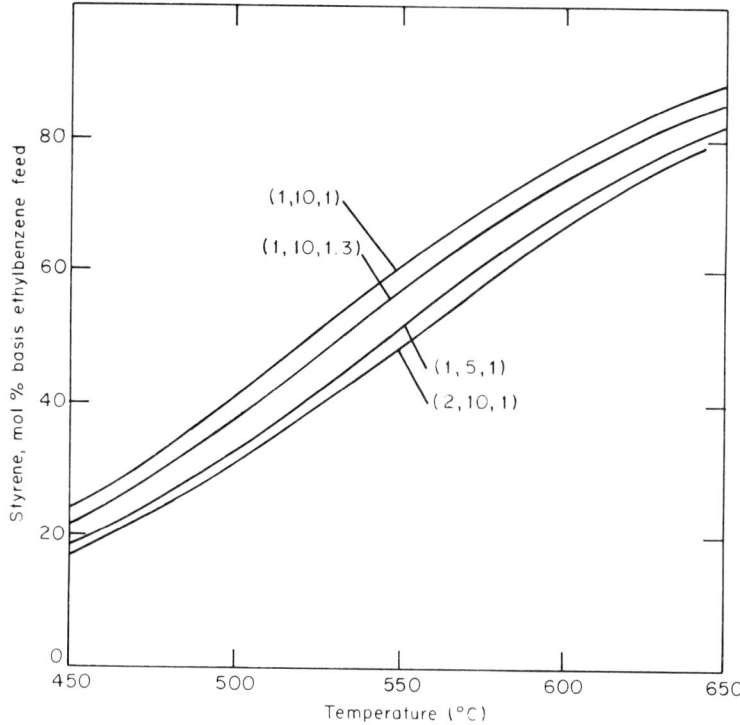

FIG. 2. Equilibrium for ethylbenzene ⇌ styrene. Parentheses show (P, n, a) where P is in atmospheres, n is moles inert diluent, and a is moles H_2 formed per mole styrene.

diminishes as conversion increases, not only because equilibrium is approached more closely, but also because in many cases reaction products act as inhibitors. The ideal temperature profile in a reactor would probably show an increase with distance, but practically attainable profiles normally are either flat or show a decline. Large adiabatic beds in which the decline is steep are often used.

The reactor pressure should be as low as possible without excessive recycle costs or equipment size. Usually it is near atmospheric, though reduced pressures have been used in the Houdry butane dehydrogenation process. In any case, the catalyst bed must be designed for a low pressure drop.

Rapid preheating of the feed is desirable to minimize cracking. Usually this is done by mixing prewarmed feed with superheated diluent just as the two streams enter the reactor. Rapid cooling or quenching at the exit of the reactor is usually necessary to prevent condensation reactions of the olefinic products. Materials of construction must be resistant to attack by hydrogen, capable of prolonged operation at high temperature, and not be unduly active for conversion of hydrocarbons to carbon. Alloy steels containing chromium are usually favored. Steels containing nickel are also used, but they can cause trouble from carbon formation. If steam is not present, traces of sulfur compounds may be needed to avoid carbonization. Both steam and sulfur compounds act to keep metal walls in a passive condition.

Dehydrogenation Processes of Industrial Importance

Ethane → Ethylene

This reaction is currently effected by pyrolysis (no catalyst), using temperatures of 800 to 900°C and steam diluent at moderate pressures of 2 to 5 atm. The selectivity to ethylene is 85 to 90%. Externally heated pyrolysis tubes are used. Details of this process are given elsewhere in this work.

Propane → Propylene

The reaction can be carried out catalytically using a chromia–alumina catalyst at about 550 to 600°C. A kinetic study of this reaction over a potassium-promoted chromia–alumina catalyst was made by Suzuki and Kanero [31]. In commercial practice it is more usual to pyrolyze propane to obtain a mixture of ethylene and propylene.

n-Butane → n-Butenes → Butadiene

This process has been much used to obtain butenes and butadiene. A chromia–alumina catalyst is used at 550 to 600°C [17, 32]. For butenes production, operation at atmospheric pressure is acceptable, but when butadiene is the main product it is preferable to operate at a reduced pressure of about one-fourth of an atmosphere. In the most prominent process for this reaction, the Houdry Process [14], the catalyst bed is diluted with inert solids which act as heat carriers, receiving heat during the regeneration periods when coke is burned off with air, and supplying heat for the endothermic reaction during process periods of 7 to 50 min duration. A process period of 15 min may be normal. A study of the kinetics of reaction, coking of catalyst, and reactor design for low-pressure dehydrogenation of 1-butene to butadiene over chromia–alumina catalyst was published by Dumez and Froment [10]. An earlier kinetic study of butane dehydrogenation, primarily to butenes, and using chromia–alumina catalyst at 1.1 to 2.7 atm pressure with 30 min process periods was that of Dodd and Watson [9]. Conversions in this process are limited to 40% or less by equilibrium.

n-Butenes → Butadiene

There are three general methods of converting n-butenes to 1,3-butadiene. The first is the Houdry process, just mentioned, in which dehydrogenation is at reduced pressure for brief process periods over a chromia–alumina catalyst. In this process the feed can be butenes, butane, or a mixture.

Second is the continuous dehydrogenation in the presence of steam over a suitable catalyst at atmospheric pressure and about 620°C. This process has been widely used, with the butenes usually coming from refinery cracking processes. Initially the Esso 1707 catalyst based on MgO was used [17] and hourly steam regenerations were needed. An excellent study of the kinetics for this system was published by Beckberger and Watson [4]. Later, catalysts based on iron oxide were introduced, and these permitted dehydrogenation with no or very infrequent regenerations. The Shell 205 catalyst (62.5% Fe_2O_3, 2.2% Cr_2O_3, and 35.3% K_2CO_3) is a prominent example. With it, about 12 mol of steam are used per mole of butenes. At a conversion of 30%, the selectivity to butadiene is 75 to 80% mol. The steam serves to carry heat, to lower the hydrocarbon partial pressure, and to maintain the catalyst in an active condition. The steam is preheated to a higher temperature than the butenes stream, the two are mixed immediately before entering the large adiabatic catalyst bed, and at the exit from the bed the products are rapidly cooled by water quench or by tubular heat exchangers. Typical operating conditions are 590 to 650°C, butenes gas hourly space velocity (for gas at NTP) 200 to 600, H_2O/C_4H_8 8 to 14, conversion per pass 25 to 35%, and selectivity 70 to 80%. A $3/16$ in. diameter catalyst pellet is normally used, but as Voge and Morgan [34] have shown, selectivity is somewhat improved with smaller sizes in the laboratory tests. However, in commercial plants the size may have a critical effect on pressure drop and bed integrity.

Another catalyst that has been used for butenes dehydrogenation in the presence of steam is the Dow Type B, which consists of calcium and nickel phosphates promoted with a small amount of chromium oxide [23]. This catalyst gives a higher selectivity to butadiene of about 90%, but requires more steam (20 mol per mole of butenes) and also requires hourly regeneration with air.

The third method of butenes dehydrogenation is oxidative dehydrogenation over a catalyst by means of added air. This has the great advantage of not being limited by equilibrium. The reaction is highly exothermic but it can be operated adiabatically if a sufficient quantity of inert gases is present. Oxidative dehydrogenation has been developed for commercial operation by Phillips as the O-X-D Process [15, 16]. Air and steam are mixed, heated, and mixed with butenes feed. The mixture is passed over the oxidative dehydrogenation catalyst in an adiabatic reactor at about 480 to 590 C. The butenes gas hourly space velocity is in the range 200 to 600, O_2/C_4H_8 is about 1, H_2O/C_4H_8 about 35, temperature rise about 100 C, conversion of butenes 70 to 80%, and selectivity 85 to 90%. Suitable catalysts for this reaction may be of the bismuth–molybdenum oxide type, or of the tin phosphate type, or still others. Variants of the oxidative dehydrogenation process have also been developed by British Petroleum [20] and by Petro-Tex [3]. There has been much recent research effort in the field of oxidative dehydrogenation, and various papers describe studies of catalysts for butenes conversion by this method [13, 26]. A kinetic study of the oxidative dehydrogenation of butenes to butadiene over ferrite catalyst was described by Sterrett and McIlvried [30].

Higher Paraffins to Olefins

Paraffins from about C_6H_{14} to about $C_{20}H_{42}$, usually segregated into four-carbon fractions, can be dehydrogenated to monoolefins at low pressure over nonacidic noble metal catalysts. A mixture of *n*-olefins is obtained since the position of the double bond shifts readily under dehydrogenation conditions. This multiplicity of olefins on the right-hand side of the chemical equation raises the equilibrium conversion relative to what it would be with only a single olefin formed. An example of such a process is the PACOL Process of UOP, which is used to prepare detergent range olefins for alkylation from *n*-paraffins [5, 33]. Conversion levels are of the order of 20 to 40%, and selectivity to monoolefins is about 90%. Diolefins, cracked products, and aromatics are undesired minor by-products, produced at selectivities of 2 to 4% each (basis feed molecules converted). A separation process is necessary to isolate unconverted *n*-paraffins for recycle to the dehydrogenation step. Selective extraction of olefins or olefin alkylation followed by distillation serves for this separation. A fixed bed of catalyst is used in which the reaction is carried out in the vapor phase in the presence of excess hydrogen. Little detail has been published.

Paraffins → Olefins, General

The lower paraffins can be dehydrogenated over chromia–alumina or over noble metal catalysts. For higher paraffins only certain noble metal catalysts can be used since formation of aromatic rings is prominent for six or more carbon atoms with a chromia–alumina catalyst. Paraffins cannot be dehydrogenated at practical conversions over the alkalized iron oxide catalysts in the presence of steam, nor can they be dehydrogenated by the oxidative processes used for butenes conversion. There has been much research interest devoted to finding an oxidative dehydrogenation process suitable for paraffins. Good conversions are possible by use of iodine or sulfur compounds as initiators [12, 18, 25]. Iodine is especially effective as a dehydrogenating agent, and when used in conjunction with an alkaline acceptor to bind the HI produced can give very extensive dehydrogenation to olefins, diolefins, or aromatics. By reaction with oxygen, the iodine can then be regenerated from the acceptor–HI combination [18]. Operating conditions for this process are quite corrosive, construction costs are high, and iodine losses must be very low.

Other promoted oxidative processes for paraffins use catalysts and smaller amounts of iodine or sulfur compounds along with added oxygen. There is much continuing research on various types of oxidative dehydrogenation, including treatment in the presence of oxygen and small amounts of halides over various catalysts [3], and treatment in the presence of oxygen and hydrogen sulfide over suitable catalysts [12]. An older review article on oxidative dehydrogenation of hydrocarbons lists 177 references [28].

Cyclohexanes → Aromatics

This is one of the easier dehydrogenations. Equilibrium conversion is greater than 80% above about 350°C, even at a pressure of 5 atm. Platinum–alumina catalysts are very effective, and this is one of the more important reactions in the platinum catalyst reforming of gasoline for octane improvement. In fact, the reaction is not usually carried out as a simple reaction of a single cyclohexane, but instead the desired aromatics are recovered from the mixed products of reforming. For details see the section on reforming, or the book of Thomas [32].

Alcohol → Aldehyde or Ketone

A typical reaction is that of isopropyl alcohol to give acetone and hydrogen. The equilibrium is favorable at temperatures of 250°C or above, but somewhat higher temperatures are needed for a good rate. Catalysts that can cause dehydration have to be avoided. Metallic copper is effective, and may be used on an inert support or as particles of brass. Supported silver is another suitable catalyst. Certain oxide catalysts, such as copper chromite, can be used. Conditions are 300 to 450°C, 1 to 5 atm pressure, and 1 to 5 LHSV. Externally heated tubes are used to avoid large temperature decline. Alternatively, an oxidative dehydrogenation with air added and partial conversion to both steam and hydrogen may be effected using silver on an inert support.

Conversion of methanol to formaldehyde is usually done oxidatively over silver catalyst or over an iron–molybdenum oxide catalyst [32]. A special type of dehydrogenation, represented by the reaction

$$CH_3OH + H_2O \rightarrow 3H_2 + CO_2$$

can be used to convert methanol to hydrogen. Manganous oxide on silica gel is a very active catalyst which operates in a water–methanol solution under pressure at about 130°C.

Ethylbenzene → Styrene

Styrene is manufactured on a very large scale for use in various polymers. Most of the styrene is made by dehydrogenation. The technology has been continually improved and has reached an advanced status. Design needs include: (1) a high temperature and a low pressure for favorable rate and equilibrium, (2) an active and selective catalyst, (3) a method of supplying the heat of reaction, (4) the avoidance of side reactions, (5) a low pressure drop through the catalyst bed, (6) the highest possible conversion to avoid costly recycle, (7) reactor materials suitable for operating conditions, (8) conservation of heat and steam, and (9) good stream factor. These needs have been adequately met. For example, large plants have demonstrated uninterrupted operation for over 18 months.

There are several competing processes, but all appear to use iron oxide catalysts and steam dilution. Prominent are: (1) Monsanto-Combustion Engineering-Lummus [1, 8], (2) Union Carbide-Cosden-Badger [2], and (3) Societe Chimique des Charbonnages [29]. Proprietary styrene processes are operated by Dow Chemical Co. and by BASF AG.

Reactors for ethylbenzene dehydrogenation usually contain large adiabatic beds of catalyst. Vapor flow is either downward or radial-outflow. Beds may be staged, with intermediate heaters or intermediate addition of hot gases to supply the heat of reaction in part. By means of such devices and possibly other undisclosed innovative reactor designs, it has been possible to increase operating conversion levels from about 40% to about 60%, with consequent savings. German practice (BASF) has been to use catalyst in externally heated tubes, a lower steam dilution, and a conversion level of about 40% [24].

In the adiabatic reactors the pressure is not much above 1 atm abs, and the pressure drop through the reactor must be kept low. Since the inlet temperatures border on those for thermal cracking of ethylbenzene, the inlet residence time must be kept low. Usually highly superheated steam is mixed at the reactor entrance with ethylbenzene vapors which are somewhat below the desired inlet temperature. Operating conditions in the catalyst beds may cover the following ranges:

Temperature	550 to 650°C
Pressure	0.7 to 3 atm abs
Inerts/hydrocarbon	10 to 20 (inerts, mainly steam)
Flow rate of ethylbenzene	0.3 to 1.5 vol./vol. catalyst h
Conversion, %	40 to 65
Selectivity, %	85 to 95 mol styrene/mol ethylbenzene converted

Catalysts for ethylbenzene dehydrogenation have undergone a continuing evolution. The earliest catalyst was bauxite (an impure aluminum oxide). This was replaced in the United States during the mid-1940s by catalysts based on MgO and Fe_2O_3 [17]. Early European history was reviewed by Ohlinger and Stadelmann [24], starting with I. G. Farben studies and continuing with the postwar work of BASF. The best BASF catalyst as of 1965 was Lu-144F, with the composition 85% Fe_2O_3, 7.7% ZnO, 0.7% Al_2O_3, 0.5% CaO, 0.5% MgO, 5% KOH, 0.3% K_2CrO_4, and 0.3% K_2SO_4. However, in most of the world the Shell 105 catalyst became the standard. This catalyst consists of 87.9% Fe_2O_3, 2.5% Cr_2O_3, and 9.6% K_2O. In use, the red Fe_2O_3 is largely converted to black Fe_3O_4 which is magnetic and electrically conductive, while the K_2O is partially converted to K_2CO_3. Lee [19] gives much information about composition and effectiveness of iron oxide catalysts for styrene manufacture, including X-ray data for fresh and used Shell 105. He notes that these catalysts are readily poisoned by chlorides.

Catalysts are used as $1/8$ in. diameter cylindrical pellets or sometimes as larger particles. The effect of particle size on selectivity for the related Shell 205 catalyst was noted in butenes dehydrogenation [34]. The effect is less in the styrene reaction, and the increased pressure drop with smaller particles may very well cancel any gain in intrinsic selectivity.

The velocity of the dehydrogenation reaction is determined by catalyst activity, temperature, vapor composition, and pressure. In adiabatic reactors, equilibrium is closely approached at the outlet, and the distance from equilibrium may govern the reaction velocity. An early kinetic analysis of reactor design for ethylbenzene dehydrogenation in externally heated tubes and in adiabatic beds was that of Wenner and Dybdal of Monsanto [35]. They give integral conversion data with 3/16 in. pellets of a catalyst similar to Shell 105 (their catalyst "B"). Later differential and integral reactor studies with 10 to 30 mesh granules of Shell 105 were reported by Carrà and Forni [6]. The reaction rate was essentially independent of ethylbenzene partial pressure at low conversions, but was strongly inhibited by styrene. Rate data are well fitted by the equations

$$dx/d(W/F) = k_1(P_E - P_H P_S/K)/(P_E + bP_S)$$
$$dy/d(W/F) = k_2 P_E/(P_E + bP_S)$$

where x is the fractional conversion to styrene, y is the fractional conversion of ethylbenzene to the by-products benzene and toluene, W is catalyst weight, F is inlet flow rate of ethylbenzene, the k's are reaction velocity constants, K is the equilibrium constant, the P's are partial pressures, and b is an adsorption constant, evaluated to be 8.0. An additional small term can be added to the denominator to account for rates at higher pressures, which were extended to 11 atm in the work of Carrà and Forni. These equations are based on presumption of a parallel reaction to form the by-products. However, selectivity to styrene does decline at high conversions, probably in part from the reaction of styrene with steam to form CO_2, toluene, and other products. To account for this, a term $-k_3 P_S$ should be entered at the right of the first equation, and a term $+k_3 P_S$ should be entered at the right of the second equation. An indication of the effect of conversion on selectivity is given by the following data:

Conversion of ethylbenzene, %	10	25	40	Refs.
Selectivity to styrene, %	97.5	93.5	90	35
H_2/styrene in product	1.1	1.25	1.4	35
Selectivity to styrene, %	96.7	95.4	92.8	24

The exact selectivity (defined as moles of styrene obtained per mole of ethylbenzene converted) is, of course, somewhat dependent on plant design, pressure, temperature, catalyst, and steam dilution as well as on conversion level.

Since a number of different reactions are involved, including dehydrogenation, cracking to ethylene and benzene, hydrocracking to toluene and methane, and reaction with steam to produce hydrogen and CO_2, there should be an optimum temperature profile in the reactor for maximum selectivity to styrene. This problem was conditionally treated by Modell [21], but a lack of well-established values for the activation energies of the various reactions prevented firm conclusions.

Alternative methods of dehydrogenation of ethylbenzene to styrene have been studied. Oxidative dehydrogenation using oxygen as a reactant is not as successful as in the case of butenes conversion to butadiene [26]. However, SO_2 as an oxidant with proper catalysts gives high conversion and selectivity to styrene [11, 26]. Small amounts of sulfur-containing heterocycles are by-products. In the experiments of Gaspar and co-workers [11], SO_2 and O_2 were used in conjunction, the O_2 being added in three successive increments in a flow system. Using mole ratios of SO_2/ethylbenzene = 0.15, total O_2/ethylbenzene = 0,45, H_2O/ethylbenzene = 4, and an alkalized alumina or titania catalyst at 635°C, 85% conversion was obtained with a selectivity to styrene of 94%.

Industrial Processes

A process developed jointly by Monsanto and the C-E Lummus division of Combustion Engineering, Inc., is offered for licensing. A styrene plant of this design with capacity of 670,000 t of styrene per year has been built at Texas City and is said to be the largest in the world. The dehydrogenation of ethylbenzene is carried out in the vapor phase in an innovative reactor design that has permitted extremely large single-train units to be built. The commercially available catalyst is replaced about every 2 years. Starting from ethylene and benzene, the net requirements per kilogram of styrene are 0.307 kg ethylene, 0.820 kg benzene, 1.4 kg of 75 lb/in.² gauge steam, 0.088 kWh electricity, and 1545 kcal fuel. Conversion is probably about 60% with a selectivity of the order of 90%. Catalyst and chemicals cost about 0.2¢/kg of styrene [1, 8].

Another integrated process, developed by Carbide-Cosden-Badger, has likewise been used for quite large units [2]. Dehydrogenation is again in the vapor phase with steam (and perhaps other) diluent. This process is said to use unique methods to improve the dehydrogenation step without adversely affecting catalyst performance or life. Conversion, selectivity, and amount of

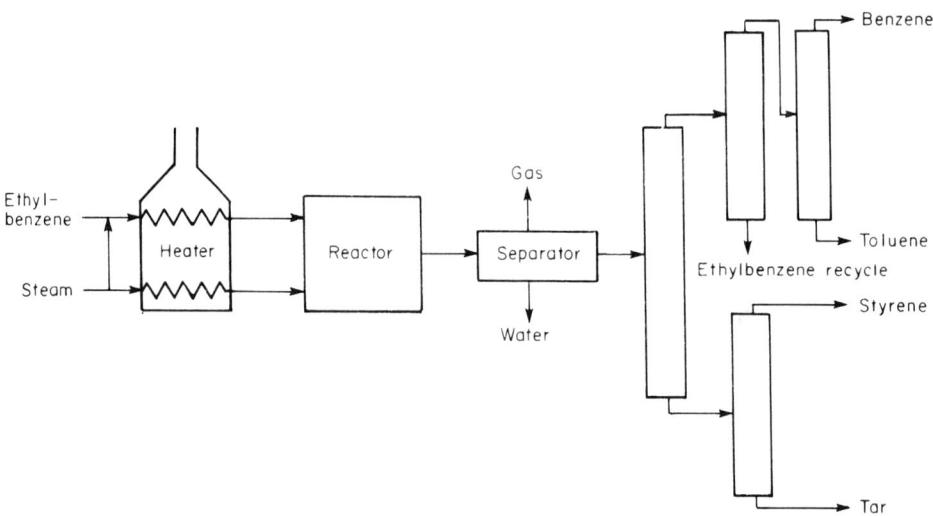

FIG. 3. Generalized flow diagram for ethylbenzene dehydrogenation.

steam are variables that must be optimized in each design. One plant using this process has completed 19 months of continuous operation, and a single charge of catalyst lasted for 27.5 months of operation, giving over 3000 kg styrene/kg catalyst.

A generalized flow diagram for styrene production by dehydrogenation in an adiabatic catalyst bed is shown in Fig. 3. Variations are possible in the sequence of distillation columns as well as in the design of the reactor itself.

Other styrene processes have been developed by Dow Chemical, by BASF (Badische) in Germany, and by Société Chimique de Charbonnages in France. Little has been published about the Dow process. BASF has used semi-isothermal reactors with the catalyst contained in externally heated tubes that are about 16 cm diameter by 300 cm long [24]. Since heat is supplied through the tube walls, it is possible to operate with lower steam dilution in the BASF process, and consequently the more expensive reactor is offset by lower steam expense. The French company offers its process for licensing. It is said to employ a multistage dehydrogenation reactor to obtain high conversion [29].

References

1. Anonymous, *Chem. Week*, p. 34 (November 17, 1976).
2. Badger Co., Inc., *Hydrocarbon Process.*, p. 204 (November 1975).
3. Bajars, L., et al., Petro-Tex Chemical Corp., various U.S. patents, including 3,308,186–3,308,198 and 3,207,805–3,207,811.
4. Beckberger, L. H., and Watson, K. M., *Chem. Eng. Prog.*, 44, 229 (1948).
5. Broughton, D. B., and Berg, R. C., *Hydrocarbon Process.*, 48(6), 115 (1969).
6. Carrà, S., and Forni, L., *Ind. Eng. Chem., Process Des. Dev.*, 4, 281 (1965).
7. Chemetron Corp., *Physical and Thermodynamic Properties of Elements and Compounds*, 1969.
8. Combustion Engineering-Lummus, *Hydrocarbon Process.*, p. 205 (November 1975).
9. Dodd, R. H., and Watson, K. M., *Trans. Am. Inst. Chem. Eng.*, April 25, 1946.
10. Dumez, F. J., and Froment, G. F., *Ind. Eng. Chem., Process Des. Dev.*, 15, 291 (1976).
11. Gaspar, N. J., Cohen, A. D., Vadekar, M., and Pasternak, I. S., *Can. J. Chem. Eng.*, 53, 79 (1975).
12. Gaspar, N. J., and Pasternak, I. S., *Can. J. Chem. Eng.*, 49, 248 (1971) and earlier papers.
13. Gibson, M. A., Cares, W. R., and Hightower, J. W., *Am. Chem. Soc., Div. Pet. Chem., Prepr.*, 22(2), 475 (1977).
14. Hornaday, G. F., Ferrell, F. M., and Mills, G. A., *Pet. Chem. Refin.*, 4(10) (1961).
15. Husen, P. C., Deel, K. R., and Peters, W. D., *Oil Gas J.*, August 2, 1971.
16. Huston, T., Jr., Skinner, R. D., and Logan, R. S., Paper presented at American Chemical Society Meeting, Los Angeles, 1973.
17. Kearby, K. K., "Dehydrogenation," in *Catalysis*, Vol. 3 (P. H. Emmett, ed.), Reinhold, New York, 1955.
18. King, R. W., *Hydrocarbon Process.*, 45, 189 (1966).
19. Lee, E. H., *Catal. Rev.*, 8, 285 (1973).
20. Newman, F. C., *Ind. Eng. Chem.*, 62, 42 (1970).

21. Modell, D. J., *Chem. Eng. Comput.*, *1*, 100 (1972).
22. Noddings, C. R., and Mullet, G. M., *Handbook of Compositions at Thermodynamic Equilibrium*, Wiley-Interscience, New York, 1965.
23. Noddings, C. R., Heath, S. B., and Corey, J. W., *Ind. Eng. Chem.*, *47*, 1373 (1955); Britton, E. C., Dietzler, A. J., and Noddings, C. R., *Ind. Eng. Chem.*, *43*, 2871 (1951).
24. Ohlinger, H., and Stadelmann, S., *Chem. Ing. Tech.*, *37*, 361 (1965).
25. Raley, J. H., Mullineaux, R. D., and Bittner, C. W., *J. Am. Chem. Soc.*, *85*, 3174 (1963).
26. Rennard, R. J., Innes, R. A., and Swift, H. E., *J. Catal.*, *30*, 128 (1973).
27. Rossini, F. D., et al., *Selected Values of Physical and Thermodynamic Properties of Hydrocarbons and Related Compounds* (API Project 44), Carnegie Press, Pittsburgh, 1953.
28. Skarchenko, *Int. Chem. Eng.*, 9(1), 1 (1969).
29. Societe Chimique des Charbonnages, *Hydrocarbon Process.*, November 1969, 121.
30. Sterrett, J. S., and McIlvried, H. G., *Ind. Eng. Chem.*, *Process Des. Dev.*, *13*, 54 (1974).
31. Suzuki, I., and Kanero, Y., *J. Catal. 47*, 239 (1977).
32. Thomas, C. L., *Catalytic Processes and Proven Catalysts*, Academic, New York, 1970.
33. Universal Oil Products Co., *Hydrocarbon Process.*, p. 145 (November 1973); p. 157 (November 1975).
34. Voge, H. H., and Morgan, C. Z., *Ind. Eng. Chem.*, *Process Des. Dev.*, *11*, 454 (1972).
35. Wenner, R. R., and Dybdal, E. C., *Chem. Eng. Prog.*, *44*, 275 (1948).

HERVEY H. VOGE

Dewaxing, Catalytic

Introduction

Historically, most commercial dewaxing operations have been based on the use of a solvent. The mixture of oil and solvent is refrigerated and the wax which separates out is removed by filtration. The dewaxed oil is then recovered by distilling off the solvent.

The solvent dewaxing process has two main disadvantages. First, the operating costs for the process are high. Second, the pour point that can be achieved is limited by the high cost of refrigerating to very low temperatures.

Because of these disadvantages, considerable research has been carried out on hydrocatalytic processes for pour point reduction. These processes are based on the ability of certain zeolites to selectively crack the paraffinic hydrocarbons which are the major constituent of wax.

This article summarizes a number of aspects of the BP Catalytic Dewaxing Process, including how the process works, applications, and economics.

Process Description

Catalyst

The key to any hydrocatalytic process for pour point reduction is the catalyst. Wax consists mainly of *n*-paraffins and slightly branched paraffins. The dewaxing catalyst needs to be able to differentiate between these components and the rest of the feedstock, and to crack them selectively. The two main criteria for a good dewaxing catalyst can be summarized as

1. That it is shape selective for paraffins.
2. That it is an active hydrocracking catalyst.

The BP Catalytic Dewaxing Process uses a catalyst based on the zeolite mordenite which has the correct pore structure to be selective for normal and slightly branched paraffins. It is also a highly active cracking material. The catalyst also contains a noble metal which acts as a hydrogenation function. Breakdown products from the cracking reaction are rapidly hydrogenated, thereby minimizing carbon deposition on the catalyst and leading to a highly stable catalyst system.

The active sites of the catalyst effectively only come into contact with paraffinic chains. Because of this, it is possible to operate the process successfully when dewaxing straight run distillates of high sulfur and nitrogen contents. For the same reason, little desulfurization or denitrogenation is observed.

Process Conditions and Flow Sheet

Catalytic dewaxing is a mild hydrocracking process and, therefore, has to be operated at elevated temperatures and hydrogen partial pressures. In general, the process can be operated within the following range of conditions.

Hydrogen partial pressure (psi)	300–1500
Liquid space velocity (v/v/h)	0.5–5
Catalyst temperature (°F)	550–750
Gas recycle rate (SCF/bbl)	2000–5000

Within the ranges set out above, the conditions for a particular dewaxing operation will depend upon the nature of the feedstock and the required product pour point. Important feedstock parameters are

Boiling range
Amount and type of wax present
Sulfur and nitrogen contents

A flow diagram for a typical dewaxing unit is shown in Fig. 1. Feed from storage is mixed with recycle gas and heat exchanged with reactor effluent before being heated to the required reactor inlet temperature by a fired heater. The reactor feed passes downflow through the reactor. Because the reaction is

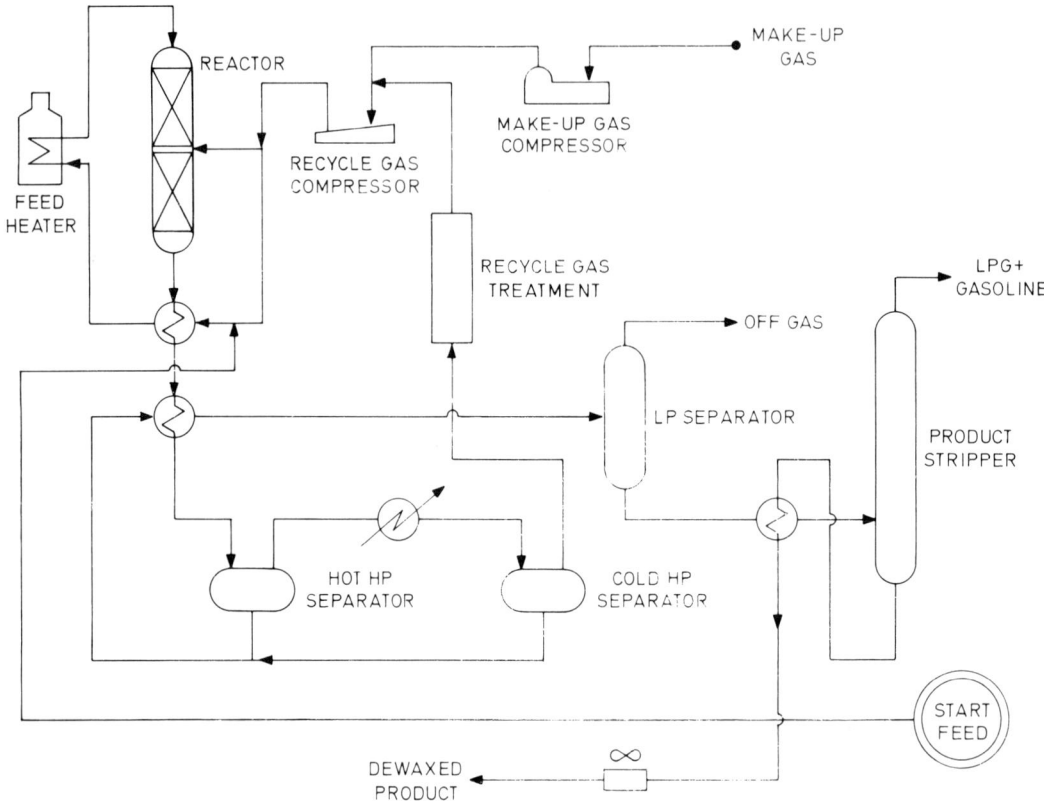

FIG. 1. Flow diagram of the BP Catalytic Dewaxing Process.

TABLE 1

	Solvent Extraction Followed by Catalytic Dewaxing	Catalytic Dewaxing Followed by Solvent Extraction	I.E.C. Class II Specification
Density at 68°F	0.850	0.850	0.895 max
Kinematic viscosity (cSt):			
at 68°F	18.3	15.75	25 max
at −22°F	550	444	1800 max
Pour point (°F)	−49	−60	−49 max
Flash point (°F)	335	295	266 min
Neutralization number (mg KOH/g)	0.01	0.01	0.03 max
Corrosive sulfur	None	None	None
Loss tangent at 90°C	3.5×10^{-4}	1.7×10^{-3}	5×10^{-3} max
Oxidation test:			
Sludge (wt.%)	0.03	0.06	0.10 max
Acidity after oxidation (mg KOH/g)	0.15	0.35	0.40 max

Dewaxing, Catalytic

exothermic, it is sometimes necessary for cold recycle gas to be injected between the catalyst beds to limit the temperature rise. The reactor effluent is cooled against the reactor feed before passing to a hot high-pressure separator. The gas from this separator is cooled and then passed to a cold high-pressure separator. The gas from the cold separator is recycled to the reactor inlet via a gas treatment section and a recycle gas compressor.

The liquid hydrocarbons from the two high pressure separators are combined and flow to a low-pressure separator where off-gas is removed. The remaining liqud passes to a product stripper where LPG and gasoline are removed as overheads fractions. The dewaxed oil is produced as a bottoms fraction.

Breakdown Products

The main breakdown products are saturated paraffins in the range C_3-C_8. Typical distributions are given later in the section dealing with applications of the process. Even when operating on high sulfur feedstocks, the yield of hydrogen sulfide is less than 0.1% weight on feed.

Applications of the Catalytic Dewaxing Process

The process can be used for a wide variety of dewaxing applications in two main areas:

1. Production of speciality oils and lubricating oils
2. Production of low pour point middle distillates

These applications will be considered separately.

Speciality Oils and Lubricating Oils

Very low pour point oils such as transformer oils, refrigerator oils, and certain hydraulic fluids have traditionally been manufactured from specially selected low wax naphthenic crudes. However, this route is becoming increasingly unattractive because of the high price and shortage of supply of such crudes. These factors led BP to develop a catalytic dewaxing process capable of producing speciality oils from paraffinic Middle East crudes.

Catalytic dewaxing alone only removes wax and gives some color improvement. A further processing step is required to make oils which meet all quality specifications. This step is normally solvent extraction although, in certain circumstances, high-pressure hydrogenation can also be used.

Table 1 shows the properties of transformer oils prepared by the two possible routes involving solvent extraction. In each case, the feed was a Kuwait light vacuum gas oil.

Both oils meet all of the International Electrotechnical Commission (I.E.C.) Class II transformer oil specifications. Generally, the route in which catalytic dewaxing follows solvent extraction is preferred for the following reasons:

Smaller dewaxing capacity needed
Reduced dewaxing severity
Slightly better finished product quality

In the preparation of refrigerator oils, catalytic dewaxing has some limitations with respect to the viscosity grades that can be produced. The higher viscosity oils tend to contain microcrystalline wax which cannot be removed by catalytic dewaxing and, therefore, the degree of pour point reduction which can be obtained is limited. However, the lower viscosity grades are particularly amenable to catalytic dewaxing, and finished oils of very high quality can be produced. Table 2 shows the properties of refrigerator oil obtained from an Iranian light vacuum gas oil by solvent extraction followed by catalytic dewaxing.

The low values for floc point and Freon insolubles show that catalytic dewaxing has successfully removed all wax from the feedstock.

Catalytic dewaxing can also be used as an alternative to solvent dewaxing for the production of low pour point lube oils of medium viscosity index. For lubricating oil expansion projects, consideration can be given to the use of catalytic dewaxing for low viscosity oils and to utilizing existing solvent dewaxing capacity for high viscosity oils.

Middle Distillates

In some refinery situations it is necessary to restrict the quantity of heavy boiling distillate used in gas oil blends in order to meet winter pour point specifications on middle distillates. Catalytic dewaxing can be used to treat these heavy distillates to produce acceptable blending components and, therefore, increase the yield of middle distillate products. As there is no practical upper limit to the boiling range of distillates that can be catalytically dewaxed, the process is able to treat any fraction from atmospheric distillation.

Table 3 gives examples of how catalytic dewaxing can be used to reduce the pour point of heavy gas oils from three different crudes of varying wax content.

In operations such as those shown in Table 3, the product pour point can be

TABLE 2

Kinematic viscosity (cSt):	
at 68°F	59.6
at 100°F	24.8
at 210°F	4.2
Pour point (°F)	−40
Flash point (°F)	385
Neutralization number (mg KOH/g)	0.01
Ash content at 1427°F (wt.%)	<0.01
Floc point (°F)	−60
Freon 12 insolubles at −40°F (wt.%)	0.01

Dewaxing, Catalytic

TABLE 3

Crude Source	A	B	C
Feed data:			
Pour point (°F)	+60	+80	+75
Sulfur (wt.%)	1.92	1.73	0.43
Viscosity at 50°C (cSt)	7.96	12.37	—
Wax content (wt.%)	11.0	16.5	20.0
Distillation (°F):			
10 wt.%	630	621	547
50 wt.%	718	766	660
90 wt.%	772	855	795
Product data:			
Pour point (°F)	−5	−5	−10
Viscosity at 50°C (cSt)	9.04	16.35	—

changed by adjusting the catalyst temperature. In the case of the feedstock from crude source A in Table 3, it was possible to reach a pour point as low as −35°F.

Table 4 illustrates the breakdown products obtained from catalytic dewaxing by giving the product distributions (output bases) for the above three operations.

Process Economics

The detailed capital cost and utility requirements for a BP Catalytic Dewaxing unit will depend upon the particular application. However, to give an indication of process economics, Table 5 gives the capital cost and utility requirements for a 10,000 BPSD dewaxing unit in a UK location (2nd quarter, 1980).

The capital cost is based on materials plus direct labor and includes all major items of equipment and piping, electrical equipment, instruments, painting and insulation, and structures, together with the cost of labor for erecting the plant on a prepared site.

TABLE 4

Crude Source (wt.%)	A	B	C
Methane + ethane	0.1	0.4	0.1
Propane	2.5	5.2	3.8
Butanes	6.4	9.4	11.0
C_{5+} gasoline	8.2	10.4	16.3
Dewaxed oil	82.8	74.6	68.8

TABLE 5

Unit Capacity	10,000 BPSD
Capital cost (£)	3,200,000
Utilities:	
Electricity (kW)	680
Heat absorbed (BTU/h)	12.5×10^6
Medium pressure steam (lb/h)	8,620
Cooling water (lb/h)	165,000

Commercial Installation

A commercial unit with a capacity of 2000 BPSD was commissioned satisfactorily in the United States in 1977. This unit has successfully dewaxed a wide variety of paraffinic and naphthenic feedstocks and, to date, process performance has been excellent with no major operational or mechanical trouble.

Bibliography

Further information about the BP Catalytic Dewaxing Process can be found in the following.

Bennett, R. N., and Elkes, G. J., *Low Pour Oils from Paraffinic Crudes by the Cat Dewaxing Process*, Paper Submitted to the NPRA Fuels and Lubricants Meeting, Houston, Texas, November 1974.

Burbidge, B. W., Keen, I. M., and Eyles, M. K., "Physical and Catalytic Properties of the Zeolite Mordenite," in *Molecular Sieve Zeolites II* (Advances in Chemistry Series No 102), 1971.

Donaldson, K., and Pout, C. R., *The Application of a Catalytic Dewaxing Process to the Production of Lubricating Oil Basestocks*, Paper Submitted to the American Chemical Society, August 1972.

Hargrove, J. D., Elkes, G. J., and Richardson, A. H., *BP Cat Dewaxing—Experience in Commercial Operation*, Paper Submitted to the NPRA Fuels and Lubricants Meeting, Houston, Texas, November 1978; Reprinted in *Oil Gas J.*, pp. 103–105 (January 15, 1979).

Weinstabl, H., and Elkes, G. J., "Catalytic Dewaxing of Middle Distillates," *Erdoel, Erdgas Z., Proceedings of the OGEW Annual Meeting 1977*, pp. 193–195.

J. D. HARGROVE

Dewaxing, Solvent

Introduction

One of the important properties of lubricating oils is their flow characteristics at low temperatures. In order to insure proper lubrication of machinery, oil must flow at temperatures well below the lowest possible working temperature. The main feedstocks used today in the fabrication of lubricating oil are mixtures of hydrocarbons produced from paraffinic crudes, but these mixtures also contain, together with hydrocarbons needed in the production of lubricating oils, variable amounts of paraffinic hydrocarbons with high melting points. At low temperatures these paraffinic hydrocarbons crystallize out of the oil and form a network in which the liquid phase is trapped and cannot be displaced. The lowest temperature observed in a standard laboratory test at which the oil still flows is called the pour point [17]. These paraffinic hydrocarbons are called waxes, and the process in which a wax is removed from oil is called dewaxing.

For high viscosity oils such as bright stock, a viscosity pour point can be observed. The oil fails to flow in the pour point test because its viscosity is too high and a deeper "dewaxing" does not result in any improvement.

The first dewaxing processes were based on cooling the waxy feed alone or diluted with naphtha and separating the wax by cold settling or, later on, by centrifuges. For the separation of wax from cold light vacuum distillates, plate and frame filter presses were used.

Although some of the old methods are still in commercial use today, three main methods are used in modern refinery technology:

1. *Solvent Dewaxing.* In these types of processes the feed oil is mixed with one or more solvents, the feed oil/solvent mixture is cooled down to allow the formation of wax crystals, and the solid phase is separated from the liquid phase by filtration.
2. *Urea Dewaxing.* Urea straight-chain paraffin adducts are formed and separated by filtration from the dewaxed oil.
3. *Catalytic Dewaxing.* Straight-chain paraffinic hydrocarbons are selectively cracked on zeolite-type catalysts, and the light reaction products are separated from the dewaxed lubricating oil by fractionation.

The most used method is solvent dewaxing. In 1973 the capacity of solvent dewaxing units in the United States and Canada was over 210,000 bbl/d [14].

Properties of Dewaxed Oil and Refinery Waxes

Dewaxed Oil

The removal of waxy constituencies from a stock used for the production of lubricating oil will change not only its pour points but other physical properties also.

The specific gravity of wax is low, ~ 0.750 to 0.780 at $210°F$, and as a consequence the specific gravity of the dewaxed oil is higher than the specific gravity of the charge stock. The viscosity of the wax in the liquid state is lower than that of the oil. The viscosity of the dewaxed oil is higher than that of the charge stock, the change in viscosity being higher for high viscosity oils.

Dewaxing also reduces the viscosity index of the oil. Starting from the same waxy oil feed, a lower pour point dewaxed oil will have a lower viscosity index than a higher pour point oil. The gradient changes with the viscosity of the feed and with the temperature range of the pour point.

Wax

Wax consists mainly of normal and low branched paraffinic hydrocarbons. It is a colorless mass, more or less translucent, and has a crystalline or semicrystalline structure. The classification of waxes is somewhat arbitrary and is related to the method of production and the molecular weight. The refinery wax produced from light viscosity distillates by chilling, pressing, and sweating is called paraffin wax. The product from solvent dewaxing is called slack wax. This slack wax is deoiled to produce hard wax or scale wax. Microcrystalline waxes are produced from the dewaxing of residual stocks.

Refinery waxes are characterized by the oil content [15], melting point [18], and penetration [19]. For a given wax the melting point and the penetration are functions of the oil content.

Depending on the molecular weight, the melting point of waxes with oil contents of less than 0.5 wt.% is between 80 and $160°F$ for distilled stocks with $C_{18}H_{38}$ to $C_{32}H_{66}$ hydrocarbons [20], and 140 to $190°F$ for microcrystalline waxes produced from bright stocks. The molecular weight of microcrystalline waxes is between 450 and 600 [20].

Solvent Dewaxing [6]

All solvent dewaxing processes are based on the same fundamental idea of cooling a mixture of feed oil and solvent to a temperature low enough to cause crystallization of the required amount of wax from the solution and subsequent separation of the solid phase. As far as the dewaxing properties are concerned, the solvent must have the following characteristics:

A high solvent power at low temperature for dewaxed oils
A very low solubility for wax
A low melting point, well below the filtration temperature

In addition, the solvent must have the same general properties desired for any solvent used in a commercial process:

Low cost
High thermal and chemical stability
Low boiling points
Low latent heat of vaporization and low specific heat
Uncorrosivity
Untoxicity

Many times, in order to optimize the dewaxing properties of the solvent, a mixture of two solvents is used. In the most used dewaxing process, known as the MEK Process, the solvent is a mixture of toluene and methyl ethyl ketone (MEK). The toluene is a good solvent for hydrocarbons and as a consequence has a quite high solubility for wax also. In order to minimize the wax solubility, MEK is added. Because the solubility of hydrocarbons in MEK is quite low, if too much MEK is present in the toluene/MEK feed oil mixture, an oily liquid phase will also separate. The effect of the MEK/toluene and the solvent/feed oil ratios on the phase separation temperature is shown in Fig. 1, using the Texaco method. It consists of determining the cloud point of a solution containing filtrate and solvent for a certain solvent/filtrate ratio by varying the MEK/toluene ratio in the solvent.

An important factor in the solvent dewaxing processes is the dewaxing temperature differential, i.e., the difference between the filtration temperature and the pour point of the dewaxed oil. It is a function of the solubility of wax in the solvent.

As it can be seen from Fig. 1, the temperature differential—the difference

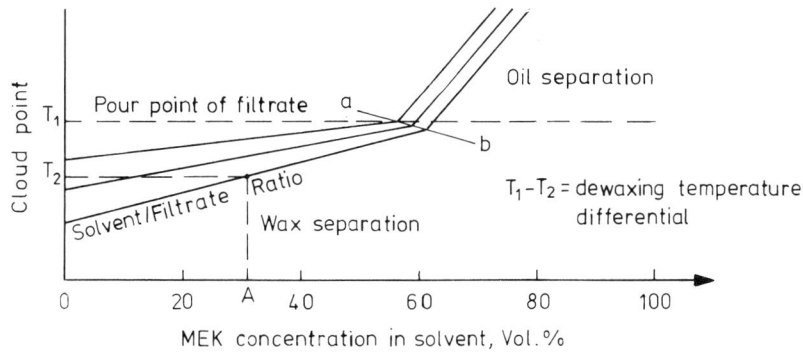

FIG. 1. Miscibility diagram of filtrate in toluene/MEK mixtures.

between the cloud point in the wax separation region and the pour point of the filtrate—decreases with MEK concentration. But, because a higher dewaxing temperature differential means a lower filtration temperature for a required pour point of dewaxed oil, the higher the MEK/toluene ratio, the higher the filtration temperature. As a consequence, lower cooling/chilling and refrigeration capacity is needed. At the same time, higher MEK/toluene ratios will produce bigger wax crystals and so a higher filtration rate. From these points of view, the maximum percentage of MEK should be used. On the other side, this maximum is defined by the miscibility diagram as a MEK/toluene ratio low enough to avoid operation in the oil phase separation region—on the right side of the line $a-b$ in Fig. 1—under normal running conditions of the unit.

The same phenomenon is observed in all solvent dewaxing processes using a mixture of two solvents like the Dichlorethane–Methylene dichloride (Di/Me) and the old Edeleanu–Benzene–SO_2 Processes. In the Di/Me Process [16] the methylene chloride (ME) is the solvent with high solvent power—equivalent to toluene in the MEK Process—and the dichlorethane (Di) is the solvent with high selectivity. A phase separation curve for MEK and Di/Me is shown in Fig. 2 using the same waxy feed.

Another solvent used in commercial units is a MEK/MIBK (methyl isobutyl ketone) mixture. Two other solvents are used as single solvents in some commercial dewaxing units—liquid propane and methyl isobutyl ketone (MIBK).

The solubility of paraffin wax in hydrocarbons increases with a decrease in the melting point of the wax and with an increase of temperature. Hydrocarbons of decreasing molecular weight, down to pentane/butane, increase in wax solubility. A further decrease of the molecular weight below pentane/butane shows a decrease of the solubility of wax [5]. On the other hand, using hydrocarbons with too high a vapor pressure (e.g., methane or ethane) will complicate the process. Propane has the advantage of having quite low solubility for wax and not too high a vapor pressure.

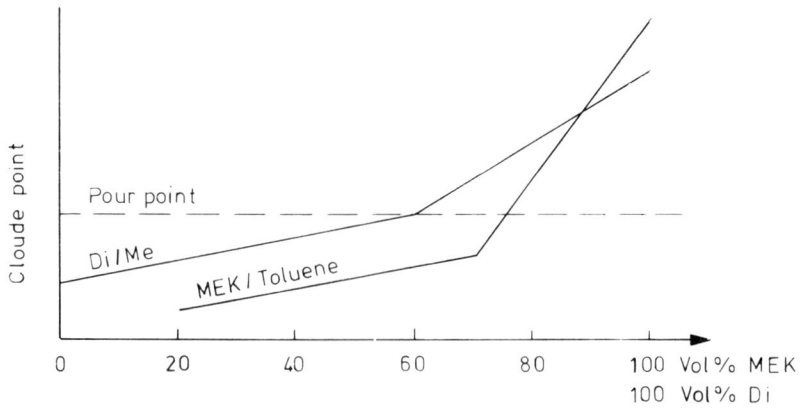

FIG. 2. Comparison between miscibility curves of MEK/toluene and Di/Me solvents using the same charge oil and the dilution ratio.

It is possible to adjust the solvent power and the selectivity by changing the ratio between the two solvents—as in the MEK and Di/Me Processes. In the one solvent dewaxing processes the characteristics of the solvent are changed by adjusting the filtration temperature. In propane dewaxing, due to the high solubility of wax in propane, a typical temperature differential is 25 to 40°F in comparison to 5 to 15°F in Di/Me, MEK/MIBK, and MEK/toluene dewaxing.

Another important variable of the Solvent Dewaxing Process is the solvent/charge oil (dilution) ratio. The dilution ratio affects the structure of the wax crystals and the filtration characteristics of the cake on the filter as well as the oil content of the wax and hence the yield of dewaxed oil. For instance, when high dilution ratios are used on low viscosity stocks, fluffy crystals are formed at the beginning of the cooling process. During the cooling process, smaller crystals are incorporated in and around this fluffy big structure. Finally, although very good filtration characteristics of the cake are obtained, the amount of oil trapped in these cages cannot be washed out and the oil content of the wax is high and dewaxed oil yields are low. Dilution ratios that are too low are even more detrimental because only very small crystals, which give poor filtration and washing rates, can be obtained.

The main parameter governing the solvent dilution technique is the viscosity of the liquid phase at various points in the cooling/chilling stage as well as on the filter. The kinetics of the solid phase formation—which are a function of the viscosity of the mother liquor among other factors—are a determining factor in the structure of the wax crystals and of the cake on the filter. The solvent dilution ratio is a function of the viscosity–temperature curve of each solvent and a function of the solvent composition for the dual solvent dewaxing processes as well. For the same reason, at a given solvent viscosity, the viscosity of the charge oil will determine the solvent dilution ratio. Heavy, high viscosity oils are diluted with more solvent than light oils.

With more wax crystallizing out of the liquid phase, the viscosity of the remaining liquid phase changes. At the same time the solvent-to-oil ratio in the liquid phase will determine the oil content of the wax. Since a given percentage of the small pores cannot be washed out with fresh solvent in the washing stage, the amount of oil trapped as the liquid phase in these pores is dependent on the oil content of the liquid phase of the slurry.

An improvement of filtration is observed if an incremental solvent dilution technique is used in the cooling/chilling stage. The feedstock is mixed with a limited amount of dilution solvent and the rest of the dilution solvent is added stagewise during the cooling. This technique is very useful for light stocks. The initial low dilution, at a point where the temperature of the solvent/charge oil solution is still high and where the crystallization process has not yet started, avoids the formation of big fluffy structures. Later on, with an increased viscosity of the solution, more solvent is added. For heavy stocks, e.g., bright stock, where the initial viscosity is high and the structure of the wax crystals is microcrystalline, this incremental dilution technique is not used.

Avoiding the formation of big crystals in the beginning stage of crystallization is accomplished in some processes by means of shock chilling and/or high turbulence. In shock chilling the fresh solvent is added incrementally at a somewhat lower temperature than the solvent/charge mixture stream. In addition, in the initial stages of wax crystal formation, a high turbulence induced in the solution will break down any fluffy structures.

The shock chilling technique is useful for dewaxing light narrow cuts. By dewaxing long cuts containing both macro- and microcrystalline waxes, shock chilling can sometimes be detrimental because the first wax crystals grown out of the solution are the high molecular weight, high melting point microcrystalline ones. Shock chilling will make these crystals even smaller.

The cooling step is accomplished in ways that differ from one process to another. The most used apparatus is the scraped-wall exchanger chiller. It is a double pipe heat exchanger where the solvent/feed charge mixture flows through the inner tube. The inside surface of the inner tube is continuously scraped by blades to avoid the deposition of wax and thus a drop in the heat transfer coefficient and an increase in the pressure drop. For scraped-wall exchangers which are located at the beginning of the cooling chain, cold filtrate solution is used as the cooling medium in the outer tube. For scraped-wall chillers the refrigeration medium (ammonia or propane) is evaporated in the outer tube. Standard sizes for the inner tubes are 6 and 8 in., although 12 in. tubes have also been used recently.

In the Propane Dewaxing Process the chiller is a horizontal tank in which propane is evaporated. At least two tanks in parallel are used, the chilling process being a batch operation. The chilling rate is controlled by the propane compressor, the mean value being approximately 1°C/min or less.

In the Dilchill Process the first cooling stage is a crystallizer provided with many mixing elements. The chilling rate is controlled by incremental injection of cold solvent and by the size of the crystallizer. Cooling rates of 0.6 to 2.8°C/min are considered to give the best results in filter rates. A minimum value for the mixing Reynolds number is required [10, 11].

Another important criteria in a Solvent Dewaxing Process is the liquid-to-solid ratio in the slurry fed to the filter. With too much solid, the agglomerates

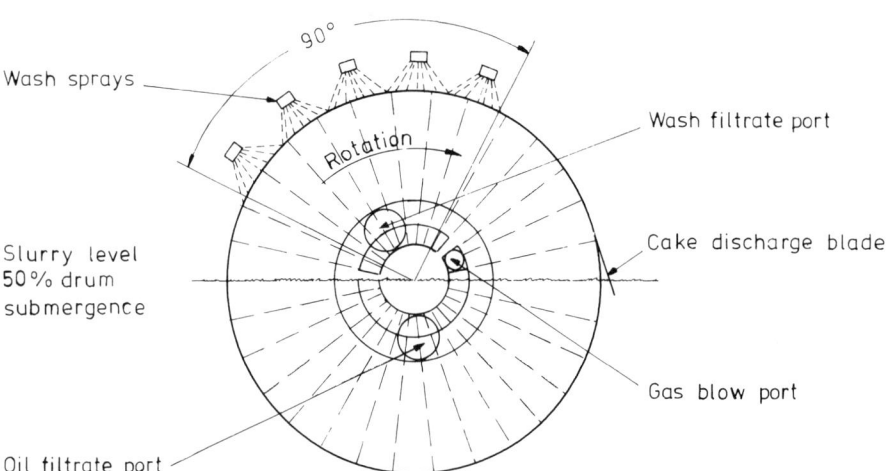

FIG. 3. Cross-sectional diagram of a rotary drum filter.

of wax crystals will stick together to give a thick, tight cake with poor filter rates and a high oil content in the wax. If the liquid-to-solid ratio is too high, the cake on the filter cloth becomes too thin with many cracks, and the wash solvent, instead of washing the cake, will flow through these cracks, leaving the cake unwashed. For adjusting the liquid-to-solid ratio in some plants where the wax content of the feed is very low, wax is recycled into the feed. Recycling of filtrate or wash filtrate into the main steam to the filter before the last chiller is used in many plants where the wax content of the feed is high.

Perhaps the most important equipment in a solvent dewaxing unit is the filter. Large rotary drum filters are used in every solvent dewaxing unit. A typical partition of the filter drum in dewaxing processes is shown in Fig. 3.

The filtration zone is at the bottom of the drum and represents about 30 to 40% of the operating cycle for most dewaxing processes (10 to 20% for the Di/Me Process).

The separation of dewaxed oil from the wax by filtration is the key step in the dewaxing process [1–4].

The basic mathematical formula used in treating the filtration is the Poiseuille equation, which for noncompressible cakes is:

$$\frac{dV}{d\phi} = \frac{PA}{\mu_l[k(V/A) + r]} \tag{1}$$

where $dV/d\phi$ = filtration rate at the moment that the quantity of filtrate V has been obtained
P = pressure difference at which filtration takes place
A = surface area of filtering medium
μ_l = viscosity of filtrate at filtration temperature
k = specific resistance of filter cake
= resistance of filter cake per unit surface area when unit quantity of filtrate has passed through it
r = resistance of filtering medium per unit of surface area

The application of Poiseuille's formula to dewaxing is complicated by many factors:

The specific resistance of a filter cake (k) is a function of the structure of the wax crystals which is by itself a function of the dilution and of the physical properties (e.g., viscosity) of the oil/solvent mixture. Beside that, the specific resistance of the filter cake changes during filtration, the cake being compressible.

The cloth resistance is a time-dependent variable due to the cloth fouling which is a function of the structure of the wax crystals (microcrystalline waxes are caught in the tissue much easier than macrocrystalline waxes), of the rotation speed (the frequency of fouling the cloth in unit time), and the frequency of warm washing of the cloth (for removing the wax crystals caught in the tissue).

Nevertheless, Eq. (1) can be used to understand how to adjust the rotary filter operation for the optimization of the filtration process. Equation (1) applies strictly to batch filtration with an uncompressible cake. To render Eq. (1) valid for compressible cakes, Mondria [7, 8] wrote it in a general form:

$$\frac{dV}{d\phi} = \frac{A f(P, \mu_l)}{k(V/A) + r} \tag{2}$$

Integrating this equation to find the quantity of filtrate V, which is obtained in time ϕ in a batch filtration under constant pressure, gives

$$V = \frac{Ar}{k}\left(\sqrt{1 + \frac{2k}{r^2}f(P, \mu_l)\phi} - 1\right) \tag{3}$$

For a continuous filtration with a rotary vacuum filter, the time ϕ from discontinuous filtration is replaced by

$$c \times t$$

where t = time units needed for carrying out a cycle consisting of filtration, washing, drying, and cake discharge
c = fraction of t used for filtration only. Equation (3) becomes

$$V_{cont} = \frac{Ar}{k}\left(\sqrt{1 + \frac{2kct}{r^2}f(P, \mu_l)} - 1\right) \tag{3a}$$

V_{cont} = the volume of filtrate obtained during one complete cycle of the drum

The volume of filtrate per unit of time, Q, will be

$$Q = \frac{V_{cont}}{t} = \frac{Ar}{kt}\left(\sqrt{1 + \frac{2kct}{r^2}f(P, \mu_l)} - 1\right) \tag{4}$$

where t is the time of revolution of the filter.
After introducing the number of revolutions n ($= 1/t$), Eq. (4) becomes

$$Q = \frac{nAr}{k}\left(\sqrt{1 + 2\frac{kc}{r^2 n}f(P, \mu_l)} - 1\right) \tag{5}$$

Mondria [7, 8] uses Eq. (5) for determining the influence of the rotational speed of the filter drum on the filtration rate by defining the ratio between the resistance of cake to the resistance of cloth

$$kq/Ar = \alpha; \quad q = Q/n$$

at a reference number of revolutions per unit time n_r

Define y as the ratio between the filtration rates $Q(n)$ and $Q(n_r)$ at n and n_r revolutions per unit time, respectively. From Eq. (5),

$$\alpha y^2 + 2xy - (\alpha + 2)x = 0 \qquad (6)$$

where

$$x = n/n_r$$

The following conclusions are drawn:

The filter rate increases continuously by increasing the rotation speeds.
For a negligible filter cloth resistance ($\alpha \to \infty$), the rate of filtration increases with the square root of x.
For a negligible cake resistance ($\alpha \to 0$), the rate of filtration is practically independent of the rotation speed.

In order to reduce the oil content of the wax and increase the yield of dewaxed oil, the cake produced in the filtration zone is washed with solvent in the washing zone.

Depending on the construction of the filter, the ratio between the washing and the filtration zones is in general between 2 to 1 and 1 to 2.

As already mentioned, the cake has a porous structure. After filtration, the pores are filled with filtrate solution and the oil content of the wax depends on the porosity of the cake and on the dilution ratio at the filter entrance. The higher the porosity and the lower the solvent/oil ratio, the higher is the content of the cake after removing the solvent. In solvent dewaxing, porosities between 0.70 and 0.90 have been measured, being somewhat higher after filtration than after washing. Increasing the washing ratio will reduce the porosity. The porosity of the cake increases with the molecular weight of the wax.

The wash solvent passing through the cake displaces the filtrate. For an ideal washing case—piston flow of wash solvent through the pores—only a volume of wash solvent equal to the pore volume will be required for complete removal of filtrate. In fact, due to laminar flow profile and diffusion, the fraction of filtrate washed out is an asymptotic function of the volume of wash solvent [9].

Based on material balances and on a flow model in capillaries, Butler and Tiedje [9] found that two washing sequences can be determined. At the beginning the washing solvent displaces the filtrate from the capillaries and the filtrate will be of the same composition as the original filtrate. After this point (breakthrough), the liquid leaving the cake will be partly original filtrate and partly wash solvent.

The following symbols will be used:

W = volume of dry solvent
K_D = volume of solvent used for dilution
K_W = volume of solvent used for washing

O = volume of oil in the feed
O_P = volume of oil in the washed cake
O_A = volume of oil in the cake after filtration
αW = volume of solvent in the cake

Knowing αW for a particular oil, the dilution ratio (K_D/O) can be calculated for the required degree of deoiling (O_P/O). By washing, the filtrate will be displaced, and up to the breakthrough point

$$O_P = O_A \left[1 - \frac{K_W}{\alpha W + O_P} \right] \qquad (7)$$

volumes of oil will remain in the cake.

The breakthrough point occurs when the cake is washed with a volume of wash solvent equal to half of the original volume of filtrate retained in the cake:

$$0 < \frac{K_W}{\alpha W + O_P} < 0.5 \qquad (8)$$

If more solvent is used for washing, i.e.,

$$\frac{K_W}{\alpha W + O_P} > 0.5 \qquad (9)$$

the volume of oil remaining in the cake will be

$$O_P = O_A \frac{\alpha W + O_P}{4 K_W} \qquad (10)$$

From Eqs. (8), (9), and (10) the total volume of solvent ($K_D + K_W$) needed for a required oil content $O_P/(K_W + O_P)$ can be calculated.

Experimental data are required for αW. Because the cake is compressible, the porosity sometimes decreases during the washing. As a consequence, αW is often a variable. Equations (8), (9), and (10) can also be used for calculating the influence of distributing the total amount of solvent between dilution (K_D) and washing (K_W) although care should be taken because the dilution ratio influences the structure of the cake itself (porosity, permeability, etc.).

Useful information about the optimization of filter operation can be obtained by applying the Poiseuille formula to the washing operation in the same manner as shown before for filtration [7, 8]. For example,

For a negligible cloth resistance, the washing efficiency is the same at any rate of rotation, assuming that the rate of washing (volume of wash solvent/volume of cake) can be kept constant.
For a finite value of cloth resistance, the washing efficiency improves somewhat as the rotational speed increases.
Assuming the cake resistance during filtration and washing is constant, the

ratio between the volume of wash solvent (q_W) and the volume of filtrate (q) is

$$x = \frac{q_W}{q} = \frac{T_W}{T} \frac{\eta}{\eta^*} \frac{\frac{1}{2}\alpha + 1}{\alpha + 1} \quad (11)$$

where T_W = time interval for washing
T = time interval for filtration
η = viscosity of filtrate
η^* = intermediate viscosity between filtrate and wash solvent
α = ratio of cake and cloth resistance

Because the spray of the wash solvent on the surface of the cake is less equally distributed than the flow of the filtrate from the immersed slurry bath, experimental values for x are somewhat lower than from Eq. (11).

The fouling of the filter cloth and hence the increase of filter cloth resistance are important factors in filtration, and thus generous allowance is made in sizing the required filter area for a dewaxing unit. The filters are periodically washed individually with warm solvent in order to remove the wax and ice crystals trapped in the cloth. The higher the molecular weight of the feed, the shorter is the time interval between two successive warm washing operations.

MEK-Dewaxing Process [23]

At the beginning of the process the feed oil is diluted with solvent. To insure that the feed oil dissolves completely in the solvent, the mixture is heated after the initial dilution (Fig. 4).

The heating temperature is a function of the characteristics of the feed oil and of the initial dilution ratio. It is about 10 to 15°F above the melting point of the wax to be separated. The charge oil–solvent mixture is cooled in a water cooler up to its cloud point and then in double-pipe heat exchangers (scraped-wall coolers) with cold filtrate. The final cooling step is achieved in scraped-wall chillers by evaporating a refrigeration medium (propane, ammonia, etc). The dilution solvent is introduced into the wax-bearing oil stream at selected points in the chilling cycle.

The slurry flows into a filter feed tank and then, by gravity, into the filters. The wax-free oil filtrate solution is drawn through the filter cloth to the filtrate receiver.

Filtration is induced by vacuum on the filtrate receiver. A closed circuit inert gas system is used, the inert gas being used for blowing the cake from the filter and for keeping a constant slight overpressure in the filter.

The wax cake deposited upon the drum during filtration is washed on the filter with cold solvent. The wash filtrate also flows into the filtrate receiver. Filtrate is recycled into the slurry to adjust the liquid-to-solid ratio of the slurry.

The filtrate solution from the filtrate receiver is pumped through the annulus of the double-pipe exchanger scraped-wall coolers to the evaporators for solvent recovery. The wax mix is heated and flows to a water settler where

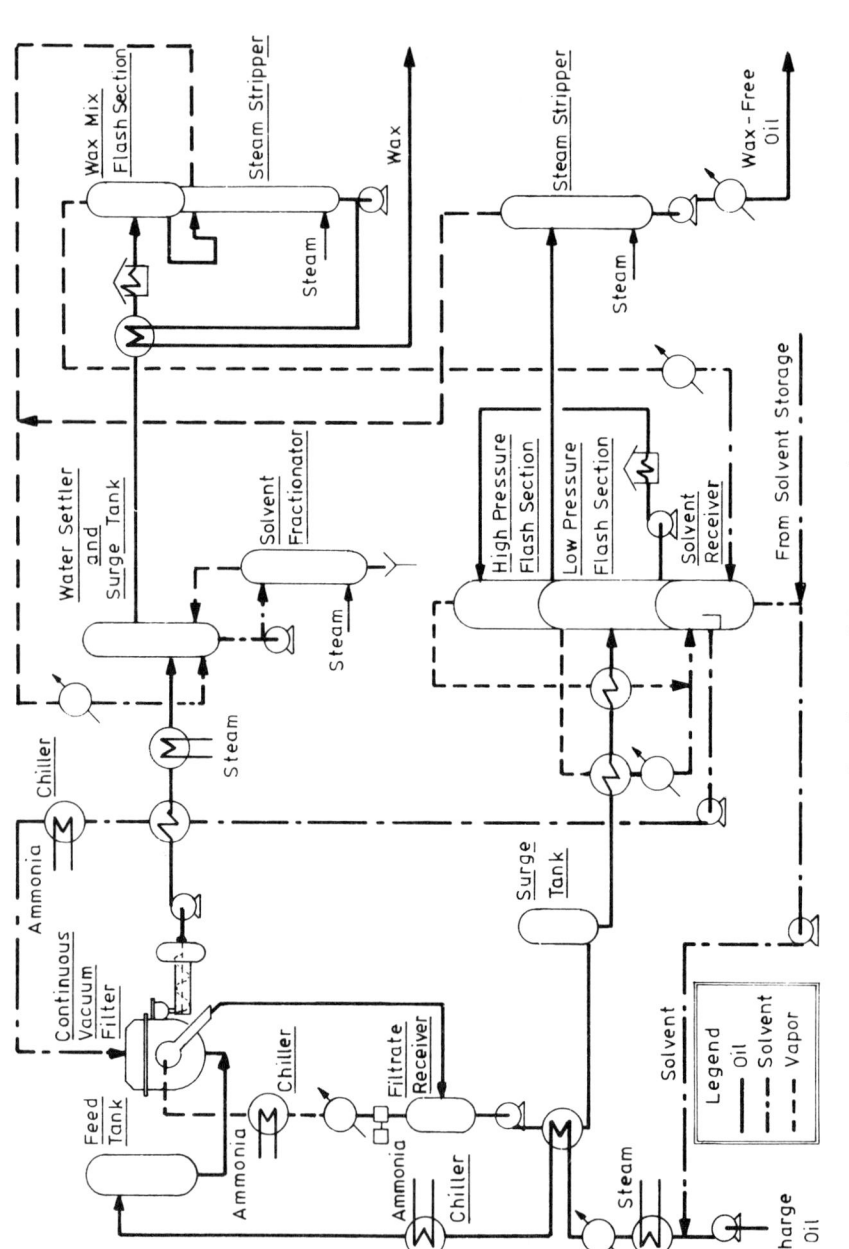

FIG. 4. Process flow diagram of the MEK Dewaxing Process.

water–solvent mixtures from the steam strippers are contacted with the wax mix. The water phase containing dissolved solvent—practically all of which is MEK—is sent to the solvent fractionator where the MEK–water azeotrope is separated as overhead and the solvent-free water as bottom product. The wax phase from the water settler is sent to evaporators for solvent recovery.

Di/Me Dewaxing Process [22]

Warm waxy feed oil is dissolved in warm Di/Me solvent (Fig. 5). The mixture is cooled with water and with filtrate in shell-and-tube coolers up to the phase separation temperature and further with cold filtrate in scraped-wall coolers where approximately 60% of the cooling is achieved. The remaining 40% of the cooling up to the filtration temperature takes place in scraped-wall chillers by means of a refrigeration system (ammonia, propane, etc.). The slurry flows to rotary drum filters from which the filtrate solution is directed into the main filtrate receiver. The wax cake is washed continuously on the filter with cold, dry solvent. The wash filtrate flows into the wash filtrate receiver and is then pumped back into the slurry. Vacuum is kept on the main filtrate and on the wash filtrate receivers. By separating the main filtrate solution from the wash filtrate solution, the oil content of the wash filtrate solution is only 3 to 5%, and by recycling it to the slurry better filtration characteristics and a lower oil content of the wax can be obtained than by recycling the mixture of wash and main filtrate.

The main filtrate solution is pumped through the scraped wall coolers and heat exchangers to the solvent recovery system. The cake consisting of wax and solvent is blown off from the filter and sent to solvent recovery. For maximum heat recovery at least two solvent evaporation stages are used for the filtrate and the wax solution. The final stage is a steam stripper. In order to avoid fouling the cold surfaces in the double-pipe heat exchangers and the filter cloth with ice, the first evaporating stages for filtrate and wax solution are fractionation columns where the evaporated solvent is used to dry the wet solvent received as overhead product from the strippers.

Propane Dewaxing Process [21]

Propane is mixed in a ratio of from 2-to-1 up to 4-to-1 vol./vol.—usually 3-to-1 vol./vol.—with waxy feed oil and brought to a temperature where all suspended solids are dissolved—normally 180°F. The propane/waxy feed oil mixture is cooled down to its cloud point temperature in water-cooled shell-and-tube heat exchangers and sent to one of the two parallel evaporative chillers which are big, cold insulated, horizontal tanks (Fig. 6). The temperature is reduced here at a controlled rate to the filtering temperature by evaporating the propane. The rate of chilling is 1 to 2°F/min. To control the viscosity of the slurry, cold liquid propane is added during the chilling.

In propane dewaxing the crystal growth and shape are very sensitive to the chilling rate and the dilution ratios. Use of dewaxing additives is very important in obtaining good filterability [13].

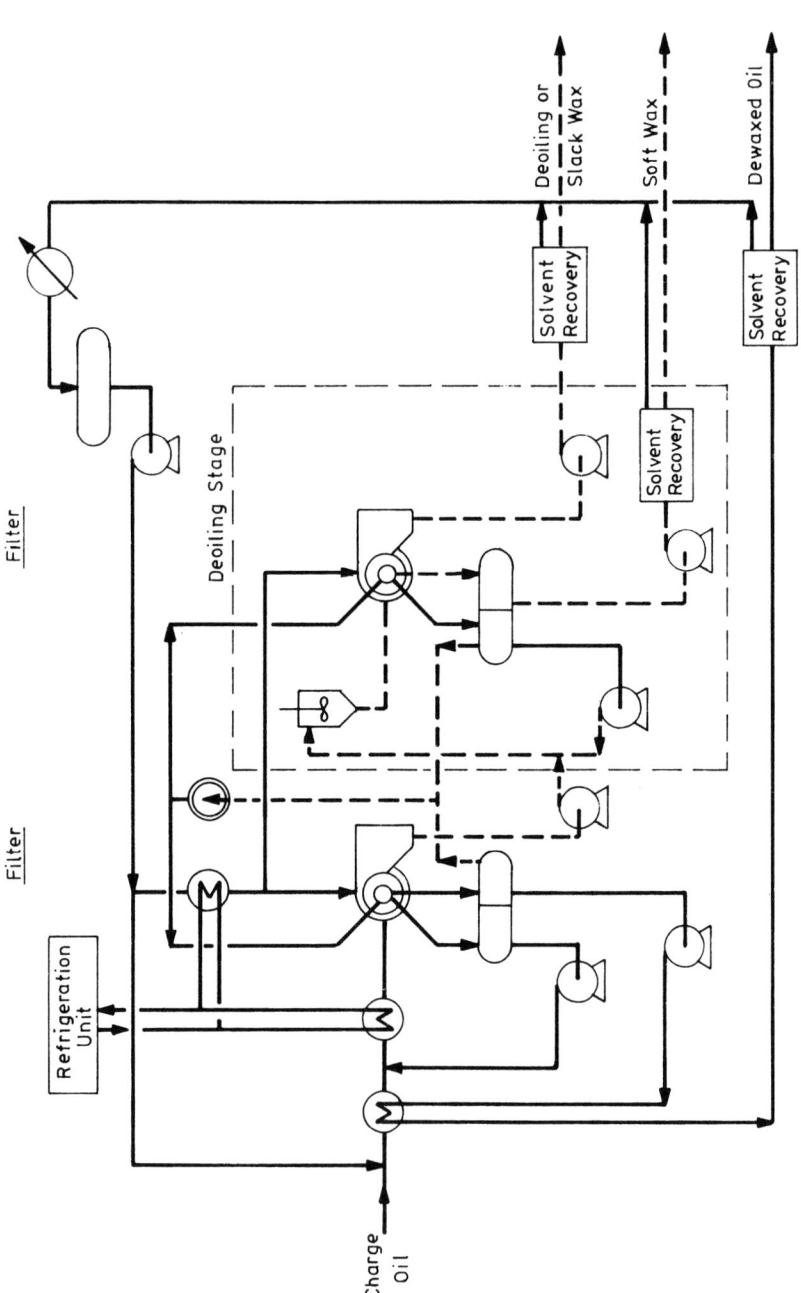

FIG. 5. Process flow diagram of the Di/Me Dewaxing Process.

Dewaxing, Solvent

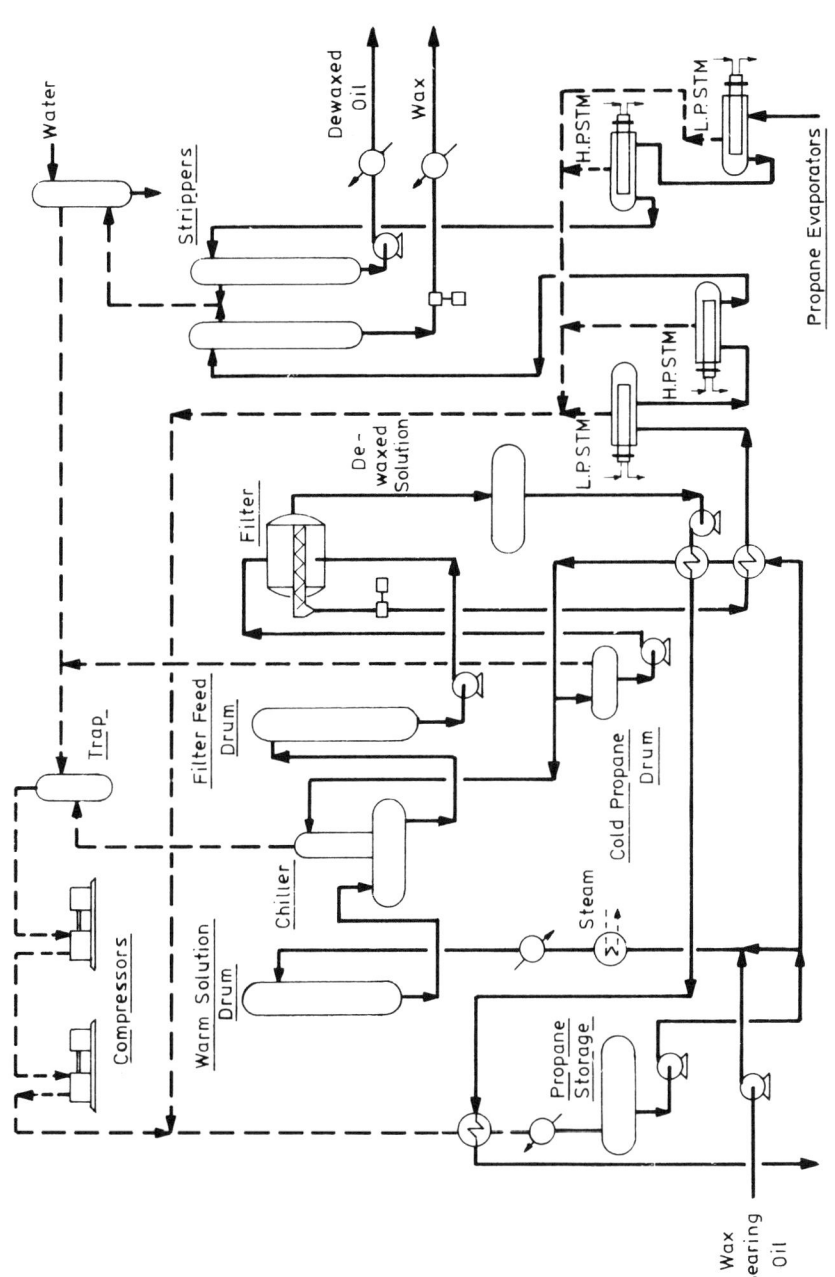

FIG. 6. Process flow diagram of the Propane Dewaxing Process.

The batch chilling used in the Propane Dewaxing Process requires big compressors sized for the fast chilling rate at low temperatures where the vapor pressure of propane is low, whereas at the beginning of chilling, where the vapor pressure of the propane is high, the compressors are underloaded. The wax is separated from the slurry in rotary pressure filters where cold propane is used for washing the cake.

Propane is used both as a blanketing gas and a filter blow-back gas. The filtrate solution and the wax mix are sent to the solvent recovery sections where most of the propane is evaporated in kettle-type reboilers under a pressure high enough to allow direct condensation. The remaining propane is recovered in stripping columns and is recompressed for condensing.

Dilchill Dewaxing Process [10–12]

This process has the advantage of tight control on the growing wax crystals during the cooling/chilling process. Incremental solvent dilution and shock chilling, together with a highly sheared environment, allows the formation of large crystals with few fine and dense crystals which occlude little oil. For distilled lube oil stocks the feed at a temperature above its cloud point is fed into the dilution tower (crystallizer), a vertical tower with injection lines for allowing the addition of cold solvent incrementally along the height. The cooling rate is between 2 to 5°F/min.

For bright stock the feed is prediluted with solvent at a ratio of about 1-to-1 by volume before entrance into the dilution tower.

The negative effect of shock chilling in the dilution tower is compensated for by a high degree of agitation. The tower is divided in mixing stages by horizontal circular plates which restrict flow between the stages to an annular opening. Each stage is provided with a mixing device, preferably a turbine-type agitator.

The required degree of agitation is defined by a modified Reynolds number of at least 10,000:

$$N_{Re} = L^2 n \frac{\gamma}{\mu}$$

where L = agitator diameter, ft
 γ = liquid density, lb/ft^3
 N = agitator speed, r/s
 μ = liquid viscosity, lb/ft · s

The last part of the cooling is achieved in scraped-wall chillers. The solvent recovery section is similar to conventional dewaxing units. The main difference is that all the solvent has to be dried.

Deoiling

Petroleum waxes are largely used in the manufacture of paper containers and wrappers, of candles, and of matches. For most uses the oil content of the wax should be less than 0.5%. The oil content of the slack wax produced in the dewaxing unit is up to 20%, depending on the feed, the process, and the operating conditions. For the production of waxes with low oil content from slack wax, an additional processing step is required. One of the oldest methods is "sweating" where melted slack wax is fed into shallow steel pans equipped midway with a screen. The pans are stacked in a heating oven. After the wax is solidified, the temperature is increased at a rate of 1 to $2°F/h$ and the different grades of oil and waxes that "sweat out" are drained from time to time. The operation is continued until only the required melting point fraction remains in the pans. Although some sweating units are still in commercial use, the main process used today is solvent deoiling.

The solvent deoiling plant can be an independent unit or integrated into a solvent dewaxing plant. For independent deoiling units, the slack wax is heated above its melting point, diluted with solvent, chilled to the filtration temperature, and filtered. The unit is similar to the solvent dewaxing unit.

In integrated solvent dewaxing-deoiling units the slack wax from the dewaxing operation, which contains 65 to 80 vol.% of solvent, is sent directly to the deoiling stage where it is diluted with additional warm solvent, cooled if necessary, and refiltered. In order to increase the yield of dewaxed oil and to have a rigorous control of the properties of the wax (oil content [15], penetration [19], and melting point [18]), an integrated solvent dewaxing-deoiling unit usually consists of three filtration stages [24]. Such a process is illustrated in Fig. 7.

Slack wax from the first dewaxing stage is repulped with cold solvent, filtered, and cold washed. The filtrate from the repulp stage is recycled to the

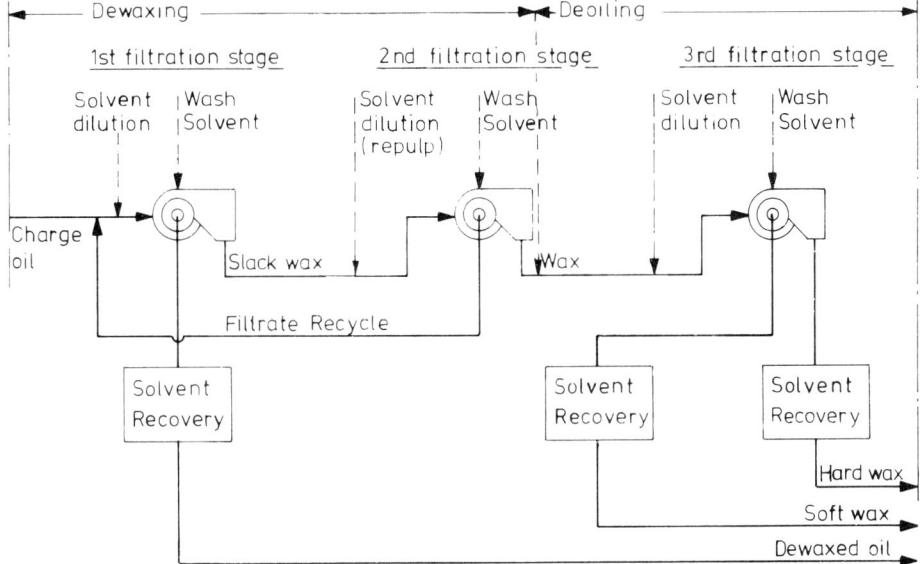

FIG. 7. Process flow diagram of an integrated solvent dewaxing/deoiling unit.

first dewaxing stage. The partially deoiled wax from the second stage is mixed with additional warm solvent, cooled, and refiltered in the third stage. The temperature and solvent ratios used in the third heated and/or fractionation stage are adjusted as required to obtain the desired oil content, melting point, and penetration on the hard wax.

References

1. E. J. Reeves, *Pet. Process.*, pp. 885–886 (August 1949).
2. E. J. Reeves, *Pet. Refiner*, 26(6), 104–105 (1947).
3. E. J. Reeves and I. E. Pattillo, *Pet. Refiner* 27(3), 80–82 (1948).
4. E. J. Reeves, *Ind. Eng. Chem.*, 39(2), 203–206 (1947).
5. V. A. Kalichevsky and K. A. Kobe, *Petroleum Refining with Chemicals*, Elsevier, Amsterdam, 1956.
6. S. J. Marple and L. J. Landry, "Modern Dewaxing Technology," *Adv. Pet. Chem. Refin.*, 10, 190–216 (1965).
7. H. Mondria, *Chem. Eng. Sci.*, 1(1), 20–35 (1951).
8. H. Mondria, *Appl. Sci. Res.*, A, 2, 165–183 (1951).
9. R. M. Butler and J. L. Tiedje, *Can. J. Technol.*, 3(1), 455–467 (1957).
10. D. A. Gudelis, J. F. Eagen, and J. B. Bushnell, *Hydrocarbon Process*, pp. 141–146 (September 1973).
11. J. D. Bushnell and J. F. Eagen, *Oil Gas J.*, pp. 80–84 (October 20, 1975).
12. F. A. Biribauer and J. D. Bushnell, U.S. Patent 3,658,688 (1972).
13. N. F. Chamberlin, J. A. Dinwiddie, and J. L. Franklin, *Ind. Eng. Chem.* 41(3), 566–570 (1949).
14. *Hydrocarbon Processing*, pp. 113–114 (June 1973).
15. ASTM D 721–56T.
16. E. Terres, *Arch. Eisenhuettenwes.*, 21(3/4), 89–95 (1950).
17. ASTM D 97–57
18. ASTM D 87–57, D 127–60
19. ASTM D 1321–61T
20. A. H. Warth, *The Chemistry and Technology of Waxes*, Reinhold, New York, 1956.
21. *Hydrocarbon Processing*, p. 196 (September 1972).
22. *Hydrocarbon Processing*, p. 186 (September 1972).
23. *Hydrocarbon Processing*, p. 200 (September 1972).
24. *Hydrocarbon Processing*, p. 207 (September 1972).

G. G. SCHOLTEN

Dewaxing, Urea

General Features

This process is based on the formation of crystalline complexes between urea and certain organic compounds [1]. The complexes are called adducts.

The principle is quite simple: When a mixture of hydrocarbons containing straight-chain components is contacted with solid urea, the N-paraffins are selectively absorbed by formation of stabile crystalline complex. In the urea adducts the urea molecules are connected into spirals by hydrogen bonds between the oxygen and the amino groups of adjacent urea molecules. The tetragonal crystal structure of urea changes to a hexagonal system. This results in a channel having a diameter of $\sim 5\text{Å}$ into which only molecules with a cross-sectional dimension equal to or less than those of the channel can fit and therefore be adducted. It seems that the urea molecules grow spiralwise around the hydrocarbons [2]. As a consequence, urea forms adducts with organic compounds with long unbranched chains.

In petroleum refining this phenomenon is used for the separation of linear aliphatic hydrocarbons—n-paraffins—from other types of hydrocarbons present in a mixture.

In the separation of long-chain unbranched hydrocarbons, one limitation should be kept in mind: the adduct formation is not very selective, and for higher molecular-weight hydrocarbons, urea adducts can be formed from hydrocarbons containing branching and even rings [2], but the molecule *must* contain a long unbranched chain. The linear chain of an isoparaffinic hydrocarbon must have 11, 12, 15, or 16 carbon atoms in order to form adducts if the methyl group is in position 2, 3, 4, and 5, respectively [8]. Cyclic hydrocarbons with a linear side-chain of at least 18 carbon atoms can also be adducted.

Adduct formation occurs under conditions similar to a reversible chemical reaction.

$$\text{Organic compound} + \text{urea} \rightleftarrows \text{adduct}$$

The first n-paraffinic hydrocarbon which forms adducts with urea is n-hexane. The higher the molecular weight of the paraffinic hydrocarbon, the easier is the formation and the stability of the adduct. Olefin hydrocarbons form adducts with urea similar to paraffinic hydrocarbons, but the adducts have a lower stability for the same molecular weight and structure. The weight ratio between urea and a normal hydrocarbon in an adduct has a constant value of ~ 3.5 g urea/g of n-paraffin. (The figure varies between 3.1 and 3.7.)

If n is the number of carbon atoms in the hydrocarbon, the mole ratio (m) of urea to hydrocarbon in an adduct can be calculated within certain limits by

$$m = 0.65n + 1.5$$

Adduct formation is an exothermic process. The heat of adduct formation can be approximated by means of the following equation [2]:

$$\Delta H = 6.5 - 2.37m \text{ kcal/mol}$$

The equilibrium of adduct formation is a function of the molecular weight of the n-paraffins (length of chain), of the temperature, and of the concentration of the urea. The normal temperature range used in urea dewaxing processes is 68 to 115°F.

The decomposition temperature of the adduct increases with the molecular weight of the hydrocarbon and with the concentration of the urea.

Decomposition is obtained by raising the temperature and/or by adding hot water or a hot hydrocarbon solvent. No urea adduct can exist above 260°F.

Adduct Formation

The urea needed for adduct formation can be made available in the reaction vessel in two ways: as solid particles [4, 5, 7a, 10, 12, 14] and/or as urea solution in water [7, 8, 9].

With solid urea particles the adduction can be a quite slow process. Dewaxing a Middle East wax distillate (490–1020°F) with solid urea, Marechal and de Radzitzki [3] found that 18 h is needed for 100% removal of wax. However, if the urea has been recovered from the adduct by treatment with hot toluene (210°F), only 30 min is needed for the same degree of dewaxing.

When working with urea solution in water, the solution is saturated at temperatures higher than the reaction temperature. By cooling the mixture of urea solution and feed oil under intensive stirring, urea is precipitated out of solution and forms adducts with the wax. In this type of adduction a higher amount of urea must be used than with solid urea. Only the difference between the urea in the initial solution and the urea left in the mother liquor is made available for the adduction. The higher the temperature difference between the saturation temperature of the urea solution and the reaction temperature, the lower is the volume of urea solution which has to be used for producing the required dewaxing effect.

With low saturation temperatures [8, 9] the volume of mother liquor is very large and in the adduction vessel an independent aqueous liquid phase will be present together with the oily liquid and adduct solid phases. In commercial units the separation of these three phases is a very difficult operation. Using high saturation temperatures [7], the volume of mother liquor is reduced to a level where the whole aqueous phase is trapped in the solid phase adduct, the separation of the oily liquid phase and the solid adduct phase being an easy problem.

Mass transfer phenomena are very important in the urea dewaxing process. Due to the high viscosity of the feed oils at the reaction temperature, solvent is used for diluting the oil. This solvent has, in fact, many functions:

Reduces the viscosity of the oily phase

Reduces the concentration of the solid phase in the reacting system, allowing better contact between the reactants; avoids agglomeration of the adduct particles; and gives a slurry with good pumping properties

Is used for washing the adduct in the filtration/washing stage of the process

Adduct formation is accelerated by the addition of activators. In order to be able to grow around the *n*-paraffin molecules, the urea molecules must form a homogeneous solution with the hydrocarbon molecules. Consequently, the activators are solvents for hydrocarbons and urea. Alcohols have been widely used as activators. In dewaxing spindle oil distillates with powdered urea, Yata [4] found that by adding 1 to 3 vol.% (on feed oil) alcohols as activators, methanol gives the highest yield of adduct whereas *n*-butanol gives the lowest.

The lower the molecular weight of alcohols, the higher is the activation. The same phenomenon was observed for ketones, with acetone and methyl ethyl ketone producing the highest yields. Activators with an excellent solvency power for urea, such as methanol, lower the activation ability by the addition of water. For higher molecular weight activators, such as butanol or methyl isobutyl ketone—activators with a poor solvency power for urea—the addition of water improves the activation ability [4].

In many processes using urea solutions in water, the activator is the solvent itself, i.e., methyl isobutyl ketone (MIBK) in the Shell/Wilmington process [9] and dichloromethane, CH_2Cl-CH_2Cl, in the Edeleanu Process [7].

Resinous substances containing carbonyl groups as well as naphthenic acids act as inhibitors in adduct formation. Sulfur compounds have the same effect [13].

By treating spindle oil with sulfuric acid, Yata [4] reduced the time required for the completion of the equilibrium in adduct formation from practically no adduction at 0% H_2SO_4 to 4 min at 10 vol.% H_2SO_4. A sample treated with 5 vol.% H_2SO_4 required 26 min for equilibrium when adduction took place immediately after treatment, and 195 min when the same sample was adducted 2 weeks after sulfuric acid treatment.

The presence of acidic compounds in the feed (carry over in vacuum distillation, oxidizing during the storage, and/or naphthenic acids) is without doubt the main reason for the very different results shown in the literature for adduct formation in the urea dewaxing processes. For tight control of the urea dewaxing process, a feed with an acid value of less than 0.01 mg KOH/g oil is required. In general, surface-active constituents in the oil inhibit adduction [2]. As a consequence, ordinary surface-active agents, ionic or nonionic, may also act as strong inhibitors. These surface-active materials are adsorbed at the surface of urea and interrupt the formation of adducts. The solubility of these surface-active materials into the solvent and/or activator will change the adsorbtion equilibrium and, as a consequence, in addition to the earlier-mentioned effects, the solvent and/or the activator used in the urea dewaxing process will also influence the rate of adduct formation through this adsorption mechanism.

Urea Dewaxing Technology

In principle, every urea dewaxing process consists of the following steps:

Contacting the feed with urea and adduct formation
Separation of adduct from the liquid phase
Decomposition of adduct
Separation of paraffins from the urea
Recovery of solvent and/or activators from every stream
Recycling the urea to the adduction vessel

Stirred tanks are used in practically every process for adduct formation. These contacting vessels are called reactors although adduct formation is not a real chemical reaction. Fixed- [15] and fluidized- bed reactors [5] have also been tried on a laboratory scale.

The degree of agitation, the residence time, and the residence time distribution will affect the degree of dewaxing as well as the purity of the adducted n-paraffins.

The degree of dewaxing is also a function of the urea/feed oil ratio. A certain excess of urea is required for deep dewaxing. On the other hand, more solid phase—from excess urea—means more dewaxed oil trapped in the solid phase. Part of this oil left in the solid particles after filtration and washing will reduce the n-paraffin content of the wax.

If molecular weight of the feed is high, isoparaffinic hydrocarbons will be adducted together with n-paraffins if a complete removing of n-paraffins from the dewaxed oil is required. These isoparaffins together with the filtrate trapped in the solid adduct after washing will reduce the n-paraffin content to a great extent. For processes using a solvent for the feed oil, the n-paraffin content of the wax can be increased up to a certain level by increasing the solvent/feed oil ratio in the reactor. In this way the original liquid left in the solid phase after washing will have a lower oil content. Above this level, the n-paraffin content of wax can be increased by increasing the adduction temperature because less isoparaffins will be adducted.

The conditions in the reactor [temperature, water content (if any), urea to feed and solvent to feed ratios, flow and mixing, type of solvent and/or activator, etc.] will determine the structure of the adduct. A very fine adduct will give almost as many problems as a soft, fluffy, and sticky one in the filtration stage. The temperature of the reactor is kept constant by indirect cooling [7a, 8, 9, 12], by undercooling the liquid streams to the reactor [14], or by evaporating part of the solvent dichloromethane (boiling point at 760 mmHg, 106°F) [7].

The separation of the adduct from the liquid phase is realized by means of continuous vacuum filters [14], rotary pressure filters [7], pan chain horizontal filters [7], or centrifuges [12]. The cake is washed on the filter with fresh solvent from the solvent recovery section. The washed adduct is decomposed at temperatures up to 220°F. In processes using solid urea, hot solvent is mixed with the adduct. The solid urea is separated and recycled to the reaction vessel, and the wax/solvent solution is sent to solvent recovery.

In processes using dilute urea solution [8, 9], the decomposition takes place at temperatures of about 140°F.

With concentrated urea solution [7] the decomposition takes place at about 220°F with hot water. The two liquid phases are separated and the diluted urea solution is concentrated and recycled to the reaction section.

In the phase separation stages—filtration and decanting—the main problem is to avoid the carry-over of one phase into another. Fine adduct particles passing into the dewaxed oil solution and/or urea dissolved in this solution will plug the solvent recovery section. Washing the oily phase with water is practiced in at least two processes [7, 8]. The same happens with the wax solution. Heavy emulsions can be formed in decanters where oil from urea solutions has to be separated.

Urea Dewaxing Processes

The urea dewaxing process was first used mainly for the production of wax with a high content of n-paraffinic hydrocarbons. With an additional purification stage—repulping [7], readduction [7], or two-stage decomposition of the adduct [7, 7a, 14]—n-paraffins of high purity (99%) can be produced. Normal paraffins in the kerosene range (C_{10} to C_{15}) are valuable raw materials in producing easily biodegradable detergents. Higher normal paraffins (C_{13} to C_{17}) are widely used in producing alkyl sulfates and plasticizers for the plastics industry. Wax-cracking plants can convert n-paraffins in the C_{20} to C_{40} range into high-purity olefins which have wide use in the chemical industry. Normal paraffins in the C_{13} to C_{17} range are used to produce proteins from paraffins.

The process is also used in the production of oils with pour points below 0°F from paraffinic crudes. Good quality transformer oils can be obtained. If the pour point requirement for dewaxed oils is 0°F or above, this can be accomplished more economically by means of solvent dewaxing processes.

The Nurex Process [14]

Nippon Mining Co. commercially developed a process for producing n-paraffins in the range C_9 to C_{30}. The purity of n-paraffins varies with the recovery rate and with the n-paraffin content of the charge oil. Purity of 98% or greater is reported for the C_{15} to C_{20} range. The main impurities are monomethyl paraffins. Less than 1000 ppm aromatics and naphthenes have been detected. The process flow diagram is shown in Fig. 1.

Feed oil diluted with deparaffined oil recycle and solvent as activator is undercooled and mixed with solid urea in the reactor. The solvent is a mixture of aromatic solvent and methanol. The solid content of the slurry is about 30 vol.%. The adduct slurry is filtered on a continuous vacuum filter where the cake is washed with more of the aromatic solvent. The filtrate solution is sent to the solvent recovery system. The washed adduct is decomposed at temperatures below 212°F in the decomposer vessels with a mixture of solvent and n-paraffins. The solid urea is separated from the liquid phase—solvent plus n-

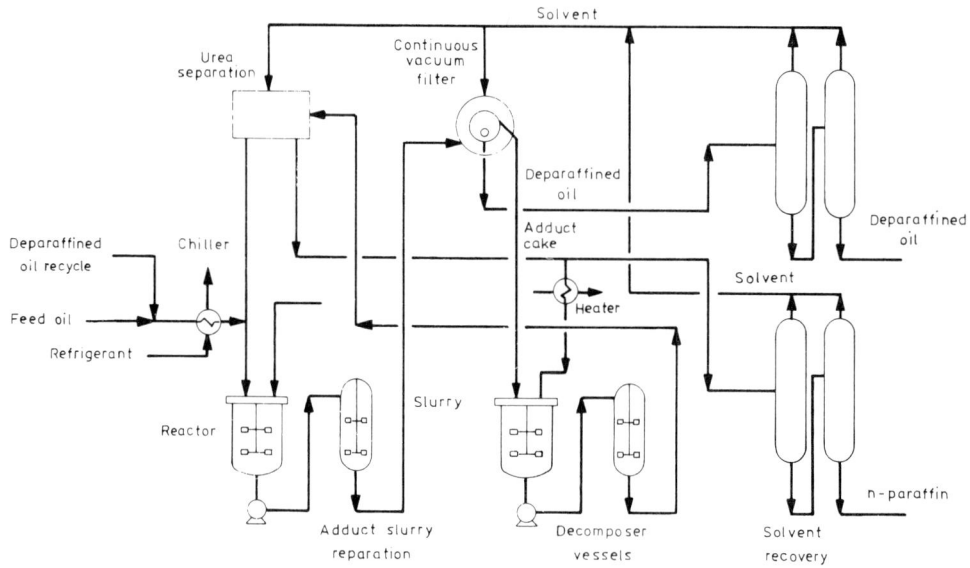

FIG. 1

paraffins—and recycled to the reaction. The solvent–n-paraffin solution is sent to solvent recovery.

The Edeleanu Process [6, 7]

The first commercial unit started in 1955 with a capacity of about 38,000 mt/yr. In 1976 the total capacity of urea dewaxing plants using the Edeleanu Process reached 3,300,000 mt/yr for one-stage operation and 670,000 mt/yr for two-stage operation. The process flow diagram is shown in Fig. 2.

A solvent—dichlormethane (DCM)—with a low oil content (coming from the washing of the adduct on the filter) is mixed with charge oil, and the aqueous urea solution is saturated at about 160°F. The concentration of urea is about 76%.

The mixture is cooled in the reactor section at the reaction temperature by means of DCM vaporization. The reaction temperature is usually between 68 and 120°F and is controlled through the suction pressure of a compressor. Intensive mixing is required in the reactor section.

By cooling down the urea solution from 160°F to the reaction temperature, the major part of urea is crystallized out of the solution and is used for adduct formation. The mother liquor representing the aqueous urea solution with a concentration corresponding to the saturation at the reaction temperature will be trapped entirely into the granular adduct. The solid content of the slurry is ~ 30 vol.%. The liquid phase consists of the mixture of unreacted hydrocarbons from the feed and the part of the solvent which was not evaporated in the reaction section.

The adduct is separated from the dewaxed oil/solvent solution by filtration. Almost every type of filter can be used for this purpose except vacuum filters because under vacuum the solvent will evaporate in the cake. The adduct is washed with clean, dry solvent on the filter to remove the dewaxed oil solution left in the cake after filtration.

If a high purity concentrate of n-paraffins is required for the wax, the adduct from the first filtration stage is repulped with solvent at temperatures equal to or higher than the reaction temperature and separated again by filtration. The liquid phase, mainly solvent, is recycled to the first filtration stage as wash solvent.

Adduct decomposition takes place at temperatures between 175 and 200°F. With higher molecular weights of n-paraffins, higher temperatures have to be used. To avoid higher decomposer temperatures, water is added in an amount corresponding to a drop of about 20°F in the saturation temperature of the urea solution. In the decomposer the major part of the solvent left in the adduct by washing is evaporated. Due to the DCM/water azeotrope, a small quantity of water will also be evaporated.

The two liquid phases flow from the decomposer into the paraffin/urea solution decanter. The paraffin phase containing a few percent DCM is sent to solvent recovery. The urea solution from the bottom of the decanter passes into the urea solution concentrator where the water added in the decomposer is evaporated under a pressure of about 3 lb/in.^{2}abs and the saturation of the urea solution is brought back to the level at 160°F.

The main contaminant of the urea solution is biuret. By avoiding high temperatures, the level of biuret in the circulated urea solution can be kept below 5%. Biuret is constantly degradated into cyanuric acid, ammelide, and

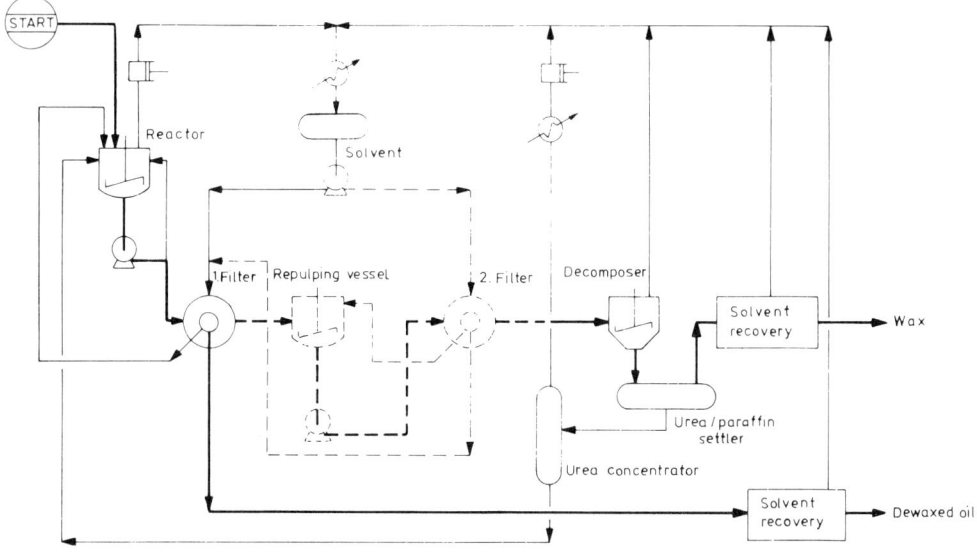

FIG. 2

TABLE 1. Typical Feed and Product Analysis in a Urea Dewaxing Plant—One Stage (Edeleanu Process)

Characteristics	Feed I		Feed II		Feed III		Feed IV	
	Charge Oil	Dewaxed Oil	Charge Oil	Dewaxed Oil	Charge Oil	Dewaxed Oil	Charge Oil	Dewaxed Oil
Distillation range (°F)	680/775	—	560/625	—	640/750	—	715/825	—
Viscosity (cSt/210°F)	3.35	—	—	1.9	3.4	3.57	7.35	9.1
Pour point (°F)	75	−15	10	−80	50	−55	90	−15
Yield (wt.% on feed)	100	76	100	90	100	85	100	85

other compounds which, being insoluble in the aqueous urea solution, are removed by filtration. For final purification purposes the urea solution passes through an activated carbon tower before it is fed to the reaction section.

The level of contaminants in the urea solution is an important parameter in the urea dewaxing process because it affects adduct formation. A high level of contaminants affects the reaction rate and the yield in the reactor section, the oil content of the wax—poor filtration and washing characteristics of the adduct—and the separation of the aqueous urea solution in the paraffin/urea settler.

Table 1 shows some typical feed and product characteristics from a commercial urea dewaxing unit.

In a one-stage operation the wax has a n-paraffin content of 75 to 85% depending on the solvent/oil ratio in the reactor, on the degree of dewaxing (pour point requirements for the dewaxed oil), and on the n-paraffin content of the feed.

The wax produced in the one-stage urea dewaxing unit has three main constituents:

TABLE 2 Chromatographic Analysis of a Oil-Free Wax from a Urea Dewaxing Unit (Edeleanu Process)

Hydrocarbons	wt.%	$(i/n) \times 100$
n-C_{15}	1.6	
i-C_{16}	0.1	<1
n-C_{16}	11.5	
i-C_{17}	0.7	3
n-C_{17}	22.6	
i-C_{18}	1.0	3.6
n-C_{18}	27.2	
i-C_{19}	1.1	4.4
n-C_{19}	24.9	
i-C_{20}	0.5	6.1
n-C_{20}	8.2	
$i + n$-C_{21}	0.6	

TABLE 3 Feedstocks Dewaxed by the Edeleanu Urea Dewaxing, Adduction, and Repulping Process [6]

	Light Gas Oil	Light Gas Oil
Feed:		
Specific gravity at 60°F	0.817	0.833
Pour point (°F)	25	−5
Paraffin content (wt.%)	35	17
Boiling range (°F)	465–625	410–610
Dewaxed oil:		
Yield (wt.%)	68	84
Pour point (°F)	−50	−80
Wax:		
Paraffin content (wt.%)	99	99

Adducted n-paraffins
Adducted i-paraffins
Oil left in the adduct cake after washing on the filter.

The oil content of the wax is between 8 and 15% depending on the solvent/oil, on the liquid/solid ratio in the slurry, on the degree of washing, and on the porosity of the adduct.

The i-paraffin content is a function of the urea/feed ratio and of the reaction temperature.

A typical analysis of the oil-free wax produced in a one-stage dewaxing unit is shown in Table 2.

The two-stage Edeleanu Process with repulping will produce wax with a n-paraffin content of about 99 wt.% together with low pour point oils.

Table 3 [6] shows the feed, dewaxed oil, and wax characteristics from a large-scale plant.

References

1. F. Bergen, German Patent 869,070.
2. R. L. McLaughlin, in *The Chemistry of Hydrocarbons*, Vol. I, Reinhold, New York, 1954.
3. J. Marechal and P. de Radzitzki, *J. Inst. Pet.* 47(434), 33 (1960).
4. N. Yata, *Bull. Jpn. Pet. Inst.*, 4, 35 (March 1962).
5. P. K. Karanth and R. W. Sinha, *Pet. Hydrocarbons*, 6(3), 215 (1972).
6. A. Hoppe, *Adv. Pet. Chem. Refin.*, Φ (1964).
7. Edeleanu Gesellschaft mbH, German Patents 1,105,090, 1,104,101, 1,104,100, 1,098,657, 1,094,390, 1,085,281, 1,020,143, 1,015,972, 1,015,168, 1,109,299,

1,156,925, 1,225,328, 1,470,547, 1,470,550, 1,645,730, 1,919,663, 1,919,664, 1,945,902, 2,039,120, 2,138,428.
7a. Deutsche Texaco AG, German Patents 2,220,594, 2,220,621, 2,364,333, 2,333,470.
8. L. N. Goldsbrough, 4th World Petroleum Congress Rome, 1955, Section III-B, Paper 6.
9. W. A. Bailey, Jr., R. A. Bonnerot, L. C. Fetterly, and A. G. Smith, *Ind. Eng. Chem.*, 43, 2125 (1951).
10. Sonneborn Sons, *Chem. Eng.*, p. 114 (1956).
11. E. J. Fuller, *Sep. Purif. Methods*, 1(2), 253 (1972).
12. J. Bathory, *Chem.-Anlagen Verfahren*, p. 63 (April 1972).
13. Y. G. Geletii, D. F. Varfolomeer, and V. V. Tyurin, *Khim. Tekhnol. Topl. Masel*, p. 16 (October 1974).
14. Nurex Process, *Oil Gas J.*, p. 142 (October 16, 1972).
15. K. P. Kukanova, A. G. Sardanashvili, and V. A. Matishev, *Khim. Tekhnol. Topl. Masel*, p. 18, (July 1972).

G. G. SCHOLTEN

Hydrocracking

Hydrocracking, which is a more recent process development compared to the long-established conversion processes of visbreaking and thermal cracking, figures prominently in plans for production of gasoline and middle distillates.

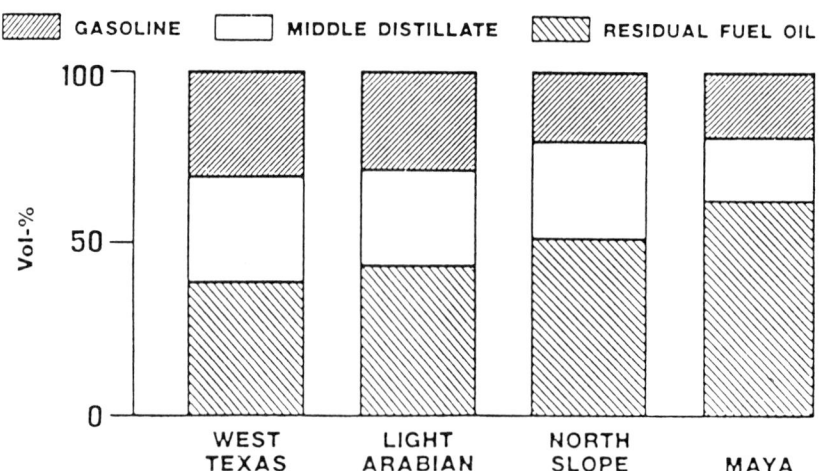

FIG. 1. Yields from some important crudes. (Source: UOP.)

Hydrocracking

Fuels Production in the United States

The ability of refiners to cope with the renewed trend toward distillate production creates renewed interest in hydrocracking—particularly as crude supplies become heavier. Without the required conversion units, heavier crudes result in lower yields of straight run gasoline and middle distillate. To maintain current gasoline and middle distillate production levels, additional conversion capacity will be required. Typically, the yield of gasoline and middle distillate is greater with lighter crudes (Fig. 1).

With the United States currently undergoing lead phase out, hydrocracking—alone or as a supplement to other refinery processed—deserves considerable attention. It adds flexibility to refinery processing and to the product slate.

Depending on the feedstock being processed and the type of plant design employed (single stage or two stage), flexibility can be provided to vary product distribution among the following principal end products:

LPG
High octane light gasoline
Heavy naphtha (catalytic reformer feed)
Jet fuel components
Middle distillate products—diesel/kero
Low sulfur fuel oil blendstocks
High viscosity index lube oil blendstocks

Background

Today's refiner is faced with the need to convert the heavier components of the crude barrel into lighter, more valuable products. This situation is a result of increasingly heavier crudes, lower demand for fuel oil, a steady to lower demand for motor gasoline, and an increasing demand for mid-distillates, particularly diesel.

In the United States the demand for fuel oil as a percentage of total crude dropped during the early 1980s from 14 to 11%. Correspondingly, the percentages going to mid-distillates have gone up. A further shift in demand, especially in the direction of diesel oil and jet fuel relative to gasoline, is expected for the coming years.

Ultimately, new conversion facilities will be built to upgrade the bottom of the petroleum barrel to light products. However, neither the available cash flow nor the project economics will support such investments today in most refineries. Redeploying existing facilities to provide a partial solution to residual fuel conversion is an immediately viable move. Mild hydrocracking is a widely applicable example of such reuse of existing facilities.

Other options include fluid catalytic cracking (FCC) and coking, but both of these require substantial investments. However, even if these units already exist on-site, some of the end products (e.g., diesel fuel) are of such poor quality (e.g., cetane shortfall) that further upgrading via hydrogenation at high pressures may still be required.

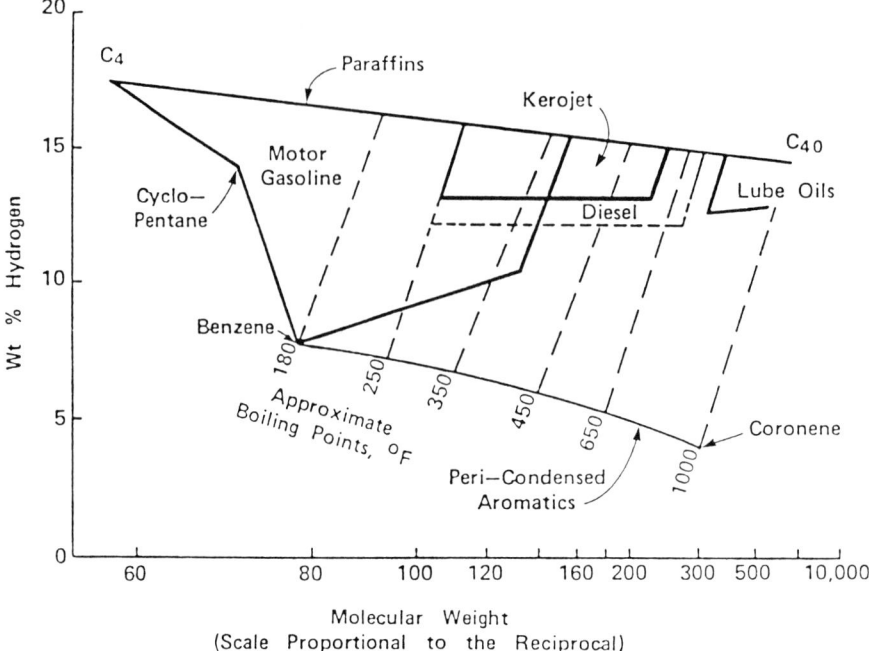

FIG. 2. Hydrogen content versus molecular weight. Delineation by pure hydrocarbons, approximate boiling points, and potential ranges for specification petroleum products. (Source: Chevron Research Co.)

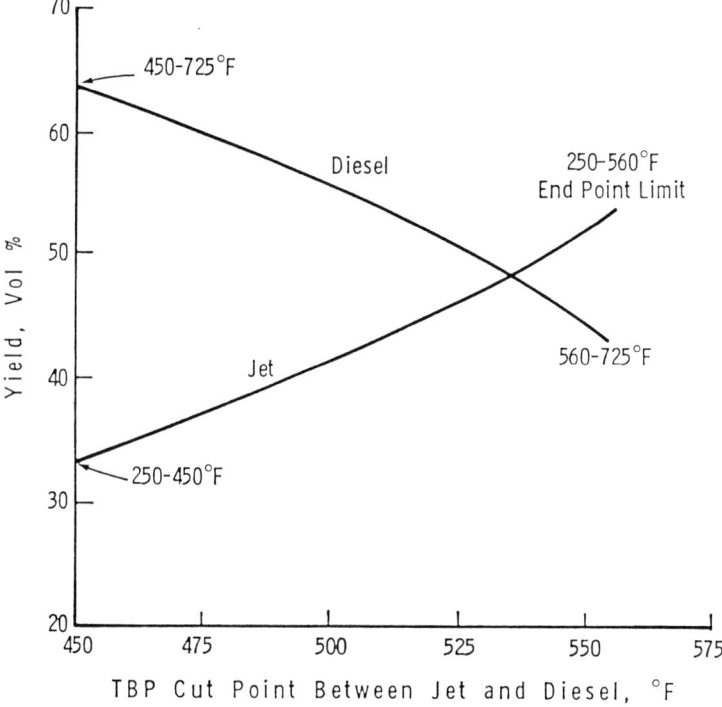

FIG. 3. Yields from hydrocracking vary with the catalyst being employed. With Chevron catalyst ICR 120, varied product slates can be achieved. The cut point between jet and diesel can be controlled while keeping the jet initial cut point at flash point specification and the RCP at 725°F. Varying the intermediate cut point from 450 to 555°F gives a range of jet to diesel production ratios of 0.5 to 1.2. (Source: Chevron Research Co.)

Hydrocracking

Fundamentally, the trend toward lower feed gravity is related to an increase in the carbon/hydrogen ratio of crudes (Fig. 2) because of higher resid content. This can be overcome by upgrading methods which lower this ratio by adding hydrogen, rejecting carbon, or using a combination of both methods.

Though the technology to upgrade heavy oils exists, selection of the optimum process units is very much dependent on each refiner's needs and goals. With this in mind, systems to "dig deeper into the barrel" should not only be cost effective and reliable, but also flexible.

Hydrocracking adds that flexibility. Hydrocracking offers the refiner a process that can handle varying feeds and operate under diverse process conditions. Utilizing different types of catalysts (Fig. 3) can modify the product slate produced. Reactor design and number of processing stages play a role in this flexibility.

Mild Hydrocracking

The objective of the mild hydrocracking process is to produce significant yields of lighter products at operating conditions similar to those of a vacuum gas oil (VGO) desulfurizer. Consequently, the flow scheme for a mild hydrocracking unit is virtually identical to that of a VGO desulfurizer as evidenced by Figs. 4, 5, and 6.

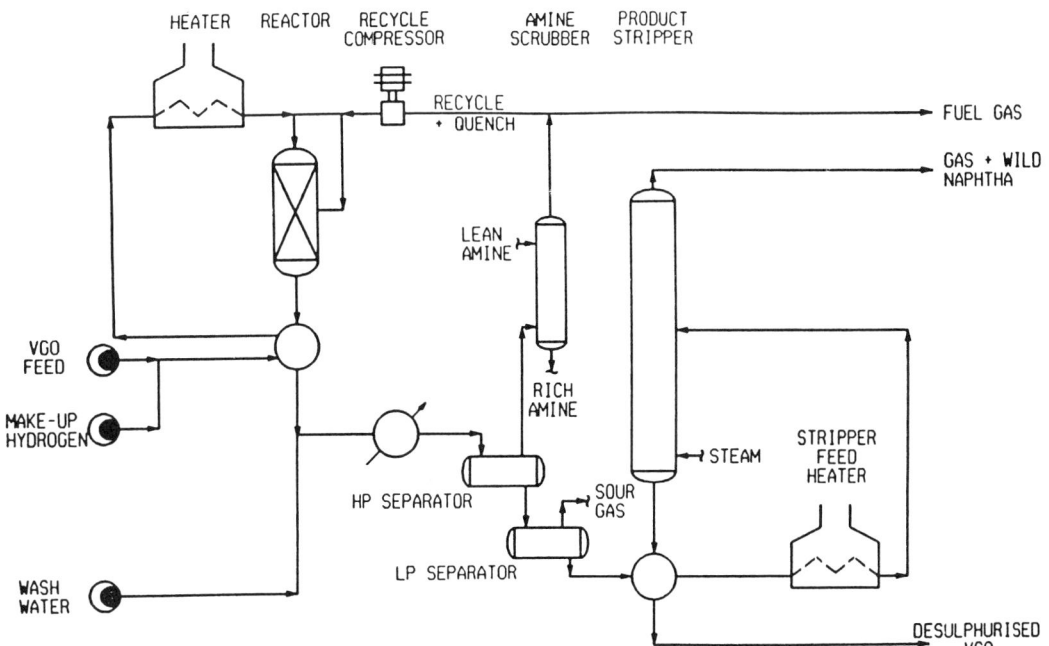

FIG. 4. VGO desulfurizer. (Source: M. W. Kellogg Co.)

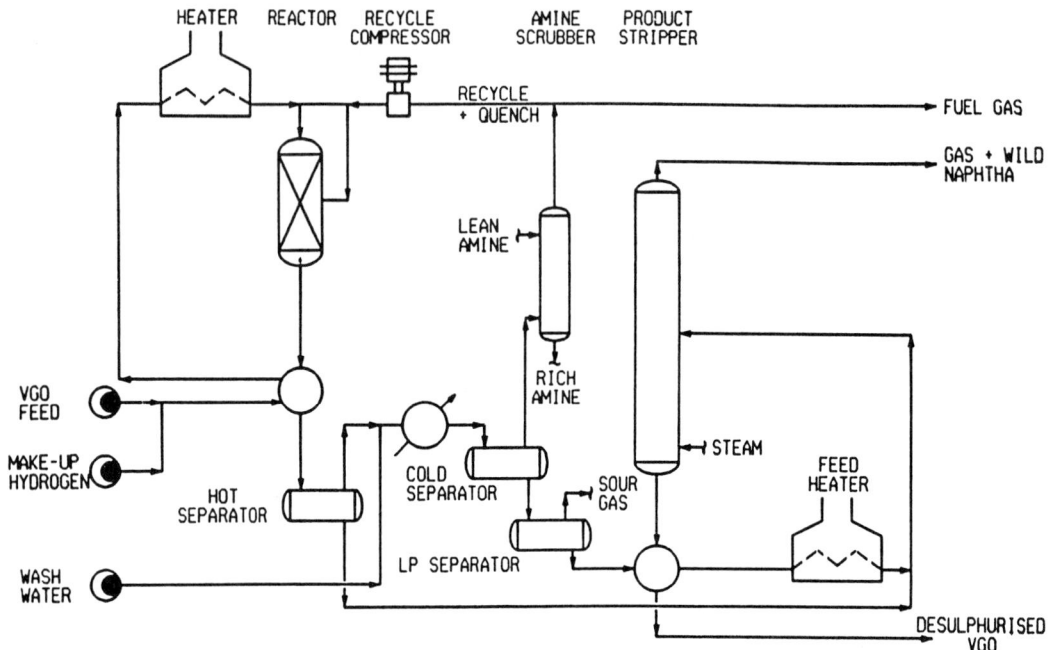

FIG. 5. VGO desulfurizer (hot separator design). (Source: M. W. Kellogg Co.)

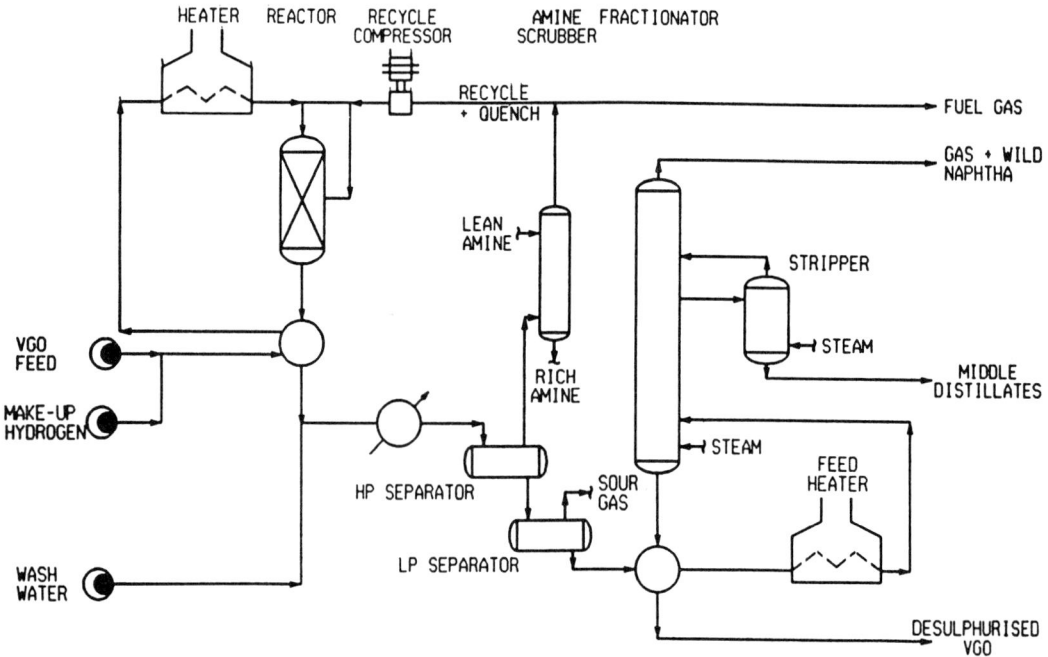

FIG. 6. VGO desulfurizer operating in the mild hydrocracking mode. (Source: M. W. Kellogg Co.)

Figure 4 illustrates a simplified process flow diagram of a vacuum gas oil desulfurizer. VGO feed is mixed with hydrogen make-up gas and preheated against reactor effluent. Further preheating to reaction temperature is accomplished in a fired heater. The hot feed is mixed with recycle gas before entering the reactor. The temperature rises across the reactor due to the exothermic heat of reaction. Catalyst bed temperatures are usually controlled by using multiple catalyst beds and introducing recycle gas as an interbed quench medium. Reactor effluent is cooled against incoming feed and air or water before entering the high-pressure separator. Vapors from this separator are scrubbed to remove H_2S before compression back to the reactor as recycle and quench. A small portion of these gases is purged to fuel gas to prevent buildup of light ends. Liquid from the HP separator is flashed into the low-pressure separator. Sour flash vapors are purged from the unit. Liquid is preheated against stripper bottoms and in a feed heater before steam stripping in a stabilizer tower.

Water wash facilities are provided upstream of the last reactor effluent cooler to remove ammonium salts produced by denitrogenation of the VGO feed.

Figure 5 is a variation of Fig. 4 and is referred to as the hot separator design. The process flow scheme is identical to that described above up to the reactor outlet. After initial reactor effluent cooling against incoming VGO feed and make-up hydrogen, a hot separator is installed. Hot liquid is routed directly to the product stabilizer. Hot vapors are further cooled by air and/or water before entering the cold separator. At this point the process is identical to that described for Fig. 4.

This arrangement reduces the stabilizer feed preheat duty and the effluent cooling duty by routing hot liquid direct to the stripper tower. This utility and equipment saving must be offset against the capital cost of the hot separator.

When considering the revamp of a VGO desulfurizer to mild hydrocracking operation, due regard must be given to the unit design in the following areas.

(1) Reactor. Desulfurization, denitrogenation, and hydrocracking reactions all occur with the liberation of heat. During mild hydrocracking operations these reactions occur to the same or to a greater extent than in a conventional VGO desulfurizer. Therefore, the process temperature rises more rapidly through the catalyst bed. To protect the catalyst from the deactivating effects of high temperatures, a multiple catalyst bed arrangement with interbed quench facilities must be employed. Cold recycle hydrogen is most commonly used as the quench medium.

Where this internal reactor arrangement does not already exist, the VGO desulfurizer must be revamped.

(2) Washwater. Washwater facilities are normally installed upstream of the last reactor effluent cooler before the high-pressure separator. Water injection is used to dissolve ammonium salts formed through denitrogenation reactions in the reactor. These salts would otherwise deposit in exchanger tubes, leading to increased system pressure drop and eventually complete plugging of the exchangers.

Very much higher levels of denitrogenation are achieved during mild hydrocracking operation than during VGO desulfurization. The incidence of salt deposition is proportional to the level of nitrogen removal, as are the wash water requirements. The washwater injection and removal facilities must therefore be checked for adequacy in the mild hydrocracking mode of operation.

Alternatively, where washwater facilities do not already exist on the VGO desulfurizer, these must be provided during the revamp.

(3) Fractionation/Stripping. Because the amount of hydrocracking occurring during conventional VGO desulfurization is small, more complex recovery schemes than simple steam stripping of the desulfurized product cannot be economically justified.

The higher conversions that are achieved during mild hydrocracking operations result in increased middle distillate yields which require a fractionator for their recovery. Recovery may be accomplished by simple modification of the stripper tower or by relocation and modification of an existing, spare fractionator or may involve purchase of a completely new tower.

(4) Hydrogen Requirements. As a result of the increased hydrocracking and denitrogenation activity of the mild hydrocracking catalyst, hydrogen consumption on the unit is correspondingly higher than on conventional VGO desulfurizers. It is a prerequisite of any revamp to mild hydrocracking operation that sufficient make-up hydrogen be available from the catalytic reformer for implementation of the project.

It is not possible to generalize about the magnitude of the change in hydrogen consumption since refinery revamp objectives differ and therefore influence the outcome. For example, if capacity of the unit is to be reduced, leading to a reduction in space velocity, then total hydrogen consumption may actually fall. Alternatively, a reduction in catalyst cycle length may be acceptable under the revamp conditions, thus permitting operation at high conversion levels, resulting in an increase in hydrogen consumption.

TABLE 1 Feedstock Qualify for VGO HDS and MHC Operations

Feedstock source	Heavy Arabian VGO
Density at 15°C (60°F)	0.938
Sulfur, wt.%	3.3
Nitrogen, wppm	1110
Boiling range, TBP–GLC, °C (°F)	
IBP	350 (662)
10%	400 (752)
30%	439 (822)
50%	472 (882)
70%	507 (945)
90%	553 (1027)
EP	614 (1137)

TABLE 2 Operating Conditions for VGO HDS vs MHC Operation

	VGO HDS	MHC
Maximum pressure, bar (lb/in.^{2}abs)	75 (1090)	75 (1090)
Maximum temperature, °C (°F)	430 (806)	430 (806)
Make-up gas flow, Nm^3/T (SCFB)	90 (500)	150 (835)
Hydrogen content, mol%	90	90
Recycle gas flow, Nm^3/T (SCFB)	500 (2780)	500 (2780)
Hydrogen content, mol%	80	80
H$_2$S content, mol%	0	0
LHSV, h^{-1}	0.5	0.5
Catalyst	S-424, dense loaded	MHC-1
Specification	0.3 wt.% S in 370°C+ (698°F+) product	0.3 wt.% S at EOR conditions
Cycle length, months	11	11

Example. An example of the possible benefits of a switch to mild hydrocracking operations follows.

Table 1 details the quality of the feedstock for both VGO HDS and mild hydrocracking conditions.

Table 2 details the unit conditions for the two modes of operation.

Table 3 details the product composition for mild hydrocracking operation.

Properties of the 370°C+ (698°F+) product are:

Density at 15°C (60°F)		0.89
Sulfur, wt.%:	SOR	0.02
	EOR	0.3
Nitrogen, wppm:	SOR	10
	EOR	900

In addition, the viscosity of the unconverted VGO is very much lower and the H/C ratio much improved compared to the original feedstock, making it an excellent fuel oil blend component or FCCU feedstock.

Middle distillate product in the gas oil boiling range has a cetane number in the range 40 to 45.

Smoke point of middle distillate product in the kerosene boiling range is feedstock dependent, but typically a value of about 18 mm can be expected.

All products are low in both sulfur and nitrogen contaminants.

TABLE 3 Product Composition for Mild Hydrocracking Operation

Composition	wt.%
C$_1$–C$_4$	1.6
C$_5$–180°C (356°F)	14.8
180 (356)–250°C (482°F)	12.9
250 (482)–300°C (572°F)	5.2
300 (572)–370°C (698°F)	9.1
370°C+ (698+)	56.4
Conversion of 370°C+ (698°F+)	42%

Catalysts and Processes

More needs to be said about hydrocracking catalysts. The process employs high-activity catalysts that produce a significant yield of light products. Catalyst selectivity to middle distillate is a function of both the conversion level and operating temperature, with values in excess of 90% being reported in commercial operation. In addition to the increased hydrocracking activity of the catalyst, percentage desulfurization and denitrogenation at start-of-run conditions are also substantially increased. End-of-cycle is reached when product sulfur has risen to the level achieved in conventional VGO HDS operation.

An important consideration, however, is that commercial hydrocracking units are often limited by design constraints of an existing VGO hydrotreating units. Thus, the proper choice of catalyst(s) is critical when searching for optimum performance. Typical commercial distillate hydrocracking (DHC) catalysts contain both the hydrogenation (metal) and cracking (acid sites) functions required for service in existing desulfurization units. UOP, for example, investigated using all-alumina-base hydrotreating catalysts in more severe service (high temperatures) to determine cracking activity relative to DHC catalysts. The results of these pilot studies are summarized in Figs. 7 and 8. In this comparison, as temperature is increased above normal hydrotreating conditions, DHC catalyst demonstrates a much higher catalytic activity for hydrocracking, while both systems approach 100% desulfurization. At 30% conversion, this cracking activity advantage corresponds to about 25°F in reactor temperature requirement.

The catalyst selected for mild hydrocracking service must also be capable of stable operation at high temperatures. As shown in Table 4, DHC catalysts are designed for high severity (100% conversation) distillate hydrocracking

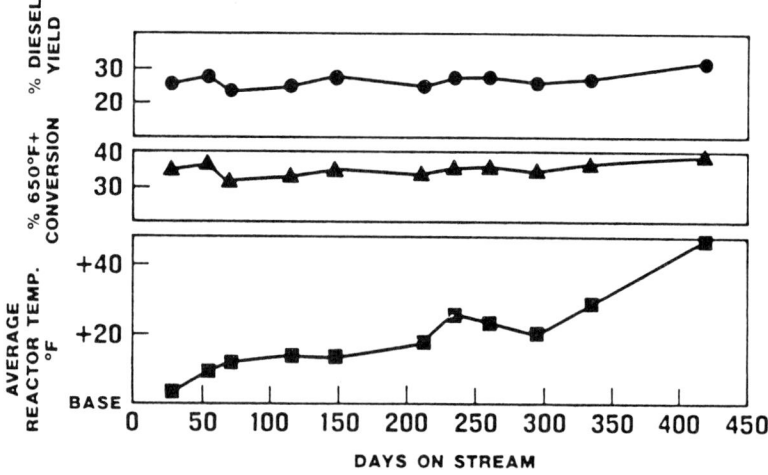

FIG. 7. Mild hydrocracking catalyst life study for Middle Eastern VGO/DHC catalyst. (Source: UOP.)

Hydrocracking

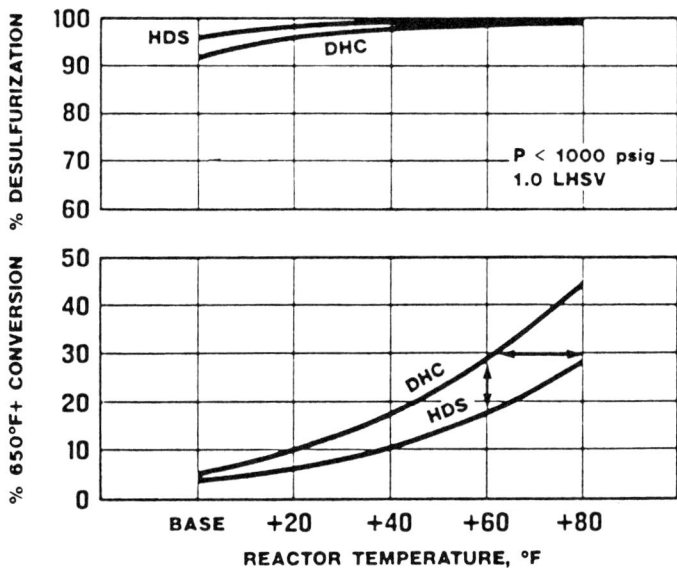

FIG. 8. DHC versus HDS catalyst conversion and desulfurization response to temperature. (Source: UOP.)

service and can operate stably at temperatures up to 850°F. This has been verified at low pressures and higher space velocities in both commercial and pilot-plant operations.

In fact, several pilot operations are nearing commercial development.

Commercializing Related Processes

Energizing hydrocracking technology, such as that developed by Petro-Canada, shows promise for some applications. Their scheme is a high conversion, high demetallization, residuum hydrocracking process which, using an additive to inhibit coke formation, achieves conversion of high boiling point hydrocarbons into lighter products. Initially developed to upgrade tar sands bitumen and heavy oils of Canadian

TABLE 4 Processing Conditions[a]

	Gas Oil Hydrotreating	Distillate Hydrocracking
Catalyst function	Promote hydrogenation reactions	Promote hydrogenation reactions Promote hydrocracking reactions
Catalyst promoters	Metal sites	Metal sites Acid sites
Operating pressure, lb/in.^{2} gauge	600–1500	1500–3000
Operating temperature, °F	650–780	720–850
LHSV, h^{-1}	0.5–3.0	0.4–1.5
H$_2$ consumption, SCFB	200–600	1000–2000
H$_2$ circulation, SCFB	1500–5000	8000–15,000

[a]Source: UOP.

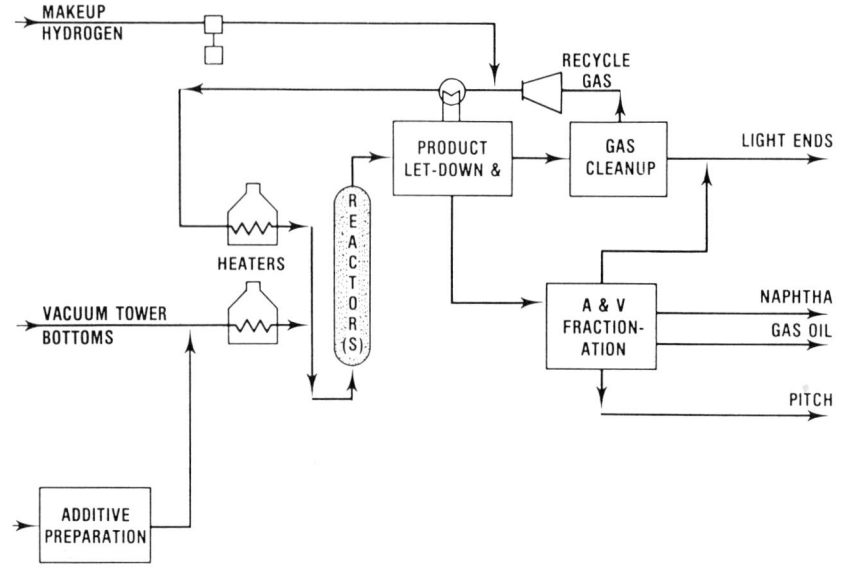

FIG. 9. CANMET unit, simplified schematic. (Source: Petro-Canada.)

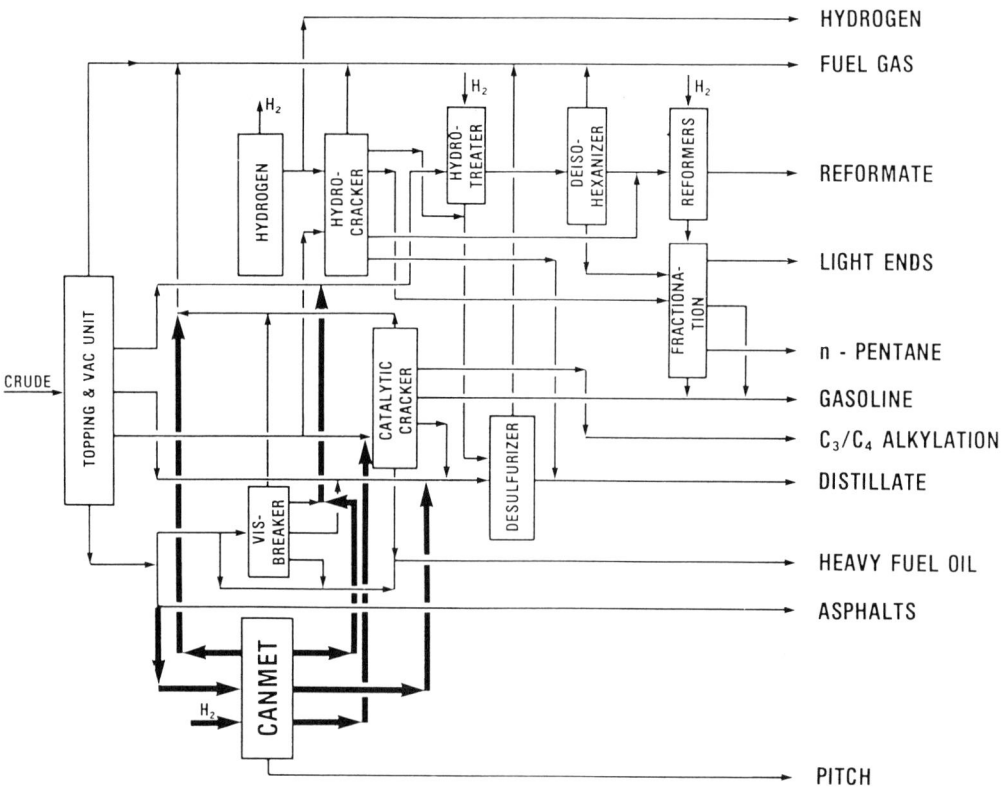

FIG. 10. The demonstration plant is part of the refinery. (Source: Petro-Canada.)

TABLE 5 Typical Hydrocracking Catalysts[a]

Single-stage isocracking catalysts:
    ICR 106:   High ratio of mid-distillate to naphtha
    ICR 120:   Very high ratio of mid-distillate to naphtha
    ICR 113:   Used for hydrocracking heavy oils like DAO
    ICR 117:   High ratio of naphtha to mid-distillate
    ICR 201:   Hydrocracking naphtha or raffinate to LPG

Two-stage isocracking catalysts:
    ICR 113:   First stage denitrification
    ICR 106:   First stage denitrification and crackling
               Second stage hydrocracking for mid-distillate
    ICR 120:   First stage denitrification and cracking
               Second stage hydrocracking for maximum mid-distillate
    ICR 117:   First stage denitrification and cracking
               Second stage hydrocracking for naphtha and mid-distillate
    ICR 201:   Second stage hydrocracking for LPG from naphtha or raffinate
    ICR 202:   Second stage hydrocracking for naphtha or jet fuel
    ICR 204:   Second stage hydrocracking for naphtha, aromatics, and jet fuel

[a]Source: Chevron Research Co.

origin, an ongoing program of development has broadened the technology to processing offshore heavy oils and the bottom of the barrel from so-called conventional crudes. A 5000 bbl/d demonstration plant is located at Petro-Canada's Montreal East refinery. Here is how the process works: Heavy feedstock is mixed with an additive, heated, contacted with hydrogen, and sent to an upflow reactor. The reactor products are withdrawn overhead and are separated into a hydrogen-rich recycle gas stream, process gas, conventional refinery cuts, and a residual pitch. The demonstration plant is fully integrated into the refinery (Figs. 9 and 10). It makes maximum use of existing equipment and the existing infrastructure. The feed will be obtained from the vacuum tower, hydrogen will be derived from the existing reformers, and the required additive will be prepared on-site.

Product naphtha will be hydrotreated and reformed, light gas oil will be hydrotreated and sent to the distillate pool, the heavy gas oil will be processed in the FCC, and the pitch will be sold.

The above discussion on catalysts should not lead the design engineer to believe that only a few hydrocracking catalysts are necessary. In fact, just the opposite is true. Whole families of catalysts (Table 5) are required depending on feed available and products required. In addition, the number of process stages are important to catalysts choice. Generally, one of three options are utilized by the refinery.

Two-Stage Design

The most common is a two-stage plant (Fig. 11). This flow scheme has been very popular in the United States since it maximizes the yield of transportation fuels and has the flexibility to produce gasoline, naphtha, jet fuel, or diesel fuel to meet seasonal swings in product demand.

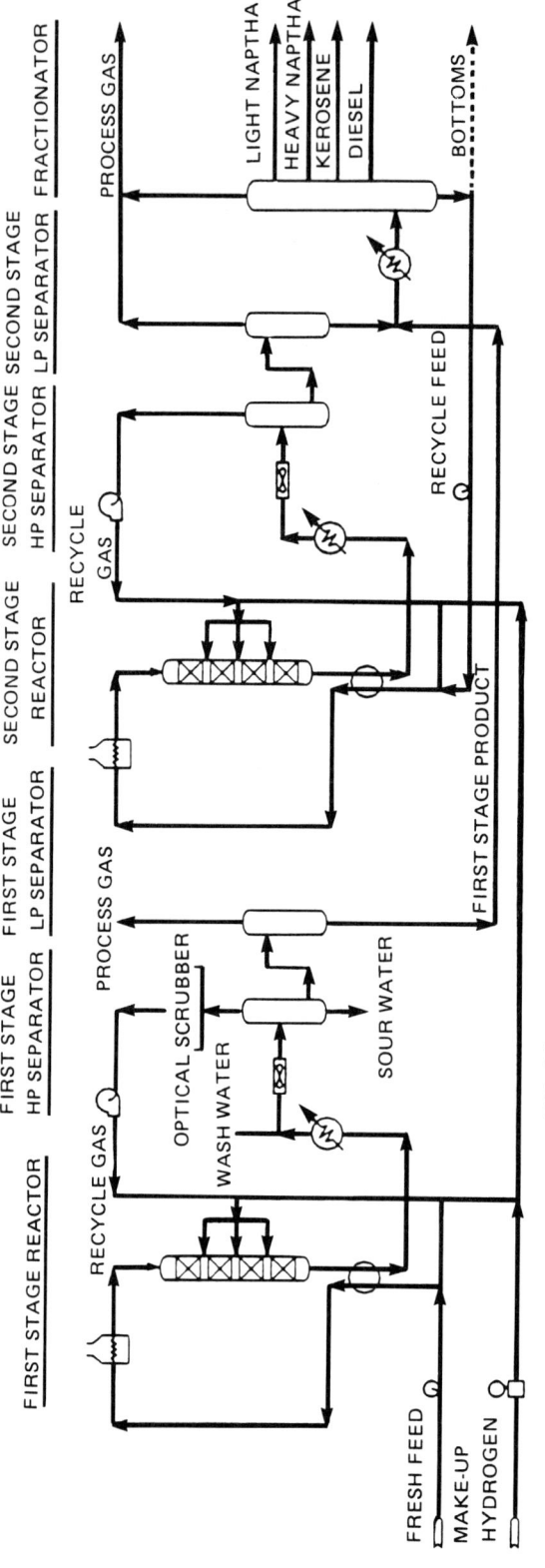

FIG. 11. Two-stage hydrocracking design. (Source: Chevron Research Co.)

Hydrocracking

This type of hydrocracker consists of two reactor stages together with a product distillation section. The choice of catalyst in each reaction stage depends on the product slate required and the character of the feedstock. In general, however, the first-stage catalyst is designed to remove nitrogen and heavy aromatics from raw petroleum stocks. The second-stage catalyst carries out a selective hydrocracking reaction on the cleaner oil produced in the first stage.

Both reactor stages have similar process flow schemes. The oil feed is combined with a preheated mixture of makeup hydrogen and hydrogen-rich recycle gas and heated to reactor inlet temperature via a feed–effluent exchanger and a reactor charge heater. The reactor charge heater design philosophy is based on many years of safe operation with such two-phase furnaces. The feed–effluent exchangers take advantage of special high-pressure exchanger design features developed by Chevron engineers to give leak-free end closures. From the charge heater, the partially vaporized feed enters to the top of the reactor. The catalyst is loaded in separate beds in the reactor with facilities between the beds for quenching the reaction mix and ensuring good flow distribution through the catalyst.

The reactor effluent is cooled through a variety of heat exchangers including the feed–effluent exchanger and one or more air coolers. Deaerated condensate is injected into the first-stage reactor effluent before the final air cooler in order to remove ammonia and some of the hydrogen sulfide. This prevents solid ammonium bisulfide from depositing in the system. A body of expertise in the field of materials selection for hydrocracker cooling trains is quite important for proper design.

The reactor effluent leaving the air cooler is separated into hydrogen-rich recycle gas, a sour water stream, and a hydrocarbon liquid stream in the high-pressure separator. The sour water effluent stream is often then sent to a Chevron WWT Process plant for ammonia recovery and for purification so water can be recycled back to the hydrocracker. The hydrocarbon rich stream is pressure reduced and fed to the distillation section after light products are flashed off in a low-pressure separator.

The hydrogen-rich gas stream from the high-pressure separator is recycled back to the reactor feed by using a recycle compressor. Sometimes with sour feeds, the first-stage recycle gas is scrubbed with an amine system to remove hydrogen. If the feed sulfur level is high, this option can improve the performance of the catalyst and result in less costly materials of construction.

The distillation section consists of a H_2S stripper and a recycle splitter. This latter column separates the product into the desired cuts. The column bottoms stream is recycled back to the second-stage feed. The recycle cut point is changed depending on the light products needed. It can be as low as 320°F if naphtha production is maximized (for aromatics) or as high as 720°F if a low pour point diesel is needed. Between these two extremes, a recycle cut point of 500 to 550°F results in high yields of high smoke point, low freeze point jet fuel.

Single-Stage Design

A single-stage once-through (SSOT) unit resembles the first stage of the two-stage plant. This type of hydrocracker usually requires the least capital investment. The feedstock is not completely converted to lighter products. For this application the refiner must have a demand for a highly refined heavy oil. In many refining situations, such an oil product can be used as lube oil plant feed or as FCC plant feed or in low sulfur oil blends or as ethylene plant feed. It also lends itself to stepwise construction of a future two-stage hydrocracker for full feed conversion.

Single-Stage Recycle

A single-stage recycle (SSREC) unit converts a heavy oil completely into light products with a flow scheme resembling the second stage of the two-stage plant. Such a unit maximizes the yield of naphtha, jet fuel, or diesel depending on the recycle cut point used in the distillation section. This type of unit is more economical than the more complex two-stage unit when plant design capacity is less than about 10,000–15,000 BPOD. Commercial SSREC plants have operated to produce low pour point diesel fuel from waxy Middle East vacuum gas oils. Recent emphasis has been placed on the upgrading of lighter gas oils into jet fuels.

Once-Through Partial Conversion

Building on the theme of one- or two-stage hydrocracking, process engineers should consider once-through partial conversion (OTPC). This concept offers the means to convert heavy vacuum gas oil feed into high quality gasoline, jet fuel, and diesel products by a partial conversion operation. The advantage is lower initial capital investment and also lower utilities consumption than a plant designed for total conversion.

Because total conversion of the higher molecular weight compounds in the feedstock is not required, once-through hydrocracking can be carried out at lower temperatures, and in most cases at lower hydrogen partial pressures than in recycle hydrocracking, where total conversion of the feedstock is normally an objective.

Proper selection of the types of catalysts employed can even permit partial conversion of heavy gas oil feeds to diesel and lighter products at the low hydrogen partial pressures for which gas oil hydrotreaters are normally designed. This so-called "mild hydrocracking" has been attracting a great deal of interest lately from refiners who have existing hydrotreaters and wish to increase their refinery's conversion of fuel oil into lower boiling, higher value products without a large capital expenditure. A wide range of products (Table 6) can be efficiently produced.

Union Oil has compared investment and operating cost of OTPC with recycle hydrocracking and believes OTPC offers advantages in both areas while maintaining product quality and selectivity.

Recycle hydrocracking plants are designed to operate at hydrogen partial pressures from about 1200 to 2300 lb/in.^{2} depending on the type of feed being processed. Hydrogen partial pressure is set in the design in part depending on required catalyst cycle length, but also to enable the catalyst to convert high molecular weight polynuclear aromatic and naphthene compounds which must be hydrogenated before they can be cracked. Hydrogen partial pressure also affects properties of the hydrocracked products which depend on hydrogen uptake, such as jet fuel aromatics content and smoke point and diesel cetane number. In general, the higher the feed endpoint, the higher the required hydrogen partial pressure necessary to achieve satisfactory performance of the plant.

Once-through partial conversion hydrocracking of a given feedstock may be carried out at hydrogen partial pressures significantly lower than required for recycle total conversion hydrocracking. The potential higher catalyst deactivation rates experienced at lower hydrogen partial pressures can be offset by using higher activity catalysts and designing the plant for lower catalyst space velocities. Catalyst deactivation is also reduced by the elimination of the recycle stream. The lower capital cost resulting from the reduction in plant operating pressure is much more significant than the increase resulting from the possible additional catalyst requirement and larger volume reactors.

Additional capital cost savings from once-through hydrocracking result from the reduced overall required hydraulic capacity of the plant for a given fresh feed rate as a result of the elimination of a recycle oil stream. Hydraulic capacity at the same fresh feed rate is 30–40% lower for a once-through plant compared to one designed for recycle.

TABLE 6 Applications of Produce Purified, Lower-Boiling Products from a Variety of Feedstocks[a]

Feeds	Products
Straight-run gas oils	LPG
Vacuum gas oils	Motor gasolines
Fluid catalytic cracking oils and decant oils	Reformer feeds
Coker gas oils	Aviation turbine fuels
Thermally cracked stocks	Diesel fuels
Solvent deasphalted residual oils	Heating oils
Straight-run naphthas	Solvents and thinners
Cracked naphthas	Lube oils
	Petrochemical feedstocks
	Ethylene feed pretreatment process (FPP)

[a]Source: UOP.

Utilities savings for a once-through versus recycle operation arise from lower pumping and compression costs as a result of the lower design pressure possible and also lower hydrogen consumption. Additional savings are realized as a result of the lower oil and gas circulation rates required, since recycle of oil from the fractionator bottoms is not necessary.

Lower capital investment and operating costs are obvious advantages of once-through hydrocracking compared to a recycle design. This type of operation may be adaptable for use in an existing gas oil hydrotreater or atmospheric resid desulfurization plant. The change from hydrotreating to hydrocracking service will require some modifications and capital expenditure, but in most cases these changes will be minimal.

The fact that unconverted oil is produced by the plant is not necessarily a disadvantage. The unconverted oil produced by once-through hydrocracking is a high quality, low sulfur and nitrogen material that is an excellent feedstock for an FCC unit or ethylene pyrolysis furnace, or a source of high viscosity index lube oil base stock. The properties of the oil are, of course, a function of the degree of conversion and other plant operating conditions.

One disadvantage of once-through hydrocracking compared to a recycle operation is a somewhat reduced flexibility for varying the ratio of gasoline to middle distillate that is produced. A greater quantity of naphtha can be produced by increasing conversion, and jet fuel plus diesel yield can also be increased in this manner. But selectivity for higher boiling products is also a function of conversion. Selectivity decreases as once-through conversion increases. If conversion is increased too much, the yield of desired product will decrease, accompanied by an increase in light ends and gas production. Higher yields of gasoline or jet fuel plus diesel are possible from a recycle than from a once-through operation.

Middle distillate products made by once-through hydrocracking are generally higher in aromatics content of poorer burning quality than those produced by recycle hydrocracking. However, the quality is generally better than produced by catalytic cracking or from straight run. Middle distillate product quality improves as the degree of conversion increases, and as hydrogen partial pressure is increased.

Feed Characteristics

For a conventional hydrotreating catalyst and an improved mild hydrocracking (MHC) catalyst, the effects of pressure, temperature, and space velocity on conversion are presented in Fig. 12. Clearly, all three variables have a strong effect on conversion. Second generation catalysts are found to be the most dependent on process conditions. In general, increasing severity of operation will increase the delta conversion between first and second generation MHC catalysts. Vis-á-vis a conventional catalyst, the effect of feedstock on version plays a key role in system operation.

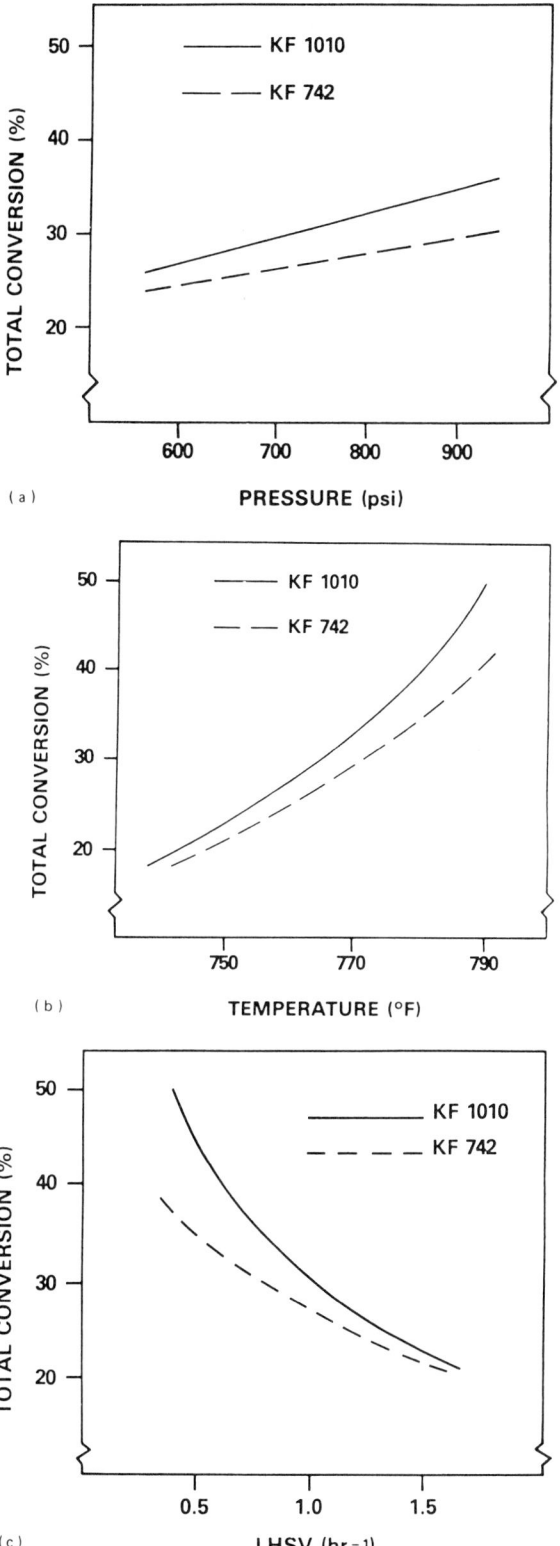

FIG. 12. Effect of pressure (a), temperature (b), and space velocity (c) on conversion. (Source: AKZO Chemie.)

TABLE 7 Properties of Feedstocks Applied in MHC Studies[a]

Type of Feedstock	API Gravity (122°F)	S (wt.%)	N (ppm)	Distillation, ASTM D-1160 (°F)			
				10%	50%	80%	90%
VGO number 1	20.2	3.03	882	752	846	934	973
VGO number 2	23.2	1.96	873	738	814	897	937
VGO number 3	20.0	2.10	2001	718	865	959	993[b]
VGO number 4	21.1	1.10	2386	667	754	826	856
CGO	19.2	3.52	1892	522	777	977	1044
Kuwait atmospheric resid	15.0	3.99	3800	680	945		

[a]Source: AKZO Chemie.
[b]EBP at 85% recovery.

Tests by AKZO Chemie on typical hydrocracking feeds (Table 7) give varying results (Table 8) depending on the type of catalyst used.

In other AKZO Chemie studies, vacuum gas oil (VGO) was blended with coker gas oil (CG)) and/or with atmospheric resid (AR). The results seem encouraging.

Although the conversion drops with blending, it remains at an acceptable level. Another important observation is the relatively better performance of the second generation MHC catalyst for these blends. Furthermore, it is expected that lowering the space velocity will improve the delta conversion between the first and second generation catalysts. In the case of multiple bed operation, one of the questions is the choice of the first-stage catalyst. In study with VGO/CGO and VGO/resid blends, the effect of a highly active HDS and a highly active HDN catalyst on the performance of the dual bed is critical.

TABLE 8 Effect of Feedstock on Total Conversion[a,b]

Feedstock	Conversion Observed for Conventional Catalyst	Increase in Conversion Compared to Base Case	
		KF-1002	KF-1011 with KF-742 as First Stage Catalyst
VGO number 1	36.1	6.9	5.5
VGO number 2	33.5	6.8	5.3
VGO number 3	34.2	3.7	1.4
VGO number 4	30.2	3.0	0.5

[a]Source: AKZO Chemie.
[b]The first column gives the conversion observed for base case, the conventional HDS catalyst. The second and third columns show the increase in conversion for the two second generation catalyst systems. The conversion measured with the conventional HDS catalyst depends mildly on the feedstock applied. For the second generation MHC catalysts, the differences are far more severe, especially for KF-1011. For heavy feeds, KF-1002 is generally preferred over KF-1011.

Hydrocracking

TABLE 9 Feedstocks

Component	Typical Range	
	SR VGO	SR Cracked Distillates to Visbroken or Coker Gas Oils
Gravity, °API	22	17–40
Boiling range (°F)	650–1050	300–1100
H_2 content, wt.%	12	>10.5
Sulfur content, wt.%	2	0.1–4.0
Nitrogen content, wt-ppm	800	200–5000
Bromine no.	3	1–30
Conradson carbon, wt.%	0.3	0.1–1.5
Asphaltenes, ppm	100	<500
Metals, ppm	1	<2

Another feedstock consideration is that the level of contaminants, particularly metals, asphaltenes, and Conradson carbon residue, in a given feed has a large impact on catalyst stability and cycle length. For this reason, some companies have set some practical limits for acceptable charge stock. UOP uses:

Nickel + vanadium, ppm	<2
Asphaltenes (C_7 insolubles), ppm	<500
Conradson carbon residue, wt.%	<2

Table 9 characterizes the range of feedstock that can be processed.

Metal and asphaltene specifications can usually be easily met with proper control of the vacuum column operation, whereas carbon residue is more dependent on crude type. API gravity, hydrogen content, and sulfur and nitrogen levels are also crude dependent and have a large effect on the quality of the converted products. In general, a more paraffinic feed produces a more paraffinic, higher quality product while consuming less hydrogen than a more aromatic feed. Thermally cracked stocks can also be processed in a mild hydrocracking operation to hydrotreat and stabilize distillates and achieve some 650°F+ conversion, but very careful consideration must be given to unit heat balance, since olefin saturation is highly exothermic (Table 10). In many cases, interbed H_2 quenching or heat exchange is required.

Before closing the book on feedstocks, one should look closely at other important feedstock characteristics.

Chevron says heavier feedstocks with a 700–1050°F boiling range will produce somewhat lower mid-distillate yields as a percent to isocracker feed than the above feedstock. The mid-distillate production rate as a percent of crude oil processed will, of course, be higher. The severity of operation is the factor affecting the process yields. The combinaton of LHSV, temperature, and pressure needed to convert the heaviest molecules in the feed will be

TABLE 10 Some Causes of Runaway Reactions or Excursions in Hydrocracking Reactors[a, b]

Cause	Comments and Cures
Furfural Co	Temperature can reach 1000°F. To stop, use wide open bed quench. Not always necessary to replace catalyst
Loss of charge pump Loss of recycle compressor Power failure in the unit	Cut fires, depressure unit, and restart.
Reduction or loss of water injection Upsets in charge heater Quench system fails to operate	High nitrogen feeds lead to high ammonia levels that deactivate catalyst
Too much feed for amount of hydrogen available	Can result in fire if pressure sags, reducing recycle and quench capacity. For any incident, notification of appropriate authorities is a must
Other	Chances of runaway reaction at mild hydrocracking or hydrotreating are less than for 2500 lb/in.^{2} cracking

[a] Source: NPRA Q&A Sessions.
[b] Depends on process being employed.

more severe, and light end yields in the C_1–C_4 range as well as light naphtha yields will increase. Conversely, a 700–950°F feed will produce a somewhat higher process yield of mid-distillates than a 700–1000°F boiling range feed. Generally, the objective for any refiner is to run the heaviest boiling range feed consistent with producing a nearly metals-free and asphaltene-free feedstock. The future differential between residuum and transportation fuel values is expected to be sufficiently high that feeding the heaviest possible VGO (giving the lowest residuum yield) will be the economic choice, even though middle distillate yield will be somewhat lower. Amorphous catalysts are especially well suited for this severe operation.

Typical Middle Distillate Hydrocracking

Figure 13 shows a typical flow scheme for a hydrocracker designed to crack gas oils to Jet A-1 fuel. Figures 14 and 15 are typical flow schemes for hydrocracking gas oils to diesel fuel. The following discussion describes

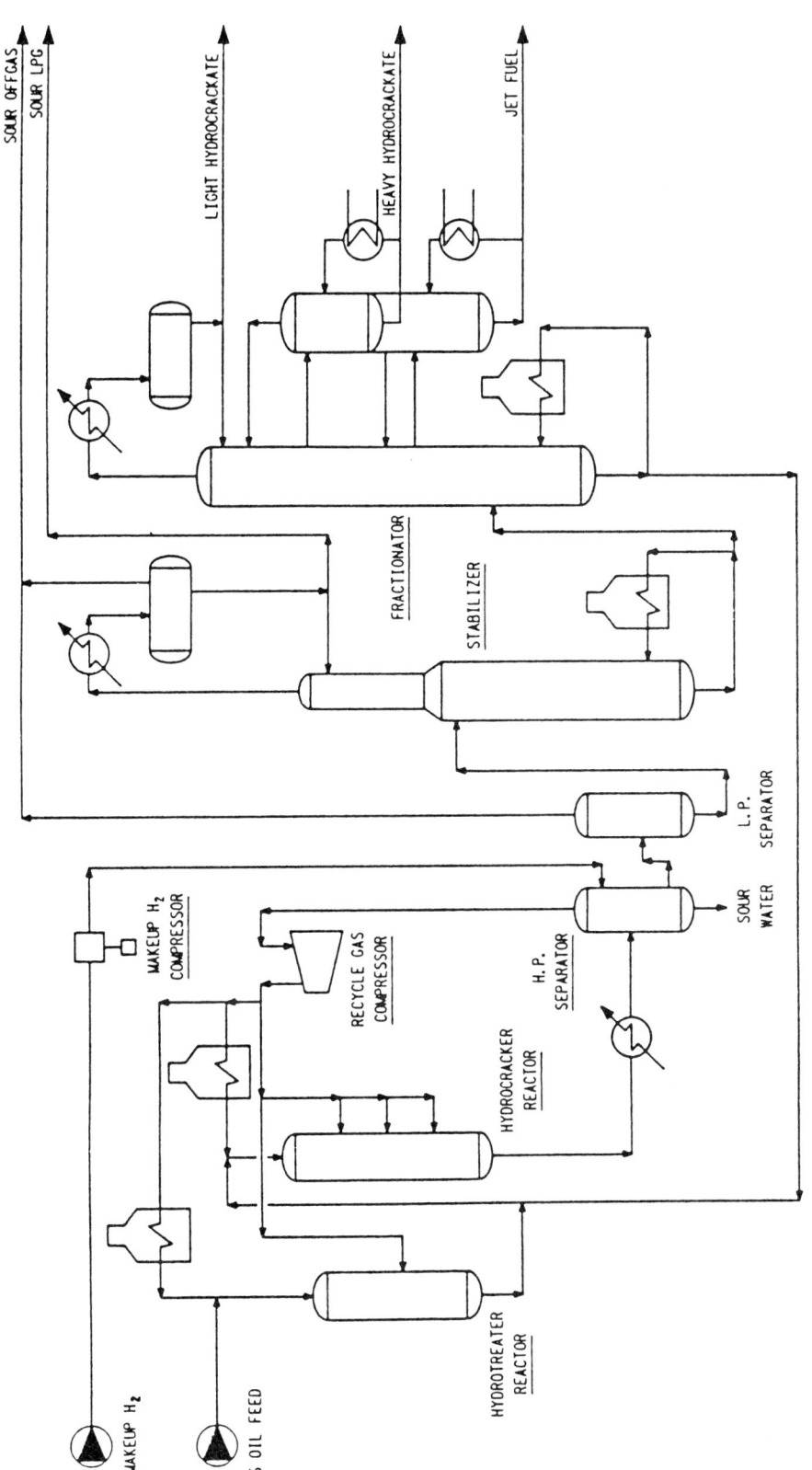

FIG. 13. Hydrocracker flow scheme for jet fuel production. (Source: Litwin Engineers.)

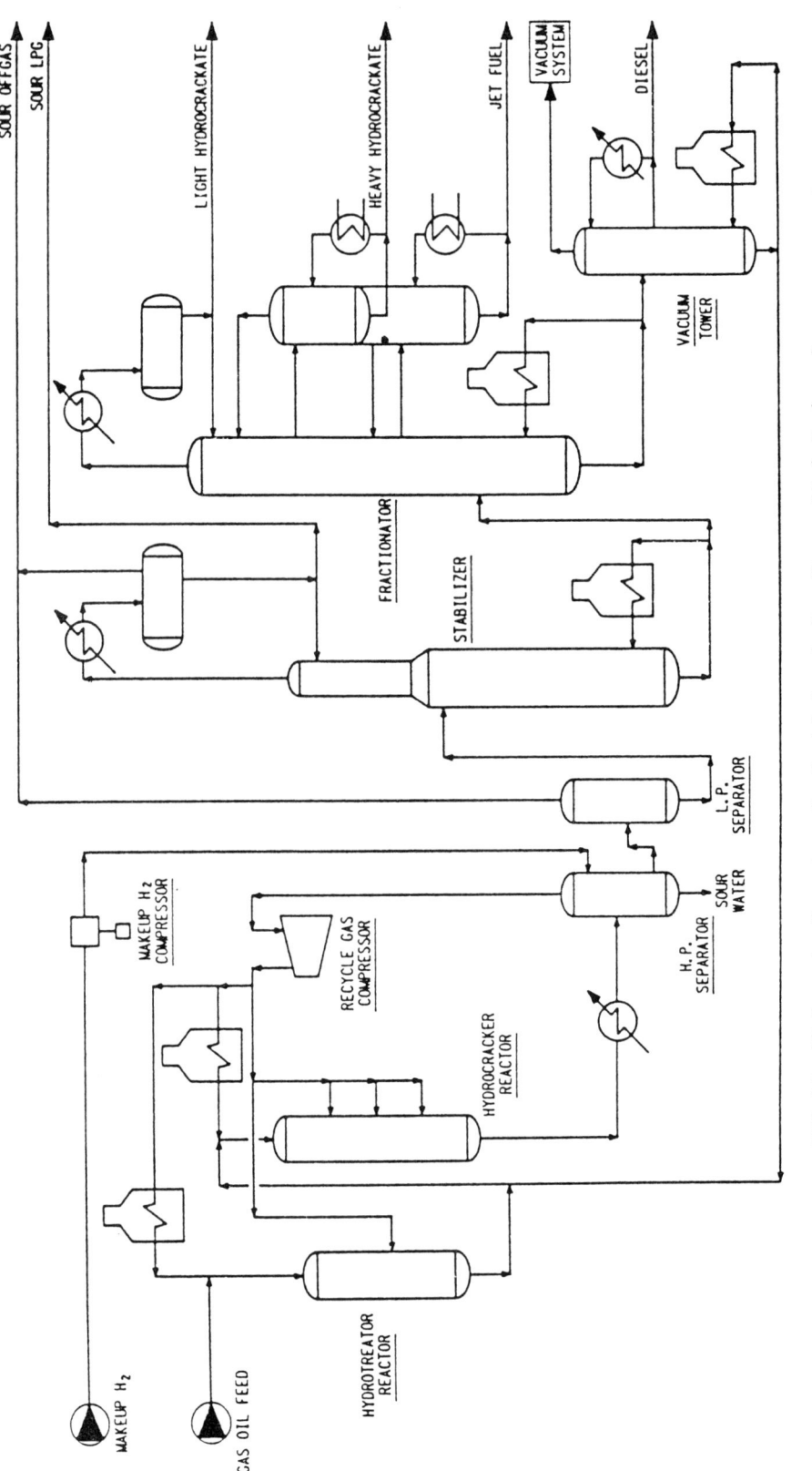

FIG. 14. Hydrocracker flow scheme for diesel fuel production. (Source: Litwin Engineers.)

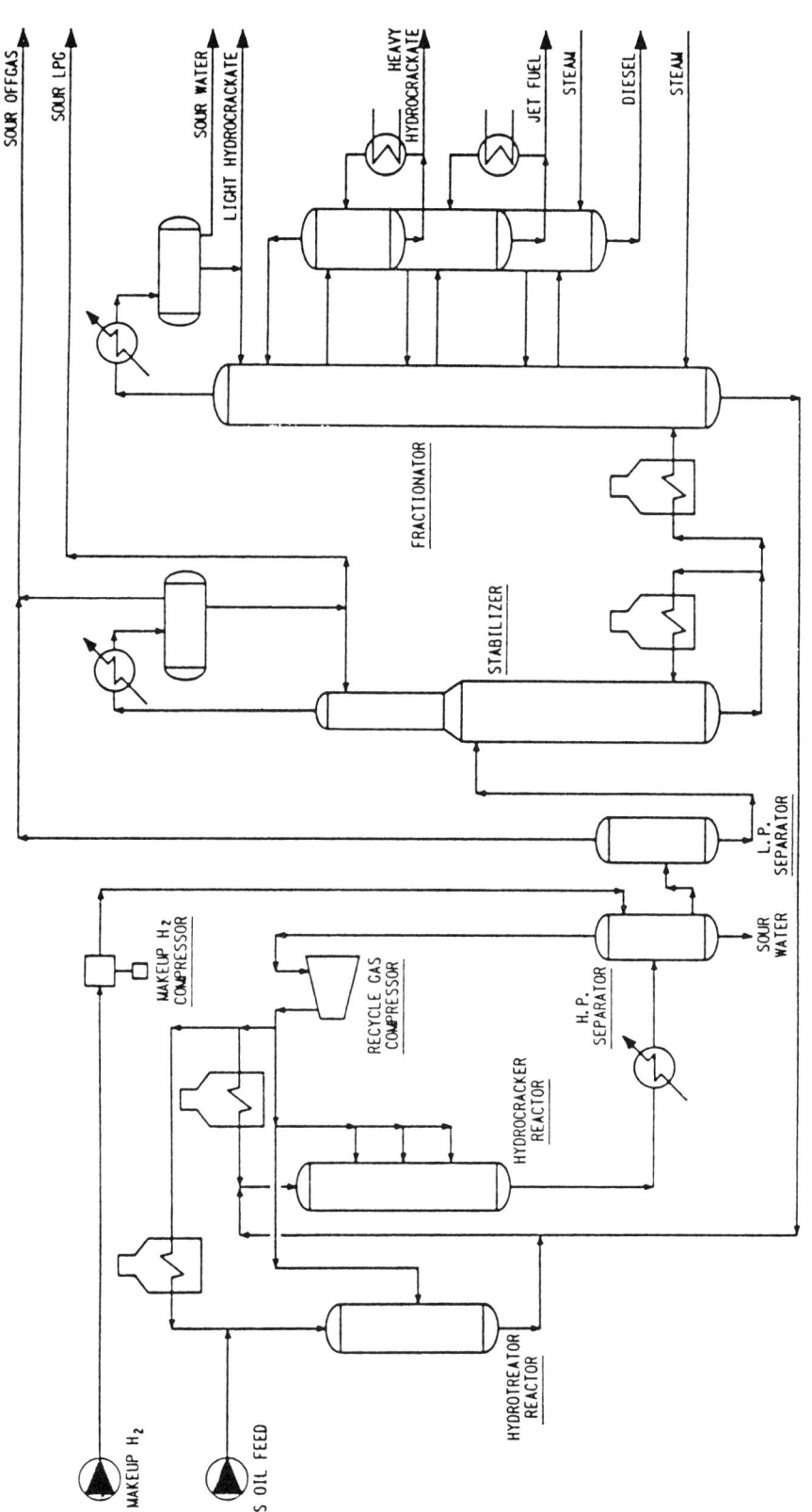

FIG. 15. Alternate hydrocracker flow scheme for diesel fuel production. (Source: Litwin Engineers.)

primary process decisions an engineering company and an operating company must consider to successfully design and construct and to properly operate a hydrocracker. These decisions concern:

1. Liquid-feed preparation
2. Hydrogen make-up preparation
3. Product fractionation

Litwin Engineers has summarized some of the key considerations regarding refinery decisions in these three areas.

Liquid Feed Preparation

There are typically one of two virgin feeds to a hydrocracker: Atmospheric gas oil (AGO) or AGO plus vacuum gas oil (VGO).

If a refiner elects to hydrocrack only AGO, the hydrocracker can be designed for lower pressure operation rather than a unit handling both AGO and VGO (i.e., 1500 lb/in.^{2} gauge versus 2200–2400 lb/in.^{2} gauge for AGO plus VGO), resulting in a 15–20% lower capital investment than a full range gas oil hydrocracker. The lower pressure design is, however, more sensitive to gas oil endpoint. At a reactor operating pressure of 1500 lb/in.^{2} gauge, for example, a gas oil feed ASTM D-86 endpoint of 800–850°F is generally the limit.

Atmospheric Gas Oil Feed Preparation. In preparing virgin AGO feed to a hydrocracker, either of two situations will normally exist: (1) AGO feed will come from a new crude distillation column, or (2) AGO feed will come from an existing crude column. From its studies, Litwin has concluded the following: To improve AGO fractionation in a new crude distillation column, installing seven real fractionating trays in the wash oil section is preferable to increasing overflash; to improve AGO fractionation in an existing crude column, using 6 ft of Koch's #2 Flexipac to replace three distillation trays offers a very short payout of 2–3 months by increasing usable AGO yield at no sacrifice in quality.

The foregoing conclusions are discussed in the following paragraphs.

If the AGO feed comes from a new crude column, steps can be taken to maximize AGO yield while minimizing entrainment of atmospheric resid into the AGO draw. This result is desirable since entrainment of atmospheric resid will increase the endpoint of the AGO feed to the hydrocracker, promote coke laydown in the hydrocracker reactors, and shorten run length. Litwin has found that by installing seven real fractionating trays in the wash oil section of the tower (i.e., that section between the AGO draw and the flash zone), a theoretical count of three trays can be used in predicting column performance. Of course, another approach to improving fractionation between feed and bottoms is to increase overflash, but the total amount of liquid vaporized in the flash zone is limited by flash zone temperature, which is typically 690–720°F, dependent upon downstream processing of tower products. In attempting to maximize AGO yield, a refiner will normally hold overflash to approximately 5.0 LV% on feed, consistent with good distillation practices. As a result, increasing overflash has limited capabilities if a refiner wishes to produce maximum AGO. Improving distillation hardware is the better solution to the problem.

In the event that AGO feed to be processed is from an existing crude column, there most likely were no special measures taken in the original design to minimize

Hydrocracking

carryover of atmospheric resid in the AGO product. A typical crude-vacuum distillation unit in which AGO is combined with VGO for further processing or for blending for sales probably has three distillation plates in the atmospheric crude fractionation section between the flash zone and the AGO draw. Steps can be taken, however, to improve both AGO quantity and quality.

Vacuum Gas Oil Feed Preparation. Care must be taken in the design of the vacuum distillation unit preceding a hydrocracker to ensure that entrainment of vacuum resid from the flash zone of the vacuum tower does not cause the levels of metals and Conradson carbon in the gas oil to exceed certain design levels. Such contaminants will cause premature deactivation of the hydrocracking catalyst.

Hydrogen Make-up Preparation

The purity of hydrogen in the make-up stream to the hydrocracker affects the hydrogen partial pressure in the reactor. This partial pressure in turn sets the operating pressure of the hydrocracker. There are two main conclusions: (1) If hydrogen is supplied from a steam reformer hydrogen plant, the addition of a pressure swing adsorption (PSA) section is economically attractive. (2) If hydrogen is supplied from a catalytic reformer unit, justification of purifying reformer off-gas must be done on a case-by-case basis.

These conclusions are discussed in the following paragraphs.

When a hydrogen plant is the supplier of make-up for a hydrocracker, Litwin has found that increased hydrogen purity is a desirable option, and that such purification by PSA is a better choice than by potassium carbonate absorption followed by methanation. The hydrogen purity for a steam reformer using PSA can be 99.99+% while for a conventional adsorption plus methanation scheme it is normally 95–97%. The resulting reduction in hydrocracker design pressure (approximately 200–400 lb/in.^{2}) translates into lower hydrocracker capital investment. The somewhat higher hydrogen plant investment attributable to the PSA unit is generally paid out rapidly by the reduced hydrocracker investment and by savings in hydrogen plant and hydrocracker utilities usage. In addition, PSA facilities require less winterization than those employing potassium carbonate solutions.

If make-up gas for a hydrocracker is from a catalytic reformer, Litwin has found that the question of whether this gas should be further purified has no clear and consistent answer. At first glance it would seem obvious that purification of gas with an 80–85 mol% H_2 content to 99+% H_2 should offer improved economics. However, the only pertinent measure of H_2 concentration is the hydrogen-to-methane ratio, since ethane and heavier components will condense out in the hydrocracker separator at its high pressure and relatively low temperature. This, of course, means that the effective hydrogen purity of reformer off-gas to a hydrocracker is above 90%. Since some of the methods for purifying hydrogen streams require large pressure drops and therefore higher compressor operating and investment costs, others such as PSA yield a low-pressure hydrocarbon off-gas by-product at 1.0–2.0 lb/in^{2}gauge which usually cannot be disposed of in conventional fired equipment burners (see Table 5). Justification of purifying reformer off-gas must be investigated for each individual project.

Product Fractionation

When a refiner operates a hydrocracker to produce maximum jet fuel, the product fractionation arrangement is different from that required to yield maximum diesel fuel. In a jet fuel operation, the liquid product is first stabilized to avoid the need for compressing the overhead gas from the fractionator to feed to the stabilizer, as would be the case if the fractionator preceded the stabilizer. Stabilizer bottoms are then distilled in a reboiled fractionator for separation of naphthas and jet fuel.

Fractionation flow patterns for a diesel fuel hydrocracker are similar to those for jet fuel, but the subsequent fractionation differs. A full range diesel fuel stream cannot be withdrawn from a reboiled fractionator, since the reboiler feed contains components heavier than the diesel fuel fraction. These components, when heated in excess of 700°F at atmospheric pressure, cause severe coking of the heater. Therefore, when a refiner operates to produce maximum diesel fuel, he must employ either a reboiled fractionation column steam stripped, or an unstripped fractionator reboiled at a lower temperature followed by a vacuum column which is needed to recover the heavier portion of the diesel fuel. The scheme of a reboiled fractionator followed by vacuum distillation is preferable due to its inherently better fractionation capabilities, ensuring the highest possible yields of product and recycle oil. Litwin has seen diesel fuel lifted by both methods in hydrocrackers, but feels that the economics of yield versus capital investment must be studied and decided with the refiner to determine the preferred fractionation method of product recovery.

Yields and Performance

Considering the number of licensors of hydrocracking technology and the variations available with each process, there is an infinite number of possible yields and performances. This section will touch on a few of the available options.

An example of a small refinery that integrated hydrocracking is USA Petrochemical Co. that operates a 30,000 BPSD refinery in Ventura, California, just north of Los Angeles. The refinery was built without conversion capacity, no FCC or hydrocracker. As designed, reduced crude was fed to vacuum fractionation with the VGO, then desulfurized and sold as low sulfur fuel oil or FCC feedstock. When mild hydrocracking technology became available in 1982, USA Petrochemical quickly recognized the potential profit in increased distillate production if their existing gas oil desulfurizer was converted to MHC Unibon operation.

Early in 1983, USA Petrochemical modified their gas oil desulfurizer for mild hydrocracking operation. The additions to the system included:

1. Additional heat exchange in the feed–product exchangers
2. Fin fan cooler for the product fractionator overhead
3. Water wash system for the reactor effluent heat exchanger train
4. Revisions to the sour water stripper to process an increased load
5. Relocation of the diesel draw tray to minimize diesel loss to fractionator bottoms

The modifications were finished in a short period of time at a total cost of approximately $250,000. USA Petrochemical estimates that the incremental

Hydrocracking

TABLE 11 Licensed Unicracking/HDS Plants[a]

Company	Plant Location	Capacity (BPSD)	Start-up Date
Maruzen Oil Company	Chiba, Japan	60,000	1976
Philips Petroleum Co.	Sweeny, Texas	75,000	1980
Kuwait Oil Co.	Ahmadi, Kuwait	66,000	1984
Kuwait Oil Co.	Ahmadi, Kuwait	66,000	1985
Kuwait National Petroleum Co.	Abdulla, Kuwait	66,000	1986
Chinese Petroleum Corp.	Taoyuan, Taiwan	30,000	1986

[a]Source: UOP.

increase in product values with UOP's MHC over hydrotreating ranges between $0.70 and $1.16/bbl of feed to the unit. This increase comes from both an increase in product volume and higher value of the product slate. Incremental utility consumption was $0.175/bbl for fuel gas (higher temperature requirement for MHC over desulfurization) and electrical power (increased pressure, water wash, and cooling requirements). Catalyst cost is estimated at about $0.20/bbl. The net result for USA Petrochemical was a payback in less than 6 months.

In the case of Union Oil's technology, the HC process is often married to HDS (hydrodesulfurization). Over the past 10 years, the company has licensed six commercial units of this type with over 360,000 BPSD capacity as shown in Table 11. The initial installation in 1976 at the Maruzen Oil Co. in Japan was designed to produce low sulfur fuel oil (LSFO) from residua. Since then, the UK/HDS process has become more flexible with the ability to upgrade feedstock for fluid catalytic cracking (FCC) or hydrocracking. Thus, for example, the Phillips' UK/HDS unit in Sweeny, Texas, provides feedstock for a heavy oil cracker (HOC). New plant installations are scheduled in Kuwait and Taiwan. The latest Union data on product yields are shown on Table 12.

The calculated overall conversion has been corrected to account for the 13 vol.% of heavy diesel in the feed. From this table it can be seen that C_5-400°F naphtha and 265–500°F jet fuel yields increase as conversion to 700°F-minus products increases. Heavy diesel yield increases to a maximum at

TABLE 12 Feed #1: Product Yields as a Function of Conversion[a]

	Conversion to 700°F-Minus (TBP), (vol.%)			
Yields (vol.% of feed)	23.9	59.3	79.2	94.2
C_5–400°F naptha	12.0	33.1	55.8	75.0
265–500°F jet fuel	12.5	30.5	40.8	46.7
500–700°F diesel	20.3	25.0	21.8	14.1
265–700°F jet fuel + diesel	32.8	55.5	62.6	60.8

[a]Conversion = $\left(1 - \dfrac{\text{vol.\% 700°F+ in product (\% of feed)}}{\text{vol.\% 700°F+ in feed}}\right) \times 100$.

about 60% conversion, and then begins decreasing as conversion is further increased. Total jet fuel plus diesel yield reaches a maximum at about 80% conversion to 700°F-minus. Jet fuel to heavy diesel ratio increases from 0.61:1 at 23.9% conversion to 3.3:1 at 94.2% conversion. This change in yield distribution with conversion can be more clearly seen in Fig. 16. This plot further illustrates that the particular catalyst used for this study gives an increasingly greater yield of naphtha at the expense of jet fuel and diesel as conversion is increased. This is expected, since it is a high activity molecular

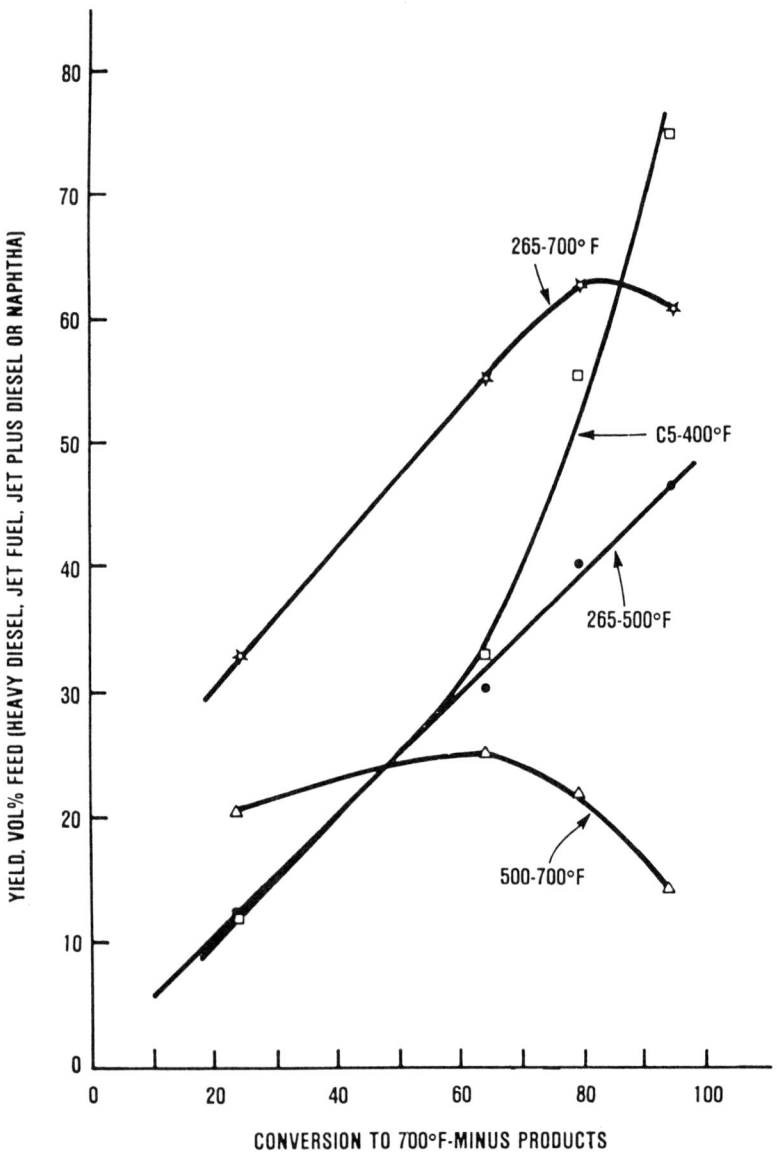

FIG. 16. Feed #1: Product yields as a function of conversion.

Hydrocracking

sieve catalyst designed for conversion to naphtha and lighter products. The relative slopes of these curves would be quite different for a catalyst designed for production of middle distillates. This will be illustrated later. The trend of decreasing selectivity for middle distillates with increasing once-through conversion is observed with all hydrocracking catalysts.

The change in conversion described above is achieved in the hydrocracker, assuming constant feed rate and feed quality, by raising or lowering the reactor average temperature.

Reactor temperature has a definite effect on yield and conversion (Figs. 16 and 17). These plots illustrate the tremendous flexibility hydrocracking gives to the refiner. By raising the average reactor temperature by about 10°F, the net conversion of feed to 700°F-minus products increases from 60 to 80%. This same 10°F change in reactor temperature increases conversion to 400°F-minus products from 25 to 50%. The relationships described by Fig. 17 are again for a high activity molecular sieve catalyst. The response of conversion

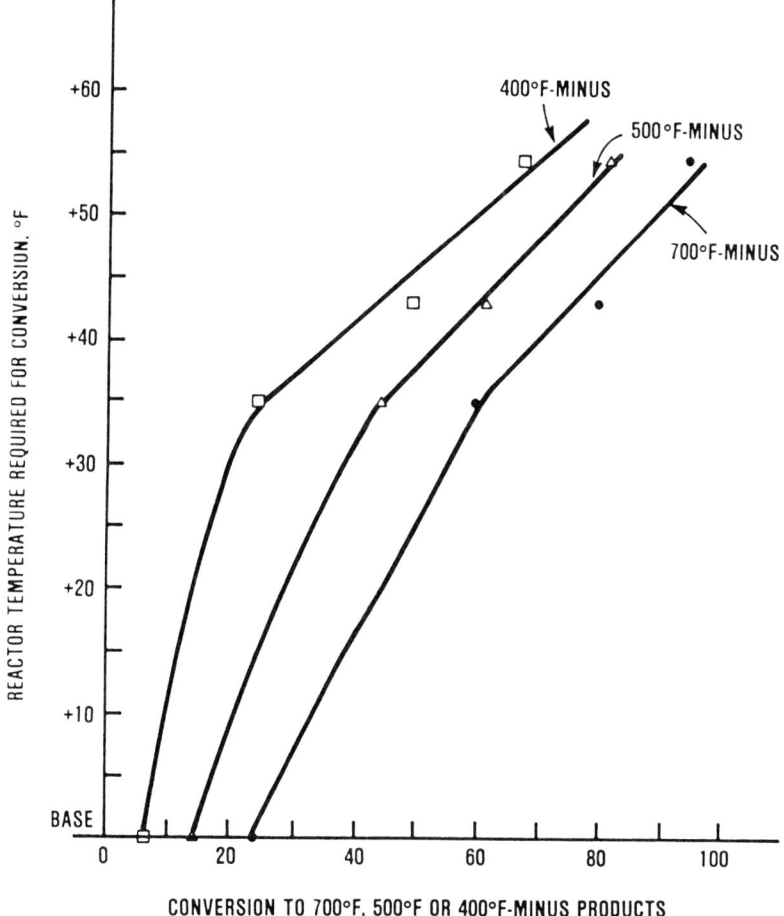

FIG. 17. Feed #1: Conversion as a function of hydrocracking temperature.

to changing temperature is much different for an amorphous hydrocracking catalyst, for example. An amorphous catalyst would require as much as a 50% greater temperature increase than a molecular sieve catalyst for the same increase in conversion, according to Union.

A number of Chevron-designed hydrocrackers are in operation (Table 13). The use of Chevron's dual catalyst mild hydrocracking technology has enabled Nippon Petroleum Refining Co. (NPRC) to produce excellent yields of good-quality mid-distillate from VGO in an existing VGO desulfurizer at its Muroran Refinery. This technology enables refiners to take advantage of underutilized desulfurization capacity for hydrocracking with minimal additional capital investment. The Muroran unit has been in this high conversion mode since June 1982 and has operated in the range of 10–30 liquid volume

TABLE 13 Chevron-Designed Hydrocracking Plants[a]

No.	Company	Location	Major Products[b]		Capacity (BPOD)
1	Sohio	Ohio	N	1962	12,000
2	Chevron	Mississippi	N	1963	28,000
3	Shell Canada	Ontario	N/D	1963	4,000
4	Tosco	California	N	1963	22,000
5	Caltex	Germany	N/K/D	1964	8,500
6	Chevron	California	N/K/F	1966	30,000
7	Chevron	California	N/K	1966	50,000
8	Sohio	Ohio	N/F	1966	25,000
9	Nippon Mining	Japan	G	1966	3,000
10	Mobil	California	N	1967	16,000
11	NIOC	Iran	K/D	1968	14,000
12	Tenneco	Louisiana	N	1968	16,000
13	KNPC	Kuwait	K/D	1969	18,000
14	Mobil	Texas	N	1969	29,000
15	EMP	Spain	G	1969	20,000
16	Chevron	California	N/K	1969	50,000
17	Sohio	Ohio	N/L	1970	20,000
18	Chevron	Mississippi	N/K	1971	32,000
19	BP Oil	Pennsylvania	N	1975	20,000
20	Irving Oil	Canada	N/D	1977	31,000
21	KNPC	Kuwait	K/D	1978	44,000
22	Hawaiian Independent	Hawaii	K	1981	12,000
23	NIOC	Iran	K/D	1981	15,000
24	NIOC	Iran	K/D	1981	15,000
25	Chevron	California	L	1984	18,500
26	Chevron	California	L	1984	12,000
27	KNPC	Kuwait	K/D	1986	38,000
28	Unannounced	Middle East	K/D	1986	44,000
29	Unannounced	Middle East	K/D	1987	39,000
	Total				686,000

[a]Source: Chevron Research Co.
[b]D = diesel, F = FCC feed, G = LPG, K = kerojet, L = lubes, N = naptha.

(LV) percent of 680°F cut point diesel and lighter products. The amount of recoverable diesel and its quality are strong functions of the feed boiling range, synthetic conversion, and plant pressure.

For commercial operation, the 40,000 BPOD at NPRC was designed to desulfurize Arabian VGO. In the early 1980s, faced with a reduction in fuel oil demand and increased light product demand, Muroran shifted to severe desulfurization with the existing hydrodesulfurization catalyst. In 1982, Chevron's mild Isocracking catalyst system was installed, and the unit has been operating in this mode since then. Feed rate to the unit, consisting primarily of VGO, varies between 20,000 and 27,000 BPOD. Conversion levels typically yield 20–40 LV and diesel at 900–1000 lb/in.^{2}abs hydrogen partial pressure.

A comparison of product yields and properties for three modes of operation are shown in Table 14. In all cases the light Isomate distillate product meets Japanese diesel specifications for sulfur, cetane index, pour point, and distillation. The heavy Isomate bottoms product is an excellent fuel oil blend stock or FCC feed. The viscosity reduction of the heavy Isomate enhances its low sulfur fuel oil blending quality by reducing the amount of cutter required. The reduced nitrogen content leads to improved FCC catalyst activity, conversion, and yields.

TABLE 14 Feed and Product Inspections, Commercial Data

Operation	Conventional Desulfurization	Severe Desulfurization	Mild Isocracking
% HDS	90.0	99.8	99.6
Yields, LV (%):			
Naptha	0.2	1.5	3.5
Light Isomate	17.2	30.8	37.1
Heavy Isomate	84.0	70.0	62.5
Feed:			
Gravity °API	22.6	22.6	23.0
Sulfur, wt.%	2.67	2.67	2.57
Nitrogen, ppm	720	720	617
Ni + V, ppm	0.2	0.2	—
Distillation, ASTM, °F	579–993	579–993	552–1031
Light Isomate:			
Gravity, °API	30.9	37.8	34.0
Sulfur, wt.%	0.07	0.002	0.005
Nitrogen, ppm	18	20	20
Pour point, °F	18	14	18
Cetane index	51.5	53.0	53.5
Distillation, ASTM, °F	433–648	298–658	311–683
Heavy Isomate:			
Gravity, °API	27.1	29.2	30.7
Sulfur, wt.%	0.26	0.009	0.013
Nitrogen, ppm	400	60	47
Viscosity, cSt at 122°F	26.2	19.8	17.2
Distillation, ASTM, °F	689–990	691–977	613–1026

Product TBP distillation curves for two mild Isocracking conversion levels are shown in Fig. 18. Due to stringent Japanese diesel distillation specifications, diesel yield is limited to ~690°F TBP cut point (Diesel Cut "A"). With less restrictive distillation specifications, the diesel TBP cut point could be raised as high as 725°F (Diesel Cut "B"), producing an incremental 9 LV percent diesel.

The effect of varying the conversion is illustrated in Fig. 19. Naphtha and synthetic diesel yields are plotted as a function of synthetic conversion. Synthetic conversion is total conversion corrected for straight-run distillate material in the feed. We have also included pilot-plant data for comparison. Although the naphtha yield increases significantly as conversion is increased, the bulk of conversion (~80%) is to diesel material. This illustrates the mid-distillate selectivity of the catalyst system. The data shown were obtained over a wide range of reactor temperatures (690–790°F). The linear relationship between the naphtha and synthetic diesel yield versus synthetic conversion indicates the catalyst selectivity to mid-distillate production is temperature independent.

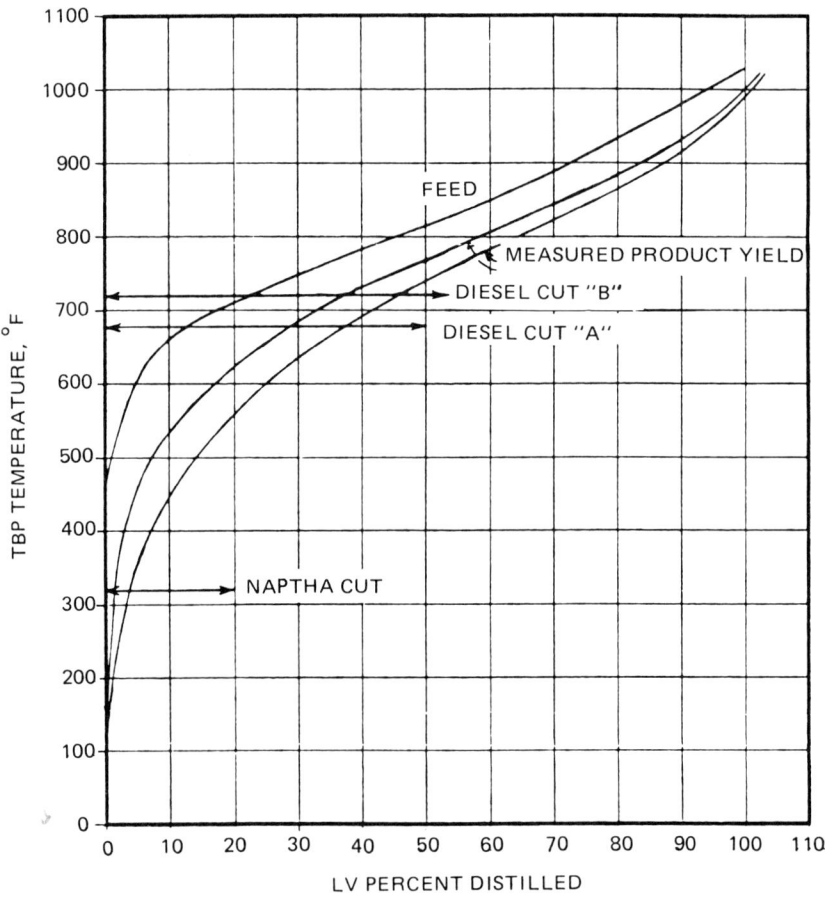

FIG. 18

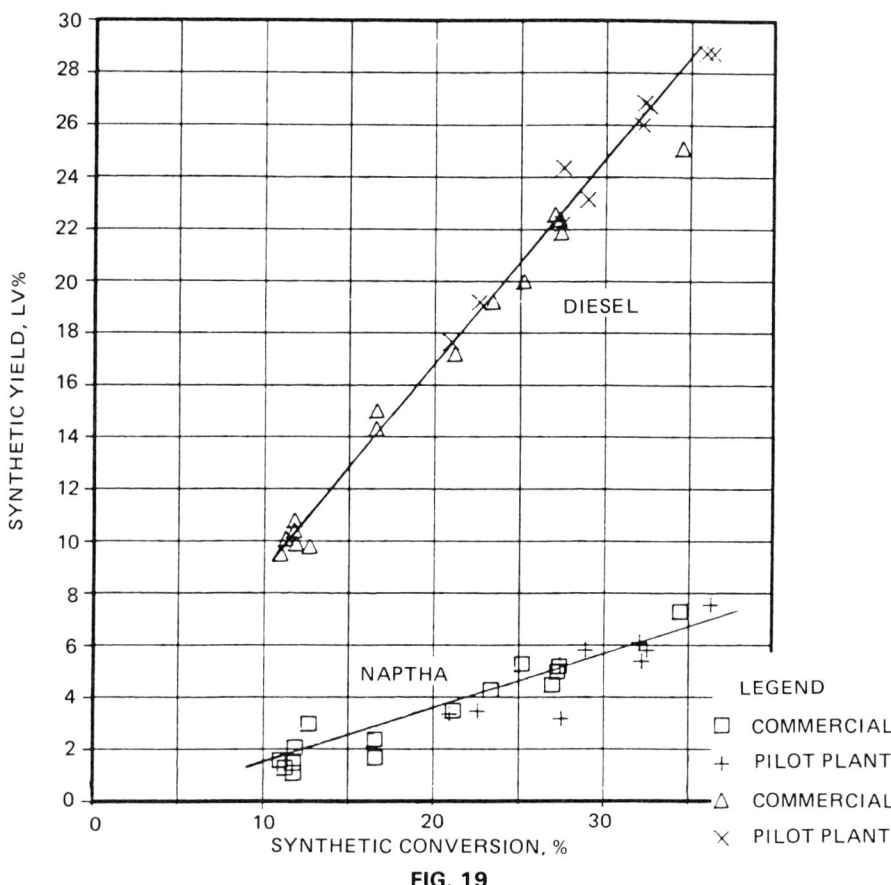

FIG. 19

Economics

One must realize that payback period or return on investment (ROI) is highly esoteric for each company and each location. Even the price of natural gas has a dramatic effect on payout (Fig. 20).

Tables 15, 16, and 17 compare overall economics for Cases 1G, 2D, 3G, and 4D. All costs are on a U.S. Gulf Coast basis and in fourth quarter 1983 dollars. Investments are based on curve-type estimates for typical on-plot and off-plot plants. Actual costs for specific plants could vary by ±25%. Initial catalyst and paid-up royalty costs are included in on-plot investment. Off-plot costs were estimated by summing the contributions of boilers, tanks, lines, buildings, and other required items.

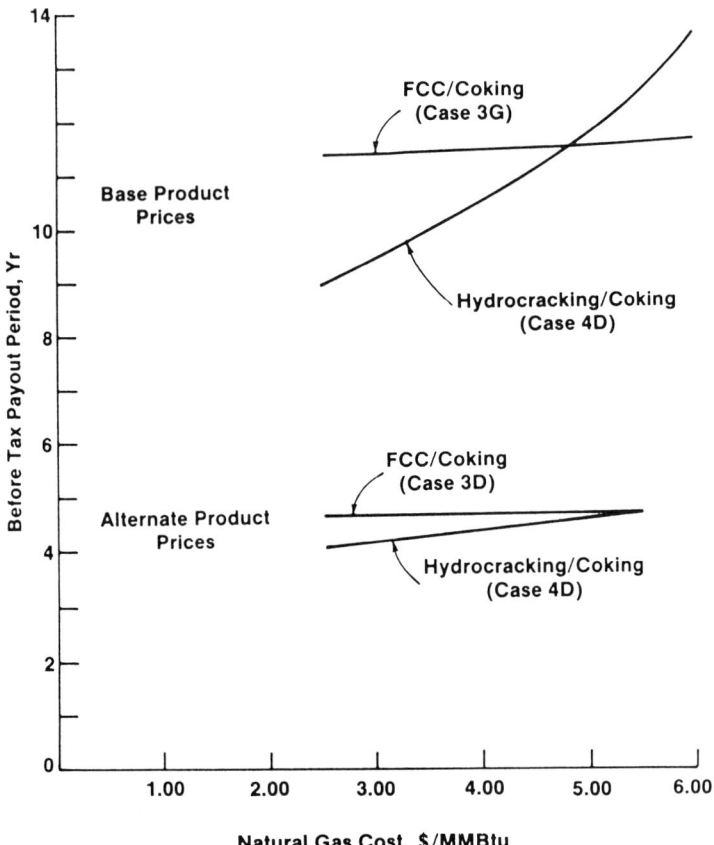

FIG. 20. Effect of natural gas cost on payout period. Base cost of $4.50/MMBtu assumed in study. (Source: Chevron Research Co.)

TABLE 15 Feed and Product Prices Used[a]

	Base ($/bbl)	Alternate ($/bbl)
Arabian Heavy crude oil	27.20[b]	
Purchased fuel	4.50[c]	
C_3 LPG	17.00	
C_4 LPG	28.00	
Chemical naptha	32.00	35.00
Unleaded gasoline	35.00	38.00
Jet fuel	34.00	37.00
Diesel fuel	33.00	36.00
Residual fuel oil	25.00	
Delayed coke	20[d]	
Sulfur	100[d]	

[a]Source: Chevron Research Co.
[b]Delivered to U.S. Gulf Coast, fourth quarter 1983.
[c]$/MMBtu.
[d]$/short ton.

TABLE 16 Economic Comparison (primary cases)[a]

Case	IG	2D	3G	4D
Refinery Type	FCC	Hydrocracking	FCC	Hydrocracking
Cracker Feed	VGO	VGO	VGO/CGO	VGO/CGO
Operating Mode	Gasoline	Mid-Distillate	Gasoline	Mid-Distillate
Investment,[b] $MM:				
On-plot	490	460	700	710
Off-plot	380	330	450	400
Total	870	790	1150	1110
Product revenue, $MM/yr	1616	1649	1696	1743
Refining cost, $MM/yr:				
Crude oil	1489	1489	1489	1489
Natural gas to H_2 plant	0	29	2	52
Purchased fuel	19	23	9	3
Utilities, catalyst, and chemicals	13	15	20	24
Labor	9	7	11	8
Maintenance, taxes, and insurance	48	44	65	67
Total	1578	1607	1596	1643
Net revenue, $MM/yr	38	42	100	100
Before-tax payout period, yr	23	19	11	11
Before tax payout period at alternate produce prices, yr	7.2	6.1	4.7	4.5

[a]Source: Chevron Research Co.
[b]Fourth quarter 1983, U.S. Gulf Coast.

TABLE 17 Economic Comparison (flexibility cases)[a]

Case	3G/3D		4G/4D	
Refinery Type	FCC		Hydrocracking	
Cracker Feed	VGO/CGO		VGO/CGO	
Investment,[b] $MM:				
On-plot	720		800	
Off-plot	450		410	
Total	1170		1210	
	Gasoline Mode	Distillate Mode	Gasoline Mode	Distillate Mode
Product revenue, $MM/yr	1696	1693	1738	1743
Refining cost, $MM/yr:				
Crude oil	1489	1489	1489	1489
Natural gas to H_2 plant	9	15	8	3
Purchased fuel	2	6	39	52
Utilities, catalyst, and chemicals	20	20	26	24
Labor	11	11	9	8
Maintenance, taxes, and insurance	65	65	67	67
Total	1596	1606	1638	1643
Net revenue, $MM/yr	100	87	100	100
Before-tax pay period, yr	12	13	12	12
Before tax payout period at alternate product prices, yr	4.8	5.1	4.9	4.9

[a]Source: Chevron Research Co.
[b]Fourth quarter 1983, U.S. Gulf Coast.

TABLE 18 Stock Balances (VGO cracking cases)[a,b]

Case	1G	2D
Refinery Type	FCC	Hydrocracker
Operating Mode	Gasoline	Mid-Distillate
Feed:		
Arabian Heavy crude oil	150.0	150.0
Natural gas to H_2 plant	—	2.9
Net products:		
LPG	3.4	5.4
Chemical naphtha	3.0	6.7
Unleaded gasoline	54.2	18.7
Jet fuel	5.2	37.1
Diesel fuel	12.7	15.6
Residual fuel oil	70.2	69.1
Total	148.7	152.6
Total light products[c]	78.4	83.5
Net-fuel purchased	1.9	2.3

[a]Source: Chevron Research Co.
[b]All numbers in MBPCD.
[c]All products except residual fuel oil.

Product revenues were calculated using the stock balances given in Tables 18 and 19 and the product prices given in Table 20. Prices for naphtha, gasoline, jet fuel, diesel fuel, and residual fuel oil are spot prices for the U.S. Gulf Coast in the fourth quarter of 1983 as reported by various industry newsletters. C_3 and C_4 LPG are priced at $18 and $7/bbl less than unleaded gasoline, respectively. The refining cost includes Arabian Heavy crude oil at $27.20/bbl delivered to the Gulf Coast. Natural gas for hydrogen plant feed and purchased refinery fuel were assumed to cost $4.50/MMBtu based on the price of low sulfur fuel oil. Electrical power was assumed to cost 6.0 ¢/kWh. Maintenance, taxes, and insurance costs were estimated as percentages of on-plot and off-plot investments.

Naturally, economics vary for NHC depending on what hardware is already on the ground. It is estimated that roughly one-third of the world's refining capacity is in the FCC refinery configuration. In the United States, where the FCC is the workhorse of the refinery, that ratio is even higher, so it is interesting to see how MHC fits into an FCC refinery.

If one converts a VGO desulfurizer ahead of an FCC to an MHC, this has the effect of backing out roughly 30% of the VGO from the FCC, thus producing more distillate and less gasoline. Many refiners may not be willing to operate the FCC at such a low capacity. It is likely that a refiner would want production of gasoline either maintained or possibly increased, except for possible seasonal variations. United States refiners tend to keep their FCC's relatively full.

The refiner has an interesting and attractive option in this case—he can fill up his FCC to either its throughput, coke, or SO_x emission capacity, whichever is reached first, with VGO purchased outside the refinery on the spot

Hydrocracking

TABLE 19 Stock Balances (VGO/CGO blend cracking cases[a,b])

Case Refinery Type Operating Mode	3G FCC Gasoline	3D FCC Mid-Distillate	4G Hydrocracking Gasoline	4D Hydrocracking Mid-Distillate
Feed:				
Arabian Heavy crude oil	150.0	150.0	150.0	150.0
Natural gas to H_2 plant	0.2	0.6	4.0	5.3
Net products:				
LPG	4.3	4.9	6.0	6.6
Chemical naphtha	2.6	3.8	1.0	6.4
Unleaded gasoline	63.0	45.5	58.3	25.3
Jet fuel	1.8	17.3	34.8	34.9
Diesel fuel	64.6	65.2	39.2	68.3
Residual fuel oil	0	0	0	0
Total	136.3	136.7	139.3	141.5
Net fuel purchased	0.9	1.5	0.8	0.3
Coke, TPCD	2660	2750	2550	2550

[a]Source: Chevron Research Co.
[b]All numbers in MBPCD unless otherwise noted.

market. Thus, while producing additional No. 2 distillate, he can maintain his gasoline production. In some cases, blending some vacuum resid into the FCC feed may also be an attractive alternative.

Table 21 summarizes the details of two operations. Yields around the FCC are shown for two cases: (1) A base case where the VGO hydrotreater is operated for desulfurization only, no conversion, and (2) the VGO hydrotreater is converted to MHC which converts 30% to No. 2 distillate, with outside VGO brought in to fill the FCC so that gasoline production is maintained equal to the base case. Product and feed prices were average values for 1983.

Except for the outside VGO and the increased No. 2 distillate, all other yields are similar. Therefore, the profitability of the MHC/FCC combination depends primarily on the delta price between No. 2 distillate and the outside VGO. In the example, for a $1.5/bbl price delta between these two materials, there is $4600/d additional revenue for operating the gas oil hydrotreater as an MHC. Obviously, this will increase as the No. 2 distillate/outside VGO price delta increases. In these cases, the FCC complex also includes HF Alkylation. Only C_4's are alkylated, no outside i-C_4's are purchased.

The FCC in itself is a relatively flexible process. It produces high gasoline yields, and by changes in process variables, catalyst activity and product cut points can also produce significant amounts of distillate. The addition of MHC can further enhance the distillate/gasoline ratio of an FCC complex and allow the refiner to take further advantage of seasonal product demands.

Figure 21 shows the swing in yields that is possible from an MHC/FCC complex. Here it is assumed that in the gasoline season the MHC is operated as a desulfurizer and the FCC is operated in the gasoline production mode. In the distillate season the MHC is operated in the mild hydrocracking mode

TABLE 20 Actual Product Properties[a]

	Specification	1G	2D	3G	3D	4G	4D
Unleaded gasoline:							
Reid vapor pressure, lb/in.^{2}	11 max	11	11	11	11	11	11
Octane, (R + M)/2	87 min	87	87	87	87	87	87
Jet:							
Sulfur, wt.%	0.3 max	0.23	0.13	0.23	0.20	0.02	0.05
Freeze point, °F	−58 max	−58	−58	−58	−58	−58	−58
Smoke point, mm	20 min	26	25	26	27	21	21
Flash point (tag, closed), °F	100 min	130	100	130	100	140	100
Diesel/No. 2 heating oil:							
Sulfur, wt.%	0.5 max	0.5	0.02	0.5	0.5	0.5	0.5
Pour point, °F	20 max	20	−5	15	15	20	10
Cetane index	40 min	52	70	49	48	52	57
D 86 90% distilled °F	680 max	645	660	630	650	640	605
Viscosity, cSt at 122°F	1.6–3.0	3.0	3.0	3.0	3.0	2.8	3.0

[a]Source: Chevron Research Co.

Hydrocracking

TABLE 21 MHC Unibon in FCC Refinery: Economic Analysis[a,b]

Case	Base	MHC	Average 1983 Prices
Mode	Desulfurize	Mild hydrocracker	$/bbl
Feeds, BPSD:			
Middle East VGO	30,000	30,000	29.0
Outside VGO	—	6,750	31.5
H_2 (MM SCFD)	13.5	15.5	($3.5/MSCF)
Products, BPSD:			
LPG	3,547	3,662	25.0
C_4's	1,505	1,586	28.0
Gasoline	22,507	22,707	35.5
No. 2 distillate	3,970	10,865	33.0
Slurry oil	1,315	1,277	27.0
Fuel gas (MM SCFD)	6.74	7.18	($5.0/MMBtu)
Gross margin, M$/d	226.7	241.4	
Operating costs, M$/d	35.3	45.3	
Margin M$/d	191.4	196.0	

[a]Source: UOP.
[b]Basis 30,000 BPSD VGO desulfurized; desulfurized bottoms to FCC; HF alkylation of C_4S from FCC; no outside butanes purchased.

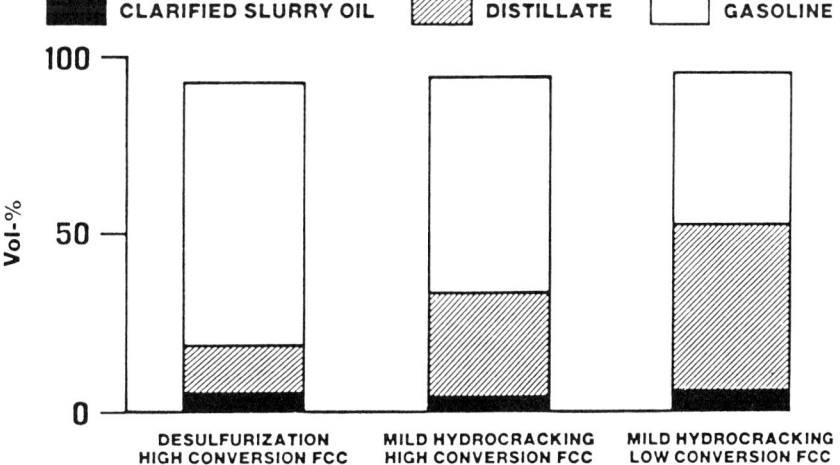

FIG. 21. Yield flexibility in FCC refinery for MHC Unibon/FCC/HF alkylation. (Source: UOP.)

and the FCC is operated in a low conversion, maximum distillate mode. Outside VGO is brought in to fill the FCC to its barrel capacity. The gasoline/distillate ratio from the FCC complex can be as high as 7:1, or when maximum distillate is required can be lowered to about 0.9.

The economic data presented are from 1983–1984. Accordingly, dollar adjustments must be made and up-to-date values considered when doing actual design and engineering.

Summary

For many refiners, the potential still exists to balance fuel oil supply and demand by adding vacuum gas oil cracking capacity. Situations where this is the case include:

Refineries with no existing vacuum gas oil cracking capacity.
Refineries with more vacuum gas oil available than vacuum gas oil conversion capacity.
Refineries where addition of vacuum residuum conversion capacity has resulted in production of additional crackable stocks boiling in the vacuum gas oil range (e.g., coker gas oil).
Refineries that have one of the two types of vacuum gas oil conversion units but could benefit from adding the second type. This might be done out of necessity to meet changing product demands or to increase overall flexibility.

Of course, in some cases a refiner might add both gas oil cracking *and* residuum conversion capacity simultaneously.

Those refiners who do choose gas oil cracking as part of their strategy for balancing residual fuel oil supply and demand must decide whether to select a hydrocracker or an FCC. Numerous papers comparing the two processes have been written over the years. However, neither process has evolved to be the universal choice for gas oil cracking. Both processes have their advantages and disadvantages, and process selection can be properly made only after careful consideration of many case-specific factors. Among the most important factors are: (1) product slate required, (2) amount of flexibility required to vary the product slate, (3) product quality (specifications) required, and (4) the need to integrate the new facilities in a logical and cost-effective way with any existing facilities.

In summary, the simplest form of hydrocracking is the so-called "mild hydrocracking." In this case, hydrotreaters designed for vacuum gas oil desulfurization and catalytic cracker feed pretreatment are converted to once-through hydrocrackers. Because existing units are being used, the hydrocracking is often carried out under nonideal hydrocracking conditions due to some inherent original design feature of the unit. Typical mild

Hydrocracking

TABLE 22 Typical Mild Hydrocracking Conditions[a]

Pressure, lb/in.^{2} gauge	600–1200
Temperature, °F	700–825
H_2: oil, SCF/bbl	1500–4000
Space velocity, v/v/h	0.4–1.0
Conversion, vol.%	5–35

[a]Source: UOP.

hydrocracking conditions are given in Table 22. These conditions are typical of many low-pressure desulfurization units which, for hydrocrackers in general, are marginal in pressure and hydrogen:oil ratio capabilities. Space velocities are often on the higher side of the range quoted. For hydrocracking, in order to obtain satisfactory run lengths (approximately 11 months), reduction in feed rate or addition of an extra reactor may be necessary. In most cases, since the product slate will be lighter than for normal desulfurization service only, changes in the fractionation system may be necessary. When

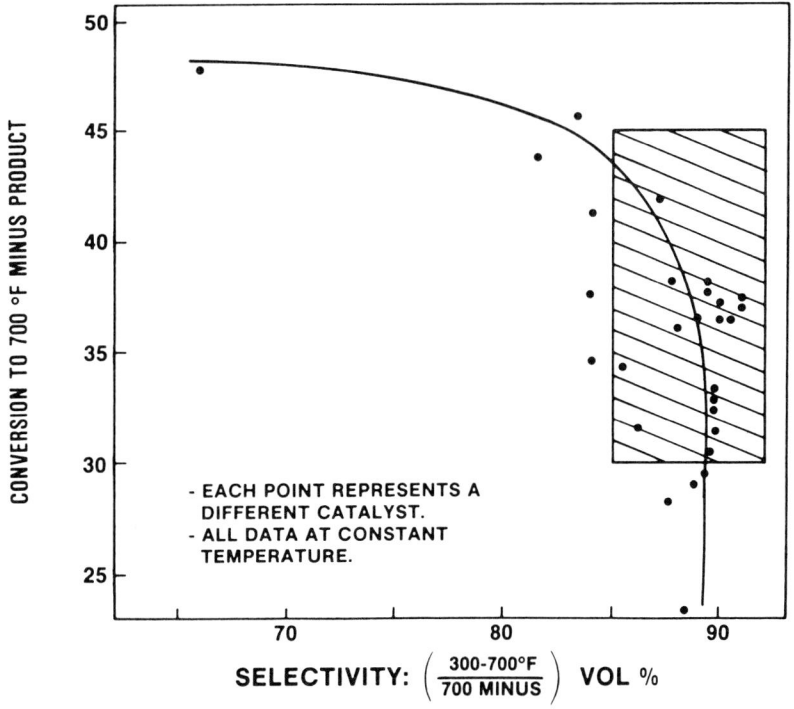

FIG. 22. Variation of selectivity with conversion and catalyst.

these limitations can be tolerated, the product value from mild hydrocracking versus desulfurization can be greatly enhanced.

As illustrated above for various forms of more conventional hydrocracking, the type of catalyst used can influence the product slate obtained. This is illustrated in Fig. 22 for a large number of catalysts. Here it is seen that for a mild hydrocracking operation at constant temperature, the selectivity of the catalyst varies from about 65 to about 90 vol.%.

$$\text{Selectivity} = \frac{300-700°F}{700°F-\text{minus vol.\%}}$$

As a result of an active research and development program, several catalytic systems have now been developed with a group of catalysts specifically for mild hydrocracking operations. Depending on the type of catalyst, they may be run as a single catalyst or in conjunction with a hydrotreating catalyst.

The author gratefully acknowledges the cooperation of the National Petroleum Refiners Association (NPRA) in preparing this article. The information presented is primarily a summary of NPRA reports, papers, and publications plus data from the author's own personal files on companies and processes.

GUY E. WEISMANTEL

Lubricating Oils: Manufacturing Processes

Introduction

The manufacture of mineral lubricating oil base stocks consists of five basic steps: (1) distillation and (2) deasphalting to prepare the feedstocks, (3) solvent or hydrogen refining to improve viscosity index and quality, (4) solvent or catalytic dewaxing to remove wax and improve the low temperature properties of paraffinic lubes, and (5) clay or hydrogen finishing to improve color, stability, and quality of the lube base stock. Figures 1 and 2 are simplified processing diagrams illustrating some of the ways these processes are combined to manufacture paraffinic and naphthenic lube base stocks and byproducts, respectively. The finished lube base stocks are blended with each other and/or with additives using batch and continuous methods to prepare fully formulated lubricants. The most common route is the mixed processing route which consists of solvent refining, solvent dewaxing, and hydrogen finishing although a few refiners still use the solvent and clay processing route

Lubricating Oils: Manufacturing Processes

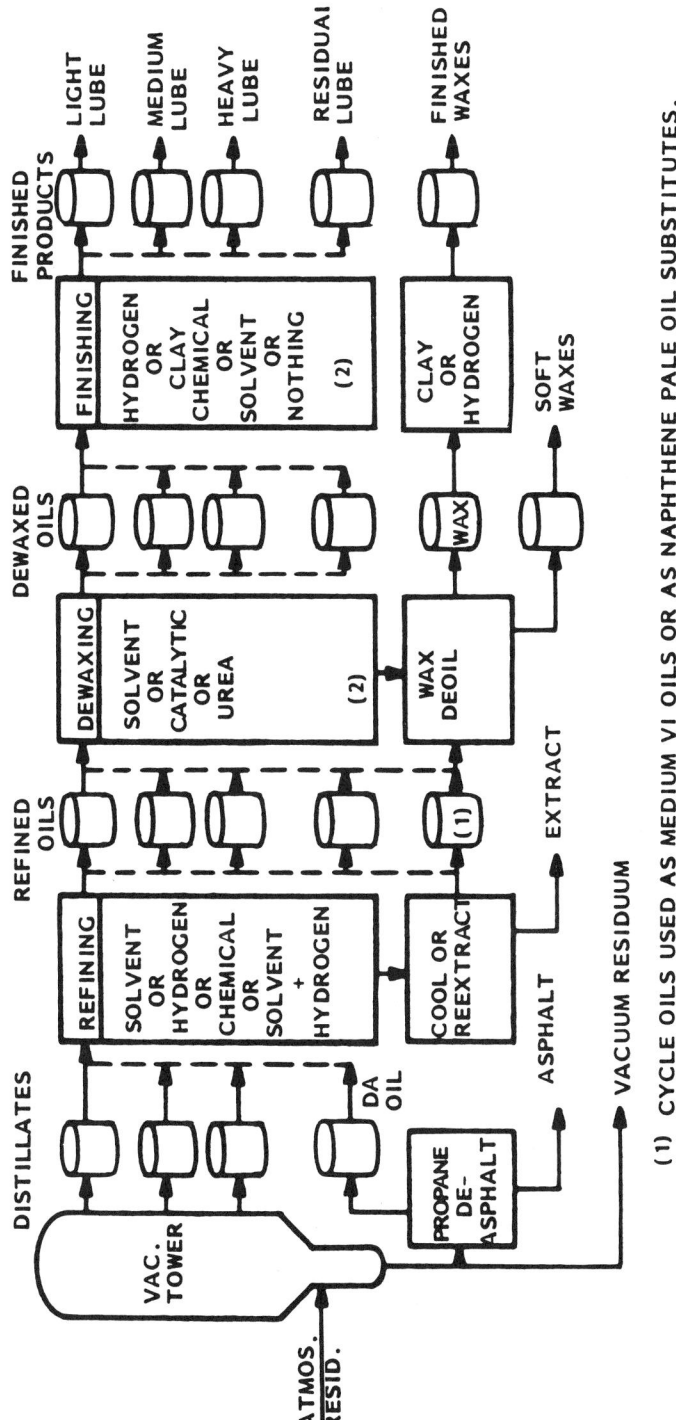

FIG. 1. Process flow for manufacture of paraffinic lube oils.

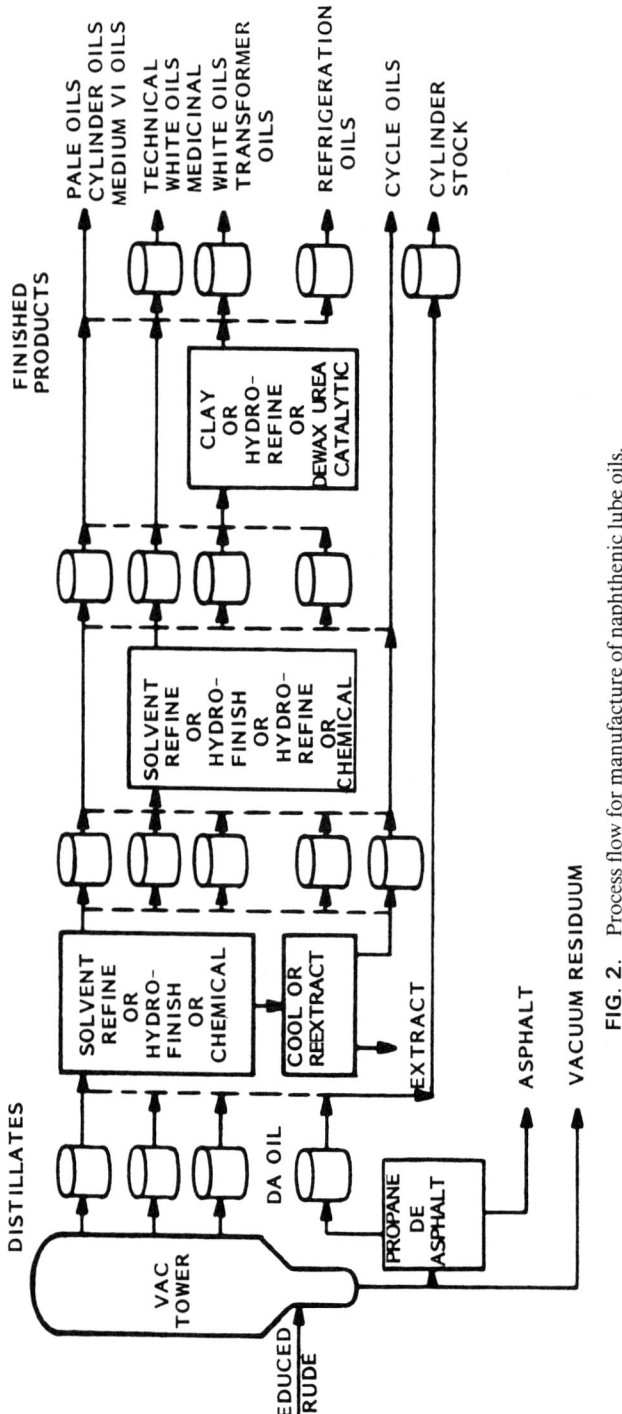

FIG. 2. Process flow for manufacture of naphthenic lube oils.

TABLE 1 Effect of Processes on Properties and Chemical Composition of Lubricating Oil Base Stocks[a]

		Refining			Process Dewaxing			Finishing		
	PDA	Solvent	Hydrogen	Chemical	Solvent	Catalyst	Urea	Hydrogen	Acid	Clay
Property:										
API gravity	+	+	+	+	–	–	–	N	N	N
Flash point	N	N	N	N	N	N	N	N	N	N
Viscosity	–	–	–	–	+	+	+	N	N	N
Viscosity index	+	+	+	+	–	–	–	N/+	N/+	N
Pour point	+/–	+	+	+	–	–	–	N/+	N/+	N
Color	–	–	–	–	+	–	+	–	–	–
Thermal stability	+	+	+/–	+/–	N	N	N	+/–	+/–	+
Oxidative stability	+	+	+/–	+/–	N	N	N	+/–	+/–	+
Inhibitor response	+	+	+	+	N	+/–	N	+	+	+
Constituent:										
Asphaltenes	–	–	–	–	+	N/–	+	N/–	–	–
Resins	–	–	–	–	+	N/–	+	N/–	–	–
Aromatics	–	–	–	–	+	+	+	N/–	–	–
Naphthenes	+	+	+	+	+	+	+	N/+	+	+
Paraffins	+	+	+	+	–	–	–	N/+	+	+
Wax	+	+	+	+	–	–	–	N/+	+	+
Nitrogen	–	–	–	–	+	N/–	+	–	–	–
Oxygen	–	–	–	–	+	–	+	–	–	–
Sulfur	–	–	–	–	+	N/–	N	–	–	–

[a]N = nil (min), + = increase, – = decrease, +/– = depends on severity, PDA = propane deasphalting.

or the hydrogen refining and solvent dewaxing route. The all-hydrogen processing (lube hydrocracking—catalytic dewaxing–hydrorefining) route is used by two refiners for the manufacture of a limited number of paraffin base oils.

The effect of lube processes on the chemical constituents and physical properties of lube base stocks are summarized in Table 1. It should be noted that each process is generally used for one purpose and that any accompanying change must be accepted to obtain the desired change in physical properties or chemical composition. For example, solvent extraction is used to increase the viscosity index and to improve the quality of lube base stocks by extracting aromatics. Accompanying this change are a reduction in sulfur and nitrogen content, an increase in API gravity, a decrease in viscosity, and improvement in color.

Crude Oils

Crude oils are the source of the feedstocks used in the manufacture of mineral lube base oils and the hydrocarbons used in the manufacture of synthetic oils. Crude oils contain considerable quantities of carbon and hydrogen and small quantities of sulfur, nitrogen, and inorganic salts, and relatively smaller quantities of metals. The hydrocarbon compounds present in crude oils range in carbon number from 1 through 100 or more and from the very simple to the very complex.

Since the quantity and chemical composition of specific lube boiling range fractions vary widely from crude to crude, entirely different processing sequences and/or severities are used to produce lube base stocks which also vary in chemical composition. The manufacture of lube base stocks is further complicated in some refineries where as many as 15 to 25 different crude oils are processed. The refiner thus finds it necessary to segregate crudes based on their suitability for lubes manufacture. This information is developed using laboratory evaluation techniques which vary from simple distillations and a few physical and chemical tests to very complex evaluations involving distillation, processing in pilot plants which simulate the refiners lube plant (processing sequence), physical and chemical tests, product formulation, and bench or field testing.

Lube Distillation

The initial step in refining a crude oil is to separate the crude into a series of narrow boiling range fractions which are the raw stocks for the various products to be produced. After the water, salts, and sediment have been removed by desalting, this initial separation is carried out on a crude distillation unit (CDU) or vacuum pipe still (VPS) which is composed of at

Lubricating Oils: Manufacturing Processes

least two distillation sections generally referred to as the atmospheric distillation tower or unit (ADT or ADU) and a vacuum distillation tower or unit (VDT or VDU). Figure 3 is a simplified flow diagram for a lube crude still. The gases, naphtha, and kerosene fractions are removed in the ADU. The residuum from the ADU is distilled in the VDU. Although some fractions boiling at 500°F may be distilled on the ADU and used for the manufacture of low viscosity base oils, the majority of the lube distillates boiling between 580 and 1000°F are distilled in the VDU to the proper viscosity and flash specifications. A solution of caustic or lime is introduced into the feed to CDU or VDU for the purpose of neutralizing the organic acids present in some crude oils. Some refiners also use a circulating caustic-oil stream to neutralize these acids in the vapor phase and to prevent caustic from entering the vacuum residuum [1]. Neutralization of crude oils containing high contents of organic acids is used to reduce or eliminate corrosion in downstream processing units. It also improves the color, stability, and refining response of lube distillates.

The design and operation of the VDU is of utmost importance in the manufacture of lube base stocks because distillate properties, particularly boiling range and purity of the feedstocks, have a significant effect on processing response in the lube refining, dewaxing, and finishing units as well

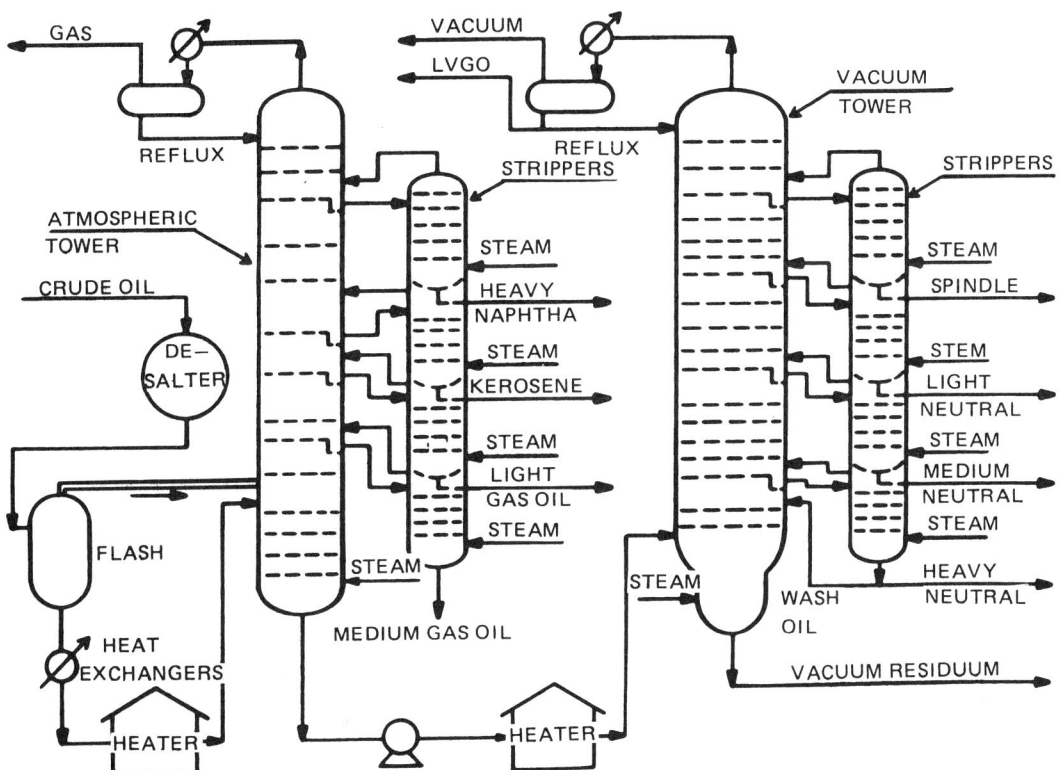

FIG. 3. Lube crude distillation unit flow diagram.

as the quality of the finished base stock. It is for these reasons that lube VDUs contain more fractionation stages, operate at lower pressure, and use more reflux than VDUs used to prepare fuels fractions for further processing. Refiners have in recent years installed a second VDT operating in series with the first VDT [2] and refurbished existing VDUs and/or installed new VDUs with high efficiency internals to reduce the flash zone pressure and improve the quality and yield of lube distillates. Information on crude oil distillation was reported by Atkins, McBride, and Owens [3] and Watkins [4].

Lube Deasphalting Processes

The vacuum residua contain a considerable quantity of high viscosity components useful in the manufacture of lubricating oils and asphaltenes and resins which contribute an undesirable dark color to the lube base stocks and form deposits of carbonaceous materials on heating. Refiners have employed many methods (adsorption, chemical treating, and precipitation with alcohols, ketones, and light hydrocarbons) to remove these undesirable constituents from vacuum residua and heavy distillates. Propane deasphalting [5–7] is most often used to remove these undesirable components which also impede the refining action of the downstream processes. Propane is preferred over the other liquefied gases used in the milder "deep" deasphalting (sometimes called decarbonizing) processes used to prepare feedstocks for fuels processing, because lube deasphalting requires removal of a greater quantity of the asphalts and resins present in the vacuum residuum.

The major process variables in deasphalting are dosage and temperature which range from about 500–1000 vol.% and 100 to 190°F, respectively. Pressure is a minor process variable which can be used to control the degree of separation. Deasphalted oil (DAO) yield is most dependent on crude source and will range from 30 to 90 wt.% basis vacuum residuum feed for most crudes currently being used. An increase in temperature decreases the yield and carbon residue of the DAO whereas an increase in solvent dosage increases the yield of deasphalted oil.

Heavy neutral distillates are sometimes produced by deasphalting the very high viscosity (wash oil) stream from vacuum fractionation of reduced crudes. The yield of DAO for use in bright stock and cylinder oil manufacture can be increased by including a higher than normal proportion of this wash oil in the vacuum residuum feed to the deasphalting unit.

Although mixer-settlers were used in the early deasphalting units and are still in use at a few locations, baffle towers of various designs and rotating disk contactors are used in modern deasphalting units. A simplified flow diagram for a propane deasphalting unit is shown in Fig. 4.

Single and dual effect evaporators are used to recover the solvent in the older deasphalting units with triple effect evaporation being used in modern or refurbished deasphalting units [5, 6]. Supercritical solvent recovery methods have also been developed [8, 9] and are being used in some deasphalting units.

Lubricating Oils: Manufacturing Processes

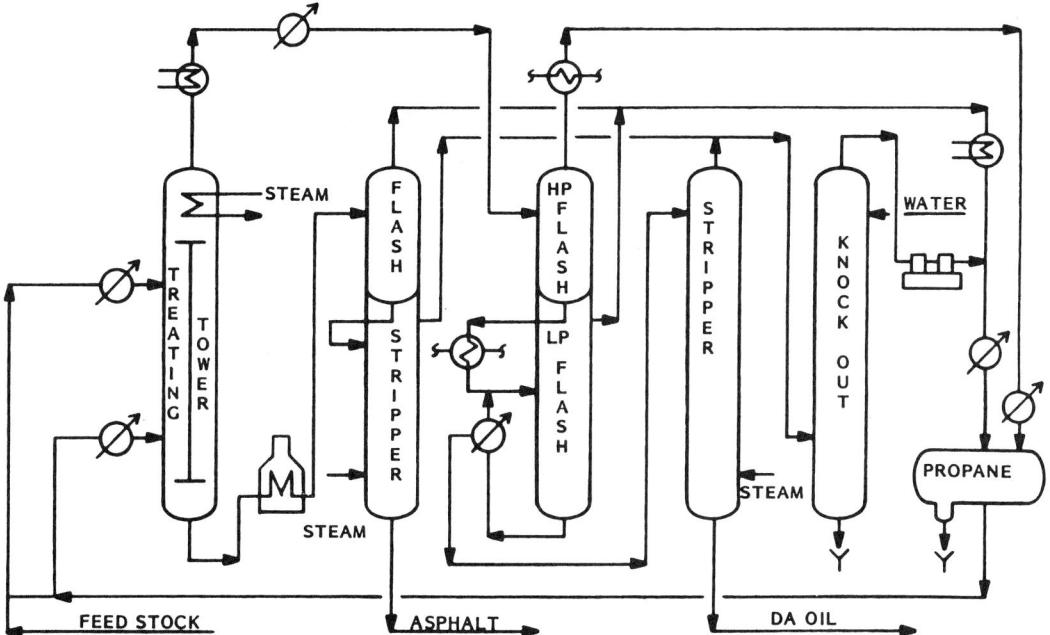

FIG. 4. Propane deasphalting unit flow diagram.

The Duo-Sol Process [1, 10, 11] is the only double-solvent process used to both deasphalt and extract lubricating oil feedstocks. Since it is both a deasphalting and extraction (refining) process, it does not conveniently fit into either the deasphalting or refining processes used for the manufacture of lube oils. Propane is used as the deasphalting or paraffinic solvent and Selecto (a mixture of phenol and cresylic acids) is used as the extraction solvent in this process. The Duo-Sol Process is most often used in the manufacture of bright stocks from vacuum residua. The process variables are temperature, solvent composition, and solvent dosage. The extraction is conducted in seven to nine batch extractors which are connected end to end followed by solvent recovery conducted in multistage flash distillation and stripping towers. The product raffinate requires no additional solvent or hydrogen refining for the manufacture of lubricating oil base stocks. Solvent dosages range from 100 to 400% Selecto and 150 to 500% propane basis feedstock. The temperature and solvent dosage used and yields of refined oil are highly dependent (like all refining processes) on the crude source being used and the quality of the base stock being manufactured. Since this process is capital intensive and phenol is toxic, no new units have been built since the mid-1950s and some units have been abandoned or replaced with deasphalting and/or solvent and hydrogen refining processes.

Lube Refining Processes

The vacuum distillates and deasphalted oils contain aromatics and other undesirable constituents which result in rapid darkening, oxidation, and sludging in service. Chemical, solvent, and hydrogen refining processes have been developed and are used to remove these undesirable components and to improve the viscosity index and quality of lube base stocks.

The classical chemical (sulfuric acid and clay refining) processes which were originally used for the refining of lube oil base stocks have been replaced by solvent extraction (solvent refining) and hydrotreating (hydrogen refining) processes because they are (1) more effective for the upgrading of feedstocks, (2) more cost effective, and (3) environmentally more acceptable. Although some refiners still use chemical refining processes, chemical refining is most often used for the reclamation of used lubricating oils or in combination with solvent or hydrogen refining processes for the manufacture of speciality lubricating oils and by-products.

Chemical Refining Processes

Acid-alkali refining [1, 10], also called "wet refining," is a chemical refining process in which lubricating oils are contacted with about 25 to 400 lb/bbl of 85–104 wt.% purity sulfuric acid followed by neutralization using aqueous or alcoholic alkali. Acid-alkali refining is conducted in a batch or continuous manner. In the batch process the oil to be treated is pumped to a treating agitator and mixed with acid of the desired strength. The oil and acid are mixed by mechanical means or by air blowing for a period of 10 to 30 min, after which time water may be added to assist in coagulation of the acid sludge. The sludge is removed or the oil decanted after settling for a period of several hours. More acid is added and the process repeated as needed. The acid or "sour" oil from this operation is then neutralized using an aqueous or alcoholic neutralizing agent followed by water washing and drying. Continuous acid refining involves the same steps as batch refining with the exception that (1) the acid, oil, "sour" oil, and neutralizing agent are mixed with pumps or static mixers; (2) excess acid, sludge, excess neutralizing agent, and soaps are removed by centrifuge or centrifugal extractors; (3) water washing is conducted using centrifugal extractors; and (4) brightening of the oil is conducted in continuous strippers. The advantages for the continuous process over the batch process are (1) higher yields of oil, (2) lower manpower requirements, (3) lower chemical consumption, (4) lower maintenance cost, (5) smaller space requirements, and (6) a reduction in air and water pollution.

Acid-clay refining [1, 10], also called "dry refining," is carried out in a manner similar to the acid-alkali refining process with the exception that clay and a neutralizing agent are used for neutralization of the "sour" oil. This process is used with oils that tend to form emulsions during neutralization, and it may be conducted in a batch or continuous manner. The neutralization products and bleaching clay are separated from the oil by using pressure filters. Continuous processes frequently employ the clay contacting methods used for the finishing of lubricating oils.

Neutralization [1] with aqueous and alcoholic caustic, soda ash lime, and other neutralizing agents is used to remove organic acids from some feedstocks. This type of treating can be conducted in a batch or continuous manner, as is done in sulfuric acid treating, or may be introduced into the crude distillation unit. Neutralization of feedstocks containing large quantities of organic acids is usually conducted to reduce organic acid corrosion in downstream units or to improve the refining response and color stability of lube feedstocks.

Hydrogen Refining Processes

Hydrogen refining, commonly called hydrotreating, was first used in the 1930s and discontinued because the then co-emerging solvent refining processes were more cost effective. A cheap source of hydrogen resulting from the use of catalytic reforming units and the need by some refiners to use "nonlube crudes" for lube oil manufacture led to reintroduction of the hydrogen refining processes in the late 1960s and early 1970s.

The hydrogen refining processes [12–26] are more severe than the mild hydrogen finishing processes (used as a replacement for the older chemical finishing processes). These processes consist of (1) the severe hydrotreating (lube hydrocracking) processes used as an alternative to solvent extraction and (2) the moderate severity hydrotreating (hydrorefining) processes used to prepare speciality products or to stabilize hydrocracked base stocks. A summary of process conditions for hydrogen refining processes is provided in Table 2.

The catalysts used in the lube hydrocracking processes are proprietary to the licensors of lube hydrocracking processes and are reported as various mixtures of cobalt, nickel, molybdenem, and tungsten on an alumina or silica–alumina-based carrier with nickel–tungsten probably used most often for manufacture of solvent neutral oils and bright stocks. Fluorides are sometimes incorporated in the catalysts or added to the feed to enhance catalyst activity. The frequency of catalyst regeneration is once every 1 to 3 years.

The catalysts used in the hydrorefining processes are proprietary but

TABLE 2 Process Conditions for Hydrogen Refining Processes

Process Variable	Hydrocracking	Hydrorefining
Pressure, lb/in.² gauge	1500–3300	200–3000
Temperature, °F	600–850	480–850
Space velocity, $V_0/V_c/h$	0.25–1.5	0.3–5
Hydrogen recycle rate, SCF/bbl	3000–8000	550–8000
Hydrogen purity, mol%	70–90	90–100
Hydrogen consumption, SCF/bbl	500–3000	100–3000

TABLE 3 Investment Cost, Utilities, and Chemical Consumption for the Solvent and Hydrogen Processing Routes—Kuwait Crude [22]

	Processing Route	
	Solvent Refining (US $)	Hydrogen Refining (US $)
Investment cost (France 1981):		
Vacuum distillation	7,500,000	5,100,000
Propane deasphalting	5,600,000	6,900,000
Furfural refining	8,700,000	—
Hydrogen refining	—	18,200,000
Solvent dewaxing	15,500,000	15,500,000
Hydrogen finishing	3,500,000	2,200,000
Steam reforming	—	6,200,000
Total Investment	40,800,000	54,100,000
Utilities consumption:		
Fuel, ton/h	5.98	5.44
Electricity kWh/h	4,465	6,455
Steam, ton/h	36.1	36.8
Cooling water, m^3/h	2,465	2,045
Chemicals consumption:		
Solvents, ton/yr	2,200	2,660
Hydrogen, ton/yr	357	See reformer
Catalysts, US $/yr	18,500	199,500
Reformer feed, ton/h	—	1.25
Steam production, ton/h	—	7

usually consist of nickel–molybdenum on alumina in the first stage and high nickel content or precious metal catalysts in the second stage.

A comparison of the investment costs and utilities and chemicals consumption for the manufacture of lube base stocks from Kuwait Crude using the solvent and hydrogen refining routes are summarized in Tables 3 and 4. Although these data [22] and other reported information [13, 15, 16] indicate that investment, utilities, catalysts, and chemicals costs are lowest for the solvent extraction route and highest for use of the hydrocracking route when processing the most widely used lube crudes, determination of the most cost-effective route requires a detailed study of the refiners available crude source and cost, available facilities, process combinations to be used, product quality level, and by-product use and value. The cost for addition of hydrogen refining to an existing lube plant will depend on the facilities available and if the existing solvent refining plant will be abandoned or used in parallel to pretreat the feed or posttreat the product from hydrogen refining.

Hydrogen refining processes are also used in combination with each other and with solvent extraction as summarized below.

1. Hydrocracking–hydrorefining
2. Hydrocracking–solvent extraction
3. Hydrogen finishing–solvent extraction (hydrogen starting)
4. Solvent extraction–hydrorefining (hydroextraction)

TABLE 4 Feedstock Requirements, Product, and By-products for the Solvent and Hydrogen Processing Routes—Kuwait Crude [22]

	Processing Route	
	Solvent Refining	Hydrogen Refining
Feedstock, ton/yr:		
Reduced crude (380°C)	1,021,989	609,480
Products, ton/yr:		
150 Solvent neutral	45,000	45,000
500 Solvent neutral	75,000	75,000
Bright stock	30,000	30,000
By-products, ton/yr:		
Vacuum gas oil	179,392	121,009
Vacuum residuum	328,860	0
Asphalts	151,602	217,889
Extracts	169,665	0
Waxes	39,415	39,415
Diesel and naphtha	2,055	74,115

The lube hydrocracking processes [12, 13, 15–18] are the most severe lube hydrogen refining processes and are most often used as a replacement for solvent refining or for viscosity reduction. These processes remove most of the nitrogen, oxygen, and sulfur present in lube oil feedstocks and convert the undesirable polynuclear aromatics and polynuclear naphthenes to mononuclear naphthenes, aromatics, and isoparaffins. The isoparaffins and mononuclear naphthenes and aromatics with multiple paraffinic side chains are generally recognized as the hydrocarbon structures in lube base stocks which have the desired viscometric properties (VI).

The advantages of lube hydrocracking as compared to solvent refining are (1) manufacture of lube base stocks from some (nonlube) crudes: (2) higher VI base stocks can be manufactured; (3) higher yield of lube base stock at a specified VI: (4) elimination of solvent extraction and finishing processes for some lube base stocks: (5) conversion of residual oils to distillate lube base stocks: (6) lube stocks of substantially better quality, in terms of VI and response to inhibition: and (7) production of high quality by-products, such as naphtha and middle distillate.

The disadvantages of lube hydrocracking as compared to solvent extraction are (1) lube base stocks darken and form sludge on exposure to air and light; (2) solvent refining, clay treating, or a second-stage hydrogenation is required to stabilize the hydrocracked base oils; (3) oils stabilized by hydrorefining are highly paraffinic and in some cases require a change in the additive package to overcome additive solubility problems; (4) severely hydrocracked lube stocks result in poorer solvent dewaxing filtration rates; (5) investment costs and operating costs are higher for hydrocracking; and (6) highly aromatic by-products suitable for use in carbon black or rubber extender oil manufacture are not produced.

The hydrocracking processes may be used alone or in series with single or dual reactors with the same or a different catalyst in each stage. Figure 5 is a

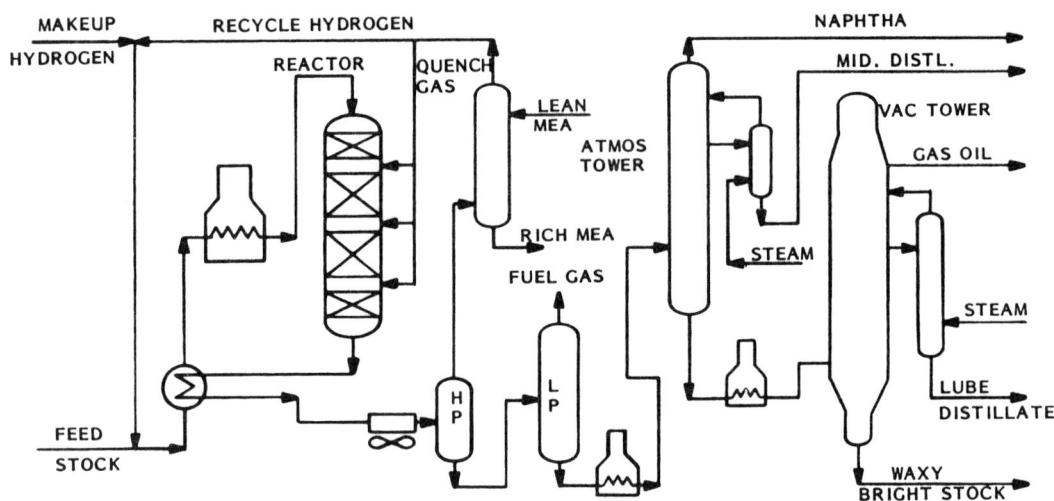

FIG. 5. Lube hydrocracker flow diagram.

simplified flow diagram for a single-stage lube hydrocracking unit in which the oil is mixed with hydrogen and processed downflow over a fixed bed of catalyst. Hydrogen quench is used to remove the heat generated and control the reaction temperature. The effluent from the reactor passes through a series of separators to remove the hydrogen and other light gases. The effluent product is then fractionated in atmospheric and vacuum distillation towers to provided the desired lube stocks and by-products. The effluent gas (mostly hydrogen) from the process is purified, if required, and recycled in the process.

These processes are generally operated in a blocked-out manner using narrow boiling range feedstocks because the VI of the low viscosity base oils is lower than desired and the VI of the high viscosity base oils are higher than desired if bulk (wide boiling range) feedstocks are used. These disadvantages can be overcome by recycle of the light lube fraction in the process or by subsequent hydrogen refining or solvent refining. The flow schemes covering the various modes of operation were reported by Bryson [15].

The feedstocks to the lube hydrocracking processes consist of (1) unrefined distillates and deasphalted oils, (2) solvent extracted distillates and deasphalted oils, (3) cycle oils, (4) hydrogen refined oils, and (5) mixtures of these hydrocarbon fractions. The use of slack wax as a feed to a lube hydrocracker for manufacture of very high VI base oils has been reported by Bull [16]. The composition and properties of the feedstocks to hydrocracking like the feedstocks to solvent extraction are dependent on the crude source and lube base stocks to be produced. Aromatic extracts from solvent refining can also be used as feedstocks for the manufacture medium VI base stocks.

Conventional VI base oil products obtained from the hydrocracking processes closely resemble those of solvent extracted oils with the exception that aromatic, sulfur, and nitrogen contents are usually lower and color is usually lighter at comparable VI levels. The composition of base oils produced by a combination of hydrocracking and solvent extraction is more like that of the solvent-refined base oils and do not require modification of the additive packages.

The yield, quality, and composition of lube base stock from hydrocracking processes are dependent on the quality of the feedstock, prior and postprocessing, catalyst, crude source, and VI of the base stock being produced. For a given feedstock, increased processing severity results in lower yield and viscosity and higher base oil viscosity index.

The operating conditions previously listed cover a wide range of temperature, pressure, and space velocity. Each of these variables has an effect on product characteristics which are also dependent on the mode of operation, catalyst used, nature of the feedstock, and products to be made.

Lube hydrocracking processes have been developed and are being used by several lube refiners. Shell France and Pennzoil currently use hydrocracking without a stabilizing process such as hydrogen, solvent, or clay treating. Pennzoil has also used a lube hydrocracker to convert surplus DA oil into distillate lube oils. The hydrocracking plant operated by EMP at Puertollano, Spain, was shut down in the late 1970s because it was less cost effective than furfural extraction for refining feedstocks derived from Arabian Light Crude. Chevron, Idemitsu Kosan, Petro Canada, and Ssangyong Oil use hydrorefining to stabilize the hydrocracked oils either immediately following hydrocracking or following fractionation and dewaxing. Quaker State used clay treating and Sun Oil uses furfural refining to stabilize the hydrocracked oils.

The hydrorefining processes [14, 20–24] operate at lower temperatures and at the same or somewhat lower pressures than the hydrocracking processes and are used to stabilize or improve the quality of lube base stocks from the lube hydrocracking processes and for manufacture of speciality oils.

Feedstocks to the hydrorefining processes are dependent on the nature of the crude source but generally consist of waxy or dewaxed solvent-extracted or hydrogen-refined paraffinic oils and refined or unrefined naphthenic and paraffinic oils from some selected crudes.

Figure 6 is a simplified flow diagram for a single-stage hydrorefining unit used for the stabilization of hydrcracked lube base stocks and manufacture of speciality lube base stocks and waxes. The two-stage process uses two single-stage hydrorefining units for the manufacture of medicinal grade white oils by removing essentially all of the sulfur, nitrogen, and aromatics in the first stage and elimination of the last traces of aromatics and impurities in the second stage.

The combination hydrogen and solvent refining processes result from the need to stabilize the hydrocracked lube base stocks and from the refiners desire to reduce the cost of lubes manufacture by elimination of the acid and clay refining of speciality products. Texaco and some of Texaco's affiliates have used and/or are using the Hydrogen Starting (mild hydrorefining to severe hydrogen finishing prior to solvent refining) route for the manufacture of lube base oils. Shell is using hydroextraction (mild furfural refining followed by severe hydrorefining) to manufacture lube base stocks [16]. Details concerning process conditions, catalysts, and economics for these processes have not been published. Vacuum gas oil, residual oil, and reduced crude desulfurizers and hydrocrackers are also sometimes used to provide feedstocks for the manufacture of lube oils using the solvent refining route.

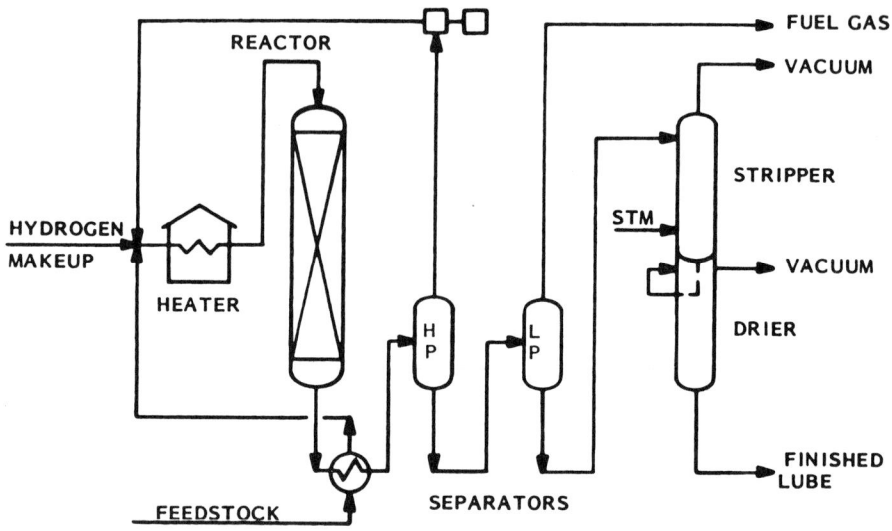

FIG. 6. Single-stage lube hydrorefining unit flow diagram.

Solvent Refining Processes

Petroleum refiners introduced solvent extraction (refining) processes in the late 1920s and early 1930s as an alternative to chemical and clay treating of lube base stocks for the removal of undesirable constituents and to improve the viscosity index of paraffinic lube base stocks [1]. Extraction has also been used to a lesser degree to improve the quality of naphthene pale oils and is now being more extensively used to reduce the toxicological aggressiveness of these base stocks.

The feedstocks to lube solvent extraction processes are (1) paraffinic and naphthenic distillates, (2) deasphalted oils, (3) hydrogen refined distillates and deasphalted oils, and to a lesser extent (4) cycle oils and (5) dewaxed oils. The products are (1) raffinates (refined oils) for further processing or (2) finished lube base stocks. The by-products are aromatic extracts which are used in the manufacture of rubber, carbon black, petrochemicals, FCCU feed, fuel oil, asphalt, etc.

Although no solvent meets all of the requirements of the ideal extraction solvent, a large number of solvents have been proposed and used commercially. The major solvents in use today are N-methyl-2-pyrrolidone (MP) and furfural, with phenol and liquid sulfur dioxide being used to a lesser extent. In addition to the quality of the crude being used, the main process variables are the processing temperature, solvent dosage, and solvent purity. These and solvent recovery methods each have an effect on process economics.

The extraction devices used in these processes are listed in Table 5.

Packed towers are generally used in the liquid sulfur dioxide process.

Since lube refining units must process a variety of different feedstocks of different viscosity and chemical composition from many crude sources in blocked operation at a variety of feed rates and processing conditions, design

TABLE 5 Extraction Devices Used in Solvent Refining Processes

Furfural	N-Methyl-2-pyrrolidone	Phenol
Centrifugal contactors	Baffle towers	Baffle towers
Packed towers	Centrifugal contactors	Centrifugal contactors
Rotating disk contactors	Packed towers	Packed towers
	Rotating disk contactors	

data for the proprietary extraction equipment is generally not available and can best be obtained from the designers of the extractors or the licensors of the processes. The report by Fiocco [27] presents an example of the long periods of time and experimentation (both laboratory and commercial) required to develop and improve the efficiency of lube oil extraction equipment.

Solvent recovery in the solvent refining processes is generally conducted in a series of multistage evaporators (commonly called flash towers) to reduce the energy requirements. Steam or inert gas strippers are used to remove the last traces of solvent, and a solvent purification section is used to remove water and other impurities. Additional information on the use of multistage evaporation and energy reductions which can be achieved from the use of additional evaporation stages or inert gas stripping is reported in the published literature [28–30].

The Edeleanu Process [1, 10], based on the use of liquid sulfur dioxide, was the first extraction process used by the petroleum industry and was introduced in 1907 to reduce the smoke point of kerosene. It was later applied to the extraction of lubricating oils and has been used in combination with benzene for the extraction and dewaxing of lubricating oil base stocks.

Although liquid SO_2 has good solvent power and selectivity for aromatic compounds, the requirement for low extraction temperatures limits its use to the extraction of naphthenic and low pour or dewaxed paraffinic feedstocks. The main disadvantages for use of liquid SO_2 are (1) toxicity, (2) air pollution control requirements, (3) moisture control to prevent corrosion, and (4) maintenance costs related to these items. No new units have been built for extraction of lube oils since the late 1950s and some units have been abandoned or replaced by solvent or hydrogen refining units. Although there are still some units in operation in Europe, currently it is being used at only one location in the United States for the extraction of naphthene oils.

The Furfural Refining Process [1, 10, 31, 35], the most widely used lube refining process on a worldwide basis, was first used commercially in the early 1930s and has been most extensively developed by Texaco. It is also used for the extraction of straight run gas oils and light and heavy cycle oils from catalytic cracking operations.

In addition to the quality of the feedstock, the main process variables are temperature, solvent dosage, purity of the solvent, and the quantity of extract recycled to the feed or below the feed in the extraction device. The yield of refined oil (selectivity) of furfural is equivalent to or better than that of the other refining solvents. The temperatures and solvent dosages used are highly dependent on the quality of the feedstock, crude source, and quality level of base stocks being produced. Although the normal ranges are 100 to 250°F and 100 to 500 vol.% furfural basis feed, higher and lower temperatures and dosages are used in some cases.

A simplified flow diagram for a furfural refining unit is shown in Fig. 7. RDCs are the currently favored extraction devices because of their excellent turn-down ratios, rapid change-over of feedstocks, and the elimination of the need to shut down for cleaning (decoking) which is experienced with packed towers.

The Phenol Refining Process [1, 33] was introduced in the early 1930s and has for the most part been replaced by the MP refining processes in the United States and to a lesser degree in Europe.

Phenol is a highly selective and highly toxic solvent which was first used for the extraction of lubricating oil feedstocks in 1930. The selectivity of phenol is good but generally lower than that of furfural or MP. In addition to the quality of the feedstock, the main process variables are temperature, dosage, the quantity of water or oil in the circulating solvent, and the dosage of wet solvent or water injected into the feed or extractor below the feed. The preferred extraction devices are packed towers and baffle towers.

N-Methyl-2-pyrrolidone (MP) Refining Processes [29, 30, 34–37] were developed by Exxon (EXOL N Extraction) and Texaco (Texaco MP Refining Process). MP has replaced furfural and phenol as the major extraction solvent in recent years and is beginning to replace these solvents on a worldwide basis. Figures 8 to 10 are representative of a few of the simplified flow diagrams for these processes. The process flow for these processes is similar to that of the furfural and phenol refining processes. The main differences in these processes are the different techniques used in the drying of the solvent, the use of wet MP in the Exxon EXOL N Process and the preference for anhydrous MP in the Texaco MP Refining Process. Cooling and/or water injection into the extract mix leaving the extractor can be used to produce cycle oils for the manufacture of medium VI base oils and to control the aromatic content of the extract. Preferred extraction devices are packed towers and trayed towers.

In addition to the quality of the feedstock, the main process variables are temperature, dosage, purity of the solvent, and the amount of water or wet solvent injected into the feed or treating solvent or into the extraction device below the feed. MP is nontoxic and has excellent solvent power and selectivity which is equivalent to that of furfural. Although the water content of the solvent should be minimized to maximize solvent power and minimize the solvent circulation, there are some feedstocks (crude source dependent) where water is used to improve the selectivity of MP. Since the limitation in most lube extraction units is solvent turnover, increases of 25 to 60% in feed rate have been obtained from the conversion of furfural and phenol refining units to MP.

The investment, operating, and energy costs are lower for the use of MP as compared to the use of furfural or phenol because the higher solvent power (lower treating dosage) of MP results in the need for smaller units and less energy consumption for a given size lube plant. Comparative investment costs and utilities consumption for the use of furfural and MP are shown in Table 6 [34].

Lubricating Oils: Manufacturing Processes

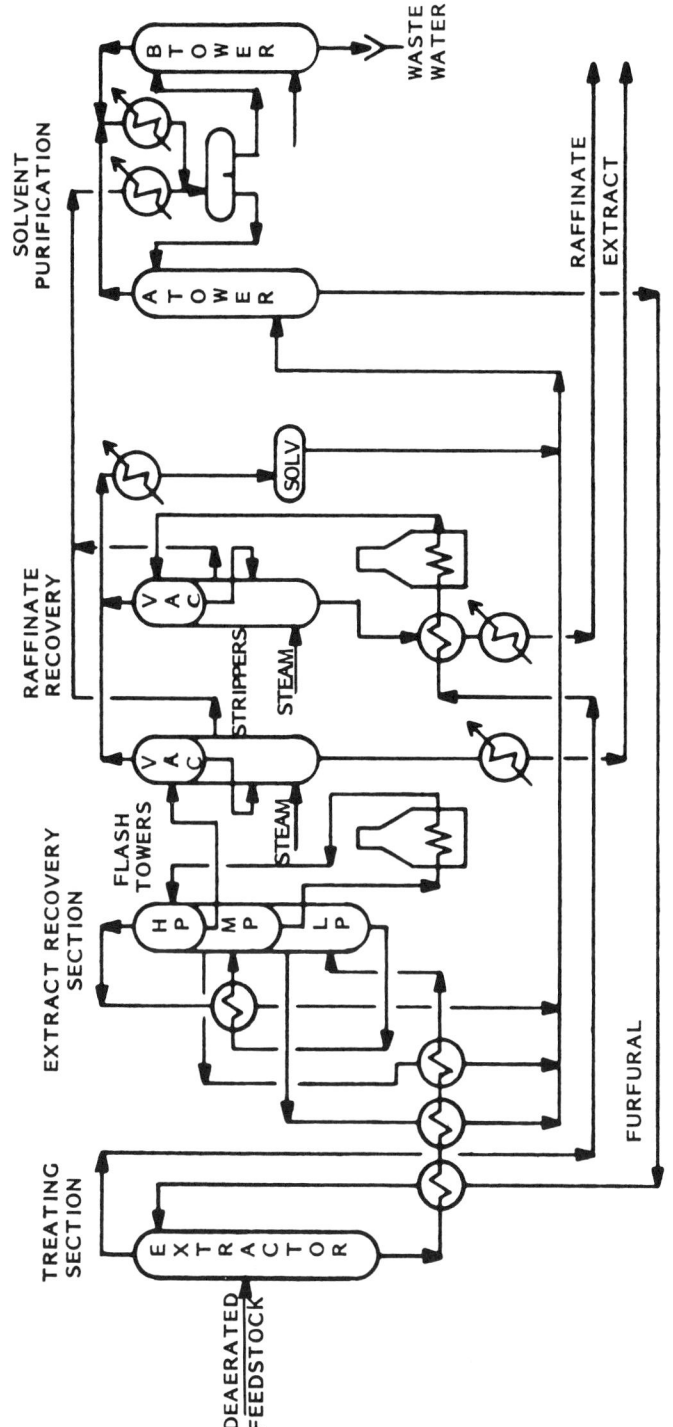

FIG. 7. Texaco furfural refining flow diagram.

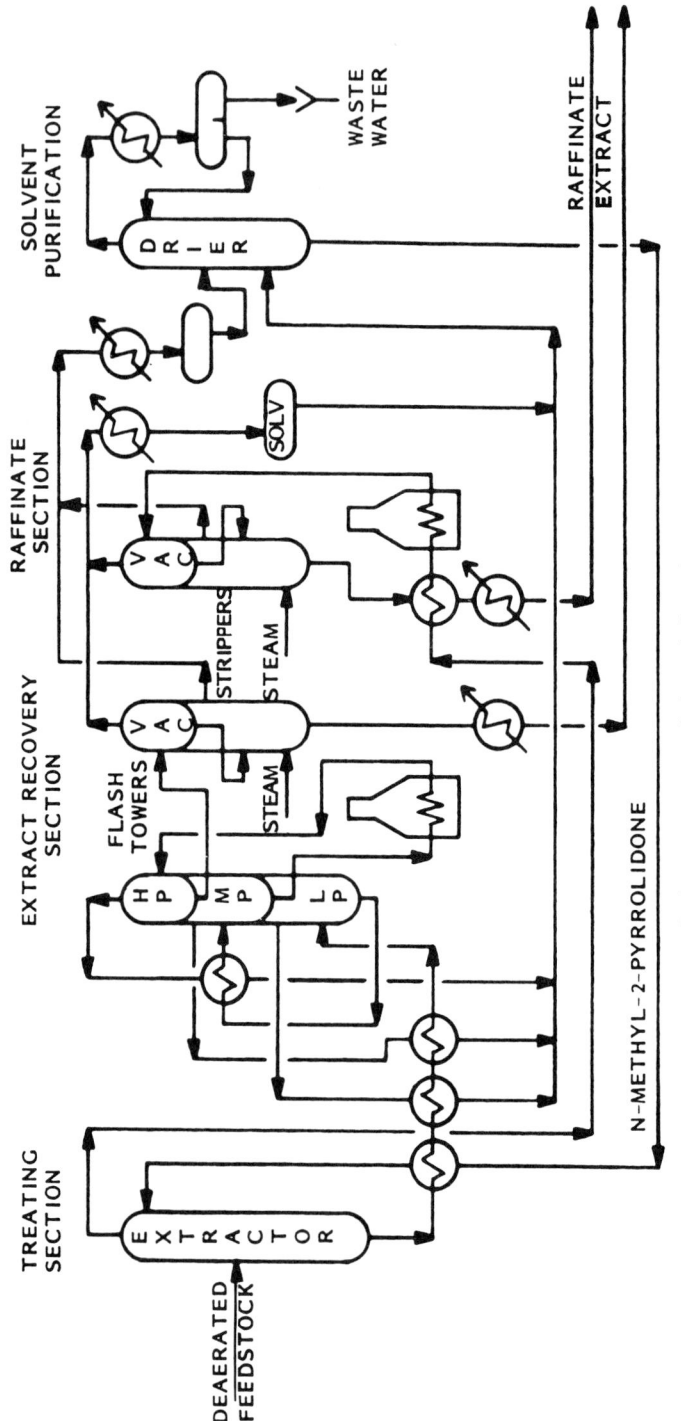

FIG. 8. Texaco MP refining unit flow diagram.

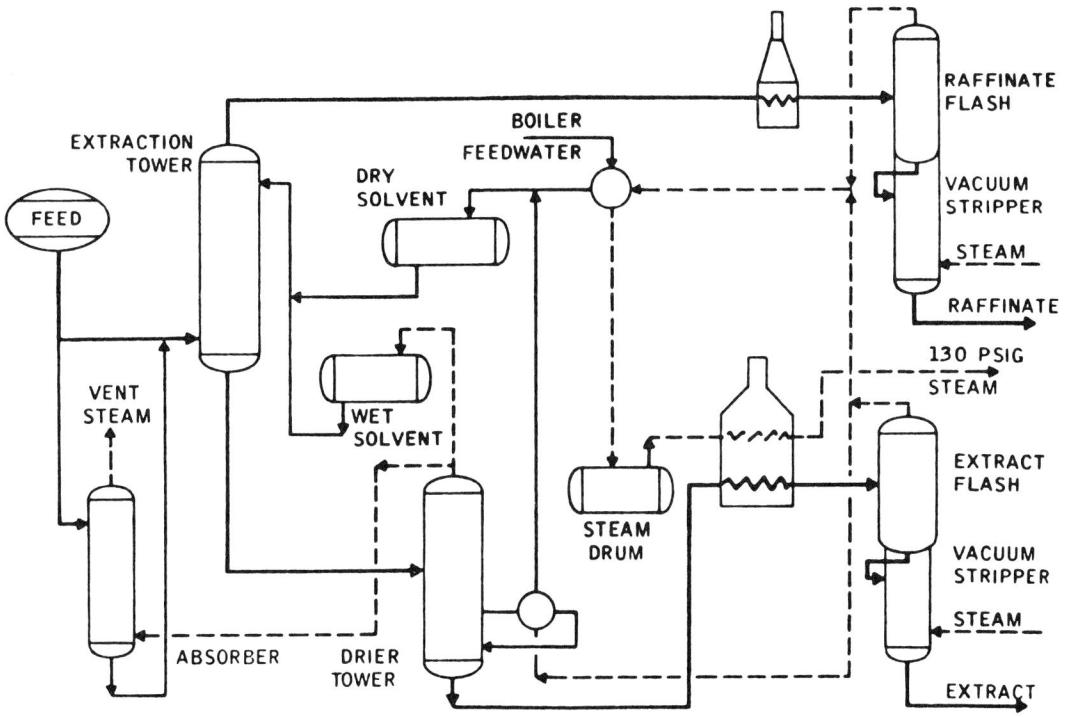

FIG. 9. EXOL N extraction unit using steam stripping.

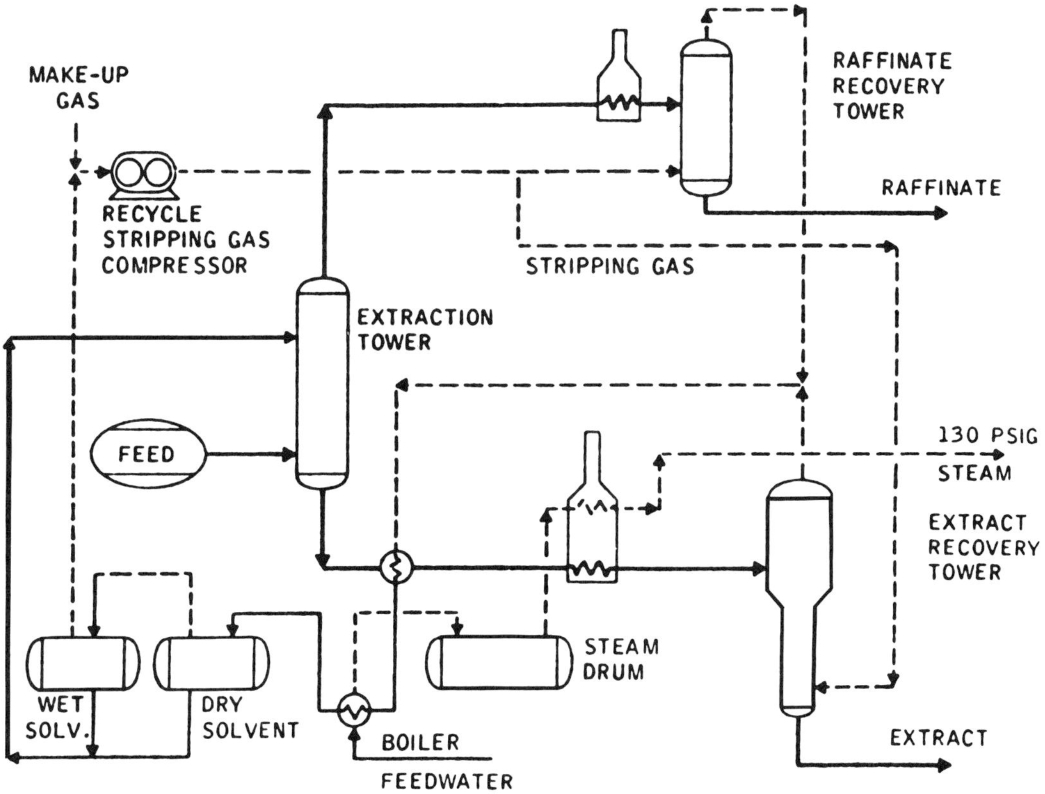

FIG. 10. EXOL N extraction unit using recycled stripping gas.

TABLE 6 Investment, Utility, and Chemical Costs for Solvent Refining Arabian Light Feedstocks [34]

		MP Refining		Furfural Refining	
Nominal Feed Rate, BPSD		10,000		10,000	
kg/h		63,000		63,000	
Investment cost (1986 US Gulf Cost), $		13,600,000		17,200,000	
Utilities and Chemicals Cost	$/Unit	Rate	Cost, $	Rate	Cost, $
Fuel (heat absorbed), MM · kcal/h	16.5	8.18	135.0	12.9	212.9
High pressure steam, kg/h	0.014	7864	110.1	5745	80.4
Low pressure steam, kg/h	0.012	1117	13.4	696	8.4
Cooling water, m^3/h	0.024	472	11.3	720	17.3
Electrical power, kW	0.05	267	13.4	402	20.1
Solvent, kg/h	2.9/1.5	2.5	7.3	38	57.0
Na_2CO_3, kg/h	—	0.2	—	1.6	—
Nitrogen, kg/h	—	8	—	8	—
Total cost, $/h			290.4		396.0
$/metric ton			4.65		6.34

Lube Dewaxing Processes

The raw paraffin distillates and residual oils leaving the crude stills contain wax and are normally solids at ambient temperature. The deasphalting and refining processes increase the wax content of lube feedstocks. Removal of wax from these fractions is necessary to permit manufacture of lubricating oils with the desired low temperature properties. Catalytic dewaxing, solvent dewaxing, and urea dewaxing processes have replaced the older cold settling, pressure filtration, and centrifuge dewaxing methods in the manufacture of lubricating oils.

Catalytic Dewaxing Processes

The removal of wax from lubricating oil base stocks by solvent dewaxing is expensive from the standpoint of investment, operating costs, and production of low pour point (below about −35°F) oils is generally not practical. Several petroleum refiners have patented various selective hydrocracking (catalytic dewaxing) processes for the manufacture of lube oil base stocks. The processes which have been commercialized to date were developed by British Petroleum [38], Chevron [19, 26], and Mobil [39–42]. Modified versions of the BP and Mobil processes are used to dewax fuels fractions. These processes are more cost effective and permit obtaining lower pour point products than the solvent dewaxing processes. Viscosity index is generally lower at the same pour point for catalytic dewaxed as compared to solvent dewaxed neutrals prepared from the same feedstock. Dewaxed oil yields are in some cases lower and in others higher than that obtained by solvent dewaxing. Products formulated from the lower VI catalytically dewaxed oils are equivalent to the solvent dewaxed oils and have demonstrated better low temperature properties [42]. However, more severe extraction or hydrogen refining of the distillate feedstocks to catalytic dewaxing is required to produce lube base stock of the same VI as that obtained by solvent dewaxing. The BP process can be used to dewax a wide range of naphthenic feedstocks and waxy or partially dewaxed feedstocks. However, it is not suitable for the manufacture of high VI paraffinic lube base oils. The Mobil MLDW process can be used to process all grades of lube base stocks. The Chevron Process has to date been used to dewax hydrogen refined 100 and 240 neutral oils and is believed suitable for dewaxing the full range of hydrogen and solvent extracted feedstocks.

The process flow of the BP process is essentially that of the hydrorefining unit shown in Fig. 6 with the exception that hydrogen quench (not needed with the Chevron or Mobil processes) is required to control the reaction temperature. Finishing, if required for purification and color improvement, requires an additional processing step unless a hydrogen finishing reactor is added. The BP Catalytic Dewaxing Process uses a proprietary synthetic mordenite containing platinum as the dewaxing catalyst.

The simplified process flow for the Mobil MLDW process is shown in Fig. 11 and consists of two reaction stages. The first reactor contains a proprietary ZSM-5 based dewaxing catalyst, and the second reactor contains an amorphous hydrogen finishing catalyst used to saturate olefins that are created by the dewaxing catalyst and to improve oxidative stability, color, and demulsibility of the finished lube oil.

The simplified process flow of the Chevron process is like that of the MLDW process. This process also uses a proprietary dewaxing catalyst in the first stage and a hydrogen finishing type catalyst in the second stage. Catalysts

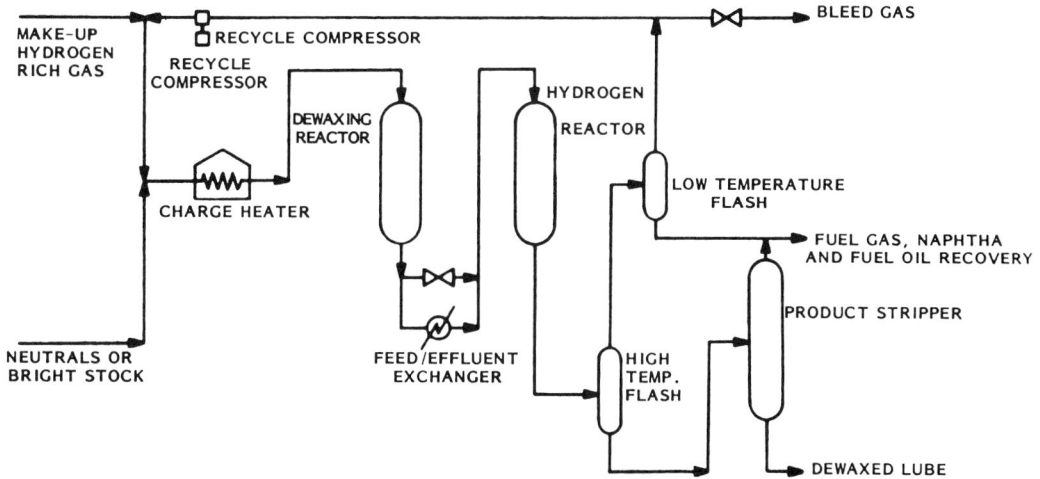

FIG. 11. MLDW process flow diagram.

TABLE 7 Process Conditions

	Process		
	British Petroleum	Mobil	
	Dewaxing	Dewaxing	Finishing
H_2 partial pressure, lb/in.^{2}	300–1500	250–3000	
Space velocity, V_0/V_c/h	0.5–5.0	Depends on feedstock	
Temperature, °F	550–750	525–675	475–550
H_2 circulation, SCF/bbl·d	2000–5000	500–5000	

used in this process have not been disclosed but are probably similar to the ZSM-5 zeolite containing a small quantity of a hydrogeneration metal in the dewaxing reactor and a conventional hydrogen finishing catalyst in the second reactor.

The processing conditions and hydrogen consumption are dependent on the (1) boiling range, (2) wax content, (3) pour point of the dewaxed oil, (4) viscosity, and (5) chemical composition of the feed. Generalized process conditions [38, 40] are summarized in Table 7.

Process conditions for the Chevron process have not been reported.

Technical information and details concerning process conditions and design for the catalytic dewaxing processes are currently only available through secrecy agreements with the licensors of the processes.

The first BP unit (a converted hydrotreating unit) was brought on stream in 1977 and a grassroots unit was brought on stream in 1983 with additional units in the planning stage. The Mobil Lube Dewaxing Process was demonstrated in 1978. Commercial MLDW units in operation include two Mobil units (one each in Australia and the United States) and two licensed units. Additional units are in the planning stage. The Chevron unit was brought on stream in 1985 at their Richmond California Refinery.

Solvent Dewaxing Processes

A considerable number of solvent-based processes have been developed over the years for the dewaxing of lubricating oils [1, 10, 43]. These processes can be divided into three basic sequential steps: (1) crystallization, the dilution and chilling of the feedstock with solvent; (2) filtration of the wax from the solution of dewaxed oil and solvent; and (3) solvent recovery from the wax cake and filtrate for recycle in the process by flash distillation and stripping. This processing sequence is depicted in Fig. 12 for a ketone dewaxing unit. The major process variables include (1) the nature of the feedstock, (2) the solvent and solvent composition, (3) the solvent dilution procedure, (4) the chilling procedure, (5) the filtration procedure, and (6) the solvent recovery method. The manner in which the above variables are controlled can have a significant effect on (1) the production rate, yield of dewaxed oil; (2) the pour point of the dewaxed oil; (3) the oil content of the wax; and (4) the investment and operating costs.

Lubricating Oils: Manufacturing Processes

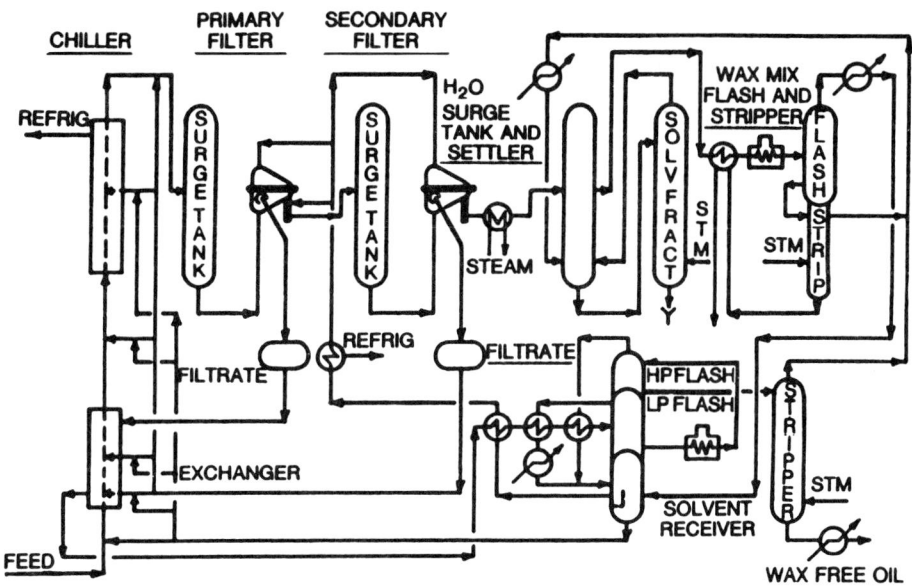

FIG. 12. Texaco solvent dewaxing unit flow diagram.

The major processes in use today are the ketone dewaxing processes. Processes used to a considerably lesser degree are the Di/Me Process and the propane dewaxing process.

The ketone dewaxing processes are based on improvements which have been made in the original solvent dewaxing process, the Acetone-Benzol Process (A-B process), which was first used commercially in 1927 [1, 10, 44, 45]. The original A-B or "classical" solvent dewaxing process was a single dilution single-stage filtration process. It has been modified to use various mixtures of acetone, benzene, methyl ethyl ketone (MEK), methyl isobutylketone (MIBK), toluene, and other aromatic hydrocarbons. The solvent mixtures used most often are MEK–toluene and MEK–MIBK with neat MIBK being used in some cases. Improvements made to the ketone dewaxing processes over the years include (1) replacement of batch chillers with scraped-surface exchangers/chillers and crystallizers; (2) rotary vacuum filters as replacements for the pressure-type filters; (3) synthetic filter cloth replaced cotton canvas for longer on-stream time; (4) controlled shock chilling, multiple dilution, and filtration techniques; (5) multiple effect evaporation for solvent recovery; (6) inert gas in place of steam for stripping; (7) cold backwashing as a replacement for hot washing of the filters; and (8) improved refrigeration plants and techniques [29, 45–47]. These processes and modifications of these processes are also used to prepare low oil content hard and finished waxes. Although some refiners use the older double-dilution single-stage filtration procedures or modifications of these processes described in the literature, the most widely used ketone dewaxing processes in use today are the Texaco Solvent Dewaxing Process and the Exxon Dilchill Process, respectively.

The Texaco Solvent Dewaxing Process [1, 7, 10, 43, 44], generally called the MEK process, normally uses a mixture of MEK and toluene as the dewaxing solvent and sometimes uses mixtures of other ketones and aromatic solvents.

Figure 12 is a simplified flow diagram for this process. The waxy feedstock is heated to 10–15°F above the cloud point of the oil and diluted with solvent while chilling at a controlled rate in double-pipe scraped-surface exchangers and chillers. Two to four volumes of solvent using an incremental dilution procedure for low to medium viscosity stocks and a double-dilution controlled shock chilling procedure for high viscosity stocks are used in the more modern MEK units with the older double-dilution single-stage filtration used in some units. About 60% of the chilling is obtained by heat exchange between the feedstock and solvent with the cold filtrate from filtration in the annulus of the double-pipe exchangers. The remainder of the refrigeration required for chilling is obtained by indirect heat exchange with a refrigerant in the annulus of the double-pipe chillers. The slurry leaving the chillers at a temperature of 5 to 20°F below the desired pour point is filtered using rotary vacuum filters and the wax cake is washed with a spray of cold solvent before being discharged by an inert gas blow back. The filtrate from these primary filters is used to prechill the feedstock and solvent mixture in the scraped-surface exchangers. The filtrates from these filters may be separated into a primary (oil-rich) filtrate used as the chilling medium and the wash (oil-lean) filtrate used as dilution solvent. The wax cake from the primary filter is diluted with additional solvent and filtered in a second (repulp) rotary vacuum filter with cold solvent washing of the wax cake to reduce the oil content of the wax. The filtrate from the repulp filter is normally used as feedstock dilution solvent in the crystallization train but may also be separated into an oil-rich filtrate for use as feedstock dilution solvent and an oil-lean filtrate for dilution of the wax from the primary filter. The solvent is recovered from the resultant dewaxed oil filtrate and wax cake by dual- or triple-effect flash vaporization and recycled in the process.

The wax recrystallization and wax warm-up deoiling procedures are used in those cases in which the lube refiner also manufactures a hard or finished wax. The Edeleanu Spray Deoiling Process [48] and older wax sweating processes [1] are also used to manufacture low oil content waxes.

The Exxon Dilchill Dewaxing Process [10, 45, 46, 49, 50] uses direct cold solvent dilution-chilling technique with a high degree of mixing in a special crystallizer in place of the scraped-surface exchangers. The high degree of mixing is used to overcome the poor filtration rates obtained with the conventional shock chilling techniques. A solvent drying step must be added to remove the last traces of water from the solvent to prevent the icing of the solvent chillers during chilling of the solvent to the low temperatures required for direct chilling of the feedstock. Indirect refrigeration and scraped-surface chillers are used to complete the chilling cycle. Filtration and multiple effect evaporation for solvent recovery are conducted in the usual manner.

Data have been reported which show that the filtration rates oil content of the slack wax produced by this process are better than those obtained using the conventional (single-dilution single-stage filtration) ketone dewaxing processes [50]. Although comparative commercial data on the same waxy feedstock using the Dilchill procedure and multidilution or controlled shock chill procedures are not available, laboratory comparisons have shown that filtrations rates are comparable when the same dilution ratios and filtration procedures are used. In addition, filtration rates comparable to those reported for the Dilchill process have been obtained in some commercial ketone dewaxing processes using the conventional incremental dilution and controlled shock chill procedures.

The warm-up wax deoiling procedure is used with the Dilchill Process in the manufacture of hard or finished wax.

The Di/Me Dewaxing Process [45, 51] uses a mixture of dichloroethane (Di) and methylene dichloride (Me) as the dewaxing solvent. This process, developed by Edeleanu Gellschaft, is used by a few refineries in Europe for the dewaxing of lubricating oil base stocks and the manufacture of low oil content wax. The process flow is essentially the same as that of the ketone dewaxing processes except that the oil-lean wash filtrate is used as dilution in the chilling train. The wax cake from the primary filter is diluted with the oil-lean wash filtrate from the second-stage filter and the repulped mixture filtered in the second-stage (repulp) filter to produce hard or finished wax. The second stage is omitted if hard wax production is not desired.

The Propane Dewaxing Process [1, 10, 45] was developed and first used in 1932 by Standard Oil Company of Indiana and further improved by the JUIC (Standard Oil of New Jersey, Union Oil Co., Standard Oil of Indiana, and M. W. Kellog patent combine). Process flow is essentially that of the ketone dewaxing processes with the exception that higher pressure equipment is required, and chilling is done in evaporative chillers by vaporizing a portion of the dewaxing solvent. Filtration is conducted using rotary filters and washing of the wax cake with cold propane. Solvent is recovered for recycle in the process using two or more evaporators and by stripping. Advantages for this process are (1) the solvent is cheap and readily available, (2) evaporative chilling greatly reduces the adhesion of wax to the crystallizer walls and eliminates the need for expensive crystallizers, and (3) wax cloud points of propane dewaxed bright stocks are usually lower as compared to ketone dewaxed bright stocks. The main disadvantages for the use of propane are that (1) the differential between the dewaxed oil and filtration temperature (25–45°F) is considerably higher than that of the ketone dewaxing or Di/Me dewaxing processes, (2) control of batch chilling is difficult because the compressors are underutilized during the initial chilling phase of the process, and (3) dewaxing aids are required to obtain good filtration rates.

The use of ketone solvents as a wax antisolvent to improve the economics of the propane/propylene dewaxing process using a mixture of propylene and acetone as the dewaxing solvent was tested commercially [49]. Although this test was reported to be successful, it is not known if the modified process is being used commercially.

Lube Urea Dewaxing Processes

The formation of crystalline complexes (adducts) between urea and straight-chain hydrocarbons was discovered by Bengen in 1940 and has been used as a basis for lube oil dewaxing processes and for the manufacture of normal paraffins. No new urea dewaxing units have been constructed and some existing units have been abandoned in recent years. Those interested in additional information should review Hoppe [52] and Scholten [53].

Lube Oil Finishing Processes

Although acid and clay are still used to some extent for the finishing of lubricating oil base stocks and wax, these processes have for the most part been replaced with hydrogen finishing processes.

Sulfuric acid treating [1, 10] is the original or "classical" refining and finishing process for the manufacture of lubricating oils. Although it is still used for this purpose by a few refiners, it finds it main use in the manufacture of specialty oils and the reclamation of used lubricating oils. The treating of lubricating oils with sulfuric acid is conducted in a batch or continuous manner as was described in the section on acid refining. The main difference in acid finishing as compared to acid refining is that the quantity of acid used is usually low (10 or less lb/bbl) in comparison to the large quantities (25 to 400 lb/bbl) used in the acid refining processes.

Clay contacting [1, 10] involves the intimate mixing of oil with fine bleaching clay at elevated temperature for a short period of time followed by separation of the oil and clay. This process is frequently combined with the acid treating process for the finishing and neutralization of lube base stocks. The process improves color and chemical, thermal, and color stability of the lube base stock. The process variables include the clay dosage and treating temperature. Clay dosages are usually relatively low (4 to 20 lb/bbl) and treating temperature ranges from about 220 to 650°F or just below the flash point of the oil being treated. Use of excessively high temperatures can result in darkening of the oil.

Clay percolation [1, 10, 54] is a static bed absorption process used to purify, decolorize, and finish lube base stocks and waxes. This process has in large part been replaced by hydrogen finishing but is still in limited use for the manufacture of refrigeration oils, transformer oils, turbine oils, white oils, and waxes. Although Attapulgus clay can be used, the most frequently used clay is Porocel, an activated bauxite. The process variables include temperature, flow rate, throughput, and type of clay.

Hydrogen finishing processes [56–60] are mild hydrogenation processes used in place of the older and more costly acid and clay finishing processes for the purpose of improving color, odor, thermal, and oxidative stability, and demulsibility of lube base stocks. They are fixed-bed catalytic hydrogenation processes used for neutralization, desulfurization, and denitrification of lube base stocks. Unlike the hydrogen refining processes, hydrogen finishing processes do not saturate aromatics nor break carbon—carbon bonds.

Process flow is essentially identical to the hydrorefining process shown in Fig. 6. The operating conditions are dependent on feedstock composition (related to crude source as well as type and severity of prior processing), catalyst, and product specifications. General operating conditions for these processes are listed in Table 8.

TABLE 8 Operating Conditions

Hydrogen partial pressure, lb/in.^{2}	200–1500
Temperature, °F	450–650
Liquid hourly space velocity, V_0/V_c/h	0.5–3.0
Hydrogen rate, SCF/bbl feed	100–5000
Hydrogen purity, mol%	50–100
Hydrogen consumption, SCF/bbl feed	50–200

Hydrogen finishing catalysts are usually cobalt–molybdenum, nickel–molybdenum, or iron–cobalt–molydenum on alumina. The catalysts are received in the oxide form and must be sulfided before use. They provide long service life and are regenerated using a controlled oxygen burn. Metal poisoning of the catalysts is usually not a problem with lube feedstocks because the metals are removed in the deasphalting and refining processes.

Feedstocks to these processes are usually naphthenic or dewaxed paraffinic oils which may be or may not have been chemically, solvent, or hydrogen refined. These processes sometimes (1) precede solvent dewaxing to improve the quality or processing respones of the wax by-product from solvent dewaxing or (2) precede solvent extraction (hydrogen starting) to improve the yield of refined oil or reduce the sulfur content of aromatic extracts. When used in this manner, the size of the units, hydrogen consumption, and operating costs are increased in comparison to hydrogen finishing following solvent dewaxing. However, the size of the extraction unit may be reduced, and installation of finishing units for the manufacture of wax or desulfurization of extracts (if needed) are eliminated or reduced in size.

The effect of hydgen finishing temperature and pressure is highly dependent on the quality of the feedstock and the type of catalyst used. An increase in temperature or pressure will normally improve neutralization, desulfurization, denitrification, product color, and product stability. However, increasing the temperature above some maximum which is related to the catalyst and feedstock quality will degrade the color, oxidation stability, and other properties of the base oil.

The hydrogen finishing processes used by most of the lube oil refiners are properietary processes such as Ferrofining (British Petroleum), Gulfinishing (Gulf), Hydrofining (Exxon), Hydrogen Finishing (Texaco), and Lube Oil Hydrotreating (IFP).

The hydrorefining processes previously discusses are sometimes used as finishing processes for the stabilization of hyrocracked oils and as a replacement for the acid and clay finishing of specialty products such as transformer oils, white oils, and waxes.

References

1. V. A. Kalichevsky and K. A. Kobe, *Petroleum Refining with Chemicals*, Elsevier, London, 1956, pp. 72–122, 319–456.

2. R. G. Brand, "Mobil's New 100,000 bbl/day Crude Distillation Unit," *Heat Eng.*, pp. 98–101 (January-February 1960).
3. G. T. Atkins, E. O. McBride and L. W. Owens, "Crude Oil Distillation," in *Encyclopedia of Chemical Processing and Design*, Vol. 13, Dekker, New York, 1981, pp. 238–260.
4. R. N. Watkins, *Petroleum Refining Distillation*, 2nd ed., Gulf Publishing, Houston, 1981.
5. J. A. Bonilla et al., "FW Solvent Deasphalting," in *Handbook of Petroleum Refining Processes*, McGraw-Hill, New York, 1986, pp. 8.19–8.51.
6. C. P. Chang and J. R. Murphy, "Deasphalting," in *Encyclopedia of Chemical Processing and Design*, Vol. 14, Dekker, New York, 1983, pp. 149–166.
7. W. P. Gee and H. H. Gross, "Dewaxing and Deasphalting," in *Advances in Chemistry Series*, No. 5 (Progress in Petroleum Technology), American Chemical Society, Washington, D.C., 1951, pp. 160–176.
8. S. R. Nelson and R. G. Roodman, *ROSE: The Energy Efficient Bottom of the Barrel Alternative*, Paper Presented at the 1985 Spring AIChE Meeting, Houston, March 24–28, 1985.
9. J. R. Salazar, "UOP Demex Process," in *Handbook of Petroleum Refining Processes*, McGraw-Hill, New York, 1986, pp. 8.61–8.70.
10. D. Klamann et al., "Production of Petroleum Based Lubricating Oils," in *Lubricants and Related Products*, Verlag Chemie, Weinheim, 1984, pp. 51–83.
11. M. H. Tuttle, "The Performance and Flexibility of the DUOSOL Process," in *Proceedings—Fifth Mid-Year Meeting, API*, Vol 16M, No. III, 1935, pp. 112–123.
12. A. Billion et al., "Consider Hydrorefining for Lubes," *Hydrocarbon Processing*, September 1975, pp. 139–144.
13. A. Billion et al., "Procede D'Hydroraffinage Pour La Production D'Huiles Lubrifiantes," in *Proceedings—Tenth World Petroleum Congress*, Vol. 4, 1980, pp. 211–220.
14. A. Billon et al., Improvements in Waxes and Special Oil Refining," in *1980 Proceedings—Refining Department, API*, Vol. 59, pp. 168–177.
15. M. C. Bryson et al., "Gulfs Lubricating Oil Hydrotreating Process," in *Proceedings—Division of Refining, API*, Vol. 49, 1969, pp. 439–443.
16. S. Bull and A. Marmin, "Lube Oil Manufacture by Severe Hydrotreatment," in *Proceedings—Tenth World Petroleum Congress*, Vol. 4, 1980, pp. 221–228.
17. V. P. Burton and L. E. Hutchings, *Production of Quality Lubricants by the HDC Unibon Process*, Paper Presented at the PetroPeru's Lube Oil Manufacturing Operations Seminar, Lima, May 8–12, 1978.
18. T. R. Farrell and J. A. Zakarian, "Lube Facility Makes High-Quality Lube Oil from Low-Quality Feed," *Oil Gas J.*, pp. 47–51 (May 19, 1986).
19. R. W. Geiser and L. E. Hutchings, "Quality Lubricants from Pennsylvania Grade Crude Oil by the Isomax Process," in *1973 Proceedings—Division of Refining, API*, Vol. 53, pp. 738–757.
20. J. B. Gilbert and R. Kartzmark, "Advances in the Hydrogen Treating of Lubricating Oils and Waxes," in *Proceedings—Seventh World Petroleum Congress*, III, 1967, pp. 193–205.
21. J. B. Gilbert et al., "Hydroprocessing for White Oils," *Chem. Eng.*, pp. 87–89 (September 15, 1975).
22. *IFP Technology for the Refining of Lube Base Oils, White Oils and Waxes*, Reference 29676, Institut Francais Du Petrole, France, November 1981.
23. H. C. Murphey et al., "High Pressure Hydrogenation—Route to Specialty Products," in *Proceedings—Division of Refining, API*, Vol. 49, 1969, pp. 817–904.
24. M. K. Rausch and G. E. Tollefsen, *DUOTREAT Process*, Presented at the 1972

National Fuels and Lubricants Meeting of the NPRA, New York, September 14–15, 1972, Paper No. F&L-72-44.
25. D. H. Shaw, "Recent Developments in Oil Refining," in *Proceeding of the Eleventh World Petroleum Congress*, Vol. 4, 1984, pp. 345–357.
26. J. A. Zakarian et al., *All Hydroprocessing Route for High-V.I. Lubes*, Paper Presented at the AIChE Spring National Meeting, New Orleans, April 6–10, 1986.
27. R. J. Fiocco, "Development of the Cascade Weir Tray for Extracton," in *AIChE New Developments in Liquid-Liquid Extractors: Selected Papers from ISEC '83, AIChE Symposium Series*, 238, Vol. 80, 1984, pp. 89–93.
28. P. Taylor, *Operating Lube Plants Efficiently*, Paper Presented at the AIChE Spring National Meeting, New Orleans, April 6–10, 1986.
29. A. Sequeira Jr. et al, "Return To Basics—How to Reduce Energy Requirements in Lube Oil Solvent Extraction and Dewaxing Processes," in *1980 Proceedings—Refining Department, API*, Vol. 59, pp. 133–150.
30. J. D. Bushnell and R. J. Fiocco, "Engineering Aspects of the EXOL N Lube Extraction Process," in *1980 Proceedings—Refining Department, API*, Vol. 59, pp. 133–150.
31. L. C. Kemp, G. B. Hamilton, and H. H. Gross, "Furfural as a Selective Solvent in Petroleum Refining," *Ind. Eng. Chem.*, 40(2), 220–227 (1948).
32. R. E. Manley et al., "Refining of Lubricating Oils with Furfural," in *Proceedings—Fourteenth Mid-Year Meeting American Petroleum Institute*, Vol. 16M, No. III, 1936, pp. 39–46.
33. R. K. Stratford et al., "The Use of Phenol as a Selective Solvent in the Production of High-Grade Lubricating Oils," in *Proceedings—Fourteenth Annual Meeting American Petroleum Institute*, Section III (Refining), Chicago, October 23–26, 1936, pp. 90–95.
34. F. C. Jahnke, *Solvent Refining of Lube Oils—The MP Advantage*, Paper Presented at the AIChE Fall National Meeting, Miami Beach, November 2–7, 1986.
35. B. M. Sankey et al., "EXOL N: New Lubricants Extraction Process," in *Proceedings of the Tenth World Petroleum Congress*, Vol. 4, 1979, pp. 407–414.
36. A. Sequeira, P. B. Sherman, J. U. Douciere, and E. O. McBride, "MP Refining of Lubes," *Hydrocarbon Processing*, pp. 155–160 (September 1979).
37. A. Sequeira et al., *Conversion of Furfural Refining Units to N-Methyl-2-pyrrolidone (MP) Refining Units*, Paper Presented at the AIChE Spring National Meeting, New Orleans, April 6–10, 1986.
38. J. D. Hargrove, "Dewaxing, Catalytic," in *Encyclopedia of Chemical Processing and Design*, Vol. 15, Dekker, New York, 1983, pp. 346–352.
39. R. G. Graven and J. R. Green, *Hydrodewaxing of Fuels and Lubricants Using ZSM-5 Type Catalysts*, Paper Presented at the Australian Institute of Petroleum 1980 Congress.
40. F. A. Smith, *Mobil Lube Oil Dewaxing (MLDW) Technology*, Paper Presented at the Texaco Lubricating Oil Manufacturing Licensee Symposium, White Plains, New York, May 18–20, 1982.
41. K. W. Smith et al., "New Process Dewaxes Lube Base Stocks: Mobil Lube Dewaxing," in *1980 Proceedings—Refining Department, API*, Vol. 59, pp. 151–158.
42. W. C. Starr and J. W. Walker, *Quality of Hydrodewaxed Base Stocks*, Presented at the 1981 National Fuels and Lubricants Meeting of the NPRA, Houston, November 5–6, 1981, Paper No. F&L-81-85.
43. S. Marple Jr. and L. J. Landry, "Modern Dewaxing Technology," in *Advances in Petroleum Chemistry and Refining*, Vol. 10, Wiley-Interscience, New York, 1965, pp. 192–216.
44. F. X. Govers and G. R. Bryant, "Solvent Dewaxing of Oils with Benzol and

Acetone," in *Proceedings of the American Petroleum Institute*, Vol. 14, No. III, May 1933, pp. 7–15.
45. G. G. Scholten, "Solvent Dewaxing," in *Encyclopedia of Chemical Processing and Design*, Vol. 15, Dekker, New York, 1983, pp. 353–370.
46. J. D. Bushnell and J. F. Egan, *Commercial Experience with Dilchill Dewaxing*, Presented at the NPRA Annual Fuels and Lubricants Meeting, Houston, 1975, Paper No. F&L-75-50.
47. J. M. Scalise et al., "Solvent Dehydration System Cuts Energy Use, Improves Dewaxed Oil Yield," *Oil Gas J.*, pp. 84–86 (August 27, 1984).
48. G. W. Wirtz, *Spray Deoiling Process*, Paper Presented at the Texaco Lubricating Oil Manufacturing Licensee Symposium, White Plains, New York, May 18–20, 1982.
49. J. F. Eagen et al., "Successful Development of Two New Lubricating Oil Dewaxing Processes," in *Proceedings—Ninth World Petroleum Congress*, Vol. 5, 1975, pp. 345–357.
50. D. A. Gudelis et al., "Improvements in Dewaxing Technology," *API Proceedings—Division of Refining*, Vol. 53, 1973, pp. 724–737.
51. S. Norbert, "German Unit Gives Dewaxing Data," *Hydrocarbon Processing Pet. Refiner*, pp. 104–106 (December 1963).
52. A. Hoppe, "Dewaxing With Urea," in *Advances in Petroleum Chemistry and Refining*, Vol. 8, Wiley-Interscience, New York, 1964, pp. 193–234.
53. G. G. Scholten, "Dewaxing, Urea," in *Encyclopedia of Chemical Processing and Design*, Vol. 15, Dekker, New York, 1983, pp. 353–370.
54. Englehard Minerals & Chemicals Corp., *Hydrocarbon Refining with Static-Bed Percolation*, EM-9349, Revised, Menlo Park, New Jersey, April 1986.
55. R. M. Butler and R. Kartzmark, "Chemical Changes in Lubricating Oil on Hydrofining," in *Proceedings—Fifth World Petroleum Congress*, Vol. III, 1959, pp. 151–160.
56. H. F. Dare and J. Demeester, *Ferrofining—First Commercial Unit on Stream*, Paper No. Tech. 62-15 Presented at the Annual Meeting of the NPRA, April 2–4, 1962, San Antonio, Texas.
57. J. B. Gilbert et al., "Hydrogen Processing of Lube Stocks," *J. Inst. Pet.*, 53(526), 317–327 (October 1967).
58. E. O. Kindschy et al., "Lubricating Oil Hydrotreatment to Improve Quality and Yields," *Preprint 35C, 55th National Meeting of AIChE*, Houston, February 1965.
59. R. L. Menzl and W. L. Webb, "Hydrotreating of Lubricating Oil Stocks for Industrial Oils," in *Proceedings—American Petroleum Institute*, Section III, Refining, 1965, pp. 48–53.
60. F. Tsuneyoshi et al., "Hydrogen Treating of Some Lubricating Oil Fractions," *Bull. Jpn. Pet. Inst.*, 6, 1–10 (June 1964).

AVILINO SEQUEIRA, Jr.

4
Treating Processes

Desalting, Crude Oil

Salt is added to crude by produced brines, and when such brines cannot be completely removed, residual salt levels may be high. For example, 1 bbl of brine can contribute 70 lb or more of salt, an amount that is unacceptable in as much as 1000 bbl oil.

The common removal technique is to dilute the original brine with fresher water so that the salt content of water that remains after separation treatment is acceptable, perhaps 10 pounds per thousand barrels (pptb) oil, or less.

In areas where fresh water supplies are limited or expensive to obtain, the economics of this technique become critical. This article presents an engineering approach to the calculation of dilution water requirements. Methods given should enable more accurate, efficient design of desalting facilities for maximum dilution water conservation.

Water Supply Problems

Crude oil desalting techniques in the field have improved tremendously in the past 15 years with the introduction of the electrostatic coalescing process. Still, local producing regions have differing problems and, as a result, require different solutions. For instance, a desalting installation in West Africa or rain forest areas of South America will have ample fresh water for dilution. But arid producing areas of the Middle East do not have fresh water in large quantities and must depend on seawater or slightly salty subsurface water prevalent in the area. Even then, preparation of either water source for dilution purposes is expensive and must be studied with care in the early planning stages of any desalting scheme.

General Requirements

Almost every area will produce water at some stage of field life; many areas produce saltwater from the beginning. But the prolific Middle East fields have produced dry, salt-free crude for many years. This is rapidly coming to an end as the reservoirs are taxed for more and more oil.

Saltwater has begun to cause operators in these areas many problems. Some wells have been shut in, and some operators have begun installation of facilities for removing and disposing of produced water. Also, most operators are in at least the planning stages of water injection. Since the primary source for this water is the sea, it too will add to salt-removal problems.

There are several acceptable methods of produced saltwater removal. Some are more efficient than others, such as electrostatic coalescing. But whatever the method used, there will be some remnant water to be *sold* with the crude.

Desalting, Crude Oil

Remnant water content will vary with efficiency of dehydration equipment, ranging generally from as low as 0.1 to 0.3%. Since this water is the salt carrier, the amount of salt will depend on water salinity and the volume remaining after dehydration.

Example Situation

Assume 1000 bbl crude are produced with 10.0% water. Water salinity is determined to be 200,000 ppm equivalent sodium chloride (NaCl). This converts to pounds of NaCl per barrel of water by:

$$\frac{200,000}{1,000,000} \times 350 = 70 \text{ ppb water}$$

See Fig. 1 for easy conversion of ppm NaCl to ppb water.

The amount of produced water initially present is (where bopd is barrels of oil per day)

$$\frac{1000 \text{ bopd}}{1 - 0.10} - 1000 = 111 \text{ bbl water}$$

Total produced salt content of crude before dehydration is (where bwpd is barrels of water per day)

$$111 \text{ bwpd} \times 70 \text{ ppb} = 7770 \text{ pptb}$$

See Table 1 for direct conversion of ppm NaCl equivalent to pptb.

For comparison, assume that two methods of dehydration-desalting are present, one slightly more efficient than the other—for instance, one electrostatic and one mechanical coalescer. Further assume total net oil produced is 100,000 bopd, split evenly between the two methods.

Effluent from the electrostatic method is 0.1% basic sediment and water (BS&W). From the mechanical method it is 0.2% BS&W. This converts to

$$\frac{1000}{1 - 0.001} - 1000 = 1 \text{ bbl water per 1000 bbl net oil}$$

for the electrostatic process effluent, and

$$\frac{1000}{1 - 0.002} - 1000 = 2 \text{ bbl water per 1000 bbl net oil}$$

for the mechanical process effluent.

As determined earlier, each barrel of produced water contains 70 lb of equivalent NaCl, and for the above cases, oil for marketing—after dehydration to very low water contents—still will contain 70 and 140 pptb with no further desalting.

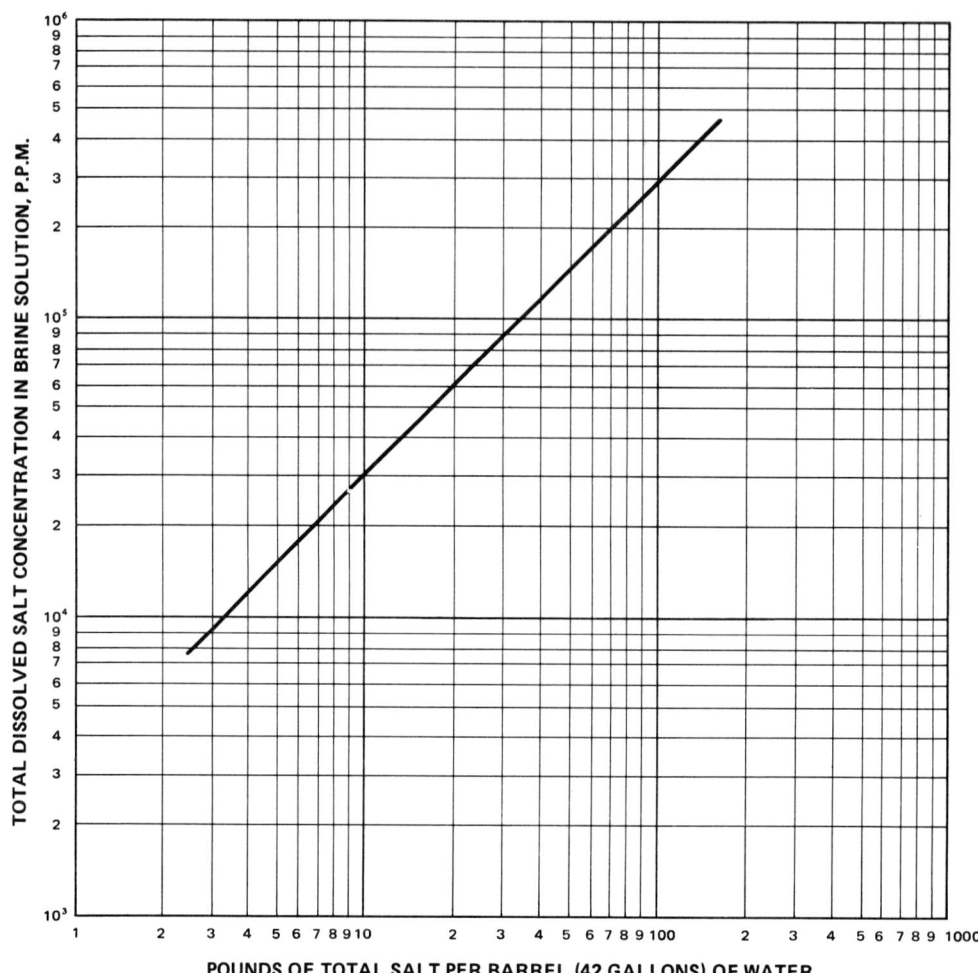

FIG. 1. This chart may be used to convert total dissolved salt concentration in a brine solution expressed in ppm to pounds of salt per barrel of water. Total dissolved salt content is based on chloride ions and calculated as equivalent NaCl.

TABLE 1 Pounds of Salt per 1000 bbl of Oil

Salt Content of Water (ppm)	Volume Percent Water Content in Oil (barrels of water)				
	1.00 (10)	0.50 (5)	0.20 (2)	0.10 (1)	0.05 (0.5)
10,000	35.00	17.50	7.00	3.50	1.75
20,000	70.00	35.00	14.00	7.00	3.50
30,000	105.00	52.50	21.00	10.00	5.25
40,000	140.00	70.00	28.00	14.00	7.00
50,000	175.00	87.50	35.00	17.50	8.75
100,000	350.00	175.00	70.00	35.00	17.50
150,000	525.00	262.50	105.00	52.50	26.25
200,000	700.00	300.00	140.00	70.00	35.00

Desalting, Crude Oil

The only practical method of further salt reduction in this example is addition of dilution water that contains less salt than the produced water. Dilution water is injected into the produced crude stream prior to the dehydration (desalting) treatment stage. Some form of agitation or shearing is required in the stream, after the water injection point, to contact and mix dilution water with dispersed droplets of produced water. The combination reduces salt concentration, followed by dehydration or water removal. And remaining BS&W in very low percentages will contain a low concentration of NaCl. Figure 2 is a flow diagram of the process using single-stage dehydration.

In concept, this seems simple and straightforward, but in fact it is somewhat difficult to achieve due to problems of water dispersion and contact and the possibility of creating a difficult emulsion. This article is not a discussion of emulsion treating, so it is noted only that some form of efficient emulsion treating and water removal are necessary, and remarks will be limited to dilution water, injection, mixing, and required ratios of dilution water to produced oil.

Calculation Methods

The practice of desalting crude is not new. Refineries have been successfully desalting crude charge stock to less than 5 pptb for many years, and mechanical and electrostatic desalting have been improved greatly. However, very little

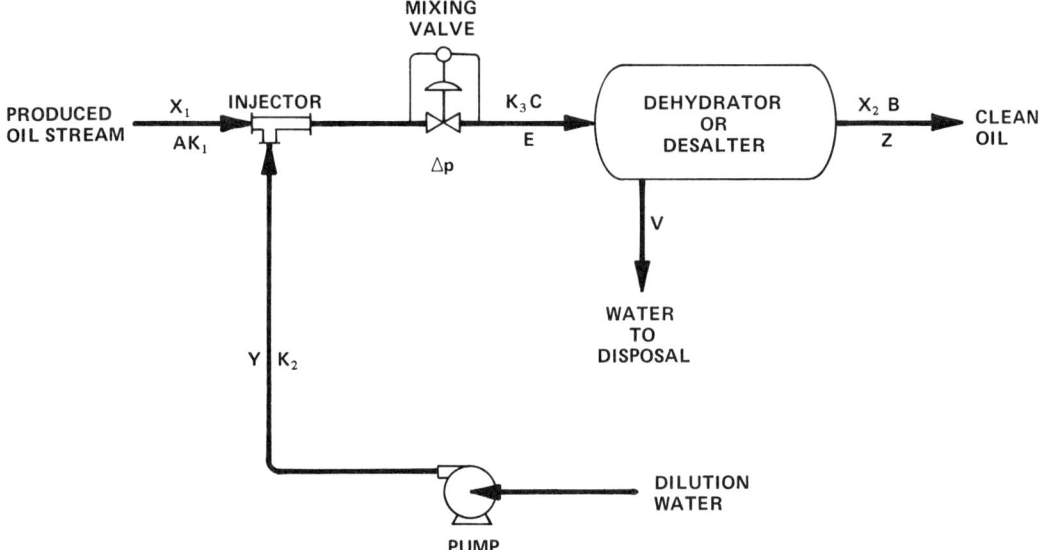

FIG. 2. In the single-stage desalting system, dilution water is injected into the crude stream prior to the dehydration stage.

attention has been given to use of dilution water, probably due to the general availability of both fresh and waste waters in and near refineries. As a result, only *rule-of-thumb* methods for dilution water have been developed. Such shortcuts are no longer satisfactory under present economics.

Requirements for dilution water ratios can be calculated easily as a material balance. By combining the arithmetic of material balance and a unique method of water injection and dispersion for contact efficiency, very low dilution water use rates can be achieved. This can be highly significant in an area where production rates of 100,000 bopd are common and fresh water supply is a problem.

Example Calculation

The previous example of a typical crude containing 10% produced saltwater contained 70 pptb for 0.1% BS&W and 140 pptb for 0.2% BS&W, and each barrel of remnant water contained 70 lb of NaCl (equivalent). To dilute this concentration enough to market a 10-pptb crude oil, the following calculations will determine the amount of dilution water required for a comparison of Cases A and B—single stage electrostatic and mechanical separation, respectively, Table 2. The calculation basis is 1000 bbl net oil, and the systems in Fig. 2 are assumed.

Basic equations for calculating required dilution water are as follows (see Fig. 2):

$$Z = BK_3 \qquad (1)$$

$$K_3 = \frac{AK_1 + YK_2E}{A + YE} \qquad (2)$$

where $A = \dfrac{1000X_1}{1 - X_1}$ = water in inlet stream, bbl

$B = \dfrac{1000x_2}{1 - X_2}$ = water in clean oil, bbl

$C = A + Y$ = water to desalter inlet, bbl

Y = injection water (varies with each problem), bbl

$V = A + Y - B$ = water to disposal, bbl

E = mixing efficiency of Y with A (as a fraction), assume 80%

K_1 = salt per barrel of water in produced oil stream, lb

K_2 = salt per barrel of dilution water, lb

$K_3 = \dfrac{AK_1 + YK_2E}{A + YE}$ = salt per barrel of water to desalter inlet, lb

X_1 = fraction of water in produced oil stream

X_2 = fraction of water in clean oil outlet

Z = salt in outlet clean oil per 1000 bbl of net oil, lb

Since it has been determined that B is 1 in Case A, and 2 in Case B, the equations for Case A reduce to

TABLE 2

	Case A	Case B
Process	Single stage	Single stage
Method	Electrostatic	Mechanical
Water inlet, %	10	10
Water outlet, %	0.1	0.2
Salt content of produced water, ppb	70	70
Salt content of dilution water, ppb[a]	2.1	2.1
Desired salt content of oil, pptb	10	10
Total net oil, bbl/d	50,000	50,000

[a] Assumed 6000 ppm NaCl.

$$K_3 = \frac{Z}{B} = \frac{10}{1} = 10$$

$$10 = \frac{(111 \times 70) + Y(2.1 \times 0.8)}{111 + Y(0.8)}$$

$$Y = 1{,}053.8 \text{ bbl water required for 1000 bbl net oil}$$

$1053.8 \times 50 \,(1000 \text{ bopd}) = 52{,}690 \text{ total bwpd}$

and for Case B

$$K_3 = \frac{10}{2} = 5$$

$$5 = \frac{(111 \times 70) + Y(2.1 \times 0.8)}{111 + Y(0.8)}$$

$$Y = 3110 \text{ bbl water required for 1000 bbl net oil}$$

$3110 \times 50 \,(1000 \text{ bopd}) = 155{,}500 \text{ total bwpd}$

It is readily seen that, even though both water requirements are high, the 52,690 bbl of dilution water in Case A are less than half of that for Case B. The reason is simply that outlet water content in A of 0.1% BS&W and 0.2% BS&W in B were assumed.

This exercise illustrates the importance of an efficient water removal system. No other factor plays as important a part in dilution water savings as does the lower level of dehydration.

These particular examples require so much dilution water that it is uneconomical to pursue them further. We must look at other methods to conserve dilution water.

Two-Stage System

The proper approach to desalting this example oil is the use of a two-stage desalting system as represented by the flow diagram of Fig. 3. In the two-stage

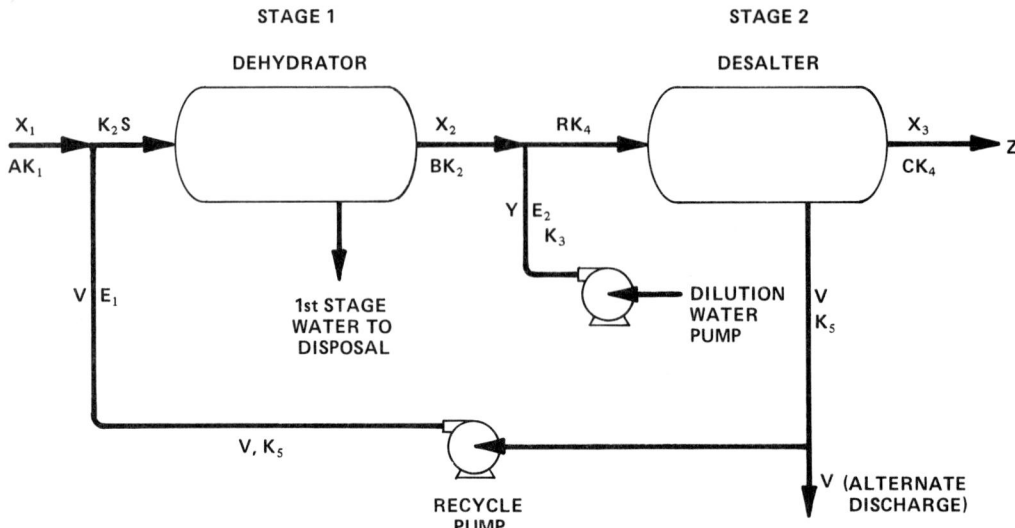

FIG. 3. In the two-stage desalting system, dilution water is injected between stages, greatly reducing the amount of dilution water required. Further reductions may be achieved by adding the second-stage recycle pump.

process, dilution water is injected between stages after stream water content has been reduced to a very low level by the first stage. This will reduce dilution water requirements to only a few thousand barrels per day.

Further reduction is achieved by adding the second-stage recycle pump. Second-stage water is much lower in NaCl than the produced stream inlet water due to addition of dilution water. By recycling this water to the first-stage inlet, both salt reduction and dehydration is obtained in the first stage. The water volume to be recycled is assumed to be the same as dilution water injection volume.

Desalter units generally will produce a dehydrated stream containing like amounts of BS&W from each stage. Therefore, BS&W can be considered *pass through volume* and the dilution water added is the amount of water to be recycled. The recycle pump, however, generally is oversized to compensate for difficult emulsion conditions and upsets in the system.

Dilution water calculations for a two-stage system using recycle are slightly more complicated than for a single stage or two stage without recycle. These problems are better suited to a computer, as hand calculations require several entries of dilution water variables in order to *zero in* on the proper balance.

Referring to Fig. 3, the basic equations for computation are established as follows:

$$VK_5 = \frac{BAK_1(R - C) + SYK_3(R - E_2C)}{SR + BE_1(C - R)} \quad (3)$$

$$V = B + Y - C \quad (4)$$

$$Z = BK_2 + YK_3 - VK_5 \quad (5)$$

Desalting, Crude Oil

where $A = \dfrac{1000X_1}{1 - X_1}$ = water in inlet stream to facility, bbl

$B = \dfrac{1000X_2}{1 - X_2}$ = water in effluent of first stage, bbl

$C = \dfrac{1000X_3}{1 - X_3}$ = water in effluent of second stage desalter, bbl

Y = injection water of lesser salinity than inlet water (A), bbl
V = recycle water to injection in first stage inlet line, bbl
E_1 = mixing efficiency of V with A (as a fraction), assumed 80%
E_2 = mixing efficiency of Y with B (as a fraction), assumed 80%
X_1 = fraction of water in inlet stream to facility
X_2 = fraction of water in first-stage outlet oil
X_3 = fraction of water in second-stage outlet oil
K_1 = salt per barrel of water to facility, lb

$K_2 = \dfrac{AK_1 + VE_1K_5}{A + VE_1}$ = salt per barrel of water to first-stage desalter, lb

K_3 = salt per barrel of dilution water, lb

$K_4 = \dfrac{BK_2 = YE_2K_3}{B + E_2Y}$ = salt per barrel of water to second-stage desalter, lb

K_5 = salt per barrel of water to recycle injection into inlet line to first stage, lb
$S = A + E_1V$ = water to first-stage desalter, bbl
$R = B + E_2Y$ = water to second-stage desalter, bbl
Z = salt in outlet per 1000 bbl of net oil, lb

Note: All water volumes are per 1000 bbl net oil, all salt contents are pounds of total dissolved salts per barrel of water, and Y varies with each individual problem.

Solutions to example Cases A and B, when two-stage desalting and recycle is applied, are 8.95 and 49.6 bbl of dilution water required, respectively. For the 50,000 bbl net oil in Case A, this is $50 \times 8.95 = 448$ bwpd, and for Case B, $50 \times 49.6 = 2,480$ bwpd dilution water. Note that the 0.2% BS&W in Case B requires more than 5 times the dilution water as does Case A, with only 0.1% BS&W effluent.

One would think that increased efficiency of mixing (standard efficiency is taken as 80%) could further improve water reduction. However, as can be proven by making variations in these calculations, efficiency does not affect water use rate to a large degree. *The two most important variables for water use rate control are BS&W effluent percentage and salt content of dilution water.*

It is not always possible to control the injection water salinity without considerable expense. Therefore, BS&W content of the clean desalted crude should be of prime importance when dilution water conservation is an economic factor.

Dilution Water/Net Oil Ratio

It has been determined through field experience that a very low BS&W content at the first-stage desalter exit requires a high percentage of dilution water to properly contact dispersed, produced water droplets and achieve desired salt concentration reduction. This percentage dilution water varies with the strength of the water/oil emulsion and oil viscosity. Empirical data show that the range is from 4.0% to as high as 10%. Obviously, this indicates that the mixing efficiency of 80% is not valid when low water contents are present.

Additional field data show that the low dilution water use rates can be maintained and still meet the required mixing efficiencies. The problem encountered with very low BS&W contents is produced water droplet size and their dispersement in oil. When 99.9% of the produced water has been removed, the remaining 0.1% consists of thousands of very small droplets more or less evenly distributed throughout the oil. To contact them would require either a large amount of dilution water dispersed in the oil or a somewhat smaller amount with better droplet dispersion. Whatever the required amount, it can be attained without exceeding the dilution water rates shown in the earlier example.

Water contained in each desalter unit is an excellent source of volume ratio makeup. Figure 4 shows the typical two-stage process with recycle, but in this case note that recycle water is routed to the second-stage inlet as well as the first-stage inlet. The amount of water recycled to the first-stage must be the same as the dilution water injection rate to maintain the water level in the second-stage unit. The volume of water recycled to the second stage inlet can be any amount, since it immediately rejoins the controlled water volume in the lower portion of the desalter unit. This is usually referred to as *internal recycle*, and field installations of this type have been operating successfully for several years.

Also note in Fig. 4 that an additional pump is shown recycling first-stage water to first-stage inlet. The first-stage internal recycle is not necessarily required for each installation. It is dependent upon amount of produced water present in the inlet stream. Its use should be evaluated for each case.

In calculating dilution water requirements, *internal recycle* may be ignored since it does not add salt or water volume to the stream process.

Injection and Mixing

Thus far developed is the need for dilution water, arithmetic for determining volumes, and a simple but unique approach to low dilution water use rates incorporating the volumes of water retained internal to the process. It is also necessary to develop the need for an economical, simple method of water injection, dispersion, and mixing.

Produced salt water, if it remains in the oil in sufficient quantities to require a dehydration-desalting facility, is carried in an emulsion state. The emulsion

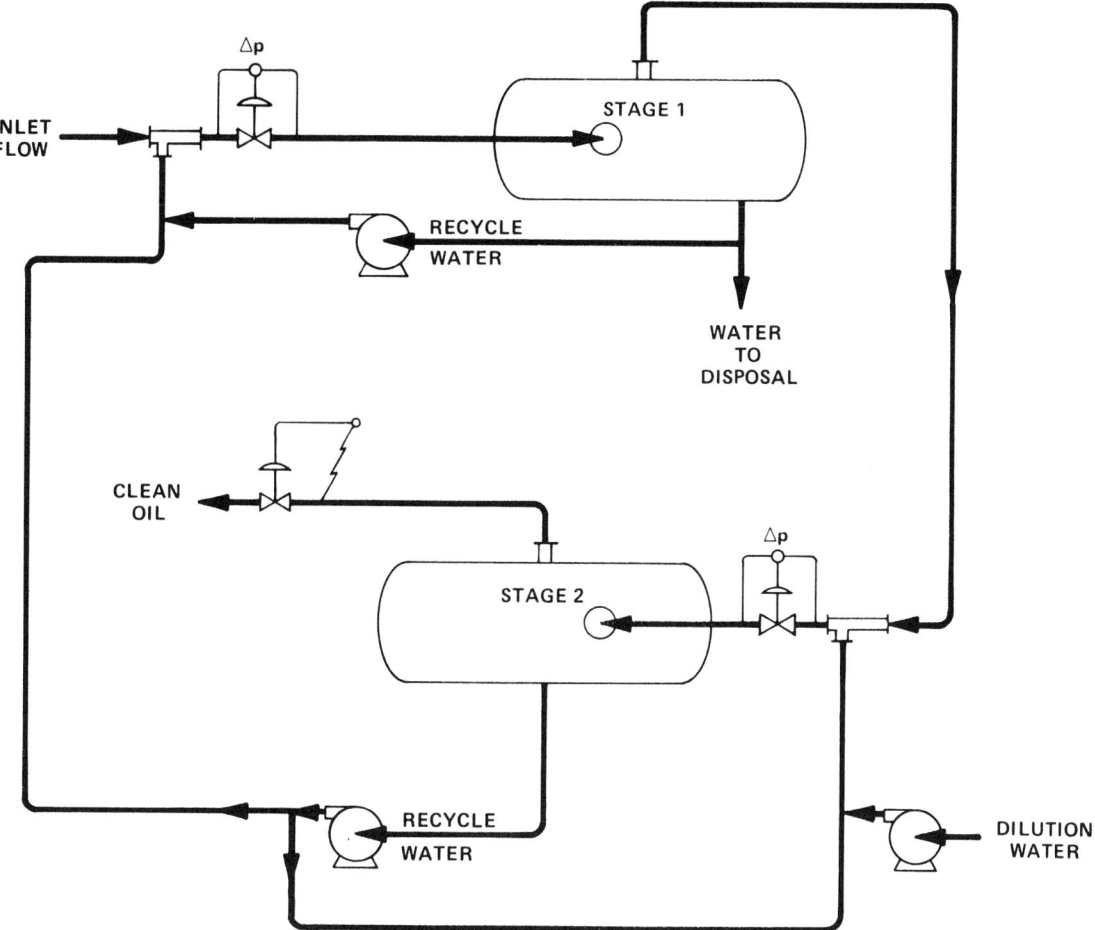

FIG. 4. This schematic shows a two-stage desalting system with recycle and *internal recycle*.

exists as a result of mechanical shearing of produced water and its dispersion into oil by turbulence. Once water is dispersed in droplets, many of them are combined with the various emulsifying agents produced in the crude, and they resist natural separation because of their size and envelopment by the emulsifying agents.

For the majority of these droplets to coalesce with dilution water, some method of dilution shearing and dispersion will be required, much the same as initial emulsifying mechanical actions. Many methods have been used for water injection and mixing over the years. These have ranged from simply pumping water through a tee connection into the oil stream to mixing oil and water together through the crude transfer pump. The former is not adequate for mixing purposes and the latter often makes an emulsion very difficult to break in the separation phase of the process.

The most successful method consists of a system of spray nozzles in the oil flow line positioned so that water pumped through the nozzles enters the oil in very fine droplets. Immediately following the spray injector is a differential control valve that provides suitable plug and seat surfaces for mechanical shearing along with controlled pressure drop to provide energy necessary for the shearing action. See Fig. 5 for a schematic representation of this method.

It is not always possible to predict accurately the pressure drop required for adequate mixing. Most field desalting installations require 15 to 25 $lb/in.^2$ differential across each mixing valve when a spray injector is not used. This large differential is required because the mixing valve must disperse and shear at the same time. When the spray injector is installed ahead of the mixing valve, the differential required for mixing will range from 5 to 15 $lb/in.^2$. This lower total pressure drop across a two-stage desalting plant can mean a significant savings in the horsepower required to move the crude, along with a possible reduction in equipment costs, since units may be designed to operate at a lower pressure.

Besides savings that may result from pressure and horsepower reduction, it is generally known that the more violent mixing of oil and water also creates an emulsion that would require more emulsion breaker chemical and a higher temperature for separation. Savings in chemical additives and fuel could amount to several hundred thousand dollars per year on an installation processing 100,000 bopd or more.

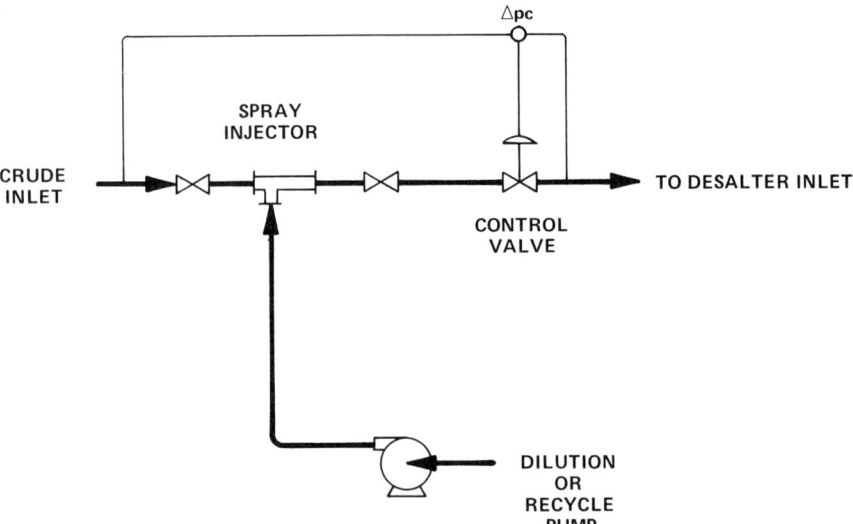

FIG. 5. The most successful method of contacting water droplets dispersed in crude is through the use of a spray injector system. Water is pumped through nozzles and enters the stream in very fine droplets. The differential control valve provides for mechanical shearing and a controlled pressure drop that gives the energy necessary for the shearing action.

Most oil producers have experienced salt water produced with crude oil at some time. Generally, removal of this unwanted, bothersome water has been fairly straightforward, involving wash tanks or a heater-treater. Removal of this same water along with salt concentration reduction presents a completely different set of problems. These can be overcome with relative ease if the operator is willing to spend many thousands of dollars each year for water, fuel, power, and chemical additives. Conversely, if some time is spent in the initial design stages to determine the best methods for water removal to achieve lower BS&W remnants, mixing, and injection, then a system can be designed that will operate at a greatly reduced annual cost. In most cases, yearly savings on operational expenses will pay for the complete installation in 1 to 3 years.

This material appeared in *World Oil Magazine*, copyright Gulf Publishing Co., pp. 150–156, June 1978.

DONALD R. BURRIS

Demetallization/Desulfurization of High Metal Content Petroleum Feedstocks

A number of methods are available for segregating metals from crude oil. Deasphalting and coking are two refinery, noncatalytic methods. Catalytic cracking and hydrodesulfurization are two catalytic methods. This article is devoted to demetallization as it relates to hydrodesulfurization.

Petroleum crude oil, which is found in almost all parts of the globe, varies in its physical characteristics from area to area, and even from field to field within a particular area. Properties such as sulfur content, gravity, and distillation can vary greatly among crudes, requiring a refiner to plan flexibility in his operations to compensate for different feedstocks.

Metals content also varies with crude type and may or may not present a problem to the refiner, depending on the amount of metals present and the downstream processing required. Vanadium and nickel, the primary metals found in petroleum, can range from less than 1 weight part per million (wppm) in some crudes to as high as 1100 wppm vanadium and 85 wppm nickel for

Venezuela's Boscan crude. Vanadium is usually present in higher concentrations than nickel for Middle East and Venezuelan crudes. However, for many domestic crudes, particularly Californian, the nickel content is higher. Other metals, such as sodium and iron, are also found in quantities up to 100 and 60 wppm, respectively, though usually much lower.

The metals are contained in the large asphaltenic structures in crude oils; those crudes having higher concentrations of heavy molecules are more often the ones with the most metals. A trend is therefore observed where the lower API gravity crudes tend to be the higher metal content crudes. Within a particular gravity range the metals content may be further broken down into the sequence where Venezuelan crudes have the most metals followed by Middle East crudes, Californian crudes, and other domestic crudes. Figure 1 shows the relationship between gravity and vanadium for several important oil-producing areas.

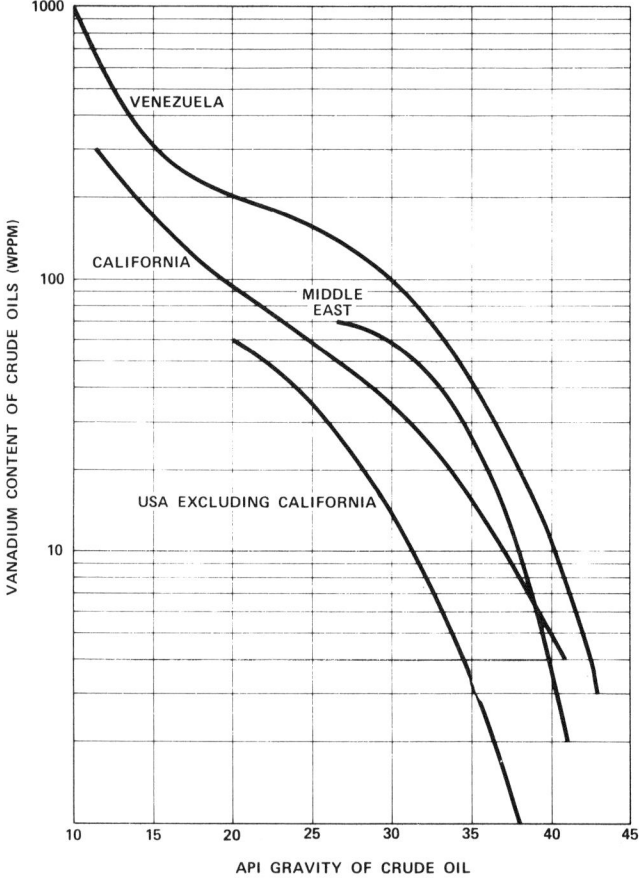

FIG. 1. Metal content of important world crude oils.

Properties of Khafji Crude Oil

Petroleum crude oils are composed of thousands of different hydrocarbon compounds. The major elements in petroleum crude oils are carbon and hydrogen. Khafji crude oil has been chosen as an example, and an analysis is shown in Table 1. Khafji crude oil contains 82.2 wt.% carbon and 11.5 wt.% hydrogen. The remaining 6.3 wt.% is composed of oxygen, sulfur, nitrogen, and trace metals such as aluminum, barium, calcium, chromium, cobalt, copper, iron, lead, magnesium, manganese, molybdenum, nickel, silicon, strontium, tin, titanium, vanadium, and zinc. The major metal contaminants are nickel and vanadium. Table 1 shows how properties vary from fraction to fraction in crude oil. Fractionation of crude oil will concentrate the heavy asphaltenes and metals into the higher boiling range fractions. Low-boiling fractions such as naphthas are low in density, viscosity, and sulfur. Naphthas contain no measurable metals content to poison desulfurization catalysts. High-boiling fractions such as vacuum residua are high in density, viscosity, and sulfur content. They also contain appreciable quantities of vanadium and nickel. The vanadium and nickel deposit on the catalyst, lowering the catalyst activity during hydrodesulfurization.

Hydrodesulfurization

Hydrotreating processes, in particular the hydrodesulfurization of petroleum residua, are catalytic processes. Hydrocarbon feedstock and hydrogen are passed through a catalyst bed at elevated temperatures and pressures. Some of the sulfur atoms attached to hydrocarbon molecules react with hydrogen on the surface of the catalyst to form hydrogen sulfide (H_2S). Thermodynamic equilibrium calculations show that these reactions can be driven to almost 100% completion, but economic considerations normally limit the commercial application of hydrodesulfurization.

Hydrodesulfurization and demetallization occur simultaneously on the active sites within the catalyst pore structure. Sulfur and nitrogen occurring in residuum are converted to H_2S and NH_3 in the catalytic reactor. These gases are scrubbed out of the reactor effluent gas stream. The metals in the residuum oil are deposited on the catalyst in the form of metal sulfides. Cracking of residuum to distillate produces carbon laydown on the catalyst. Both carbon and metal deposition on the catalyst poison catalyst activity. Carbon deposition is a fast reaction which soon equilibrates to a particular carbon level. Carbon deposition is controlled by hydrogen partial pressure within the reactors. Metal deposition is a slow reaction which is directly proportional to the amount of feedstock passed over the catalyst.

TABLE 1 Properties of Khafji Crude Oil

Boiling Range (°F)	Vol.%	Gravity[a] (°API)	Kinematic Viscosity		Carbon Residue Ramsbottom (wt.%)	Sulfur (wt.%)	Vanadium (wppm)	Nickel (wppm)
			Centistokes at 100°F	Centistokes at 210°F				
Straight run light naphtha IBP-185	7.3	92.3	—	—	—	0.03	—	—
Naphtha, 185–360	14.1	59.8	—	—	—	0.04	—	—
Kerosene, 360–500	10.9	45.3	1.5	—	—	0.32	—	—
Light gas oil, 500–700	17.0	33.3	4.6	1.6	0.08	1.65	0.1	0.2
Vacuum gas oil, 700–1000	23.7	21.0	87.0	8.0	0.16	2.60	0.2	0.2
Vacuum residuum, 1000-EP	27.0	3.5	—	4,940	21.7	5.71	128	47
Whole crude	100.0	29.0	19.2	13.2	6.3	2.90	41	15
Atmospheric residuum, 700-EP	50.7	11.2	10,360	147.4	12.6	4.74	71	26

[a]Specific gravity = $\dfrac{141.5}{°API + 131.5}$

TABLE 2 Hydrogenation

Saturation:

$$\text{Benzene} + 3H_2 \longrightarrow \text{Cyclohexane}$$

Hydrodesulfurization:

$$R-SH + H_2 \longrightarrow RH + H_2S$$
$$R-S-S-R' + 3H_2 \longrightarrow RH + R'H + 2H_2S$$

Hydrocracking:

$$R-R' + H_2 \longrightarrow RH + R'H$$

Definition of Terms

The reaction of molecular hydrogen with a hydrocarbon such as petroleum is called *hydrogenation*. As shown in Table 2, the field of hydrogenation is further subdivided into more specific types of reactions. *Saturation* is the addition of hydrogen to aromatic compounds or compounds with double or triple bonds without bond cleavage. When hydrogen reacts with bond cleavage, the reaction is called *hydrogenolysis*. Hydrogenolysis reactions with bond cleavage of carbon-sulfur or sulfur-sulfur bonds are called *hydrodesulfurization*. Hydrogenolysis reactions with cleavage of carbon—carbon bonds are referred to as *hydrocracking*. Hydrogenolysis reactions are applied in petroleum processes for the purpose of hydrodesulfurization, hydrodenitrification, hydrocracking, and demetallization. These processes differ in the choice of catalyst and process conditions selected to achieve the desired result.

Examples of Petroleum Sulfur Compounds

Sulfur compounds have different structures and molecular weights. They react at various rates of reaction to produce H_2S and different hydrocarbon products. During the process of hydrodesulfurization, nonsulfur-containing molecules also undergo reaction. Oxygenated molecules are the most reactive, sulfur molecules the next reactive, and nitrogen molecules the least reactive.

Table 3 contains a partial list of some of the types of sulfur compounds which exist in crude oils. Besides those on this list, crude oils also contain elemental sulfur, dissolved hydrogen sulfide, benzonaphthothiophenes, and many other classes of sulfur compounds. Petroleum crude oils contain hundreds of sulfur compounds of which only 200+ compounds have been identified. The easier to remove (more reactive) sulfur compounds are found in the distillate fractions. The less easy to remove (less reactive) cyclic sulfur compounds are found in the resid portion of crude oils.

In hydrodesulfurization the metals will also react with hydrogen. Generally, it is easier to crack a bond between carbon and a heavy metal than a carbon—sulfur bond. Consequently, the rate of metals conversion from organometals to metallic sulfides is faster than the rate of organosulfur conversion to hydrogen sulfide. The free metals, or metal sulfides, are solids and deposit on the catalyst surface. The metals cover the catalyst surface, filling the pore volume or blocking the pores, and this action "poisons" the catalyst, at least in the sense that active sites are made inaccessible to the oil and hydrogen. It has been observed that after a certain fraction of the metals are removed from the oil, the remaining metals are much more difficult to remove. This experimental observation has led to the development of methods for demetallizing heavy oils prior to desulfurization.

Discussion of Organometallic Compounds

The structure of asphaltenes is theorized to be a micellar structure of various molecules which are stacked and aligned similar to graphite. Molecular weights of the organometallic compounds are as low as 450. The majority of the organometallic compounds are in vacuum residuum boiling above 1050°F. The micellar intermolecular sheets are not chemically bound since high tempera-

TABLE 3 Examples of Sulfur-Containing Hydrocarbons in Petroleum Crude Oil

Name	Structure	Typical Reaction
Thiols (mercaptans)	R—SH	R—SH + H_2 ⟶ RH + H_2S
Disulfides	R—S—S—R'	R—S—S—R' + $3H_2$ ⟶ RH + R'H + $2H_2S$
Sulfides	R—S—R'	R—S—R' + $2H_2$ ⟶ RH + R'H + H_2S
Thiophenes	(thiophene with R)	(thiophene) + $4H_2$ ⟶ n-C_4H_{10} + H_2S
Benzothiophenes	(benzothiophene with R)	(benzothiophene) + $3H_2$ ⟶ CH_3CH_2—(benzene) + H_2S
Dibenzothiophenes	(dibenzothiophene with R)	(dibenzothiophene) + $2H_2$ ⟶ (biphenyl) + H_2S

Demetallization/Desulfurization of Petroleum Feedstocks

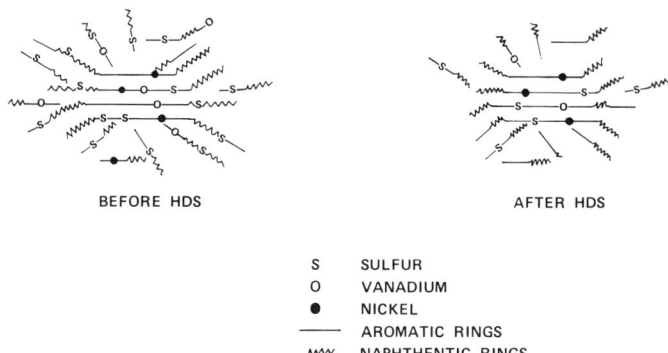

S	SULFUR
O	VANADIUM
●	NICKEL
——	AROMATIC RINGS
⩗⩗⩗	NAPHTHENTIC RINGS

FIG. 2. Demetallization of asphaltenes and surrounding resins during hydrodesulfurization.

tures and solvents can break the micelles apart. Analyses of an asphaltene compound isolated from a Kuwait crude indicate four aromatic sheets with a 12-Å distance across the aromatic center (Fig. 2). Note the greater concentration of the vanadium compounds on the periphery of the micelle and the location of the nickel compounds.

The tendency of vanadium atoms to project from asphaltenes makes the vanadium more accessible to the catalyst surface. At a given set of process conditions, vanadium removal is greater than nickel removal. Figure 2 illustrates an asphaltene structure after hydrodesulfurization (HDS). Note the significant reduction of vanadium on the periphery of the micelle and the insignificant change of nickel within the micelle.

The relative reactivity of organometallic compounds also affects where the metals deposit within a catalyst pellet. It has been experimentally determined that the organometallic compounds decompose at different depths in catalyst pores. Nickel organometallic compounds decompose deeper in the pores than do vanadium organometallic compounds. This has been experimentally observed by the pore mouth blockage caused by vanadium compounds. The vanadium deposits form a shell in the pores on the external parts of the catalyst pellets.

Reaction Kinetics

The life of a catalyst used to hydrotreat petroleum residuum is dependent on the rate of carbon deposition and the rate at which organometallic compounds decompose and form metal sulfides on the surface. Several different metal complexes exist in the asphaltene fraction of the residuum. Therefore, an explicit reaction mechanism of decomposition which describes all the compounds is not possible. In general, however, this reaction could be described as hydrogen (A) dissolved in the residual oil contacting an organometallic compound (B) at the surface of the hydrotreating catalyst and producing a metal sulfide (C) and a hydrocarbon (D):

$$A + B \xrightarrow{\text{Catalyst}} C + D$$

Different rates of reaction may occur with various types and concentrations of metallic compounds. Medium metals content Venezuelan resid, for example, has a lower rate of demetallization compared to high metals content Venezuelan Crude (Boscan). Individual organometallic compounds decompose according to both first- and second-order rate expressions. For reactor design a second-order rate expression is applicable to the decomposition of residuum as a whole.

Typical Hydrodesulfurization Process Configuration

All hydrodesulfurization processes react hydrogen with a hydrocarbon feedstock to produce H_2S and a desulfurized hydrocarbon product. The basic process flow sheet for a typical hydrodesulfurization process is shown in Fig. 3. Liquid petroleum feedstock is preheated and mixed with hot recycle gas containing hydrogen. The hot feedstock and recycle gas are passed over catalyst in the reactor section at temperatures between 550 and 850°F and pressures

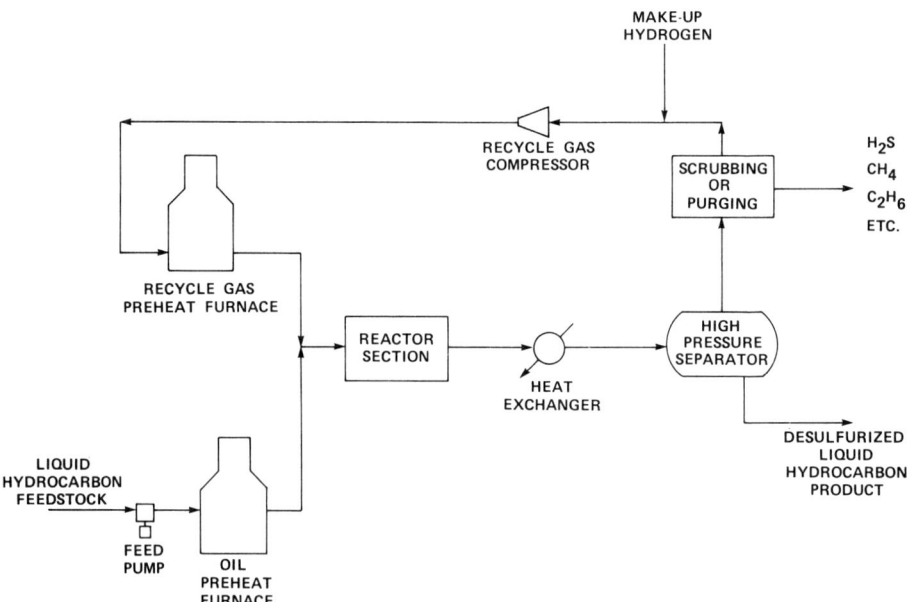

FIG. 3. Typical hydrodesulfurization process.

between 150 and 3000 lb/in.^{2} gauge. The reactor effluent is then cooled by heat exchange, and desulfurized liquid hydrocarbon product and recycle gas are separated at essentially the same pressure as used in the reactor. The recycle gas is then scrubbed and/or purged of the H_2S and light hydrocarbon gases, mixed with fresh hydrogen makeup, and preheated prior to mixing with hot hydrocarbon feedstock.

The recycle gas scheme is used in the hydrodesulfurization process to minimize physical losses of expensive hydrogen. Hydrodesulfurization reactions require a high hydrogen partial pressure in the gas phase to maintain high desulfurization reaction rates and to suppress carbon laydown (catalyst deactivation). The high hydrogen partial pressure is maintained by supplying hydrogen to the reactors at several times the chemical hydrogen consumption rate. The majority of the unreacted hydrogen is cooled to remove hydrocarbons, recovered in the separator, and recycled for further utilization. Hydrogen is physically lost in the process by solubility in the desulfurized liquid hydrocarbon product, and from losses during the scrubbing or purging of H_2S and light hydrocarbon gases from the recycle gas.

Downflow Fixed-Bed Reactor Design

The reactor design commonly used in hydrodesulfurization of distillates is the fixed-bed reactor design. Feed enters at the top of the reactor and the product leaves at the bottom of the reactor (Fig. 4). The catalyst remains in a stationary position, with hydrogen and petroleum feedstock passing in a downflow direction through the bed of catalyst. The hydrodesulfurization reaction is exothermic and the temperature rises from the inlet to the outlet of each catalyst bed. With a high hydrogen consumption and subsequent large temperature rise, the reaction mixture can be quenched with cold recycled gas at intermediate points in the reactor system. This is done by dividing the catalyst charge into a series of catalyst beds. The effluent from each catalyst bed is quenched to the inlet temperature of the next catalyst bed. This sequence continues from catalyst bed to catalyst bed until the oil passes through the reactor or, in some cases, a series of reactors.

The extent of desulfurization is controlled by raising the inlet temperature to each catalyst bed to maintain constant catalyst activity over the course of the run. Fixed-bed reactors are mathematically modeled as plug-flow reactors with very little backmixing in the catalyst beds. The first catalyst bed is poisoned with vanadium and nickel at the inlet to the bed. As the catalyst is poisoned in the front of the bed, the temperature exotherm moves down the bed. Also, the activity of the entire catalyst charge declines. This is the reason why it is necessary to raise the reactor temperature over the course of the run. After catalyst regeneration, the reactors are opened and inspected, and the high metal content catalyst layer at the inlet to the first bed may be discarded and replaced with fresh catalyst. The catalyst loses activity after a series of regenerations. Consequently, after a series of regenerations it is necessary to replace the

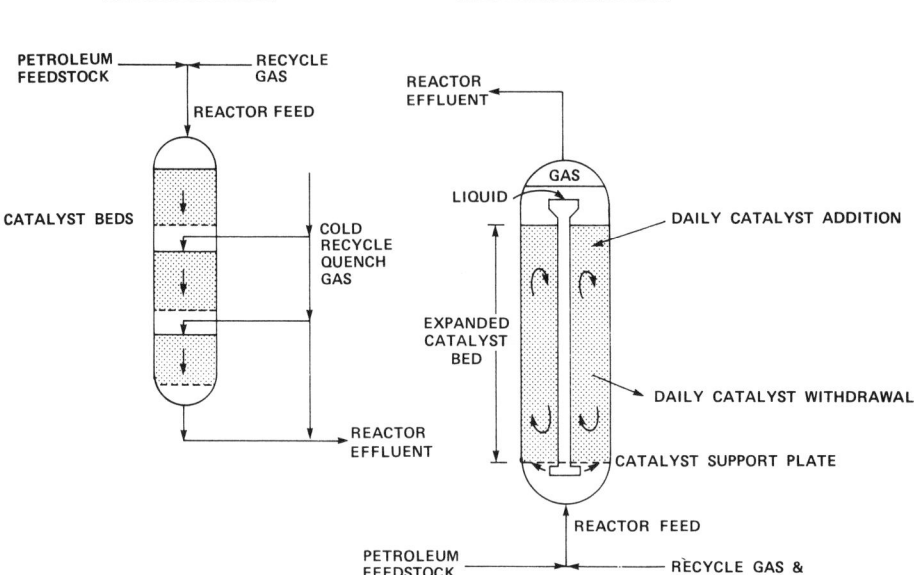

FIG. 4. Hydrosulfurization reactor design.

complete catalyst charge. In the case of very high metal content feedstocks such as residua, it is necessary to replace the entire catalyst charge rather than to regenerate it. This is due to the fact that the metal contaminants cannot be removed by economical means during rapid regeneration. Moreover, the heavy metals have been reported to interfere with the combustion of carbon and sulfur. The metals catalyze sulfur dioxide to sulfates which have a permanent poisoning effect on the catalyst.

Fixed-bed hydrodesulfurization units are generally used for distillate hydrodesulfurization. They are also used for residuum hydrodesulfurization, but require special precautions in processing. The residuum must undergo two-stage electrostatic desalting so that salt deposits do not plug the inlet to the first catalyst bed. The residuum must be low in vanadium and nickel content to avoid plugging the beds with metal deposits.

During the operation of a fixed-bed reactor, contaminants entering with fresh feed are filtered out and fill the voids between catalyst particles in the bed. The buildup of contaminants in the bed can result in the channeling of reactants through the bed, thus reducing hydrodesulfurization efficiency. As the flow pattern becomes distorted or restricted, the pressure drop throughout the catalyst bed increases. If the pressure drop becomes high enough, physical damage to the reactor internals can result. When high-pressure drops are observed throughout any portion of the reactor, the unit is shut down and the catalyst bed is skimmed and refilled.

With fixed-bed reactors a balance must be reached between reaction rate and pressure drop across the catalyst bed. As catalyst particle size is decreased, the desulfurization reaction rate increases, but so does the pressure drop across the catalyst bed. Expanded-bed reactors do not have this limitation, and small $\frac{1}{32}$ in. extrudate catalysts or fine catalysts may be used without increasing the pressure drop.

Upflow Expanded-Bed Reactor Design

Expanded-bed reactors are applicable to distillates, but are commercially used for very heavy, high metals, and/or dirty feedstocks having extraneous fine solids material. They operate in such a way that the catalyst is in an expanded state so that the extraneous solids pass through the catalyst bed without plugging. They are isothermal, which conveniently handles the high-temperature exotherms associated with high hydrogen consumptions. Since the catalyst is in an expanded state of motion, it is possible to treat the catalyst as a fluid and to withdraw and add catalyst during operation.

Expanded beds of catalyst (Fig. 4) are referred to as particulate fluidized. This means that the petroleum feedstock and hydrogen flow upward through an expanded bed of catalyst with each catalyst particle in independent motion. The catalyst migrates throughout the entire reactor bed. Expanded-bed reactors are mathematically modeled as backmix reactors with the entire catalyst bed at one uniform temperature. Spent catalyst may be withdrawn and replaced with fresh catalyst on a daily basis. Daily catalyst addition and withdrawal eliminate the need for costly shutdowns to change out catalyst and also result in a constant equilibrium catalyst activity and product quality. The catalyst is withdrawn daily with a vanadium, nickel, and carbon content which is representative on a macroscale of what is found throughout the entire reactor. On a microscale, individual catalyst particles have ages from that of fresh catalyst to as old as the initial catalyst charge to the unit, but the catalyst particles of each age group are so well dispersed in the reactor that the reactor contents appear uniform.

Referring again to the expanded-bed reactor in Fig. 4, petroleum feedstock and recycle gas enter the bottom of the reactor, pass up through the expanded catalyst bed, and leave from the top of the reactor. Commercial expanded-bed reactors normally operate with $\frac{1}{32}$ in. extrudate catalysts which provide a higher rate of desulfurization than the larger catalyst particles used in fixed-bed reactors. With extrudate catalysts of this size, the upward liquid velocity based on fresh feedstock is not sufficient to keep the catalyst particles in an expanded state. Therefore, for each part of the fresh feed, several parts of product oil are taken from the top of the reactor, recycled internally through a large vertical pipe to the bottom of the reactor, and pumped back up through the expanded catalyst bed. The amount of catalyst bed expansion is controlled by the recycle of product oil back up through the catalyst bed.

The expansion and turbulence of gas and oil passing upward through the expanded catalyst bed are sufficient to cause almost complete random motion in the bed (particulate fluidized). This effect produces the isothermal operation.

It also causes almost complete backmixing. Consequently, in order to effect near complete sulfur removal (over 75%), it is necessary to operate with two or more reactors in series. The ability to operate at a single temperature throughout the reactor or reactors, and to operate at a selected optimum temperature rather than an increasing temperature from the start to the end of the run, results in more effective use of the reactor and catalyst contents. When all these factors are put together, i.e., use of a smaller catalyst particle size, isothermal, fixed temperature throughout run, backmixing, daily catalyst addition, and constant product quality, the reactor size required for an expanded bed is often smaller than that required for a fixed bed to achieve the same product goals. This is generally true when the feeds have high initial boiling points and/or the hydrogen consumption is very high.

Expanded Bed Reactor Design with Fine Catalyst

It was previously mentioned that $1/32$ in. extrudate catalyst could not be completely expanded by the velocity of fresh feedstock. The catalyst bed was expanded by recycling liquid product oil to create a high enough liquid velocity to expand the catalyst bed. By using fine catalyst in the 50 to 200-μm size, it is possible to operate an expanded bed without recycle of liquid product oil. The velocity of gas and fresh oil feedstock is sufficient to expand the bed of fine catalyst.

In addition to eliminating the cost of an expanded bed pump, the fine catalyst system has several other advantages over extrudate catalyst. The average diffusional path in a catalyst particle is shorter. The relative number of pores on the external surface of the fine catalyst is larger. Therefore, the fine catalyst is less subject to vanadium pore-mouth blockage of external pores. Since the fine catalyst has a shorter diffusional length and less chance of pore blockage, the sulfur removal is greater for fine catalyst for a given set of process conditions, i.e., pressure, temperature, reactor volume, and catalyst usage. Offsetting these demonstrated advantages are restrictions on reactor size, configuration, and turndown of the fresh feed rate. Work is underway to develop an intermediate size range between fine catalyst and $1/32$ in. extrudate.

Separate Demetallization Reactors

One method of controlling demetallization is to employ separate smaller "guard" reactors just ahead of the fixed-bed hydrodesulfurization reactor section. The preheated feed and hydrogen pass through the guard reactors which are filled with an appropriate catalyst for demetallization which is often the same as the catalyst used in the hydrodesulfurization section. The advantage of this system is that it enables the refiner to replace the most

Demetallization/Desulfurization of Petroleum Feedstocks

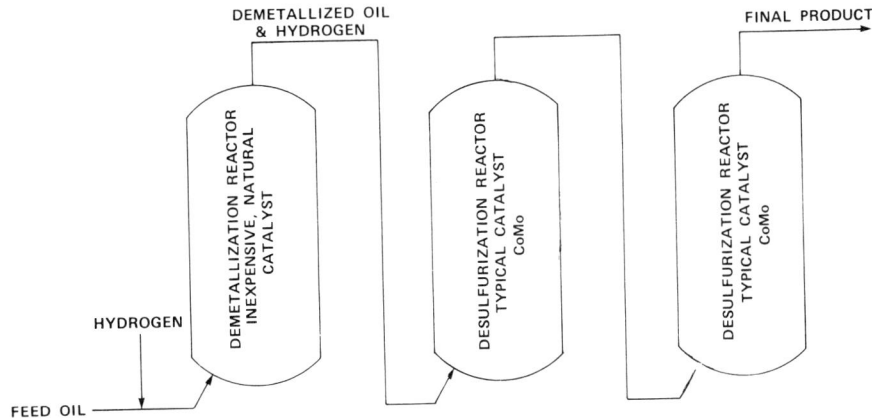

FIG. 5. Reactor configuration for demetallization/desulfurization.

contaminated catalyst, where pressure drop is highest, without having to replace the entire inventory or bring down the unit. The feedstock is alternated between guard reactors while catalyst in the idle guard reactor is being replaced.

When the expanded-bed design is used, the first reactor could employ a low-cost catalyst (5% of the cost of Co/Mo catalyst) to remove the metals, and subsequent reactors can use the more selective hydrodesulfurization catalyst (Fig. 5). The demetallization catalyst can be added continuously without taking the reactor out of service. Spent demetallization catalyst can be loaded to more than 30% vanadium, which makes it a valuable source of vanadium ore.

Hydrodesulfurization Catalysts

Hydrodesulfurization catalysts consist of metals impregnated on a porous alumina support. Almost all of the surface area is found in the pores of the alumina (200 to 300 m^2/g). The metals are dispersed in a thin layer over the entire alumina surface within the pores. This type of catalyst does display a huge catalytic surface for a small weight of catalyst.

Cobalt (Co), molybdenum (Mo), and nickel (Ni) are the most commonly used metals for desulfurization catalysts. Co/Mo and Ni/Mo catalysts resist poisoning and are the most universally applied catalysts for hydrodesulfurization of everything from naphthas to residuums.

Co/Mo and Ni/Mo catalysts promote both demetallization and desulfurization. The vanadium deposition rate at a given desulfurization level is a function of the pore structure of the alumina support and the types of metals on the support. A catalyst support having large pores preferentially demetallizes with a low degree of desulfurization (Fig. 6). A catalyst support having small pores preferentially desulfurizes with a low degree of demetallization. This relationship for vanadium removal is also observed for nickel removal (Fig. 7).

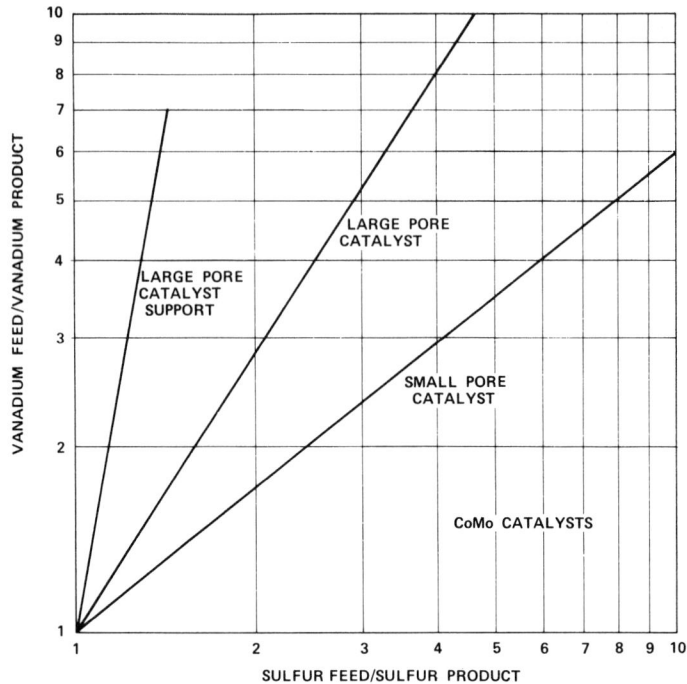

FIG. 6. Relationship between vanadium and sulfur removal with different catalysts.

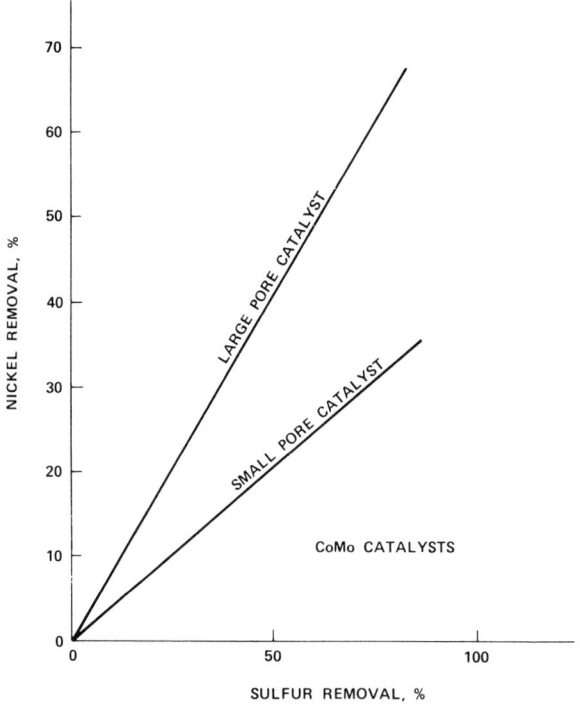

FIG. 7. Relationship between nickel and sulfur removal.

Demetallization/Desulfurization of Petroleum Feedstocks

Processing Vacuum Residuals in Expanded-Bed Reactors

Table 4 presents results from commercial and pilot plant units for processing vacuum residuals with varying metal contents. The commercial data are from steady-state performance with continuous Co/Mo catalyst addition. These data do not represent any upper limit on desulfurization, but were selected to illustrate the effect of feed metals content on process conditions and Co/Mo catalyst addition rates.

Case 1 data are from the Lake Charles commercial LC-Fining Unit. The unit has two expanded-bed reactors in series. The feedstock is a West Texas atmospheric residuum. This feedstock has a low metals content of 23 wppm

TABLE 4 Desulfurization of Residuums with Co/Mo Catalysts

	Case 1				Case 2				Case 3		
Data source	Commercial				Commercial				Pilot plant		
Location	Lake Charles, Louisiana				Salamanca, Mexico				—		
Crude type	West Texas				Mexican blend				Gach Saran		
Feed rate, bbl/std. d	2,100				16,300				—		
Number of reactors in series	Two				One				Two		
Feedstock data:											
Gravity, °API	17.1				10.7				9.9		
Sulfur, wt.%	2.52				3.22				3.24		
V/Ni, ppm	23/24				147/35				249/80		
975°F+, vol.%	45.0				74.5				77.3		
Type Co/Mo catalyst	Small pores				Large pores				Large pores		
Process conditions:											
Relative space velocity	3.0				1.6				1.0		
Product Yields, Qualities	Yield	Sulfur (wt.%)	V (ppm)		Yield	Sulfur (wt.%)	V (ppm)		Yield	Sulfur (wt.%)	V (ppm)
H_2S, NH_3, wt.%	2.3				2.6				3.0		
C_1–C_3, wt.%	1.0				2.2				1.7		
C_4–400°F, vol.%	7.7	<0.1			13.1	<0.1			9.7	<0.1	
400–650°F, vol.%	13.3	0.1			23.2	0.2			21.4	0.1	
650–975°F, vol.%	57.1	0.2			29.0	0.5			32.7	0.4	
975°F+, vol.%	25.5	1.2	14		39.8	1.9	186		41.0	1.4	68
C_4+	103.6	0.4	3		105.1	1.0	78		104.8	0.7	29
Chemical H_2 consumption, std.ft³/bbl	680				940				880		
Desulfurization, %	84				69				77		
Vanadium removal, %	87				49				89		
Catalyst addition, lb/bbl	0.05				0.13				0.31		

vanadium and 24 wppm nickel. The desulfurization level is 84% and the demetallization as measured by vanadium removal is 87%. The space velocity is given a relative value of 3.0 compared to a relative space velocity of 1.0 for Case 3. The relative space velocities of Cases 1 to 8 are all calculated relative to Case 3. The Co/Mo catalyst addition rate required to maintain constant product qualities and yields is 0.05 lb of catalyst/bbl of feedstock.

Case 2 data are from the Pemex commercial LC-Fining Unit in Salamanca, Mexico. The feedstock is a mixture of vacuum bottoms from Mexican crudes blended with gas oil. This unit also has a single-stage reactor and was designed for conversion rather than desulfurization. The feed contains 147 ppm vanadium. The desulfurization rate is 69% and the demetallization rate is 49%. The catalyst addition rate to maintain constant catalyst activity is 0.13 lb of catalyst/bbl of feedstock.

Case 3 data are from a pilot plant run on Gach Saran vacuum bottoms. This vacuum bottoms was cut from commercially available Iranian heavy crude. Although it is referred to as Gach Saran, it is reported to contain a small amount of lighter crude. The feedstock is a blend of 80 vol.% vacuum bottoms and 20 vol.% heavy gas oil and contains 249 ppm vanadium.

The pilot plant run was carried out in a two-stage expanded-bed unit with continuous catalyst replacement. The desulfurization rate was 81%. However, the data were adjusted to 77% desulfurization to allow a better comparison with the other cases. The adjusted demetallization rate is 89%. The catalyst replacement rate is 0.31 lb of catalyst/bbl of feedstock.

By comparison of the three cases it is possible to identify a general relationship between the cost of desulfurization and the vanadium content of the feedstock. As vanadium content increases, the cost of desulfurization increases. The higher cost is for increased catalyst replacement rates and for larger reactor volumes required to obtain a lower space velocity.

Processing residual oils with vanadium contents below 200 to 300 wppm can be handled economically with Co/Mo catalysts. With vanadium contents above this level, the economics favor a separate demetallization reactor followed by one or more desulfurization reactors of Co/Mo catalyst.

Demetallization/Desulfurization in Expanded-Bed Reactors

The stocks that have relatively high metals contents, 300 wppm and higher, substantially increase catalyst consumption. The actual catalyst consumption is high because the metals poison the catalyst, and the catalyst must be replaced often. The usual desulfurization catalysts are relatively expensive for these consumption rates. There are proprietary catalysts which are very inexpensive and can be used in the first reactor to remove a large percentage of the metals. Subsequent reactors downstream of the first reactor would use normal hydrodesulfurization catalysts. Since the catalyst materials are proprietary, it is not possible to identify them here. However, it is understood that such catalysts contain little or no metal promoters, i.e., nickel, cobalt, molybdenum. Metals removal on the order of 90% has been observed with these materials.

Our studies have shown that the most economic method of desulfurizing very high metal content oils is to demetallize the oil first in an expanded-bed reactor using an inexpensive, natural catalyst and then to pass the effluent directly through two expanded-bed reactors in series with Co/Mo catalyst.

The demetallization catalyst is a commercially available material. At steady-state conditions it has a slightly better activity for vanadium removal than Co/Mo catalyst. Its main advantage, however, is that its cost is only 5% of the cost for Co/Mo catalyst. It has been used successfully in expanded-bed pilot plant units.

As with most other properties of residual oil, there is a large variation in the rate of reaction of the vanadium compounds from one crude to another. Demetallization data have been collected on a considerable number of crudes. Results are presented for a medium metals content Venezuelan residuum which exhibits a low demetallization rate and a high metals content Venezuelan crude oil which exhibits a high demetallization rate.

Table 5 shows the advantages of the demetallization/desulfurization scheme on a medium metals content atmospheric residuum which contains about 375 to 400 ppm vanadium. Considerable work has been done with this feedstock in the pilot plants. Case 4 is a desulfurization run with two reactors of Co/Mo catalyst in series. The run was made in the pilot plants with continuous catalyst replacement.

Operating conditions have been adjusted to 77% desulfurization to allow comparison with Cases 1 to 3. The demetallization rate is 56%. The catalyst addition rate is high at 0.42 lb of catalyst/bbl of feedstock.

Case 5 is an equivalent demetallization/desulfurization scheme. The data are from two consecutive runs made in the pilot plant. The feed was first demetallized using an inexpensive catalyst to an optimum vanadium level which had been established in previous screening work. The demetallized oil was then processed over Co/Mo catalyst. The required space velocity for 77% desulfurization is similar to Case 3. The demetallization rate is 79%. The use of a first-stage demetallization catalyst has substantially reduced the Co/Mo catalyst consumption rate to 0.13 of catalyst/bbl of feedstock.

Table 6 shows the advantage of the demetallization/desulfurization scheme with Boscan crude. Because Boscan is such a heavy crude, most of our work has been done on desulfurization of whole crude. The whole crude contains 1100 ppm vanadium, and 56 vol.% boils above 975°F. The feedstock is similar in boiling range to a normal atmospheric residuum.

Case 6 shows desulfurization with Co/Mo catalyst. This case was estimated from a pilot plant run. For 77% desulfurization and 72% demetallization, the catalyst replacement cost for constant catalyst activity is 0.62 lb of catalyst/bbl and the required space velocity is 1.6 relative to Case 3.

Case 7 shows demetallization/desulfurization of Boscan crude. This case was estimated from two separate pilot plant runs. The crude was first demetallized to the proper levels with an inexpensive catalyst and then desulfurized with Co/Mo catalyst. The space velocity for 77% desulfurization is 1.5 relative to Case 3. The overall demetallization rate is 91%. The Co/Mo

TABLE 5 Demetallization and Desulfurization of Venezuelan Medium Metals Atmospheric Residuum

	Case 4			Case 5		
Data source	Pilot plant			Pilot plant		
Type of operation	Desulfurization			Demetallization/desulfurization		
Crude type	Laguna			Laguna		
Number of reactors in series	Two			Three		
Feedstock data:						
Gravity, °API	11.8			12.6		
Sulfur, wt.%	2.8			2.8		
V/Ni, ppm	375/55			398/57		
975°F+, vol.%	55			55		
Type Co/Mo catalyst	Small pores			Small pores		
Process conditions:						
Relative space velocity	1.5			0.9		
Product Yields, Qualities	Yield	Sulfur (wt.%)	V (ppm)	Yield	Sulfur (wt.%)	V (ppm)
H_2S, NH_3, wt.%	2.4			2.4		
C_1–C_3, wt.%	2.4			1.9		
C_4–400°F, vol.%	13.0	0.1		11.4	<0.1	
400–650°F, vol.%	31.1	0.2		29.5	<0.1	
650–975°F, vol.%	32.6	0.4		34.1	0.3	
975°F+, vol.%	26.9	1.6	590	28.3	1.8	290
C_4+	103.6	0.6	170	103.3	0.6	88
Chemical H_2 consumption, std.ft³/bbl	720			680		
Desulfurization, %[a]	77			77		
Vanadium removal, %	56			79		
Catalyst addition, lb/bbl	0.42			0.13		

[a]Desulfurization for two Co/Mo reactors in series.

catalyst addition rate is substantially reduced to 0.11 lb of catalyst/bbl of feedstock.

Case 8 is included to show that 90% desulfurization is feasible with demetallized Boscan crude. The required space velocity is not excessively low and is in the same range as commercial hydrodesulfurization plants.

Conclusions

The world's supply of low sulfur, low metals crude oil is being depleted. Substantial reserves of heavy, high metals crude oil have been discovered and

TABLE 6 Demetallization and Desulfurization of Venezuelan High Metals Whole Crude

	Case 6			Case 7			Case 8		
Data source	Pilot plant			Pilot plant			Pilot plant		
Type operation	Desulfurization			Demetallization/desulfurization			Demetallization/desulfurization		
Crude type	Boscan			Boscan			Boscan		
Number of reactors in series	Two			Three			Three		
Feedstock data									
Gravity, °API	10.4			10.4			10.4		
Sulfur, wt.%	5.6			5.6			5.6		
V/Ni, ppm	1100/85			1100/85			1100/85		
975°F+, vol.%	56			56			56		
Type Co/Mo catalyst	Large pores			Small pores			Small pores		
Process conditions									
Relative space velocity	1.6			1.5			0.9		
Product Yields, Qualities	Yield	Sulfur (wt.%)	V (ppm)	Yield	Sulfur (wt.%)	V (ppm)	Yield	Sulfur (wt.%)	V (ppm)
H_2S, NH_3, wt.%	4.9			4.9			5.3		
C_1–C_3, wt.%	3.4			2.7			3.9		
C_4–400°F, vol.%	15.1	0.2		12.5	<0.1		16.4	<0.1	
400–650°F, vol.%	27.5	0.5		24.0	<0.1		25.1	<0.1	
650–975°F, vol.%	37.8	1.1		38.9	0.6		41.9	0.3	
975°F+, vol.%	24.6	2.8	1270	29.0	3.3	360	22.4	1.8	290
C_4+	105.0	1.3	333	104.4	1.3	110	105.8	0.6	70
Chemical H_2 consumption, std.ft³/bbl	1140			1030			1320		
Desulfurization, %[a]	77			77			90		
Vanadium removal, %	72			91			94		
Catalyst addition, lb/bbl	0.62			0.11			0.18		

[a]Desulfurization based on two Co/Mo reactors in series.

are awaiting production. Future refineries will have to process higher sulfur, higher metal content feedstocks. Technology does exist for demetallizing high metal content residual oils prior to hydrodesulfurization. Deasphalting and coking are means to remove the metals; however, these processes reduce the volume of oil produced. Demetallization of residua combined with hydrodesulfurization allows removal of metal compounds with an increase in the volume of oil produced.

Residuum hydrotreating for metals removal can be accomplished through the use of a variety of reactors to include either downflow fixed-bed or expanded-bed designs. Of these reactors, the expanded bed offers the advantages of catalyst addition and removal at constant operating conditions, isothermal operation, and elimination of bed pressure drop.

Two challenges exist for further improvement in residuum demetallization/desulfurization technology: (1) development of a catalyst or catalyst treating agent which would reduce the amount of surface poisoning by organometallic compounds would be of value, and (2) an inexpensive method of regenerating poisoned catalyst by removing deposited metal sulfides could substantially reduce net catalyst cost.

Bibliography

Beuther, H., and Schmid, B. K., "Reaction Mechanisms and Rates in Residue Hydrodesulfurization," in *Sixth World Petroleum Congress Proceedings*, Section III, 1963, Paper 20, pp. 297–310.

Celestinos, J. A., Zermeno, R. G., Van Driesen, R. P., and Wysocki, E. D., "Ebullated Bed Hydrocracker Gives High Yields from Vacuum Resids," *Oil Gas J.*, *73*(48), 127–134 (December 1, 1975).

Considine, D. M., *Energy Technology Handbook*, McGraw-Hill, New York, 1977, Chap. 3, pp. 251–264.

Hastings, K. E., James, L. C., and Mounce, W. R., "Demetallization Cuts Desulfurization Costs," *Oil Gas J.*, *73*(26), 122–130 (June 30, 1975).

Hiemenz, W., in *Sixth World Petroleum Congress Proceedings*, Section III, 1963, Paper 20, p. 307.

Nelson, W. L., "Questions on Technology," *Oil Gas J.*, *64*(11), 128 (March 14, 1966).

Schuit, G. C., and Gates, B. C., "Chemistry and Engineering of Catalytic Hydrodesulfurization," *AIChE J.*, *19*(3), 417–438 (1973).

RICHARD A. BAUSELL
JOHN CASPERS
KENNETH E. HASTINGS
JOHN D. POTTS
ROGER P. VAN DRIESEN

Desulfurization, Liquids, Petroleum Fractions

Desulfurization—General

Definition

Desulfurization is the removal of sulfur or sulfur compounds from hydrocarbons or hydrocarbon mixtures derived from petroleum or natural gas. Desulfurization processes can also be used to treat hydrocarbons from other sources such as oil shale or coal.

Three general techniques are used for sulfur removal [1].

1. Catalytic hydrodesulfurization involves the reaction of sulfur compounds with hydrogen in the presence of a catalyst, resulting in their decomposition into hydrogen sulfide and hydrocarbon remnants of the original sulfur compound. This is the most broadly applicable process for desulfurization and is the principal topic of this article.
2. Solvent extraction removes entire sulfur-bearing molecules from hydrocarbon streams. Absorption is widely employed for removing acid gases (H_2S and CO_2) from natural gas and light petroleum gases. Popular solvents for this purpose are aqueous or alcohol solutions of alkanolamines, and aqueous solutions of alkali carbonates.

    Caustic treating, usually with aqueous sodium hydroxide, is widely practiced to remove mercaptans and organic acids from gasoline and occasionally from light distillates. Some processes combine caustic treating with air oxidation of mercaptans to convert them to disulfides. Such "sweetening" processes eliminate the bad odor and corrosiveness associated with mercaptans without necessarily removing sulfur from the hydrocarbon stream being treated.

TABLE 1 Examples of Sulfur Compounds in Petroleum

H_2S	S_8	R—SH
Hydrogen sulfide	Elemental sulfur	Mercaptans
R—S—R^1	R—S—S—R^1	Cyclic sulfides
Sulfides	Disulfides	(tetrahydrothiophene)
Benzothiophenes	Dibenzothiophenes	

Extraction is not economically feasible for removing sulfur from heavier fractions of petroleum because of high yield losses. However, during extraction of heavy gas oils with furfural or phenol to manufacture lubricating oil-base stocks, sulfur compounds are concentrated in the extracts. This is observed since sulfur is predominantly contained in the aromatic structures which are preferentially removed as extract.

3. Adsorption on molecular sieves is widely used to remove sulfur compounds from natural gas and from light hydrocarbon streams such as propane, butane, pentane, and even light gasolines. This process is effective for removing hydrogen sulfide and light mercaptans, sulfides, and disulfides. It simultaneously dries the hydrocarbon and will also remove carbon dioxide from natural gas streams.

Sulfur in Petroleum

Crude petroleum contains a variety of nonhydrocarbons, principally organic compounds of sulfur, nitrogen, and oxygen, but also organometallic compounds containing nickel and vanadium and occasionally iron and arsenic. Of these, sulfur compounds are generally the most prevalent and of most concern in refining.

The sulfur contents of crude oils range from virtually nil to as high as 7 or 8 wt.% in extreme cases, depending upon the source of the crude [2]. In moderately high sulfur crudes containing 2 to 3% sulfur, organic sulfur compounds could theoretically represent as many as half of the total molecules present. Higher boiling fractions of such crudes may contain, on the average, two or more sulfur atoms per molecule. This explains why separation of sulfur compounds from petroleum by physical means like extraction is not practical for higher boiling fractions.

Not only do the sulfur contents of crude oils vary, but the relative amounts of various classes of sulfur compounds (Table 1) vary from crude to crude. Some generalizations, however, can be made.

Both hydrogen sulfide and elemental sulfur are found dissolved in crude oils. These materials can be responsible for corrosion problems in processing of the crude. Most of the hydrogen sulfide is normally flashed off with the light hydrocarbon gases (methane, ethane, propane) in gas separation units at the oil field. Elemental sulfur, when present in the crude oil as produced, is low in concentration and because of its high boiling point remains in the heavier fractions upon distillation. Another source of elemental sulfur is the oxidation of hydrogen sulfide in liquefied petroleum gases (propane, butane) and in natural and straight-run gasolines subsequent to production. The elemental sulfur thus formed in these low-boiling fractions can react with mercaptans to produce hydrogen sulfide which creates the corrosion problem.

The organic sulfur in petroleum fractions ranges from very simple identifiable compounds in the lightest fractions to much more complicated compounds, which are very difficult to identify, in heavier fractions. In light straight-run fractions, sulfur is typically present as aliphatic compounds. As the boiling range increases, the sulfur compounds tend to be more cyclic, and in the heaviest fractions the sulfur is almost exclusively present in complex ring structures.

In gasolines the principal sulfur compounds include mercaptans, aliphatic sulfides, aliphatic disulfides, and five- and six-membered ring cyclic sulfides. Thiophenes are found in gasoline made by cracking heavier fractions, but are not usually present in straight-run gasolines.

Few mercaptans occur above the gasoline boiling range. The predominant sulfur structures in kerosines and gas oils are sulfides, cyclic sulfides, benzothiophenes, and dibenzothiophenes.

In petroleum residues [3], sulfur structures similar to those in gas oils account for most of the sulfur present. However, the most difficult sulfur in residues to remove by hydrodesulfurization is associated with asphaltenes. Asphaltenes (defined as the n-pentane insoluble components of petroleum) typically have molecular weights averaging about 10,000. They are generally high in nitrogen content as well as sulfur and contain much of the trace metals found in crude oil, particularly nickel and vanadium. They may be viewed as polymers formed from simpler organic sulfur, oxygen, and nitrogen compounds found in petroleum.

Purposes of Desulfurization

In light petroleum products, such as liquefied petroleum gases, naphtha, gasoline, and even kerosine, sulfur compounds are objectionable because of their potential to cause corrosion, the bad odor of light sulfur compounds, and the toxic character of hydrogen sulfide [4]. The presence of sulfur compounds in gasoline tends to lower the octane number and also decreases the susceptibility of gasolines to octane improvement by addition of tetraethyl or tetramethyl lead. Furthermore, sulfur poisons noble metal catalysts. In catalytic reforming of naphthas and in isomerization of C_4 to C_6 paraffins, it is necessary to pretreat the feedstocks to sulfur contents of a few parts per million or lower to avoid catalyst deactivation.

As indicated above, amine or caustic treating or molecular sieve adsorption is generally used to remove sulfur compounds from light hydrocarbon streams. Caustic treating or sweetening processes are also useful for naphtha and kerosines. However, the very low sulfur levels required in feedstocks for catalytic reforming can only be achieved by hydrodesulfurization.

Sulfur compounds are not a particularly serious problem in most middle distillate fuels. The exceptions are middle distillates from some crudes which contain mercaptans and are potentially corrosive unless sweetened or hydrodesulfurized. Nevertheless, desulfurization may be required with middle distillate fuels from high sulfur crudes in order to meet product specifications set for either environmental or competitive reasons.

Most hydrotreating of distillate fuels, however, is not done solely for desulfurization. The fuels are frequently mixtures of virgin and catalytically or thermally cracked distillates containing olefins and diolefins, nitrogen and oxygen compounds, and acidic and basic materials in addition to sulfur. Hydrotreating such fuels improves storage stability, color, odor, corrosion properties, and burning properties. High quality kerosines, jet fuels, diesel fuels, and home heating oils can be prepared this way.

Desulfurization of heavier oils, such as atmospheric or vacuum gas oils and deasphalted oil, is widely practiced in order to meet environmental restrictions. Often the desulfurized heavy oils are blended back with undesulfurized residue, thus indirectly preparing a moderate sulfur level residual fuel.

A fairly recent application, which is being practiced by more and more refiners, is desulfurization of gas oil which is subsequently fed to fluid catalytic cracking units. This allows gas oils from high sulfur crudes to be catalytically cracked without excessive sulfur oxide emissions in the regenerator flue gas. There are additional advantages: higher conversions, greater yields of gasoline, reduced coke make, and reduced cracking catalyst consumption. Most of the metals in the catalytic cracking feedstock are removed during desulfurization, eliminating a principal source of cracking catalyst deactivation.

Direct desulfurization of residues, to make lower sulfur residual fuels than can be prepared by indirect desulfurization, is a relatively recent development in hydrodesulfurization technology [5]. This process is in fairly widespread use in Japan to make low sulfur residual fuels from high sulfur Middle East crudes.

A proposed use of residual desulfurization is to prepare a feedstock suitable for catalytic cracking [6]. This is an extension of the gas oil operation described above. Both metals and sulfur removal would be important in such an application; thus the combination of residual desulfurization and catalytic cracking is a possible route to complete conversion of high sulfur crude oils to clean distillate products.

Hydrodesulfurization Catalysts

The most commonly used hydrodesulfurization catalysts consist of cobalt oxide and molybdenum oxide (CoMo), or nickel oxide and molybdenum oxide (NiMo), dispersed on high surface area alumina supports [7]. Molybdenum is generally regarded as the active desulfurization component, with cobalt or nickel acting as a promoter which increases catalytic activity.

The catalysts are manufactured with the metals in an oxide state. In the active form they are in the sulfide state, which is obtained by sulfiding the catalyst either prior to use or with the feed during actual use. Any catalyst which exhibits hydrogenation activity will catalyze hydrodesulfurization to some extent. However, the Group VIB metals (chromium, molybdenum, and tungsten) are particularly active for desulfurization, especially when promoted with metals from the iron group (iron, cobalt, nickel).

CoMo catalysts are by far the most popular choice for desulfurization, particularly for straight-run petroleum fractions. NiMo is often chosen instead of CoMo when higher activity for polyaromatics saturation or nitrogen removal is required, or when more refractory sulfur compounds such as those in cracked stocks must be desulfurized. In some applications, at least, nickel–cobalt–molybdenum (NiCoMo) catalysts appear to offer a useful balance of hydrotreating activity [8]. Nickel–tungsten (NiW) is usually chosen only when very high activity for aromatics saturation is required along with

TABLE 2 Typical Catalyst Metals Compositions

	CoMo	NiMo	NiCoMo	NiW
Cobalt, wt.%	2.5	—	1.5	—
Nickel, wt.%	—	2.5	2.3	4.0
Molybdenum, wt.%	10.0	10.0	11.0	—
Tungsten, wt.%	—	—	—	16.0

activity for sulfur and nitrogen removal. Table 2 shows some typical metals compositions for commercially available catalysts.

Naphtha Desulfurization

The relative importance of the various desulfurization methods has shifted significantly over the years. From the beginning of the petroleum industry up through the early 1950s, caustic, amine, clay, and acid treating were the main methods of desulfurization for light hydrocarbon streams. However, the advent of catalytic reforming in the 1940s and 1950s made hydrogen cheaper and more available. Since that time, catalytic hydrotreating has continually increased in importance until today it is by far the most widely used desulfurization technique in the petroleum industry. The methods of desulfurization are now discussed in more detail.

Catalytic Hydrotreating

Catalytic hydrotreating [9] (sometimes called catalytic hydrodesulfurization) of naphtha is widely applied to prepare charge for catalytic reforming and isomerization processes. It is generally accomplished by passing a feedstock together with hydrogen over a fixed catalyst bed at elevated temperature and pressure. The main purpose is to remove sulfur, but denitrogenation, deoxygenation, and olefin saturation reactions occur simultaneously with desulfurization. These reactions are also beneficial since the noble metal catalysts used in reforming and isomerization can be poisoned by olefins, oxygen, and nitrogen compounds as well as by sulfur in the feedstocks. Sulfur and nitrogen limitations can in some cases be 0.5 ppm or less. Failure to adequately remove these contaminants can lead to poor yields, low catalyst activity, and brief catalyst life.

A typical flow diagram of a hydrotreating unit is shown in Fig. 1 [10]. A hydrogen-rich gas (usually above 70% H_2) joins the feed. This mixture is then heated and, in almost all cases, totally vaporized. The feed–hydrogen mixture then passes over a fixed catalyst bed at conditions which depend on the feedstock properties and the desired product specifications. In the most common light hydrocarbon application of catalytic hydrotreating (naphtha pretreating), the reactor effluent is cooled and sent to a separator where most of the hydrogen, methane, ethane, and some of the hydrogen sulfide and ammonia are flashed off as gas. This gas can be scrubbed and either recycled or used for other purposes. The liquid phase from the separator is sent to a stripper (a pressured distillation column) in order to remove any residual H_2S or ammonia that may be dissolved in the liquid. The stripper is also used to separate any C_3 and C_4 material from the pretreated naphtha product. In the pretreatment of naphthas for catalytic reforming (or other processes which employ noble metal catalysts), the stripper operation is important since residual dissolved H_2S in the naphtha could easily cause sulfur limitations to be exceeded.

Typical yields and product specifications are shown in Table 3 for the catalytic hydrotreatment of a virgin naphtha for a catalytic reforming operation. Such data for the pretreatment of cracked naphthas or gasolines would be similar with one exception. Hydrogen consumption for cracked stocks would be much higher and would depend heavily on the olefin content of the charge.

Chemistry of Catalytic Hydrotreating

As noted above, catalytic hydrotreating can accomplish more than just desulfurization. Denitrogenation and olefin saturation can occur as well as the removal of oxygen in certain applications. But what are these reactions and what are their relative rates?

The different types of sulfur compounds found in light hydrocarbon streams (boiling below 400°F) and the basic desulfurization reaction for each type are listed in Table 4 [7, 11, 12]. These reaction equations illustrate that de-

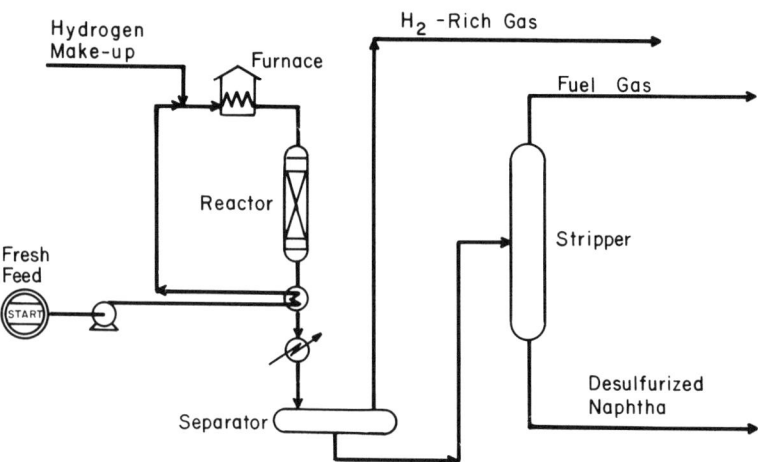

FIG. 1. Typical naphtha hydrotreater.

TABLE 3 Typical Yields and Product Specifications for Naphtha Hydrotreating

Feed: Typical virgin naphtha (sulfur content 100–700 ppm
nitrogen content ~0.3–2 ppm)

Product stream to be used for catalytic reforming

Operating conditions: 7–15 LHSV; 200–400 lb/in.² gauge total pressure
400 SCF/bbl gas rate (85% H_2 purity)
450–600°F average catalyst bed temperature

Yields:	
H_2S, wt.%	0.01–0.07
C_1–C_2, wt.%	0.2
C_3–C_4, vol.%	0.1–0.3
C_5+, vol.%	99.5–100.0
Hydrogen consumption, SCF/bbl	10–20
Product inspections:	
Sulfur, ppm	<1
Nitrogen, ppm	<0.5
Olefins, vol.%	<0.5
Oxygen, ppm	<5
API	Same as feed
Distillation	
Hydrocarbon analysis	Essentially same as feed

sulfurization is accomplished by the cleavage of carbon—sulfur bonds. These bonds are replaced by C—H and S—H bonds.

In a similar manner, nitrogen and oxygen are removed in catalytic hydrotreating by the cleavage of C—N and C—O bonds. The basic reactions are:

TABLE 4

Type of Sulfur Compound	Desulfurization Reaction
1. H_2S	1. None—usually separated by stripping
2. Mercaptans (R—SH)	2. R—SH + H_2 → R—H + H_2S
3. Sulfides (R—S—R′)	3. R—S—R′ + $2H_2$ → R—H + R′—H + H_2S
4. Disulfides (R—S—S—R′)	4. R—S—S—R′ + $3H_2$ → R—H + R′—H + $2H_2S$
5. Cyclic sulfides (thiophene)	5. (thiophene) + $2H_2$ → C_5H_{12} + H_2S
6. Benzothiophene	6. (benzothiophene) + $3H_2$ → (ethylbenzene) CH$_2$—CH$_3$ + H_2S

Nitrogen-containing compound + $H_2 \rightarrow$ nitrogen-free compound + NH_3

Oxygen-containing compound + $H_2 \rightarrow$ oxygen-free compound + H_2O

These C—N and C—O bond cleavages are much more difficult to achieve than the C—S bond cleavage. Consequently, denitrogenation and deoxygenation occur to a much lesser extent than does desulfurization. Fortunately, nitrogen and oxygen are not usually significant problems in virgin naphthas, although in cracked stocks or in synthetic naphthas (such as those from coal or shale) concentrations of nitrogen and oxygen can be quite high (see Table 5). In such cases, very severe hydrotreating conditions are required to reduce the concentrations to levels acceptable for reforming or isomerization (<1 ppm).

Finally, olefin saturation is effected simply by the addition of hydrogen to an unsaturated hydrocarbon molecule (a molecule with one or more carbon—carbon double bonds) to produce a saturated product. For example, R—CH = CH—R' is saturated to form R—CH_2—CH_2—R'. The olefin saturation reaction is very exothermic and proceeds relatively easily. Straight-chain monoolefins are easy to hydrogenate while branched cyclic diolefins are somewhat more difficult. Aromatic rings are not hydrogenated except under the most severe hydrotreating conditions. As with nitrogen and oxygen compounds, olefins are not found to any great extent in virgin naphthas, but concentrations can get as high as 40 or 50 vol.% in cracked or synthetic stocks.

More detailed discussions of the reaction mechanisms and the catalysis of

TABLE 5 Ranges of Conditions Used for Various Types of Light Oil Hydrotreating

Case	I	II	III
General feedstock type	Virgin	Cracked or blends of virgin and cracked	Synthetic[a] or blends of virgin and synthetic
Severity	Mild	Intermediate	Severe
Feedstock sulfur, ppm	100–700	100–3000	500–8000
Feedstock nitrogen, ppm	<2	2–30	500–6000
Feedstock oxygen, ppm	<10	10–100	100–5000
Feedstock olefins, vol.%	<1.0	5–40	<1[b]–45[c]
Operating conditions:			
Space velocity, LVHSV	7.0–15.0	3.0–6.0	0.8–2.0
Total pressure, lb/in.² gauge	200–400	400–700	800–1500
H_2 partial pressure, lb/in.² gauge	60–150	150–400	500–1000
Hydrogen purity in reactor gas, mol%	70–90	70–90	70–90
Gas rate, SCF/bbl	200–400	400–1000	1000–6000
Average temperature, °F-SOR	450–600	550–680	650–730
Cycle life, months	6–24	6–18	3–12

[a] For example: Naphthas from coal or shale oil.
[b] For naphtha from coal.
[c] For naphtha from shale oil.

hydrotreating have been undertaken by Schuman and Shalit [11] and by Schuit and Gates [12].

Catalysts Used in Hydrotreating

Most hydrotreating catalysts that are used for light hydrocarbon treating consist of various combinations of nickel oxide, cobalt oxide, and molybdenum oxide supported on high surface area alumina. Although the metals are initially in the oxide state, they are activated to the sulfide state by sulfiding the catalyst. This is often done by adding carbon disulfide or some other light, sulfur-containing hydrocarbon to the feed for the first few hours of operation.

Cobalt–molybdenum (CoMo) catalysts are widely used throughout the industry for naphtha hydrotreating although nickel–molybdenum (NiMo) or nickel–cobalt–molybdenum (NiCoMo) catalysts are often chosen when better nitrogen removal activity is desired. In extreme cases where feedstock nitrogen levels are very high, nickel–tungsten (NiW) catalysts are sometimes considered. NiW catalysts have not, however, had wide application in naphtha hydrotreating because of their relatively high cost.

Hydrotreating catalysts are generally either in the form of spheres or cylindrical extrudates, though currently some catalyst manufacturers are reporting advantages for shaped extrudates. Some typical ranges of catalyst properties are shown in Table 6.

Processing Considerations

By far the two most important factors in determining the operating conditions needed for a given hydrotreater are (1) the feedstock properties and (2) the desired product properties. These two factors set the general severity that is required in the hydrotreating operation although other considerations (such as catalyst type, heater limitations, or the amount, pressure, and purity of the hydrogen available) will have a bearing on the actual conditions that are used. The major processing considerations are now discussed.

Feedstock Properties. As the concentrations of sulfur, nitrogen, oxygen,

TABLE 6 Typical Ranges of Hydrotreating Catalyst Properties

Nominal catalyst diameter, in.	1/8 to 1/20
Average length for extrudates, in.	~1/4
Packed catalyst density, lb/ft^3	30 to 50
Surface area, m^2/g	200 to 300
Nickel content, wt.%	0 to 6
Cobalt content, wt.%	0 to 6
Molybdenum content, wt.%	5 to 15

and/or olefins increase, more severe hydrotreating conditions will be needed to bring the concentrations down to any given level. The rate of hydrodesulfurization reactions of single sulfur compounds follows first-order kinetics; however, when present in combinations, the compounds taken as a group exhibit apparent reaction orders of up to 1.6.

There are three very broad categories of light hydrocarbon feedstocks that can be hydrotreated: (1) virgin, (2) cracked (e.g., coker naphtha), and (3) synthetic (e.g., coal or shale liquids). The conditions typically used to hydrotreat virgin stocks are mild, whereas treating cracked feeds (or blends of cracked and virgin feeds) requires more severe conditions. Hydrotreating synthetic stocks represents the most severe type of operation. Table 5 presents typical ranges of conditions for three hydrotreating cases. Case I is mild hydrotreating of a virgin naphtha, Case II represents relatively severe treating of a blend of virgin naphtha and cracked (coker) naphtha, and Case III is the severe pretreating of shale oil or coal naphtha or a blend of these feeds with petroleum naphthas. In these cases it is assumed that the product is to be charged to a reformer. This means that the product sulfur and nitrogen levels must be reduced to below 1 and 0.5 ppm, respectively, and the olefin content must be less than 0.5 vol.%.

Space Velocity. When holding all other conditions constant, increasing space velocity will cause product sulfur (and/or other contaminant concentrations) to increase. This effect can in some cases be offset by increasing either the reactor temperature and/or the hydrogen partial pressure. Table 5 shows typical space velocity ranges for the various types of hydrotreating. In mild operations, space velocities range from 7 to 10 or even as high as 15. In more severe operations, space velocities are usually between about 3 and 6, and in the most severe cases space velocities typically vary between 0.8 and 2.

Hydrogen Partial Pressure. Increasing hydrogen partial pressure increases the rates of desulfurization, denitrogenation, olefin saturation, and deoxygenation, if all other conditions are held constant. Studies [13, 14] have shown that desulfurization reactions are first order in hydrogen partial pressure at least for pressures below 420 lb/in.^{2} gauge. It should be noted that hydrogen partial pressure can be altered by changing (1) total pressure, (2) gas circulation rate, or (3) the hydrogen purity of the circulating gas. Typical ranges of these variables are presented in Table 5.

Temperature. Higher temperatures increase the rates of desulfurization and the other desired reactions which remove contaminants. In the design of hydrotreaters, the initial (start-of-cycle) temperature is set by the design throughput, the operating pressure, and other economic factors. Enough temperature flexibility is built into the unit so that temperatures may be increased throughout the cycle to offset the losses in catalyst activity that occur as the catalyst ages. Table 5 shows that hydrotreater temperatures are normally kept between 500 and 700°F, with a maximum of about 800°F. Above 800°F, hydrocracking reactions and coke deposition become too prominent for economical operation.

Catalyst Aging and Hydrotreater Cycle Life. The major factor in determining naphtha hydrotreater cycle lives and catalyst aging rates is the feedstock character. If the charge stock is high in unstable olefins or organometallic compounds and other heavy contaminants which can deposit on the catalyst, then catalyst aging will be relatively fast and pressure drop problems could force an early unit shutdown. These contaminants are usually only a problem when treating synthetic feeds or high concentrations of cracked stocks. In general, catalyst aging due to coke laydown increases with increasing temperature or with decreasing hydrogen partial pressure. When hydrotreating a virgin naphtha, catalyst aging due to coke laydown generally occurs very slowly and, in fact, does not usually determine cycle life. Most commercial hydrotreaters that are in mild naphtha pretreating service are shut down when it is convenient and often for some purpose other than a needed catalyst regeneration such as general cleaning and maintainance or a refinery turnaround. Catalyst regenerations are done at such times as a matter of expedience. Typical cycle lives for the different types of light oil hydrotreating are presented in Table 5.

Economics of Hydrotreating

Although hydrotreating of naphthas and other light hydrocarbons is not inexpensive, it is much less costly than accepting the consequences of inadequate contaminant removal, especially when the treated material is to be used in a process which employs a noble metal catalyst. Hydrotreating is generally recognized to be the only economically feasible process that is available for such applications. Brief investment and utility requirements for a typical mild hydrogenation unit (see Table 5 for operation conditions) are presented in Table 7 [10]. It should be mentioned, however, that the cost of a hydrotreating unit is not only dependent on the unit size but is also very sensitive to the design operating pressure. Consequently, the cost of a unit to be used for severe hydrotreating service could be considerably higher than that

TABLE 7 Investment and Operating Requirements for a Medium-Sized Mild Naphtha Hydrotreater (application: preparation of reformer feed)

Unit capacity, bbl/d	20,000
Investment, $M (including catalyst and engineering fee, erected cost)[a]	7,400
Utilities (average):	
Fuel, MM Btu/h	27
Steam, M lb/h	11
Power, kW	1,000
Cooling water, gal/min	106

[a]U.S. Gulf Coast, 2nd quarter, 1981, NRC Index = 890.

TABLE 8 Major Current Naphtha Pretreating Processes

Distillate Hydrodesulfurization	Institute Francais Du Petrole
Unibon	Universal Oil Products Co.
Hydrofining	Esso Research and Eng. Co.
Hydrogenation	M. W. Kellogg Co.
Naphfining	Gulf Research and Development Co.
Ultrafining	Standard Oil Co. (Indiana)
Unionfining	Union Oil Co. of California
Vapor-Phase Hydrotreating	Shell Internationale Research Maatschappij N.V. or Shell Development Co.

shown in Table 7. Table 8 lists some widely used naphtha pretreating processes which are available by license [15].

Acid Gas Absorption

Amine treating is an absorption process that is widely used to remove H_2S, CO_2, and other acid gases from natural and light petroleum gases. The process employs an organic amine (mono-, di-, or triethanolamine) which will absorb H_2S at temperatures up to 180°F.

The process is relatively simple. The untreated gas is passed upflow through a packed or bubble tray tower countercurrent to the amine solution at approximately 100°F and operating line pressure. Treated gas is removed at the top of the tower and the spent solution is regenerated by steam stripping at higher temperatures (200 to 300°F).

Other processes [1] have been developed that are basically very similar to amine treating but which employ a different type of reagent. Some popular reagents are solutions of potassium phosphates, amino acids, or potassium carbonates.

Caustic Treating

Caustic treating is basically an extraction process that removes organic acids, mercaptan sulfur, H_2S, and phenolic sulfur compounds from petroleum fractions [1]. The process is usually applied to light distillates because yield losses are too great with heavier stocks [16]. The major use of caustic treating today is the removal of mercaptans from light straight-run gasoline.

Caustic treating is accomplished by contacting a 5 to 20 wt.% caustic solution countercurrently with the untreated stream in a packed tower. The treating is carried out at pressures ranging from 10 to 50 lb/in.² gauge and at temperatures between about 70 and 120°F. The ratio of product to caustic can vary up to 10 to 1 [1].

The spent caustic can be regenerated either by steam stripping or air contacting. The regeneration conditions depend on the concentration of impurities in the caustic solution. The caustic is not 100% regenerable: thus make up solution must also be supplied.

There are a number of licensed caustic treating processes that have been adequately described in the literature [1, 16]. Many of these processes employ promoters which, when added to the caustic, either improve mercaptan removal or provide some other useful service. Popular solubility promoters are methanol, cresols, cresylates, phenol, phenolates, and ferrocyanides. One other useful variation is a process which combines caustic treating with oxidation sweetening. The caustic treating removes H_2S and most of the mercaptans while the oxidation sweetening step converts the remaining mercaptans to less objectionable disulfides. The general oxidation sweetening reaction is

$$4RSH + O_2 \rightarrow 2RSSR + 2H_2O$$

As the reaction equation indicates, oxidation sweetening does not reduce the sulfur content of the stream, it only converts the sulfur to a more acceptable form. Disulfides, however, reduce the lead susceptibility of gasoline and, for this reason, processes which remove the sulfur altogether are more popular than sweetening. Still, sweetening is a useful process for certain applications. Some sweetening processes that have enjoyed popularity are the doctor, copper chloride, hypochlorite, Merox, and lead sulfide processes [16]. Probably the most widely used sweetening process today is inhibitor sweetening which employs phenylene diamine-type compounds. This method is popular because of its low cost.

Molecular Sieve Adsorption

Molecular sieves are commonly used to selectively adsorb water and sulfur compounds from light hydrocarbon streams such as LPG, propane, butane, pentane, light olefins, and alkylation feed. Sulfur compounds that can be removed are H_2S, sulfides, disulfides, and mercaptans.

In the process [17], the sour feed usually passes through a bed of sieves at ambient temperature. The operating pressure must be high enough to keep the feed in the liquid phase. The operation is cyclic in that the adsorption step is stopped at a predetermined time before sulfur breakthrough occurs. Sulfur and water are removed from the sieves by purging with fuel gas at 400 to 600°F.

There are various types of molecular sieves, and the choice of sieve can depend on the particular application. Types 4A, 5A, and 13X sieves are alkali metal alumino silicates which are quite similar to some natural clays and feldspars [18]. All of these sieves can adsorb sulfur and water compounds, but the 13X sieves have a somewhat higher capacity. In addition, 13X sieves have a larger effective pore diameter and can adsorb molecules of larger critical diameters such as aromatics.

Molecular sieve processes have increased in importance over the years, and sieves are now used for a variety of purposes [18] in the petroleum and petrochemical industries.

Middle Distillate Desulfurization

Early desulfurization processes, such as caustic, amine, and clay treating, were not as successful with middle distillates as they were with naphthas and lighter feedstocks [1]. More sophisticated extraction processes utilizing hydrogen fluoride or sulfur dioxide contacting were somewhat more applicable to the removal of sulfur from diesel, gas oils, and cycle oils [1]. Such processes did suffice in producing satisfactory distillate fuels although substantial yield losses were inherent.

The availability of cheaper hydrogen from catalytic reforming led to the development of many middle distillate hydrotreating (hydrodesulfurization) processes in the 1950s. These processes result in superior treatment of distillates to improve color, odor, corrosion properties, thermal stability, and burning characteristics in addition to accomplishing essentially complete sulfur removal. This section deals only with the catalytic processes since they are used almost exclusively for treatment of middle distillates in the petroleum industry today.

Process Description

A typical flow diagram for distillate hydrodesulfurization is shown in Fig. 2 [10]. The oil feed is mixed with fresh and recycled hydrogen and heated under pressure to the proper reactor temperature. The oil–hydrogen mixture is charged to the reactor, passing downflow through the catalyst. In the reactor,

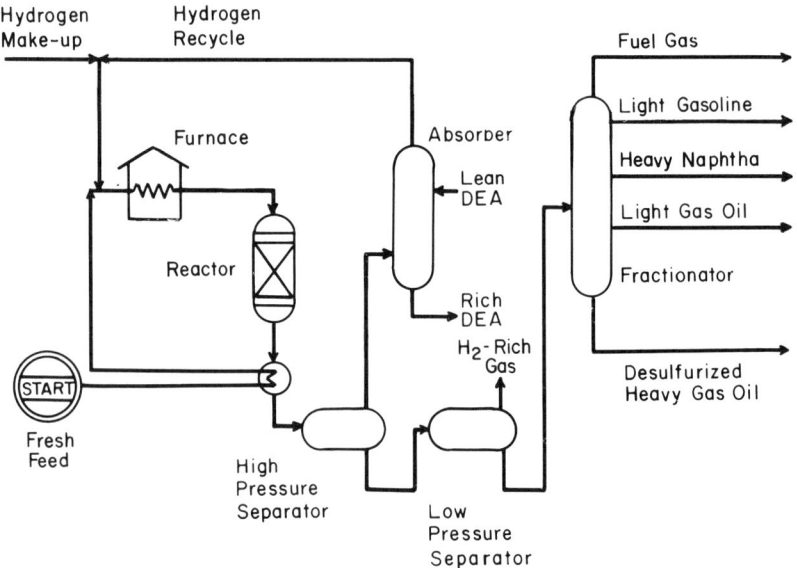

FIG. 2. Typical heavy gas oil desulfurizer.

fresh feed is hydrotreated and a limited amount of hydrogenation, isomerization, and cracking occurs to produce a small amount of C_1 through C_5 paraffins. In addition, sulfur compounds are converted to hydrogen sulfide and nitrogen compounds are converted to ammonia. Olefins are also saturated. These reactions are exothermic and, in the cases of vacuum gas oils or unsaturated feedstocks, reactor temperature rise is regulated by the use of cold recycle gas quench.

Reactor effluent is cooled and enters the high-pressure separator where the oil is separated from the hydrogen sulfide and hydrogen-rich gas. Hydrogen sulfide is scrubbed from the gas (optional for light distillate units) and the hydrogen-rich gas is recycled. The liquid is passed through a low-pressure separator and stripper to remove the remaining light ends and dissolved hydrogen sulfide. Fractionation of the liquid product is sometimes employed, especially on heavier charge stocks.

Typical yields and product properties are shown in Tables 9 and 10 for the desulfurization of West Texas virgin furnace oil and Kuwait vacuum gas oil, respectively. Cracked stocks would show similar upgrading but would entail much greater hydrogen consumptions.

Chemistry of Process

All of the reactions described in the naphtha section of this article also apply to the middle distillate operations so they will not be repeated here. In addition, there are much more complex molecules involved, especially in vacuum gas oil feedstocks. While formulas for the complex molecules cannot be readily

TABLE 9 Typical Yields and Properties for Desulfurization of West Texas Furnace Oil (duty: 97% desulfurization)

Yields, % of charge (average for cycle):			
H_2S, wt.%			0.81
NH_3, wt.%			0.01
C_1–C_4, wt.%			0.10
C_5–375°F, vol.%			0.25
375°F + furnace oil, vol.%			99.87
Hydrogen consumption, SCF/bbl			100
Properties	Feed	C_5–375°F	375°F +
Gravity, °API	41.8	51.0	43.3
Sulfur, wt.%	0.78	<0.01	0.02
Nitrogen, ppm	53	<2	25
Distillation, °F:			
10%	363	—	358
50%	471	—	464
90%	612	—	600

TABLE 10 Typical Yields and Properties for Desulfurization of Kuwait Vacuum Gas Oil (duty: 95.5% desulfurization)

Yields, % of charge (average for cycle):				
H_2S, wt.%				2.64
NH_3, wt.%				0.03
C_1–C_4, wt.%				0.70
C_5–375°F, vol.%				1.08
375–650°F, vol.%				12.05
650°F + gas oil, vol.%				87.74
Hydrogen consumption, SCF/bbl				·380
Properties	Feed	C_5–375°F	375–650°F	650°F +
Gravity, °API	24.1	51.0	35.0	29.0
Sulfur, wt.%	2.60	<0.01	0.02	0.13
Nitrogen, wt.%	0.07	—	0.005	0.05
Aniline point, °F	174.4	—	—	186.3
Distillation, °F:				
10%	682	—	358	659
50%	810	—	464	791
90%	961	—	600	954

defined, a simple example would be dibenzothiophenes which would react as follows:

$$\text{R-dibenzothiophene} + 3H_2 \rightarrow \text{R-dihydrodibenzothiophene} + H_2 \rightarrow \text{R-cyclohexylbenzene} + H_2S$$

As in naphtha hydrotreating, nitrogen and oxygen are removed in distillate hydrodesulfurization but to a lesser extent than sulfur removal. Olefin saturation is readily accomplished at the normal severities employed. Partial saturation of polynuclear aromatics is normally experienced, especially when the charge stocks are previously cracked materials, i.e., FCC cycle oils and thermal distillates.

Detailed discussions of the reaction mechanisms involved in distillate treating have been presented by Schuman and Shalit [11], Schuit and Gates [12], and Hoog [19].

Catalysts

All of the information shown in the previous section for naphtha hydrotreating catalysts apply to distillate hydrodesulfurization catalysts and will therefore not be repeated here.

Processing Considerations

The operating conditions in distillate hydrodesulfurization are dependent upon the stock to be charged as well as the desired degree of desulfurization or quality improvement. Kerosines and light gas oils are generally processed at mild severity and high throughput. Light catalytic cycle oils and thermal distillates require slightly more severe conditions. Higher boiling distillates, such as vacuum gas oils and lube oil extracts, require the most severe conditions. Typical ranges of conditions are listed in Table 11.

The principal variables affecting the required severity in distillate desulfurizers are hydrogen partial pressure, space velocity, reaction temperature, and feed properties.

Hydrogen Partial Pressure. The important effect of hydrogen partial pressure is the minimization of coking reactions. If hydrogen pressure is too low for the required duty at any position within the reaction system, premature aging of the remaining portion of catalyst will be encountered. The effect of hydrogen pressure on desulfurization varies with feed boiling range. For a given feed there exists a threshold level above which hydrogen pressure is beneficial to the desired desulfurization reaction. Below this level, desulfurization drops off rapidly as hydrogen pressure is reduced.

Space Velocity. As space velocity is increased, desulfurization is decreased. However, increasing hydrogen partial pressure and/or the reactor temperature can offset the detrimental effect of increasing space velocity.

The space velocity–desulfurization relationship has been evaluated using basic kinetic relationships [12, 19–21]. The presence of a complex mixture of sulfur-bearing compounds results in an apparent reaction order between first and second. First-order behavior has been shown if either liquid holdup or effective catalyst wetting is accounted for [21].

Reaction Temperature. Higher temperature increases the rate of desulfurization. The start-of-run temperature is set by the design desulfurization level, space velocity, and hydrogen partial pressure. Capability to increase temperature as the catalyst deactivates is built into the unit design. Temperatures of 780°F and above result in excessive coking reactions and higher than normal catalyst aging rates. Therefore, units are designed to avoid the use of such temperatures for any significant part of the cycle life.

TABLE 11 Typical Operating Conditions for Distillate Hydrodesulfurization

	Kerosine and Light Gas Oils	Heavy Gas Oils and Lube Oil Extracts
Total pressure, lb/in.^{2} gauge	100–1000	500–1500
Reactor temperature, °F	450–800	650–800
Space velocity, V/h/V	2–10	1–3

Feedstock Properties. The feed boiling range has the greatest effect on the ultimate design of the unit. The reaction rate constant in the kinetic relationships decreases rapidly with increasing average boiling point in the kerosine and light gas oil range but much more slowly in the heavy gas oil range. This is attributed to the difficulty in removing sulfur from ring structures present in the entire heavy gas oil boiling range.

Catalyst Aging and Cycle Life. Catalyst life depends on the charge stock properties and the degree of desulfurization desired. The only permanent poisons to the catalyst are metals included in the feed. The metals deposit on the catalyst, usually quantitatively, causing permanent deactivation as they accumulate. However, this is usually of little concern except when deasphalted oils are charged, since most distillate stocks contain low amounts of metals. Nitrogen compounds are a temporary poison to the catalyst but there is essentially no effect on catalyst aging except that caused by a higher temperature requirement to achieve the desired desulfurization. Hydrogen sulfide can be a temporary poison in the reactor gas. Recycle gas scrubbing is employed to counteract this condition.

Providing that pressure drop buildup is avoided, cycles of 1 year or more and ultimate catalyst life of 3 years or more can be expected. The catalyst employed can be regenerated by normal steam–air or recycle combustion gas–air procedures. The catalyst is restored to near fresh activity by regeneration during the early part of its ultimate life. However, permanent deactivation of the catalyst occurs slowly during usage and repeated regenerations, so replacement becomes necessary.

Economics

Investment and utility requirements for distillate desulfurizers will vary widely, depending on feedstock properties and degree of treating required. A representative set of values is shown in Table 12 for a vacuum gas oil unit. Representative economics for the various processes available by license, listed in Table 13, are reported elsewhere on a biannual basis [22].

Residue Desulfurization

Residue hydrodesulfurization evolved over a period of at least 25 years to become a commercial reality in 1963. Since then, a significant number of commercial plants has been constructed. Total capacity of the units now operational amounts to about 800,000 bbl/stream day. Many more units are in various stages of planning, engineering, or construction. However, shortage of capital funds as well as uncertainties of crude supply and ultimate sulfur regulations in various locations have been instrumental in slowing the installation of additional units.

TABLE 12 Economics for Vacuum Gas Oil Desulfurizer

Charge Stock	Light Arabian VGO
Rate, bbl/stream day	37,840
Desulfurization duty, %	90
Investment, $M[a]	24,000
Operating requirements (average):	
Fuel, MM Btu/h	180
Power, shaft kW	3,700
Cooling water, gal/min	2,800
Deaerated condensate, gal/min	16
Boiler feed water, gal/min	35
Steam produced, M lb/h	115

[a] U.S. Gulf Coast, 2nd quarter, 1981, NRC Index = 890. Total erected cost for all major processing equipment including contractor's fees. Includes first catalyst charge cost and royalty. Does not include sulfur and hydrogen units.

Much of the driving force behind the building of residual hydrodesulfurizers was provided by legislative action on the part of some industrialized nations, primarily Japan and the United States. Requirements for 0.3% sulfur fuels were established in many congested areas and, in some cases, 0.1% sulfur fuels were projected in forecasts for the future. Implementation moved faster in Japan due to their aggressive legislative action and high degree of dependence on imported Middle East crudes. Construction in the

TABLE 13 Available Middle Distillate Desulfurization Processes

Process	Licensor
Distillate Hydrodesulfurization	Institute Francais du Pétrole
Fuel Hydrodesulfurization	Badische Anilin- und Soda Fabrik AG and Institute Francais du Pétrole
GO-fining	Exxon Research and Engineering Co.
Gulfining	Gulf Research and Development Co. and Houdry Division of Air Products and Chemicals, Inc.
Trickle Flow Hydrodesulfurization	Shell Internationale Research Maatschappij B.V. or Shell Development Co.
Hydrofining	BP Trading Ltd.
Hydrofining	Exxon Research and Engineering Co.
Ultrafining	Standard Oil Co. (Indiana)
Unionfining	Union Oil of California
VGO and DAO Hydrotreating	Chevron Research Co.
Unibon	Universal Oil Products Co.

United States has been much slower due to the availability of low sulfur crudes, allowing production of low sulfur residual fuels directly or indirectly by desulfurization of vacuum gas oil and backblending with the residue. This is not expected to last, however, since 60% of the projected reserves of recoverable crude oil worldwide are of the "high sulfur" type (greater than 1% sulfur in crude oil). Thus it would seem that residue desulfurization will be necessary in the future to maximize the utilization of the crude oil barrel.

Process Description

Direct residual desulfurization processes are available from a number of licensors, as listed in Table 14. Most of these processes are similar in design, with distinguishing features being catalyst compositions and shapes employed. The different catalysts allow other minor differences in operating conditions and peripheral equipment. A typical flow diagram is shown in Fig. 3. The flow is quite similar to the distillate hydrodesulfurization unit described previously. Primary differences include the use of higher purity hydrogen makeup gas (usually 95% or greater), inclusion of filtration equipment in most cases, and facilities to upgrade the off-gases to maintain higher concentration of hydrogen in the recycle gas. Most of the processes listed utilize downflow operation over fixed-bed catalyst systems. Exceptions to this are the H-Oil and LC-fining Processes, which employ upflow designs and ebullating catalyst systems with continuous catalyst removal capability, and the Shell Process, which may involve the use of a "bunker flow" reactor ahead of the main reactors to allow periodic changeover of catalyst [23]. Table 15 lists those units now in service or well into construction together with their design capacities and start-up dates.

TABLE 14 Available Residue Desulfurization Processes

Process	Licensor
Resid HDS	Gulf Research and Development Co.
Fuel Hydrodesulfurization	Badische Anilin- und Soda Fabrik AG and Institute Francais du Pétrole
RESIDfining	Exxon Research and Engineering Co.
H-Oil	Hydrocarbon Research, Inc.
Residual Oil Hydrodesulfurization	Shell Internationale Research Maatschappij B.V. or Shell Development Co.
LC-fining	C-E Lummus and Cities Service Research and Development Co.
RCD Unibon	Universal Oil Products
RDS and VRDS Hydrotreating	Chevron Research Co.
Resid Hydroprocessing	Standard Oil Co. (Indiana)
Residue Desulfurization	BP Trading Ltd.
Unicracking/HDS	Union Oil Co. of California

Desulfurization, Liquids, Petroleum Fractions

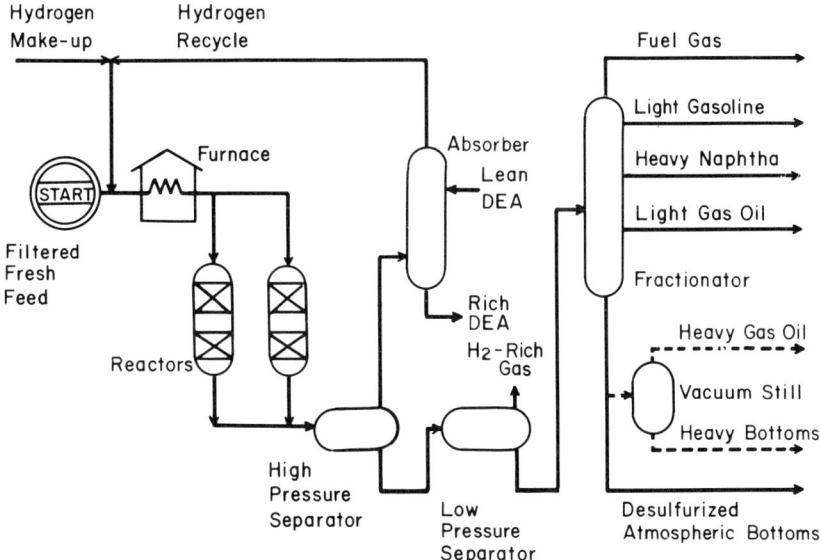

FIG. 3. Typical residual desulfurizer.

Chemistry of Process

The primary objective in most of the residue desulfurization processes is to remove sulfur with minimum consumption of hydrogen. However, complete elimination of undesired side reactions is not feasible. Thus substantial percentages of nitrogen, oxygen, and metals are also removed from the feedstock. In addition, hydrocracking, thermal cracking, and aromatic saturation reactions occur to some extent. Certain processes, i.e., H-Oil and LC-fining, can be designed to accomplish greater amounts of hydrocracking to yield larger quantities of lighter distillates at the expense of desulfurization.

Reactions of typical sulfur, nitrogen, and oxygen compounds as described in previous sections of this article will also take place during residue desulfurization. However, many compounds are much more complex than the basic structures shown previously. A typical residual molecule is shown in Fig. 4. The thiophenic sulfur in this case is much more difficult to remove because the aliphatic groups interfere with the adsorption of the sulfur compound on the reactive sites of the catalyst. The size of the molecule itself can inhibit adsorption on the catalyst. This helps explain why residues are more difficult to desulfurize than lighter oils.

Removal of nitrogen is much more difficult than removal of sulfur. Nitrogen removal is about 25 to 30% when sulfur removal is 75 to 80%. Hydrocracking and thermal cracking occur to a limited extent at start-of-cycle but become more important as temperatures are raised to maintain catalyst activity. True olefins are not normally present to any extent in residues, but olefinic side groups may be present and minor quantities of true olefins may be formed during primary distillation. Any olefins present will be readily saturated

TABLE 15 Residue Hydrodesulfurization Units in Operation or Well into Construction

Company	Location	Capacity (bbl/stream day)	Licensor	Start-up Date
Cities Service	Lake Charles, Louisiana	6,000	Cities Service/HRI	1963
Standard Oil of California	Richmond, California	25,000	Chevron	1966
Idemitsu Kosan	Chiba, Japan	40,000	UOP	1967
Kuwait National Petroleum	Shuaiba, Kuwait	28,800	Cities Service/HRI	1968
American Independent Oil	Mena Abdullah, Kuwait	35,000	UOP	1969
Nippon Mining	Mizushima, Japan	31,000	GR & DC	1969
Kashima Oil	Kashima, Japan	45,000	UOP	1970
Idemitsu Kosan	Hyogo, Japan	40,000	GR & DC	1972
Okinawa Sekiyu Seisei	Okinawa, Japan	38,000	GR & DC	1972
Petroleos Mexicanos	Salamanca, Mexico	18,500	Cities Service/HRI	1973
Shell	Gothenburg, Sweden	3,000	Shell	1973
Mitsubishi Oil	Mizushima, Japan	45,000	GR & DC	1974
Idemitsu Kosan	Aichi, Japan	50,000	GR & DC	1975
Asia Oil	Yokohama, Japan	30,000	UOP	1976
Seibu Oil	Yamaguchi, Japan	45,000	Shell	1976
Maruzen Oil	Chiba, Japan	60,000	Union	1976
Exxon	Baytown, Texas	75,000	Exxon	1977
Standard Oil of California	El Segundo, California	24,000	Chevron	1977
Dow Chemical	Freeport, Texas	63,000	Exxon	1980
Asia Oil	Sakaide, Japan	28,000	GR & DC	1980
Phillips Petroleum	Sweeny, Texas	75,000	Union	1981
Phillips Petroleum	Borger, Texas	50,000	GR & DC	1982

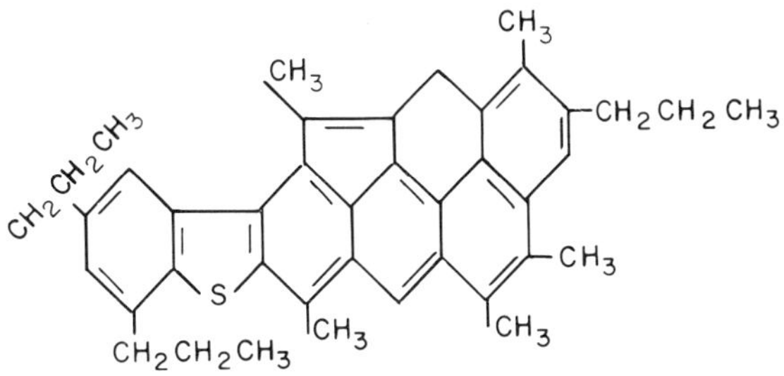

FIG. 4. Typical residual molecule.

at the operating conditions used. Saturation of aromatic rings occurs to only a limited extent. It is most likely to occur at low temperatures and long residence times.

Metals are removed from the feed in substantial quantities. They are mainly deposited on the catalyst surface and exist as metal sulfides at processing conditions. As these deposits accumulate, the catalyst pores eventually become blocked and inaccessible, thus catalyst activity is lost.

Desulfurization of residues is considerably more difficult than lighter oil desulfurization because many more contaminants are present and very large, complex molecules are involved. The most difficult portion of feed in residue desulfurization is the asphaltene fraction. The asphaltenes have a strong tendency to associate themselves so that they exist as nuclear clusters of micelles of weakly bound molecules as shown in Fig. 5. The apparent molecular weight of such clusters can be as high as 100,000. The micelles are apparently not chemically bonded since violent agitation, high temperature, and solvents can break them apart. At normal operating conditions they probably do not exist. However, the aromatic nucleus of the asphaltene molecule is strongly adsorbed on the catalysts employed. It is essential that these large molecules be prevented from condensing with each other to form coke, which deactivates the catalyst. This is accomplished by selection of proper catalysts, use of adequate hydrogen partial pressure, and assuring intimate contact of the hydrogen-rich gases and oil molecules in the process design.

Catalysts

Exact catalyst metals formulations and support characteristics are considered as proprietary information by the various process licensors. However, some generalities can be derived from various literature and patent sources.

The general types of catalyst used in residue desulfurization are combinations of metal oxides on alumina or silica-stabilized alumina supports. Molybdenum always seems to be one of the metals, with cobalt and/or nickel being used in combination with the molybdenum in many cases. Patent literature indicates that other metals may be used as substitutes for the cobalt and/or nickel or as additional promoters or stabilizers to the base catalyst.

Germanium, magnesium, sodium, phosphorus, titanium, and zinc are some examples.

The supports are usually tailored to accomplish those objectives which are deemed most important or to arrive at the best cost effectiveness in the overall economics of the process. Different support selections can be made to accomplish particular goals of a specific unit. For example, smaller pored

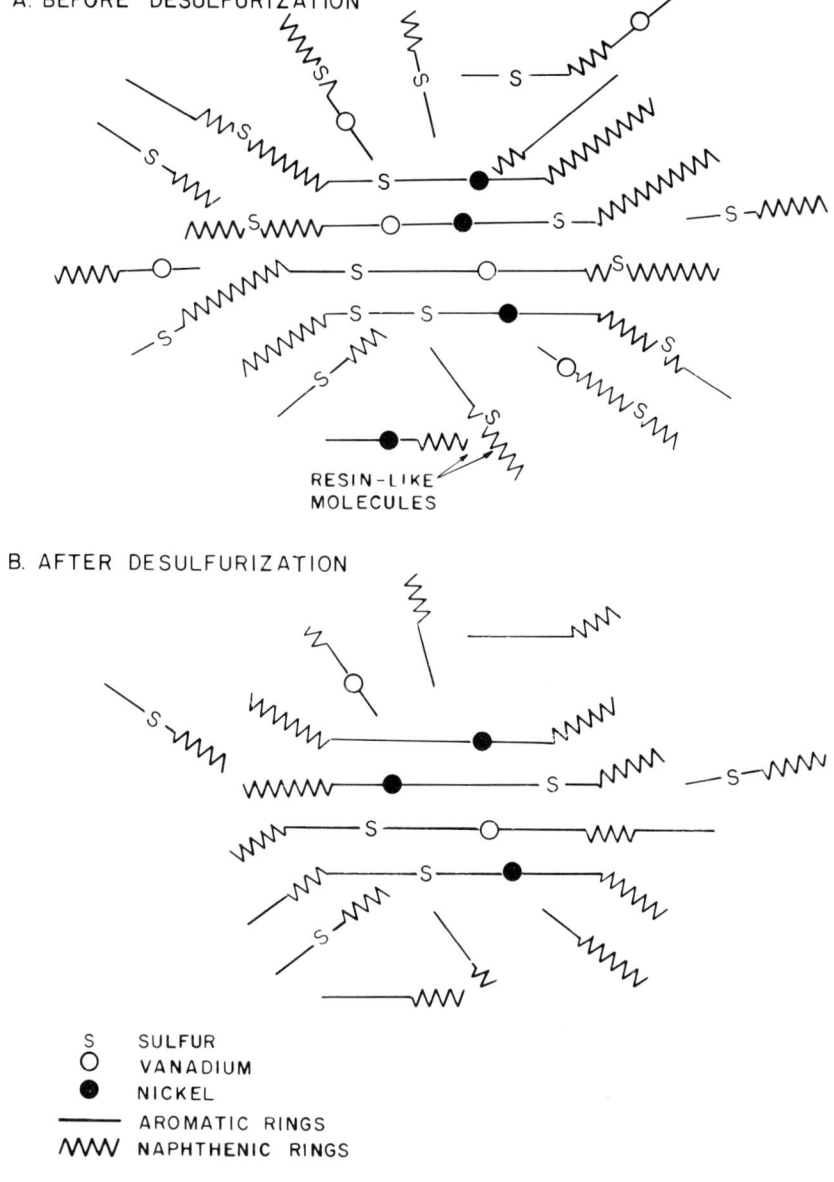

FIG. 5. Qualitative changes in asphaltenes and suurounding resins before and after desulfurization.

catalysts will tend to remove less metals than larger pored catalysts and be more active for desulfurization reactions. However, the metal-holding capacity of the small-pored catalyst will also be less than the large-pored catalyst, which results in a sacrifice of catalyst life. The better catalyst selection would depend on the combination of these two characteristics which will allow the best life–activity relationship for a given application. In some cases, critical combinations of large and small pore sizes are used to arrive at the best catalyst for a given feedstock and operating conditions. The objectives of the process unit are also important since metals removal would be more critical in an application to produce residual feedstock for catalytic cracking versus an application to produce residual fuel.

Catalyst size and shape are also important factors in residue desulfurization processes. Smaller size contributes to improved desulfurization and demetallization; however, pressure drop considerations become more important. Recent experience with shaped catalyst has been touted as responsible for improved performance and lower pressure drop [24].

Design Criteria

Metals Deposits on Catalyst

While many of the conventional design criteria in distillate desulfurization (hydrogen partial pressure, degree of desulfurization, gas circulation rates) must be considered in residue desulfurization, an additional important criteria is the effect of metals accumulation on the catalyst. The effective life of a particular catalyst will vary depending on its pore structure and total pore volume. It is also dependent upon the particular feedstock being processed and the operating conditions employed. As mentioned previously, many of the competitive processes use different catalyst characteristics which have been tailored to achieve the objectives of most concern to the individual licensor. Thus some processes will remove more metals from the feedstock while others will reject metals to a greater extent.

In general terms, catalysts which show better selectivity for metals removal will also hold more total metals before they become inoperable for the required desulfurization duty. This holding capacity for metals has been defined previously [25] as a "saturation level." This saturation level increases with decreasing size for a given catalyst. However, the selectivity for demetallization over desulfurization reactions also increases with decreasing size. The combination of these effects results in an optimum particle size to maximize the cycle life.

Estimation of Catalyst Requirements

Accepting the existence of a saturation level for a given catalyst, the estimation of catalyst volume requirements in the metals-limited portion of the reaction system is greatly simplified. In effect, one starts with an end-of-run metals on

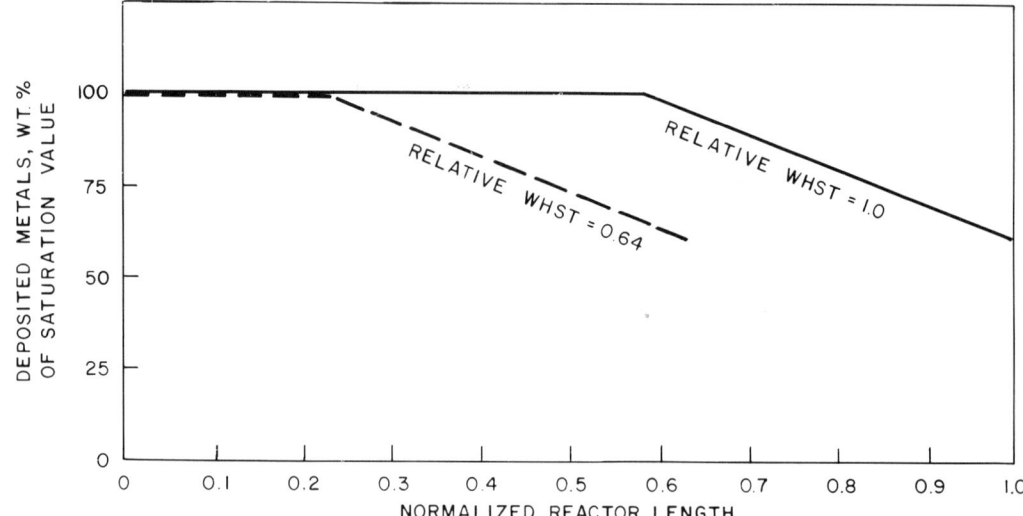

FIG. 6. EOR metal deposits profile for two pilot plants aging runs—HDS of Kuwait atmospheric tower bottoms at 1% sulfur, 650°F + fuel oil.

catalyst profile as shown in Fig. 6. The length of the saturated zone and the unsaturated zone are calculated to give sufficient catalyst volume to store the metals which will be removed from the feedstock over the required cycle life. Catalyst requirements for the unsaturated zone result from an analysis of the activity and active space time required to produce the required product sulfur content at the end-of-run temperature. Extensive use is made of desulfurization–demetallization relationships for different stocks as well as metals on catalyst–activity relationships determined experimentally.

For deeper desulfurization duty the catalyst requirements for the remainder of the system are generated from conventional kinetic relationships. These are determined through experimental evaluations and fitting of the data to kinetic equations.

Many investigators [21, 26–37] have analyzed the kinetic relationships of residue desulfurization. In a number of cases, these evaluations have shown apparent second-order relationships to hold up to moderate desulfurization levels (about 80%). After this point, a declining or lower reaction order appears to prevail. Some investigators [21] have been able to show that if either liquid holdup or effective catalyst wetting are accounted for, results will follow apparent first-order kinetics.

Important Operating Parameters

Desalting, Filtration, and Metallurgy

Efficient desalting and filtration are critical to proper operation of the fixed-bed residue desulfurization processes. Double desalting is a common practice to eliminate potential deposits of sodium and potassium in the catalyst bed. Also, chloride ions are minimized in the system, and this will alleviate plugging of low-temperature areas of the recovery systems by ammonium chloride.

Suitable filter systems are available for handling the quantity of solids present in conventional petroleum residues. Special advanced designs may be needed to handle the quantities and size distributions of solids which are expected in some anticipated charge stocks, such as tar sands and shale oil.

Metallurgy selection and proper turnaround maintenance are also of great importance because of the inherent corrosion potential of the system. The presence of hydrogen sulfide, chlorides, ammonia, organic acids, and water in the same system calls for preventive measures which are included in detailed procedures from the process licensor.

Catalyst Aging

A gradual loss of catalyst activity occurs during normal operation of the residue process. Therefore a gradual increase in catalyst temperature is required through the cycle to maintain the desired product sulfur content. This loss in activity is caused by deposition of coke and metals (from Ni and V in the feedstock) on the catalyst surface and in the catalyst pores.

Ultimate catalyst life is directly related to the total metals tolerance of the catalyst, which is a function of particle size, shape, and pore size and volume. Metals deposited cause permanent deactivation of the catalyst and preclude the restoration of catalyst activity by normal regeneration procedures. Spent catalysts are either discarded or returned to reclaimers for recovery of the various metals of value.

The amount of coke deposited on the catalyst depends primarily on hydrogen partial pressure but is also influenced by the asphaltene content of the feedstock. Higher hydrogen pressure decreases coking while higher asphaltene content increases coking.

Hydrogen Partial Pressure, Circulation, and Purity

Besides the effect of hydrogen partial pressure on catalyst aging, maintenance of adequate amounts of hydrogen within the system is required. Circulation rates and purity requirements are set to avoid a shortage of hydrogen anywhere in the reaction system in order to prevent undesirable side reactions.

Catalyst Temperature and Space Velocity

These variables are very important to proper operation of the process. In a given system a reduction in space velocity without an appropriate reduction in temperature will result in overtreating of the feedstock. This will lead to

TABLE 16 Typical Yields and Properties for Desulfurization of Kuwait Atmospheric Tower Bottoms

Yields, % of HDS Charge	Feed	1% Sulfur 650°F + Fuel[a]	0.3% Sulfur 375°F + Fuel[a]	0.1% Sulfur 375°F + Fuel[a]
H$_2$S, wt.%		3.07–3.14 (3.10)	3.73–3.74 (3.73)	3.93–3.94 (3.93)
NH$_3$, wt.%		0.08–0.07 (0.08)	0.13–0.12 (0.12)	0.17–0.17 (0.17)
C$_1$–C$_4$, wt.%		0.27–1.10 (0.62)	0.33–1.67 (0.89)	0.40–2.07 (1.14)
C$_5$–375°F naphtha, vol.%		1.4–4.1 (2.6)	2.0–6.4 (3.8)	2.5–7.6 (4.6)
375–650°F distillate, vol.%		6.6–11.2 (8.5)	8.9–17.4 (12.5)	9.1–20.8 (14.0)
650°F + residue, vol.%		92.7–86.5 (90.1)	90.4–78.9 (85.6)	89.9–74.6 (83.5)
Chemical hydrogen consumption, SCF/bbl		(497)	(650)	(725)
Residue product properties:				
Gravity, °API	16.6	20.5–22.1 (21.2)	22.0–23.9 (22.8)	22.5–24.4 (23.3)
Sulfur, wt.%	3.8	1.0	0.32–0.35 (0.33)	0.10–0.11 (0.11)
Carbon residue, wt.%	9.0	5.6–6.1 (5.8)	3.6–4.1 (3.8)	3.0–3.6 (3.2)
Nitrogen, wt.%	0.22	0.17–0.19 (0.18)	0.13–0.15 (0.14)	0.09–0.11 (0.10)
Pour point, °F	+60	+65 – +55 (+60)	+70 – +20 (+35)	+60 – 0 (+15)
Nickel, ppm	15	4.5–4.8 (4.6)	1.3–1.6 (1.5)	0.4–0.5 (0.5)
Vanadium, ppm	45	7.9–8.6 (8.2)	2.4–2.8 (2.6)	1.0–1.3 (1.2)
Viscosity, SUS at 210°F	250	122–88 (108)	104–70 (90)	94–64 (81)

[a] Values in parentheses indicate average values for cycle.

Desulfurization, Liquids, Petroleum Fractions

TABLE 17 Typical Cost and Utility Requirements for Desulfurization of Kuwait Atmospheric Tower Bottoms

	1% Sulfur 650 °F+	0.3% Sulfur 375 °F+ Fuel	0.1% Sulfur 375 °F+ Fuel
Investment, $MM[a]	58.8	77.7	97.5
Chemical hydrogen consumption, SCF/bbl (average)	497	650	725
Utility requirements (average demand):			
Power, shaft kW	13,300	15,100	18,600
Fuel, MM Btu/h	163	174	174
Cooling water, gal/min	4,900	6,300	7,600
Steam, M lb/h (net)	36	73	87
Condensate, gal/min	45	60	82
Boiler feedwater, gal/min	45	48	48

[a]50,000 bbl/stream day, U.S. Gulf Coast, 2nd quarter, 1981, NRC Index = 890. Battery limits cost including first catalyst fill but excluding royalty, sulfur, and hydrogen plants.

premature aging of the catalyst by virtue of increased coking and incremental metals deposition. Such premature aging effects are essentially irreversible.

Yields and Product Properties

Since catalyst temperatures are increased over the length of an operating cycle, both yields of lighter materials and properties of the remaining residue are affected. Distillate properties are affected to a much lesser extent. The magnitude of the variations will depend on the catalyst selected for the operation and the operating conditions employed. As mentioned earlier, catalysts can be tailored to accomplish different objectives. Typical values for Gulf hydrosulfurization are shown in Table 16 for three levels of desulfurization of Kuwait 650°F+ residue. The product properties affected to the greatest extent are viscosity and pour point. These changes, as well as the changes in distillate yields, indicate that cracking reactions are increasing as the run progresses. This is due to temperature increases which lead to a greater degree of thermal cracking as described previously [25].

Economics

Investment, manufacturing costs, and utility requirements for residue desulfurization will vary widely, depending on charge stock properties and product quality required. Values for the same three cases for Kuwait 650°F+ residue shown in Table 16 are listed in Table 17. Representative economics for the various processes are reported elsewhere on a biannual basis [22].

References

1. W. F. Bland and R. L. Davison, *Petroleum Processing Handbook*, McGraw-Hill, New York, 1967, Chap. 3.
2. H. T. Rall, C. J. Thompson, H. J. Coleman, and R. L. Hopkins, *Sulfur Compounds in Crude Oil*, U.S. Department of the Interior, Bureau of Mines, Washington, D.C., 1972.
3. D. M. Jewell, R. G. Ruberto, E. W. Albaugh, and R. C. Query, *Ind. Eng. Chem., Fundam.*, *15*(3), 206 (1976).
4. W. A. Gruse and D. R. Stevens, *Chemical Technology of Petroleum*, McGraw-Hill, New York, 1960, Chap. 7.
5. H. Taylor, *Hydrocarbon Process.*, *52*(5), 86 (1973).
6. G. F. Ondish, J. A. Frayer, and J. D. McKinney, *Hydrocarbon Process.*, *55*(7), 105 (1976).
7. J. B. McKinley, in *Catalysis*, Vol. 5, Reinhold, New York, 1957, p. 405.
8. R. A. Flinn and J. B. McKinley, U.S. Patent 2,880,171 (March 31, 1959).
9. W. A. Horne and J. McAfee, *Adv. Pet. Chem. Refin.*, *3*, 193 (1960).
10. Gulf Research & Development Company, *Gulf Process Brochure—Hydrocarbon Processes and Catalysts*, Pittsburgh, Pennsylvania, 1975, pp. 4–7.
11. S. C. Schuman and H. Shalit, *Catal. Rev.* *4*(2), 245 (1970).
12. G. C. A. Schuit and B. C. Gates, *AIChE J.*, *19*(3), 417 (1973).
13. C. G. Frye and J. F. Mosby, "Kinetics of Hydrodesulfurization," *Chem. Engr. Prog.*, *63*(9), 66 (1967).
14. C. N. Satterfield, *Mass Transfer in Heterogeneous Catalysts*, MIT Press, Cambridge, Massachusetts, 1970.
15. A. A. Montagna, J. F. Marmo, and S. W. Chun, *Sweetening and Naphtha Pretreating*, Paper Prepared for American Chemical Society New York Meeting, August 27, 1972.
16. W. L. Nelson, *Petroleum Refinery Engineering*, McGraw-Hill, New York, 1969, Chap. 10.
17. Anonymous, *Hydrocarbon Process.*, *51*(9), 213 (1972).
18. J. J. Collins, "Molecular Sieves in the Process Industry," *Chem. Eng. Prog.*, (August 1968).
19. H. Hoog, *J. Inst. Pet.*, *36*, 738 (1950).
20. T. B. Metcalfe, *Hydrogen Sulfided Inhibition of Hydrodesulfurization*, Presented at AIChE 64th National Meeting, New Orleans, Louisiana, March 16–20, 1969.
21. J. A. Paraskos, J. A. Frayer, and Y. T. Shah, *Ind. Eng. Chem., Process Des. Dev.*, *14*, 315 (July 1975).
22. *Hydrocarbon Processing*, *55*(9), (1976).
23. A. J. J. van Ginneken, M. M. van Kessel, and G. Renstrom, *Oil Gas J.* (April 28, 1975).
24. H. R. Jackson and K. E. Whitehead, *Oil Gas J.* (February 7, 1977).
25. J. A. Paraskos, A. A. Montagna, and L. W. Brunn, *AIChE Symp. Ser.* *72*, 127 (1976).
26. H. Beuther and B. K. Schmid, Sixth World Petroleum Congress, June 1963.
27. J. J. van Deemter, *Chemical Reaction Engineering*, Pergamon, New York, 1965, p. 215.
28. Y. Shimizu et al, *Bull. Jpn. Pet. Inst.*, *12*, 10 (May 1970).
29. T. Ohtsuka, Y. Hasegawa, and N. Takanari, *Bull. Jpn. Pet. Inst.* *10*(14), (1968).
30. T. Ohtsuka, Y. Hasegawa, M. Koizumi, and T. Ono, *Bull. Jpn. Pet. Inst.*, *9*(1), (1967).
31. H. Nakamura et al., *Int. Chem. Eng.* *10*(3), (July 1970).

32. D. E. Mears, *Chem. Eng. Sci.*, 26, 1361 (1971).
33. E. S. Parkin, J. A. Paraskos, and J. A. Frayer, *AIChE Symp. Ser.*, 71(148), 241, (1975).
34. L. E. Ross, *Chem. Eng. Prog.* 61(10), 77 (1965).
35. H. C. Henry and J. B. Gilbert, *Ind. Eng. Chem., Process Des. Dev.*, 12, 328 (1973).
36. J. M. Hochman and E. Effron, *Ind. Eng. Chem., Fundam.*, 8, 63 (1969).
37. C. N. Satterfield, A. A. Pelessof, and T. K. Sherwood, *AIChE J.*, 15, 226 (1969).

ROBERT J. CAMPAGNA
JAMES A. FRAYER
RAYNOR T. SEBULSKY

Desulfurizing Cracked Gasoline and Other Hydrocarbon Liquids by Caustic Soda Treating

Caustic soda treating of gasoline components is a long-standing process practice of the petroleum refining industry. Virgin gasolines produced by crude oil distillation and cracked gasolines from catalytic cracking and delayed coking contain hydrogen sulfide, aliphatic mercaptans, and aromatic mercaptans—all of which are detrimental to antiknock compound susceptibility, odor, and cleanliness. Gasolines containing these components are corrosive and objectionably odoriferous (termed "sour" by the industry). Aromatic mercaptans have an unfavorable influence on the oxidation stability of gasoline and are notorious gum formers. Gasolines are treated with caustic soda solutions to remove hydrogen sulfide, while mercaptans are either extracted or converted to less objectionable disulfides.

Converting mercaptan sulfur (RSH-S) to disulfide sulfur improves the odor, corrosiveness, and stability of cracked gasoline. However, the octane performance and antiknock compound susceptibility are only slightly improved. "Sweet" gasolines have no mercaptan odor and pass the industry's doctor test

TABLE 1 Sensitivity of Mercaptans to Doctor Test

Mercaptan	Minimum Detectable Quantity of RSH-S (ppm wt.)
Ethyl	6.0
n-Propyl	3.0
n-Butyl	1.5
n-Amyl	2.0
Isoamyl	1.0
n-Heptyl	1.0
Phenyl	10.0

TABLE 2 Chemistry of Caustic Extraction and Sweetening

Hydrogen sulfide extraction:	$H_2S + 2NaOH \rightarrow Na_2S + 2H_2O$
Mercaptan extraction:	$RSH + NaOH \rightarrow NaSR + H_2O$
Mercaptide oxidation:	$2NaSR + \frac{1}{2}O_2 + H_2O \rightarrow RSSR + 2NaOH$
Cresylic acids (phenolics) Extraction:	$ROH + NaOH \rightarrow RONa + H_2O$

[1]. Various mercaptans have different sensitivities to the doctor test as shown in Table 1.

Generally, light cracked gasolines treated to a mercaptan sulfur content of 5.0 ppm by weight pass the doctor test, whereas heavy cracked gasoline requires treatment to 3.0 ppm in order to pass. Table 2 shows the basic chemical reactions that take place during caustic treating of cracked gasolines.

Hydrogen sulfide, aromatic mercaptans, and cresylic acid are all readily extracted by caustic of 15 wt.% NaOH and higher. However, aliphatic mercaptans have such low distribution coefficients into caustic of any strength that extraction is difficult and air oxidation to disulfide sulfur is usually required to pass the doctor test. The distribution coefficients [2, 3] for various mercaptans in caustic are shown in Table 3.

As indicated in Table 3, the C_4 and heavier aliphatic mercaptans are difficult to extract; however, the solutizing effect of cresylic acids enhances the distribution of mercaptans into the caustic phase. Other solutizers, including methanol [4], sodium isobutyrate [4], and tannin [5], have also been used. The presence of solutizers is especially beneficial where air sweetening is practiced because the oxidation reaction occurs principally in the caustic phase. Currently, the most widely used solutizer is a mixture of caustic and acid oils that have been extracted by caustic treatment of heavy cracked gasoline at the refinery in question. These acid oils are predominantly cresylic acid (75%) containing a significant amount (25%) of thiophenols.

Typical Two-Stage Treating

A continuous two-stage, 25 wt.% caustic treating/sweetening system is shown schematically in Fig. 1. Note that the first stage is an extraction stage whereas the second stage is a combined extraction/air-inhibitor-sweetening step. The fresh 25 wt.% caustic is added to the second stage of this countercurrent treater on flow control. The rate of fresh caustic addition is set to give an acid oil content of at least 15 vol.% in the second stage recirculating caustic to provide adequate solutizer.

Caustic is transferred from the second stage to the first stage and is withdrawn from the first stage on level control. Air and a sweetening inhibitor are added to the hydrocarbon stream between stages. The quantity of air added is approximately 30 SCF/lb of mercaptan sulfur to be converted.

Some gasolines contain a large amount of nonextractable mercaptan sulfur which makes inhibitor sweetening costly and difficult. In some cases an

TABLE 3 Distribution Coefficients for Various Mercaptans

	Distribution Coefficient	
Mercaptan	In 10 wt.% Caustic	In 30 wt.% Caustic with 20% Cresylic Acid
Thiophenols	~510,000.0	~510,000.0
Cresylic acids	~510,000.0	~510,000.0
Propyl mercaptan	30.0	175.0
Butyl mercaptan	5.0	77.0
Amyl mercaptan	0.9	51.0
Heptyl mercaptan	0.03	36.0

oxidation catalyst is added to the sweetening stage to enhance the rate of conversion of mercaptans to disulfides.

For conventional mixer-settler treaters, the recycle caustic rate is typically 15 to 20% of the gasoline rate with the mix valve pressure drop 5 to 15 lb/in². Since the smaller droplets formed by the mix valve often do not coalesce and settle completely in the settler, this leads to carryover into the rundown tanks. Sometimes carryover is so severe that a coalescer and/or water wash are utilized to minimize caustic entrainment to storage.

The design of a mixer-settler stage for caustic treating involves essentially a tradeoff between stage efficiency (related to intensity of mixing) and settling velocity (related through Stokes law to size of settler vessel). Consequently, the viscosities of the two phases are primary parameters, particularly the viscosity

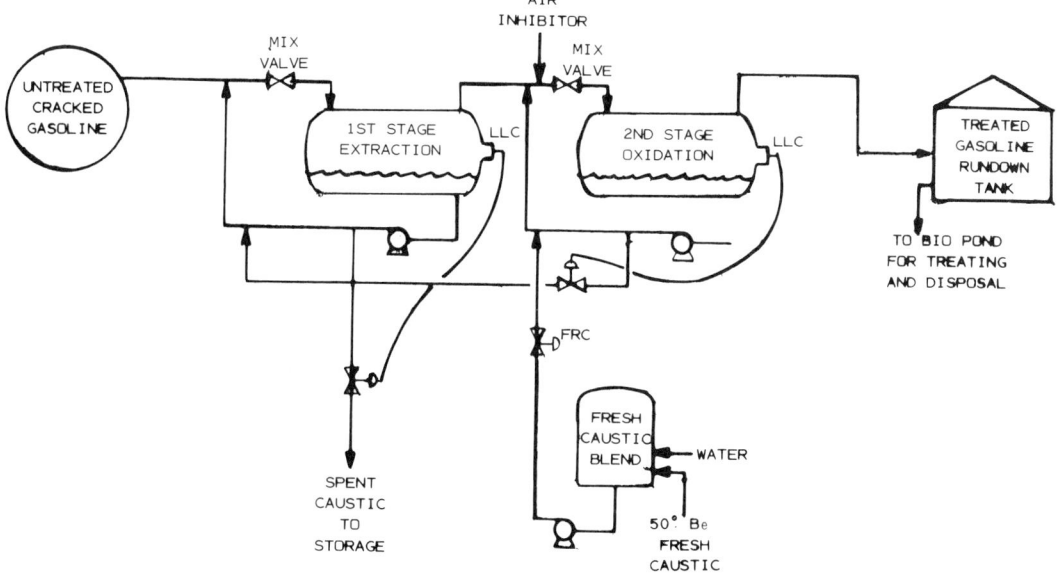

FIG. 1. Two-stage caustic treating/sweetening.

of the caustic phase. As caustic becomes "spent" by absorbing acid oils, its viscosity increases markedly. The viscosities of 20 and 25 wt.% caustic at various percent spent levels are shown in Fig. 2. Settler design is normally keyed to the upward terminal velocity of gasoline droplets through the caustic phase, with a minimum droplet size of 0.005 to 0.010 in. diameter as criterion. The effects on settler size of temperature, caustic strength, precent spent, and droplet size criteria are dramatic. The effects of these variables on settler size (for horizontal settlers of $L/D=4$) are illustrated in Table 4.

The most important operating variables of caustic treating/sweetening follow.

Treating Temperature. Extraction of mercaptans is increased at low temperatures, whereas air-inhibitor sweetening is more rapid at higher temperatures. The normal range of treating temperatures is 90 to 125°F.

Degree of Mixing. As the mixing intensity is increased, treating effectiveness improves, but so does the tendency for caustic to carry out of the settler. For treating units already in operation, the maximum acceptable mixing level should be determined by observing the amount of caustic entrainment at different recycle caustic rates and mix valve pressure drop settings. The nuisance or actual cost of accumulating and disposing of the caustic entrainment must then be weighed against the treating advantages.

Amount of Air Added. Stoichiometrically, it requires 14.8 SCF of air/lb of mercaptan sufur converted. Since 100% efficient utilization of the air is unlikely,

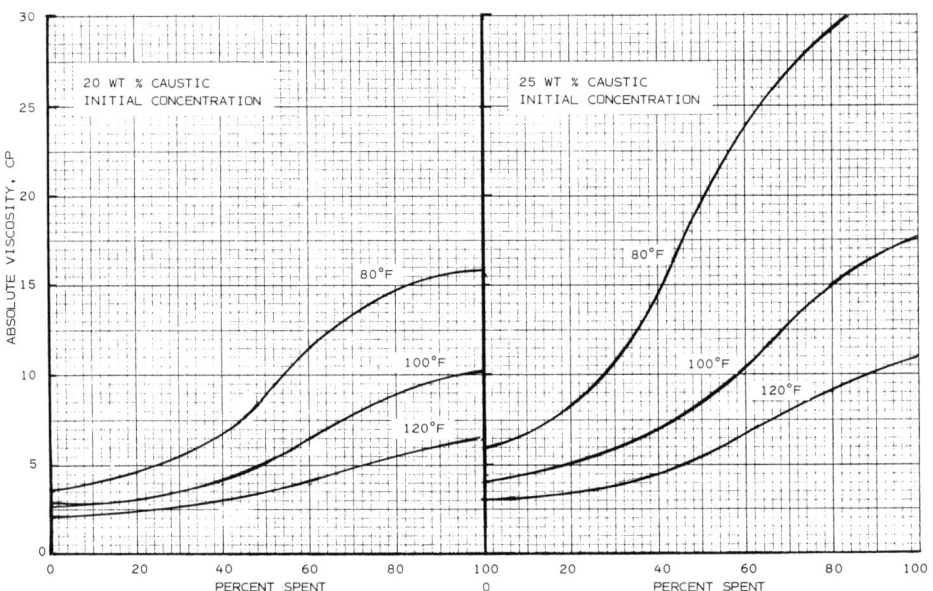

FIG. 2. Viscosity of cresylate solutions.

TABLE 4 Effects of Variables on Settler Size

Wt.% Fresh Caustic	Temperature (°F)	% Spent	Droplet Size (in.)	Size Settler for 10,000 bbl/d Naphtha (15% caustic recycle)
20	80	100	0.005	17' 01" × 68' 04"
20	100	100	0.005	13' 08" × 54' 08"
20	120	100	0.005	10' 10" × 43' 04"
20	100	33	0.005	7' 10" × 31' 04"
20	100	100	0.010	6' 10" × 27' 04"
25	80	100	0.005	24' 01" × 96' 04"
25	100	100	0.005	17' 08" × 70' 08"
25	120	100	0.005	13' 11" × 55' 08"
25	100	33	0.005	11' 08" × 46' 08"
25	100	100	0.010	8' 10" × 35' 04"

excess air should be used. The amount of air should be at least 30 SCF/lb of mercaptan, as long as this does not exceed the solubility of air in the stream being treated. Table 5 shows the approximate amounts of air that cracked gasoline can absorb.

Amount and Type of Sweetening Inhibitor. The sweetening rate is increased by increasing the inhibitor addition rate. Phenylenediamine sweetening inhibitors of various chemical structures are generally used. The active-ingredient content varies from supplier to supplier and must be considered by the refiner to determine inhibitor cost-effectiveness comparisons. In Table 6, performances of several cracked gasoline treating plants are summarized. Note that the treated product directly from the treater does not always pass the doctor test. Some refiners rely upon residual sweetening to occur in downstream lines and tanks.

TABLE 5 Solubility of Air in Cracked Gasoline at 100°F

Pressure (lb/in.² gauge)	SCF Soluble Air/ bbl Gasoline[a]	RSH-S[b] Equivalent (ppm wt.)
0	1.29	329
30	3.91	998
45	5.23	1337
60	6.54	1673
75	7.85	2009

[a] Air at 60°F and 14.7 lb/in.² gauge.
[b] Gasoline gravity of 6.30 lb/gal.

TABLE 6 Performance Summary of Commercial Treating Units

Inlet RSH-S (ppm wt.)	Outlet RSH-S (ppm wt.)	Air (SCF/bbl)	Temperature (°F)	Pounds Active Inhibitor/1000 bbl Gasoline	Passes Doctor Test
60.0	5.0	0.5	105	0.8	Yes
291.0	6.6	0.6	115	7.5	Yes
390.0	117.0	1.5	90	3.5	No
805.0	195.0	1.5	90	3.5	No

Caustic Strength Employed. Stronger caustics provide better sulfur extraction; however, to prevent severe caustic entrainment with conventional gravity settlers, it is recommended that no stronger than 25 wt.% caustic be used. The viscosity of the spent caustic (see Fig. 2) increases rapidly as the caustic becomes saturated with acid oils.

Acid Oil Content of the Treating Caustic. As mentioned earlier, the solutizing effect of acid oils in the treating caustic improves the extraction and sweetening of mercaptans by increasing their solubility in the caustic environment. Each fresh caustic strength has its own equilibrium acid oil content. Figure 3 shows equilibrium acid oil content versus fresh caustic strength, assuming that the gasoline stream being treated is saturated with water at the treating tempera-

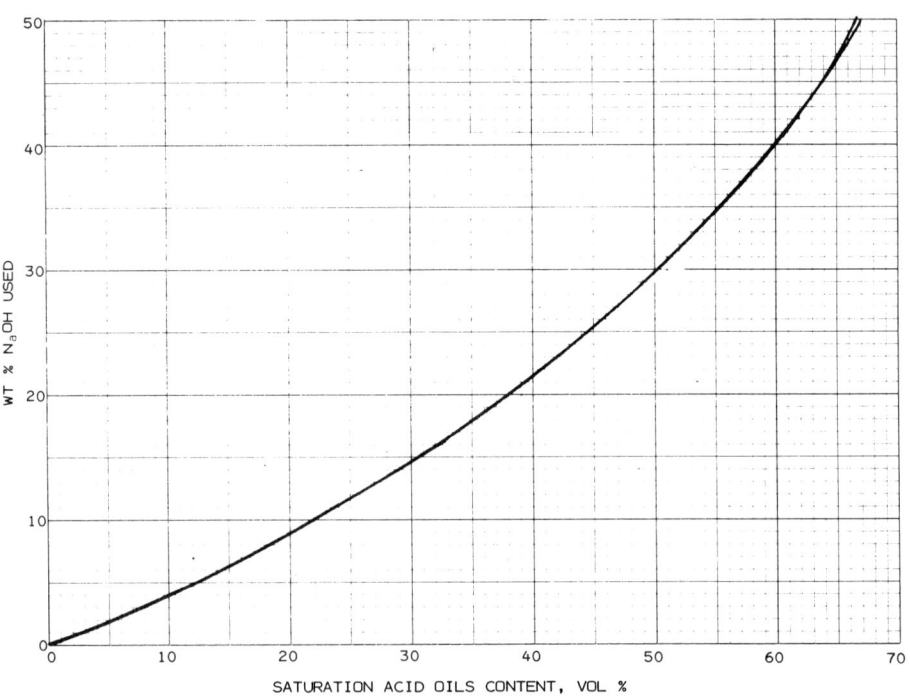

FIG. 3. Saturation limit for acid oils as a function of caustic strength.

ture. If the stream is dry, the gasoline dehydrates the caustic and the acid oils concentration builds above the curve shown on Fig. 3, whereas a wet stream prevents the acid oils from reaching the level shown in the curve.

Number of Treating Stages. Increasing the number of treating stages increases the treating efficiency by adding mixing, reaction time, and the opportunity for caustic containing less mercaptides to work on the cracked gasoline. The staging of the caustic and gasoline is countercurrent, providing a system that allows the freshest caustic to contact gasoline containing the least amount of mercaptan sulfur.

The design/operating variables relationships for mixer-settler caustic treating are illustrated in Fig. 4a and 4b. For a given settler design (superficial hydrocarbon upflow velocity of 0.01 ft/s), the overall performance of a mixer-settler stage as mixing intensity is increased is shown in Fig. 4a for the case where the hydrocarbon/caustic phase ratio is 20:1. As the mixing intensity is increased, the acidic material extracted increases, ultimately reaching virtually complete extraction. As the extraction efficiency reaches ~75%, however, the removal efficiency begins to suffer due to entrainment. Although an acidic

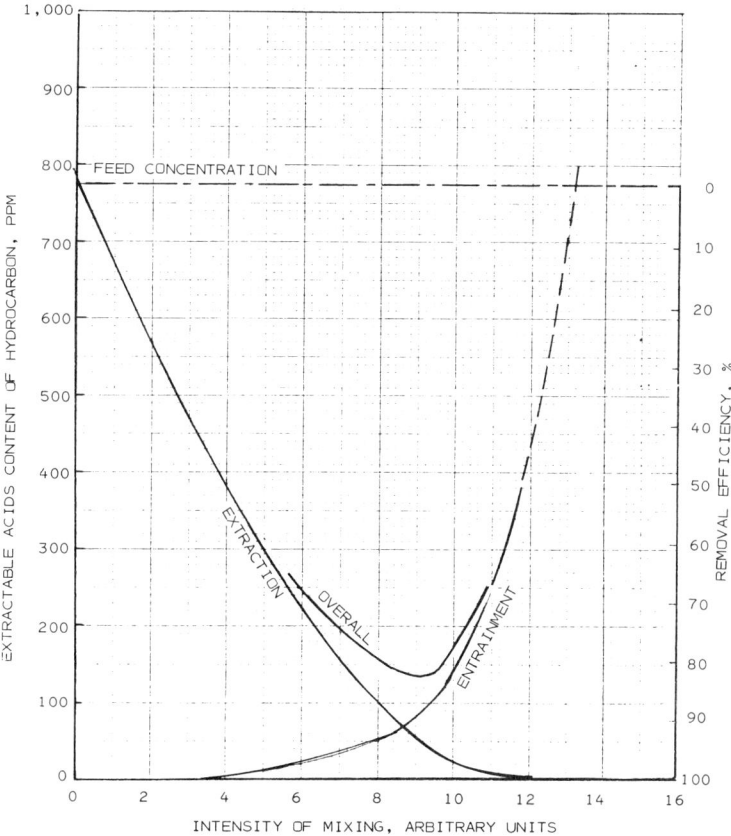

FIG. 4a. Performance of mixer/settler at a 20:1 phase ratio.

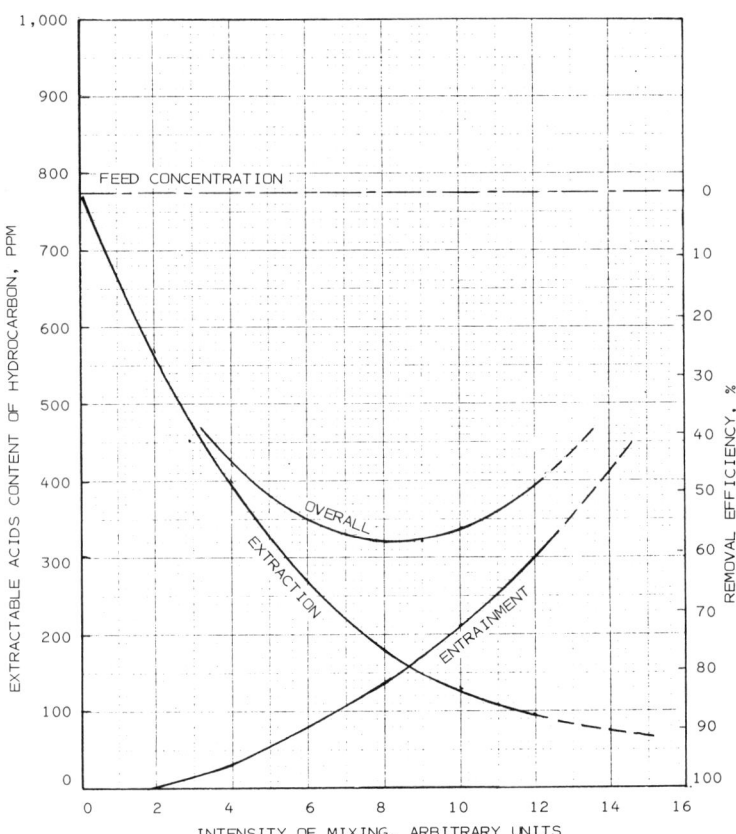

FIG. 4b. Performance of mixer/settler at a 120:1 phase ratio.

specie may be successfully extracted from hydrocarbon to the aqueous caustic phase, a portion of the aqueous caustic phase is entrained in the exit hydrocarbon phase and, hence, is not removed. Figure 4b shows the comparable case for a phase ratio of 120:1. At this smaller ratio the limit of extraction before entrainment becomes significant is only about 40%, and the optimum is only about 60% removal.

Technological Improvements

With the widespread use of caustic treating of gasolines, numerous improvements have been introduced. Electrostatic precipitation has been employed to minimize carryover of caustic in treaters [6].

Since the introduction of "static mixers" in the late 1960s [7], there has been considerable activity in utilizing these devices as the mixing element instead of mix valves. With these devices the size distribution of the dispersed droplets generated can be relatively narrow. Consequently, for a given extraction efficiency, the carryover of caustic is relatively smaller than with mix valves or other devices which utilize turbulent energy for mixing.

Desulfurizing Cracked Gasoline

A novel approach to liquid–liquid contacting in which phase dispersion is virtually eliminated was introduced by Merichem Co. in 1974 [8, 9]. As applied to caustic treatment of hydrocarbons, this technology is available under the service mark Merifining.

A contactor stage in this art (called a Fiber-Film contactor) consists of a bundle of generally aligned, long metal fibers of small diameter placed in a shroud and extending into a vessel as indicated in Fig. 5. The caustic solution is introduced upon and coats the surface of the metal fibers. The hydrocarbon stream flows as a continuous phase through the bundle, intimately contacts the caustic on the fiber surfaces, and pulls the caustic phase along the metal fibers by viscous-shear (drag) forces. Upon entering the vessel the hydrocarbon phase disengages from the bundle and is withdrawn while the caustic phase collects in the bottom of the vessel and is withdrawn. A very large interfacial surface area between the two liquid phases is thus generated for mass transfer, yet the caustic is never freely dispersed. Since settling of dispersed droplets is unnecessary, this allows operation without entrainment and eliminates the need for settlers, coalescing, water washing, and/or large rundown tanks.

A typical contactor handling 10,000 bbl/d of hydrocarbon consists of an 18-in. diameter shroud, is 8 ft in shrouded length, and provides approximately 8500 ft^2 of caustic/hydrocarbon interfacial surface. The bundle of fibers extends beyond the shroud and into the caustic collection zone of the disengagement vessel. The pressure drop through such a stage is only 2 to 5 lb/in.^{2} and the total hydrocarbon residence time can be 5 to 10 min.

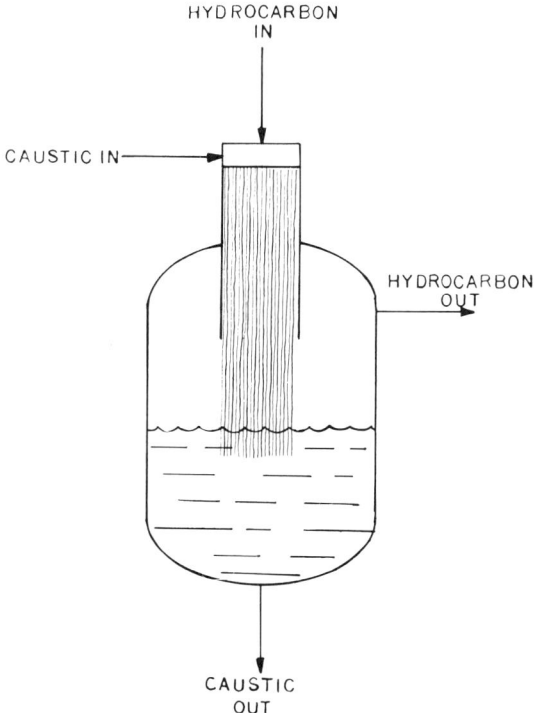

FIG. 5. Schematic diagram of Fiber-Film contactor stage.

References

1. F. J. Welcher (ed.), *Standard Methods of Chemical Analysis*, Vol. 2, Part B, 6th ed., Van Nostrand Reinhold, New York, 1963, p. 2005.
2. D. L. Yabroff, "Extraction of Mercaptans with Alkaline Solutions," *Ind. Eng. Chem.*, *32*, 257–262 (1940).
3. J. Happel and D. W. Robertson, "Removal of Mercaptans from Naphthas by Caustics," *27*, 941–943 (1935).
4. D. L. Yabroff and D. W. White, "Action of Solutizers in Mercaptan Extraction," *Ind. Eng. Chem.*, *32*, 950–953 (1940).
5. Anonymous, "Regeneration Steps Cut Cost of Removing Mercaptans," *Pet. Process.*, *3*, 435–438 (May 1948).
6. L. C. Waterman, "Electrical Coalescers," *Chem. Eng. Prog.*, *61*, 51–57 (October 1965).
7. Anonymous, *Business Week*, p. 156 (June 21, 1969).
8. Anonymous, "New Mass Transfer Method," *Chem. Eng.*, *81*, 54 (September 20, 1974).
9. B. E. Norris, "Mass Transfer without Mixing," *Hydrocarbon Process.*, *54* (September 1975).

K. E. CLONTS
RALPH E. MAPLE

Doctor Sweetening

Introduction

The Doctor Sweetening Process, or to use a more technically correct term, the Plumbite Sweetening Process, is one of the oldest petroleum refining processes extant. It survived the demise of the batch-shell stills with which the refining industry began, and it preceded the introduction of the continuous pipe still, which is a name only the older refinery operators will recognize as being the descriptive term for modern crude distillation units. Acid treating of petroleum distillates was once a common refining technique, particularly for thermally cracked gasoline to remove gum-forming olefins and diolefins, and was certainly contemporaneous with doctor treating. There could be arguments as to which was first used, but acid treating has disappeared as a means of refining gasoline and kerosine fractions. There are still some doctor sweetening units in operation and there exists a modern version of the principles of doctor treating—the Bender Process—that uses litharge deposited on a bed of blast furnace slag over which the feed, caustic soda, oil, air, and elemental sulfur are simultaneously percolated. This process finds its principal use in treating virgin distillate in the kerosine and furnace oil boiling range at the present time.

There appears to be no definite information on when the Doctor Sweetening Process was first used or what function it was originally intended to serve. It is

© 1982 UOP Inc., All Rights Reserved.

reported to have been first employed about 1910 [1]. It seems almost certain that the process evolved from "rule of thumb" practices, perhaps related to metallurgical practices as applied to problems of the infant petroleum industry for removal of hydrogen sulfide and elemental sulfur and improving odor associated with both H_2S and mercaptan. There is some doubt that those early refiners were aware that mercaptan was responsible for the sickening odor they perceived.

In reviewing the early literature on chemical treatment of petroleum products, it is obvious that chemists of distinction such as J. C. Morrell and Gustav Egloff as late as 1923 [2] were not certain of the function or chemistry of the process. For example, Morrell and Egloff appear to have been under the impression that the function of the process was to remove elemental sulfur and they gave the following erroneous equation for the reaction:

$$2Na_2PbO_2 + 4S^0 \rightarrow Na_2S + Na_2SO_4 + PbS \qquad (1)$$

It is interesting to note that these authors did not mention the word "mercaptan" in their discussion of the "Interpretation of the 'Doctor Test.'"

As understanding of the chemistry of the process improved, it became generally recognized that the principal purpose of the Doctor Sweetening Process is to eliminate malodorous mercaptans:

$$RSH + \frac{1}{4}O_2 \rightarrow RSSR + \frac{1}{2}H_2O \qquad (2)$$

Originally it may have been used for the removal of H_2S, which also has a nauseous odor, and, as Morrell and Egloff indicated, for the removal of elemental sulfur which is a cause of corrosiveness and which is the reaction product of H_2S and atmospheric oxygen. In the light of later knowledge of the chemistry involved, it is possible that the process worked for the removal of malodorous mercaptan and elemental sulfur in those early days only when both happened to be present, which may have been the usual case.

Wendt and Diggs [3] in 1924 appear to be the first authors to explain the basic chemistry of the process. When oil containing mercaptan is brought into contact with a concentrated aqueous solution of sodium plumbite (doctor solution) containing an excess of sodium hydroxide, the followiof reaction takes place:

$$Na_2PbO_2 + 2RSH \rightarrow Pb(SR)_2 + 2NaOH \qquad (3)$$

where R is a hydrocarbon radical.

The lead mercaptide may or may not be oil-soluble depending upon its molecular weight and structure. For the most part the mercaptans found in gasoline and kerosine form oil-soluble lead mercaptides.

Upon completion of Reaction (3), addition of elemental sulfur to the reaction mixtures causes the following reaction to take place:

$$Pb(SR)_2 + S \rightarrow RSSR + PbS \qquad (4)$$

The disulfide is oil-soluble and remains in the oil. The lead sulfide is a heavy precipitate not soluble in either the oil or the spent doctor solution. While disulfide has a somewhat disagreeable odor, it is much, much less offensive than mercaptans.

At later dates, Ott and Reid [4], Schulze and Buell [5], and Birch and Stanfield [6] showed that the chemistry of the process is much more complex than indicated by Wendt and Diggs, with several complex competing side reactions taking place. The most important of these side reactions are those that lead to the formation of polysulfides, $RSSS_xR$, which are quite undesirable for reasons discussed below:

$$2(RS)_2Pb + (x + 2)S \to RSSR + RSSS_xR + 2PbS \tag{5}$$

Wendt and Diggs did not speak of regeneration of the spent doctor solution to recover the litharge from the precipitated lead sulfide, but it is known that recovery of lead was practiced. Originally, the lead sulfide was converted, after settling and decanting the spent solution, to lead oxide by roasting on a heated steel plate open to the atmosphere. This is akin to ore roasting and resulted in evolution of residual hydrocarbons, carbon monoxide, sulfur dioxide, and much smoke, probably containing particles of lead sulfide, lead oxide, sodium hydroxide, and sodium carbonate. The dominant reaction for reconstitution of lead oxide by this means is

$$PbS + \frac{3}{2}O_2 \to PbO + SO_2 \tag{6}$$

In the light of modern air pollution concerns, it is shocking to think that such practices existed. Roasting was still in use in some places as late as 1945.

At some point in the evolution of the process, and this discovery is apparently not documented, it was learned that the sodium plumbite could be reconstituted merely by blowing the spent slurry of lead sulfide and excess caustic soda solution with air at elevated temperature. The reaction is usually depicted as

$$PbS + 4NaOH + 2O_2 \to Na_2PbO_2 + Na_2SO_4 + 2H_2O \tag{7}$$

This reaction, too, is more complex than indicated, with evidence showing several alternative routes. Evidence indicates that the principal reaction is

$$PbS + O_2 + 3NaOH \to Na_2PbO_2 + \frac{1}{2}Na_2S_2O_3 + \frac{3}{2}H_2O \tag{8}$$

with perhaps 15% of the total sulfoxide salts being in the form of sulfate and 85% in the form of thiosulfate. When regeneration of the spent solution is conducted in the liquid phase, there is an inherent net chemical consumption of caustic soda as a consequence of the formation of sodium sulfate and sodium thiosulfate.

With the advent of tetraethyl lead (TEL) as an antiknock agent in gasoline, and with the pioneering work of Universal Oil Products (UOP) Co. in the commercial use of oxidation inhibitors in the early 1930s to prevent gum formation that occurs through oxidation of olefins in cracked gasoline, certain disadvantages of doctor treating that had been ignored, or not previously recognized, demanded attention.

Endo [7] and Schulze and Buell [5] observed that sulfur compounds in gasoline had an antagonistic effect on the susceptibility of gasoline to octane number improvement by the use of TEL. Ryan [8] developed a correlation showing the quantitative relationship between concentration and type of sulfur compounds on the suppression of TEL susceptibility of gasoline. Ryan showed that polysulfide sulfur, one of the products of the doctor sweetening reaction that cannot be completely avoided, is 2.3 times more harmful than either mercaptan sulfur or disulfide sulfur.

Lowry, Dryer, and Wirth [9], in their pioneering efforts to apply oxidation inhibitors to thermally cracked gasoline to prevent gum formation to eliminate the 3 to 10% gasoline losses suffered during acid treatment, as then commonly used, recognized the need for improved doctor sweetening if their efforts were to be successful. They undertook the first comprehensive, practical study to determine the effect of doctor sweetening operating variables on product quality. In addition, as a companion work, Wirth, Lowry, and Strong [10] studied the operating variables of doctor solution regeneration. The Lowry et al. work was carried out in the laboratory, in a 500-bbl/d full-scale experimental treating plant, and in treating plants of several petroleum refineries.

The design and operating criteria discussed below are based primarily on the Lowry et al. work, together with a feedback of field experience by the UOP treating engineers up to about 1958 when doctor sweetening was rendered obsolete by the successful introduction of the UOP Merox (registered trademark of UOP Inc.) Process [11]. The UOP Merox Process accomplished the same objective as doctor sweetening at a much, much lower cost, with greatly reduced pollution potential, and with no possibility of harming the quality of the product. Moreover, the Merox Process could also be used, in one of its operating modes, to extract mercaptans and thereby reduce sulfur content. A good deal of the early success and rapid acceptance of the Merox Process was because of the possibility of converting existing doctor sweetening equipment to Merox application.

Chemical Requirements

Despite the complexity of the chemistry involved, the simple equations of Wendt and Diggs, (3) and (4), plus the regeneration equations (7) and (8) can be used to approximate the theoretical chemical requirements for doctor sweetening and solution regeneration.

Based on 1 mol of mercaptan, the doctor sweetening reactions can be written as

$$RSH + \frac{1}{2}Na_2PbO_2 \rightarrow \frac{1}{2}Pb(SR)_2 + NaOH \qquad (9)$$

$$\tfrac{1}{2}\text{Pb(SR)}_2 + \tfrac{1}{2}\text{S} \to \tfrac{1}{2}\text{PbS} + \tfrac{1}{2}\text{RSSR} \tag{10}$$

and, on the assumption that the molar ratio of $\text{Na}_2\text{S}_2\text{O}_3/\text{Na}_2\text{SO}_4$ in the regenerated solution is 85/15, the regeneration reactions can be combined and written as

$$0.500\text{PbS} + 1.540\text{NaOH} + 0.540\text{O}_2 \to$$
$$0.500\text{Na}_2\text{PbO}_2 + 0.230\text{Na}_2\text{S}_2\text{O}_3 + 0.0406\text{Na}_2\text{SO}_4 + 0.772\text{H}_2\text{O} \tag{11}$$

The chemical requirements in Table 1 are calculated from these equations. Also indicated in Table 1 is the chemical consumption actually expected on the basis of operating experience for an efficient doctor sweetening unit.

Process Flow and Effect of Operating Variables

The work of Lowry et al. showed that the variables having the greatest effect on doctor sweetening results are

1. Proper mixing of feed and doctor solution
2. Quantity of sulfur used
3. Point of sulfur injection
4. Quantity and condition of doctor solution used
5. Operating temperature

and that the variables having the greatest effect on the rate of regeneration of the solution to keep it in the proper condition are

TABLE 1 Chemical Requirements for Doctor Sweetening

	Theoretical Requirements		Actual Consumption for Design Purposes (lb/lb, RSH—S)
	mol/mol RSH—S	lb/lb RSH—S	
PbO	0.5	3.5	0.35[a]
S	0.5	0.5	0.65
NaOH	1.54	1.92	3.84[b]
O_2	0.54	0.54	—
Air (oxygen source)	—	30.5 (SCF)	60[c]

[a] About 90% of the litharge required can be recovered in the regeneration step.
[b] Doctor solution becomes ineffective when about $\frac{1}{2}$ of the NaOH has been converted to equivalent sodium thiosulfate, so about 2 times the theoretical amount of NaOH is consumed.
[c] Two to 3 times the theoretical air requirement is needed depending upon mixing efficiency in the regeneration column, the temperature, and how thoroughly hydrocarbons are stripped from the spent solution prior to regeneration. Another factor affecting air requirements is the amount of oxidizable phenolic materials that are present in the spent solution to consume oxygen.

Doctor Sweetening

1. Temperature
2. Mixing intensity of air and doctor solution
3. Quantity of air used
4. Quantity of litharge in the solution at any time
5. Quantity of dissolved salts (Na_2SO_4 and $Na_2S_2O_3$) in the solution
6. Thoroughness of oil removal from the spent solution before regeneration

Mixing

To bring about Reaction (3), a doctor solution consisting of plumbite (PbO) dissolved in an excess of aqueous sodium hydroxide solution is intimately mixed with the sour feedstock consisting of a petroleum fraction containing mercaptan. When elemental sulfur is added to this mixture, usually in the form of sulfur dissolved in a portion of the feedstock, Reaction (4) proceeds.

Sufficient sulfur must be used, of course, to complete the reaction. When only the theoretical amount of sulfur is added, Reaction (4) proceeds slowly. The reaction can be speeded up by adding an excess of sulfur; however, excess sulfur results in leaving unreacted elemental sulfur in the product or in the formation of polysulfide as shown in Reaction (5), or both. The point in the mixing sequence at which the sulfur is added also affects the amount of excess sulfur needed and consequently the amount of polysulfide formed.

Lowry found that if sufficient litharge is supplied to react with all of the mercaptan present, and Reaction (3) is completed before sulfur is added, that the amount of polysulfide formed is minimized and the best possible product quality is achieved.

Referring to the primary mixer arrangement shown in Fig. 1 or 2, it will be noted that doctor solution and feed are premixed for a short time before the sulfur–oil is introduced. After the sulfur is introduced, mixing is continued.

Mixing must be sufficiently intense to keep the two phases in intimate contact during the course of Reaction (4), but it must not be so intense that a stable emulsion is formed that will not separate in the settler provided. Residence time in the mixer should be sufficient to complete Reaction (4). In this way the lead sulfide that forms a molecule at a time in the oil phase, and is, therefore, very finely divided, is coagulated and washed out of suspension by the doctor solution. If the reaction goes to completion in the settler, the lead sulfide which then forms settles from the oil very slowly and may be entrained in the product leaving the settler.

It is to overcome the problem of lead sulfide entrainment from the primary settler that the secondary mixer-settler system is sometimes added, wherein regenerated doctor solution is used to recover lead sulfide entrained from the first settler. The solution used in the secondary mixer-settler suffers no consumption of litharge and after picking up lead sulfide can be used in the primary mixers. The secondary mixer-settler is not a necessity, but it enables the operator to use the very minimum of excess sulfur without running the risk of lead sulfide entrainment if he should momentarily use too little excess sulfur and thereby not get a complete reaction in the primary mixer.

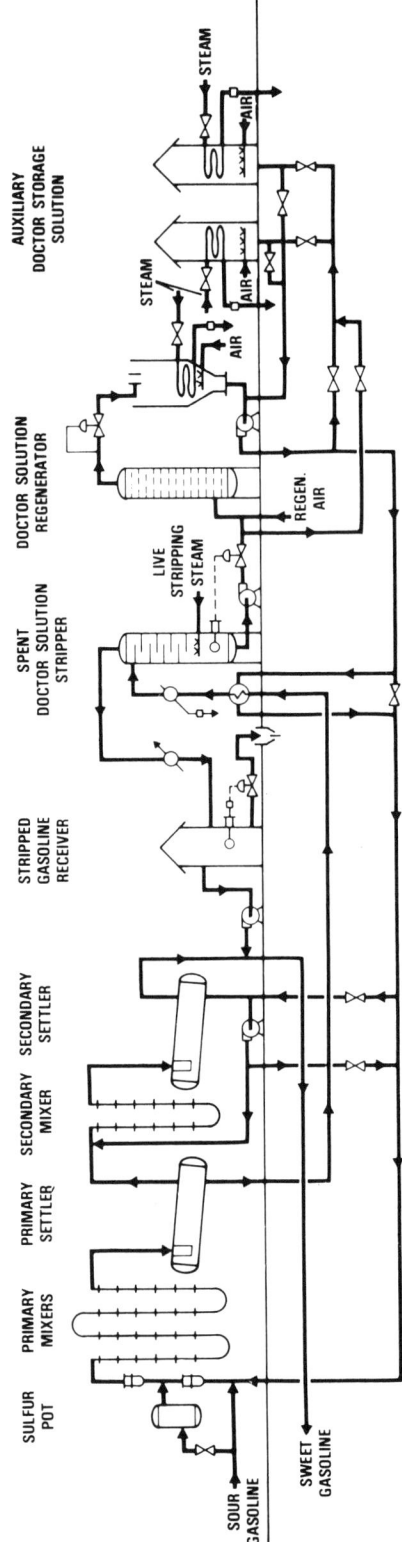

FIG. 1. Simplified flow chart of doctor sweetening unit. Once through doctor solution on sweetening; once through doctor solution on regeneration.

Doctor Sweetening

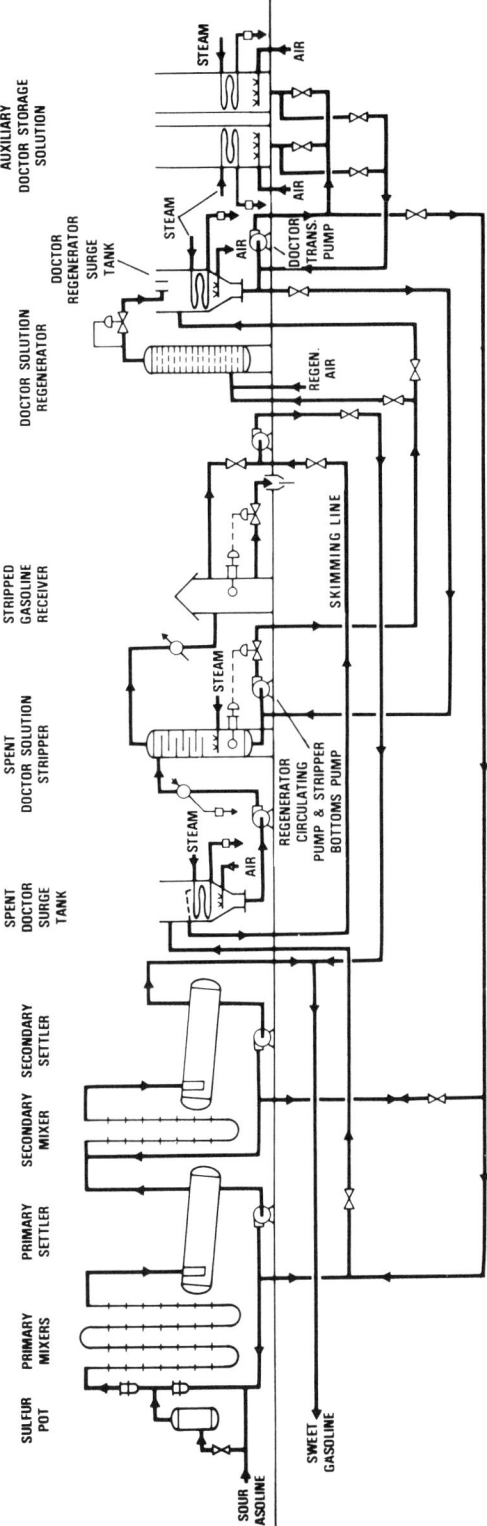

FIG. 2. Simplified flow chart of doctor sweetening unit. Recirculating doctor solution on sweetening. Batch stripping: 16 h/d. Batch regeneration: 8 h/d.

As shown in the regeneration Reactions (7) and (8), sodium sulfate and thiosulfate are formed at the expense of sodium hydroxide and sulfur. It is important to maintain the alkalinity of the solution by periodically adding caustic soda to replace that consumed. Ultimately, the solution must be discarded because it will no longer be possible to keep the plumbite in solution because of salt accumulation. This becomes evident either by a failure to bring about oxidation of lead sulfide when conditions are otherwise proper for regeneration, or by a tendency of the nearly saturated sulfate salt solution to form emulsions in the mixers and to cause foam in the regenerator. In some extreme cases, salt crystals form and the solution solidifies. Recovery of lead from spent solution before discard is discussed below.

Treating temperature is important. For gasoline and lighter stocks, sweetening will take place readily at temperatures about 80°F; for kerosine and stove oils it is best to use a temperature of 120 to 140°F. When the treating temperature is too high, say, above 140°F, other oxidation reactions that are unimportant at lower temperatures can take place and lead to product deterioration; for example, phenols and nitrogen compounds can oxidize to form colored compounds undesirable in the product from an aesthetic standpoint. Elemental sulfur can react directly with hydrocarbons to increase the sulfur content, and some peroxidation of cracked stocks might take place which would greatly reduce the effectiveness of oxidation inhibitors.

After separation of the spent doctor solution from the treated hydrocarbon, regeneration is usually carried out by intimately mixing it with air at 180 to 200°F. Simply sparging air through doctor solution in a tank will effect regeneration, but recirculation through a regeneration column such as shown in Fig. 3 greatly speeds the reaction and reduces the amount of air needed. The higher the temperature the faster the reaction, of course, and the limiting temperature is usually set by the temperature of available steam. Aside from temperature, the most important variable is mixing intensity. Excess air above the theoretical requirements is not beneficial per se except to the extent that it increases mixing intensity.

The rate of oxidation of lead sulfide is affected by the amount of lead oxide present. Thus the regeneration rate of a PbS slurry containing no PbO increases considerably after 1 to 2 lb of PbO per barrel have formed. For this reason it is advisable to provide for enough solution in the primary mixer so that the spent solution withdrawn from the settler to regenerator still contains 1 to 2 lb of PbO per barrel.

As pointed out above, it is important that all of the oil in the spent solution is thoroughly removed. The presence of oil with the solution in the regenerator can lead to coating of the lead sulfide particles with a varnishlike deposit and thus slow regeneration. This is particularly true when treating cracked products. In some cases such varnishlike coatings have been known to make regeneration impossible. Thorough settling of the spent solution to separate out suspended oil, or preferably stripping with steam to remove oil, is a very important step in a doctor sweetening operation. In some cases an emulsion of gasoline and doctor solution, which is so stable that it will not separate in several days, may be present in the spent solution. In such cases steam stripping is a necessity.

In some cases, for a reason never fully explained, the precipitated lead

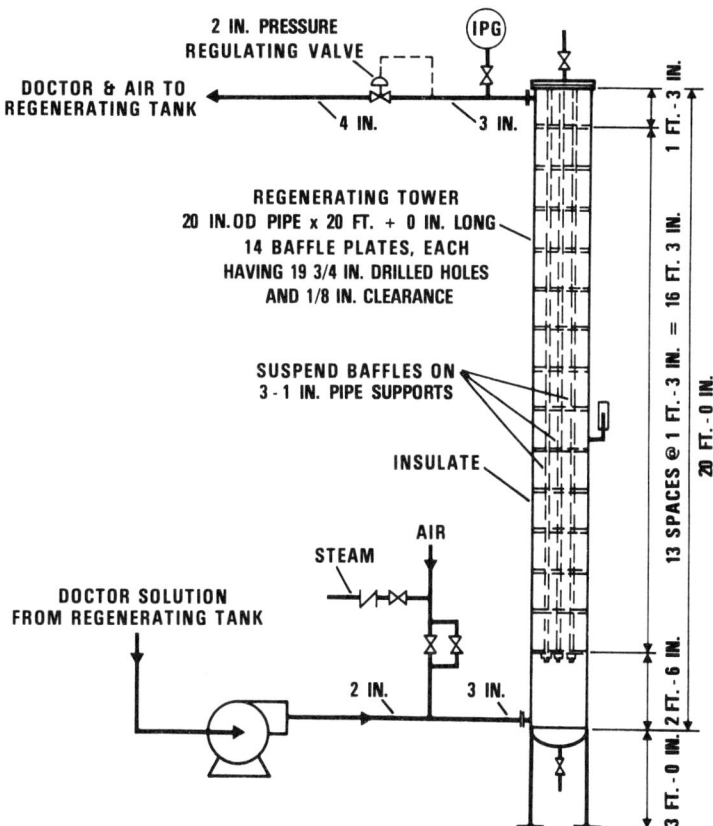

FIG. 3. Doctor solution regenerating tower.

sulfide is preferentially wetted by the feed rather than by the doctor solution. This phenomenon is often observed when treating stocks that have a higher than normal aromatic content and also have a high naphthenic acid content. When the lead sulfide is not wetted by the doctor solution, it floats at the interface between the oil and the solution in the settler. It is impossible to recover the lead sulfide from the interface in such cases without withdrawing considerable oil with the spent solution. In those cases use of a steam stripper is a necessity.

The effect of accumulated dissolved salts in used doctor solution was discussed above—if PbO cannot dissolve in the doctor solution, none will be formed from the PbS.

Effect of Nonmercaptan Acidic Components in the Feed

Since doctor solution is highly alkaline, it will extract acidic materials from the feed. Thus hydrogen sulfide will be extracted.

Some naphthenic acids will react with lead to form oil-soluble plumbous naphthenate. If this should remain in the treated solution, it can oxidize in storage to the plumbic salt which is not oil soluble, thus giving rise to development of a finely divided precipitate in an oil that went to storage free of precipitate.

Not only will the H_2S neutralize the alkaline solution, but the sulfide ion will also precipitate the lead as sulfides. A small amount of H_2S can be tolerated, but to the extent that it is present, it increases NaOH consumption and increases the regeneration load.

It is more efficient to remove hydrogen sulfide with caustic soda in a prewash where greater utilization of sodium hydroxide can be achieved than to allow it to be removed by the doctor solution.

Naphthenic acids and sulfonic acids are excellent stabilizers of oil–caustic solution emulsions. It is impossible to operate a doctor plant if there is more than a trace of naphthenic acid in the feed. The same is true to a lesser extent of aliphatic acids. Naphthenic acids are always present to some extent in virgin stocks boiling above about 300°F but are not found in lower boiling stocks. Aliphatic acids are found in all boiling ranges of thermally cracked stocks derived from naphthenic acid-bearing crudes. Apparently the aliphatic acids are fragments of naphthenic acids.

Phenol and its homologs appear in both virgin and cracked stocks (both thermal and catalytic) boiling above about 300°F. The concentration in virgin gasolines is low and does not interfere. In higher boiling virgin stocks (kerosine) the phenolic materials are not sufficiently extracted by the doctor solution to cause sweetening problems, but they often cause color problems. It appears that the phenols that are extracted become oxidized to some extent in the regeneration step and are then redissolved by the kerosine when the solution is reused. It is only where such oxidized phenols are colored that color loss is experienced. Color loss in sweetening kerosine can usually be avoided if the solution is not reused. This is only possible, of course, if there are other doctor treaters in operation that can use the once-used solution from the kerosine sweetener.

Phenols in cracked gasoline are extracted and accumulate in the doctor solution. They are not particularly harmful. They may cause some discoloration of the gasoline for the reasons mentioned above. They have some tendency to stabilize emulsions between gasoline and doctor solution, probably because in the concentrations to which they accumulate they tend to increase the viscosity of the solution.

Design of a Doctor Sweetening Plant

There are probably as many design techniques used in doctor sweeteners as there are designers who have tried their hand at it. The basic objectives have been discussed above. The technique used by one designer for accomplishing these objectives is outlined below for a feed pretreated to remove hydrogen sulfide, naphthenic acid, and sulfonic acid.

Doctor Sweetening

Preparation of Doctor Solution. Provide a means of adding litharge to heated caustic soda solution. A chart showing the solubility of PbO in a NaOH solution is given as Fig. 4. A convenient solution that will give good results can be made up of 10 lb of litharge added per barrel of a 20°Bé NaOH solution (14.4 wt.% NaOH or 58.4 lb of NaOH/bbl).

A convenient method of adding litharge is to dump it manually from the shipping drums directly into a tank of hot caustic solution while gently agitating the solution using an air sparger in the bottom of a suitable tank.

Doctor Solution Requirements. Provide a doctor solution circulating pump of sufficient capacity to mix not less than 10 volumes of regenerated doctor solution with 100 volumes of the oil to be treated. This minimum volumetric ratio is to assure good contact in the mixer. The net quantity of doctor solution should be such that after completion of Reaction (3) there should still be about 2 lb of unreacted PbO per barrel of solution; in other words, 8 lb of litharge converted to PbS per barrel of solution. If the feedstock is of low mercaptan content, it may be necessary to recirculate solution from the settler back to the mixer to maintain the $\frac{1}{10}$ volumetric ratio and to react 8 lb of litharge per net barrel of solution provided to the mixer. This recirculation arrangement is shown in Fig. 1.

Mixer. A typical mixing system is shown in Fig. 5. There have been many other mixing systems in use over the years, but the one shown is effective, easily built, and easily cleaned if plugging occurs.

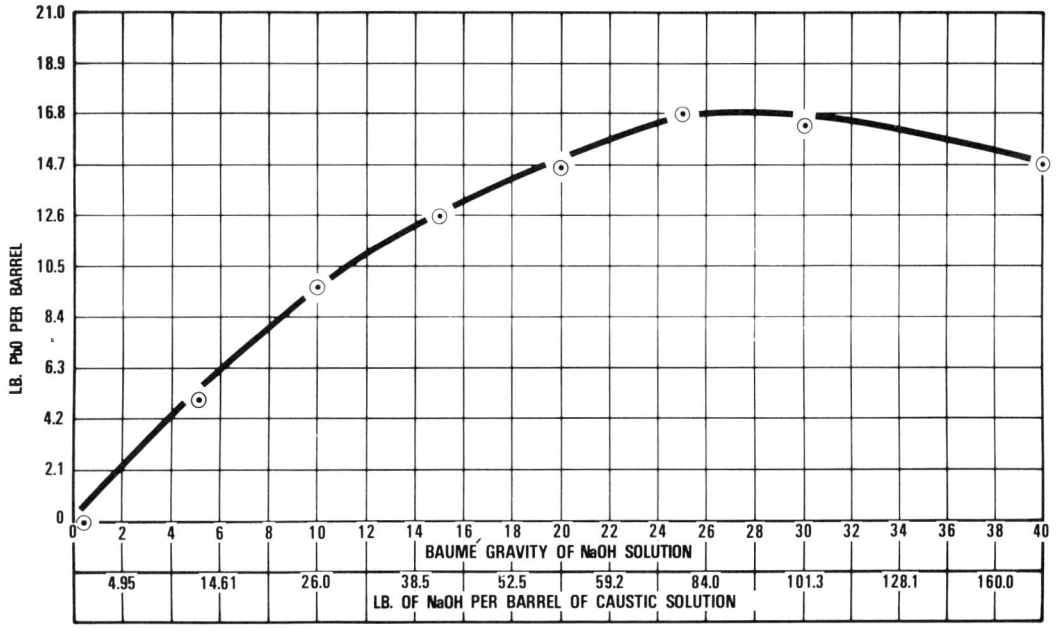

FIG. 4. Solubility of PbO in NaOH solution at 90°F.

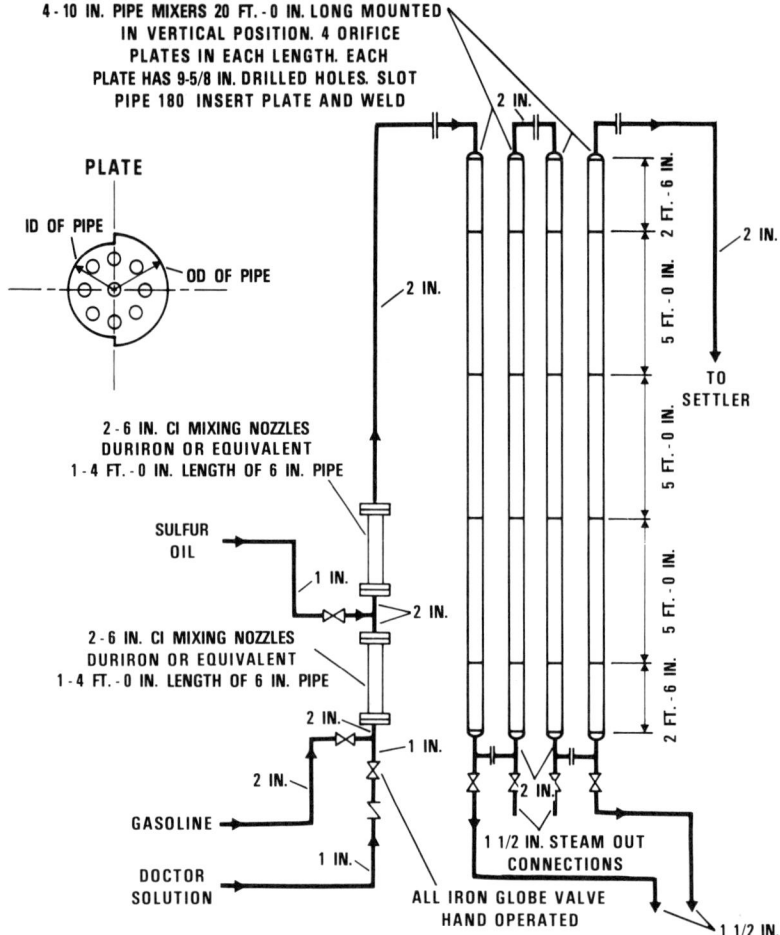

FIG. 5. Gasoline and doctor mixing equipment: 100 bbl/h gasoline, 4 min mixing time.

Premixing. Provide a mixer which can be of the cast Durion type, or a set of orifice plates that will provide 20 to 30 s residence time for the doctor solution and the oil to be sweetened with about a 3 to 5 lb pressure drop in which to complete Reaction (3). If a set of orifice plate mixers is to be used, provide at least three plates in series with a multiplicity of holes in each plate and 2-ft spacing between plates. Calculate the pressure drop using ordinary orifice formulae and an average density for the oil–doctor mixture. Great accuracy in the calculation is not necessary.

Sulfur Oil Injection. Bypass a portion of the feed around the premixer and allow it to flow upward through a vessel filled with lump or stick sulfur. The quantity that must be bypassed will depend upon the temperature and the mercaptan content of the feed. The amount can be estimated from the sulfur

Doctor Sweetening

solubility chart, Fig. 6. In Table 1, for design purposes, the amount of sulfur needed is given as 0.65 lb/lb mercaptan sulfur to be converted, but in operation the quantity of sulfur used must be continually modified by the operator by manual adjustment of the sulfur oil rate depending upon results of tests made on samples withdrawn from the outlet of the primary mixer. This is a most important control for successful doctor sweetening operation, and careful attention must be given to providing good temperature control and fine adjustment of the sulfur oil rate.

The sulfur pots, usually two, connected for both parallel and series flow with either one first in line, should be designed with a sufficient volume of sulfur so that they may be topped up with sulfur at convenient, say, 2-week intervals.

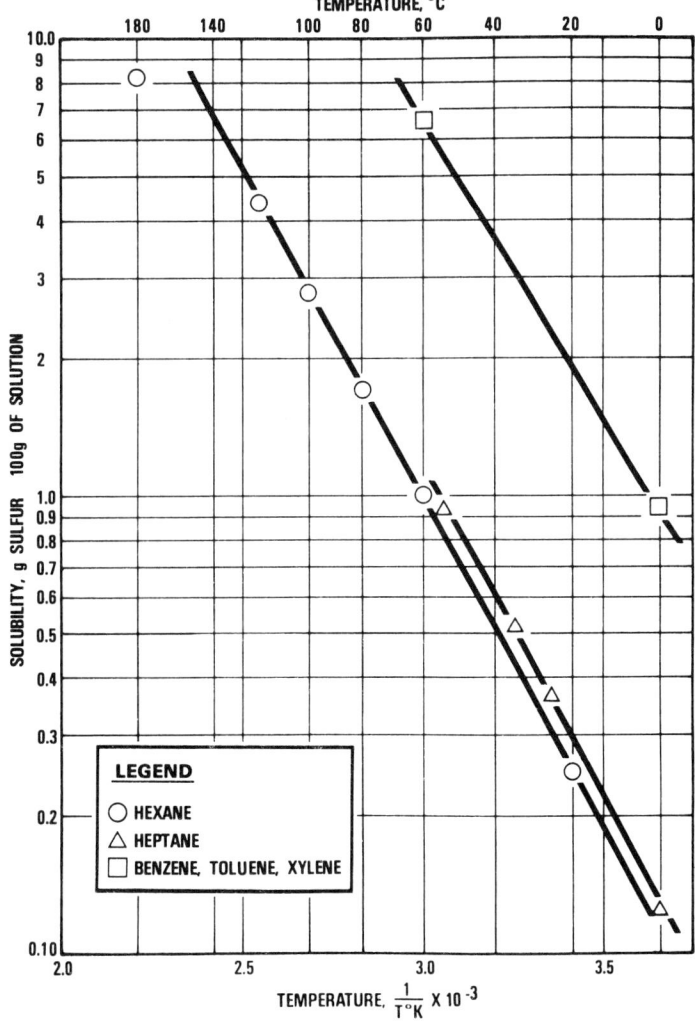

FIG. 6. Solubility of sulfur in hydrocarbons. (Based on A. Seidel, *Solubility of Inorganic and Metal Organic Compounds*, Vol. 1, 3rd ed., Van Nostrand, New York, 1940.)

Primary Mixer. Residence time in the primary mixer may vary from about 2 min for low boiling, easily sweetened materials to as much as 20 min for a high boiling, high mercaptan, difficultly sweetened kerosine or furnace oil. For an unknown stock a batch laboratory test is needed to estimate the mixing time. Details of the laboratory test are given in the Appendix. With a mixer of the type shown in Fig. 5, a pressure drop of 10 to 15 lb/in.^{2} gauge will provide adequate mixing without resulting in stable emulsions. Several connecting flanges and steam-out connections should be provided so that plugging can be easily cleared up if it should occur.

Primary Settler. The usual practice is to provide for 1-h hydrocarbon residence time in the settler for gasoline and 2 h for kerosine. The settler should be sloped at 5° or more from the horizontal so as to minimize the amount of settled PbS cake that accumulates in the bottom. A baffle of some sort should be provided at the inlet to reduce turbulence. The doctor solution outlet should be provided with a vortex spoiler to minimize the amount of hydrocarbon withdrawn with the doctor solution. Judiciously located manways and collector basins should be provided for flushing accumulated lead sulfide out of the settler when the unit is shut down.

Secondary Mixer and Settler. The secondary mixer is not a necessity but is considered worthwhile, especially when treating higher mercaptan cracked gasoline or kerosine where product quality (TEL susceptibility and inhibitor susceptibility for the gasoline and color loss and corrosion test for the kerosine) is critical. Only 2 min mixing time and 5 to 10 lb/in.^{2} pressure drop are needed in the secondary mixer. The secondary settler should be identical to the primary settler.

Doctor Solution Regeneration. Both batch and continuous doctor solution regeneration have been commonly used. The choice between the two usually depends upon whether the treating plant is to be staffed for 8 or 24 h. If only one experienced operator is to be employed, batch regeneration is used. Large plants will usually regenerate continuously when equipment cost savings offset the extra cost for 24 h operator coverage. For those plants practicing batch regeneration, an unskilled operator or an operator from another nearby unit will usually be used to monitor the sweetening section of the plant which is preferably run on a 24-h schedule. (Batch operation of the sweetening plant as well as batch regeneration is sometimes used for small plants when intermediate storage tanks are available.)

Batch Regeneration. In the case of batch regeneration, the only added piece of equipment not found in continuous regeneration is the spent doctor surge tank. This should be a cone bottom tank of sufficient size to hold at least a 1-d supply of spent doctor solution plus freeboard for emulsion accumulation. It should be equipped with a steam coil to keep the solution hot for better separation of emulsified oil. It should also be fitted with an air sparger to be used to put settled lead sulfide into suspension before transferring the spent solution to the regeneration system. A skimming line should be provided to remove separated oil floating on top of the doctor solution.

Doctor Sweetening

Spent Doctor Solution Stripper. Not all doctor plants will be equipped with a stripper. If a unit is not equipped with a stripper, it should always have a spent doctor surge tank to remove entrained and emulsified oil from the solution before going to the regenerator, whether or not regeneration is continuous or batch.

A doctor solution stripper is shown in Fig. 7. The quantities and sizes shown can be prorated for other capacities. The stripper is for the purpose of removing emulsified and suspended oil from the doctor solution prior to regeneration. Since it is impossible to predict exactly how much oil may be present, the quantity of stripping steam needed can only be roughly estimated. A stripper to remove kerosine needs more steam per unit of oil stripped than one removing gasoline. The stripped doctor solution should flow from the stripper directly to the regenerator to avoid the need for reheating.

Frequently a solution stripper will be mounted on top of a doctor regenerator surge tank so that the stripped solution can flow by gravity into the regeneration tank.

In those units not using a stripper, the solution coming from the surge tank is heated to 200°F and flows directly to the regeneration system.

Regeneration. As stated above, regeneration can be accomplished merely by air blowing in a tank, but a regeneration column is much more efficient. In operation, spent doctor solution containing suspended lead sulfide and an

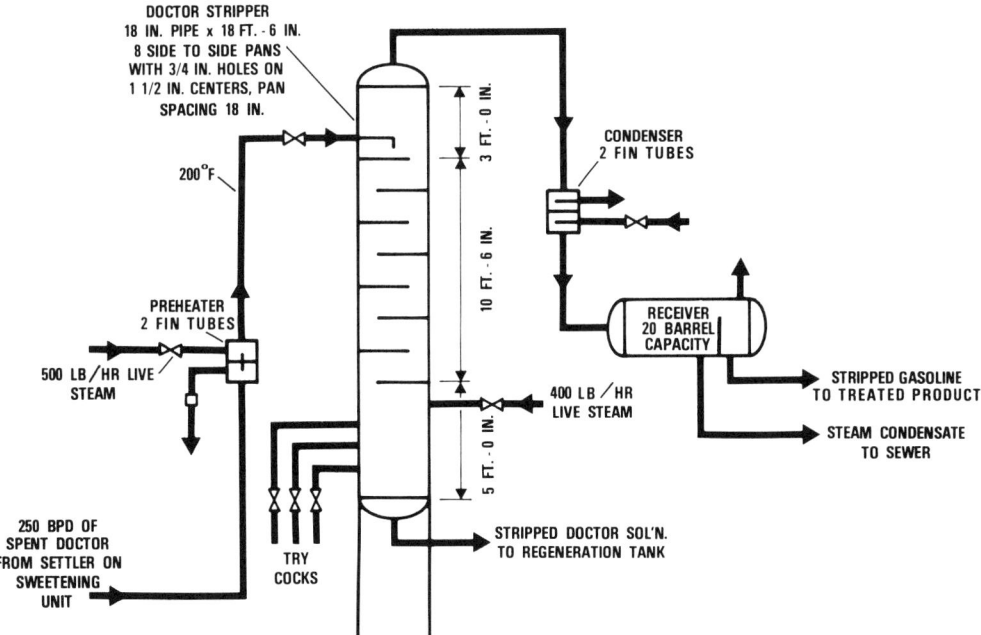

FIG. 7. Doctor solution stripping column: 250 bbl/d spent doctor containing 12% (vol.) gasoline in suspension.

excess of sodium hydroxide, and preferably containing 1 to 2 lb of unreacted litharge per barrel, is circulated from the bottom of a cone bottom regeneration surge tank upward through a perforated plate mixing column. Air is blown into the bottom of the column and flows upward with the solution. A steam coil in the bottom of the regeneration tank keeps the preheated solution hot. An air spider in the bottom of the regeneration tank keeps the unreacted lead sulfide in suspension. In batch regeneration, circulation through the tank is continued until all of the lead sulfide has been oxidized. In continuous regeneration the equipment is sized to give complete oxidation of all the lead sulfide in one pass.

A typical doctor regeneration column is shown in Fig. 3, and design criteria are given in Table 2.

The regenerated doctor solution must be cooled to treating temperature before it is reused. Doctor solution coolers could be used for this purpose, but such coolers are always troublesome. Common practice is to provide auxiliary or regeneration solution surge tanks in which to store the solution until it is cool enough for reuse. Addition of water and continued air blowing can be used to speed up cooling. It is at this point also that alkalinity adjustments are made to compensate for expenditure of NaOH as sulfate salts and adjustments in water content by additional air blowing in the auxiliary tank or water addition as needed.

Special Design Precautions

Throughout this discussion mention has been made of precipitated lead sulfide and lead sulfide slurry. If lead sulfide is allowed to settle out of suspension, it forms a dense cake that is difficult to resuspend. Plugging of pipes and mixers and accumulated deposits of lead sulfide are nuisance problems in operating doctor sweetening plants, and designers should give special consideration to these problems.

The various sections of the mixers should be flanged to permit easy opening. There should be an abundance of air blasting and steam-out connections for clearing plugged lines. The settlers should have judiciously located manholes to permit flushing out accumulated lead sulfide with a hydraulic lance without the need for entering the vessel as it is difficult to make a vessel safe to enter as long as it contains a lead sulfide deposit. Each of the various tanks should be fitted with a steam coil and with an air sparger below it on the very bottom of the tank to put settled lead sulfide into suspension. Even the regenerated solution tanks should have this equipment because it is certain that lead sulfide will get into these tanks at some time.

Although caustic embrittlement or stress-corrosion cracking might be expected in doctor sweeteners that use concentrated caustic soda solution, which may be heated to 180 to 220°F in the regenerator, caustic embrittlement has not been a problem.

Fire and explosion in doctor treaters, especially in the solution regeneration tank, is a matter for special consideration. The partially regenerated solution

Doctor Sweetening

TABLE 2 Empirical Design Factors for Orifice Column Doctor Solution Regeneration

Basis: Minimum concentration of PbO in spent solution, 2.0 lb/bbl
Not more than a trace of emulsified gasoline
Salt content to be not greater than 35 lb/bbl
Sixteen orifice plates at 12-in. spacing

Temperature, °F	180–200
Pressure at outlet, lb/in.² gauge	25
Pressure at inlet, lb/in.² gauge	40–50
Solution circulation rate, bbl/h/ft² of tower cross section	30–35
Air rate, SCF/bbl circulated	90–95
PbO concentration differential across tower (attainable under above conditions), lb/bbl	1.5

Calculate the number and size of hole in orifice plate as follows:

1. Calculate average specific gravity of air and doctor solution at tower conditions (midpoint) disregarding water vapor
2. From this calculate feet of liquid head equivalent to 1 lb/in.²
3. Calculate number of barrels of air plus doctor solution flowing per hour
4. From this information calculate diameter of a single orifice to give desired pressure drop per plate using:

$$Q_B = 18.26 D^2 \sqrt{H}$$
$$= \text{Flow of air plus doctor solution, } B/H$$
$$D^2 = \text{Diameter of orifice, in.}^2$$
$$H = \text{Pressure drop per plate, ft of fluid}$$

5. From the diameter of the single orifice, calculate the number of holes of some smaller diameter that will give equal orifice area. Usually $\frac{1}{2}$, $\frac{3}{4}$, or 1 in. holes are used. The larger holes are used when drilling of a larger number of smaller holes would be tedious and expensive.

Sample Calculation

It is desired to regenerate 60 bbl of 20°Bé spent doctor solution containing 2 lb of residual litharge per barrel to a solution containing 10 lb/bbl. It is understood that the spent solution has had emulsified hydrocarbon removed, and that it contains enough suspended lead sulfide to form the desired litharge content. This regeneration is to be carried out in 6 h.

Regeneration rate, lb PbO/h: $\dfrac{60 \times (10 - 2)}{6} = 80$

Circulation rate, bbl/h: $\dfrac{80}{1.5} = 53.3$, say 53

Tower diameter, ft: $\sqrt{\dfrac{53}{32.5 \times 0.785}} = 1.49$, say 18-in. pipe

(continued)

TABLE 2 (continued)

Air rate, SCF/h: $92.5 \times 53 = 4,900$

Density of doctor-air mixture:

    Basis: 1 bbl doctor (20°Bé)
             92.5 ft³ (standard conditions) air

    Total weight, lb:

Doctor: $350 \times 1.16 =$	406.0
Air: $92.5 \times 29/359 =$	7.5
Total	413.5, say 414

    Total volume, ft³

Doctor: $1 \times 5.6 \text{ ft}^3/\text{bbl} =$	5.6
Air: $92.5 \times \dfrac{14.7}{\left(\dfrac{50+25}{2} + 14.7\right)} \times \dfrac{(460+180)}{460} =$	36.2
Total	41.8

Density, $\text{lb/ft}^3 = \dfrac{413.5}{41.8} =$ 9.9

Feet of fluid equivalent to 1 lb/in.²: $\dfrac{144}{9.9} =$ 14.6

Pressure drop per plate, ft of fluid $= \dfrac{25}{16} \times 14.6 =$ 22.8

Volume of doctor + air, bbl/h:

$$\dfrac{41.8 \text{ ft}^3/\text{bbl of doctor solution}}{5.6 \text{ ft}^3/\text{bbl}} \times 53 \text{ bbl/h} = 396$$

Diameter of single (orifice) hole:

$$Q_B = 396 = 18.26 D^2 \sqrt{22.8}$$

$$D^2 = \dfrac{396}{18.26 \times 4.78} = 4.54 \text{ in.}^2; \quad D = 2.42$$

Equivalent to 23, ½ in. holes in each plate

---

plus excess air issuing from the top of the regeneration column passes to the top of the regeneration tank where excess air, evaporated hydrocarbon, and steam separate and vent to atmosphere. (In a modern design, air pollution regulations would require control of the vapors issuing from this tank.)

Static electrical discharges frequently ignite these vapors, particularly in those plants that do not have steam strippers to remove hydrocarbon prior to regeneration. The static charge is generated by the liquid dropping through air into the tank. If a grounding chain or pipe is placed so that the liquid is in contact with it as it drops into the tank, a static charge will not develop. Steam snuffing should be provided to put out any fire that might occur. If the regeneration tank is covered, as it would be in a modern design to control atmospheric venting, explosion doors should be provided in the cover to protect the tank in case of an explosion. (NOTE: Doctor regeneration can get rather exciting at times with explosions and explosion doors opening and slamming shut.)

Foaming of the doctor solution issuing from the regeneration column is a frequent problem. Foaming occurs because of: (1) Accumulation of excessive sodium sulfate and thiosulfate in the solution as it reaches the end of its useful life, and (2) accumulation of sulfonic, naphthenic, or aliphatic acids, or all three, extracted from certain feedstocks high in these materials. Pretreating for removal of organic acids and judicious replacement of solution can be practiced to avoid foaming, but provision should be made to control foaming and to cope with it if it occurs.

Silicone antifoam agents have proven to be effective in preventing foaming and suppressing it if it occurs. A supply of antifoam agent and a means of quickly adding it should be provided. The regeneration tank should be enclosed in a diked area that will prevent foam from spreading if the tank should foam over.

Operating Controls

Aside from the usual attention to operating levels, temperatures, pressures, and flow rates, there is one critical control in operation of a doctor sweetening unit that demands constant operator attention. This control involves sulfur addition. Unless the quantity of sulfur is held within a narrow acceptable range, either a sour product that contains lead mercaptide in solution is obtained from the settler or a product that is corrosive to the copper strip will be obtained. Even before it has so much sulfur or polysulfide as to be corrosive, the sweetened product may not have optimum properties with respect to inhibitor response or TEL response, or in the case of kerosine, to color or sediment stability in storage. Sulfur addition should be minimized so that the longest possible butyl mercaptan test can be achieved. (NOTE: This is really a doctor test run in reverse with mercaptan added to test for excess sulfur and polysulfide. It is called a reverse doctor test by some. A butyl mercaptan test time of 20 to 30 min is indicative of satisfactory operation.)

Constant variations in feed mercaptan content, even in the smoothest operating refineries, together with the lack of temperature and flow control precision needed to assure constant sulfur addition, makes it necessary for the operator to run tests on the product from the mixer every 30 to 60 min to check product quality with the doctor test and butyl mercaptan test, and to adjust the sulfur–oil rate accordingly.

The regenerated doctor solution should be checked daily for gravity and total alkalinity. Laboratory equipment for these tests as well as the doctor test and butyl mercaptan test should be provided at the unit for the operator's use. Caustic soda solution should be added to keep the total alkalinity at the desired level. If the gravity gets too high, water can be added. If the gravity becomes too low, a higher regeneration temperature can be used to evaporate some water, or a more concentrated NaOH solution can be used for alkalinity adjustment. Also, excess air can be used in the regenerator to evaporate the excess water.

Litharge content and salt content of regenerated solution should be determined at not less than weekly intervals. When the total salt content exceeds 35 lb per barrel of solution, some spent solution should be discarded and replaced with freshly prepared solution.

In doctor plants that operate to contain 1 or 2 lb of litharge per barrel in the spent solution, steps should be taken to remove all the litharge before discarding. This can be done in one of two ways:

1. For that part of the solution that is to be discarded, continue to recirculate it in the primary mixer system until the litharge has all been consumed before withdrawing it to disposal.
2. After the spent solution is withdrawn to a spent solution tank or to the spent doctor surge tank, the residual lead oxide can be precipitated with sulfide either by use of a gas stream containing H_2S or by addition of a caustic solution containing sodium sulfide. After the lead sulfide thus formed settles out, the spent doctor solution can be decanted and withdrawn for disposal. The settled lead sulfide can then be put into suspension in fresh caustic solution and added to the system.

Analytical Methods

Analytical methods used in controlling doctor sweetening plants are identified in Table 3.

Density determination of doctor solution and total salts in doctor solution are not published methods.

Density is readily determined with sufficient accuracy for doctor sweetener control by use of simple, heavier than water, hydrometer calibrated in Baumé degrees.

Total salts is determined by the method given in the Appendix.

Appendix

Determination of Mixing Time

Primary mixing time for stock on which there is no previous experience can be determined readily in the laboratory by a simple procedure.

1. Determine the mercaptan concentration accurately by using any reliable test method.

TABLE 3

Test Required	Test Method		
	UOP[a]	ASTM	Other
Mercaptan by			
Silver nitrate potentiometric titration	163	D-3227 or D-1323	—
Mercury bromide method	198	—	—
Copper sulfate titration (Cu no.)	197	—	—
Litharge in doctor solution	145	—	—
Doctor test for petroleum distillates	41	D-484 (Section 4.2.6.1)	—
Organic acids in petroleum distillates	587	D-974 or D-664	—
Phenols in petroleum products by direct titration	729	—	—
Phenols and thiophenols in petroleum products and spent caustic by spectrophotometer	262	—	—
Density	—	—	Bé hydrometer
Total alkalinity	210 or 209	—	—
Phenols in used caustic solution	211	—	—
Total salts in used doctor solution	—	—	See Appendix
Butyl mercaptan test	143	—	—
Copper strip corrosion test	—	D-130	—
Elemental sulfur, mercury number	286	—	—
Elemental sulfur by polarograph	700	—	—

[a] Available through subscription to *UOP Laboratory Test Methods for Petroleum and Its Products.*

2. Weigh out sufficient flowers of sulfur to provide 5 to 10% excess over the amount theoretically needed to react with the mercaptan present (see Table 1).
3. To a 4-liter flask fitted with a power stirrer, add 3 liters of the oil to be treated and 0.3 liters of doctor solution made from 20°Bé caustic solution containing 10 lb PbO per barrel. (Note: If 0.3 liters of doctor solution does not provide at least 10% excess of the theoretical PbO required for the mercaptan in the feedstock, add a sufficient volume to provide at least 10% excess PbO.) The flask should be provided with a heating mantle so that the test can be run at different temperatures, as mixing time also depends on temperature.
4. Start the stirrer and mix for 30 s.
5. Stop the stirrer; add the measured amount of sulfur. Start the stirrer and measure the time elapsed until a dark brown precipitate of lead sulfide can be seen in the mixture.
6. Stop the mixer and allow the bulk of the doctor solution to settle. Filter a sample of the oil to remove suspended lead sulfide and run a doctor test, a butyl mercaptan test, and a copper strip corrosion test. If the doctor test fails, repeat the experiment using more sulfur. If the butyl mercaptan or corrosion test fails, repeat the experiment using less sulfur.

7. If it is impossible to get a "break"; that is, lead sulfide precipitate formation in a reasonable time, say, 20 min, without using so much sulfur that a failing butyl mercaptan test or a failing copper strip test results, repeat the experiment at a higher temperature.
8. Design the primary mixer to provide a residence time 50% in excess of the mixing time determined by the above procedure.

Total Salts Determination

Place 5 cc of the solution in a 140-cc porcelain dish covered with a watch glass. Add 1 cc of 50% H_2SO_4 and 2 cc of concentrated HNO_3. Heat over a low flame until charring begins. Cool, add 5 cc of concentrated HNO_3, and heat over low flame in a hood until completely charred. Wipe off the watch glass with quantitative filter paper, char the entire material in a furnace at low temperature, and determine the weight of the residue.

Calculations

A. Grams of ignited sulfated residue × 70.1 = total sulfates in lb/bbl
B. NaOH present as sulfate = (142.06/80.02) × lb NaOH/bbl in original doctor solution
C. PbO as sulfate = (303.26/223.20) × lb PbO/bbl in original doctor solution
D. Salts present, expressed as sulfate, lb/bbl = A − (B + C)
E. Salts expressed as thiosulfates, lb/bbl = D × (158.12/142.06)

References

1. *The Science of Petroleum*, Vol. 3, Oxford University Press, 1938.
2. J. C. Morrell and G. Egloff, *Refiner Nat. Gasoline Manuf.*, 2, 7 (1923).
3. G. L. Wendt and S. H. Diggs, *Ind. Eng. Chem.*, 16, 11 (1924).
4. E. Ott and E. Reid, *Ind. Eng. Chem.*, 22, 878, 884(1930).
5. W. E. Schulze and A. E. Buell, *Natl. Pet. News*, 27(4), 25–26, 28, 30–31 (1935).
6. S. F. Birch and R. Stanfield, *Ind. Eng. Chem.*, 28, 668 (1936).
7. E. Endo, *J. Fuels Soc. Jpn.*, 13, 292–308 (1934).
8. J. G. Ryan, *Ind. Eng. Chem.*, 34(7), 824–832 (1942).
9. C. D. Lowry, Jr., C. G. Dryer, and C. Wirth III, *Ind. Eng. Chem.*, 30, 1275 (1938).
10. C. Wirth III, J. R. Strong, and C. D. Lowry, Jr., *Studies in Doctor Sweetening*, Presented at the American Chemical Society Milwaukee Meeting, September 5–9, 1938.
11. K. M. Brown, W. K. T. Gleim, and P. Urban, *Oil Gas J.*, 57(44), 73 (1959).

KENNETH M. BROWN

Permission for the publication herein of the paper entitled *Doctor Sweetening* has been granted, and all rights reserved, by UOP Inc.

Index

Abu Dhabi:
    operating refineries, capacities, 216
    refinery locations, capacities, and types of processing, 219
Acid gas absorption, 708
Africa:
    crude oil consumption, 4
    petroleum production and demand, 69
Air toxicity, land forming of hazardous wastes and, 187-188
Albania, refinery locations, capacities, and types of processing, 219
Algeria:
    operating refineries, capacities, 216
    refinery locations, capacities, and types of processing, 219
Alkylate octane, increasing, 56-57
Alkylation, 86, 101, 102
Alternate feedstocks for refineries, 68
Amorphous Al_2O_3-based hydrotreating, economics of, 167-168
Angola:
    operating refineries, capacities, 216
    refinery locations, capacities, and types of processing, 219
API gravity, 77
API separator sludges for management of wastes, 184
Argentina:
    operating refineries, capacities, 216
    refinery locations, capacities, and types of processing, 219
Aromatics, 88, 104, 290
    catalytic cracking compared to thermal cracking of, 368
Asia, crude oil consumption, 4
Asphalt, 13
Asphalt residual treatment (ART) process, 109-128, 513-515
ASTM distillation, 78
Atmospheric residue:
    BEV formula for, 19
    production costs, 22-24

Australasia, crude oil consumption, 4
Australia:
    operating refineries, capacities, 216
    refinery locations, capacities, and types of processing, 219-220
Austria:
    operating refineries, capacities, 216
    refinery locations, capacities, and types of processing, 220
Aviation fuels, 10

Bahrain:
    operating refineries, capacities, 216
    refinery locations, capacities, and types of processing, 220
Bangladesh:
    operating refineries, capacities, 216
    refinery locations, capacities, and types of processing, 220
Barbados:
    operating refineries, capacities, 216
    refinery locations, capacities, and types of processing, 220
Belgium:
    operating refineries, capacities, 216
    refinery locations, capacities, and types of processing, 220
Best demonstrated available technologies (BDAT) for treatment of toxic wastes, 192
    examples of, 193-194
    land treatment of refinery wastes and, 195-198
    standards for refinery wastes, 193, 194
Boiling range, 5
Bolivia:
    operating refineries, capacities, 216
    refinery locations, capacities, and types of processing, 220
Brazil:
    operating refineries, capacities, 216
    refinery locations, capacities, and types of processing, 221

Breakeven value (BEV) formula for petroleum products, 13, 15, 19
Brunei:
  operating refineries, capacities, 216
  refinery locations, capacities, and types of processing, 221
Building materials, 2
Bulgaria, refinery locations, capacities, and types of processing, 221
Burma:
  operating refineries, capacities, 216
  refinery locations, capacities, and types of processing, 221

Cameroon:
  operating refineries, capacities, 216
  refinery locations, capacities, and types of processing, 221
Canada:
  crude oil consumption, 4
  operating refineries, capacities, 216
  petroleum production and demand, 69
  refinery locations, capacities, and types of processing, 221-223
Catalyst economics, 155-169
  catalytic reforming, advances in, 157-158
  fluid catalytic cracking (FCC), 158-160
  functions and costs, 156-157
  future developments, 168-169
  general hydrotreating (HT) and hydrodesulfurization (HDS) catalysts, 167-168
  hydrocracking, 163-166
  mild hydrocracking, 160-162, 163
  residue hydroprocessing, 166-167
Catalysts:
  for HOC units, 488-490
    effect of metal on, 512
  for hydrodesulfurization, 689-690, 700-701
  for sulfur recovery, 148-149
  See also Octane catalysts
Catalyst usage, 130-155
  Claus unit tail gas treatment catalysts, 151-152
  combustion promotors, 153-154
  dimerization catalysts, 133
  fluid catalytic cracking, 133-138

[Catalyst usage]
  hydrocracking, 138-140, 600-603
    mild, 140-141
  hydrorefining, 145-148
  hydrotreating/hydrogenation, 141-145
  isomerization C_4 catalysts, 133
  isomerization (C_5 and C_6) catalysts, 133
  isomerization (xylene) catalyst, 133
  methyl tertiary butyl ether (MTBE) catalysts, 153
  for naphtha reforming, 131-133
  other refining catalysts, 152-153
  polymerization, 148
  saturation catalysts, 141-145
  steam hydrocarbon reforming catalysts, 149-151
  sulfur oxide reduction catalysts, 154
  sulfur (elemental) recovery, 148-149
  sweetening catalysts, 151
Catalytic cracking, 85, 95-97, 349-480
  cracking catalysts, 414-423
    activity tests, 420-421
    catalyst deactivation, 421-422
    catalyst manufacture, 417-419
    catalyst selectivity characteristics, 419-420
    clay catalysts, 415-416
    high alumina catalysts, 416
    mixed oxide catalysts, 416
    physical/chemical testing, 422-423
    semisynthetic cracking, 416-417
    synthetic catalysts, 416
    zeolite catalysts, 417
  engineering aspects of fluid processing, 423-476
    application of fluidized solids principle to catalytic cracking, 436-462
    cat cracker control, 428-432
    catalyst stripping, 436
    energy recovery, 462-464
    heat balance, 432-436
    inspection and maintenance, 474-476
    material selection and design aspects, 464-472
    mechanical considerations, 464-476
    regeneration, 423-427
    special design factors, 472-473

Index

[Catalytic cracking]
  [engineering aspects of fluid processing]
    suggested operation precautions, 473-474
    troubleshooting,, 436, 437-447
  Exxon Research and Engineering Co. system, 358-360
  fixed-bed catalytic cracking, 350-351
  fluidized solids units, 355-358
  Gulf Research and Development Co. system, 363, 364
  M. W. Kellogg Co. system, 361-362
  moving-bed catalytic cracking, 352-354
  optimization and control, 516-527
  reactions and kinetics, 367-414
    chemistry, 367-374
    feedstock quality effects, 390-407
    kinetics, 374-379
    operating variable effects, 380-390
    piloting, 413-414
    process models, 407-413
  residuum cracking, 364-366
  Texaco Development Co. system, 363
  Universal Oil Products Co. system, 360-361
Catalytic dewaxing, 558-564, 565
  applications of the process, 560-563
    middle distillates, 562-563
    specialty oils and lubricating oils, 560-562
  commercial installation, 564
  lubricating oil processes, 654-656
  process description, 559-560
    catalyst, 559
    process conditions and flow sheet, 559-560
  process economics, 563-564
Catalytic hydrotreating, 701-702
  chemistry of, 702-705
Catalytic naphtha reforming, 131-133
Catalytic reforming, 82, 93-95
  economics of, 157-158
Caustic soda treatment, 708-709
  desulfurizing cracked gasoline by, 727-736
    technological improvements, 734-736
    typical two-stage treating, 728-734

Caustic wash, 81
Characterization factor, 5-6
Chemical refining processes for lubricating oil, 642-643
Chemicals derived from petroleum, 2, 3
Chemistry of catalytic cracking, 367-374
Chile:
  operating refineries, capacities, 216
  refinery locations, capacities, and types of processing, 223
China:
  crude oil consumption, 4
  operating refineries, capacities, 216
  petroleum production and demand, 69
  refinery locations, capacities, and types of processing, 223
Claus unit tail gas treatment catalysts, 151-152
Clay catalysts, 415-416
Cleanup of wastes, 182
Coal, direct combustion of, 174-175
  limitations, 176-177
Coastal Eagle Point Co., octane boosting program of, 25, 26-31
Coke, direct combustion of, 174-175
  limitations, 176-177
Coking, 85, 97-99, 245
  delayed, 97, 98, 245-249
    coke drums and ancillary equipment, 249
    design features, 248
    heater, 248-249
    installed capacities for (U.S.), 245
  fluid, 97, 245, 249-251, 253-281
    applications of the process, 258-263
    coke properties and uses, 251-252
    coker charge stocks, 245
    Flexicoking process, 256-257
    installed capacities for (U.S.), 245
    operations, 278-279
    process, 255-256
    process calculations, 263-271
    process design of equipment, 271-278
Colombia:
  operating refineries, capacities, 216
  refinery locations, capacities, and types of processing, 224

Combustion promotors, 153-154
Commercial synthesis gas technology, 172, 173
Comprehensive Environmental Response Compensation and Liability Act (Cercla), 179
Congo:
  operating refineries, capacities, 216
  refinery locations, capacities, and types of processing, 224
Costa Rica:
  operating refineries, capacities, 216
  refinery locations, capacities, and types of processing, 224
Cracking, 85
  catalytic, 85, 95-97, 349-480
    cracking catalysts, 414-423
    engineering aspects of fluid processing, 423-476
    Exxon Research and Engineering Co. system, 358-360
    fixed-bed catalytic cracking, 350-351
    fluidized solids units, 355-358
    Gulf Research and Development Co. system, 363, 364
    M. W. Kellogg Co. system, 361-362
    moving-bed catalytic cracking, 352-354
    optimization and control, 516-527
    reactions and kinetics, 367-414
    residuum cracking, 364-366
    Texaco Development Co. system, 363
    Universal Oil Products Co. system, 360-361
  heavy oil, 480-515
    asphalt residual treating (ART) process, 513-515
    background, 480-481
    catalysts, 488-490
    commercial performance of an HOC unit, 504-511
    effect of metal on catalyst, 512
    environmental and energy factors, 503-504
    equipment features, 495-496
    feedstocks, 481-485
    heat removal system, 487-488
    operating characteristics, 493-495

[Cracking]
  [heavy oil]
    operations and maintenance, 512-513
    process information, 490-493
    retrofitting a refinery to include HOC, 500-503
    unit design, 485-487
    yields and utilities, 496-500
  thermal, 281-349
    chemistry of, 289-290
    commercial applications, 286-289
    definition of thermal processes, 282-283
    economics, 346-348
    equipment design, 339-346
    history and development, 283-286
    yield and quality correlations, 291-338
  See also Hydrocracking
Crude oil desalting, 79, 88-89, 666-677
  calculation methods, 669-673
    two-stage system, 671-673
  dilution water/net oil ratio, 674
  general requirements, 666-669
  injection and mixing, 674-677
  water supply problems, 666
Crude oil distillation, 80, 89-91
Crude oils:
  in lubricating oil manufacture, 638
  pretreatment of, 78-79
  world proven reserves, 67-68
Cuba, refinery locations, capacities, and types of processing, 224
Cyprus:
  operating refineries, capacities, 216
  refinery locations, capacities, and types of processing, 224
Czechoslovakia, refinery locations, capacities, and types of processing, 224

Dealuminated Y (DY) zeolite, 26
Deasphalting, 527-544
  deasphalted oil (DAO) and solvent asphalt, 528-530
  lubricating oil deasphalting process, 634, 640-641
  operating problems, 541-542
  process applications, 530-533

Index

[Deasphalting]
  process description, 533-535
  solvent extraction, 535-540
    extraction temperature and pressure, 537-538
    extraction tower, 538-540
    selection of a solvent, 535-537
    solvent/oil ratio, 537
  solvent recovery system, 540-541
  utilities, 542-543
Dehydrogenation, 544-558
  design principles, 548-549
  enthalpies and equilibria, 545-548
  industrial processes, 556-557
  processes of industrial importance, 550-556
    alcohol to aldehyde or ketones, 553
    n-butane to n-butenes to butadiene, 550
    n-butenes to butadiene, 550-551
    cyclohexanes to aromatics, 552-553
    ethane to ethylene, 550
    ethylbenzene to styrene, 553-556
    higher paraffins to olefins, 551-552
    paraffins to olefins, general, 552
    propane to propylene, 550
Delayed coking, 97, 98, 24-249
  coke drums and ancillary equipment, 249
  design features, 248
  heater, 248-249
  installed capacities for (U.S.), 245
Demetallization/desulfurization of feedstocks, 677-696
  definition of terms, 681
  discussion of organometallic compounds, 682-683
  downflow fixed-bed reactor design, 685-687
  examples of petroleum sulfur compounds, 681-682
  in expanded-bed reactors, 692-693
  hydrodesulfurization, 679, 681
    catalysts of, 689-690, 700-701
  properties of Khafji crude oil, 679, 680
  reaction kinetics, 683-684
  separate demetallization reactors, 688-689
  typical hydrodesulfurization process configuration, 684-685

[Demetallization/desulfurization of feedstocks]
  upflow expanded-bed reactor design, 687-688
    with fine catalyst, 688
    processing vacuum residuals in, 691-692
Denmark:
  operating refineries, capacities, 216
  refinery locations, capacities, and types of processing, 224
Density, 5
Deoiling, 580-582
Desalting crude oil, 79, 88-89, 666-677
  calculation methods, 669-673
  dilution water/net oil ratio, 674
  general requirements, 666-669
  injection and mixing, 674-677
  water supply problems, 666
Desulfurization, 696-727
  of cracked gasoline, 727-736
    technological improvements, 734-736
    typical two-stage treating, 728-734
  definition, 697-698
  hydrodesulfurization catalysts, 689-690, 700-701
  middle distillate, 710-714
    catalysts, 712
    chemistry of process, 711-712
    economics, 714
    processing considerations, 713-714
    process description, 710-711
  naphtha, 701-709
    acid gas absorption, 708
    catalysts used in hydrotreating, 705
    catalytic hydrotreating, 701-702
    caustic treating, 708-709
    chemistry of catalytic hydrotreating, 702-705
    economics of hydrotreating, 707-708
    molecular sieve adsorption, 709
    processing considerations, 705-707
  residue, 714-725
    catalysts, 719-721
    chemistry of process, 717-719
    design criteria, 721-722

[Desulfurization]
[residue]
economics, 725
important operating parameters, 722-725
process description, 716
yields and product properties, 725
sulfur in petroleum, 697, 698-699
Developmental synthesis gas technology, 172, 173
Dewaxing:
catalytic, 558-564, 565
applications of the process, 560-563
commercial installation, 564
lubricating oil processes, 654-656
process description, 559-560
process economics, 563-564
solvent, 565-582
deoiling, 580-582
dewaxed oil, 566
Di/Me dewaxing process, 568, 569, 577, 578, 659
Dilchill dewaxing process, 570, 580, 658-659
lubricating oil processes, 656-659
MEK-dewaxing process, 567-569, 575-577
propane dewaxing process, 570, 577-580, 659
wax, 566
urea, 565, 583-592
adduct formation, 584-585
Edenleanu process, 588-591
Nurex process, 587-588
technology of, 586-591
Diesel fuels, 10-12
Di/Me dewaxing process, 568, 569, 577, 578, 659
Dimerization catalysts, 133
Discretionary specifications for petroleum products, 6
Disposal of wastes, 181-182
Doctor sweetening, 736-758
analytical methods, 756
background, 736-739
chemical requirements, 739-740
design of sweetening plant, 746-752
determination of mixing time, 756-758

[Doctor sweetening]
effect of nonmercaptan acidic components in the feed, 745-746
mixing, 741-745
operating controls, 755-756
process flow and effect of operating variables, 740-741
special design precautions, 752-755
total salts determination, 758
Dominican Republic:
operating refineries, capacities, 216
refinery locations, capacities, and types of processing, 224

Eastern Europe, petroleum production and demand, 69
Ecuador:
operating refineries, capacities, 216
refinery locations, capacities, and types of processing, 224
Edenleanu process:
for lubricating oils, 649
for urea dewaxing, 588-591
Egypt:
operating refineries, capacities, 216
refinery locations, capacities, and types of processing, 225
El Salvador:
operating refineries, capacities, 216
refinery locations, capacities, and types of processing, 225
Engelhard ART process (*see* Asphalt residual treatment process)
Engelhard ESR process, 109-110, 126-127
Environmental and energy factors in HOC units, 503-504
Ethiopia:
operating refineries, capacities, 216
refinery locations, capacities, and types of processing, 225
Ethylene, 87-88, 103-104
Exempted wastes, management of, 183-184
Exxon Research and Engineering Co.:
catalytic cracking system of, 358-360
Dilchill dewaxing process of, 570, 580, 658-659

Far East, petroleum production and demand, 69

Index

Feedstocks, 13
  alternate feedstocks for refineries, 68
  charged to cat cracking units, quality effects of, 390-407
  demetallization/desulfurization of, 677-696
    definition of terms, 681
    discussion of organometallic compounds, 682-683
    downflow fixed-bed reactor design, 685-687
    examples of petroleum sulfur compounds, 681-682
    in expanded-bed reactors, 692-693
    hydrodesulfurization, 679, 681
    hydrodesulfurization catalysts, 689-690, 700-701
    properties of Khaji crude oil, 679, 680
    reaction kinetics, 683-684
    separate demetallization reactors, 688-689
    typical hydrodesulfurization process configuration, 684-685
    upflow expanded-bed reactor design, 687-688
  from heavy oils, 481-485
  identification of, 76-78
  for refineries, 68
Field separation, 79
Finishing process for lubrication oil, 660-661
Finland:
  operating refineries, capacities, 216
  refinery locations, capacities, and types of processing, 225
Fixed-bed catalytic cracking, 350-351
Flexicoker, 256-257
  heat balance for, 270-271
Flexicoker heater, 274-275
Flexicoking, 97, 98, 256-257
Fluid catalytic cracking (FCC), 133-138
  economics of, 158-160
FCC gasoline, upgrading, 55-56
Fluid coking, 97, 245, 249-251, 253-281
  applications of the process, 258-263
    investment and operating costs, 263
    process variations, 262-263
    product mix flexibility, 258-262

[Fluid coking]
  coke properties and uses, 251-252
  coker charge stocks, 245
  Flexicoking process, 256-257
  installed capacities for (U.S.), 245
  operations, 278-279
  process of, 255-256
  process calculations, 263-271
    Flexicoker heat balance, 270-271
    fluid coker heat balance, 269-270
    operating conditions and process control, 263
    reactor holdup requirement, 264-269
    reactor yields and product qualities, 264
  process design of equipment, 271-278
    coking reactor design, 271-273
    Flexicoker heater, 274-275
    fluid coker burner design, 273
    gasifier design, 275-276
    heater overhead cooling and dust recovery, 276
    particle size control, 273
    quench elutriator design, 273-274, 275
    transfer line design, 276-278
Fluidized solids units, 355-358
  application of principles of, to catalytic cracking, 436-462
France:
  operating refineries, capacities, 216
  refinery locations, capacities, and types of processing, 225-226
Fuel gas, 7
Fuel oil blendstocks, 593
Fuel oils, 12-13
Fuels, 2
Furfural refining process for lubricating oils, 649-650

Gabon:
  operating refineries, capacities, 216
  refinery locations, capacities, and types of processing, 226
Gaseous products, 6-13
  diesel fuelds, 10-12
  fuel gas, 7
  fuel oils, 12-13
  gasoline, 7-10
  LPG, 7

[Gaseous products]
  natural gas, 6
  nonfuel products, 13
  still gas, 7
Gases, reconstituting, 86
Gas oil:
  BEV formula for, 19, 20
  production costs, 20
Gasoline, 7-10
  BEV formula for, 19
  FCC, upgrading, 55-56
  high octane light, 593
  poly, upgrading, 56-57
  production costs, 21-22
  refining, 80-82
    octane number, 81-82
    sulfur content, 81
    volatility, 80
  See also Thermal reforming of gasoline stocks
Gasoline desulfurization, 727-736
  technological improvements, 734-736
  typical two-stage treating, 728-734
Germany:
  operating refineries, capacities, 216
  refinery locations, capacities, and types of processing, 226-227
Ghana:
  operating refineries, capacities, 217
  refinery locations, capacities, and types of processing, 227
Greece:
  operating refineries, capacities, 217
  refinery locations, capacities, and types of processing, 227
Guatemala:
  operating refineries, capacities, 217
  refinery locations, capacities, and types of processing, 227
Gulf Research and Development Co., catalytic cracking system of, 363, 364

Hazardous and Solid Waste Amendments of 1980 (HSWA), 179, 180, 181-182
  corrective action for release of hazardous wastes into the environment, 189
  land treatment of wastes and, 185-186
  surface impoundment of wastes and, 185

Hazardous waste liquids, land disposal ban on, 186-187
Heat balance in a catalytic cracking unit, 432-436
Heavy naphtha, 593
Heavy oil cracking (HOC), 480-515
  asphalt residual treating process, 513-515
  background, 480-481
  catalysts, 488-490
  commercial performance of an HOC unit, 504-511
  effect of metal on catalyst, 512
  environmental and energy factors in, 503-504
  equipment features, 495-496
  feedstocks, 481-485
  heat removal system, 487-488
  operating characteristics, 493-495
  operations and maintenance, 512-513
  process information, 490-493
  retrofitting a refinery to include an HOC, 500-503
  unit design, 485-486
  yields and utilities, 496-500
High alumina catalysts, 416
High octane light gasoline, 593
High viscosity index lube oil blendstocks, 593
Honduras:
  operating refineries, capacities, 217
  refinery locations, capacities, and types of processing, 227
Hungary:
  operating refineries, capacities, 217
  refinery locations, capacities, and types of processing, 227
Hydrocarbons, 4-6
  light, 60-64
Hydrocracking, 85, 99-100, 592-634, 681
  background, 593-595
  catalysts for, 138-140, 600-603
  economics of, 163-166, 625-632
  feed characteristics, 608-612
  middle distillate hydrocracking, 612-618
    hydrogen make-up preparation, 617
    liquid feed preparation, 616-617
    product fractionation, 617-618

Index

[Hydrocracking]
  mild hydrocracking, 595-599
    catalysts for, 140-141
    economics of, 160-162, 163
    once-through partial conversion, 606-608
    single-stage design plant, 606
    single-stage recycle unit, 606
    two-stage design plant, 603-605
    yields and performance, 618-624
Hydrodesulfurization, 679, 681
  catalysts for, 689-690, 700-701
  economics of, 167-168
Hydrogenation, 681
  catalysts for, 141-145
Hydrogenolysis, 681
Hydrogen refining process for lubricating oil, 634, 643-648
  catalysts for, 145-148
Hydrotreating, 81, 82, 91-93
  catalysts for, 141-145
  economics of, 167-168, 707-708

India:
  operating refineries, capacities, 217
  refinery locations, capacities, and types of processing, 227-228
Indonesia:
  operating refineries, capacities, 217
  refinery locations, capacities, and types of processing, 227
International Refining Catalyst Compilation, 130-155
Iran:
  operating refineries, capacities, 217
  refinery locations, capacities, and types of processing, 228
Iraq:
  operating refineries, capacities, 217
  refinery locations, capacities, and types of processing, 228
Ireland:
  operating refineries, capacities, 217
  refinery locations, capacities, and types of processing, 229
Isomerization C_4 catalysts, 133
Isomerization C_5 and C_6 catalysts, 133
Isomerization (xylenes) catalysts, 133
Isoparaffins, catalytic cracking compared to thermal cracking of, 368

Israel:
  operating refineries, capacities, 217
  refinery locations, capacities, and types of processing, 229
Italy:
  operating refineries, capacities, 217
  refinery locations, capacities, and types of processing, 229
Ivory Coast:
  operating refineries, capacities, 217
  refinery locations, capacities, and types of processing, 230
Jamaica:
  operating refineries, capacities, 217
  refinery locations, capacities, and types of processing, 230
Japan:
  crude oil consumption, 4
  operating refineries, capacities, 217
  petroleum production and demand, 69
  refinery locations, capacities, and types of processing, 230-232
Jet fuels, 10, 593
Jordan:
  operating refineries, capacities, 217
  refinery locations, capacities, and types of processing, 232

Kellogg, M. W., Co., catalytic cracking system of, 361-362
Kenya:
  operating refineries, capacities, 217
  refinery locations, capacities, and types of processing, 232
Ketone dewaxing processes, 657
Khafji crude oil, properties of, 679, 680
Kinetics of catalytic cracking, 374-379
Korea:
  operating refineries, capacities, 217
  refinery locations, capacities, and types of processing, 232
Kuwait:
  operating refineries, capacities, 217
  refinery locations, capacities, and types of processing, 232-233

Land disposal ban on hazardous
    wastes, 186-189
  air toxicity and, 187-188
  corrective action, 189
  exemptions to bans, 188
  liquids, 186-187
  notification, 187
  toxicity measurement, 188-189
Land treatment of refinery wastes,
    190-191
  BDAT and, 195-198
  HSWA and, 185-186
Leaded gasoline, ,9
Lebanon:
  operating refineries, capacities, 217
  refinery locations, capacities, and
    types of processing, 233
Liberia:
  operating refineries, capacities, 217
  refinery locations, capacities, and
    types of processing, 233
Libya:
  operating refineries, capacities, 217
  refinery locations, capacities, and
    types of processing, 233
Light gasoline, high octane, 593
Light hydrocarbons, 60-64
Light naphtha, upgrading, 53-54
Linear programming (LP), 14
Liquids, hazardous waste, land disposal ban on, 186-187
Low sulfur fuel oil blendstocks, 593
LPG (liquid petroleum gas), 7, 593
Lubricants, 13
Lubricating oil manufacture, 634-664
  crude oils, 638
  deasphalting processes, 634, 640-641
  dewaxing processes, 654-659
    catalytic dewaxing, 654-656
    solvent dewaxing, 656-659
  distillation, 634, 638-640
  finishing processes, 660-661
  five basic steps in, 634-638
  refining processes, 642-654
    chemical refining, 642-643
    hydrogen refining, 634, 643-648
    solvent refining, 634, 648-654
  urea dewaxing processes, 659-660

Madagascar:
  operating refineries, capacities, 217

[Madagascar]
  refinery locations, capacities, and
    types of processing, 233
Malaysia:
  operating refineries, capacities, 217
  refinery locations, capacities, and
    types of processing, 233
Martinique:
  operating refineries, capacities, 217
  refinery locations, capacities, and
    types of processing, 233
MEK-dewaxing process, 567-569, 575-577
Methanol-based oxygenates, 57-60
Methyl tertiary butyl ether (MTBE)
  catalysts, 153
N-Methyl-2-pyrrolidone (MP) refining processes for lubrication
    oils, 650
Mexico:
  operating refineries, capacities, 217
  petroleum production and demand,
    69
  refinery locations, capacities, and
    types of processing, 233-234
Middle distillate desulfurization,
    710-714
  catalysts, 712
  chemistry of process, 711-712
  economics, 714
  process description, 710-711
  processing considerations, 713-714
Middle distillate hydrocracking, 612-618
  hydrogen make-up preparation, 617
  liquid feed preparation, 616-617
  product fractionation, 617-618
Middle distillates, 593
  catalytic dewaxing process for,
    562-563
Middle East:
  crude oil consumption, 4
  petroleum production and demand,
    69
Mild hydrocracking, 595-599
  catalysts for, 140-141
  economics of, 160-162, 163
Mixed oxide catalysts, 416
Molecular sieve adsorption, 709
Morocco:
  operating refineries, capacities, 217
  refinery locations, capacities, and
    types of processing, 234

Index

Moving-bed catalytic cracking, 352-354

Naphtha:
  BEV formula for, 19, 20
  heavy, 593
  light, upgrading, 53-54
  production costs, 20
Naphtha desulfurization, 701-709
  acid gas absorption, 708
  catalytic hydrotreating, 701-702
  catalysts used in hydrotreating, 705
  caustic treating, 708-709
  chemistry of catalytic hydrotreating, 702-705
  economics of hydrotreating, 707-708
  molecular sieve adsorption, 709
  processing considerations, 705-707
Naphtha reforming, catalytic, 131-133
Naphthenes, 290
Natural gas, 6
Netherlands:
  operating refineries, capacities, 217
  refinery locations, capacities, and types of processing, 234
Netherlands Antilles:
  operating refineries, capacities, 217
  refinery locations, capacities, and types of processing, 234
New Zealand:
  operating refineries, capacities, 217
  refinery locations, capacities, and types of processing, 235
Nicaragua:
  operating refineries, capacities, 217
  refinery locations, capacities, and types of processing, 235
Nigeria:
  operating refineries, capacities, 217
  refinery locations, capacities, and types of processing, 235
"No-migration petitions," 195
Nonfuel products, 13
Norway:
  operating refineries, capacities, 217
  refinery locations, capacities, and types of processing, 235
Nurex process for urea dewaxing, 587-588

Octane boosting, 25-31
  background, 25-26
  Coastal's Eagle Point refinery, program of, 25, 26-31
  ultrastable zeolites, 26
Octane catalysts, 31-50
  chemistry and characteristics of framework silicon-enriched zeolites, 32-36
    chemistry, 32-34
    hydrothermal properties, 34-36
  evaluation of high stability zone catalysts containing LZ-210 zeolites, 36-48
  microactivity and pilot plant evaluation, 37-48
Octane number, 9, 81-82
Octane options, 50-64
  alkylate and poly gasoline upgrading, 56-57
  best choice, 52, 53
  FCC gasoline upgrading, 55-56
  future choices, 52-53
  gasoline octane upgrading costs, 62
  gasoline upgrading costs, 63
  geographical distribution of upgrading options, 64
  light hydrocarbons, 60-64
  light naphtha upgrading, 53-54
  oxygenates, 57-60
  reformate upgrading, 54-55
Oily wastewater, management of, 184
Olefins, 290
  catalytic cracking compared to thermal cracking of, 368
Oman:
  operating refineries, capacities, 217
  refinery locations, capacities, and types of processing, 235
Once-through partial conversion (OTPC) hydrocracking design, 606-608
OPEC countries, petroleum production and demand, 69
Oxygenates (oxygenated hydrocarbons), 57-60

Pakistan:
  operating refineries, capacities, 217
  refinery locations, capacities, and types of processing, 235

Panama:
  operating refineries, capacities, 217
  refinery locations, capacities, and types of processing, 235
Paraffins, 290
  catalytic cracking compared to thermal cracking of, 368
Paraguay:
  operating refineries, capacities, 217
  refinery locations, capacities, and types of processing, 235
Peru:
  operating refineries, capacities, 217
  refinery locations, capacities, and types of processing, 236
Petrochemical feedstocks, 13
Petrochemicals, 2, 75-76, 87-88
  partial list of uses for, 3
Petroleum products, 2-13
  gaseous products, 6-13
  product identity, 4-6
  production costs, 13-24
    atmospheric residue, 22-24
    gasoline, 21-22
    methodology, 14-15
    naphtha and gas oil, 20
    white distillate, 16-19
  world consumption, 2-4
Petroleum refining, 67-107
  distillates, 82-83
  gasoline, 80-82
    octane number, 81-82
    sulfur content, 81
    volatility, 80
  modern refinery, 86-88
  processing details, 88-104
    alkylation, 101, 102
    aromatics, 104
    catalytic cracking, 95-97
    catalytic reforming, 93-95
    coking, 97-99
    crude desalting, 88-89
    crude distillation, 89-91
    ethylene, 103-104
    hydrocracking, 99-100
    hydrotreating, 91-93
    polymerization, 100
  producing more light products, 83-86
    cracking, 85
    reconstituting gases, 86
    vacuum distillation, 85-86

[Petroleum refining]
  product names, 70-76
    petrochemicals, 75-76
    product specifications, 72-73
    product yields, 74-75
    refined products, 72
  refinery size and cost, 104-106
  refining schemes, 76-80
    crude oil fractions, 80
    crude oil pretreatment, 78-79
    feedstock identification, 76-78
  residuals, 83
  from well to refinery, 68-70
  See also Refinery of the future
Phenol refining process for lubricating oils, 650
Philippines:
  operating refineries, capacities, 218
  refinery locations, capacities, and types of processing, 236
Poland:
  operating refineries, capacities, 218
  refinery locations, capacities, and types of processing, 236
Poly gasoline, upgrading, 56-57
Polymerization, 86, 100
  catalysts for, 148
Portugal:
  operating refineries, capacities, 218
  refinery locations, capacities, and types of processing, 236
Premium grade of gasoline, 8-9
Primary energy sources, 67
Propane dewaxing process, 570, 577-580, 659
Puerto Rico:
  operating refineries, capacities, 218
  refinery locations, capacities, and types of processing, 236

Qatar:
  operating refineries, capacities, 218
  refinery locations, capacities, and types of processing, 236

Refinery of the future, 108-129
  Engelhard ART process, 109-128
  Engelhard ESR process, 109-110, 126-127
  maximize refinery flexibility, 120-123

Index

[Refinery of the future]
  maximize use of existing equipment, 111-120
    advanced FCC-ART design technology, 113-115
    FCCU/coker revamps, 112-113
    hydrotreater/hydrocracker integration, 115-120
  maximize yield of transportation fuels, 124
  minimize investment costs, 110
  minimize operating costs, 124-126
Refinery wastes, land treatment of, 190-191
  BDAT and, 195-198
  HSWA and, 185-186
Refinery waxes, 566
Regular and premium grades of gasoline, 8-9
Research and Motor octane number, 9
Residuals, 83
Residue desulfurization, 714-725
  catalysts, 719-721
  chemistry of process, 717-719
  design criteria, 721-722
  economics, 725
  important operating parameters, 722-725
    catalyst aging, 723
    catalyst temperature and space velocity, 723-725
    desalting, filtration, and metallurgy, 722-723
    hydrogen partial pressure, 723
  process description, 716
  yields and product properties, 725
Residue hydroprocessing, economics of, 166-167
Residuum cracking, 364-366
Resource Conservation and Recovery Act (RCRA) of 1976, 179, 180
  prevention of wastes toxicity and, 191-193
Retrofitting a refinery to include heavy oil cracking, 500-503
Riser cracking, 96-97
Romania, refinery locations, capacities, and types of processing, 236

Saturation, 681
Saturation catalysts, 141-145

Saudi Arabia:
  operating refineries, capacities, 218
  refinery locations, capacities, and types of processing, 236-237
Semisynthetic catalysts, 416-417
Senegal:
  operating refineries, capacities, 218
  refinery locations, capacities, and types of processing, 237
Sierra Leone:
  operating refineries, capacities, 218
  refinery locations, capacities, and types of processing, 237
Silicon-enriched zeolites, 32-36
Singapore:
  operating refineries, capacities, 218
  refinery locations, capacities, and types of processing, 237
Single-stage-once-through (SSOT) hydrocracking design, 606
Single-stage recycle (SSREC) hydrocracking design, 606
Solvent dewaxing, 565-582
  deoiling, 580-582
  dewaxed oil, 566
  Dichill dewaxing process, 570, 580, 658-659
  Di/Me dewaxing process, 568, 569, 577, 578, 659
  lubricating oil processes, 656-659
  MEK-dewaxing process, 567-569, 575-577
  propane dewaxing process, 570, 577-580, 659
  wax, 566
Solvent refining processes for lubricating oils, 634, 648-654
Somalia:
  operating refineries, capacities, 218
  refinery locations, capacities, and types of processing, 237
Sour, 77
South Africa:
  operating refineries, capacities, 218
  refinery locations, capacities, and types of processing, 237-238
South America, petroleum production and demand, 69
Spain:
  operating refineries, capacities, 218
  refinery locations, capacities, and types of processing, 238

Specialty oils, catalytic dewaxing
process for, 560-562
Sri Lanka:
   operating refineries, capacities, 218
   refinery locations, capacities, and
      types of processing, 238
Steam hydrocarbon reforming catalysts, 149-151
Still gas, 7
Straight-run residual fuel oil, see
   Atmospheric residue
Sudan:
   operating refineries, capacities, 218
   refinery locations, capacities, and
      types of processing, 238
Sulfur:
   diesel fuels and, 11-12
   recovery of, catalysts for, 148-149
Sulfur compounds in petroleum, 697, 698-699
Sulfur content of gasoline, 81
Sulfur oxides reduction catalysts, 154
Surface impoundments of wastes, 185
Sweden:
   operating refineries, capacities, 218
   refinery locations, capacities, and
      types of processing, 238
Sweet, 77
Sweetening, see Doctor sweetening
Sweetening catalysts, 151
Switzerland:
   operating refineries, capacities, 218
   refinery locations, capacities, and
      types of processing, 238
Synthetic catalysts, 416
Synthetic fuels, yield improvement
   of, 175-176
Synthesis gas technology, commercial and developmental, 172, 173
Syria:
   operating refineries, capacities, 218
   refinery locations, capacities, and
      types of processing, 239

Tanzania:
   operating refineries, capacities, 218
   refinery locations, capacities, and
      types of processing, 239
Texaco Development Co.:
   catalytic cracking system of, 363
   solvent dewaxing process of, 657-658

Thailand:
   operating refineries, capacities, 218
   refinery locations, capacities, and
      types of processing, 239
Thermal cracking, 281-349
   chemistry of, 289-290
   commercial applications, 286-289
   definition of thermal processes, 282-283
   economics of, 346-348
   equipment design, 339-346
      heaters, 339-345
      miscellaneous equipment, 345-346
      pumps, 345
      soaking drums, 345
   history and development, 283-286
   yield and quality correlations, 291-338
      thermal cracking, 291-309
      thermal reforming, 323-338
      visbreaking, 309-323
Thermal reforming (of gasoline
   stocks), 283, 323-338
   process correlations, 325-333
   process flow, 324-325
   reaction kinetics, 334-338
Toxic Characteristic Leaching Procedures (TCLP) test, 188-189, 193
Trinidad:
   operating refineries, capacities, 218
   refinery locations, capacities, and
      types of processing, 239
Tunisia:
   operating refineries, capacities, 218
   refinery locations, capacities, and
      types of processing, 239
Turkey:
   operating refineries, capacities, 218
   refinery locations, capacities, and
      types of processing, 239-240
Two-stage hydrocracking design, 603-605

Ultrastable zeolites, 26
United Kingdom:
   operating refineries, capacities, 218
   refinery locations, capacities, and
      types of processing, 240
United States:
   crude oil consumption, 4

Index

[United States]
  operating refineries, capacities, 202, 218
  petroleum production and demand, 69
  refinery locations, capacities, and types of processing, 203-213
Universal Oil Products Co., catalytic cracking system of, 360-361
Unleaded gasoline, 9
Upflow expanded-bed reactor design, 687-688
  with fine catalyst, 688
  processing vacuum residuals in, 691-692
Urea dewaxing, 565, 583-592
  adduct formation, 584-585
  lubricating oil processes, 659-660
  technology of, 586-591
    Edenleanu process, 588-591
    Nurex process, 587-588
Uruguay:
  operating refineries, capacities, 218
  refinery locations, capacities, and types of processing, 240
Used Oil Recycling Act of 1980, 179
USSR:
  crude oil consumption, 4
  petroleum production and demand, 69
  refinery locations, capacities, and types of processing, 240

Vacuum distillation, 85-86
Venezuela:
  operating refineries, capacities, 218
  refinery locations, capacities, and types of processing, 241
Virgin Islands:
  operating refineries, capacities, 218
  refinery locations, capacities, and types of processing, 241
Visbreaking (viscosity-breaking), 85, 282-283, 309-323
  process flow, 309-311
  reaction kinetics, 320-323
  severity limits, 311-312
  sulfur distribution, 318-320
  yield correlations, 313-318
Volatility, 5
  of gasoline, 9, 80

Waste regulations, 179-189
  land disposal ban, 186-189
    air toxicity, 187-188
    corrective action, 189
    examptions to bans, 188
    liquids, 186-187
    notification, 187
    toxicity measurement, 188-189
  management of wastes, 183-186
    API separator sludge, 184
    exempted wastes, 183-184
    land treatment, 185-186
    land treatment and storage, 184-185
    surface impoundment, 185
  regulatory background, 179-183
    regulatory authority, 182-183
    requirements, 179-180
    submittals, 180-181
    waste cleanup, 182
    waste disposal, 181-182
Waste toxicity, prevention of, 190-198
  best demonstrated available technologies (BDAT), 192, 193-194
  land treatment and BDAT, 195-198
  "no-migration petitions," 195
  regulatory framework, 191-193
Waxes, 13
  refinery, 566
  See also Dewaxing
Western Europe:
  crude oil consumption, 4
  petroleum production and demand, 69
White distillate:
  BEV formula for, 19
  production costs, 16-19
World consumption of petroleum products, 2-4
World proven crude oil reserves, 67-68
Worldwide refining capacities, 214-242
  operating refineries, capacities, 216-218
  refinery locations, capacities, and types of processing, 219-242

Xylene catalysts, 133

Yemen:
  operating refineries, capacities, 218

[Yemen]
  refinery locations, capacities, and types of processing, 241
Yields improvement, 170-178
  better refinery gas utilization, 171-173
  direct combusion of coal or coke, 174-175
  limitations to coal and coke utilization, 176-177
  synthetic fuels, 175-176
Yugoslavia:
  operating refineries, capacities, 218
  refinery locations, capacities, and types of processing, 241

Zaire:
  operating refineries, capacities, 218
  refinery locations, capacities, and types of processing, 241
Zambia:
  operating refineries, capacities, 218
  refinery locations, capacities, and types of processing, 241
Zeolites:
  silicon-enriched, 32-36
  ultrastable, 26
Zeolitic catalysts, 168-169, 417